Giacomo Bergamini • Serena Silvi
Editors

Applied Photochemistry

When Light Meets Molecules

Springer

Editors
Giacomo Bergamini
Dipartimento di Chimica
"Giacomo Ciamician"
Università di Bologna
Bologna, Italy

Serena Silvi
Dipartimento di Chimica
"Giacomo Ciamician"
Università di Bologna
Bologna, Italy

ISSN 0342-4901 ISSN 2192-6603 (electronic)
Lecture Notes in Chemistry
ISBN 978-3-319-31669-7 ISBN 978-3-319-31671-0 (eBook)
DOI 10.1007/978-3-319-31671-0

Library of Congress Control Number: 2016947420

Printed on acid-free paper

This Springer imprint is published by Springer Nature
The registered company is Springer International Publishing AG Switzerland

Contents

Chapter 1
Supramolecular Artificial Photosynthesis

Mirco Natali and Franco Scandola

Abstract The conversion of light energy into chemical fuels by artificial means is a challenging goal of modern science, of great potential impact on long-term energy and environmental problems. As such, Artificial Photosynthesis is one of the most active research areas in applied photochemistry. In this tutorial review the basic ingredients of a biomimetic, supramolecular approach to Artificial Photosynthesis are outlined. First, a brief summary of the relevant structural-functional aspects of natural photosynthesis is provided, as a guide to plausible artificial architectures. Then, candidate energy converting reactions are examined, focusing attention on water splitting. The main functional units of an artificial photosynthetic system are dealt with in some detail, namely, charge separation systems, light harvesting antenna systems, water oxidation catalysts, and hydrogen evolving catalysts. For each type of system, design principles and mechanistic aspects are highlighted with specifically selected examples. Some attempts at integrating the various units into light-to-fuels converting devices are finally discussed. Throughout the review, the emphasis is on systems of molecular and supramolecular nature.

1.1 Introduction

Boosted by the rapid economic development of a growing world population, the global energy demand (now about 15 TW) is expected to double by 2050 and to triple by 2100. How to satisfy this enormous energy demand is the most pressing challenge facing society today. The main present energy source, fossil fuels, cannot be considered as a viable solution, not only for their limited availability, but because atmospheric CO_2 levels (already at more than 50 % above the pre-industrial values) cannot increase further without catastrophic consequences

M. Natali (✉) • F. Scandola
Department of Chemical and Pharmaceutical Sciences, University of Ferrara, and Centro Interuniversitario per la Conversione Chimica dell'Energia Solare (SOLARCHEM), sezione di Ferrara, Via Fossato di Mortara 17-19, 44121 Ferrara, Italy
e-mail: mirco.natali@unife.it; snf@unife.it

G. Bergamini, S. Silvi (eds.), *Applied Photochemistry*,
Lecture Notes in Chemistry 92, DOI 10.1007/978-3-319-31671-0_1

on climate change. It is clear that any long-term solution to the energy problem relies on the identification and exploitation of alternative energy sources that must be abundant, inexpensive, environmentally clean, and geographically widespread. In terms of such requirements, solar power (120,000 TW average irradiation at earth surface) presents itself as the most promising source of renewable energy available. In Nature, massive utilization of solar energy to sustain biological life is performed by a variety of photosynthetic organisms (plants, algae, cyanobacteria) that have evolved, along ca 2 billion years, to convert CO_2 and water into carbohydrates and oxygen. The development of artificial systems capable of efficiently converting light energy into practical fuels (Artificial Photosynthesis) can thus be envisioned as a research field of great potential, in principle able to provide a definitive solution of our energy problem.

In this tutorial review the basic ingredients of a biomimetic, supramolecular approach to Artificial Photosynthesis are outlined. First, a brief summary is given of the relevant structural-functional aspects of natural photosynthesis, both anoxygenic (Sect. 1.2.1) and oxygenic (Sect. 1.2.2). Then, candidate energy converting reactions are examined, focusing attention on water splitting (Sect. 1.3.1). The main functional units of artificial photosynthetic systems are dealt with in some detail, namely charge separation systems (Sect. 1.3.2), light harvesting antenna systems (Sect. 1.3.3), water oxidation catalysts (Sect. 1.3.4), and hydrogen evolving catalysts (Sect. 1.3.5). Some attempts at integrating the various units into light-to-fuels converting devices are finally discussed (Sect. 1.3.6). Throughout the review, the emphasis is on systems of molecular and supramolecular nature. In the discussion of the various topics, rather than aiming at comprehensive literature coverage, we have attempted to highlight design principles and mechanistic aspects with specifically selected examples.

1.2 Natural Photosynthesis

In Nature, photosynthesis, i.e., the conversion of light energy into chemical energy, is performed by a variety of organisms, ranging from plants to bacteria. The simplest form of photosynthesis is probably that performed by non-sulfur purple bacteria, such as *Rhodobacter sphaeroides* and *Rhodopseudomonas viridis*, where light energy is simply used to perform photophosphorylation, i.e., the conversion of adenosine diphosphate (ADP) into adenosine triphosphate (ATP), the molecular "energy currency" of the cell. At the other extreme of complexity is the oxygenic photosynthesis, performed by cyanobacteria, algae, and higher plants, whereby light energy is used to power the conversion of water and carbon dioxide into oxygen and sugars. A detailed description of the vast field of natural photosynthesis [1] is clearly outside of the scope of this tutorial review. A brief outline of the main features of the bacterial and oxygenic photosynthetic apparatuses will be given here, however, as these systems represent a powerful source of inspiration for research on artificial photosynthesis.

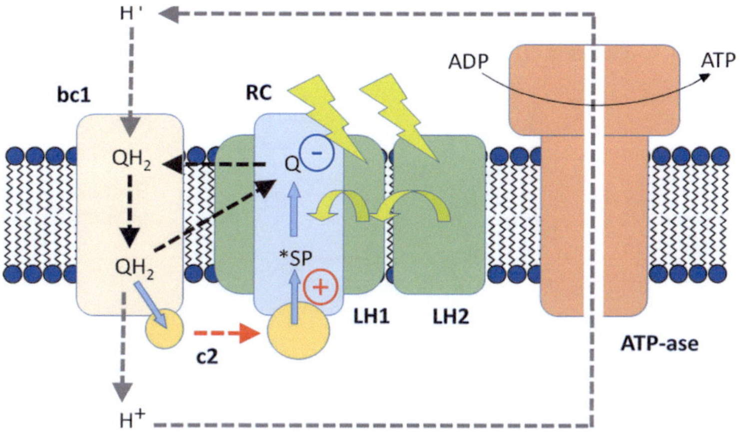

Fig. 1.1 Schematic representation of the relevant subunits involved in bacterial photosynthesis: the reaction center (*RC*), the LH1 and LH2 antenna systems, the bc1 complex, a small mobile cytochrome c2, and the ATP-ase enzyme

1.2.1 Bacterial

A schematic block-diagram picture of a bacterial photosynthetic membrane is shown in Fig. 1.1.

In simple terms, the overall function of this device is as follows: The light energy is primarily captured by light harvesting ("antenna") units, i.e., pigment protein complexes containing a large density of chromophoric molecules. Two types of such antenna units are usually present in the membrane, smaller LH2 units and the larger LH1 unit directly surrounding the reaction center. The energy is rapidly channeled, by means of exciton diffusion within and energy transfer between the antenna units to the reaction center, where a "special pair" of bacteriochlorophylls (SP) is excited. This triggers in the reaction center a series of electron transfer steps that lead to separate a positive and negative charge across the width of the membrane, the electron being localized on a quinone and the hole at the special pair. The reduced quinone picks up protons from the aqueous cytoplasmic phase, diffuses within the membrane to the cytochrome bc1 complex, where it is eventually reoxidized by the mobile cytochrome c2, releasing protons in the periplasmic aqueous phase. The cytochrome c2 shuttles from bc1 to RC giving back the electrons, thus closing the photoinduced electron transfer cycle. The net effect, therefore, is simply the building-up of a proton concentration gradient across the membrane. This proton electromotive force is then used by the ATP-ase enzyme for phosphorylation of ADP to ATP. The overall process is thus that of a cyclic photophosphorylation (Eq. 1.1, where P is inorganic phosphate)

$$ADP^{3-} + P + H^{+} + light \rightarrow ATP^{4-} + H_2O \tag{1.1}$$

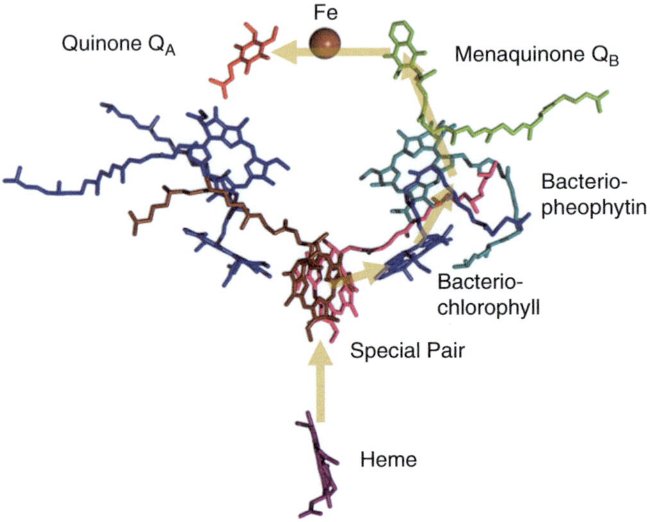

Fig. 1.2 Spatial arrangement of the cofactors within the protein matrix (not shown) of the reaction center of *Rhodopseudomonas viridis*. They are arranged in two quasi-symmetrical branches consisting of "special pair" of bacteriochlorophylls, bacteriochlorophyll, bacteriopheophytin, and quinone. A heme group of the four-heme reaction center cytochrome is also shown. The arrows point out the electron transfer chain that leads to charge separation following excitation of the special pair

In recent years, great progress has been achieved in the characterization by diffraction methods of the various functional subunits of the bacterial photosynthetic membranes, shining light on important aspects of the structure-function relationship. For their relevance to the biomimetic approach outlined in subsequent sections, attention is focused here on the structures and function of reaction center and on the antenna units. A milestone in the field, awarded with the Nobel Prize in Chemistry 1988, has been the determination of the structure of the reaction center of *Rhodopseudomonas viridis* by Diesenhofer, Huber, and Michel [2]. The spatial arrangement of the relevant cofactors, as shown in Fig. 1.2, is crucial to the function of the RC as a photoinduced charge separating device. Following selective excitation (see later) of the special pair of bacteriochlorophylls, a sequence of electron transfer steps takes place along one of the two branches of cofactors: [3] to accessory bacteriochlorophyll (3 ps), to bacteriopheophytin (1 ps), to menaquinone Q_A (150 ps), to quinone Q_B (200 μs). The hole on the special pair is eventually scavenged by electron transfer from the reduced reaction center cytochrome (0.3 μs). The long-range charge separation so obtained (22.4 Å from special pair to Q_A) is highly efficient (98 %) and long-lived (milliseconds). The key to such a performance lies in the multi-step nature of the process, where all the forward steps are kinetically optimized relative to the potentially detrimental back electron transfer steps (for a discussion of the factors that determine electron transfer kinetics, see later, Sect. 1.3.2).

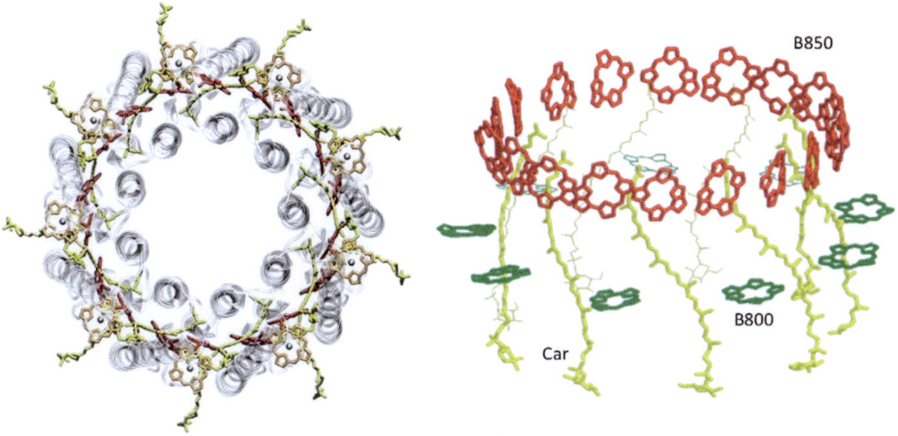

Fig. 1.3 Structure of the LH2 antenna unit of *Rhodopseudomonas acidophila*. *Left*: azimuthal view (perpendicular to membrane plane) showing the circular nonameric double-pillared protein structure. *Right* lateral view after deletion of the protein structure, showing the arrangement of three different types of chromophores: bacteriochlorophylls (B800), bacteriochlorophylls (B850), and carotenoids (Car)

Excitation of the reaction center takes place only in negligible amounts by direct light absorption. In fact, the light incident on the photosynthetic membrane is almost totally absorbed by the chromophores in the antenna units. Detailed structural and spectroscopic studies have provided clear pictures of the function of these units. The structure of the LH2 type of antenna, isolated from *Rhodopseudomonas acidophila*, [4] is shown in Fig. 1.3. In LH2, a circular double-pillared structure, made of nine heterodimers of α- and β-apoproteins spanning the membrane, acts as container of a dense ensamble of chromophores: 27 bacteriochlorophylls and 9 carotenoids. The bacteriochlorophylls are of two types, differing in wavelength of maximum absorption and position: 9 B800 molecules, lying parallel to the membrane surface, forming a ring on the cytoplasmic side; 18 B850 molecules form a tightly coupled perpendicular ring with slipped cofacial arrangement near the periplasmic side of the complex. The carotenoids, with strong visible absorption, have an extended all-trans conformation and span the entire depth of the complex, coming into van der Waals contact with both groups of BChls. Upon light absorption by the LH2 complex, a series of ultrafast energy transfer processes takes place: Car to B800–850 (140 fs), [5] B800 to B850 (0.9 ps), within the B850 ring (50–100 fs) [6]. Due to the strong interchromophore coupling, excitation in the B850 ring is partially delocalized, with an exciton delocalization length of ca four bacteriochlorophyll units [7].

The LH1 antenna units bear some structural similarity to the LH2 ones, being again based on a double-pillared cyclic protein structure, holding a large number of chromophores. The dimensions of the LH1 ring are larger, however, being composed of 16–15 heterodimers depending on the bacterium. Each heterodimer holds a pair of bateriochlorophyll molecules, forming again a large, tightly coupled ring

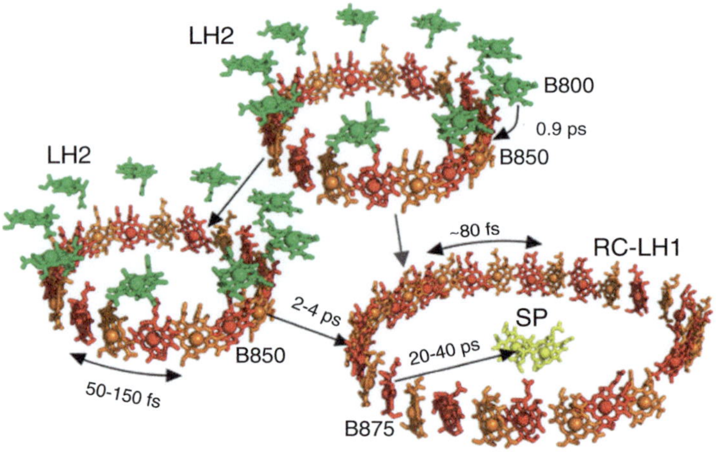

Fig. 1.4 Schematic of the processes conveying light excitation energy from the LH2 and LH1 antenna systems to the special pair of the reaction center of bacterial photosynthesis

with slipped cofacial arrangement, with maximum absorption at 875 nm. The dimensions of LH1 are appropriate to host the reaction center in its central cavity, and indeed a number of structures of LH1-reaction center complex are now available [8]. In such structures, the B875 ring of LH1 and the special-pair bacteriochloropylls of RC are aligned on the same level in the transmembrane region, in an optimal arrangement for selective excitation of the special pair by energy transfer from LH1. In the bacterial photosynthetic membrane, LH2 units and LH1-RC complexes are assembled in densely packed arrays, [9] where efficient energy transduction takes place. A summary of the time resolved energy transfer processes [6, 7] taking place in the bacterial photosynthetic membrane is given in Fig. 1.4.

In summary, the bacterial photosynthetic membrane is a perfect device performing the following light-induced functions: (a) efficient light energy harvesting, (b) fast and efficient transfer of the excitation energy to the reaction center, (c) efficient, long-lived trans-membrane charge separation. The dark processes that convert this charge separation into trans-membrane proton gradient and ultimately lead to generation of ATP fuel are of course extremely important from the biological viewpoint, but less interesting in the context of possible extensions to artificial photosynthesis.

1.2.2 Oxygenic

Unlike bacterial photosynthesis, which is a cyclic (photophosphorylation) process, the photosynthesis carried out by higher plants, algae, and cyanobacteria is characterized by a net chemical reaction consisting in the splitting of water into

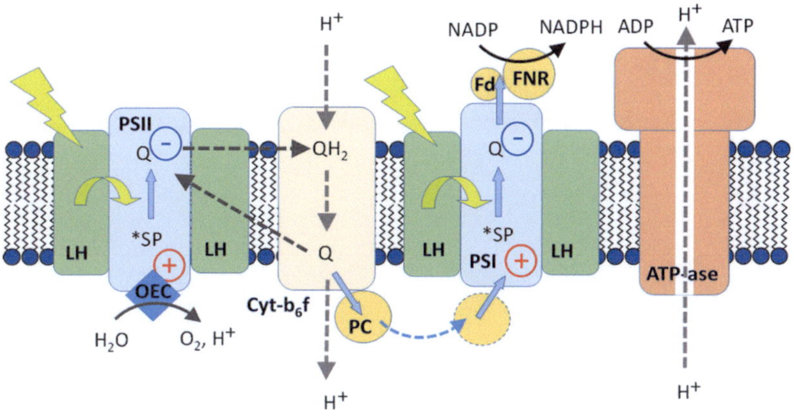

Fig. 1.5 Schematic diagram of the main functional subunits of an oxygenic photosynthetic membrane (for details, see text)

molecular oxygen, protons and electrons. The protons are used, as in bacterial photosynthesis, for the ATP synthesis and the electrons for the reduction of carbon dioxide to carbohydrates. Thus, the net reaction is shown in Eq. 1.2.

$$CO_2 + H_2O + \text{light} \rightarrow (CH_2O) + O_2 \tag{1.2}$$

A schematic block diagram of an oxygenic photosynthetic membrane is given in Fig. 1.5. A number of subunits are similar, at least from a functional viewpoint, to those of bacterial photosynthesis, e.g., reaction centers for charge separation (here called "photosystems, PS"), light harvesting antenna systems, ATP-ase. The main difference, however, lies in the intrinsic two-photon architecture of the process that requires the operation of two charge separating reaction centers, photosystems I and II. If we focus on the left part of Fig. 1.5, starting from photon absorption by the antenna (LH), energy transduction to the reaction center (PSII), charge separation in PSII, diffusion of reduced quinone with proton uptake on the stroma side and discharge on the lumen side mediated by the cytochrome b6f complex (Cyt-b_6f), all these processes bear a clear resemblance to those of bacterial photosynthesis. But now, rather than being short-circuited (as in the bacterial case) on the original reaction center, the electron transfer chain goes on, by means of a diffusing plastocyanin (PC), to the next reaction center, PSI, where a second photoinduced charge separation process takes place. The overall function is that of a device where two photosystems are chemically connected "in series". The oxidation potential of the "positive" end of PSII is used, with the help of a specific oxygen evolving catalyst (OEC, vide infra), for the oxidation of water on the periplasmic side, and the reducing power at the "negative" end of PSI is used, with the intervention of ferredoxin (Fd) and ferredoxin-NADP$^{(+)}$ reductase (FNR), for the reduction of NADP^{+} to NADPH. The primary reduced product NADPH, together with the ATP generated as usual taking advantage of the proton electromotive force, is

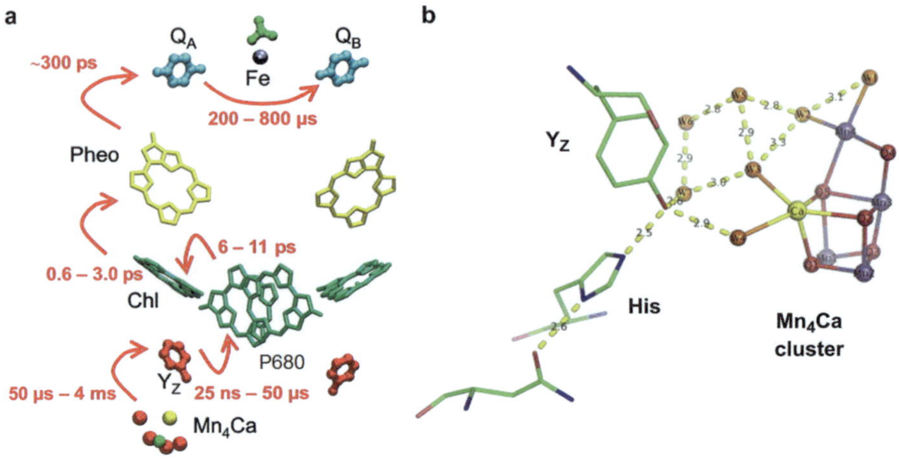

Fig. 1.6 (**a**) Arrangement of cofactors and electron transfer chain in Photosystem II (adapted with permission from ref. 10, copyright © 2012 Elsevier). (**b**) Detailed view of the OEC (adapted with permission from ref. 15c, copyright © 2011 Elsevier), showing the Mn_4Ca cluster, the primary electron acceptor tyrosine Y_Z and the nearby histidine, and the hydrogen-bond network of water molecules

used downstream in a complex thermal cycle (Calvin cycle) for CO_2 fixation and carbohydrate synthesis. Any detailed description of structure and function of the subunits of such a complex system is clearly outside the scope of this review. Let us simply point out a few aspects that may be relevant to the discussion in the subsequent sections.

Despite the wide differences in their organisms and functions, the charge separation devices involved in all oxygenic and anoxygenic (bacterial) photosynthetic systems are remarkably similar. A picture of the cofactors of PSII from *Thermosynechococcus elongatus* is shown in Fig. 1.6 [10].

Aside from the presence of chlorophylls rather than bacteriochlorophylls, the similarity to the structure of the bacterial reaction center of Fig. 1.2 is evident, with the two-branched structure involving special pair (P680), monomeric chlorophyll, pheophytin, and quinone. The photoinduced electron transfer sequence from special pair to quinones is again very similar. The main difference lies in the "positive" end of the chain, where in this case the hole on the special pair is filled with electrons coming, with a tyrosine residue (Y_Z) acting as a relay, from the Mn_4Ca cluster (Oxygen Evolving Complex, OEC) that is the actual catalyst for water oxidation. The ability of the OEC to perform such a complex process (four-electron oxidation of two water molecules, formation of a new O-O bond, and release of four protons) relies on its very specific structure, revealed by a recent high-resolution X-ray structure [15b]. Figure 1.6b shows the chair-like shape of the OEC, with the distorted cubane structure including three Mn, one Ca, and four bridging oxygen atoms, and the isolated fourth Mn and one bridging oxygen atom. It also shows four water molecules directly coordinated to the metal centers and other water molecules

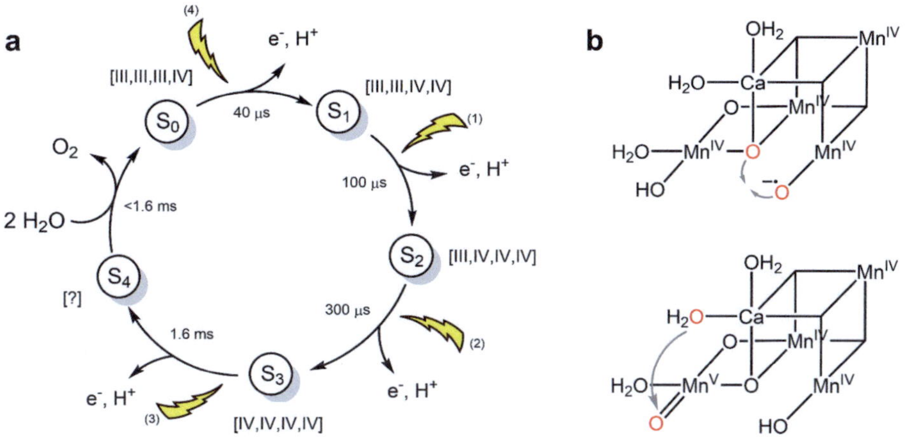

Fig. 1.7 (**a**) Kok cycle of photosynthetic water oxidation. Starting from the dark-stable S_1 state, photon absorption causes sequential electron abstraction from the OEC, accompanied by charge-compensating deprotonation steps. Oxidation-state combinations of the Mn ions are shown for the various S_0-S_3 states. Typical time constants of the ET steps are also indicated [12]. (**b**) Two plausible structures of the S_4 state, with corresponding mechanisms of O-O bond formation: oxyl radical coupling (*upper*) or water nucleophilic attack (*lower*) [13]

participating to a hydrogen-bond network linking the Mn_4Ca-cluster and Y_Z, and further from Y_Z to a nearby histidine.

The stepwise oxidation of the OEC by sequential photoinduced electron abstraction is described by the Kok cycle, [11] where four oxidation states of the OEC are termed S_i (i = 0–4), S_0 being the most reduced state and S_4 the most oxidized state in the catalytic cycle (Fig. 1.7a).

As shown by the flash number dependent oscillating pattern in O_2 evolution, the dark-stable state is S_1. From there, three photons are needed to effect the $S_1 \rightarrow S_2$, $S_2 \rightarrow S_3$, and $S_3 \rightarrow S_0$ transitions. In the last transition, likely with the intervention of an S_4 state as an intermediate, oxygen is evolved. Then, a fourth photon is needed to regenerate the dark-stable state S_1. As to the oxidation states of the four Mn centers of the OEC in the various S_i states, there is a general agreement that the dark-stable S_1 state of the OEC has two Mn(III) and two Mn(IV) centers. The oxidation states of the S_0, S_2, and S_3 states can be inferred by one-electron reduction or oxidation steps. The detailed nature of the elusive S_4 state, where oxygen evolution takes place is unknown. Among several hypotheses, [12] two plausible ones are (a) an all-Mn(IV) structure with a coordinated oxyl radical or (b) a structure with three Mn(IV) and one Mn(V)-oxo group. These two structures are related to two plausible mechanisms for O-O-bond formation (Fig. 1.7b): (a) oxyl radical coupling, (b) water nucleophilic attack [13]. A point worth of mention is that in the successive steps of the Kok cycle the OEC accumulates four oxidizing equivalents delivered by the Y_Z radical, an oxidant with an approximately constant potential (close to +1.1 V). This would be unfeasible without a charge-

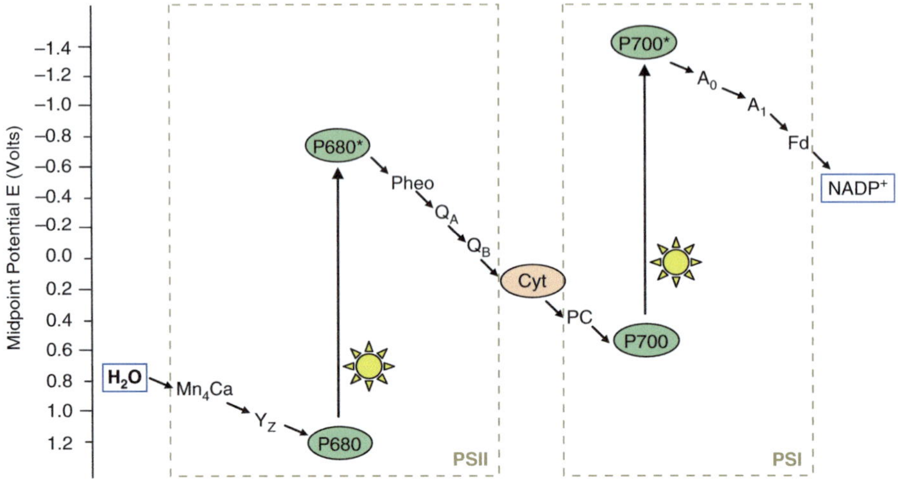

Fig. 1.8 Z-scheme of the energetics of oxygenic photosynthesis, pointing out the relationships between absorbed photon energy, chemically converted energy, and energy price paid for charge separation

compensating mechanism, by which protons are removed from acid-base sites of the OEC and relocated towards the aqueous phase.

As to the portion of the photosynthetic mechanism taking part downstream of PSII, without going into any detail, let us just point out that the structure and function of PSI [14] are again rather similar to those of PSII, with a similar two-branched chain of co-factors, involving special pair (P700), monomer chlorophyll, a second monomer chlorophyll, and quinone, and similar electron transfer chain events. The related cofactors in the two photosystems, however, have different redox properties, as shown in Fig. 1.8, where the chain of electron transfer events taking place along the whole photosynthetic process is depicted on a potential scale. This electrochemical energy diagram (known as Z-scheme) shows clearly the very reason for the use of two photosystems in series by oxygenic photosynthesis. In fact, the potential difference required for water oxidation and NADP reduction (ca 1.2 V) does not exceed the energy of a single photon (ca 2.0 eV). A large amount of the photon energy is lost, however, by each photosystem as driving force for charge separation. Therefore, two photosystems operating in series are actually required.

Although many other aspects of the complex machinery of oxygenic photosynthesis are interesting, let us only mention a few points about the light harvesting antenna systems. First, there are differences in the chromophoric units of the two types of photosyntheses, mainly dictated by the different habitats of the organisms: in plant photosynthesis the main chromophores are chlorophylls, absorbing at shorter wavelengths than their bacterio-analogues. Moreover, with respect to the LH2 and LH1 antennae of bacterial photosynthesis, the antenna units of plant photosynthesis consist of smaller, less symmetric and apparently less highly ordered subunits [15, 16]. This reflects a difference in function, which in this case requires, for optimal performance, not only efficient energy transfer to the reaction center but also a balance of the excitation pressure on the two photosystems working in series [17]. Since the absorption spectra of PSI and PSII are different, variations in light quality may drive both photosystems to a different extent, leading to imbalances. In such a case, a process called "state transition", regulated in a complex manner by phosphorylation, [18] redistributes the amount of excitations between PSI and PSII. In practice, a mobile pool of antenna complexes (LHCII) is able to reversibly associate with PSI or PSII, depending on which is preferably excited.

1.3 Artificial Photosynthesis

1.3.1 Functional Units and Candidate Reactions

With the general term "Artificial Photosynthesis" a process is commonly defined whereby light energy is exploited for the conversion of suitable substrates into chemicals with high energy content, namely fuels. Inspired by the complex machinery of natural photosynthetic organisms (Sect. 1.2), an artificial system capable of achieving this goal must be composed of several functional units, as shown in Fig. 1.9, which include: (i) an antenna system based on a series of chromophores which are responsible for the light-harvesting and the excitation energy funneling towards the reaction center, (ii) a charge separation system where the excitation energy is converted through a series of photoinduced and thermal electron transfer processes into an electrochemical potential residing in a hole and an electron kept

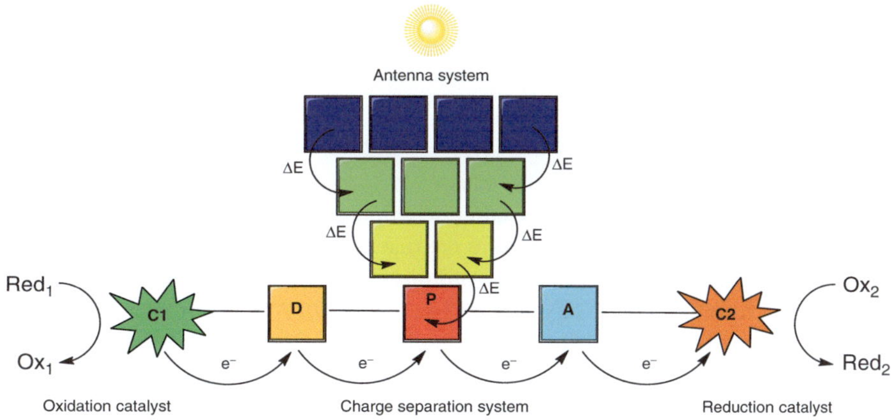

Fig. 1.9 Schematic representation of an artificial photosynthetic system comprising antennae, charge separation, and catalysts. Abbreviations used: *P* photosensitizer, *D* donor, *A* acceptor, *C1* oxidation catalyst, *C2* reduction catalyst, Red_1, Ox_2 substrates, Ox_1, Red_2 products

far apart one to each other, and (iii) two catalytic units capable of stepwise collecting and storing electrons and holes from the charge separation system to drive multi-electron redox processes on the substrates at low activation energy.

Several attempts have been performed towards the construction of either charge separation or antenna systems and towards the integration of both functional units into a single molecular device by adopting different molecular design and synthetic strategies. The basic guidelines towards this goal and some of the most relevant results will be outlined in the following sections (Sects. 1.3.2 and 1.3.3 for charge separation systems and antennae, respectively).

As far as the chemical redox reactions are concerned, several schemes can be in principle drawn for application into an artificial photosynthetic system, all sharing the common feature that the ideal substrates are naturally abundant molecules while the related products should be high energy content chemicals to be exploited either directly (e.g., combustion, fuel cells, etc.) or as intermediates for industrially relevant reaction processes. Some examples include processes such as water splitting (Eq. 1.3), carbon dioxide reduction (Eqs. 1.4, 1.5, 1.6, 1.7, 1.8), and nitrogen reduction (Eq. 1.9).

$$2\,H_2O \rightarrow 2\,H_2 + O_2 \tag{1.3}$$

$$CO_2 \rightarrow CO + 1/2\,O_2 \tag{1.4}$$

$$CO_2 + H_2O \rightarrow HCO_2H + 1/2\,O_2 \tag{1.5}$$

$$CO_2 + H_2O \rightarrow H_2CO + O_2 \tag{1.6}$$

$$CO_2 + 2\,H_2O \rightarrow CH_3OH + 3/2\,O_2 \tag{1.7}$$

$$CO_2 + 2\,H_2O \rightarrow CH_4 + 2\,O_2 \tag{1.8}$$

$$N_2 + 3\,H_2O \rightarrow 2\,NH_3 + 3/2\,O_2 \tag{1.9}$$

Notably, all these processes have in common water oxidation to dioxygen (Eq. 1.10) as the anodic half-reaction and a large extent of the energy required to power the overall reactions (Eqs. 1.3, 1.4, 1.5, 1.6, 1.7, 1.8, and 1.9) is deserved to this oxidative step.

$$2\,H_2O \rightarrow O_2 + 4\,H^+ + 4\,e^- \quad \left(E^0 = +1.23\text{ V vs NHE at pH }0\right) \tag{1.10}$$

Apart from thermodynamics, this latter reaction is also hampered by important kinetic hurdles, intimately connected to the reaction mechanism which requires the abstraction of four electrons and four protons from two water molecules with the concomitant formation of an oxygen-oxygen bond [19]. As a result, huge amount of studies are aimed at solving this quite complicated issue (see below Sect. 1.3.4).

$$N_2 + 6\,H^+ + 6\,e^- \rightarrow 2\,NH_3 \quad \left(E^0 = +0.06\text{ V vs NHE}\right) \tag{1.11}$$

As regarding the cathodic half-reaction, conversion of nitrogen into ammonia (Eq. 1.11) is a quite demanding process which is accomplished in certain natural photosynthetic organisms by nitrogenase enzymes. In these natural systems strongly reducing agents are not sufficient to promote the reaction which requires additional energy inputs (e.g., ATP). This is mainly dictated by the high activation barrier of the nitrogen-to-ammonia transformation since the mechanism requires the accumulation of six electrons, the rupture of a triple nitrogen-nitrogen bond, and the contemporary formation of six nitrogen-hydrogen bonds. As a result this chemical reaction is very difficult to afford in an artificial manner and only few examples are indeed reported which, however, perform modestly and under quite harsh reducing conditions [20]. Therefore employing nitrogen reduction as the cathodic reaction in an artificial photosynthetic system is far from a viable solution.

Reduction of carbon dioxide is, on the other hand, a potential cathodic reaction in artificial photosynthesis providing a direct way to the production of liquid fuels in a carbon-neutral fashion. The one-electron reduction of CO_2 (Eq. 1.12) is, however, highly disfavored from a thermodynamic viewpoint whereas the proton-assisted reductions significantly lower the thermodynamic barrier (Eqs. 1.13, 1.14, 1.15, 1.16, and 1.17).

$$CO_2 + e^- \rightarrow CO_2^{\bullet -} \quad \left(E^0 = -1.90 \text{ V vs NHE}\right) \tag{1.12}$$

$$CO_2 + 2\,H^+ + 2\,e^- \rightarrow CO + H_2O \quad \left(E^0 = -0.53 \text{ V vs NHE}\right) \tag{1.13}$$

$$CO_2 + 2\,H^+ + 2\,e^- \rightarrow HCO_2H \quad \left(E^0 = -0.61 \text{ V vs NHE}\right) \tag{1.14}$$

$$CO_2 + 4\,H^+ + 4\,e^- \rightarrow H_2CO + H_2O \quad \left(E^0 = -0.48 \text{ V vs NHE}\right) \tag{1.15}$$

$$CO_2 + 6\,H^+ + 6\,e^- \rightarrow CH_3OH + H_2O \quad \left(E^0 = -0.38 \text{ V vs NHE}\right) \tag{1.16}$$

$$CO_2 + 8\,H^+ + 8\,e^- \rightarrow CH_4 + H_2O \quad \left(E^0 = -0.24 \text{ V vs NHE}\right) \tag{1.17}$$

This, however, introduces non-trivial kinetic hurdles since the reduction mechanism has to deal with multi proton-coupled electron-transfer (PCET) steps. Moreover, to manage such proton electron transfer reactions a suitable catalyst must directly bind the CO_2 substrate and, since carbon dioxide is a linear and somewhat inert molecule, this process usually requires additional energetic efforts [21]. Taken together, these intrinsic restrictions significantly affect the results attainable from catalytic CO_2 reduction, usually achieving CO and/or formic acid as the main products, while rarely methanol or methane, and also showing, with few exceptions, [22] poor selectivity and/or parasite reactions such as hydrogen evolution [23].

With respect to both nitrogen and carbon dioxide reduction, proton reduction to dihydrogen (Eq. 1.18)

$$2H^+ + 2e^- \rightarrow H_2 \quad \left(E^0 = 0 \text{ V vs NHE at pH } 0\right) \tag{1.18}$$

is a much easier solution. Indeed, although a catalyst is anyway required to enable hydrogen evolution, the overall reaction is less affected by parasite processes and competitive pathways. Moreover, hydrogen is a clean fuel since its combustion in

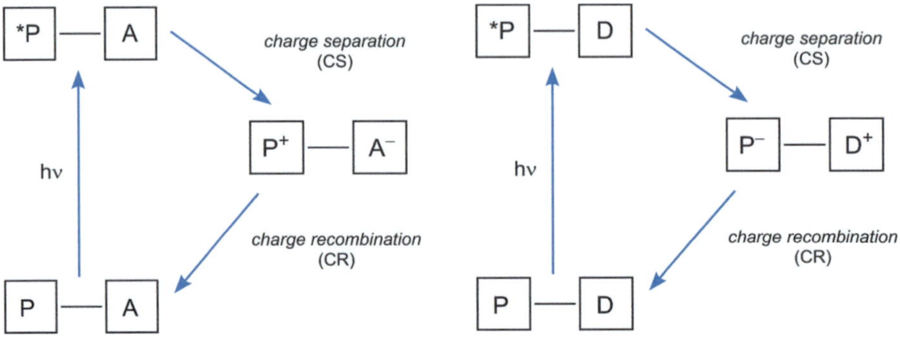

Fig. 1.10 Photoinduced electron transfer processes: oxidative quenching by an acceptor A (*left*) and reductive quenching by a donor D (*right*) of the light-harvesting photosensitizer P

the presence of oxygen produces heat, its combination with oxygen in a fuel cell generates electricity and heat, and the only byproduct of such energy-producing processes is water. More importantly, once hydrogen is produced via water splitting several application can be envisaged in addition to its direct consumption as a fuel. Indeed, it may also be exploited for catalytic hydrogenation of CO_2 to produce formic acid [24] which can be used as a preservative agent or synthetic precursor, or even as a liquid hydrogen storage material. Moreover, hydrogen produced from photochemical water splitting can be used in combination with CO_2 in the presence of suitable bacteria to renewably generate either biomass or liquid fuels such as alcohols [25]. Altogether, these evidences clearly demonstrate the actual potential arising from proton reduction catalysis and thus explain the enormous efforts made in the last years to prepare and characterize artificial systems capable of promoting this reaction (see below Sect. 1.3.5).

1.3.2 Charge Separation Systems

In natural photosynthesis, "reaction centers" (bacteria) and "photosystems" (plants) are the subunits that play the key role of converting the excitation energy, via photoinduced charge separation, into chemically exploitable redox potential. Any conceivable artificial photosynthetic system must include at least one subsystem with this functional role, that can be shortly named "charge separation system". In its simplest terms, a charge separation system is a supramolecular device where light absorption by a photoexcitable molecular component (photosensitizer, P) is followed by electron transfer to an acceptor A (Fig. 1.10, left) or from a donor D (Fig. 1.10, right), leading to a pair of one-electron oxidized and reduced transient species. As shown later, many variations of this basic scheme can be envisioned, with various degrees of complexity in both supramolecular architecture and electron transfer pathways. A simple two-component ("dyad") system, such as, e.g., those of Fig. 1.10 can be used, however, to introduce some basic concepts. The key requisites for a good charge separation system are as follows: the one-electron oxidized and reduced products (i) should be formed in high quantum yield, (ii) should store a large fraction of the original excitation energy in the form of redox potential, and (iii) should be long-lived towards charge recombination. The achievement of such requisites, relying on a complex interplay of kinetic and thermodynamic factors, is not trivial and constitutes the main challenge in the design of artificial charge separation systems.

A general theoretical framework for understanding electron transfer processes in supramolecular systems is provided by the Marcus non-adiabatic classical electron transfer theory, [26, 27] that relates the electron transfer rate constant to a number of relevant parameters (Eq. 1.19).

$$k = \left(\frac{4\pi^3}{h^2\lambda k_B T}\right)^{\frac{1}{2}} V^2 \exp\left[-\frac{(\Delta G^0 + \lambda)^2}{4\lambda k_B T}\right] \tag{1.19}$$

In Eq. 1.19 V is the electronic matrix element, h is Planck's constant, k_B is Boltzmann's constant, T is the temperature, ΔG^o is the standard free energy change of the process, and λ is the so-called reorganization energy, i.e., the energy required to reorganize the nuclear geometry of donor and acceptor as well as their solvation environments upon electron transfer. The reorganization energy has inner (molecular geometry) and outer (solvent) contributions (Eq. 1.20)

$$\lambda = \lambda_i + \lambda_o \tag{1.20}$$

that in classical terms are given as in Eqs. 1.21 and 1.22

$$\lambda_o = e^2\left(\frac{1}{2r_1} + \frac{1}{2r_2} - \frac{1}{r_{12}}\right)\left(\frac{1}{D_{op}} - \frac{1}{D_s}\right) \tag{1.21}$$

$$\lambda_i = \frac{1}{2} k \Delta Q^2 \tag{1.22}$$

where D_{op} and D_s are the optical and static solvent dielectric constants respectively, r_1 and r_2 are the radii of the two molecular components, r_{12} is the inter-component distance, e is the electron charge, ΔQ represents the nuclear displacement accompanying electron transfer along a relevant nuclear coordinate Q, and k is an appropriate force constant for that motion.

In order to produce the charge separated state with high quantum yield [requisite (i)], the forward electron transfer reaction (charge separation) must be as fast as possible, so as to be able to beat the excited-state lifetime of the sensitizer (which for most organic chromophore singlet states is of the order of few nanoseconds). For the charge separated state surviving long enough to permit utilization of the stored redox power [requisite (iii)], the back electron transfer reaction (charge recombination) must be as slow as possible. A solution to this double constraint is provided by the Marcus theory (Eq. 1.19) whereby, assuming for the sake of simplicity a constant electronic matrix element V, [28, 29] the main source of

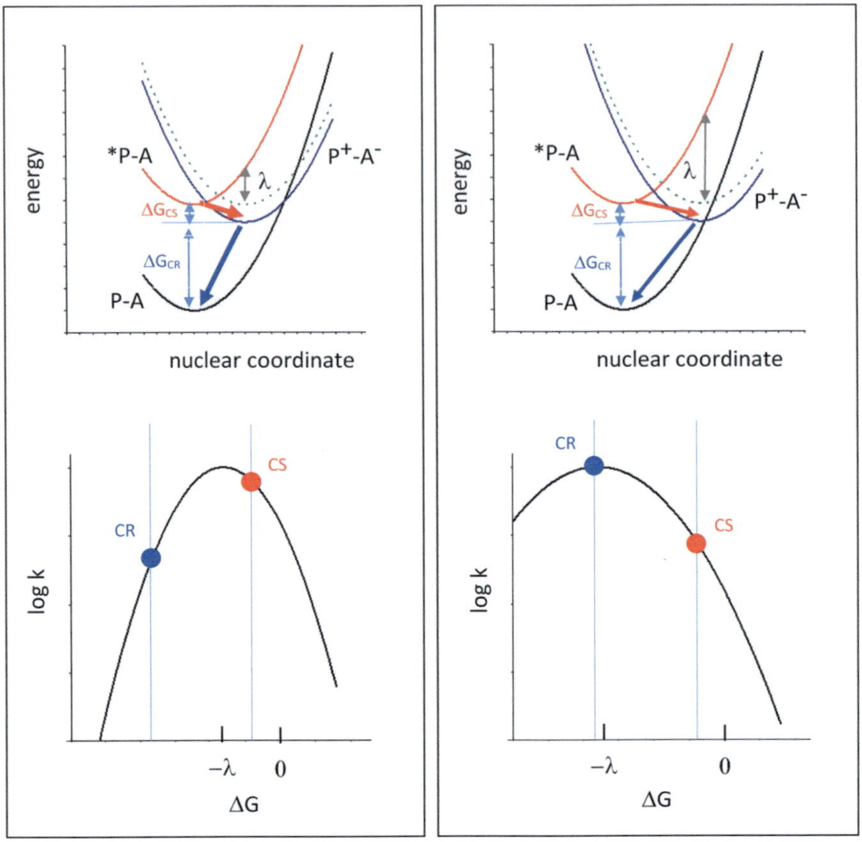

Fig. 1.11 Effects of small (*left panel*) or large (*right panel*) reorganization energy λ on photoinduced charge separation (*CS*) and recombination (*CR*). *Upper part*: potential energy curves for ground state (P-A), excited state (*P-A) and charge separated state (P^+-A^-) of a photosensitizer-acceptor dyad. *Lower part*: charge separation and recombination rates as predicted by Eq. 1.19

difference between forward and back ET rates is to be looked in their driving forces, ΔG. According to Eq. 1.19, the dependence of *log k* on driving force is quadratic, with values first increasing for increasing exergonicity ("normal" region), reaching a maximum value for $-\Delta G = \lambda$ (activationless regime), and then decreasing for more exergonic reactions (Marcus "inverted" region). It is clear that the key rate-determining factor is not the driving force as such, but rather its value relative to the reorganization energy λ. Two hypothetical situations of small and large λ, with the corresponding predictions of Eq. 1.19, are sketched in Fig. 1.11. Clearly, the ideal situation in order to obtain fast charge separation and slow charge recombination is to have, as in Fig. 1.11 (left panel), the first process in the "normal" free energy region (possibly as close as possible to the activationless regime) and the latter as a much more exergonic process lying deep into the "inverted region". The comparison between left and right panels of Fig. 1.11 clearly shows that having a small λ is

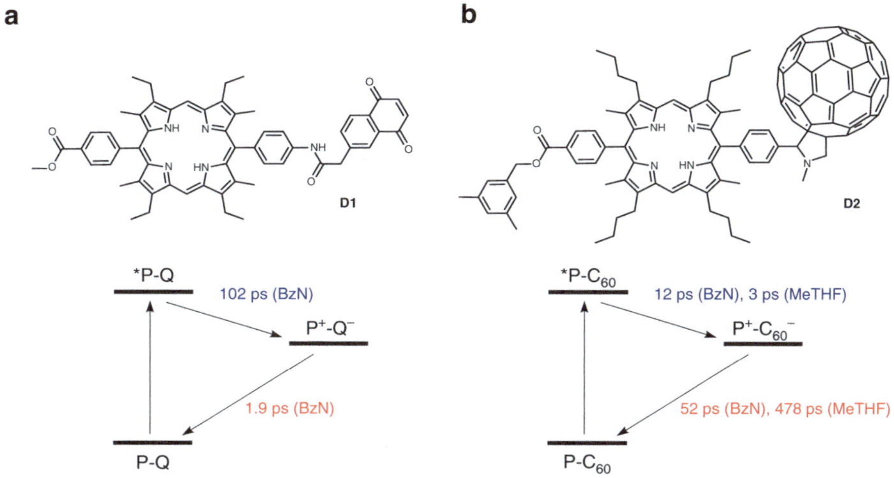

Fig. 1.12 Photoinduced charge separation and recombination in dyads (**a**) **D1** and (**b**) **D2** involving a porphyrin as photoexcitable chromophore (P) and a quinone (Q) [30] or a fullerene (C_{60}) [31] as acceptor unit, respectively

instrumental towards the achievement of this ideal kinetic condition. Furthermore, it is obvious that the smaller λ the larger the fraction of the excitation energy that can be converted, in the kinetically optimal situation ($-\Delta G_{CS} = \lambda$), into redox energy of the charge separated products (condition (ii) above).

In the light of the above arguments, the design of artificial charge separation systems should preferably use molecular components that, besides having the appropriate excited-state and redox properties, could warrant small reorganization energy values. Typically, relatively large molecules with highly delocalized electronic systems are expected to induce little solvent repolarization (large r_1 and r_2 in Eq. 1.21) and to undergo small internal geometry changes (small ΔQ in Eq. 1.22) upon ET. As to the environment to be used, a low-polarity solvent (small D_s in Eq. 1.21) will be beneficial to reduce λ_o. The comparison between the behavior of the two dyads **D1** and **D2** in Fig. 1.12 is instructive in this regard. They are made of the same photosensitizer, a porphyrin, and two different acceptors with similar redox properties, namely, a quinone and a fullerene. Despite the very similar energetics, the two dyads behave (in the same solvent, e.g., benzonitrile) quite differently: in the fullerene-based system **D2** (Fig. 1.12b) fast charge separation is followed by slow charge recombination, but the opposite situation occurs in the quinone-based **D1** (Fig. 1.12a).

This difference can be traced back, at least qualitatively, to differences in reorganization energies: [32] the fullerene A/A$^-$ couple, with its larger dimensions and electronic delocalization, provides a smaller λ than the quinone-based one. A further reduction in λ, and thus a further gain in charge separation vs recombination rates, can be obtained by lowering the solvent polarity, as shown by the comparison between the time constants obtained for dyad **D2** in benzonitrile and in 2-methyltetrahydrofurane. With reference to the schemes of Fig. 1.11, the large-λ situation (right panel) could be used to describe the porphyrin-quinone dyad, while for the porphyrin-fullerene one (and particularly so in MeTHF) the small-λ situation (left panel) very likely applies.

As discussed above, a clever choice of molecular components and medium, providing the right combination of driving forces and reorganization energies in the exponential term of Eq. 1.19, is essential to optimize the balance between charge separation and charge recombination in a dyad. Even so, however, the lifetime of the charge separated state in dyads is usually too short to permit eventual charge accumulation and utilization (see below, however, for some purported exceptions). In order to overcome this limitation, a commonly used, biomimetic strategy has been that of increasing the supramolecular complexity of the system (from dyads to triads, tetrads, etc.) [33–36]. In such systems, multiple, sequential charge separation steps can take place and the expected exponential decrease of the electronic coupling V (Eq. 1.19) with distance [29, 37] can be exploited to slow down charge recombination. An example of such strategy is provided in Fig. 1.13, where the behavior of a simple porphyrin-fullerene dyad **D3** (Fig. 1.13a) is compared with that of a tetrad **T1** obtained by adding to the same dyad two further electron-donor units, namely zinc porphyrin and ferrocene [38]. The sequence of events taking place in **T1** is schematically shown in Fig. 1.13b.

Excitation of the free-base porphyrin (either by direct light absorption or following energy transfer from the zinc porphyrin) gives a primary charge separation process (144 ps), quite similar to that obtained in **D3**. In the tetrad, however, in competition with primary charge recombination (expected to take place, as in the dyad, in some 80 ns), a fast charge shift process takes place (400 ps) whereby the hole in the free-base porphyrin is filled by an electron from the zinc porphyrin. Now, in competition with secondary charge recombination (20 μs), a further charge shift process takes place (5.9 ns) leading to the final charge separated state where the positive charge resides on the ferrocene unit. The fully charge-separated state Fc^+-ZnP-H_2P-$C_{60}{}^-$ so obtained has a remarkable lifetime of 0.34 s, with a gain factor $> 10^6$ relative to the corresponding dyad system [38]. The key to this result is evident from the reaction scheme, namely, the strong (actually, exponential) decrease in the charge recombination rates, as the distance between the charges increases going from the dyad, to the triad, and finally to the tetrad. It is important to realize, however, that the remarkable elongation in lifetime achieved by the strategy of stepwise charge separation is obtained at the expenses of the energy of the CS state, and thus of the amount of light energy converted in redox potential. Therefore, in the design of artificial charge separating systems an appropriate balance has to be sought between the lifetime and the energy of the final charge separated state.

A sharply different strategy has been advocated by Fukuzumi [39] claiming that long-lived charge separation can be obtained even in very short dyads, provided that (i) the reorganization energy is very small, (ii) the charge separated state has high energy, and (iii) no locally excited triplet states of donor and acceptor are energetically accessible from the charge separated state. As an example, the dyad **D4** (Fig. 1.14), where the acridinium unit acts as a photoexcitable acceptor and the mesityl group as the donor, was reported to have an extremely long-lived electron

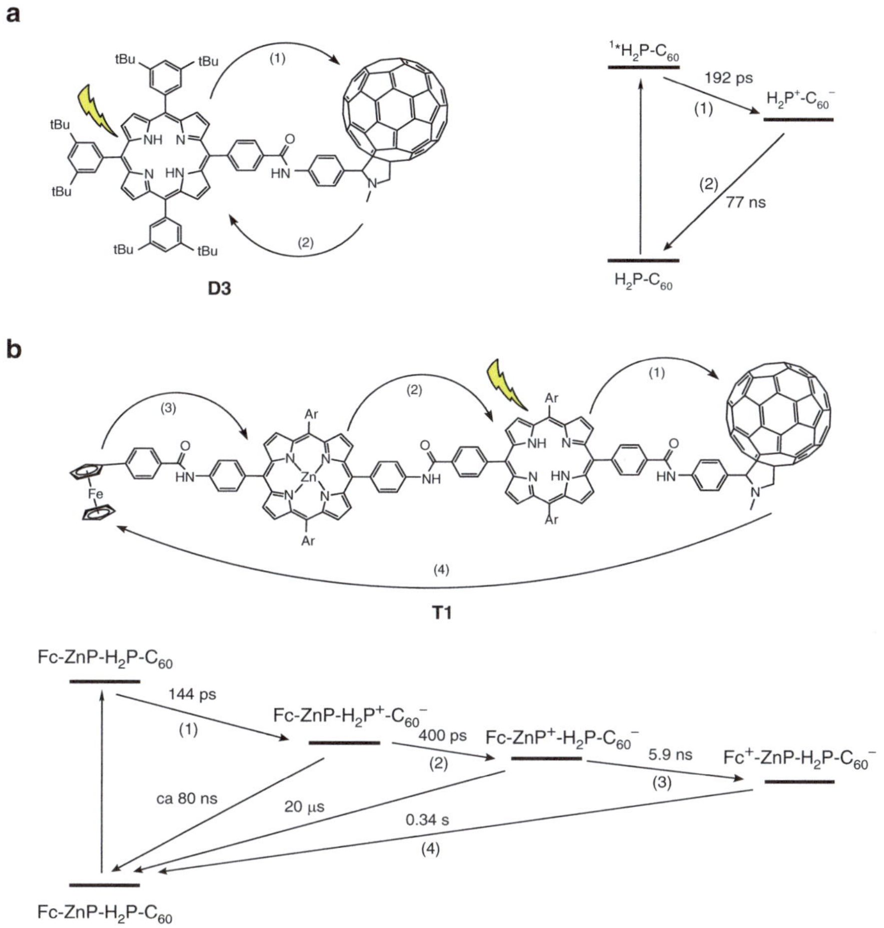

Fig. 1.13 Structure and photophysical behavior of (**a**) a porphyrin-fullerene dyad **D3** and (**b**) a ferrocene-zinc porphyrin-porphyrin-fullerene tetrad **T1**

Fig. 1.14 Molecular structure of the Acr^{+}-Mes dyad **D4** described by Fukuzumi and coworkers

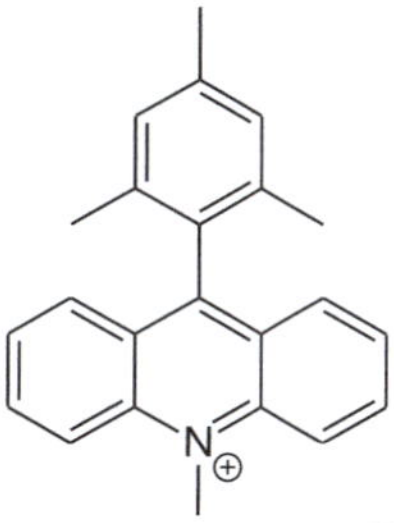

D4

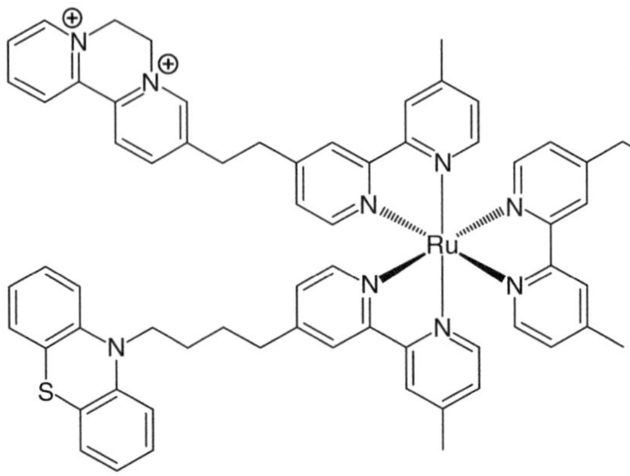

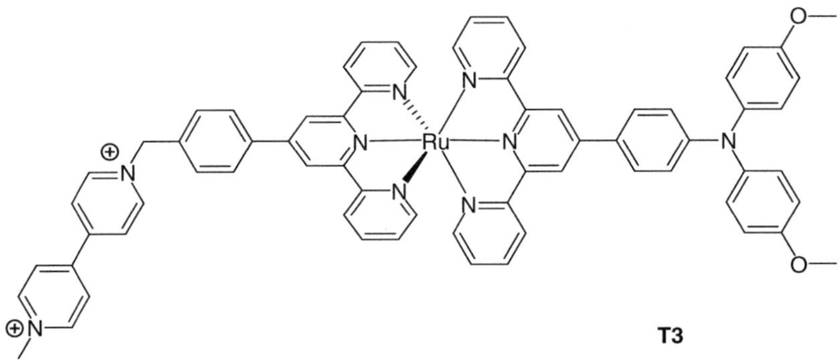

Fig. 1.15 Molecular structures of the tris(bipyridine)-based pseudo-triad **T2** and bis(terpyridine)-based triad **T3**

transfer (pseudo-CS) state $Acr^{\bullet}$-$Mes^{\bullet+}$ (2 h at 203 K). Such long lifetimes have been questioned by other authors, [40] suggesting attribution of the long-lived transient to a locally excited triplet state of the acridinium moiety rather than to an electron transfer state. However, additional experimental results in support of the electron transfer nature of the long-lived state in this type of dyads have been produced [41].

As in the examples discussed above, the photosensitizer, donor, and acceptor molecular components used to construct charge separating devices have most often been of organic nature. A substantial amount of work has also been performed, however, using transition metal coordination complexes, particularly in the role of photosensitizer units [42–46]. The reason lies essentially in the facile tunability of spectroscopic and redox properties characteristic of such species. Out of many examples available, a couple of donor-sensitizer-acceptor triads **T2** and **T3** based on two different Ru(II) polypyridine sensitizers is shown in Fig. 1.15. As far as the sensitizers are concerned, the tris(bipyridine) one has the advantage of a longer room-temperature excited state lifetime, while the bis(terpyridine) one is much shorter-lived but allows for a better stereochemistry (trans functionalization, lack of isomers). In the DQ-Ru-PTZ_2 pseudo-triad **T2** [42] the sensitizer is a tris (bipyridine) complex (Ru), the acceptor is a cyclo-quaternarized 2,2′-bipyridine (DQ) and the donor a phenothiazine (PTZ). In this system, photoinduced electron transfer from the sensitizer to the acceptor (ca 1 ns) is followed by fast hole transfer to the donor, with formation of a fully charge separated state that lives for 100–300 ns (dichloroethane, room temperature). In the MV-Ru-TAA triad **T3** [44d] the sensitizer is a bis(terpyridine) complex (Ru), the acceptor a methylviologen (MV), and the donor is a triarylamine (TAA). Here again, primary electron transfer to the acceptor (15 ns) is followed by fast hole transfer to the donor with formation of a fully charge separated state that lives for 27 ns (butyronitrile, 150 K, required for a manageable sensitizer excited-state lifetime).

All the examples of charge separating systems described so far have the sensitizer, donor, and acceptor units connected by organic covalent linkages. Though providing chemical robustness, the use of covalent organic techniques is often plagued by long synthetic routines and low yields. Inspired by the natural systems, where the cofactors are held in their positions by weak non-covalent interactions, some attempts have been made to obtain charge separating systems by spontaneous self-assembling of molecular components.

For such intrinsically non-symmetric systems, the problem is not obvious as it requires molecular recognition between the subunits to be assembled. In the example shown in Fig. 1.16, [47] the molecular building units designed to that purpose are: (i) an aluminium porphyrin (AlMPyP) as the central photoexcitable sensitizer, (ii) a naphthalenediimide (NDI) as electron acceptor unit, and (iii) a ruthenium porphyrin (RuP) as electron donor unit. The self-assembling ability is implemented by introduction of a carboxylic function (with great affinity for Al coordination) on the naphthalenediimide unit and of a pyridyl appended ligand (with strong affinity for Ru coordination) on the aluminium porphyrin. In fact, a single three-component adduct **T4**, NDI-AlMPyP-RuP, quantitatively self-assembles from a 1: 1: 1 mixture of the three starting units in dichloromethane. As demonstrated by ultrafast spectroscopy, excitation of the aluminium porphyrin is followed by stepwise electron (3 ps) and hole transfer (35 ps), leading to a charge separated state with reduced acceptor and oxidized donor (NDI^--AlMPyP-RuP^+), with a lifetime in the few nanoseconds range [47].

The few examples shown above are meant to illustrate the main principles that underlie the design of artificial systems for charge separation. Implementing such strategies, with an appropriate choice of molecular components and supramolecular architecture, a variety of molecular devices featuring efficient and long-lived photoinduced charge separation have been produced in the last decades.

1.3.3 Antenna Systems

In natural photosynthesis, the antenna units are large arrays of chromophores that play the role of efficiently absorbing the incoming photons and to convey, by means

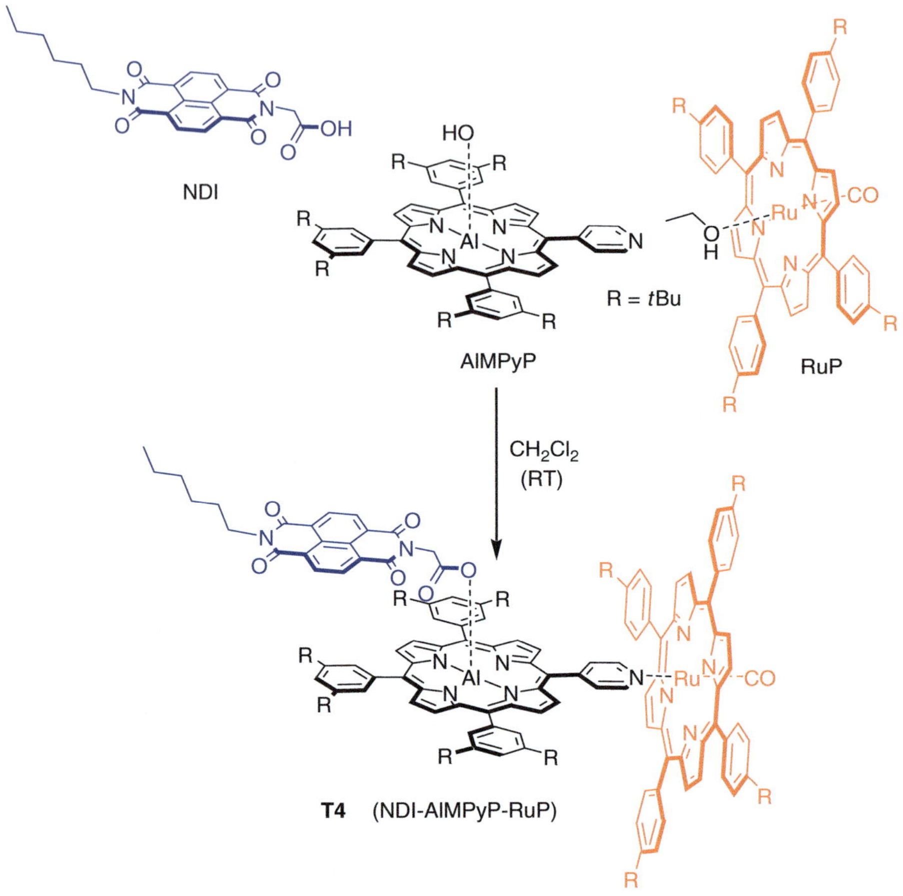

Fig. 1.16 Schematic representation of the self assembling process to obtain triad **T4**

of multiple energy transfer steps, the absorbed energy to the reaction centers (or photosystems) where charge separation takes place. Although in artificial photosynthetic systems the two functions, light absorption and charge separation, might not necessarily have to be as drastically distinguished as in Nature (see Sect. 1.2), substantial efforts have been devoted in recent years to the design and development of bioinspired artificial antenna systems.

Because of their similarity to the natural chlorophyll pigments, porphyrins have played the major role as chromophores for the construction of artificial antenna systems [48–52]. Other chromophores, such as, e.g., bodipy and perylene bisimides have been also frequently used, however. Out of the vast literature on this subject, [53] a few examples will be chosen so as to illustrate architecture strategies and functions.

A very interesting series of multichromophoric arrays has been developed and studied by Osuka and coworkers, based on zinc porphyrins directly linked in the

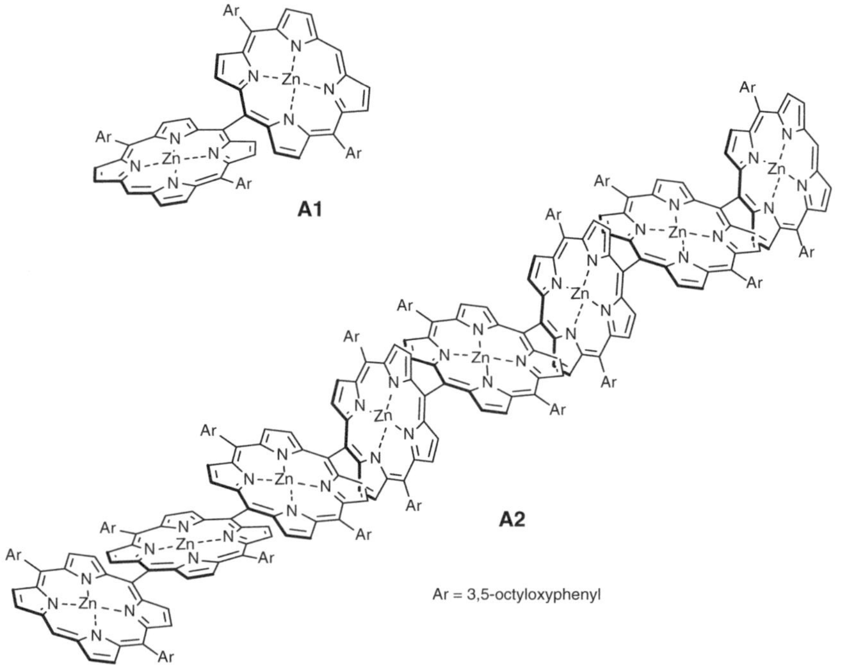

Fig. 1.17 Examples of the family of *meso-meso* linked linear porphyrin oligomers developed by Osuka and coworkers: dimer **A1** and octamer **A2**

meso positions via Ag-promoted coupling. Linear oligomers going from a simple dimer (**A1**) to tetramer, octamer (**A2**, Fig. 1.17), etc., up to a 128-mer of ca 106 nm (!) were obtained and studied [54]. In these systems, the orthogonality of the porphyrin rings imposed by steric hindrance prevents extensive π conjugation. However, the arrays exhibit dipole exciton coupling, with an exciton delocalization length (coherent length) extending over ca 4 adjacent porphyrin units. This favors fast singlet electronic energy transfer along the arrays (2.5–108 ps for arrays with $n = 2$–25 units), as measured by time-resolved fluorescence in a series of linear arrays of the type shown in Fig. 1.17 in which a terminal energy acceptor unit (5,15-bisphenylethynylated porphyrin) is added to the array [55]. Inspired by the cylindrical shape of the bacterial photosynthetic antennas (Sect. 1.2.1), a series of related porphyrin arrays with cyclic structures was synthesized by connecting *meso-meso* linked dimers or tetramers via 1.3-phenylene bridges (**A3** and **A4**, respectively, Fig. 1.18) [52]. In these arrays, the excited state is considered to be delocalized over the *meso-meso* linked dimeric or tetrameric linear porphyrin subunit, and a Forster-type model can be used to interpret the fast electronic energy transfer rates (3.6 ps for **A3**, 35 ps for **A4**) observed in these cyclic systems [56].

Aside from directly linked systems, other families of arrays can be obtained by linking the individual porphyrin chromophores by means of organic bridging groups, usually in the *meso* positions. With ethyne or butadiyne as bridging groups,

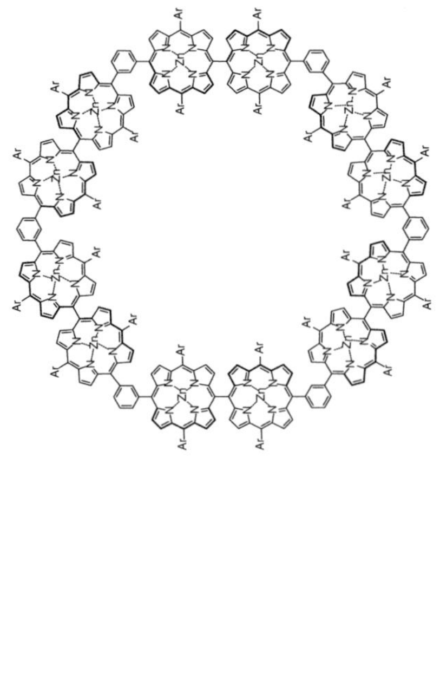

A3

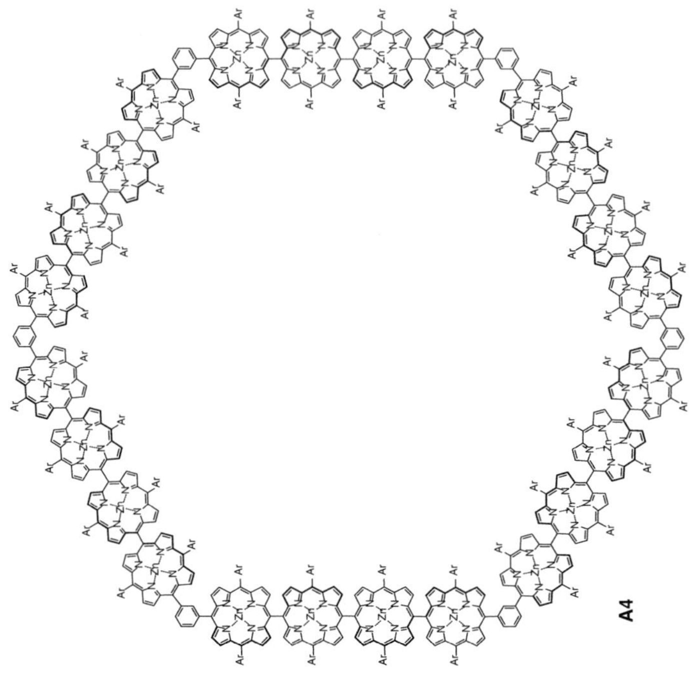

A4

Fig. 1.18 Examples of the family of *meso-meso* linked cyclic porphyrin oligomers developed by Osuka and coworkers: 12-mer **A3** and 24-mer **A4**

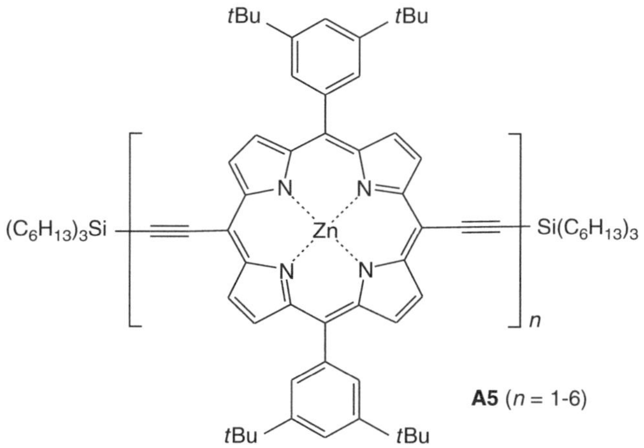

A5 (n = 1-6)

Fig. 1.19 General formula of the family of butadyine-bridged porphyrin oligomers **A5** (n = 1–6) developed by Anderson and coworkers

rotation of the porphyrin rings around the bridging linkers is essentially free, and the consequence is a high degree of π conjugation [57]. Thus, the spectroscopic behavior of the arrays is strongly different from that of the monomers. For instance, in the series of butadiyne-bridged oligomers **A5** developed by Anderson (Fig. 1.19) Q-bands shift from 650 nm in an isolated porphyrin to beyond 850 nm in the hexamer [58].

The strong conjugation makes this type of porphyrin arrays interesting not only as antenna systems but also as molecular wires. Also with such types of linkers, the possibility of having cyclic porphyrin arrays has attracted considerable attention. To that purpose, the relative flexibility of the butadyine bridge has been exploited by Anderson, [59] who developed with these oligomers a powerful template-assisted cyclization strategy. A templating molecule, possessing external pyridyl groups in appropriate number and geometry, is used to axially coordinate the zinc porphyrin subunits of linear oligomers, so as to bend them in a geometry suitable for cyclization. Addition of a Lewis base (e.g., pyridine) is then used to release the cyclic array from the template. Figure 1.20 shows schematically two types of such strategy used to afford a 12-membered cyclic oligomer, by cyclization of three tetrameric units. The two templating strategies differ by the use of a single 12-branched template ("classical") or two 6-branched ones ("Vernier") [60].

Aside from the cyclic antenna systems inspired by the light-harvesting units of bacterial photosynthesis, other covalent multiporphyrin arrays have been synthesized following a dendrimeric design, appropriate to obtain convergent energy funneling from the periphery to the center of the array. An example is antenna **A6** in Fig. 1.21 where a dendrimeric structure of 20 Zn-porphyrin units has a free-base porphyrin in the central position [61]. The diarylethyne linkers provide relatively modest interchromophoric electronic coupling, yet of sufficient magnitude to support fast excited-state energy transfer. The array behaves

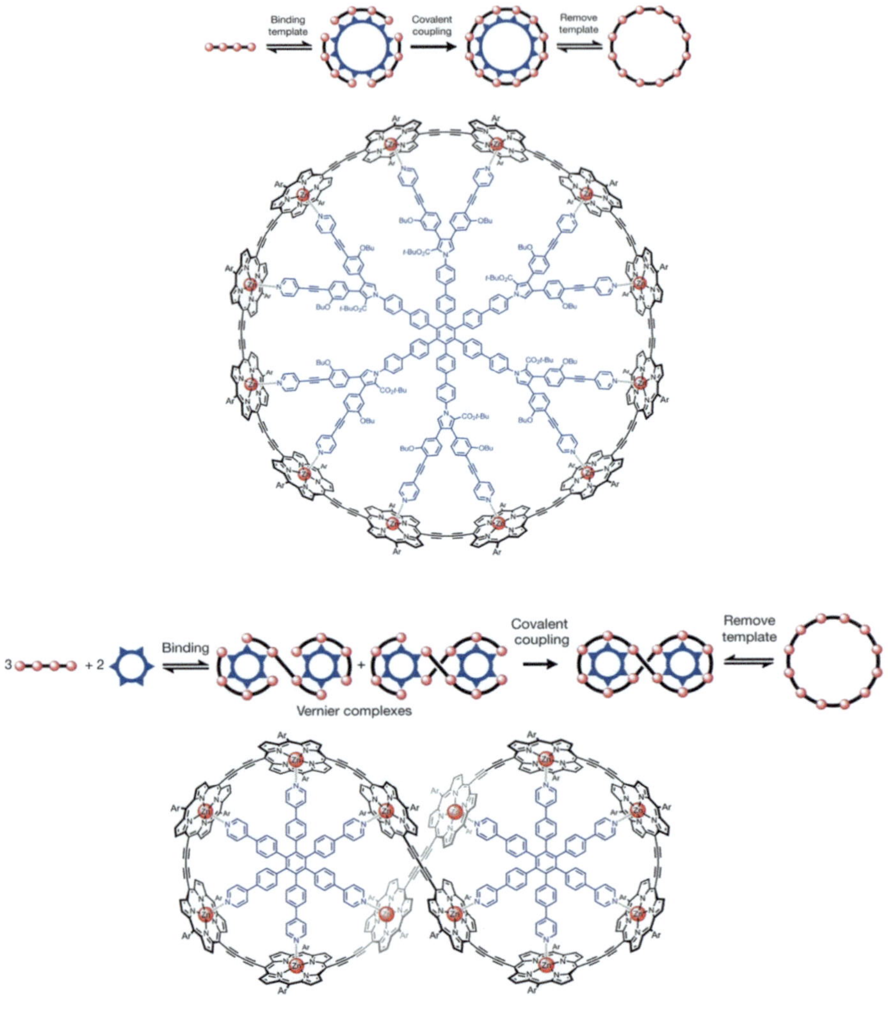

Fig. 1.20 Alternative strategies used by Anderson and coworkers to afford a 12-membered cyclic oligomer by template-assisted cyclization of three tetrameric units. The two strategies differ in the use of a single 12-branched template (*upper*) or two 6-branched ones (*lower*) (Adapted with permission from ref. [60], copyright © 2011 Nature Publishing Group)

indeed as an efficient artificial antenna, where the excitation energy absorbed by the ensemble of zinc porphyrins is transferred with high efficiency (92 %) and in a very short time scale (220 ps) to the central free-base unit.

As in the examples discussed above, the molecular units used to construct light harvesting antenna arrays have most often been of organic nature. Some work has also been performed, however, using transition metal coordination complexes as antenna chromophores. An example is the family of dendrimeric metal

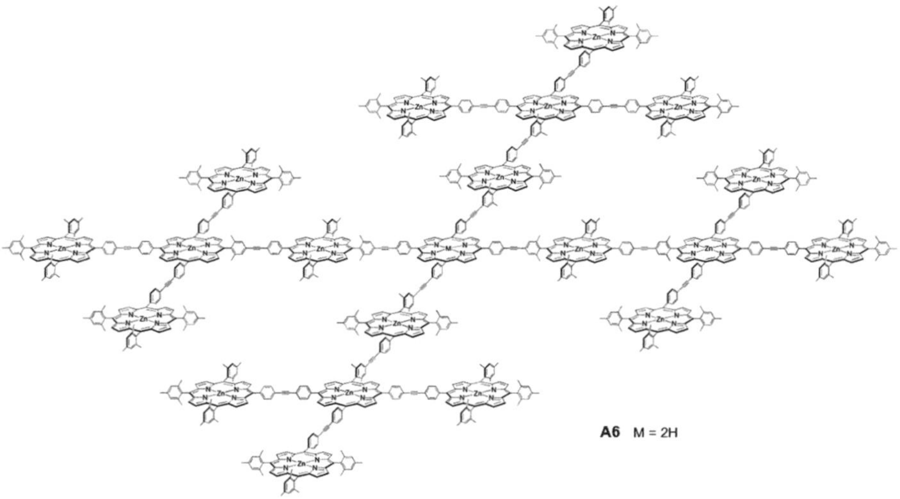

Fig. 1.21 An example of the dendrimeric antenna systems (**A6**) developed by Lindsey and coworkers, with 20 zinc porphyrins and a central free-base unit

polypyridine complexes based on the 2,3-dpp bridging ligand (where 2,3-dpp = 2,3-bis(2-pyridyl)pyrazine) [62]. In these dendrimers the relevant excited states are of metal-to-ligand charge transfer (MLCT) type involving the bridging ligands and the excitation energy is rapidly transferred ($\leq$200 fs) among adjacent chromophores to the units with excited states of lowest energy [63]. In an all-Ru system such as the decanuclear species **A7** of Fig. 1.22, the excited states of lowest energy are the $(bpy)_2Ru \rightarrow$ 2,3-dpp MLCT states of the peripheral units and the excitation energy is thus rapidly conveyed from the inner core to the periphery of the dendrimer. However, the direction of energy flow in this type of antenna systems can be conveniently controlled by synthesis, e.g., by introducing Os centers as low-energy sinks in lieu of Ru ones [62].

Covalent strategies towards multichromophoric arrays such as those outlined above, while providing structurally wonderful and kinetically stable architectures, are very expensive in synthetic work and usually modest in terms of chemical yields. Alternative strategies could be envisioned, however, taking examples from Nature, where the chromophores and cofactors of the photosynthetic machinery are held in place by weak intermolecular interactions. Some attempts to obtain multichromophoric antenna systems by soft, self-assembling strategies have been performed. The key to such strategies is to have the chromophores carrying some peripheral ligand groups and to exploit their coordination ability to metal centers [51, 64]. Two examples are shown in Figs. 1.23 and 1.24. The 12-membered cyclic array **A8** of Fig. 1.23 is representative of a general strategy mainly developed by Kobuke [51]. It is composed of six modules, each made of two Zn porphyrins carrying imidazolyl substituents connected by a *m*-phenylene bridge. The six bis (imidazolyl-Zn-porphyrin) modules self-assemble by inter-module axial

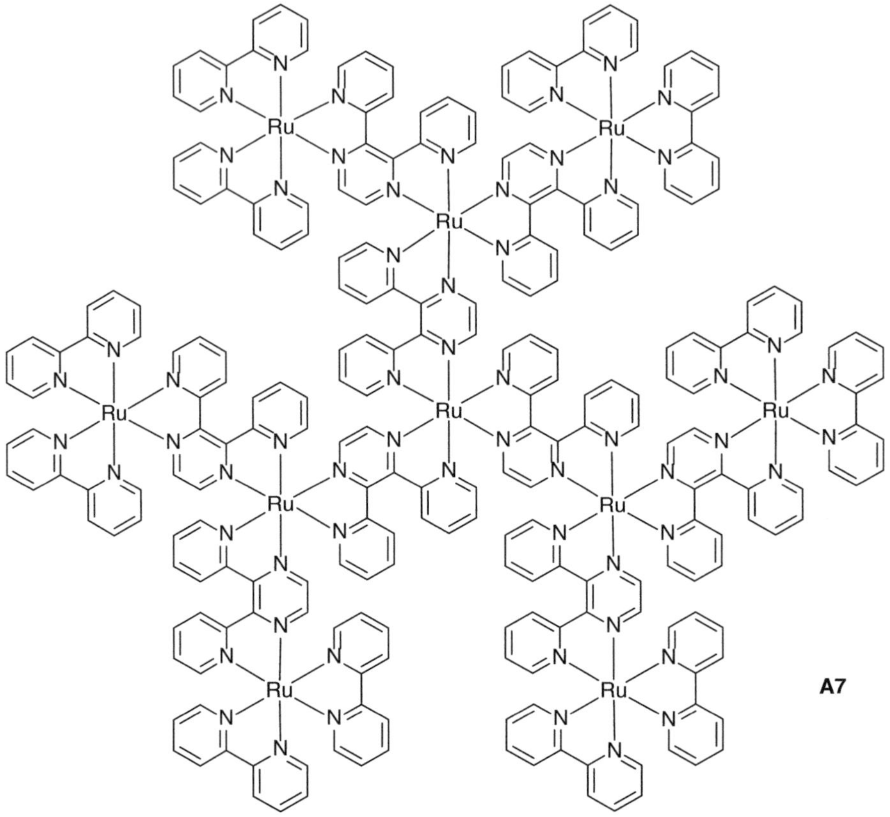

Fig. 1.22 An example of dendrimeric antenna system (**A7**) developed by Campagna, with 10 ruthenium polypyridyne units

coordination of the imidazolyl groups [65]. Excitation energy hopping through the coordination dimer units of this ring (as determined by anisotropy depolarization measurements) has a time constant of 5.3 ps, i.e., takes place ca 400 times during the 2.2 ns excited-state lifetime [66].

The six-porphyrin array **A9** of Fig. 1.24 comprises two metallacycles, each made of two *cis* dipyridyl zinc porphyrins held together by *trans*,*cis*-$RuCl_2(CO)_2$ corners, and two *trans* dipyridyl free-base porphyrins. The array self assembles in solution by side-to-face axial coordination of the pyridyl groups of the free-base "pillars" to the four zinc porphyrin units of the metallacycles [67]. In **A9**, the energy absorbed by the zinc porphyrin units is transferred to the free-base ones in 36 ps with an efficiency of ca 50 % (because of intercomponent competing electron transfer processes) [68]. Related structures can also be assembled following the same strategy, using perylene bisimide pillars instead of porphyrin-based ones [69].

Finally, a number of studies should be mentioned where some of the above-described multichromophoric light harvesting systems have been coupled with

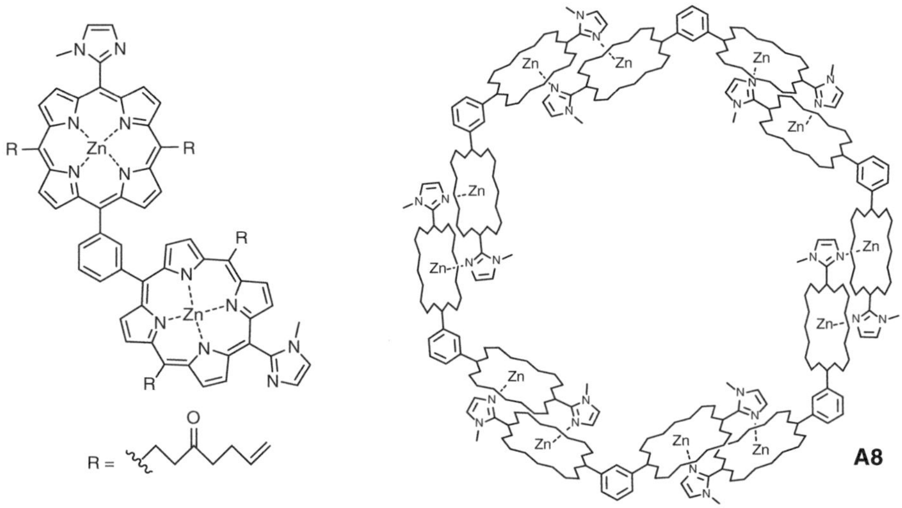

A8

Fig. 1.23 An example of cyclic porphyrin hexameric array **A8** obtained by the self-assembling strategy developed by Kobuke and coworkers, using bis(imidazolyl-Zn-porphyrin) units (*left*)

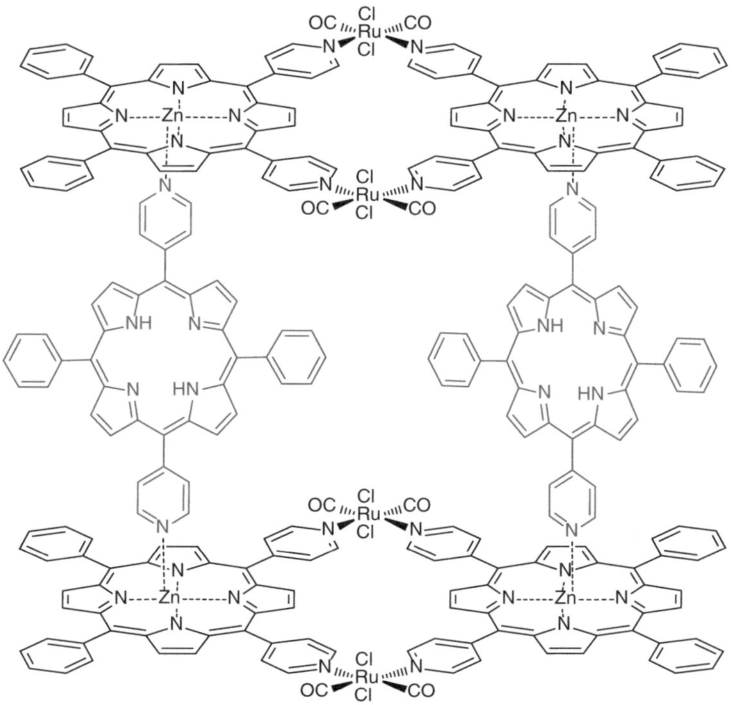

A9

Fig. 1.24 The six-porphyrin "molecular box" **A9** obtained by assembling two zinc-porphyrin metallacycles and two free-base pillars

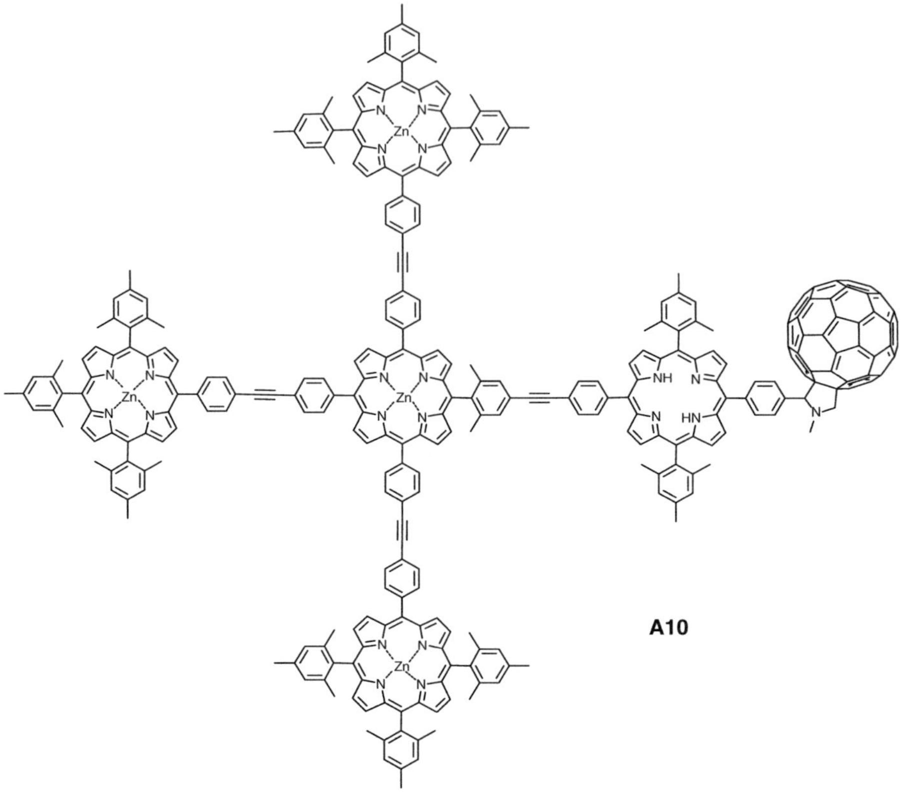

Fig. 1.25 The covalently linked six-membered antenna-reaction center mimicking system **A10**

some kind of charge separation devices, so as to obtain what could be considered antenna-reaction center mimics. A few representative examples are given below.

An early example developed by Lindsey, Moore and coworkers is **A10** shown in Fig. 1.25, featuring an antenna moiety of four Zn(II)tetrarylporphyrins attached to a free base porphyrin-fullerene reaction center [70]. In this construct, excitation of a peripheral zinc porphyrin moiety is followed by rapid (50 ps) energy transfer to the central zinc porphyrin and then (32 ps) to the free-base porphyrin. Then, a two-step charge separation occurs whereby the electron is transferred to the fullerene (25 ps) and the positive hole migrates to the zinc porphyrin array (380 ps). The final charge separated state is produced with a high quantum yield (90 %) and has a long lifetime (240 ns, data in MeTHF solvent).

A more recent and elaborated system is **A11** shown in Fig. 1.26, involving a dendrimer-type antenna system with six different chromophores (two bis (phenylethynyl)anthracene, two borondipyrromethene, and two Zn(II) tetraarylporphyrin dyes), so as to have convenient coverage across the visible spectrum. The charge separation function is added into the system by coordination of a dipyridyl functionalized fullerene to the two adjacent Zn(II)porphyrins of the

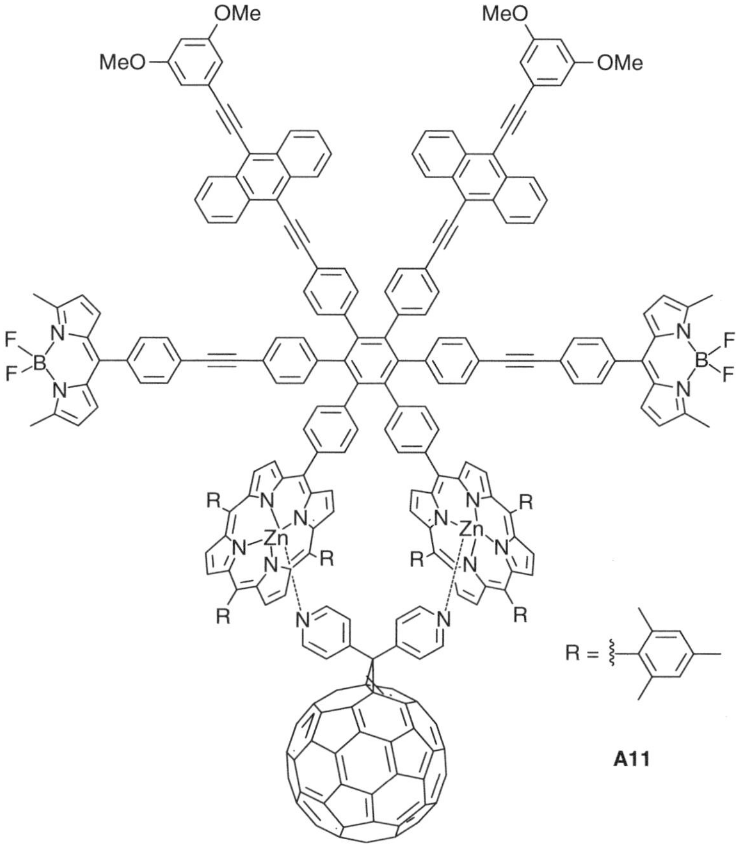

A11

Fig. 1.26 The multi-antenna reaction center complex **A11** designed by Moore, Moore, and Gust

antenna complex. We have here a hybrid (partly covalently and partly coordinatively linked) multiantenna-reaction center complex [71]. In this system, excitation of any of the antenna pigments is transferred in few ps to the zinc porphyrins and is followed by rapid (3 ps) light-induced electron transfer to the fullerene. The overall quantum yield for formation of the charge separated state (lifetime 230 ps) is close to unity.

Another example of hybrid antenna-reaction center system is **A12**, developed by Kobuke (Fig. 1.27) by coupling one of his cyclic porphyrin antenna systems (in this case an hexameric one) to a porphyrin-fullerene charge separating dyad equipped with a tripodal pyridine ligand of appropriate geometry [72]. In this composite system the excitation energy absorbed by the hexameric macrocycle is rapidly and efficiently transferred to the zinc porphyrin moiety of the tripodal ligand (ca 200 ps, 91 % efficiency), after that faster photoinduced electron transfer to the fullerene electron acceptor takes place (ca 120 ps, 94 % efficiency). The so formed charge separated state then recombines to the ground state in the ns time scale (ca 200 ns).

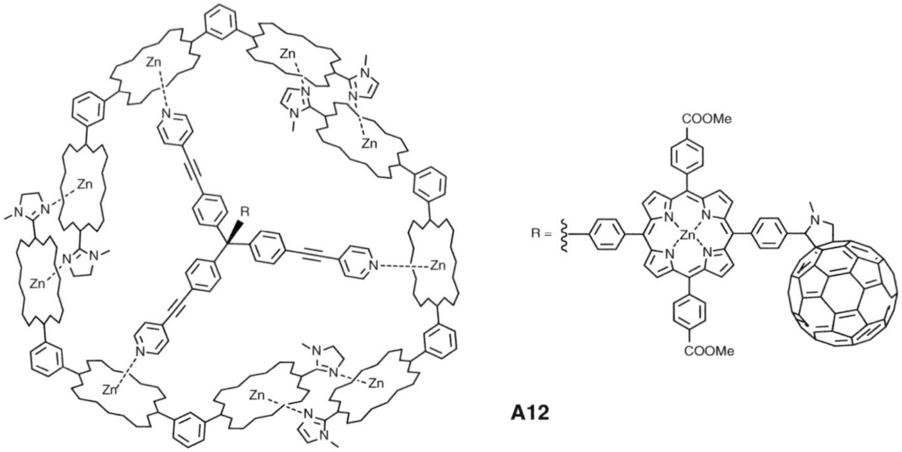

A12

Fig. 1.27 The antenna/reaction center complex **A12** designed by Kobuke and coworkers

A more complex example of coupled system is that shown in Fig. 1.28, [73] coupling an antenna with electron transfer and catalysis for photochemical hydrogen production. The system is composed of a fourth generation polyamidoamine (PAMAM) dendrimer decorated with 32 $Ru(bpy)_3^{2+}$ complexes at the periphery that act as the light-harvesting units and of small, monodisperse platinum nanoparticles (average diameter of 1.3 nm) grown inside the core of the dendrimer that act as the hydrogen evolving catalyst. Upon irradiation of the composite system in a pH 5 aqueous solution containing ascorbic acid as the sacrificial electron donor hydrogen is produced catalytically (see below Sect. 1.3.5 for additional details concerning photocatalytic hydrogen evolution). Light is efficiently absorbed by the dendrimeric $Ru(bpy)_3^{2+}$-based antenna system, whose chromophoric subunits undergo bimolecular reductive quenching by the sacrificial donor, the reduced form of the sensitizers then transfers the electrons to the platinum catalyst thus performing hydrogen evolution (Fig. 1.28).

1.3.4 Water Oxidation Catalysts

As previously mentioned in Sect. 1.3.1, water oxidation to dioxygen is a very complex reaction which is hampered by both thermodynamic and kinetic issues. Heterogeneous metal oxides have been proven to be capable of catalyzing this reaction at remarkably low overpotentials with quite good stability under oxidizing conditions [74–76]. However, their catalytic mechanism is often not very well understood; moreover they usually undergo slow electron transfer processes, in particular when interfaced with molecular sensitizers [76a].

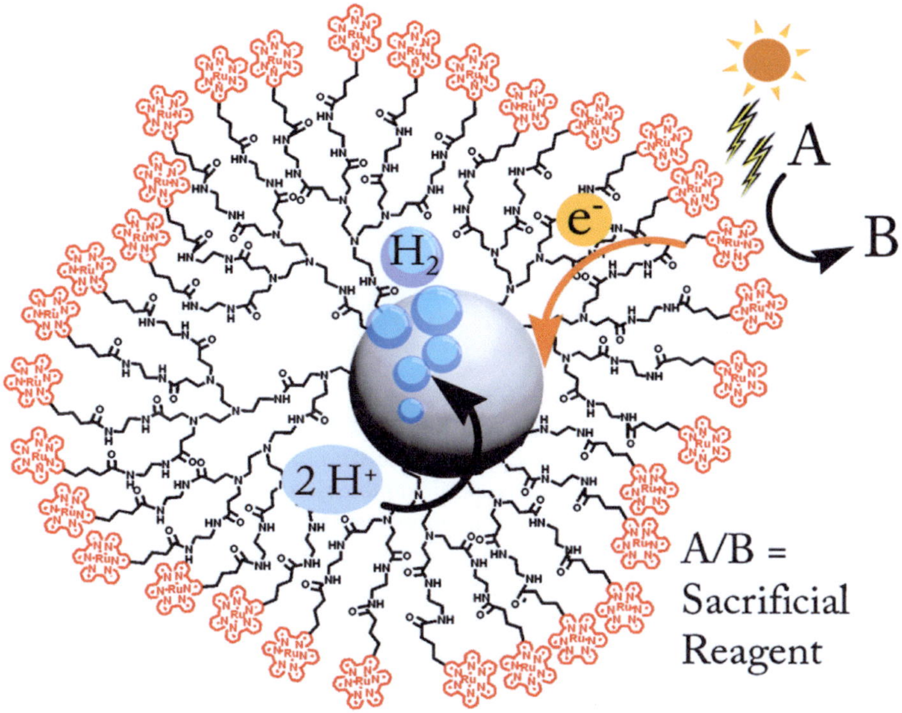

Fig. 1.28 An antenna/reaction center photocatalytic system for hydrogen production (Reprinted with permission from Ref. [73], copyright © 2014 American Chemical Society)

Molecular catalysts are, on the other hand, well defined entities whose activity can be modified and fine-tuned through molecular synthesis. Moreover, they can also be taken as discrete models of heterogeneous metal oxide phases, thus allowing for a better understanding of the catalytic mechanism of the latter [77]. As a general common feature a molecular catalyst should feature one or more metal centers stabilized by a suitable ligand framework. The main requirements of such a molecular catalyst comprise: (i) the easy accessibility of the metal center(s) to high oxidation states with sufficient oxidation power to enable water oxidation (i.e., the reduction potential of the activated catalyst must be more positive than the reduction potential of the O_2/H_2O redox couple), (ii) the possibility of the catalyst to stabilize high oxidation states via proton-coupled electron-transfer (PCET) processes thus avoiding highly charged intermediates (to this purpose the catalyst must display an aquo ligand or some anchoring site for binding a water molecule), [78] (iii) the robustness of the ligand pool towards oxidative stress.

Importantly, the water oxidation mechanism not only involves electron removal steps (each one accompanied by a proton loss) but requires a fundamental chemical step, namely the formation of the oxygen-oxygen bond. Once a metal-oxo species is formed by subsequent PCET steps from, e.g., a metal-aquo starting complex, the

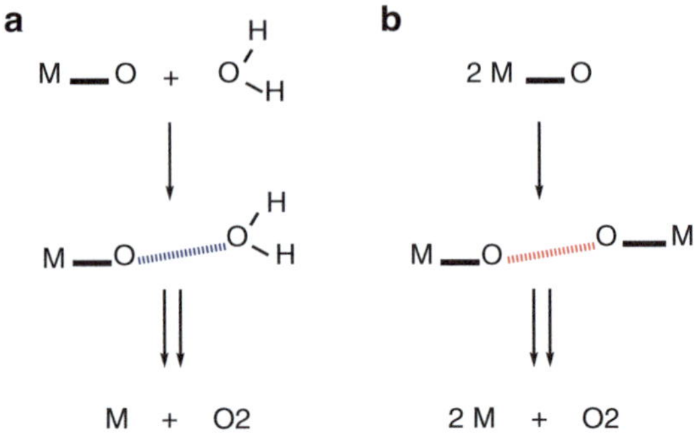

Fig. 1.29 Mechanisms for oxygen-oxygen bond formation: (**a**) water nucleophilic attack, (**b**) coupling between two M-O radicals

O-O bond formation process may occur by two different mechanisms (Fig. 1.29), depending on whether a water solvent molecule takes part to the process: (i) water nucleophilic attack, whereby a nucleophilic attack by a free water molecule to the highly electrophilic metal-oxo (M-O) intermediate is envisioned (Fig. 1.29a), after that the so formed metal-hydroperoxo species undergoes oxygen elimination upon subsequent PCET steps; (ii) coupling of two oxyl radicals, where two M-O species interact, either *intra-* or *inter*-molecularly, yielding a M-O-O-M complex which then eliminates the dioxygen molecule (Fig. 1.29b). The metal centers involved, the structure and the nuclearity of the complex as well as the electronic character of the ligand framework play a pivotal role in determining the actual water oxidation pathway [79].

Photochemical water oxidation takes advantage of light as the driving force for the conversion of the water substrate into oxygen. The minimum set of components required to foster water oxidation by light is composed of a photosensitizer (P), an electron acceptor (A), and the water oxidation catalyst (WOC). In a three-component system of this kind light is first absorbed by the sensitizer. Its excited state, with its high energy content, may in principle enable either reduction of the electron acceptor (oxidative quenching, Fig. 1.30a) or oxidation of the catalyst (reductive quenching, Fig. 1.30b). In the first case the oxidized sensitizer is responsible for oxidation of the catalyst, whereas in the second instance the reduced sensitizer undergoes electron transfer to the acceptor. Being reminiscent of the electron transfer cascade occurring in PSII, most of the artificial water oxidation schemes have been working through an oxidative quenching pathway, whereas the reductive mechanism have been less explored. Regardless of the photochemical mechanism, in both cases the sequence of photoinduced and thermal electron transfer processes has to be repeated four times in order to store four oxidizing equivalents into the catalytic platform thus leading to water oxidation and oxygen release. On the other hand, the electrons stored by the acceptor moiety should be ideally transferred to a suitable proton reduction catalyst and collected, thus powering hydrogen evolution. As pointed out above in several instances (Sects. 1.2 and 1.3.1), problems towards the achievement of this ideal behavior are represented by the possible occurrence of short-circuit processes, in this case between the

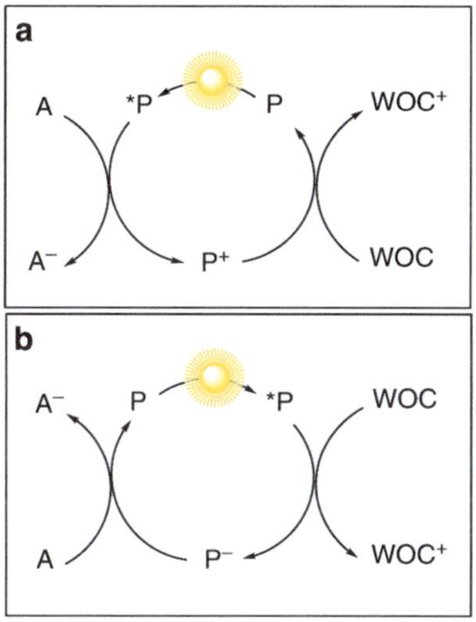

Fig. 1.30 Schematic representation of the two possible photochemical pathways leading to catalyst activation and eventually water oxidation: (**a**) oxidative quenching by the acceptor followed by hole transfer to the catalyst, (**b**) reductive quenching by the catalyst followed by electron shift to the acceptor

oxidized sensitizer or catalyst and the reduced acceptor. To get rid of such a problem, the most convenient way to study the water oxidation half-reaction and to optimize sensitizer/catalyst couples is that of exploiting a "sacrificial" electron acceptor, i.e., a molecule that undergoes decomposition upon reduction. Some examples of sacrificial agents commonly used in photochemical water oxidation schemes are the persulfate anion ($S_2O_8^{2-}$) and cobalt(III) amino complexes (e.g., $[Co(NH)_5Cl]^{2+}$).

A large amount of work is available on molecular water oxidation catalysts and their application. Many of these studies involve water oxidation catalysis under dark conditions, either electrochemical or taking advantage of chemical oxidants such as cerium salts. Herein we will concentrate only on molecular water oxidation catalysts which have been coupled to a photochemical system to power light-driven water oxidation to dioxygen. For a thorough and more general overview on molecular water oxidation catalysis the reader should refer to the following references [80–83].

Inspired by the structure of the natural oxygen evolving complex (OEC) in PSII (Sect. 1.2.2), whose outstanding activity resides in the interplay among four redox active manganese centers, several approaches towards artificial water oxidation have been devoted to the preparation and characterization of tetrametallic molecular complexes capable of performing this reaction. The main advantages arising from the exploitation of a multimetallic platform is the possibility of distributing the oxidation equivalents required for water oxidation over several metal centers, thus lowering the energetic barrier of the overall catalytic process.

Almost simultaneously Hill [84] and Bonchio [85] discovered the water oxidation ability of a tetraruthenium polyoxometalate complex $\left[Ru_4(\mu-O)_4(\mu-OH)_2(H_2O)_4(\gamma-SiW_{10}O_{36})_2\right]^{10-}$ (**WOC1**, Fig. 1.31). It features

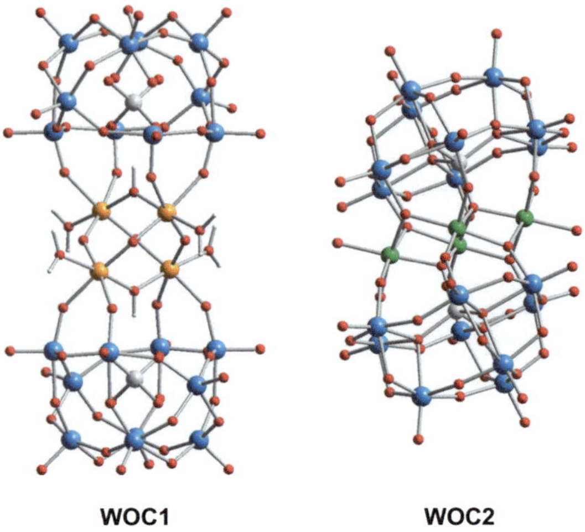

Fig. 1.31 Structural representations of the tetrametallic catalysts **WOC1-2**. Color code: ruthenium, *orange*; cobalt, *green*; tungsten, *blue*; oxygen, *red*; silicon (*left*) or phosphorus (*right*), *gray*

a $\left[Ru_4(\mu - O)_4(\mu - OH)_2(H_2O)_4\right]^{6+}$ adamantane-like core embedded between two divacant $[\gamma\text{-}SiW_{10}O_{36}]^{8-}$ tungstosilicate ligands. The presence of water molecules coordinated to the tetraruthenate core is of fundamental importance to facilitate the PCET steps required for water oxidation; moreover the all-inorganic nature of the polyoxometalate stabilizing ligands gives enhanced robustness towards the harsh conditions required to oxidize water. As a consequence, **WOC1** is extremely active as a water oxidation catalyst both under electrocatalytic conditions, either as a homogeneous species [84] or supported onto carbon nanomaterials such as nanotubes or graphene, [86, 87] and using suitable chemical oxidants such as Ce(IV) [85, 88] or $Ru(bpy)_3^{3+}$ [84, 89]. More importantly, complex **WOC1** promotes oxygen evolution at neutral pH under light-driven conditions in the presence of persulfate as the sacrificial electron acceptor and either $Ru(bpy)_3^{2+}$ or a tetranuclear dendrimer $\left[Ru\{(\mu\text{-}2,3\text{-}dpp)Ru(bpy)_2\}_3\right]^{8+}$ (where μ-2,3-dpp = 2,3-bis(2-pyridyl)pyrazine) as photosensitizers, achieving quantum yields up to 0.045 and 0.30, respectively [90, 91]. Interestingly, two different photochemical mechanisms were observed depending on the chromophore used. In particular, in the case of $Ru(bpy)_3^{2+}$ as the sensitizer, activation of the catalyst takes place through an oxidative quenching mechanism (Fig. 1.30a), i.e., the first step is the photoinduced oxidation of $Ru(bpy)_3^{2+}$ by persulfate which is followed by hole transfer from $Ru(bpy)_3^{3+}$ to the catalyst, with this latter step occurring at diffusion-controlled rates [92]. On the other hand, in the case of the tetranuclear dendrimer a reverse mechanism is observed, which involves reductive quenching of the sensitizer by the catalyst as the primary step followed by electron shift to the sacrificial acceptor (Fig. 1.30b) [93]. Importantly, in both cases complementary charge interactions between the positively charged sensitizers, either $Ru(bpy)_3^{2+}$ or $\left[Ru\{(\mu\text{-}2,3\text{-}dpp)Ru(bpy)_2\}_3\right]^{8+}$, and the negatively charged catalyst

WOC1 were observed, leading to strong quenching of the sensitizer emission. In the case of $Ru(bpy)_3^{2+}$ the ion-pairing quenching had to be avoided for an efficient photocatalytic activity (e.g., working under high ionic strength conditions or by lowering catalyst concentration), being oxidative in nature and thus detrimental for water oxidation, [92] whereas in the case of the tetranuclear dendrimer the strong ion-pairing was of fundamental importance to drive the ultrafast reductive electron transfer quenching and the subsequent electron shift to the persulfate anion [93].

Hill and coworkers also described the water oxidation activity of a tetracobalt polyoxometalate complex $[Co_4(H_2O)_2(\alpha\text{-}PW_9O_{34})_2]^{10-}$ (**WOC2**, Fig. 1.31), [94] featuring a belt of four cobalt(II) ions sandwiched between two $[\alpha\text{-}PW_9O_{34}]^{9-}$ Keggin-type ligands and representing a noble-metal-free alternative with respect to the tetraruthenium catalyst. This complex was reported to efficiently function as a water oxidation catalyst even in light-driven conditions using $Ru(bpy)_3^{2+}$ and persulfate as the photosensitizer and sacrificial acceptor, respectively, with enhanced performances with respect to the parental tetraruthenium analog **WOC1** [95]. Though several lines were reported to confirm that the catalytic process was homogenous in nature and the catalyst did not behave as a precursor of heterogeous cobalt oxide species, [96] some uncertainties still remained as far as the actual behavior of **WOC2** in the photocatalytic water oxidation scheme is concerned [97, 98].

Although "four" is the number of choice by nature, the stabilization of highly positive redox states can be also afforded by low-nuclearity clusters that combine good catalytic activity with less synthetic efforts. A particular interest towards this direction arose from the first evidences on the water oxidation activity by the so-called "blue dimer", a binuclear complex with formula $[(bpy)_2(H_2O)Ru\text{-}O\text{-}Ru(H_2O)(bpy)_2]^{4+}$ (where bpy = 2,2′-bipyridine) [99] studied by T. J. Meyer and coworkers in the early 1980s.

From these seminal works, subsequent studies by T. J. Meyer and coworkers dealt with single-site molecular catalysts since the discovery that Ru(tpy)(bpm)(OH_2) (**WOC3**, Fig. 1.32, where tpy = 2,2′:6′,2″-terpyridine and bpm = 2,2-′-bipyrimidine), Ru(tpy)(bpz)(OH_2) (**WOC4**, where tpy = 2,2′:6′,2″-terpyridine and bpz = 2,2′-bipyrazine), and Ru(Mebimpy)(bpy)(OH_2) (**WOC5**, where Mebimpy = 2,6-bis(1-methylbenzimidazol-2-yl)pyridine) were capable of

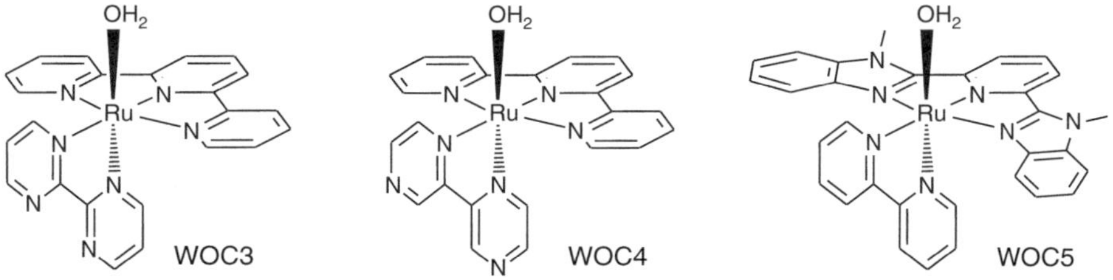

Fig. 1.32 Molecular formulae of single-site ruthenium catalysts **WOC3-5** studied by T. J. Meyer and coworkers

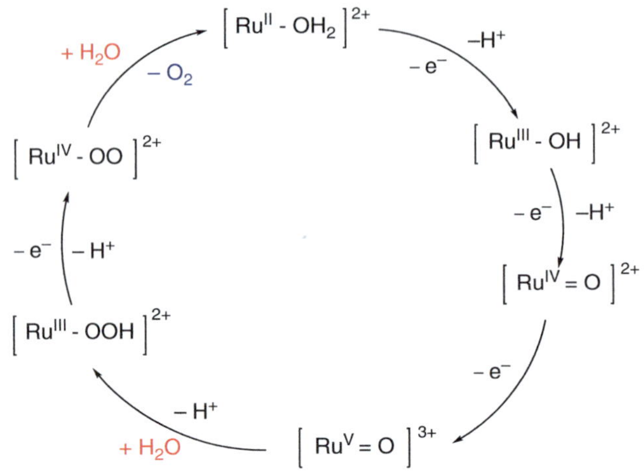

Fig. 1.33 Water oxidation catalytic mechanism by a single-site ruthenium catalyst as depicted by T. J. Meyer and coworkers

catalyzing water oxidation from acidic solutions in the presence of Ce(IV) chemical oxidant [100, 101].

Careful experimental studies were in particular undertaken to disclose the catalytic mechanism which now represents a guideline for the understanding of the reactivity towards water oxidation of single-site molecular catalysts either made of ruthenium or not (Fig. 1.33). Three PCET steps from the starting Ru^{II}–OH_2 species with formation of a Ru^{V}=O intermediate are followed by the water nucleophilic attack with formation of a Ru^{III}–OOH "hydroperoxo" intermediate which, after a subsequent proton-coupled oxidation to a Ru^{IV}–OO "peroxo" species, undergoes oxygen release.

A more recent breakthrough in the field of single-site ruthenium catalysts was the discovery of **WOC6** by Sun and coworkers (Fig. 1.34), highly active under cerium-driven water oxidation [102]. Second-order kinetics with respect to catalyst concentration were observed, suggesting that a bimolecular radical coupling mechanism (Fig. 1.29b) was operating with **WOC6** as the catalyst. The most interesting feature is the presence of the 2,2′-bipyridine-6,6′-dicarboxylate equatorial ligand which has a double function in fostering efficient water oxidation: (i) the strong electron donating character of the carboxylate groups sensitively shifts the oxidation potential of the ruthenium center to less positive values thus lowering the energetic requirement for the achievement of the Ru(V)-oxo catalytic intermediate, (ii) the wide O-Ru-O bite angle (122.99°) [102] is such that the ruthenium center may accommodate a seventh ligand (e.g., a substrate water molecule) without altering the remaining coordination sphere. The latter point was indeed demonstrated with the X-ray structure of an isolated crystal obtained upon addition of two equivalents of Ce(IV) to an acidic solution of **WOC6**, where a dimeric complex featuring seven-coordinated ruthenium(IV) centers bridged by a [HOHOH] group was observed [102].

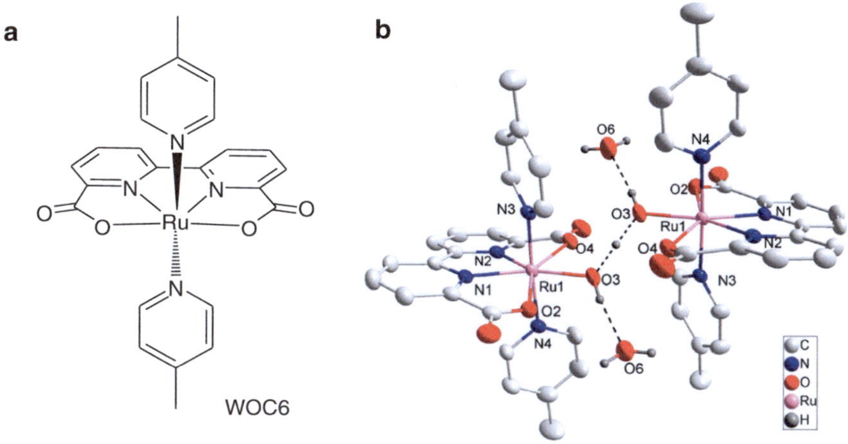

Fig. 1.34 (**a**) Molecular formula of the single-site ruthenium catalyst **WOC6** studied by Sun and coworkers; (**b**) X-ray structure of the seven-coordinated Ru(IV) dimer (Reprinted with permission from ref. [102], copyright © 2009 American Chemical Society)

WOC6 was used as the catalyst for light-driven water oxidation in sacrificial systems involving $Ru(bpy)_3^{2+}$ as the light-harvesting sensitizer and either $Co^{III}(NH_3)Cl^{2+}$ or $S_2O_8^{2-}$ as the sacrificial electron acceptor [103]. Partial depletion of both the sensitizing and catalytic moieties were detected as the main sources of photocatalysis deactivation and the reason was mainly attributed to the ability of high-valent Ru(V)-oxo species to promote parasite self-oxidation reactions. Interestingly, **WOC6** was even active when $Ru(dmb)_3^{2+}$ (where dmb = 4,4′-dimethyl-2,2′-bipyridine) was used as the sensitizer, whose oxidized species $Ru(dmb)_3^{3+}$ (obtained upon irradiation of $Ru(dmb)_3^{2+}$ in the presence of the sacrificial acceptor) is a milder oxidant than $Ru(bpy)_3^{3+}$, thus emphasizing the low operating potential of this class of catalysts.

Supramolecular triad systems (**WOC7-8**, Fig. 1.35) coupling either a $Ru(bpy)_3^{2+}$- or $Ru(tpy)_2^{2+}$-type chromophores to a **WOC6**-type catalyst were also studied as photochemical devices for light-driven water oxidation in the presence of persulfate [104]. In these systems the photoreaction was thought to involve *inter*molecular reaction of the ruthenium sensitizer with the sacrificial acceptor followed by *intra*-molecular hole transfer to the ruthenium catalyst. According to this scenario, the $Ru(bpy)_3^{2+}$-containing triad **WOC7** was five-times more active in terms of maximum turnover number (TON) than the bimolecular $Ru(bpy)_3^{2+}$/**WOC6** two-component system, most likely benefiting from the sensitizer-catalyst linkage. On the other hand, the $Ru(tpy)_3^{2+}$-containing triad **WOC8** was inactive toward water oxidation since the lifetime of the $Ru(tpy)_3^{2+}$ complex was too short to enable efficient oxidative quenching with the persulfate sacrificial acceptor.

Photochemical studies have also dealt with water oxidation catalysts containing first-row transition metals, as more viable alternatives with respect to noble metal containing catalysts in particular because of their larger abundance in nature. To

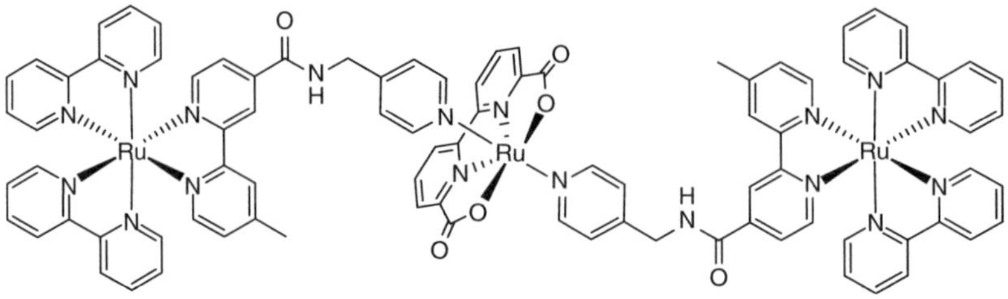

WOC7

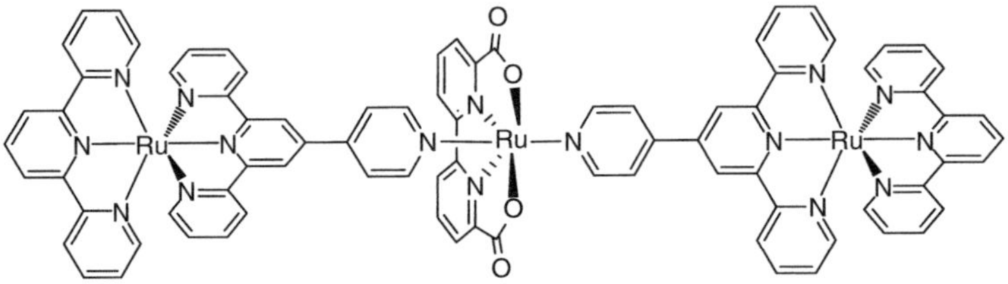

WOC8

Fig. 1.35 Molecular formulae of triads **WOC7-8** studied by Sun and coworkers

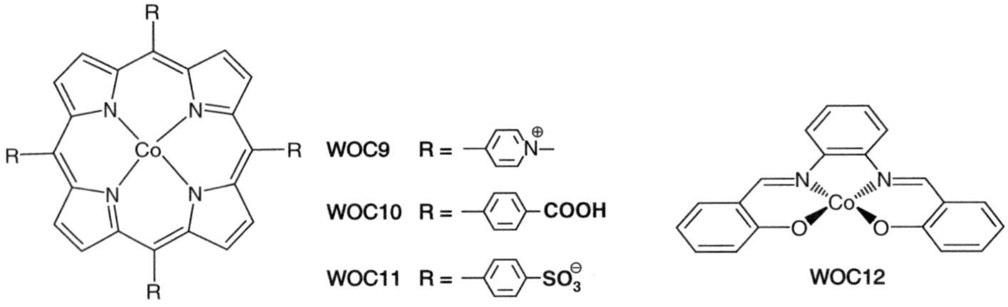

Fig. 1.36 Molecular formulae of the single-site cobalt catalysts **WOC9-11** studied by Sakai and coworkers, and **WOC12** studied by Sartorel and coworkers

cite some examples, three differently functionalized cobalt porphyrins (**WOC9-11**, Fig. 1.36) were used by Sakai and coworkers to promote water oxidation under light-activated conditions from aqueous solutions taking advantage of $Ru(bpy)_3^{2+}$ as the photosensitizer and $S_2O_8^{2-}$ as the sacrificial electron acceptor [105]. These systems showed enhanced activity at pH 11 with maximum TONs depending on the substituent in the phenyl ring in the order **WOC11** > **WOC10** > **WOC9**. Second-order kinetic dependences on catalyst concentration were also observed, suggesting that the operating water oxidation mechanism was involving coupling of two cobalt-oxyl radicals. A cobalt(II) salophen complex (**WOC12**, Fig. 1.36) was also employed by Sartorel and coworkers for photocatalytic water oxidation using the same set of reagents, namely $Ru(bpy)_3^{2+}$ and $S_2O_8^{2-}$, at neutral pH, achieving quantum yields up to 7.9 % [106].

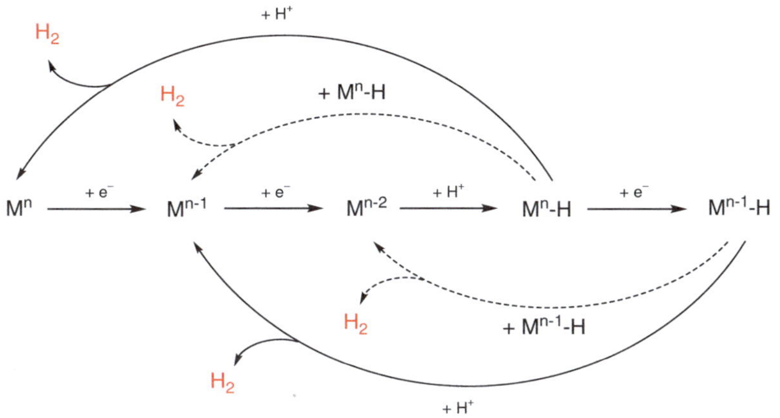

Fig. 1.37 Hydrogen evolution mechanism by a metal complex: heterolythic pathway (*solid lines*), homolythic pathway (*dashed lines*)

1.3.5 Hydrogen Evolution Catalysts

As previously outlined in Sect. 1.3.1, proton reduction to molecular hydrogen is the target reaction for the generation of an inexhaustible and sustainable fuel supply. Hydrogen formation being a two-electron two-proton process, transition metal complexes can be suitable candidates for catalyzing the hydrogen evolving reaction (HER) since redox active metal centers can in principle accommodate and store electrons thus lowering the energetic barrier required to hydrogen evolution.

As to the mechanism, [107] a general schematic representation is given in Fig. 1.37 and involves several key processes: (i) two reduction steps from a M^n to a M^{n-2} state are first required, (ii) at this stage the M^{n-2} species can be protonated yielding a M^n-H hydrido intermediate, (iii) the M^n-H species may then release hydrogen after protonation (heterolythic pathway) or by bimolecular reaction with another hydrido complex (homolythic pathway). In some instances the M^n-H could not be nucleophilic enough to be protonated and thus an additional reduction step to a M^{n-1}-H species is required, after that hydrogen may be released similarly via either a heterolytic (protonation) or a homolythic (bimolecular reaction) pathway. According to this general picture the ideal HER catalyst should display (i) a metal center (or metal centers) with several accessible reduction states, (ii) a free coordination site to favor proton binding thus enabling formation of the metal-hydrido intermediate, and preferentially (iii) some proton anchoring sites in the second coordination sphere to facilitate the formation of both the metal-hydrido species and the H-H bond.

For photochemical hydrogen generation three molecular components are required, which include a photosensitizer (P), a hydrogen evolving catalyst (HEC), and an electron donor (D). Light is first absorbed by the sensitizer, which, at the excited state level may react either with the electron donor (reductive quenching, Fig. 1.38a) or with the catalyst (oxidative quenching, Fig. 1.38b).

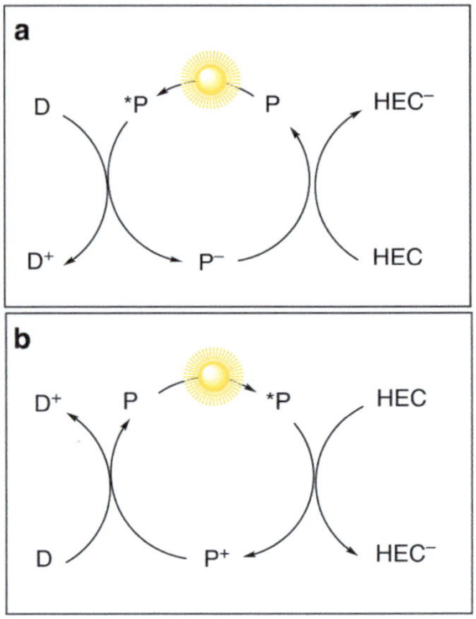

Fig. 1.38 Schematic representation of the two possible photochemical pathways leading to catalyst activation and eventually hydrogen evolution: (**a**) reductive quenching by the donor followed by electron transfer to the catalyst, (**b**) oxidative quenching by the catalyst followed by hole shift to the acceptor

In the first case the reduced sensitizer is used for the reduction of the catalyst whereas in the second case the oxidized sensitizer undergoes hole shift to the donor. Regardless of the photochemical mechanism, with such a sequence of photochemical and thermal processes two electrons are stepwise accumulated by the catalyst eventually leading to proton reduction and hydrogen evolution according to the mechanism reported above (Fig. 1.37). In a symmetric fashion with respect to photochemical water oxidation, the holes stored by the donor can be transferred to a suitable oxygen evolving catalyst thus performing water oxidation.

To detailedly study the hydrogen evolving half-reaction without kinetic complications from the oxidative side (i.e., water oxidation), one possibility is that of using a "sacrificial" electron donor, namely a molecule that undergoes decomposition upon oxidation. Examples of sacrificial reductants, to cite some of them, include aliphatic amines (e.g., triethylamine or triethanolamine), EDTA, thiols, and ascorbic acid. Huge amount of reports can be found in the literature concerning (photo)catalytic hydrogen evolution involving either noble-metal containing (e.g., palladium, platinum, rhodium) [108–110] or first-row transition (e.g., cobalt, iron, nickel) [111–113] molecular catalysts. Herein we will focus only on some representative molecular catalytic systems which have been employed together with photochemically-active components to power light-driven proton reduction to dihydrogen. For complete and detailed accounts on molecular catalysts for hydrogen evolution and their applications the reader should preferably refer to the following reports [114–119].

In Nature, certain bacteria exploit the metabolism of hydrogen, i.e., the reversible interconversion of electrons and protons into dihydrogen, as a way to either obtain chemical energy (dihydrogen oxidation) or to generate a high-energy containing chemical (i.e., hydrogen) as a substrate for fermentation processes

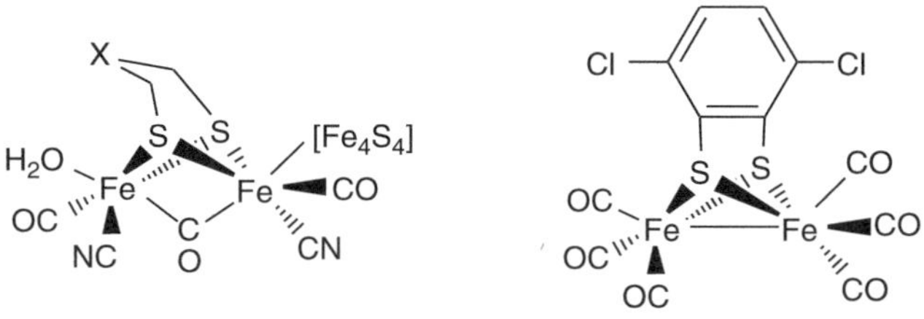

[FeFe]-H_2ase *Active site*
X= CH_2, NH, or O

HEC1

Fig. 1.39 Molecular formulae of the active site in bacterial [FeFe] hydrogenases and of the artificial analog **HEC1** studied by Ott and coworkers. In the natural system oxidation states of the iron atoms are Fe(I)-Fe(II), whereas in the artificial mimic **HEC1** are Fe(I)-Fe(I)

(proton reduction). Both reactions are catalyzed by metalloenzymes named hydrogenases, such as the [FeFe], [NiFe], and [Fe]-only H_2ase enzymes. Inspired by the natural enzymes, huge amounts of efforts have been devoted to the synthesis and preparation of artificial analogs of the active site of the natural hydrogenases with particular attention towards the [FeFe] type, the most active enzyme within the hydrogenase classes. The active site of the latter (Fig. 1.39) is composed of a binuclear iron-iron core held by a dithiolate bridge, coordinated to carbonyl and cyanide ligands, and connected to a $\{Fe_4S_4\}$-cluster (a one-electron redox mediator) through a cysteinyl bridge. Detachment of the water molecule is the key process activating the catalytic unit, leaving a free coordination site for the proton substrate.

Several synthetic artificial analogs of the [FeFe] hydrogenase active site were synthesized, usually with some chemical modifications with respect to the natural cluster (e.g., absence of carbonyl bridging ligand, different dithiolate bridge, introduction of non-biological ligands such as phosphines or carbenes, etc.). The results, however, were usually not so satisfactory as expected mainly due to the highly negative operating potentials to enable hydrogen evolution catalysis, the poor stability under (photo)catalytic conditions, and the inability to perform the reverse hydrogen oxidation reaction [113].

Nevertheless, the synthetic H_2ase-type complex **HEC1** was used by Ott and coworkers as proton reduction catalyst in light-driven photocatalytic experiments in the presence of $Ru(bpy)_3^{2+}$ as the photosensitizer and ascorbic acid as the sacrificial electron donor [120]. Importantly, the key finding in the design of **HEC1** was the introduction of electron-withdrawing chloro substituents in the phenyl ring which shifted to less negative values the potential required for proton reduction catalysis by the diiron core. A remarkable activity with maximum TONs up to 200 (after 150 min) and a quantum yield of 1 % was indeed observed upon irradiation of a 50/50 H_2O/DMF (pH 5) solution containing the donor/sensitizer/catalyst three-component system. Catalyst inactivation, however, was recognized as the limiting factor affecting the photocatalytic activity. The photochemical mechanism was of reductive type (Fig. 1.38a), with excitation triggering bimolecular photoreduction of $Ru(bpy)_3^{2+}$ by ascorbic acid, then followed by electron transfer from the Ru

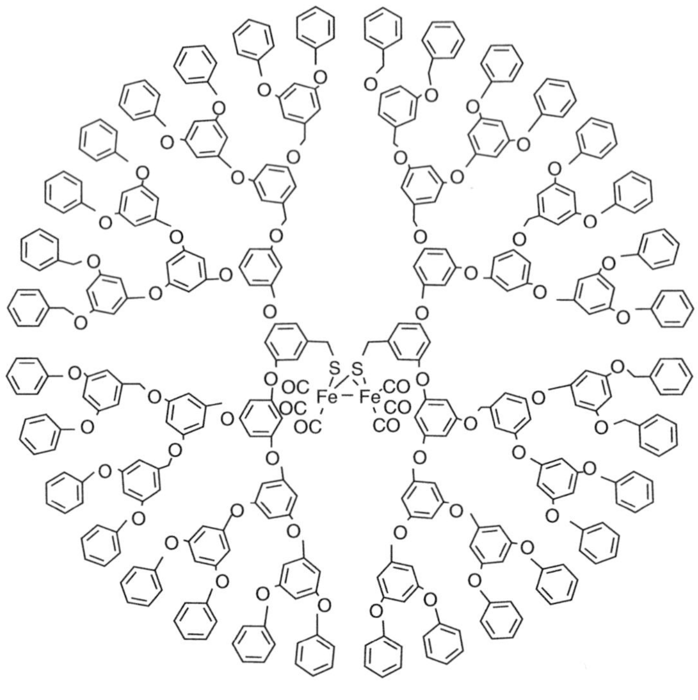

HEC2

Fig. 1.40 Molecular formula of the dendrimeric hydrogenase mimic **HEC2** studied by Li and coworkers. Oxidation states for iron atoms are Fe(I)-Fe(I)

$(bpy)_3^+$ to **HEC1**. Along the same years, tentative tests were done to prepare sensitizer/catalyst supramolecular dyads by suitably assembling artificial [FeFe] hydrogenase mimicking complexes to sensitizers such as ruthenium tris(bipyridine) [121] or porphyrins, [122] though with scarce results in terms of overall amounts of hydrogen produced upon irradiation in the presence of sacrificial donors.

More interestingly, a new approach was adopted by Li and coworkers for the preparation of more stable [FeFe] hydrogenase mimics, based on the insertion of a $\{(\mu\text{-S})_2Fe_2(CO)_6\}$ subunit inside the core of a dendrimer (Fig. 1.40) [123]. In fact, the active sites of natural hydrogenases are surrounded by the protein matrix that modulates the reactivity and protects the catalytic center so as to favor catalysis. The presence of the surrounding dendrimeric structure may thus help to better preserve the catalytic core within the artificial hydrogenase system. As a matter of fact, this approach translated into a very active and stable system for hydrogen evolution. The fourth generation dendrimeric hydrogenase mimic **HEC2** (Fig. 1.40) produced hydrogen very efficiently in a photocatalytic system involving a cyclometalated dye, namely $[Ir(ppy)_2(bpy)]^+$ (where ppy = 2-phenylpyridine), as the light-harvesting sensitizer and triethylamine as the sacrificial donor, showing

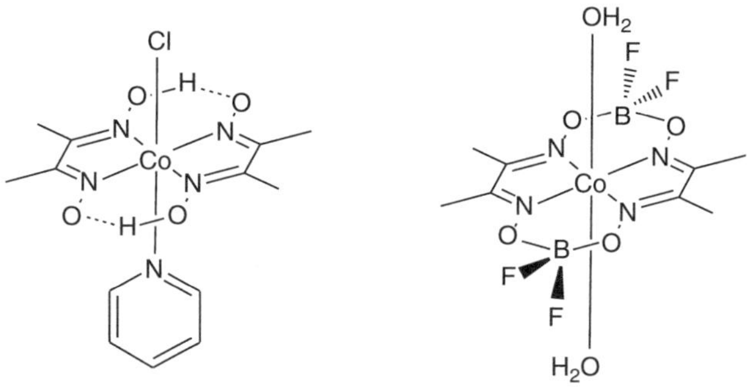

Fig. 1.41 Molecular formulae of cobaloxime type catalysts **HEC3-4**

TONs up to 22,200 after 8 h irradiation, which is the highest value ever reported with hydrogenase type catalysts and one of the highest ever observed in hydrogen evolution photocatalysis. As to the mechanism, excitation of the $[Ir(ppy)_2(bpy)]^+$ complex is followed by reductive quenching by TEA and then the reduced sensitizer undergoes electron transfer to the dendrimeric catalyst (Fig. 1.38a).

An advance in the field of molecular hydrogen evolution catalysis was the discovery by Espenson and coworkers of the ability of hydrido-cobaloxime complexes to promote hydrogen elimination [124]. Subsequently, Artero and coworkers demonstrated that cobaloxime-type complexes (e.g., **HEC3-4** in Fig. 1.41), successfully catalyze electrocatalytic hydrogen evolution from organic solvents in the presence of suitable proton sources [125]. In these complexes reduction of the cobalt center is followed by de-coordination of an axial ligand (either chloride or water), leaving a free coordination site. Upon subsequent electron transfer to the catalyst a proton can bind thus triggering hydrogen elimination either via a homolythic or a heterolythic route [107]. These results stimulated large amounts of efforts in developing sacrificial photochemical systems taking advantage of cobaloximes as the hydrogen evolving catalysts, although the catalytic activity was limited to organic solvents or organic/water mixtures. These examples included completely bimolecular systems using sensitizers such as ruthenium(II) polypyridine, [126] rhenium (I) tricarbonyl, [127] platinum(II) terpyridyl acetilide, [128] copper (I) phenantroline [129] complexes, or triplet organic chromophores like xanthene dyes [130] or porphyrins [131] in the presence of suitable sacrificial donors, usually aliphatic amines.

In most of these systems the hydrogen evolving mechanism was thought to involve reductive quenching of the excited dye by the sacrificial donor as the primary photochemical step followed by electron transfer from the reduced sensitizer to the cobaloxime catalyst, eventually leading to hydrogen evolution. Importantly, in all these examples the use of triplet sensitizers was mandatory in order to favor bimolecular processes.

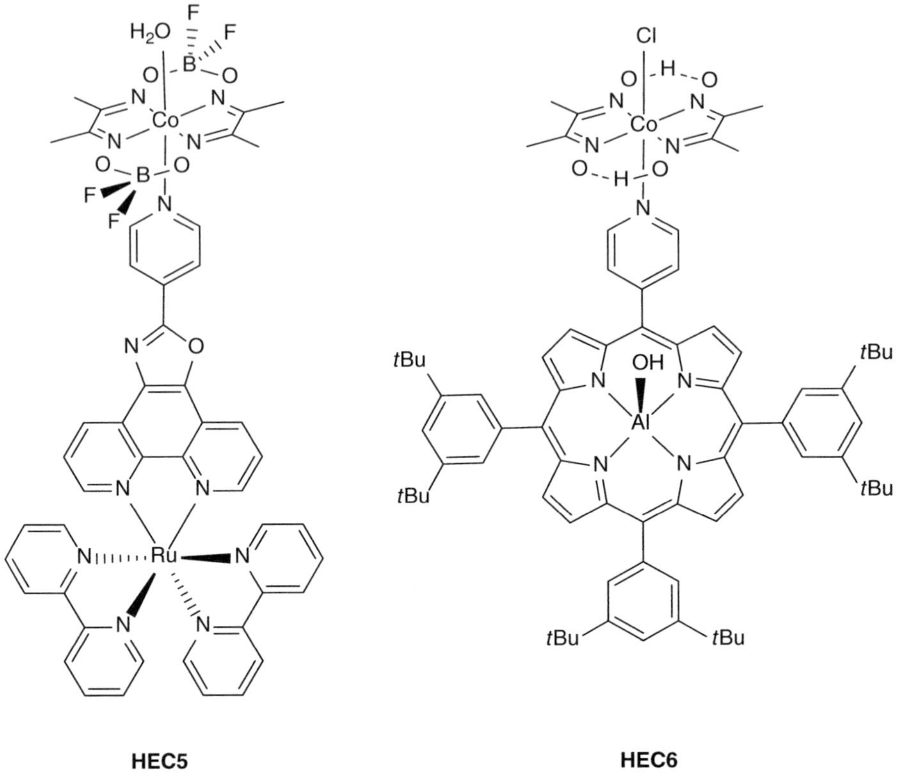

Fig. 1.42 Molecular formulae of sensitizer/cobaloxime dyads **HEC5-6**

Artero and coworkers combined the cobaloxime catalytic moiety to a ruthenium polypyridine complex in a supramolecular dyad (**HEC5**, Fig. 1.42) [132] in order to favor electron transfer processes and improve catalysis. When dyad **HEC5** was irradiated in the presence of TEA as sacrificial electron donor, hydrogen was produced and to a larger extent when compared to the purely bimolecular system [132]. Activation of the catalyst, however, still proceeded via a reductive quenching mechanism since electron transfer from the excited triplet state of the sensitizer to the catalyst (oxidative quenching) was too slow to efficiently compete with the unimolecular deactivation. On the other hand the intramolecular connection between the chromophore and the catalyst was important to facilitate the electron transfer from the reduced sensitizer, thus explaining the larger performance observed with respect to the bimolecular system.

Supramolecular dyads involving a cobaloxime catalyst and pyridyl-porphyrins as the light-harvesting unit were also studied [133, 134]. In this case, the supramolecular approach not only allows for a close proximity between the reactants, which is important for speeding up electron transfer processes and eventually catalysis, but it is also useful for taking advantage of higher energy singlet excited states which are usually short-lived to permit bimolecular processes to occur.

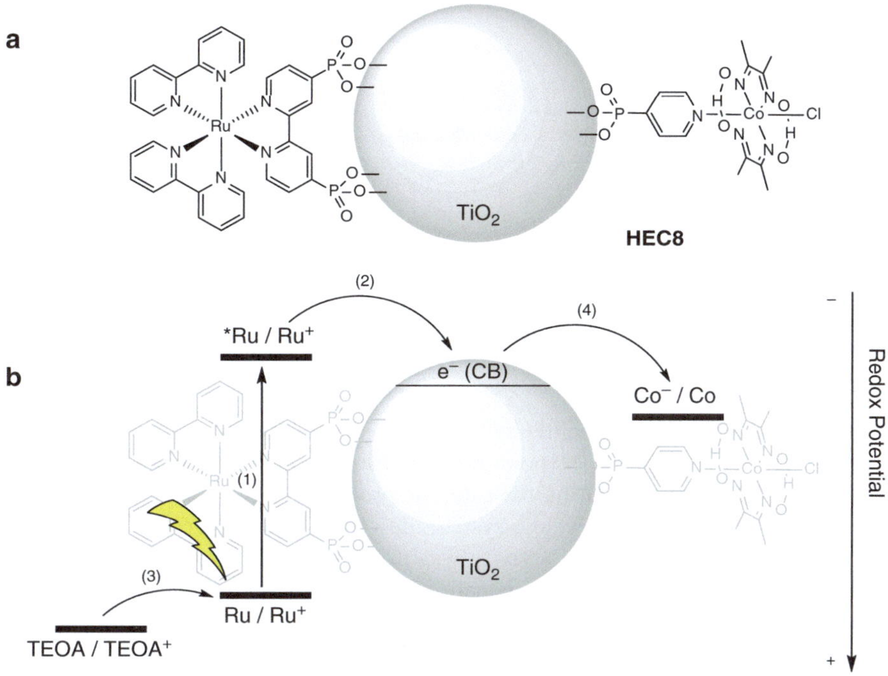

Fig. 1.43 (**a**) Structural representation of the sensitizer/TiO_2 nanoparticle/cobaloxime assembly **HEC7** studied by Reisner and coworkers; (**b**) schematic representation of the photochemical mechanism for the first catalyst reduction (for additional details see text)

In the case of **HEC6** (Fig. 1.42), [133] formation of the dyad system was, however, practically useless. Indeed singlet excited state quenching by the cobaloxime was efficient but largely reversible thus preventing hole transfer to the donor according to an oxidative quenching mechanism (Fig. 1.38b). Moreover, upon continuous irradiation of **HEC6** in the presence of ascorbic acid as the electron donor a rapid sensitizer/catalyst detachment was observed, generating a purely bimolecular system capable of hydrogen production but via a reductive photochemical pathway at the triplet level.

An interesting approach for the construction of a photochemical hydrogen evolving device is the one adopted by Reisner (**HEC7**, Fig. 1.43) [135]. A $Ru(bpy)_3^{2+}$ sensitizer and a cobaloxime catalyst were indeed anchored onto the surface of TiO_2 nanoparticles via phosphonic acid functional groups to give a hybrid system capable of efficient photocatalytic hydrogen production from neutral aqueous solutions in the presence of TEOA as the sacrificial electron donor. In this hybrid triad system (Fig. 1.43b), (1) excitation of the ruthenium(II) sensitizer is followed by (2) electron injection into the conduction band (CB) of TiO_2, (3) the TEA sacrificial donor subsequently recovers the oxidized dye and finally (4) the electron residing in the TiO_2 CB is transferred to the cobaloxime complex. When the overall photochemical process is repeated a second electron can be stored by the

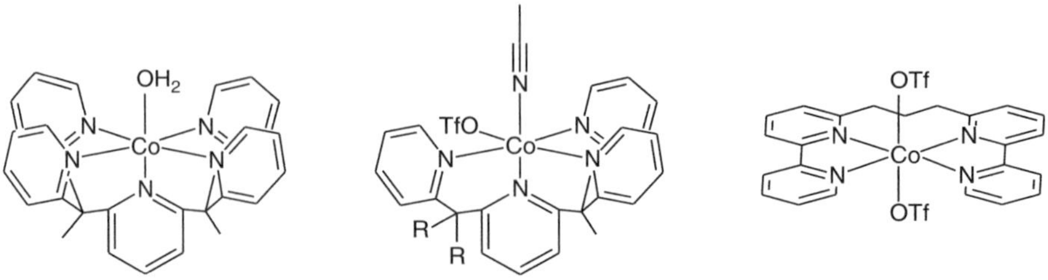

Fig. 1.44 Molecular formulae of polypyridine cobalt complexes **HEC9-12** studied by Castellano, Long, Chang, and coworkers

cobaloxime catalyst thus enabling hydrogen evolution. The main advantage arising from this hybrid approach is the possibility of efficient and multiple electron storage on the TiO_2 CB thus facilitating the catalytic process by the cobaloxime center.

In all the cobalt-based molecular systems described, however, either bimolecular or not, the hydrogen evolving activity was usually limited by degradation of the catalyst, occurring through hydrogenation of the dimethylglyoximate ligand in competition with hydrogen evolution. To circumvent stability issues, the attention was recently moved towards cobalt polypyridine complexes combining both improved stability under catalytic conditions and activity in purely aqueous environments [119]. Some examples of cobalt polypyridine catalysts (**HEC9-12**) used for hydrogen evolution catalysis are those reported in Fig. 1.44, recently studied by Castellano, Long, Chang, and coworkers, [136] which included tetradentate ligands and pentadentate ones. All these complexes were tested in standard photochemical cycles in purely aqueous solutions (usually pH 4 or 7) in the presence of $Ru(bpy)_3^{2+}$ as the sensitizer and ascorbic acid as the sacrificial electron donor, thus allowing interesting comparisons.

In particular, (i) cobalt complexes **HEC10-11** with tetradentate ligands were found to perform significantly better than the corresponding complex with pentadentate ligand (**HEC9**); (ii) complexes **HEC10-11** with tetradentate ligands capable of leaving two open *cis* coordination sites were considerably better catalysts than **HEC12** with a tetradentate ligand enforcing free *trans* coordination sites; (iii) catalyst **HEC10** with a less sterically hindering ligand was more active than the congener **HEC11**. Under photocatalytic conditions, activity is mainly limited by decomposition of the sensitizer, suggesting a net improvement in catalyst stability with respect to other cobalt complexes, e.g., cobaloximes.

Almost simultaneously to the development of cobalt-based HECs, vivid interest arose towards nickel-based catalysts for mediating the hydrogen evolving reaction since the discovery of the outstanding electrocatalytic properties of synthetic nickel complexes featuring phosphine ligands with pendent amine functionalities, e.g., the $Ni(P^{Ph}{}_2N^{Ph})_2]$ $(BF_4)_2$ (with $P^{Ph}{}_2N^{Ph}$ = 1,3,6-triphenyl-1-aza-3,6-diphosphacycloheptane) catalyst studied by DuBois and coworkers (**HEC13**, Fig. 1.45) [137]. The key feature of these catalysts is the presence of amine

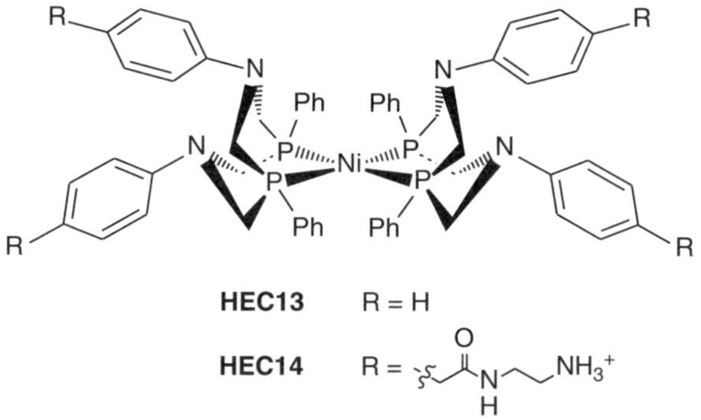

Fig. 1.45 Molecular formulae of nickel complexes **HEC13-14**

functional groups in the second coordination sphere which play an active role in the catalysis by assisting the "chemical" proton transfer steps.

Several studies were undertaken on this catalytic platform and very rapidly nickel phosphine complexes became benchmark catalysts for the construction of hydrogen evolution photochemical systems. **HEC13** was first tested as hydrogen evolving catalyst in sacrificial photocatalytic experiments in a 50/50 water/acetonitrile mixture involving $Ru(bpy)_3^{2+}$ as the photosensitizer and ascorbic acid as the sacrificial electron donor, showing enhanced and stable activity up to 150 h irradiation achieving 2,700 turnovers [138].

More interestingly, the functionalized nickel phosphine catalyst **HEC14** (Fig. 1.45) was used by Wasielewski, Stupp, and coworkers to develop a bioinspired soft-material integrating light-harvesting, charge transport, and catalysis for the production of hydrogen [139]. As the light-harvesting unit, a perylene monoimide chromophore was used, suitably functionalized at the imide position with a five-carbon chain terminated with a carboxylic group. These amphiphilic chromophores self-assembled when dissolved in aqueous solution and underwent gelation upon addition of salts such as $CaCl_2$ or NaCl. From a morphological viewpoint, the so-formed hydrogel (99 % water content) consisted of a network of flat sheet-like structures comprising stacked perylene moieties resulting in H-aggregates, a situation that increased the electronic coupling between the light-harvesting components throughout the soft-material. As to the catalytic moiety, a DuBois type catalyst was suitably functionalized with cationic ammonium groups (yielding complex **HEC14**) that permitted solubility in water solution and promoted a controlled spatial localization of the catalytic units with respect to the sensitizers through electrostatic interactions. As a result, hydrogen was efficiently produced upon visible irradiation of the mixture containing **HEC14** and the light-harvesting hydrogel in the presence of ascorbic acid as the sacrificial electron donor. The photochemical mechanism was thought to involve reductive quenching of the exciton (i.e., the delocalized excited state within the stacked perylene hydrogel) by ascorbic acid as the primary step followed by electron transfer to the nickel catalyst eventually leading to hydrogen production. The performances were dependent on the donor concentration and the electrolyte used for the gel preparation and considerably higher than those of a solid precipitate of the three components, suggesting the important role of a porous hydrated gel architecture which facilitates diffusion processes thus fostering efficient hydrogen evolution.

1.3.6 Integrated Systems

In principle, antennae, charge separation systems, and catalytic units could be assembled into an artificial photosynthetic system (see above Sect. 1.3.1, Fig. 1.9) for photochemical water splitting in which light excitation triggers a series of photoinduced energy and electron transfer processes which are ultimately responsible for water oxidation at the anodic side and hydrogen evolution at the cathodic side. In such a construct, a proton exchange membrane must be also introduced between the two compartments in order to let the protons produced via water oxidation move towards the reductive side to enable hydrogen production. The construction of a purely molecular system such as the one depicted in Fig. 1.9 is, however, intrinsically complicated. Difficulties mainly arise from the coupling of the charge separation steps, which are fast one-photon one-electron processes, with the catalytic steps, namely the formation of oxygen and hydrogen, which are slow multi-electron processes. This indeed may result in detrimental charge recombination processes inactivating the whole artificial photosynthetic system. As a consequence, a perfect arrangement of the molecular components within a suitable matrix is imperative in order to avoid such disadvantageous electron transfer processes. This is why in natural photosynthetic organisms the molecular cofactors which are responsible for light-harvesting, electron transfer, and catalysis are incorporated within suitable organized protein matrices. To achieve this goal in an artificial manner is, however, not trivial and requires a certain degree of molecular (or even biomolecular) engineering which is not easily accessible from a chemical viewpoint.

To get rid of this issue, one possibility is that of a hybrid approach, taking advantage of semiconductor (SC) materials, which are also particularly useful in view of the preparation of a solid state device for photo-assisted water splitting. As depicted in Fig. 1.46a, band gap excitation of a SC induces the promotion of an electron from the valence band (VB) into the conduction band (CB): albeit strictly dependent on the SC material, the electron in the CB has in principle a sufficient thermodynamic power to perform reduction reactions (e.g., proton reduction to hydrogen), whereas the hole left in the VB may accomplish oxidation processes (e.g., water oxidation).

In 1972 Fujishima and Honda [140] demonstrated that UV irradiation of a TiO_2 anode coupled to a platinum cathode was sufficient for promoting water splitting. The applicability of such a system was, however, limited to the use of UV light, since light absorption by TiO_2 occurs at $\lambda < 400$ nm ($E_g = 3.2$ eV), i.e., in a spectral region which has a small overlap with the solar radiation profile at Earth's surface. To shift the absorption towards the visible spectrum, sensitization of the SC surface with a molecular dye (P) is a viable solution (Fig. 1.46b): visible light is absorbed

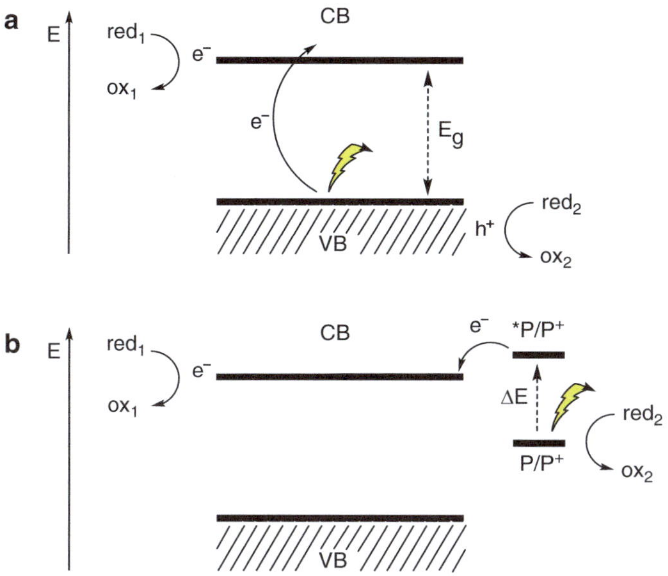

Fig. 1.46 Schematic representation of the potential processes occurring upon photoexcitation of (**a**) a SC material and (**b**) a sensitized *n*-type SC material. Abbreviations: *VB* valence band, *CB* conduction band, e^- electron, h^+ electron hole, E_g band gap energy, *ox* oxidized species, *red* reduced species, *P* photosensitizer. Note that $\Delta E < E_g$

by the chromophore which then undergoes electron injection into the CB of the SC (in other words, the sensitizer gets oxidized by the SC). The electron in the CB can be used for a reduction process (e.g., proton reduction to hydrogen), whereas the oxidized dye may perform oxidation reactions (e.g., water oxidation).

However, since absorption and electron injection are fast one-photon one-electron processes and both proton reduction and water oxidation are slow multi-electron processes, the use of suitable catalysts, capable of stepwise storing photogenerated electrons and holes, is mandatory. These few concepts translate into the photoelectrochemical cell (PEC) prototype for light-assisted water splitting depicted in Fig. 1.47a (usually called dye-sensitized photoelectrochemical cells or dye-sensitized photoelectrosynthesis cells, DS-PEC). It features a photoanode based on a wide-band-gap *n*-type nanocrystalline semiconductor of high surface area (usually TiO_2) onto which a dye photosensitizer and a water oxidation catalyst (WOC) are anchored. A platinum cathode is connected via an external circuit and the compartments are separated by a proton exchange membrane. In this artificial photosynthetic device the sensitizer acts both as the light-harvesting antenna and, together with the semiconductor, as the charge separation system. In this configuration, excitation of the sensitizer P1 is followed by electron injection into the CB of the *n*-type SC, the oxidized dye is regenerated by the WOC which acts as an electron donor, when four holes are transferred to the WOC water oxidation can take place. The electrons in the CB diffuse through the bulk of the SC, enter the external circuit, and then arrive at the platinum counter-electrode to perform the

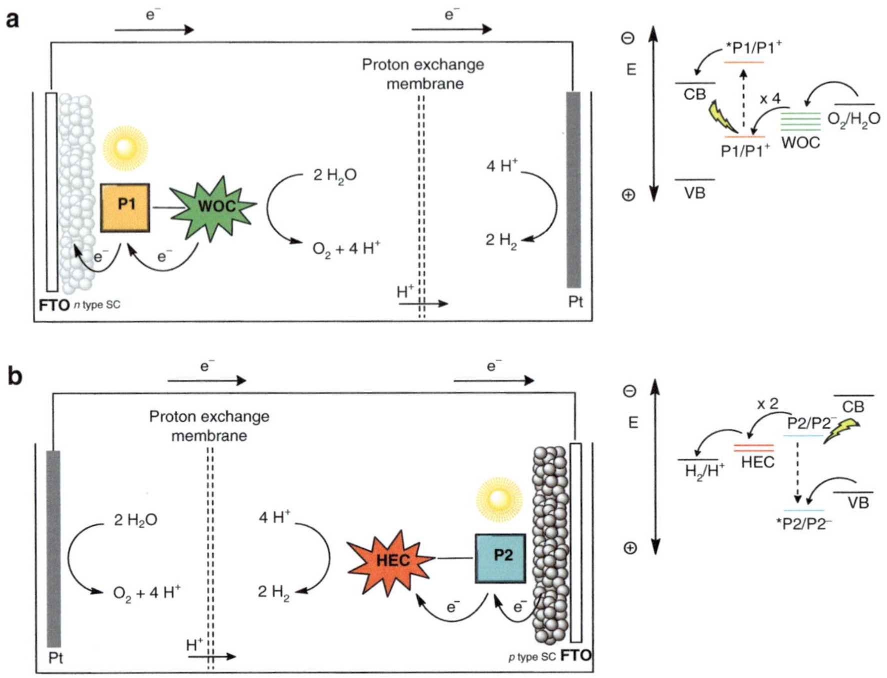

Fig. 1.47 Schematic representation of (**a**) *n*-type PEC based on a sensitized photoanode and a Pt cathode and (**b**) *p*-type PEC based on a sensitized photocathode and a Pt anode. Inserts: thermodynamics of the photoinduced electron transfer processes responsible for (**a**) oxidation and (**b**) reduction of water (*solid lines*, electron transfer processes; *dashed lines*, excitation). Abbreviations: *P1*,*P2* photosensitizer, *WOC* water oxidation catalyst, *HEC* hydrogen evolving catalyst

proton reduction reaction (in this case platinum acts both as the electron collector and the HEC). Importantly, fundamental requirements for the molecular components to be used are (see insert in Fig. 1.47a): (i) the reduction potential of the excited sensitizer (*P1) must be more negative than the CB edge potential of the *n*-type SC and (ii) the redox potential of the $P1^+/P1$ redox couple should be more positive than the oxidation potentials of the WOC. The approach adopted herein is similar to the one used in *n*-type dye-sensitized solar cells (*n*-DSSCs), [141, 142] in which a redox mediator, instead of the WOC, is used to regenerate the oxidized dye and the electrons in the CB are employed at a suitable cathode for the reduction of the oxidized component of the redox mediator thus completing a whole circuit. In this latter case, however, light-excitation results only in electrical work rather than chemical work, as obtained instead with the PEC system.

A possible alternative to the PEC configuration of Fig. 1.47a is represented by the symmetric one depicted in Fig. 1.47b, featuring a dye-sensitized wide-band-gap *p*-type SC photocathode (usually NiO) with a hydrogen evolving catalyst (HEC) and a platinum anode, both connected via an external circuit and whose compartments are separated by a proton exchange membrane. In this symmetric configuration excitation of the sensitizer is responsible for hole injection into the VB of the *p*-type SC, the reduced sensitizer then undergoes electron transfer to the HEC thus regenerating the dye, when two electrons are transferred to the HEC hydrogen evolution may occur. The holes accumulated in the VB diffuse through the bulk of the SC, enter the external circuit, and reach the platinum anode to perform water oxidation. Even in this case, this approach is reminiscent to the one adopted for *p*-type dye-sensitized solar cells (*p*-DSSCs) [143]. Main requirements for the molecular components to be used in this type of PEC are (see insert in Fig. 1.47b): (i) the oxidation potential of the excited sensitizer (*P2) must be more positive than the VB edge potential of the *p*-type SC and (ii) the redox potential of the ground state P2/P2$^-$ couple must be more negative than the reduction potentials of the HEC.

It should be remarked, however, that the energy of a single photon may not be powerful enough to drive the water splitting process without an external energy supply, because of unfavorable semiconductor redox energetics and/or excessive energy losses in charge separation steps. This implies that an external bias must be applied to enable water splitting by PECs such as those of Fig. 1.47. This problem can be, however, overcome with the design of *tandem* dye-sensitized photoelectrochemical cells as shown in Fig. 1.48.

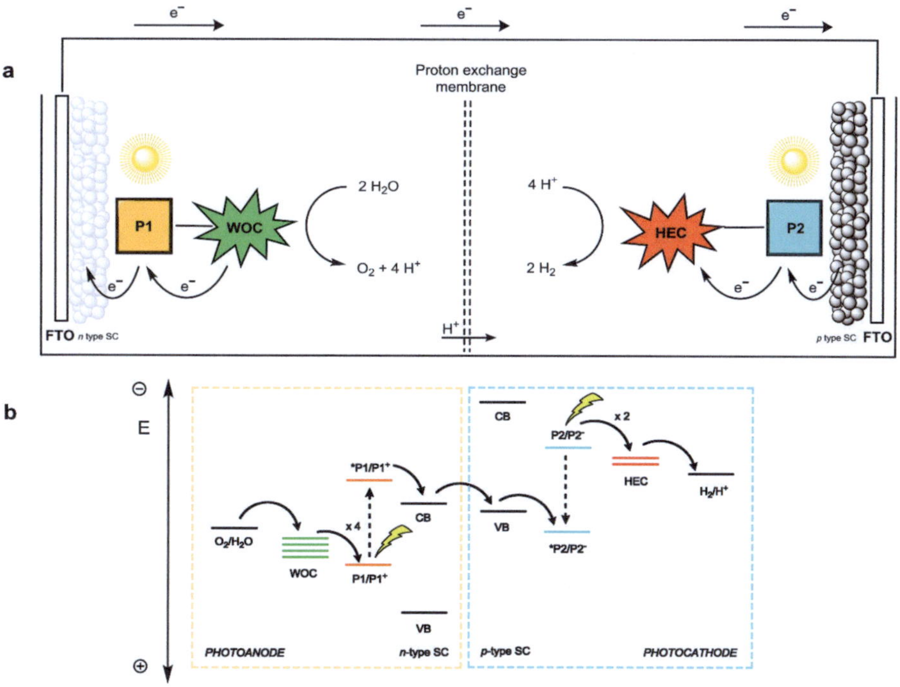

Fig. 1.48 (**a**) Schematic representation of a *tandem* PEC based on a sensitized photoanode and sensitized photocathode. (**b**) Z-scheme of photochemical water splitting in a dye-sensitized *tandem* PEC. Abbreviations: *P1,P2* photosensitizer, *WOC* water oxidation catalyst, *HEC* hydrogen evolving catalyst

Two photoelectrodes are coupled through an external circuit, namely a photoanode based on a dye-sensitized wide-band-gap *n*-type SC comprising a WOC and a dye-sensitized wide-band-gap *p*-type SC photocathode comprising a HEC. At the anodic side light absorption by the sensitizer is responsible for electron injection into the CB of the *n*-type SC and the oxidized dye enables water oxidation mediated by the WOC, while simultaneously at the cathode light absorption by the sensitizer is followed by hole injection in the VB of the *p*-type SC and the reduced chromophore triggers hydrogen evolution via the HEC. The electrons in the CB of the *n*-type SC and the holes in the VB of the *p*-type SC then recombine through the external circuit. Interestingly, the photoinduced electron transfer processes occurring in the tandem PEC configuration (Fig. 1.48b) shows strong similarities with the Z-scheme of natural photosynthesis (Sect. 1.2.2, Fig. 1.8), since two photons are needed to perform the cascade of electron transfer processes eventually leading to water splitting into oxygen and hydrogen.

The first example of dye-sensitized PEC was reported by Mallouk and coworkers [144]. It featured a photoanode based on mesoporous TiO_2 sensitized with a ruthenium tris(bipyridine) dye functionalized with phosphonic acid anchoring groups for the covalent attachment to the SC surface and a malonic acid group for the stabilization of colloidal hydrated iridium oxide nanoparticles ($IrO_2 \cdot nH_2O$), chosen as the WOC (Fig. 1.49a). The photoanode was then connected to a platinum cathode and the electrodes were immersed in a Na_2SiF_6/$NaHCO_3$ buffered solution at pH 5.8. Upon visible light irradiation the cell reached a steady-state photocurrent density of 30 μAcm^{-2} at an applied bias of +0.1 V vs Ag/AgCl, which was accompanied by oxygen evolution. A quantum yield of 0.9 % was also measured, mainly limited by the faster back electron transfer from the CB of TiO_2 to the oxidized dye molecules with respect to the forward electron transfer from $IrO_2 \cdot nH_2O$ nanoparticles to the oxidized dye. To overcome this restrictions, Mallouk and coworkers incorporated in the PEC configuration a benzimidazol-phenol (BIP) electron transfer mediator, mimetic of the tyrosine-histidine mediator in Photosystem II, and covalently bound to the $IrO_2 \cdot nH_2O$ WOC. The latter was coadsorbed onto a porous TiO_2 electrode together with a ruthenium polypyridyl sensitizer (Fig. 1.49b) [145]. The introduction of the redox mediator is of particular importance since it promotes fast reduction of the oxidized sensitizer thus preventing charge recombination processes with the injected electrons. Moreover, the oxidation of the BIP moiety is a proton-coupled electron-transfer (PCET) process involving intramolecular proton shift from the phenol unit to the benzimidazol. This reaction stabilizes the oxidized form of the redox relay thus increasing charge separation lifetime. This situation indeed positively mediates the coupling of the fast charge separation processes with the slow catalysis by the iridium oxide nanoparticles. As a result, upon visible irradiation the PEC involving the BIP electron transfer mediator (Fig. 1.49b) showed enhanced photocurrent densities by a factor of ca. 3 (80 μAcm^{-2} at +0.1 V vs Ag/AgCl) when compared to the PEC containing only the sensitizer and the catalyst (30 μAcm^{-2} at the same applied bias), with an also increased quantum yield of 2.3 %.

After these studies, important efforts have been devoted to the preparation of PECs taking advantage of molecular catalysts instead of metal oxide nanoparticles.

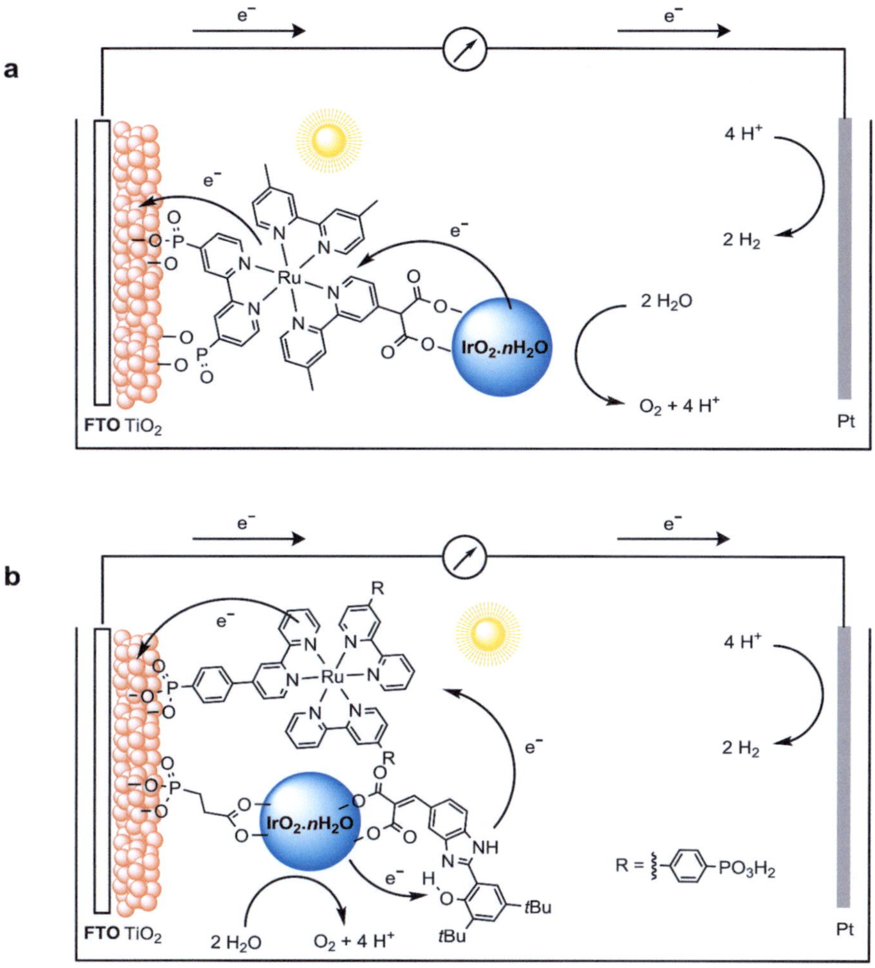

Fig. 1.49 Configurations of the *n*-type dye-sensitized PECs studied by Mallouk and coworkers: (**a**) standard chromophore-catalyst configuration and (**b**) improved cell with addition of the benzimidazol-phenol (BIP) electron transfer mediator

After several attempts, T. J. Meyer and coworkers succeeded in the preparation of a covalent sensitizer-catalyst dyad, possessing phosphonic acid anchoring groups for the covalent attachment onto a SC surface. A photoanode was thus prepared by anchoring the sensitizer-catalyst dyad onto a thin layer of TiO_2 coated onto conductive indium tin oxide (ITO) nanoparticles (Fig. 1.50). Upon 445-nm light irradiation and in the presence of an external bias of +0.2 V vs. NHE, the photoanode in combination with a platinum cathode resulted in water splitting into hydrogen and oxygen in a pH 4.6 buffer solution, yielding an average stabilized photocurrent density of ca. 100 μA cm^{-2} [146].

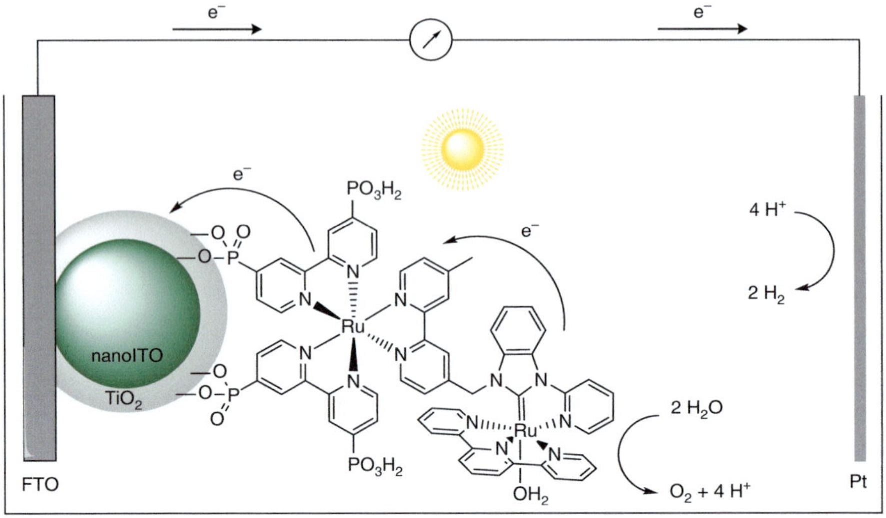

Fig. 1.50 Configuration of the *n*-type dye-sensitized PECs studied by T. J. Meyer and coworkers based on a molecular sensitizer-catalyst dyad

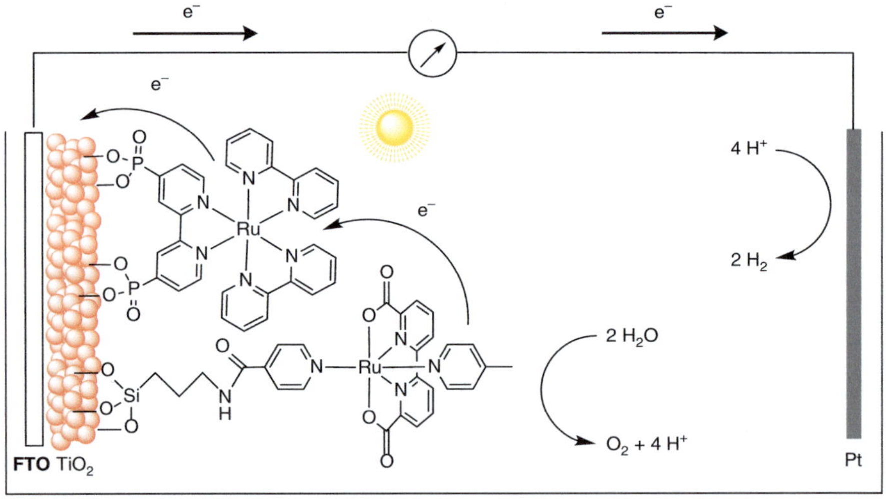

Fig. 1.51 Configuration of the *n*-type dye-sensitized PECs studied by Sun and coworkers based on the co-adsorption of ruthenium sensitizer and catalyst

Another approach was adopted by Sun and coworkers for the construction of an *n*-type PEC based on the co-adsorption of a ruthenium WOC and a ruthenium tris (bipyridine) sensitizer onto a mesoporous TiO_2 photoanode (Fig. 1.51) [147]. Both the catalytic and sensitizing moieties were suitably functionalized with anchoring groups to favor attachment onto the SC surface. When connected to a platinum cathode and immersed in a phosphate buffer solution at pH 6.8, visible irradiation (300 W xenon lamp, $\lambda > 400$ nm) of the photoanode at an applied external bias of +0.2 V vs. NHE was accompanied by oxygen and hydrogen evolution at a photocurrent density of 1.7 $mAcm^{-2}$ after 10 s, which reached a more stable value of ca. 0.7 $mAcm^{-2}$ after 100 s.

These values are the highest ever observed with PECs based on purely molecular components. Interestingly, a careful evaluation was also performed as regarding the effect of the length of the alkyl chain connecting the catalyst and the SC surface, showing improved performances in the presence of longer chains [148]. These evidences were attributed to the retardation of the back electron transfer from the CB of the TiO_2 to the oxidized WOC attainable at larger distances.

Although the mechanistic principles of a *p*-type dye-sensitized PEC are perfectly symmetric to the ones adopted in the *n*-type dye-sensitized PEC configuration (see above), the construction of such PECs have been unfortunately less explored. The reason can be mainly attributed to the lack of a good *p*-type SC. Indeed, the most widely used *p*-type SC material is NiO which suffers from several drawbacks such as competitive absorption of light and poor hole mobility. These properties dramatically limit the application of such a SC material, in particular when compared to the *n*-type counterpart like TiO_2 which shows good capability in terms of both charge transport and charge collection efficiencies.

The first example of *p*-type dye-sensitized PEC was reported by Sun and coworkers based on a NiO photocathode sensitized with a push-pull organic dye in combination with a cobaloxime HEC in solution (Fig. 1.52a) [149]. When connected to a passive platinum anode and irradiated, the PEC produced an initial photocurrent density of ca. 20 μAcm^{-2} at an applied bias of −0.4 V vs. Ag/AgCl in a pH 7.0 phosphate buffer solution, which, however, rapidly decayed mostly due to dye desorption and/or catalyst decomposition.

Subsequently, Wu and coworkers described a supramolecular sensitizer/catalyst dyad for the sensitization of an alumina coated NiO photocathode based on a bifunctional cyclometalated ruthenium chromophore and a cobaloxime HEC (Fig. 1.52b) [150]. The photocathode was then connected to a platinum anode and irradiated with a 300 W xenon lamp ($\lambda > 420$ nm) resulting in a stable photocurrent density of 9 μAcm^{-2} over a period of hours at an applied bias of +0.1 V vs. NHE in neutral solution.

As previously denoted, due to the thermodynamic restrictions related to the characteristics of the SC materials employed, both the *n*-type and *p*-type dye-sensitized PECs need an external bias in order to produce sufficient photocurrent densities under photoirradiation. The use of a *tandem* configuration by coupling a photoanode and a photocathode is a practical solution to avoid such an undesired bias. A *tandem* cell, however, is not a simple superposition of a photoanode and a photocathode, since the two compartments of the cell have to work in a concerted fashion. Indeed, the different properties of the SC materials used (different charge mobility, different charge collection efficiency, etc.) as well as the different rates of the catalytic water splitting processes may introduce not trivial kinetic hurdles which might dramatically reduce the whole photosynthetic activity. This notwithstanding some results on *tandem* PECs have been recently reported.

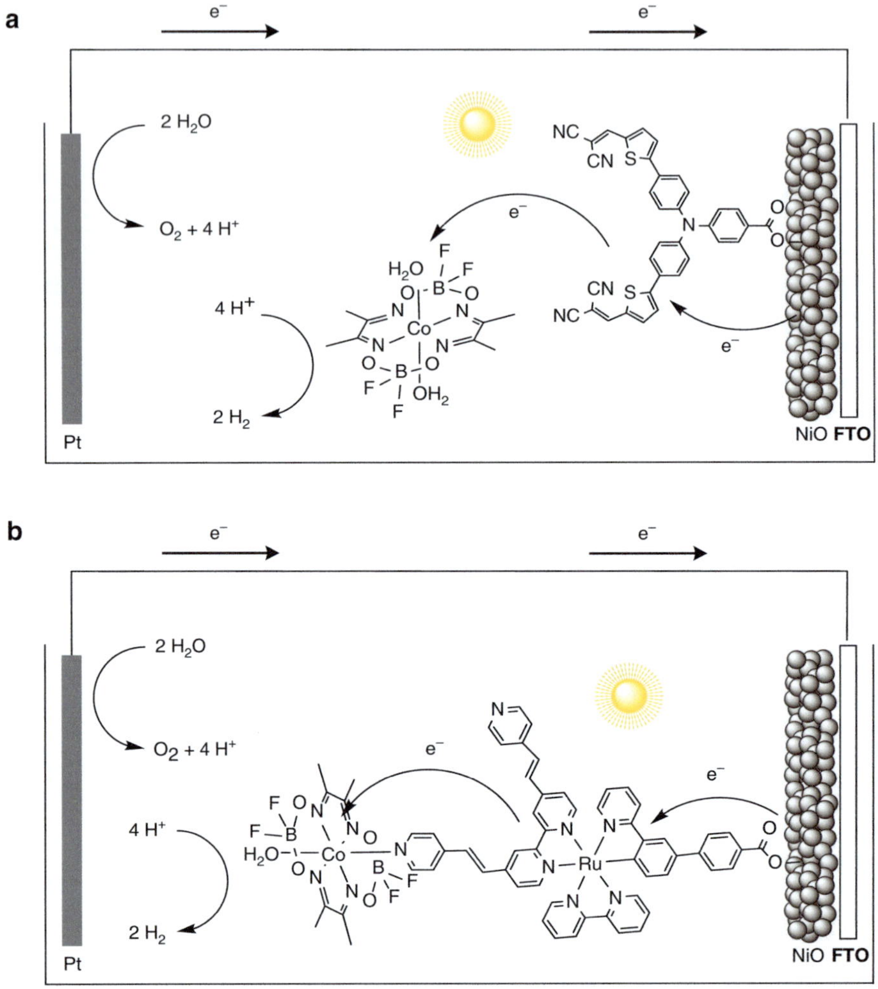

Fig. 1.52 Configuration of the *p*-type dye-sensitized PECs studied (**a**) by Sun and coworkers with the cobaloxime HEC in solution and (**b**) by Wu and coworkers based on a supramolecular sensitizer/catalyst dyad

The first example was recently shown by Sun and coworkers featuring a TiO_2 photoanode with co-adsorbed a ruthenium tris(bipyridine) sensitizer and a ruthenium WOC and a NiO photocathode with co-grafted a ruthenium sensitizer and a cobaloxime HEC (Fig. 1.53a) [151]. Upon visible light irradiation (300 W xenon lamp, $\lambda > 400$ nm) on both the anodic and cathodic side the *tandem* PEC was capable of promoting water splitting from a neutral aqueous solution without any applied external bias, yielding a stable photocurrent density of 25 μAcm^{-2}. An improvement of the photoelectrochemical response of the *tandem* PEC was subsequently performed by substituting the ruthenium polypyridine chromophores with

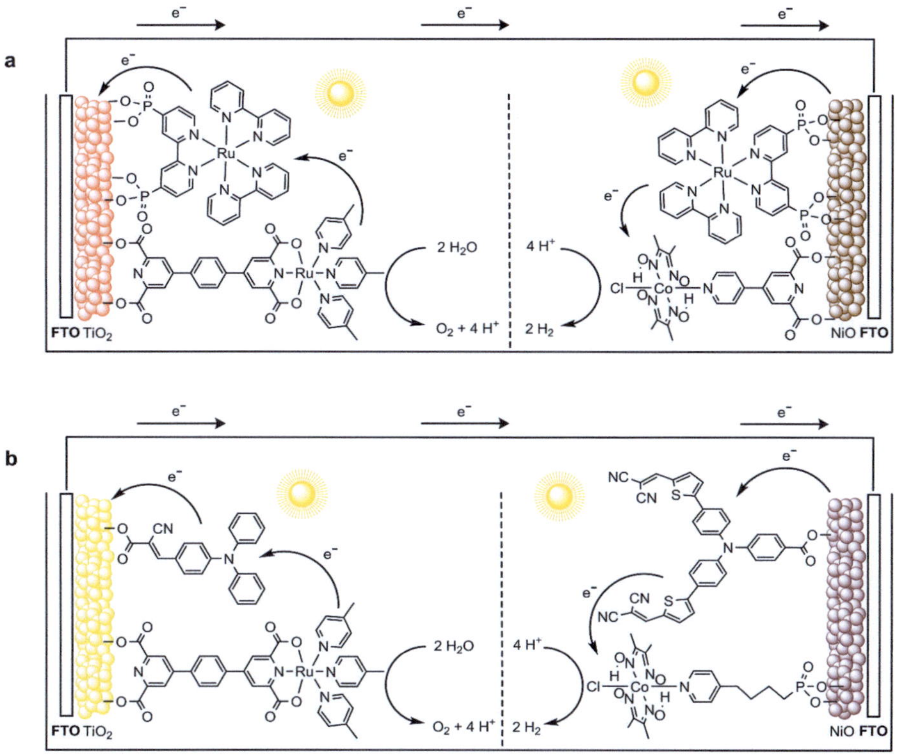

Fig. 1.53 Configuration of the *tandem* dye-sensitized PECs studied by Sun and coworkers: (**a**) with ruthenium tris(bipyridine) sensitizers and (**b**) with organic dyes

donor-acceptor organic dyes (Fig. 1.53b), achieving steady-state photocurrent densities of up to ca. 70 μAcm^{-2} without external bias applied [152]. Interestingly, in both cases, photoelectrochemical studies on the two separated photoelectrodes suggested that the kinetic bottleneck limiting the actual photocurrent values was represented by the NiO photocathode.

The presence of only few reports on functioning dye-sensitized PECs as well as the relatively modest performances of the reported systems strongly suggest that there is large room for improvement.

Main challenges in the design of dye-sensitized PECs are (i) the optimization of the interfacial electron transfer processes in order to suitably couple the fast electron transfer processes (e.g., electron injection) with the slow catalysis, (ii) the research of more effective *p*-type SC materials in order to replace the commonly used NiO, and finally (iii) the long-term durability of the systems which is intimately connected with the optimization of the immobilization process of the sensitizing and catalytic moieties onto the SC surface and their stability under operating conditions.

1.4 Conclusions

The approach outlined in this review to the generation of solar fuels is clearly inspired by natural photosynthesis, and is supramolecular in nature, being molecular species most of the light absorbing, charge transporting, and catalytic units. The virtue of a molecular approach, as demonstrated by the examples discussed in Sects. 1.3.2, 1.3.3, 1.3.4, and 1.3.5, is that the various components can be designed, functionalized, tuned, and eventually optimized by chemical synthesis. Their behavior is also amenable to characterization, at the molecular level, by a variety of physico-chemical techniques. Less obvious and still largely unaccomplished is the next step (Sect. 1.3.6), i.e., the control of the supramolecular assembling of the molecular components and their integration in working devices. A further, general open problem lies in the intrinsic chemical lability of organic systems operating under severe oxidative or reductive conditions. The stability issue, overcome by the natural systems owing to their living nature, is certainly going to be a major one in the field of artificial photosynthesis.

At the other extreme lie, at least traditionally, photovoltaic-inspired approaches. They are based on semiconductors for light absorption and charge generation, and on inorganic catalysts for oxygen and hydrogen evolution. To this domain, materials science, rather than chemistry, is central. With respect to the molecular systems, a lesser ease of fine tuning is certainly compensated by a greater robustness. Some of the most efficient devices produced to date indeed belong to this class [153]. In this regard, the photoelectrochemical cell systems discussed in Sect. 1.3.6 represent an intermediate case, making use of molecular components (as light absorbers and catalysts) but also of semiconductors (both as supports for supramolecular assembling and as electron/hole acceptors for charge separation). In perspective, these systems may be interesting to the extent to which they may gather the strengths (and perhaps mitigate some of the weaknesses) of the two approaches.

Acknowledgments Financial support from the Italian MIUR (FIRB RBAP11C58Y "NanoSolar", PRIN 2010 "Hi-Phuture"), and COST action CM1202 "PERSPECT-H2O" is gratefully acknowledged.

References

1. Blankenship RA (2002) Molecular mechanisms of photosynthesis. Blackwell, Oxford
2. (a) Deisenhofer J, Epp O, Miki K, Huber R, Michel H (1985) Nature 318:618; (b) Deisenhofer J, Epp O, Sinning I, Michel H (1995) J Mol Biol 246:429
3. Gunner MR (1991) The reaction center protein from purple bacteria: structure and function. In: Lee CP (ed) Current topics in bioenergetics, vol 16. Academic, San Diego
4. McDermott G, Prince SM, Freer AA, Hawthornthwaite-Lawless AM, Papiz M, Cogdell RJ, Isaacs NW (1995) Nature 374:517

5. Polli D, Cerullo G, Lanzani G, De Silvestri S, Hashimoto H, Cogdell RJ (2006) Biophys J 90:2486
6. Cogdell RJ, Gardiner AT, Roszak AW, Law CJ, Southall J, Isaacs NW (2004) Photosynth Res 81:207
7. Sundström V, Pullerits T, van Grondelle R (1999) J Phys Chem B 103:2327
8. (a) Roszak AW, Howard TD, Southall J, Gardiner AT, Law CJ, Isaacs NW, Cogdell RJ (2003) Science 302:1969; (b) Niwa S, Yu LJ, Takeda K, Hirano Y, Kawakami T, Wang-Otomo ZY, Miki K (2014) Nature 508:228
9. Bahatyrova S, Frese RN, Siebert CA, Olsen JD, van der Werf KO, van Grondelle R, Niederman RA, Bullough PA, Otto C, Hunter CN (2004) Nature 430:1058
10. Müh F, Glöckner C, Hellmich J, Zouni A (2012) Biochim Biophys Acta 1817:44
11. Kok B, Forbush B, McGloin M (1970) Photochem Photobiol 11:457
12. Grundmeier A, Dau H (2012) Biophys Biochim Acta 1817:88
13. Vinyard DJ, Khan S, Brudvig GW (2015) Faraday Disc 185:37
14. (a) Jordan P, Fromme P, Witt HT, Klukas O, Saenger W, Krau N (2001) Nature 411:909; (b) Amunts A, Drory O, Nelson N (2007) Nature 447:58
15. (a) Nield J, Barber J (2006) Biochim Biophys Acta 1757:353; (b) Umena Y, Kawakami K, Shen JR, Kamiya N (2011) Nature 473:55; (c) Kawakami K, Umena Y, Kamiya N, Shen JR (2011) J Photochem Photobiol B 104:9
16. Qin X, Suga M, Kuang T, Shen JR (2015) Science 348:989
17. Wientjes E, van Amerongen H, Croce R (2013) Biochim Biophys Acta 1827:420
18. Minagawa J (2011) Biochim Biophys Acta 1807:897
19. (a) Huynh MHV, Meyer TJ (2007) Chem Rev 107:5004; (b) Gust D, Moore TA, Moore AL (2009) Acc Chem Res 42:1890.
20. (a) Yandulov DV, Schrock RR (2003) Science 301:76; (b) Ritleng V, Yandulov DV, Weare WW, Schrock RR, Hock AS, Davis WM (2004) J Am Chem Soc 126:6150
21. Schneider J, Jia H, Muckerman JT, Fujita E (2012) Chem Soc Rev 41:2036
22. Sahara G, Ishitani O (2015) Inorg Chem 54:5096
23. Morris AJ, Meyer GJ, Fujita E (2009) Acc Chem Res 42:1983
24. Wang WH, Hull JF, Muckerman JT, Fujita E, Himeda Y (2012) Energy Environ Sci 5:7923
25. Torella JP, Gagliardi CJ, Chen JS, Bediako DK, Colon B, Way JC, Silver PA, Nocera DG (2015) Proc Natl Acad Sci U S A 112:2337
26. Marcus RA, Sutin N (1985) Biochem Biophys Acta 811:265
27. More elaborate models, that treat the nuclear motions quantum mechanically, are available [e.g., (a) Kestner NR, Logan J, Jortner J (1974) J Phys Chem 78:2148; (b) Jortner J (1976) J Chem Phys 64:4860; (c) Ulstrup J, Jortner J (1975) J Chem Phys 63:4358]. The main predictions on the dependence of rates on driving force and reorganization energies are, however, qualitatively the same as those of the classical Marcus model
28. Electronic factors depend on the nature and energy of the orbitals involved in donor, acceptor, and bridging linkages. Therefore, they can be different, in principle, for charge separation and charge recombination. For a more detailed treatment of this complex topic see ref. [29]
29. Natali M, Campagna S, Scandola F (2014) Chem Soc Rev 43:4005
30. Kuciauskas D, Liddell PA, Hung SC, Lin S, Stone S, Seely GR, Moore AL, Moore TA, Gust D (1997) J Phys Chem B 101:429
31. Kuciauskas D, Liddell PA, Lin S, Stone S, Moore AL, Moore TA, Gust D (2000) J Phys Chem B 104:4307
32. Imahori H, Hagiwara K, Aoki M, Akiyama T, Taniguchi S, Okada T, Shirakawa M, Sakata Y (1996) J Am Chem Soc 118:11771
33. Moore TA, Gust D, Mathis P, Mialocq JC, Chachaty C, Bensasson RV, Land EJ, Doizi D, Liddell PA, Lehman WR, Nemeth GA, Moore AL (1984) Nature 307:630
34. Gust D, Moore TA, Moore AL (2001) Acc Chem Res 34:40
35. Wasielewski MR (1992) Chem Rev 92:435
36. Wasielewski MR (2006) J Org Chem 71:5051

37. (a) Paddon-Row MN (2001) Electron Transfer in Chemistry, vol III. In: Balzani V (ed). Wiley-VCH, Weinheim. Chapter 2.1. p 179; (b) Albinsson B, Martensson J (2008) J Photochem Photobiol C 9:138
38. Imahori H, Guldi DM, Tamaki K, Yoshida Y, Luo C, Sakata Y, Fukuzumi S (2001) J Am Chem Soc 123:6617
39. Fukuzumi S, Kotani H, Ohkubo K, Ogo S, Tkachenko NV, Lemmetyinen H (2004) J Am Chem Soc 126:1600
40. (a) Benniston AC, Harriman A, Li P, Rostron JP, Verhoeven JW (2005) Chem Commun 2701; (b) Benniston AC, Harriman A, Li P, Rostron JP, van Ramesdonk HJ, Groeneveld MM, Zhang H, Verhoeven JW (2005) J Am Chem Soc 127:16054
41. Fukuzumi S, Ohkubo K, Suenobu T (2014) Acc Chem Res 47:1455
42. (a) Danielson E, Elliott CM, Merkert JW, Meyer TJ (1987) J Am Chem Soc 109:2519; (b) Larson SL, Elliott CM, Kelley DF (1995) J Phys Chem 99:6530; (c) Rawls MT, Kollmannsberger G, Elliott CM, Steiner UE (2007) J Phys Chem A 111:3485
43. Kalyanasundaram K (1992) Photochemistry of polypyridine and porphyrin complexes. Academic, London
44. (a) Sauvage JP, Collin JP, Chambron JC, Guilleraz S, Coudret C, Balzani V, Barigelletti F, De Cola L, Flamigni L (1994) Chem Rev 94:993; (b) Harriman A, Odobel F, Sauvage JP (1994) J Am Chem Soc 116:5481; (c) Flamigni L, Barigelletti F, Armaroli N, Collin JP, Dixon IM, Sauvage JP, Williams JAG (1999) Coord Chem Rev 190–192:671; (d) Baranoff E, Collin JP, Flamigni L, Sauvage JP (2004) Chem Soc Rev 33:147
45. Borgström M, Shaikh N, Johansson O, Anderlund MF, Styring S, Akermark B, Magnuson A, Hammarström L (2005) J Am Chem Soc 127:17504
46. Campagna S, Puntoriero F, Nastasi F, Bergamini G, Balzani V (2007) Top Curr Chem 280:117
47. Iengo E, Pantoş DG, Sanders JKM, Orlandi M, Chiorboli C, Fracasso S, Scandola F (2011) Chem Sci 2:676
48. Anderson S, Anderson HL, Bashall A, McPartlin M, Sanders JKM (1995) Angew Chem Int Ed Engl 34:1096
49. Imahori H (2004) J Phys Chem B 108:6130
50. Scandola F, Chiorboli C, Prodi A, Iengo E, Alessio E (2006) Coord Chem Rev 250:1471
51. Kobuke Y (2006) Eur J Inorg Chem 2006: 2333
52. Aratani N, Kim D, Osuka A (2009) Acc Chem Res 42:1922
53. Kim D (2012) Multiporphyrin arrays: fundamentals and applications, Ed. CRC Press, Boca Raton
54. Aratani N, Osuka A, Kim YH, Jeong DH, Kim D (2000) Angew Chem Int Ed 39:1458
55. Aratani N, Cho HS, Ahn TK, Cho S, Kim D, Sumi H, Osuka A (2003) J Am Chem Soc 125:9668
56. (a) Hwang IW, Ko DM, Ahn TK, Yoon ZS, Kim D, Peng X, Aratani N, Osuka A (2005) J Phys Chem B 109:8643. (b) Hori T, Aratani N, Takagi A, Matsumoto T, Kawai T, Yoon MC, Yoon ZS, Cho S, Kim D, Osuka A (2006) Chem Eur J 12:1319
57. Lin VSY, Therien MJ (1995) Chem Eur J 1:645
58. Taylor PN, Huuskonen J, Rumbles G, Aplin RT, Williams E, Anderson HL (1998) Chem Commun 909
59. (a) Hoffmann M, Wilson CJ, Odell B, Anderson HL (2007) Angew Chem Int Ed 46:3122. (b) Hoffmann M, Kärnbratt J, Chang MH, Herz LM, Albinsson B, Anderson HL (2008) Angew Chem Int Ed 47:4993
60. O'Sullivan MC, Sprafke JK, Kondratuk DV, Rinfray C, Claridge TDW, Saywell A, Blunt MO, O'Shea JN, Beton PH, Malfois M, Anderson HL (2011) Nature 469:72
61. Benites MR, Johnson TE, Weghorn S, Yu L, Rao PD, Diers JR, Yang SI, Kirmaier C, Bocian DF, Holten D, Lindsey JS (2002) J Mater Chem 12:65

62. (a) Serroni S, Campagna S, Puntoriero F, Di Pietro C, McClenaghan ND, Loiseau F (2001) Chem Soc Rev 30:367; (b) Balzani V, Ceroni P, Juris A, Venturi M, Campagna S, Puntoriero F, Serroni S (2001) Coord Chem Rev 219–221:545
63. Berglund-Baudin H, Davidsson J, Serroni S, Juris A, Balzani V, Campagna S, Hammarström L (2002) J Phys Chem A 106:4312
64. Iengo E, Scandola F, Alessio E (2006) Struct Bond 121:105
65. Takahashi R, Kobuke Y (2003) J Am Chem Soc 125:2372
66. Hwang IW, Ko DM, Ahn TK, Kim D, Ito F, Ishibashi Y, Khan SR, Nagasawa Y, Miyasaka H, Ikeda C, Takahashi R, Ogawa K, Satake A, Kobuke Y (2005) Chem Eur J 11:3753
67. Iengo E, Zangrando E, Minatel R, Alessio E (2002) J Am Chem Soc 124:1003
68. Prodi A, Chiorboli C, Scandola F, Iengo E, Alessio E (2006) ChemPhysChem 7:1514
69. Indelli MT, Chiorboli C, Scandola F, Iengo E, Osswald P, Würthner F (2010) J Phys Chem B 114:14495
70. Kodis G, Liddell PA, de la Garza L, Clausen PC, Lindsey JS, Moore AL, Moore TA, Gust D (2002) J Phys Chem A 106:2036
71. Terazono Y, Kodis G, Liddell PA, Garg V, Moore TA, Moore AL, Gust D (2009) J Phys Chem B 113:7147
72. Kuramochi Y, Sandanayaka ASD, Satake A, Araki Y, Ogawa K, Ito O, Kobuke Y (2009) Chem Eur J 15:2317
73. Ravotto L, Mazzaro R, Natali M, Ortolani L, Morandi V, Ceroni P, Bergamini G (2014) J Phys Chem Lett 5:798
74. (a) Kanan M, Nocera DG (2008) Science 321:1072; (b) Dinca M, Surendranath Y, Nocera DG (2010) Proc Natl Acad Sci USA 107:10337; c) Huynh M, Bediako DK, Nocera DG (2014) J Am Chem Soc 136:6002
75. Smith RDL, Prévot MS, Fagan RD, Zhang Z, Sedach PA, Siu MKJ, Trudel S, Berlinguette CP (2013) Science 340:60
76. (a) Morris ND, Suzuki M, Mallouk TE (2004) J Phys Chem A 108:9115; (b) Hoertz PG, Kim YI, Youngblood WJ, Mallouk TE (2007) J Phys Chem B 111:6845
77. Piccinin S, Sartorel A, Aquilanti G, Goldoni A, Bonchio M, Fabris S (2013) Proc Natl Acad Sci U S A 110:4917
78. Gagliardi AK, Vannucci A, Concepcion JJ, Chen Z, Meyer TJ (2012) Energy Environ Sci 5:7704
79. Betley TA, Wu Q, van Voohris T, Nocera DG (2008) Inorg Chem 47:1849
80. Sartorel A, Bonchio M, Campagna S, Scandola F (2013) Chem Soc Rev 42:2262
81. (a) Berardi S, Drouet S, Francàs L, Gimbert-Suriñach C, Guttentag M, Richmond C, Stoll T, Llobet A (2014) Chem Soc Rev 43:7501; (b) Sala X, Maji S, Bofill R, Garcia-Anton J, Escriche L, Llobet A (2014) Acc Chem Res 47:504
82. Kärkäs MD, Verho O, Johnston EV, Åkermark B (2014) Chem Rev 114:11863
83. Blakemore JD, Crabtree RH, Brudvig GW (2015) Chem Rev 115:12974
84. Geletii YV, Botar B, Kögerler P, Hillesheim DA, Musaev DG, Hill CL (2008) Angew Chem Int Ed 47:3896
85. Sartorel A, Carraro M, Scorrano G, De Zorzi R, Geremia S, McDaniel ND, Bernhard S, Bonchio M (2008) J Am Chem Soc 130:5006
86. Toma FM, Sartorel A, Iurlo M, Carraro M, Parisse P, Maccato C, Rapino S, Rodriguez Gonzalez B, Amenitsch H, Da Ros T, Casalis L, Goldoni A, Marcaccio M, Scorrano G, Scoles G, Paolucci F, Prato M, Bonchio M (2010) Nat Chem 2:826
87. Quintana M, Montallano Lopez A, Rapino S, Toma FM, Iurlo M, Carraro M, Sartorel A, Maccato C, Ke X, Bittencourt G, Da Ros T, Van Tendeloo G, Marcaccio M, Paolucci F, Prato M, Bonchio M (2013) ACS Nano 7:811
88. Sartorel A, Mirò P, Salvadori E, Romain S, Carraro M, Scorrano G, Di Valentin M, Llobet A, Bo C, Bonchio M (2009) J Am Chem Soc 131:16051
89. Geletii YV, Besson C, Hou Y, Yin Q, Musaev DG, Quinoñero D, Cao R, Hardcastle KI, Proust A, Kögerler P, Hill CL (2009) J Am Chem Soc 131:17360

90. Geletii YV, Huang Z, Hou Y, Musaev DG, Lian T, Hill CL (2009) J Am Chem Soc 131:7522
91. Puntoriero F, La Ganga G, Sartorel A, Carraro M, Scorrano G, Bonchio M, Campagna S (2010) Chem Commun 46:4725
92. Natali M, Orlandi M, Berardi S, Campagna S, Bonchio M, Sartorel A, Scandola F (2012) Inorg Chem 51:7324
93. Natali M, Puntoriero F, Chiorboli C, La Ganga G, Sartorel A, Bonchio M, Campagna S, Scandola F (2015) J Phys Chem C 119:2371
94. Yin Q, Tan JM, Besson C, Geletii YV, Musaev DG, Kuznetsov AE, Luo Z, Hardcastle KI, Hill CL (2010) Science 328:342
95. Huang Z, Luo Z, Geletii YV, Vickers JM, Yin Q, Wu D, Hou Y, Ding Y, Song J, Musaev DG, Hill CL, Lian T (2011) J Am Chem Soc 133:2068
96. Vickers JM, Lv H, Sumliner JM, Zhu G, Luo Z, Musaev DG, Geletii YV, Hill CL (2013) J Am Chem Soc 135:14110
97. (a) Stracke JJ, Finke RG (2011) J Am Chem Soc 133:14872; (b) Stracke JJ, Finke RG (2013) ACS Catal 3:1209; c) Stracke JJ, Finke RG (2014) ACS Catal 4:79
98. Natali M, Berardi S, Sartorel A, Bonchio M, Campagna S, Scandola F (2012) Chem Commun 48:8808
99. (a) Gersten SW, Samuels GJ, Meyer TJ (1982) J Am Chem Soc 104:4029; (b) Gilbert JA, Egglestone DS, Murphy Jr WR, Geselowitz DA, Gersten SW, Hodgson DJ, Meyer TJ (1985) J Am Chem Soc 107:3855
100. Concepcion JJ, Jurss JW, Templeton JL, Meyer TJ (2008) J Am Chem Soc 130:16462
101. Concepcion JJ, Jurss JW, Norris MR, Chen Z, Templeton JL, Meyer TJ (2010) Inorg Chem 49:1277
102. Duan L, Fischer A, Xu Y, Sun L (2009) J Am Chem Soc 131:10397
103. Duan L, Xu Y, Zhang P, Wang M, Sun L (2010) Inorg Chem 49:209
104. Li F, Jiang Y, Zhang B, Huang F, Gao Y, Sun L (2012) Angew Chem Int Ed 51:2417
105. Nakazono T, Parent AR, Sakai K (2013) Chem Commun 49:6325
106. Pizzolato E, Natali M, Posocco B, Montellano Lòpez A, Bazzan I, Di Valentin M, Galloni P, Conte V, Bonchio M, Scandola F, Sartorel A (2013) Chem Commun 49:9941
107. Dempsey JL, Brunschwig BS, Winkler JR, Gray HB (2009) Acc Chem Res 42:1995
108. Ozawa H, Sakai K (2011) Chem Commun 47:2227
109. (a) Schulz M, Karnahl M, Schwalbe M, Vos JG (2012) Coord Chem Rev 256:1682; (b) Alpin Y, Pryce MT, Rau S, Dini D, Vos JG (2013) Dalton Trans 42:16243
110. Stoll T, Gennari M, Serrano I, Fortage J, Chauvin J, Odobel F, Rebarz M, Poizat O, Sliwa M, Deronzier A, Collomb MN (2013) Chem Eur J 19:782
111. (a) Artero V, Chavarot-Kerlidou M, Fontecave M (2011) Angew Chem Int Ed 50:7238; (b) Eckenhoff WT, McNamara WR, Du P, Eisenberg R (2013) Biochim Biophys Acta 1827:958
112. DuBois DL (2014) Inorg Chem 53:3935
113. (a) Tard C, Pickett CJ (2009) Chem Rev 109:2245; (b) Lomoth R, Ott S (2009) Dalton Trans 9952
114. Andreiadis ES, Chavarot-Kerlidou M, Fontecave M, Artero V (2011) Photochem Photobiol 87:946
115. (a) Eckenhoff WT, Eisenberg R (2012) Dalton Trans 41:13004; (b) Du P, Eisenberg R (2012) Energy Environ Sci 5:6012; c) Han Z, Eisenberg R (2014) Acc Chem Res 47:2537
116. (a) McKone JR, Marinescu SC, Brunschwig BS, Winkler JR, Gray HB (2014) Chem Sci 5:965; (b) Thoi VS, Sun Y, Long JR, Chang CJ (2013) Chem Soc Rev 42:2388
117. Losse S, Vos JG, Rau S (2010) Coord Chem Rev 254:2049
118. Ladomenou K, Natali M, Iengo E, Charalambidis G, Scandola F, Coutsolelos AG (2015) Coord Chem Rev 304–305:38

119. (a) Zee DZ, Chantarojsiri T, Long JR, Chang CJ (2015) Acc Chem Res 48:2027; (b) Queyriaux N, Jane RT, Massin J, Artero V, Chavarot-Kerlidou M (2015) Coord Chem Rev 304–305:3
120. Streich D, Astuti Y, Orlandi M, Schwartz L, Lomoth R, Hammarström L, Ott S (2010) Chem Eur J 16:60
121. (a) Ott S, Borgström M, Kritikos M, Lomoth R, Bergquist J, Åkermark B, Hammarström L, Sun L (2004) Inorg Chem 43:4683; (b) Ekström J, Abrahamsson M, Olson C, Bergquist J, Kaynak FB, Eriksson L, Sun L, Becker HC, Åkermark B, Hammarström L, Ott S (2006) Dalton Trans 4599
122. (a) Li X, Wang M, Zhang S, Pan J, Na Y, Liu J, Åkermark B, Sun L (2008) J Phys Chem B 112:8198; (b) Song LC, Wang LX, Tang MY, Li CG, Song HB, Hu QM (2009) Organometallics 28:3834
123. Yu T, Zeng Y, Chen J, Li YY, Yang G, Li Y (2013) Angew Chem Int Ed 52:5631
124. (a) Chao TH, Espenson JH (1978) J Am Chem Soc 100:129; (b) Connolly P, Espenson JH (1986) Inorg Chem 25:2684
125. Razavet M, Artero V, Fontecave M (2005) Inorg Chem 44:4786
126. (a) Hawecker J, Lehn JM, Ziessel R (1983) New J Chem 7:271; (b) Deponti E, Natali M (2016) Dalton Trans doi:10.1039/c6dt01221c
127. Probst B, Roderberg A, Guttentag M, Hamm P, Alberto R (2011) Inorg Chem 49:6453
128. (a) Du P, Knowles K, Eisenberg R (2008) J Am Chem Soc 130:12576; (b) Wang X, Goeb S, Ji Z, Pogulaichenko NA, Castellano FN (2011) Inorg Chem 50:705
129. Khnayzer RS, McCusker CE, Olaiya BS, Castellano FN (2013) J Am Chem Soc 135:14068
130. (a) Lazarides T, McCormick TM, Du P, Luo G, Lindley B, Eisenberg R (2009) J Am Chem Soc 131:9192; (b) Zhang P, Wang M, Dong J, Li X, Wang F, Wu L, Sun L (2010) J Phys Chem C 114:15868
131. Lazarides T, Delor M, Sazanovich IV, McCormick TM, Georgakaki I, Charalambidis G, Weinstein JA, Coutsolelos AG (2014) Chem Commun 50:521
132. Fihri A, Artero V, Razavet M, Baffert C, Leibl W, Fontecave M (2008) Angew Chem Int Ed 47:564
133. (a) Natali M, Argazzi R, Chiorboli C, Iengo E, Scandola F (2013) Chem Eur J 19:9261; (b) Natali M, Orlandi M, Chiorboli C, Iengo E, Bertolasi V, Scandola F (2013) Photochem Photobiol Sci 12:1749
134. (a) Zhang P, Wang M, Li C, Li X, Dong J, Sun L (2010) Chem Commun 46:8806; (b) Peuntinger K, Lazarides T, Dafnomili D, Charalambidis G, Landrou G, Kahnt A, Sabatini RP, McCamant DW, Gryko DT, Coutsolelos AG, Guldi DM (2013) J Phys Chem C 117:1647; (c) Manton JC, Long C, Vos JG, Pryce MT (2014) Dalton Trans 43:3576
135. (a) Lakadamyali F, Reisner E (2011) Chem Commun 47:1695; (b) Lakadamyali F, Reynal A, Kato M, Durrant JR, Reisner E (2012) Chem Eur J 18:15464
136. Khnayzer RS, Thoi VS, Nippe M, King AE, Jurss JW, El Roz KA, Long JR, Chang CJ, Castellano FN (2014) Energy Environ Sci 7:1477
137. Helm ML, Stewart MP, Bullock RM, Rakowski DuBois M, DuBois DL (2011) Science 333:863
138. McLaughlin MP, McCormick TM, Eisenberg R, Holland PL (2011) Chem Commun 47:7989
139. Weingarten AS, Kazantsev RV, Palmer LC, McClendon M, Koltonow AR, Samuel APS, Kiebala DJ, Wasielewski MR, Stupp SI (2014) Nat Chem 6:964
140. Fujishima A, Honda K (1972) Nature 238:37
141. O'Regan B, Grätzel M (1991) Nature 353:737
142. (a) Hagfeldt A, Boschloo G, Sun L, Kloo L, Pettersson H (2010) Chem Rev 110:6595; (b) Bignozzi CA, Argazzi R, Kleverlaan CJ (2000) Chem Soc Rev 29:87
143. Odobel F, Le Pleux L, Pellegrin Y, Blart E (2010) Acc Chem Res 43:1063
144. Youngblood WJ, Lee SHA, Kobayashi Y, Hernandez-Pagan EA, Hoertz PG, Moore TA, Moore AL, Gust D, Mallouk TE (2009) J Am Chem Soc 131:926

145. Zhao Y, Swierk JR, Megiatto JD, Sherman B, Youngblood WJ, Qin D, Lentz DM, Moore AL, Moore TA, Gust D, Mallouk TE (2012) Proc Natl Acad Sci U S A 109:15612
146. (a) Alibabaei L, Brennaman MK, Norris MR, Kalanyan B, Song M, Losego MD, Concepcion JJ, Binstead RA, Parsons GN, Meyer TJ (2013) Proc Natl Acad Sci U S A 110:20008; (b) Ashford DL, Gish MK, Vannucci AK, Brennaman MK, Templeton JL, Papanikolas JM, Meyer TJ (2015) Chem Rev 115:13006
147. Gao Y, Ding X, Liu J, Wang L, Lu Z, Li L, Sun L (2013) J Am Chem Soc 135:4219
148. Gao Y, Zhang L, Ding X, Sun L (2014) Phys Chem Chem Phys 16:12008
149. Li L, Duan L, Wen F, Li C, Wang M, Hagfeldt A, Sun L (2012) Chem Commun 48:988
150. Ji Z, He M, Huang Z, Ozkan U, Wu Y (2013) J Am Chem Soc 135:11696
151. Fan K, Li F, Wang L, Daniel Q, Gabrielsson E, Sun L (2014) Phys Chem Chem Phys 16:25234
152. Li F, Fan K, Xu B, Gabrielsson E, Daniel Q, Li L, Sun L (2015) J Am Chem Soc 137:9153
153. Ager JW, Shaner MR, Walczak KA, Sharp ID, Ardo S (2015) Energy Environ Sci 8:2811, and references therein

Chapter 2
Solar Energy Conversion in Photoelectrochemical Systems

Stefano Caramori, Federico Ronconi, Roberto Argazzi, Stefano Carli, Rita Boaretto, Eva Busatto, and Carlo Alberto Bignozzi

Abstract The organization of photoresponsive molecular systems and nanomaterials on semiconductor surface holds great potential in the building of solar energy conversion devices where efficient energy conversion results from the optimized cooperation of several subsystems (semiconductor, dye sensitizers, redox mediator, hole transport medium), whose properties can be finely tuned through rational synthetic design. This chapter will review the fundamentals of semiconductor sensitization, a process relying on the quenching by charge transfer of molecular excited states coupled to semiconductor surfaces, and will move on by describing the structural and electronic properties of some of the most successful dye designs, used in conjunction with new electron transfer mediators in liquid electrolytes. From liquid electrolytes, a step forward is made by developing solid state hole conductors, which found their best employment in hybrid junctions with organo-halide lead perovskites, representing, at present, the most promising materials for solar-to-electric power conversion in mesoscopic solar cells. Finally, one of the most challenging tasks which can find solution by exploiting molecular level sensitized materials is discussed in detail through meaningful case studies: the production of solar fuels by photoelectrochemical water splitting.

2.1 Introduction

Meeting the demands of an energy-thirsty world while avoiding environmental damage poses a major scientific and societal challenge for our future. The technological development of our society will be driven by increasing energy demand, cost and environmental impact. At the global scale, the combination of various hydrocarbons sources will continue to play the major role on the next few decades,

---

S. Caramori (✉) • F. Ronconi • R. Argazzi • S. Carli • R. Boaretto • E. Busatto • C.A. Bignozzi
Department of Chemistry and Pharmaceutical Sciences, University of Ferrara, Via Fossato di Mortara 17, Ferrara 44121, Italy
e-mail: stefano.caramori@unife.it

G. Bergamini, S. Silvi (eds.), *Applied Photochemistry*,
Lecture Notes in Chemistry 92, DOI 10.1007/978-3-319-31671-0_2

but their costs and issues related to pollution, impact on human health, geo-political instability unavoidably connected to the inhomogeneous distribution of fossil resources, are doomed to increase. Besides fossil fuels, there are many energy sources like nuclear fission, wind, biomass, and geothermal but, in the long term, the most appealing and widely distributed is solar. The sun is the ultimate sustainable energy source, providing ~10,000 times the energy that is currently consumed worldwide. Solar energy is free of greenhouse gases and pollutants and is secure, without geo-political constraints. Solar energy conversion systems fall in three categories according to their primary energy product: solar electricity, solar fuels and solar thermal systems. Chemistry has played a fundamental role in the design of materials and molecular systems for solar energy conversion to electricity. The technological development of single-crystal solar cells demonstrates the practicality of photovoltaics, while novel approaches exploiting thin films, organic semiconductors and dye sensitisation offer new opportunities for cheaper, long lasting systems.

Recently, considerable advances have been made in the production of hydrogen with photoelectrochemical and photoelectrosynthetic solar cells. This chapter will cover the basic principles and some of the relevant contributions of Chemistry to the design of molecular systems for solar energy conversion to electricity and of photoelectrosynthetic devices for hydrogen production.

2.2 Semiconductor Materials

Semiconductor/liquid junctions provide the best known wet chemical method of converting solar energy into electrical energy or chemical fuels [1, 2]. Semiconductor electrodes typically operate with a quantum yield close to unity for all electromagnetic frequencies above the absorption threshold of the electrode, which can vary from the near infrared throughout the visible region of the spectrum.

Due to the essentially infinite number of atoms that must be considered, the electronic structure of solids is typically discussed in terms of energy bands which are made up of atomic orbitals of the individual atoms: [3] as a result of the large numbers of interacting orbitals the spacing of electronic energies within a band, arising from a given quantum state becomes so small that the band can be effectively considered a continuum of energy levels, however the energy gap between the groups of levels corresponding to different atomic quantum states is preserved. Thus the allowed electronic energies fall into energy bands of closely spaced levels, with forbidden gaps between these bands. As with molecular orbitals, often the energy levels of interest are the highest occupied (called the valence band, VB) and the lowest unoccuppied (called the conduction band CB) bands (Fig. 2.1).

It is the energy gap (the band-gap) between these bands (i.e., the difference in energy between the upper edge of the valence band and the lower edge of the conduction band) that determines the properties of the material. In insulators the valence band is full, the conduction band is empty and no net motion of charge

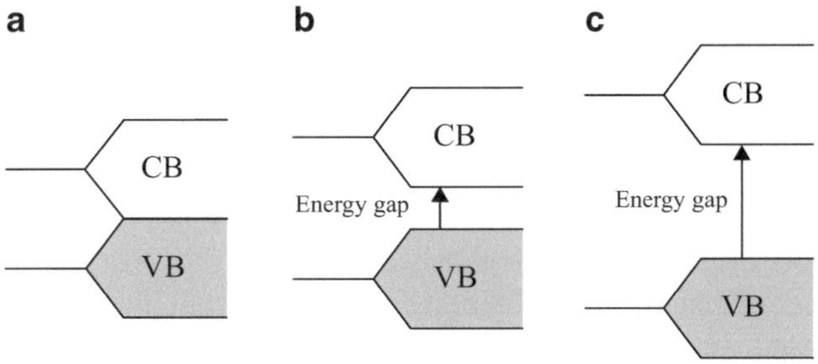

Fig. 2.1 Representation of Conduction Band (*CB*) and Valence Band (*VB*) in terms of band theory for a Metal (**a**) a Semiconductor (**b**) and an Insulator (**c**)

results from application of an electric field. In a metal the uppermost energy band containing electrons is only partially filled, or a filled band overlaps an empty band. These electrons are free to move in a field, and since they are present in large numbers they result in very high conductivities (10^4–10^6 Ohm^{-1} cm^{-1}). In semiconductors the situation is similar to that in insulators, except that the band-gap is smaller and electrons can be thermally or optically promoted to the conduction band, resulting in an electrical conductivity which is smaller than that of metals because of the smaller number of current carriers.

The promotion of electrons from the VB to the CB leaves a positively charged vacancy in the VB, which is referred to as a hole. These holes can be moved through space by the transfer of one electron to the vacancy, therefore holes are considered to be mobile.

Another important concept in the discussion of solid state materials is the Fermi Level. The Fermi Dirac distribution is given by

$$f(E) = \frac{1}{1 + e^{\frac{(E\text{-}E_F)}{kT}}} \tag{2.1}$$

where $f(E)$ is the probability that a state of energy E is occupied, E_F is a parameter called the Fermi energy, k is the Boltzmann's constant and T is the absolute temperature. It can be seen that E_F is the energy level for which $f(E) = 1/2$, therefore the Fermi level (E_F) is a virtual energy level which has the 50 % of probability to be occupied by the electrons.

The Fermi energy is just the electrochemical potential [3], or partial molar free energy per electron. This result is physically understandable considering two equilibrating systems: thermodynamically we expect them to exchange particles (electrons in this case) until their electrochemical potentials are equal; from the point of view of Fermi statistics, the transfer of particles from one system to another fills states of lower energy in one system emptying higher lying filled states in the other system and proceeds until the distributions over energy in the two systems match, i.e., the Fermi levels are equal. For an intrinsic semiconductor the Fermi level lies at the midpoint of the band-gap.

Si and Ge are prototypical examples for clarifying the basic principles of semiconductor energetics and conduction mechanism. The introduction of electron rich or electron deficient impurities into the silicon lattice (doping) has a profound influence on the resulting electronic properties of the semiconductor. For example, a small level of As in a Si semiconductor introduces an extra electron per As atom. Since the occupied molecular orbitals of Si are filled, these excess electrons fill new electronic states constituting the so called donor band (Fig. 2.2a) energetically close to the conduction band edge. Thermal excitation from the donor band to the conduction band, results in improved conductivity thanks to the presence of an excess of negative charge carriers (electrons). Such semiconductor is known as *n-type* (negative) semiconductor. Similarly, impurities of Ga in a Si semiconductor lead to the formation of band of electron acceptors in the proximity of the valence band edge. The excitation of an electron from the valence band to the acceptor band leaves a positive vacancy, or hole, into the valence band (Fig. 2.2b). Thus electronic conduction takes place by movement of holes (positive charge carriers) and materials characterized by this conduction mechanism are referred to as *p-type* (positive) semiconductors. For many solid-state electronic devices, impurities such as Ga or As are deliberately introduced at controlled levels (at ratios of 1 to 10^9) into ultrapure Si or Ge to produce *n-* and *p-type* extrinsic semiconductors which are extremely important for their semiconducting properties in solid-state electronic devices.

Sunlight contains photons with energies that reflect the Sun's surface temperature; in energy units of electron volts (eV), the solar photons reaching the Earth surface range in energy (hν) from about 3.5 eV (350 nm) to 0.5 eV (2500 nm) (Fig. 2.3). The energy of the visible region ranges from 3.0 (violet) to 1.8 eV (red); the peak power of the Sun occurs in the yellow region of the visible region, at about

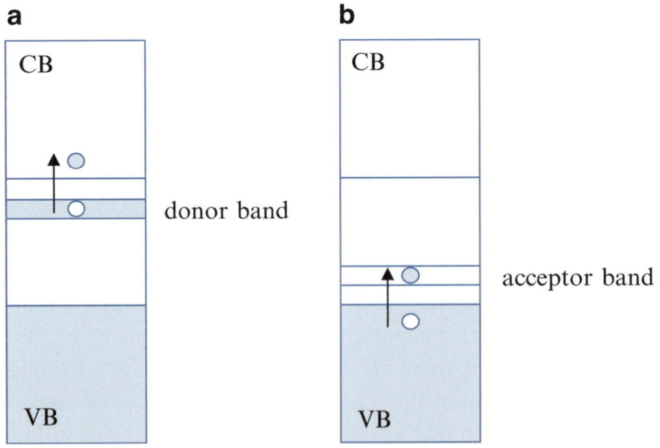

Fig. 2.2 Band structure in (**a**) an n-type semiconductor and (**b**) a p-type semiconductor

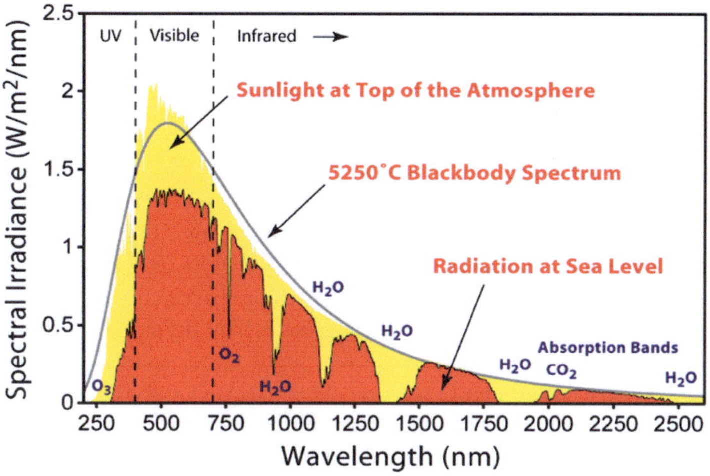

Fig. 2.3 Spectral distribution of solar irradiance

2.5 eV. At high noon on a cloudless day, the surface of the Earth at average latitudes receives 1000 watts of solar power per square meter (1 kW/m^2). Such standard irradiance is expressed as Air Mass 1.5 (AM 1.5) condition.

Photovoltaic cells generally consist of a light absorber that will only absorb solar photons above a certain minimum photon energy. This minimum threshold energy is called the "energy gap" or "band-gap" (Eg); photons with energies below the band-gap are transmitted, while photons with energies above the band-gap are absorbed. The electrons and holes formed upon absorption of light separate and move to opposite sides of the cell structure, following electrochemical potential gradients, where they are collected by the ohmic contacts. The net effect of this process is the generation of a photovoltage and a photocurrent which translates into the production of electrical power.

A French physicist, Edmund Becquerel, discovered the photovoltaic (PV) effect in 1839 [4], when he observed that a voltage and a current were produced when a silver chloride electrode immersed in an electrolytic solution and connected to a counter metal electrode was illuminated with white light. However, the birth of the modern era of PV solar cells occurred in 1954, when D. Chapin, C. Fuller, and G. Pearson at Bell Labs demonstrated solar cells based on p-n junctions in single crystal Si with efficiencies of 5–6 %.

From the mid 1950s to the early 1970s, PV research and development (R&D) was directed primarily toward space applications and satellite power. Total global PV (or solar) cell production increased from less than 10 MWp/year in 1980 to about 1200 MWp/year in 2004; the current total global PV installed capacity is about 200 GW with a further increase foreseen for the next years.

In organic molecular structures, the energy of the photons also must first exceed a certain threshold to be absorbed. This absorption creates an energetic state of the molecular system, called an excited state. These excited molecular states can also generate separated electrons and holes. Furthermore, certain organic polymers and other molecular structures can form organic semiconductors that provide the basis for organic PV devices. One difference between inorganic and organic PV cells is that in organic cells, the electrons and holes are initially bound to each other in pairs called excitons; these excitons must be broken apart in order to separate the electrons and holes to generate electricity. In inorganic PV cells, the electrons and holes created by the absorption of light are not bound together and are free to move independently in the semiconductor. The PV cells deliver maximum power P_{max} when operating at a point on the current-voltage (I–V) curve where the product IV is maximum. The efficiency (η) of a solar cell is defined as the power P_{max} supplied by the cell at the maximum power point under standard test conditions, divided by the power of incident radiation. Most frequent conditions are: irradiance 100 mW/cm^2, standard reference spectrum, and temperature 25 °C. The use of this standard irradiance value is particularly convenient since the cell efficiency in percent is then numerically equal to the power output from the cell in mW/cm^2.

PV cells can be formally divided into three categories: (1) inorganic cells, based on solid-state inorganic semiconductors; (2) organic cells (excitonic cells), based on organic semiconductors blended with molecular species; and (3) photoelectrochemical (PEC) cells, based on interfaces between semiconductors and molecules. PEC cells, relying on the interaction of molecules with extended systems, typically wide band-gap semiconductors, will be the focus of this book chapter.

2.3 Photoelectrochemical Solar Cells

Since their appearance in the late 1980s by the work of Grätzel [5, 6], photoelectrochemical solar cells based on the principle of sensitization of wide band-gap mesoporous semiconductors, have attracted the interest of the scientific community due to their outstanding performances which started the envision of a promising alternative to conventional junction-based photovoltaic devices. For the first time a solar energy device operating on a molecular level showed the stability and the efficiency required for potential practical applications.

The principle of dye sensitization of semiconductors can be traced back to the end of 1960s [7], however the efficiencies obtained with single crystal substrates were too low due to the poor light absorption of the adsorbed monolayer of dye molecules. The breakthrough in the field was brought by the introduction of mesoscopic films made of sintered nanoparticles of a semiconductor metal oxide with a large surface area which allowed the adsorption, at monolayer coverage, of a much larger number of sensitizer molecules leading to absorbance values, of thin films of a few microns, well above unity [5]. The research field grown around Dye-Sensitized Solar Cells (DSSCs or DSCs) is strongly multidisciplinary, involving several areas such as nanotechnology, materials science, molecular engineering, photochemistry and electrochemistry among the foremost.

Wide band-gap semiconductor materials such as TiO_2, SnO_2, $SrTiO_3$, ZnO, etc. have a separation between the energy levels of the valence band and the conduction band of the order of 3 eV, which means that electron-hole pairs can be generated, upon photoexcitation, by a radiation of wavelength below 400 nm, at the limit between the visible and the UV spectra. In order to increase the spectral sensitivity of these materials towards low energy photons, two general approaches have been devised: doping and molecular sensitization. Through a suitable choice of doping elements, the bulk properties of a semiconductor material can be modified in such a way that donor levels are created just below the conduction band (n-type doping) or acceptor levels are created just beyond the valence band (p-type doping). These intermediate energy levels reduce the amount of energy required to promote electrons to the conduction band, or holes to the valence band, by means of photon absorption, shifting towards red the useful range for light action. Although this approach is the preferred choice for conventional p-n photovoltaic devices, in the case of wide band-gap semiconductors it has been successfully applied only in the field of photocatalysis. Molecular sensitization is obtained when a photoexcited molecule, called sensitizer, in its electronically excited state is capable of promoting an heterogeneous charge transfer process with the semiconductor. When the excited state energy level of the sensitizer is higher with respect to the bottom of the conduction band, an electron can be injected with no thermal activation barrier into the semiconductor, leaving the sensitizer in its one electron oxidized form (Fig. 2.4a). When the excited state is lower in energy with respect to the top of the valence band, an electron transfer (formally a hole transfer) between the semiconductor and the sensitizer can take place leaving the molecule in its one electron reduced form (Fig. 2.4b) [8].

The sensitizer can be freely diffusing in solution or firmly attached to the surface of the semiconductor by virtue of suitable anchoring groups. This latter case is preferred since it can ensure a better donor-acceptor interaction between dye molecules and semiconductor substrate, increasing the electron transfer efficiency.

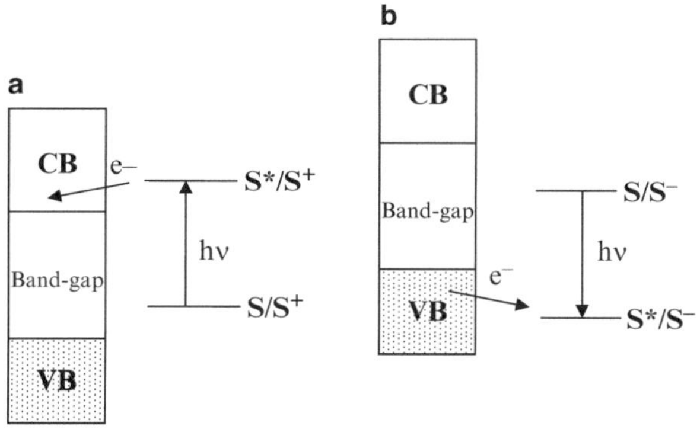

Fig. 2.4 Sensitized charge injection: (**a**) electron injection (oxidative quenching); (**b**) hole injection (reductive quenching)

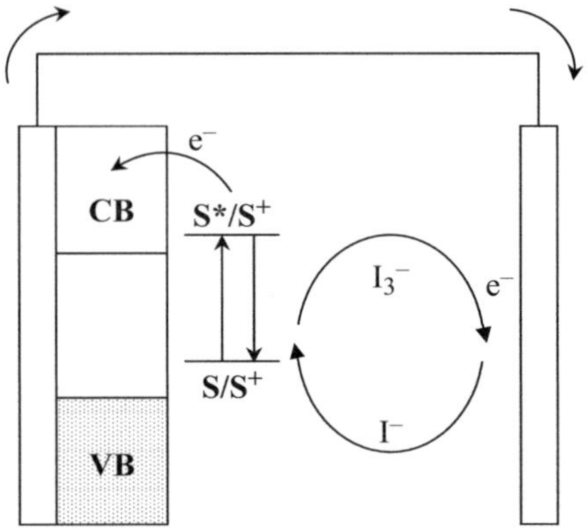

Fig. 2.5 Schematic working principle of a dye-sensitized solar cell

Electron injection sensitization is the fundamental mechanism of photoelectrochemical solar cells based on titanium dioxide and will be discussed in further details.

The working principle of a dye-sensitized solar cell is schematized in Fig. 2.5 [9]. The device is comprised of two faced electrodes, a photoanode and a counter electrode, with an electrolyte in between. Both electrodes are usually made from a sheet of common float glass coated on one side with a thin transparent conductive layer of fluorine doped tin dioxide (FTO) or indium tin oxide (ITO). Surface electrical resistivities of a few Ω-square can be easily obtained while preserving a good optical transmission over the whole visible and near infrared spectrum. Several efforts have been made recently in order to substitute the glass substrate with a less expensive polymer which have also the advantage of permitting the fabrication of flexible devices. In the photoanode, the transparent conductive electrode is covered with a thin film (7–10 μm) of a mesoporous semiconductor oxide obtained via a sol-gel procedure. This oxide layer can be deposited, starting from a colloidal suspension, by means of a variety of methods including spraying, screen printing, dip coating, spin coating or the simple doctor blade technique. After deposition, the layer is sintered in oven at a temperature of about 500 °C, a treatment which has the scope of creating a network of interconnected particles through which electrons can percolate. Dye coverage of semiconductor nanoparticles is generally obtained from alcoholic solutions of the sensitizer in which the sintered film is left immersed for a few hours. Sensitizers are usually designed to have functional groups such as –COOH, $–PO_3H_2$, –R(CN)(COOH) for stable adsorption onto the semiconductor substrate. The dye covered film is in intimate contact with an electrolytic solution containing a redox couple dissolved in a suitable solvent. The electron donor of the redox couple must reduce quickly and quantitatively the oxidized sensitizer. A variety of solvents with different viscosity and of redox mediators has been the object of intense researches, the most common being the couple I_3^-/I^- in acetonitrile or methoxy propionitrile solution.

The counter electrode is a conductive glass covered with a few clusters of metallic platinum which has a catalytic effect in the regeneration process of the electronic mediator.

The complete photoelectrochemical cycle of the device can be outlined as follows: the adsorbed sensitizer molecules (S) are brought into their excited state (S^*) by photon absorption and inject one electron in the empty conduction band of the semiconductor in a time scale of femtoseconds. Injected electrons percolate through the nanoparticle network and are collected by the conductive layer of the photoanode, while the oxidized sensitizer (S^+) in its ground state is rapidly reduced by I^- ions in solution. Photoinjected electrons flow in the external circuit where useful electric work is produced and are available at the counter electrode for the reduction of the oxidized form of the electron mediator (I_3^-). The entire cycle consists in the quantum conversion of photons to electrons.

$$\begin{array}{cc} S + h\nu \rightarrow S^* & \text{photoexcitation} \\ S^* + TiO_2 \rightarrow S^+ + (e^-, TiO_2) & \text{electron injection} \\ 2S^+ + 3I^- \rightarrow 2S + I_3^- & \text{sensitizer regeneration} \\ I_3^- + 2e^- \rightarrow 3I^- & \text{electron donor regeneration} \end{array}$$

Photoinjected electrons should escape from any recombination process in order to have a unit charge collection efficiency at the photoelectrode back contact. The two major waste processes in a DSC are due to (i) back electron transfer, at the semiconductor-electrolyte interface, between electrons in the conduction band and the oxidized dye molecules, and (ii) reduction of the electron mediator, I_3^-, at the semiconductor nanoparticles surface

$$\begin{array}{cc} S^+ + (e^-, TiO_2) \rightarrow S & \text{back electron transfer} \\ I_3^- + 2(e^-, TiO_2) \rightarrow 3I^- & \text{electron capture from mediator} \end{array}$$

A detailed knowledge of all the kinetic mechanisms occurring in a DSC under irradiation is an essential feature of its optimization process.

2.3.1 *Solar Cell Efficiency*

Since a DSC is a quantum converter, the efficiency of the conversion process is expressed in terms of the ratio between the number of electrons flowing in the external circuit and the number of photons, at a particular wavelength, striking the photoanode. This monochromatic quantum efficiency is generally referred to as Incident Photon-to-Current Conversion Efficiency (IPCE). In terms of measurable quantities the IPCE value is given by the formula:

$$IPCE(\lambda) = \frac{n_e}{n_{h\nu}} = \frac{hc}{e} \frac{J}{\lambda\ P} \tag{2.2a}$$

where h is the Planck's constant (J s), c is the speed of light in vacuo (ms^{-1}), e is the electron charge (C), J is the photocurrent density ($A\ m^{-2}$), λ is the photon wavelength (m) and P is the radiant power density ($W\ m^{-2}$) at the wavelength λ. A plot of IPCE values as a function of wavelength is called action spectrum. This quantum efficiency can be expressed as the product of three terms related to the key steps of the device working principle, as follows:

$$IPCE(\lambda) = LHE(\lambda)\ \phi_{inj}\ \eta \tag{2.2b}$$

where LHE is the light harvesting efficiency, ϕ_{inj} is the electron injection quantum yield and η is the efficiency with which electrons are collected at the photoanode. Considering their importance these three terms will be discussed in detail.

The Light Harvesting Efficiency (LHE) is the fraction of incident light that is absorbed by the dye-coated mesoporous film. Due to the small size of the semiconductor nanoparticles (≈20 nm), light scattering from a 10 μm film is almost negligible and the transparency is sufficiently high that Beer-Lambert law can be applied.

$$LHE = \frac{P_a}{P} = 1 - 10^{-A} \tag{2.3}$$

where A is the absorbance of the film due to dye molecules, given by

$$A = 1000\ \varepsilon\ c\ t \tag{2.4}$$

with ε molar absorptivity of the dye ($mol^{-1}\ l\ cm^{-1}$); c, dye concentration ($mol\ cm^{-3}$); t, film thickness (cm) and 1000 ($cm^3\ l^{-1}$), volume conversion factor. Typical values for the volume concentration of dye molecules are of the order of 10^{-4} $mol\ cm^{-3}$ and the maximum extinction coefficient for a large class of sensitizers is in the range (1–2) 10^4 $mol^{-1}\ l\ cm^{-1}$, this means that for a 10 μm (10^{-3} cm) film the term LHE is in excess of 90 %. The search for a "black" dye whose molar absorptivity is high and constant throughout the useful solar spectrum can be easily rationalized and understood in terms of the LHE factor. Recently, several strategies for increasing the LHE term have been devised bearing on the general principle that an augmented light scattering in the semiconductor film should increase the probability of photon absorption by the sensitizer molecules.

To ensure a good interfacial charge separation, the electron injection process must compete with all the excited state deactivation pathways of the sensitizer. The associated charge injection efficiency is given by

$$\phi_{inj} = \frac{k_{inj}}{k_{inj} + k_r + k_{nr} + k_q[Q]} = \frac{k_{inj}}{k_{inj} + \frac{1}{\tau}} \tag{2.5}$$

where k_{inj} is the charge injection rate constant, k_r and k_{nr} are the radiative and non-radiative rate constants and k_q is the rate constant for bimolecular quenching due to the quencher Q. Typical values of τ, the sensitizer lifetime in the absence of charge injection, lie in the ns-μs time domain. To achieve a satisfactory value of ϕ_{inj} the value of k_{inj} should be at least 100 times the sum of deactivation rate constants, which implies that electron injection should occur in picoseconds or faster times. With ultrafast transient absorption spectroscopy, the kinetics of heterogeneous charge injection for ruthenium based MLCT sensitizers has been resolved on a femtosecond time scale leading to an estimate value for $k_{inj} < 5\ 10^{13}\ s^{-1}$ for the N719 dye adsorbed on TiO_2 [10]. The reason for such a high rate can be ascribed to the optimum electronic coupling, existing between the sensitizer and the semiconductor, provided by the carboxylic functions of the dye, and to the large density of acceptor states in the TiO_2 substrate. For a large class of sensitizers the interfacial charge separation is so fast that is thought to occur from vibrationally hot excited states.

As stated before, photoinjected electrons must escape from any charge recombination or reduction process at the photoanode to ensure a 100 % electron collection efficiency. In order to reach a good charge separation after electron injection, the recombination between electrons in the semiconductor conduction band and oxidized dye molecules should be unfavorable. Several factors contribute to the kinetics of this process. Contrary to charge injection, which for MLCT sensitizers occurs from ligand π^* orbitals well coupled to the semiconductor conduction band, back electron transfer to the oxidized dye involves metal centered d orbitals with a far less overlap with the orbitals contributing to the conduction band. This implies that the electronic coupling element for back electron transfer is lower causing a retardation of the process. Another important factor comes from the large driving force and small reorganization energy (1.5 eV and 0.3 eV respectively for N719) of the back reaction which places the process in the Marcus inverted region as confirmed by the observation that the recombination rate is almost temperature independent and insensitive to the environment surrounding the coated film [11].

The presence of a strong electric field at the semiconductor-electrolyte boundary also have a remarkable effect on charge recombination. This field is created partly by proton transfer from the carboxylic functions of the dye, or from protic solvents, to the oxide surface resulting in the formation of a positively charged sheath which stabilizes electrons in the conduction band. In the presence of high charge density cations such as Li^+ or Mg^{2+} which intercalates between nanoparticles, the effect is further reinforced giving back electron transfer rates in the range 10^5–$10^6\ s^{-1}$. Finally, the existence of trapping surface states on the semiconductor have a strong influence on the kinetic regime of this interfacial electron transfer.

The second important waste process is the reduction at the photoanode of the acceptor member of the redox mediator couple. It has been shown [12–14] that in the case of the I_3^-/I^- redox couple, the relevant species is the radical anion $I_2^{\bullet -}$ which is formed by the reactions

$$S^+ + 2I^- \rightarrow S + I_2^{\bullet -} \tag{2.6}$$

$$I_3^- + (e^-, TiO_2) \rightarrow I_2^{\bullet -} + I^- \tag{2.7}$$

and decays through the conduction band electron capture (2.8) and dismutation reaction (2.9)

$$I_2^{\bullet -} + (e^-, TiO_2) \rightarrow 2I^- \tag{2.8}$$

$$2\,I_2^{\bullet -} \rightarrow I^- + I_3^- \tag{2.9}$$

The temporal decay of $I_2^{\bullet -}$ has been followed by means of nanosecond transient spectroscopy monitoring the absorption of the radical anion at 760 nm. The fast component of the decay trace (~100 ns) has been attributed to reaction (2.8) while the long-lived component is related to the dismutation reaction (2.9). Two kinetic regime have been postulated depending on the number of injected electrons per nanoparticle (N_e): when $N_e > 1$ reaction (2.8) is dominant and represents the main electron loss pathway; when N_e is lower than unity reaction (2.9) starts to compete and accounts for the long lived component. The conclusion is that at high light intensity the reduction of $I_2^{\bullet -}$ by photoinjected electrons is the only one which contributes to the quantum efficiency losses of a DSC [42].

As discussed above, the IPCE parameter reflects the performance of the entire device taking into account the efficiencies of all the processes involved and has been the basis of comparison in a variety of studies on different aspects of the DSC improvement. A complete photovoltaic characterization of a DSC requires the evaluation of parameters which contribute to the so-called global efficiency, expressed by the ratio between the maximum electric power delivered by the cell at optimum load and the optical power of the irradiating source which is normally referred to the standard AM 1.5 solar spectral irradiance distribution. The global efficiency is given by the formula

$$\eta = \frac{V_{oc}\ J_{ph}\ FF}{P_s} \tag{2.10}$$

where V_{oc} is the open circuit photovoltage (V), J_{ph} is the short circuit integrated photocurrent density (A m^{-2}), FF is the cell fill factor and P_s is the total optical power density at the cell surface (W m^{-2}). The photocurrent density is a measured quantity but can also be calculated by integration of Eq. (2.11) with respect to λ

$$J_{ph} = \frac{e}{hc} \int^{+\infty} IPCE(\lambda)\lambda\ P(\lambda)\ d\lambda \tag{2.11}$$

Fill factor values are obtained by measuring the J-V characteristics of a cell irradiated under simulated solar light and taking the ratio between the maximum electric work produced at optimum load and the product of open circuit voltage and short circuit current. Typical values for J_{sc} are in the range 16–22 mA cm^{-2}, while open circuit voltages of 0.7–0.86 V are easily obtained with fill factor values of 0.65–0.8. A global efficiency of 14 % has been recently reported for a state of the art DSC, making these devices truly competitive with solar cells based on amorphous silicon [15].

It should be noted that the maximum open-circuit photovoltage attainable in devices of this type is the energetic difference between the Fermi level of the solid under illumination and the Nernst potential of the redox couple in the electrolyte. However, for these devices this limitation has not been realized and V_{oc}, is in general much smaller. It appears that V_{oc}, is kinetically limited by electron tunneling through the solid to acceptors at the interface or in the electrolyte. For an n-type semiconductor in a regenerative cell, Eq. (2.12) holds [16].

$$V_{oc} = \frac{kT}{e_s} \ln \left(Iinj/n \sum ki[A]i \right) \tag{2.12}$$

where *Iinj*, is the electron injection flux, *n* is the concentration of electrons in TiO_2, and the summation is for all electron transfer rates to acceptors A. This fact has a marked influence on solar cell efficiency, since, as it will be described in the following sections, charge recombination between conduction band electrons and the oxidized electron mediators is one of main factors limiting V_{oc} and as a consequence the overall efficiency of the solar device.

2.4 Molecular Sensitizers

DSSC functioning is based on the interlacing of subsystems working cooperatively: the photoanode on which the dye sensitizer is adsorbed, the electron mediator and the catalytic counter electrode [6, 17, 18]. The development of new dyes and electron mediators, as well as the understanding of their interaction and of the interfacial electron transfer kinetics in which they are involved is central for the further progress in this field and is currently polarizing most of the scientific efforts of the DSSCs community. Indeed, considerable work is now directed towards the optimization of organic sensitizers [19] and of natural sensitizers [20, 21] extracted from plants, though transition metal complexes are still playing a major role in DSSC development, since many of the most efficient and most stable dye sensitizers are constituted by Ru(II) complexes displaying intense metal to ligand charge transfer (MLCT) transitions. The synthetic design of the dye sensitizer as well as the understanding of their properties has gained a significant benefit by theoretical calculations which have reached a considerable degree of sophistication, allowing to simulate TiO_2 clusters of large size (>100 Ti^{4+} ions) in order to reliably evaluate the positioning of the relevant energy levels of the dye with respect to the bands of the semiconductor [22]. Extensive studies, corroborated by theoretical calculations, revealed that the electron photoinjection observed in TiO_2/Ru-bipyridyl assemblies in contact with an inert solvent is commonly characterized by sub-picosecond dynamics, much faster than the competing process of excited state decay [23]. Thus, electron injection has not generally been considered to be a key factor limiting the device performance. Since, in principle, a moderate exoergonicity results in an activation barrier defined by $\Delta G^* = (E^*_{ox} + \lambda - E^F)$, where E^*_{ox} is the excited state potential of the dye (expressed in eV), λ is the total reorganization energy accompanying the process and E^F is the Fermi energy of the semiconductor, which is sufficiently low to allow for a very fast charge injection.

The development of new red sensitive dyes with extended light harvesting into the NIR region is an interesting possibility for improving DSSC efficiency. Zn (II) porphyrins, well documented for their intense absorption bands in the red part of the visible region, have recently received considerable attention because of their possible applications in this type of solar cells [24]. Panchromatic DSSC can be achieved by using co-sensitization with two or more dyes within a single photoelectrode [19, 25]. Ru-complexes, such as the black dye (BD) [26], and push-pull dyes reached an efficiency of 11.4 % certified by a public research center (AIST), and the co-sensitization with a zinc-porphyrin plus an organic push-pull dye led to the record value of 12.3 % in the presence of a non corrosive electrolyte [27]. The use of other dye cocktails based on push pull designs allowed to surpass 13 % when used in conjunction with a cobalt (II) electrolyte [15].

Another interesting strategy that allows, at least in theory, to improve the cell efficiencies is to use a tandem cell that integrates a n-DSSC and a p-DSSC. Not only the latter can be as efficient as the former but tandem devices may prove to be the only one capable of bettering the efficiency of 15 %. However, up to now the study of p-type dyes, which are usually coupled to NiO photoelectrodes, has represented a negligible fraction of the research on photosensitizers [28, 29]. This type of research is challenging, since energy requirements of the dyes for p-type sensitization are different from those of the n-type sensitizers and require the HOMO lying below the upper edge of the valence band. In addition the absorption spectrum of the sensitizer chemisorbed on the p-type semiconductor must possess complementary features with respect to the sensitizer of the n-type counterpart in order to increase the photocurrent output in the tandem cell.

2.4.1 Ruthenium Dye Sensitizers

Since the O'Regan and Grätzel report in 1991 [6], metal complexes based on Ruthenium have been intensively investigated for DSSC applications and hundreds of dyes have so far been tested. Many research groups have attempted to modify the structures of coordinated chromophoric and ancillary ligands with the goal of improving the photovoltaic performance though only in few cases the results have been better than those observed with the N3 dye, one of the first dyes reported by the Grätzel's group [30]. The well known N3 complex $[Ru(H_2DCB)_2(NCS)_2]$, where H_2DCB is the 4,4′ dicarboxy 2,2′ bipyridine (Fig. 2.5), is a rare example of a molecule satisfying several of the requirements that an efficient sensitizer has to fulfil: [31–37] (i) a broad and strong absorption, preferably extending from the visible to the near-infrared; (ii) minimal deactivation of its excited state through the emission of light or heat; (iii) a firm, irreversible adsorption (chemisorption) to the semiconductor's surface and a strong electronic coupling between its excited state and the semiconductor conduction band; (iv) chemical stability in the ground as well as in the excited and oxidized states; (v) negative enough excited state oxidation potential in order to bring about an effective electron injection; and (vi) ground state oxidation potential positive enough to provide the driving force for oxidizing the electron mediator. Despite irreversible oxidative electrochemistry, thermal and photochemical instability [38–41] of the N3 complex in solution, solar devices based on the partially deprotonated form and containing 0.5 M I^- perform extremely well, giving an overall efficiency in the range 8.5–10 % [13, 26] and a certified stability of 10 years [42].

In about 20 years of work, carried out in different laboratories, the most significant improvements have been obtained with dyes similar to N3 by modifying the ancillary ligands or the degree of protonation at the polypyridine ligands which provide the binding site for TiO_2 (Fig. 2.6, N719 [43], BD [26], 2 [44]). A comprehensive review on this subject has recently been reported [45].

Chemical strategies for avoiding the labile Ru-NCS bond have been recently realized. In 2009 Grätzel reported an efficiency of 10.1 % for a cyclometallated Ru dye (Fig. 2.6, dye 1) [46] sensitizer displaying an IPCE of 83 % at 570 nm and

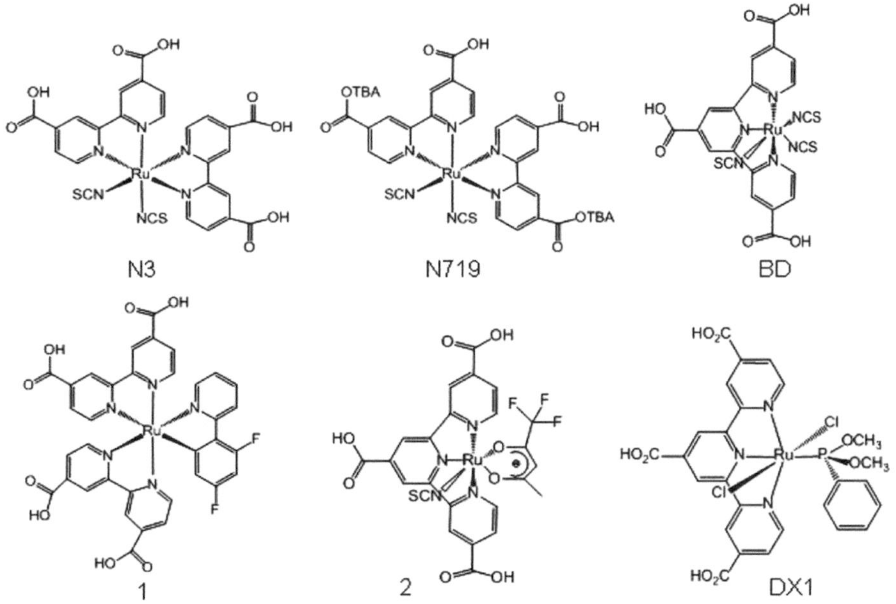

Fig. 2.6 Significant examples of n-type sensitizers based on Ru (II) polypyridine complexes

producing, at that time, the highest efficiency for any sensitizer without NCS groups. These results show unprecedented control of the HOMO energy level for Ru dyestuff, and therefore represent a new paradigm of development for the DSSCs. Although N3 has been widely used, the lack of absorption in the red region of the visible spectrum, in addition to the low molar extinction coefficient, has prompted research to synthesize panchromatic sensitizers that absorb visible light of all colours.

The complex (tri(isothiocyanato)(2,2′;6′,2″-terpyridyl-4,4′,4″-tricarboxylic acid)ruthenium(II) known as black dye (BD) has been found to have near – IR photo response up to 920 nm. It is important to note that both N3 and BD are considered among the most efficient ruthenium based sensitizers for light conversion to electricity and are composed of three fragments, namely: (i) the Ru transition metal, (ii) the polypyridine ligand and (iii) the NCS ligand. The use for BD of the terpyridyl ligand, with its low-lying π^* molecular orbitals, gives an impressive panchromatic absorption associated with the stabilization of the "hole" at the Ru(III) due to electron-donating ligands, resulting in an overall efficiency of 10.4 %.

However, the extension of the spectral response to the near IR is gained at the expense of shifting the LUMO orbital to lower energy from where injection into TiO_2 conduction band can no longer occur and the challenge to further optimize DSSC devices still remains. The complex trans-dichloro-(phenyldimethoxyphosphine) (2,2′;6′,2″-terpyridyl-4,4′,4″-tricarboxylic acid) ruthenium(II), with a phosphine-coordinated to the ruthenium metal centre [47] (Fig. 2.6, DX1), behaves as a panchromatic harvester with visible absorption extending up to 1000 nm and was found capable of higher efficiency compared to BD and N719 dyes. This result is achieved through a fine tuning of the HOMO level of the complex, thanks to the combination of electron donating (Cl^-) and electron accepting (dimethyl-phenyl-phosphine) ancillary ligands.

2.4.2 *Donor-Acceptor Organic Dyes*

An increasing attention is devoted to donor-acceptor organic dyes of the type reported in Fig. 2.7 [48, 49]. In general, the photoconversion efficiencies observed with these dyes are somewhat lower ($\leq$10 %) in comparison with the ruthenium based sensitizers, but they have advantageous features such as high molar extinction coefficient and tunable absorptions achieved through variation of the molecular structure which typically consists of the donor – π(spacer)–acceptor (D-π-A) architecture. Most of the D-π-A dyes have di-alkylamine [50] or diphenylamine moieties [50, 51] as electron donors and carboxylic acid [52, 53] or cyanoacrylic acid moieties [50] as electron acceptors, serving also as anchoring groups for attachment to the TiO_2 surface. The charge transfer character of the excited dye facilitates rapid electron injection from the dye molecule into the CB of the TiO_2, leaving the cationic charge effectively decoupled from the TiO_2 surface. The most

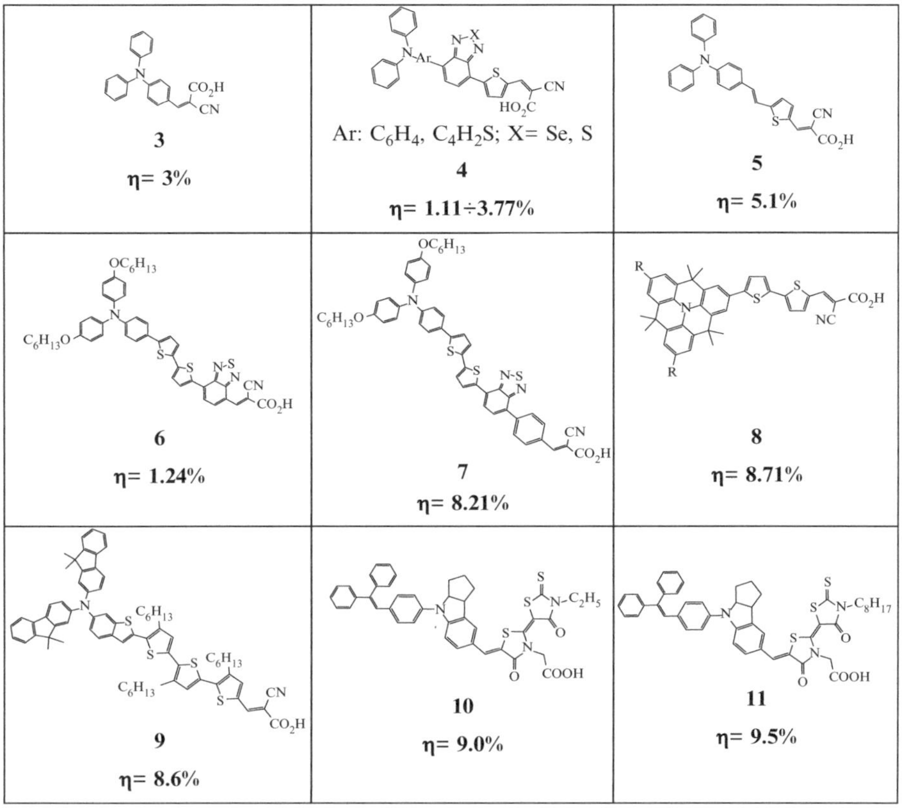

Fig. 2.7 Examples of Donor-π bridge-acceptor used for TiO_2 sensitization

successful donor units are represented by functionalized triarylamine moieties [51], which are characterized by chemical and photochemical stability due to charge delocalization in the aryl substituents. The role of small variations in the molecular design and its influence on the final performances of the solar device is exemplified by the comparison between dye 6 and 7 [54]. The insertion of a phenyl ring in the acceptor unit of dye 7 results in a ca. sevenfold increase of the efficiency (1.24 for dye 6, 8.21 for dye 7) This effect was explained by the increased twisting angle (48° according to UHF calculations) between the phenyl ring of the acceptor unit and the π -conjugated donor in the oxidized dye, decoupling the photogenerated hole from the TiO_2 surface and slowing down over five times the charge recombination process. The parallel inhibition of charge recombination between conduction band electrons and the oxidized electron mediator is also expected to enhance the electron collection efficiency. This has been elegantly demonstrated by the Ko group [55] which realized the new type of organic sensitizer 9, incorporating a planar amine with long alkyl chains units which screen the semiconductor surface from the electron acceptors present in the electrolyte. This dye yielded an overall conversion efficiency of 8.71 % under standard global air mass 1.5 solar conditions as a consequence of an increased open circuit photovoltage resulting from inhibition of the electron recapture. In general the functionalization of the donor part and of the π bridge has profound repercussions on the performance of organic dyes, since often planar or less hindered structures result in the formation of molecular aggregates on the TiO_2 surface which, by quenching the excited state, are one of the major reasons of the reduced performance of organic sensitizers compared to Ru (II) complexes. As an example of successful structural design to counter this negative effect, dye 9 [51] resulted in very high efficiencies (8.6 %) thanks to the incorporation of an extended π bridge constituted by head-to-tail oligo3-hexylthiophene which limited aggregation. Besides contributing to reduce stacking phenomena along with the hexyl chains, the bulky difluorenylphenyl-amine donor group of 9 [56] was also effective in passivating the titania surface against back recombination reactions involving I_3^-. The same motivations inspired Uchida and coworkers to the design of indoline-based dyes 10 [57] and 11 [58] which generated efficiencies of respectively 9 and 9.5 % in the presence of an antireflection coating. Although the dyes based on the indoline core have a single carboxylic group to anchor to the titania surface, maximum IPCEs exceeding 90 % were obtained, indicating nearly quantitative light absorption, charge injection and collection. These examples underline how a careful design of these dye structures can lead to the complete control over the interfacial charge separation processes, boosting cell efficiencies.

2.4.3 Push-Pull Zn-Porphyrins

Among organic sensitizers, considerable interest has been attracted by tetrapyrrolic macrocycles whose basic design may lead to a class of simple, robust, efficient and economically viable sensitizers. Porphyrins are an obvious choice when it comes to think to photosensitizing pigments, due to their well known role as light absorbers and charge separators in natural photosynthetic systems. The excited state (ca. -1 V vs. SCE) and ground state redox potentials (ca. 1 V vs. SCE) are also thermodynamically suitable to perform, respectively, electron transfer to TiO_2 and hole transfer to most redox shuttles commonly employed in DSSCs. However, in order to operate successfully as DSSC sensitizers, the basic porphyrin structure has to undergo important modifications oriented to tune and optimize the excited state characteristics. In particular, in order to obtain almost quantitative quantum yields of charge injection it is important: (1) to establish a strong electronic coupling between the excited state and the acceptor band of the TiO_2 (2) to achieve a satisfactory excited state directionality by shifting the electron density of the excited state on the electron withdrawing group linked to the titania surface. The introduction of a substantial charge transfer character which satisfies both point 1 and 2 could be obtained by chemical modification of the basic porphyrin macrocycle with appropriate donor and acceptor groups. In addition, the elongation of the π-delocalized system of the porphyrin results in a broadening and in a red shift of the absorption bands, which lead to an improved light harvesting [59].

To achieve point 1, the linkage group, usually constituted by either a carboxylic or a cyanoacetic moiety, has to be conjugated with the macrocycle, either through a direct bond or a rigid conjugated spacer, like the phenyl-ethynyl (PE) group. The functionalisation with anchoring groups could be obtained either at the meso position or at the β position of the pyrrole, without largely affecting the resulting properties of the excited state, as demonstrated by the quite similar photoelectrochemical performances that were obtained with meso and beta substituted dyes ($\eta = 2.4$–$2.7\,\%$ respectively) [60]. Beta substituted designs can also offer some advantages over meso substituted porphyrin, resulting in a better shielding of the TiO_2 surface, reducing back recombination [61].

In meso-substituted dyes donor groups are usually attached to the opposite, with respect to the location of the acceptor position of the tetrapyrrolic ring. Either aromatic or aliphatic amines, alkyl and alkoxy groups are usually privileged as donor groups. Obviously, the electron donation of the amines is stronger, leading to a marked increase in the charge transfer character, which is beneficial to increase excited state directionality. The presence of an amine donor group which allows for an effective hole localization on a portion of the molecule which favor hole transfer to the electrolyte and retard charge recombination seems to be a key feature to achieve high photon to electron conversion with porphyirin dyes. Indeed, the porphyrin YD2 (Fig. 2.8), in which the alkyl amino group is directly linked to the meso position of the tetrapyrrolic ring, without the phenyl spacer, exhibit improved charge transfer characteristics, and it is at the basis of the most successful porphyrin designs (YD2-O-C8) which achieved record efficiencies. The stronger influence of the electron donor group in YD2 has the feature of further extending the absorption spectrum to the near infrared region, to increase the electron injection rate and to suppress interfacial recombination involving I_3^-, thanks to the blocking effect of the hexylchains [62]. As a result, YD2 operates with efficiencies very similar to those of N719, reaching values higher than 7 %.

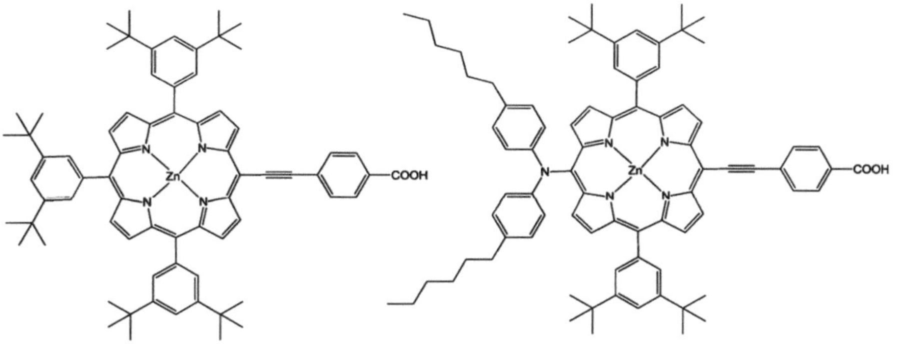

YD0 **YD2**

Fig. 2.8 Structure of YD0 and YD2 push pull porphyrins

2.5 Electron Transfer Mediators

In DSSCs light is absorbed directly at the solid/liquid interface by a monolayer of adsorbed dye and initial charge separation occurs without the need of exciton transport [63, 64]. Following the initial charge separation, electrons and holes are usually confined in two different chemical phases: electrons in the nanocrystalline semiconductor and holes (oxidized redox species) in the electrolyte solution that permeates the solid phase. According to the widely accepted kinetic model, electron transport to the charge collector occurs by diffusion in an electric field-free regime [64]. No electric field is created, essentially due to the fact that the sintered TiO_2 nanoparticles are too small and too lightly doped to support a significant space charge, and also due to a screening effect by the surrounding electrolytic solution. It is a chemical potential gradient that drives the electrons to the back contact, while the opposite electrical potential gradient, developed between the photoinjected electrons and the positive charges (either oxidized dye or oxidized electrolyte species) is screened by the electrolyte [65]. Thus, to contribute to the photocurrent the photoinjected electrons must diffuse through hundreds of nanoparticles in close proximity to the oxidized redox species (electron acceptors) in solution and, in the absence of an electrostatic barrier, interfacial recombination may result as a major energy loss mechanism. To this respect triiodide/iodide electrolytes have been the most efficient and commonly used redox mediators, due to the fact that, in the presence of a reasonable driving force (ca. 0.2 eV), I^- allows for a fast regeneration of most dyes, intercepting them on a sub-microsecond time scale [66, 67], whereas the reduction of I_2 and I_3^- is a complex multi-step reaction which, according to some authors, involves a first dissociation of I_3^- in I^- and I_2, a subsequent reduction of I_2 to I_2^- followed by the rate limiting dismutation of two I_2^- to give I^- and I_3^- [68, 69]. The same authors suggested that the reduction of I_2 occurs only with adsorbed I_2 molecules and the overall process is further slowed down by its relatively low concentration compared to I_3^- in solution. In summary, the electronic recapture involving I_3^- is kinetically so slow on both TiO_2 and SnO_2 surfaces that, under short circuit conditions, most of electrons survive the transit through the mesoporous titania film and the SnO_2 surface and appear in the external circuit. Thus the I_3^-/I^- couple appears to have ideal kinetic properties which lead to an "asymmetric behaviour" at the basis of the efficient functioning of the DSSC: the forward electron donation by I^- is a facile monoelectronic process which ensures an efficient dye recovery, while the reduction of I_3^- appears to be largely inefficient allowing for a minimization of the interfacial back recombination. In other respects the I_3^-/I^- redox couple has a list of regrettable characteristics which limits its practical use as a relay electrolyte: (1) I_2 in equilibrium with I_3^- is volatile, complicating long term cell sealing; (2) I_3^- is darkly colored and limits the light harvesting efficiency of the dye, (3) DSSC cathodes require platinum coatings to obtain the best catalysis of I_3^- reduction at the counter electrode which also exhibit instability issues [70] (4) large photovoltage loss due to the non optimal matching with the oxidised dye redox potential (5) I_3^-/I^- is corrosive and will corrode most metals, posing a serious problem for the use of metal grid collectors necessary for scaling up the solar cells to large areas. Indeed, the sheet resistance of the conductive glass is relatively high, leading to a serious ohmic loss in cells of area exceeding few cm^2 [71].

Due to all these limitations the search for new electron transfer mediators potentially capable of replacing the triiodide/iodide couple is attractive for at least two fundamental reasons: first, the investigation of new redox systems may improve our knowledge about the basic interfacial charge separation events and their dependence from the inherent structural and electrochemical properties of new mediators, second, the discovery of new efficient electron mediators may open the way to new strategies for changing, optimizing and improving DSSC design and performances.

An efficient electron mediator must simultaneously fulfill at least three strict requirements: it has to intercept efficiently the oxidized form of the dye, recombine slowly with photoinjected electrons on both TiO_2 and SnO_2 substrates and allow for an efficient mass transport in solution and into the TiO_2 mesopores. As explained above, a sluggish electron recapture onto semiconductor oxide surfaces is crucial for the correct operation of the cell. To our knowledge, there is no certain a priori indication that such requirement would be satisfied by a given chemical species on a specific substrate; however, in principle, every system characterized by an high reorganization energy associated to the electron transfer should exhibit a slow electron recapture kinetic and represent a possible candidate as electron shuttle for DSSCs. In the following we report on the most significant results made by several international research groups to develop and improve new redox systems based on coordination compounds, which, differently from monolithic molecules, offer unique possibilities in the tuning of their electrochemical properties through a rational choice of the metal and an appropriate design of the coordination sphere. These features guarantee the necessary flexibility to project an electron transfer mediator capable of meeting the kinetic requirements at the basis of the functioning of the DSSC.

2.5.1 *Cobalt Complexes*

The relatively high concentration of electrolyte employed in photoelectrochemical cells (0.1–0.5 M in redox active species) requires the use of substantial amounts of electron mediators: as a consequence the choice and the design of the redox couple has to be done with consideration towards inexpensive and available metals like the elements of the first transition row and easily synthesizable ligands. To date, octahedral Co(II)/(III) polypyridine complexes represent the most successful examples of electron transfer mediators based on coordination compounds. The Co(II)/(III) couple is usually characterized by an high inner sphere reorganization energy associated to the electron transfer, essentially due to the involvement of a metal centred e_g redox orbital with antibonding characteristics. The spin forbidden change from high spin Co(II) $t_{2g}^5 e_g^2$ (T_{1g}^4) to low spin Co(III) t_{2g}^6 (A_{1g}^1) [72] may also contribute to explain the typically low self-exchange rate constants (i.e. 4×10^{-2} M $^{-1}$s^{-1} for $[Co(Phen)_3]^{3+/2+}$) [72] observed in the case of many cobalt complexes.

The first promising report about the use of Co(II) complexes dates back to years 2001–2002 when the compound Co(II)-bis-[2,6-bis(1′-butylbenzimidazol-2′-yl) pyridine] [73] (dbbip) was evaluated by the Grätzel group as electron shuttle in conjunction with the N719 dye. Almost contemporarily to the efforts made at the EPFL (Ecole Polytechnique Federalle de Lausanne) it was found that a wide class of cobalt polypyridine complexes formed from structurally simple ligands did function as efficient electron-transfer mediators in DSSCs [74]. All of these complexes exhibit similar UV-Vis absorption spectra with a weak absorption band centred at ca. 430–450 nm. In general the polypyridine Co(II)/(III) complexes exhibit a quasi-reversible electrochemical behavior dependent on the electrode material which is typical of kinetically slow redox couples. Considering as a meaningful example the $Co(DTB)_3^{2+}$ (DTB = 4,4′ di-tert-butyl-2,2′ bipyridine) complex, the most reversible waves, with a peak separation of less than 200 mV, are usually observed on gold and on carbon based materials, like graphite or glassy carbon [75] which represent the selected materials for catalyzing the Co(III) reduction at the counter electrode of the DSSC (Fig. 2.9).

In general the performances of the first photoelectrochemical cells sensitized by the N3 dye were found strongly dependent on the composition of the electrolyte solution. A maximum monochromatic IPCE of ca. 80 %, in correspondence to the metal-to-ligand charge transfer absorption maximum of N3 was obtained in the presence of 0.25 M LiI/0.025 M I_2, whereas, with the cobalt based mediators, the best performances (ca. 50–55 % of IPCE) were observed when solutions of $Co(DTB)_3^{2+/3+}$ and $Co(TTT)_2^{2+/3+}$ were used. In the other investigated cases,

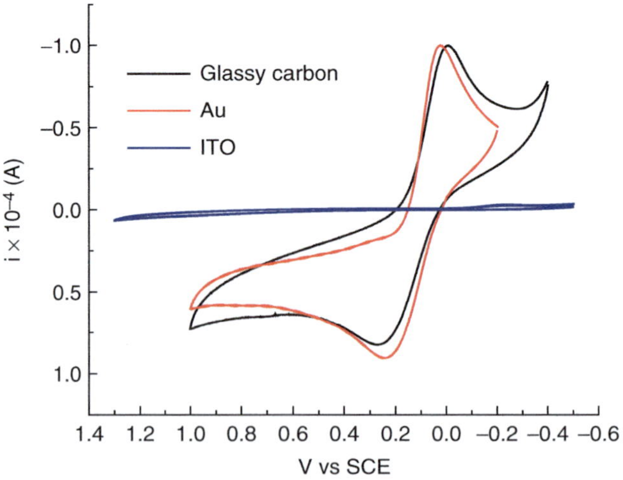

Fig. 2.9 Cyclic voltammetry of ca. 1×10^{-3}M $[Co(DTB)_3]^{2+}$ recorded in ACN/TBAPF$_6$ 0.1 M at glassy carbon (*black line*), gold (*red line*) and on ITO (*blue line*) electrodes

$Co(phen)_3^{2+/3+}$ $Co(TET)_2^{2+/3+}$ (TET = 4,4′,4″ tri-ethyl 2,2′,6′,2″ terpyridine) and $Co(DMB)_3^{2+/3+}$ (DMB = 4,4′ dimethyl-2,2′ bipyridine) mediators exhibited much lower conversion efficiencies, with maximum IPCE values in the range of 10–20 %. Electrochemical Impedance Spectroscopy (EIS) [76], confirmed the substantial amount of recombination in the presence of the less sterically hindered cobalt mediator. Thus, being a relatively efficient electron recapture one of the main limitations of Co(III)/(II) shuttles it was recognized the benefit of introducing hydrophobic chains on the bipyridine ligands, creating heteroleptic complexes of the type shown in Fig. 2.10 [77, 78]. The use of heteroleptic dyes brings two main advantages: (a) reduction of the ion pairing effect (the electrostatic interaction of negatively charged TiO_2 with Co(III) cations), due to a smaller negative ζ potential, and (b) suppression of the electron tunnelling to Co(III) acceptors caused by the steric hindrance of the long alkyl chains which limit the access of the cobalt complex to the exposed TiO_2 surface. The overall effect is a reduction of the probability of electron recapture, the longer the alkyl chain the lower the probability of electron recombination with acceptor states of the electrolyte. For similar reasons the post treatment of the dyed photoanodes with commercially available short chain tri-alkoxy propyl silanes bearing a positive charge was shown to represent a viable successful general strategy for improving the electron collection efficiency in DSSCs based on common Ru(II) complexes and kinetically fast Cobalt redox mediators, like the $[Co(bpy)_3]^{3+/2+}$ [79]. In the best cases, the siloxane treatment resulted in overall improvements of the cell efficiency of the order, respectively, of 500–600 % by comparison with the untreated photoanode. The observed enhancement in cell efficiency could be mostly ascribed to suppression of electron back recombination involving Co(III) centers resulting in a 10–20-fold increase of the electron lifetime in the TiO_2 under open circuit conditions. The steric interactions between chemisorbed dyes, silanes, and Co(III) were rationalized through molecular modeling combining the Molecular Mechanics description of the titania surface with the equilibrium geometries of the molecular entities computed at the DFT (Density Functional Theory) level (Fig. 2.10b). The modeling

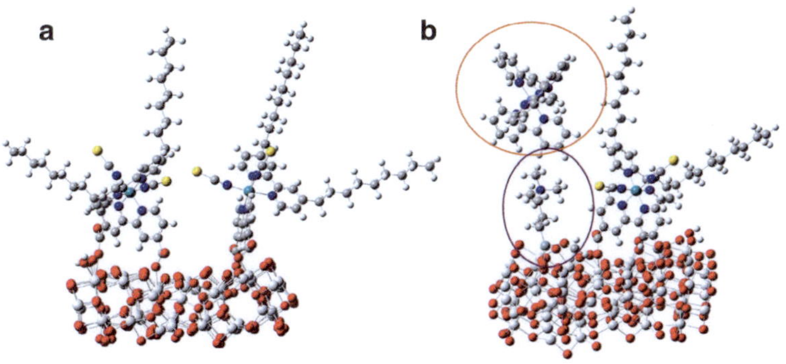

Fig. 2.10 (**a**) Two Z907 complexes interacting with the (101) surface of TiO_2; (**b**) Co-adsorbed amino-propyl-siloxane prevents electron recombination involving a $[Co(III)(bpy)_3]^{3+}$ (evidenced by the *orange circle*)

pointed out the benefit of using robust coadsorbates, displaying a strong and stable interaction with the TiO_2 surface and having enough steric bulk to protect the surface against approaching Co(III), but maintaining, at the same time, the ability to fill the exposed pockets between bulky dye molecules, where recombination may potentially occur.

The recombination process involving Co(III) centres could be further controlled by modifying the photoanode with insulating layers made by inert metal oxides. The effect of the alumina overlayer in cells based on $Co(DTB)_3^{2+}$ and sensitized by Z907 was studied in detail by employing EIS [80], which clearly showed an increased interfacial capacitance and a reduction of the charge transfer resistance (Fig. 2.11) consistent with an improved charge accumulation at the interface, originated by a suppression of residual recombination phenomena.

As discussed above, bulky substituents are usually needed to control recombination, but this type of approach increases the hydrodynamical radius and lowers the diffusion coefficient of the Co(II)/(III) couple which easily undergoes mass transport limitations, leading to a reduced J_{SC}. A detailed study by Elliott and coworkers [81] clearly addressed this point, indicating a diffusion coefficient of 6.7×10^{-6} cm^2s^{-1} for $Co(DTB)_3^{2+}$. To make a comparison, iodide was found to have a substantially larger diffusion coefficient, of the order of 1.6×10^{-5} cm^2/s^{-1}. It must be noted that the diffusion coefficients have been obtained in different conditions (i.e. diluted Co(II) solutions) from those actually met in the concentrated electrolyte employed for the DSSC. Under working conditions a large increase (11×) in solvent (acetonitrile) viscosity was found for the cobalt based mediators, whereas triiodide/iodide solutions accounted only for a ×5 increase. Considering this effect, the bulk diffusion of I^- and I_3^- can be evaluated as about six times faster than $Co(DTB)_3^{3+/2+}$. Futhermore, into the pores of the TiO_2 the slow diffusion of the cobalt mediator can be exacerbated by channel constrictivity effects on the large

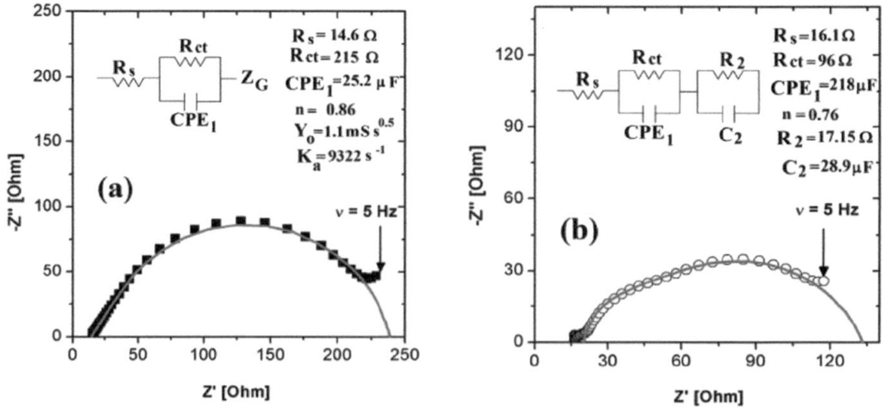

Fig. 2.11 Nyquist diagrams recorded at V_{oc} under 0.03 W/cm^2 solar illumination. (**a**) standard; (**b**) alumina coated. Note that in b a further RC parallel element was introduced to account for the new Al_2O_3 interface. Cells equipped with gold sputtered (80 nm) counter electrode

cations and by electrostatic interactions with the surface, if the dye adsorption imparts a strong negative ζ potential to the surface. Diffusional limitations were evidenced by a non linear behavior of the J_{SC} versus incident irradiance plot and by the typical shape of photocurrent transients which revealed (Fig. 2.12) an initially high photocurrent peak which quickly dropped to a steady value caused by the fact that the diffusion of the electron mediator is not fast enough to supply new reduced mediator to the TiO_2/dye interface, from which, under irradiation, is constantly depleted.

In order to circumvent these diffusional limitations caused by the steric hindrance, the most recent and promising approaches were focused on the development of custom tailored hindered organic dyes (Fig. 2.12) designed to achieve a very good passivation of the surface in order to eliminate dark current processes and allow for the use of simpler and smaller Co(II) complexes like $Co(bpy)_3{}^{2+}$ and Co $(phen)_3{}^{2+}$. The use of these basic structures which are generally characterised by a faster electron transfer with respect to their bulky analogues brings also advantages in terms of both dye regeneration, since the electron donation by Co(II) is more facile, and of faster Co(III) regeneration at the counter electrode, both processes contributing to enhance cell performance. In the following section a detailed discussion of this approach, which recently led to absolute record efficiencies, will be provided.

The substitution of Ru(II) complexes with organic dyes having high extinction coefficient (~30,000 $M^{-1}cm^{-1}$) also allows for the use of thinner titania layers, which, by shortening the electron transit time in the titania thin films, help to further reduce the likelihood of electron recapture events by photogenerated Co(III) centres.

The Dye D35 [82] (Fig. 2.13) was studied in conjunction with a series of Co(II) mediators with a variable steric bulk ($[Co(bpy)_3]^{3+/2+}$, $[Co(DMB)_3]^{3+/2+}$,

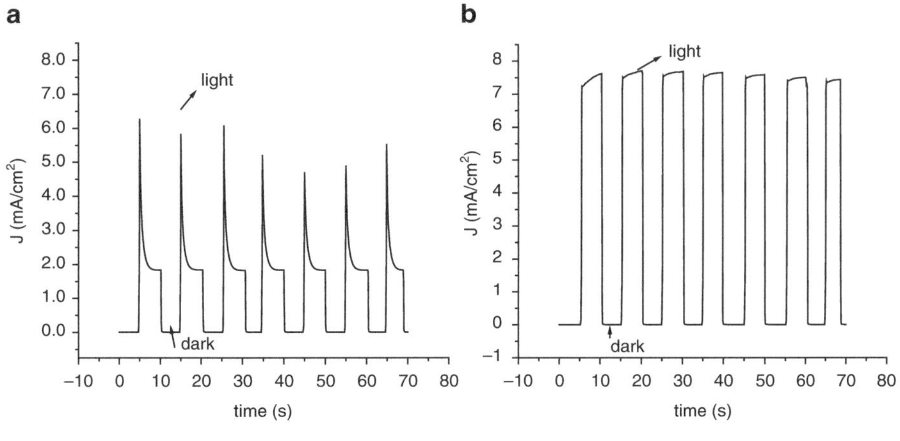

Fig. 2.12 Photocurrent transients obtained with: (**a**) spacer equipped cells; (**b**) without spacer. Electrolyte composition: 0.15 M $Co(II)(DTB)_3{}^{2+}$ + 0.5 Li^+ + 0.1 M Tbpy in acetonitrile

$[Co(DTB)_3]^{3+/2+}$) and resulted in maximum IPCE close to 90 % indicating that nearly all of the absorbed photons were converted into electrons, and negligible losses due to electron recapture by Co(III). The open circuit photovoltage obtained from the combination of D35 with $[Co(bpy)_3]^{3+/2+}$ was also remarkably high (0.92 V) and the fill factor was better than that obtained with $Co(DTB)_3^{2+}$ which, as explained above, suffers from mass transport limitations, which contribute to increase the serial resistance of the cell. The best efficiencies with D35 reached 6.7 % and were better than those obtained with the I_3^-/I^- couple (5.5 %) which

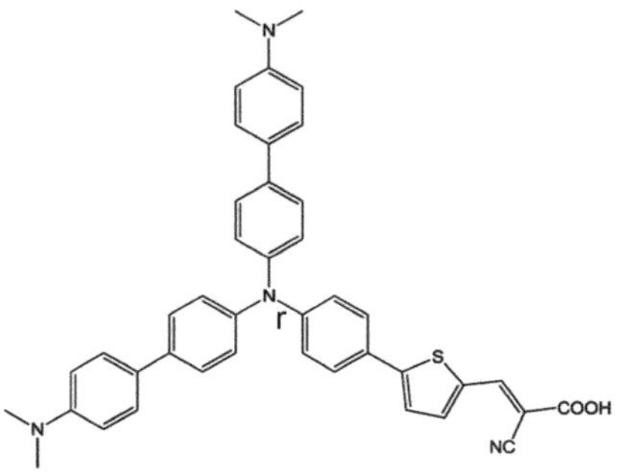

D29

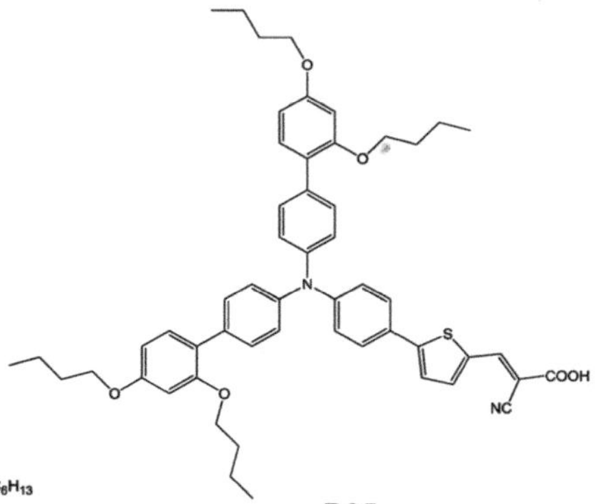

D35

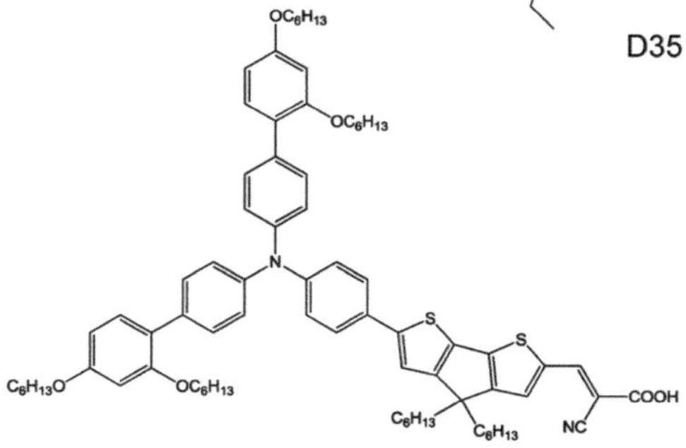

Y123

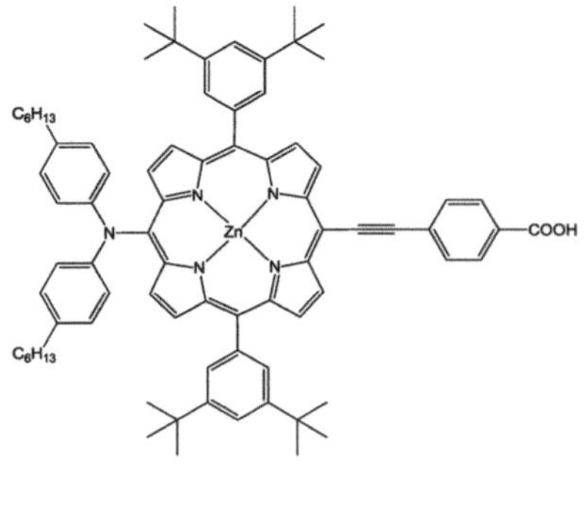

YD2

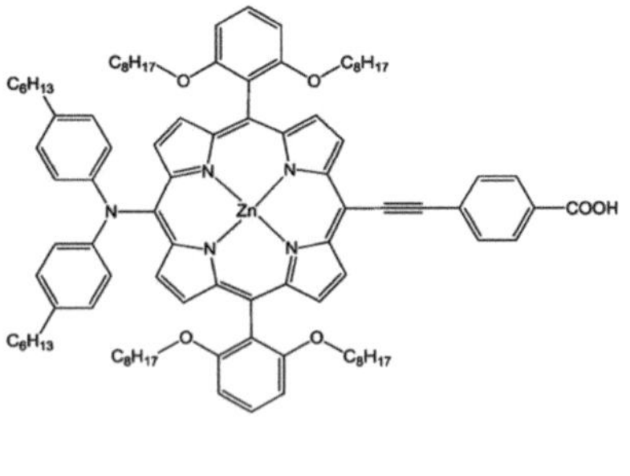

YD2-o-C8

Fig. 2.13 Some of the most successful sensitizers for Co(II)/(III) mediated solar cells affects the dye with competitive light absorption. These first studies were soon extended by considering the dye Y123 (Fig. 2.13) which, while containing the same structural features of the D35 dye has an extended π delocalized unit consisting of two rigidly bridged thiophene rings designed for improving the spectral response at low wavelengths (IPCE onset at ~700 nm) [83].

The Y123 dye served to reach a very high V_{oc}, exceeding 1000 mV, thanks to the use of the bipyridine-pyrazole Co(II) mediator [84] with $E_{1/2} = 0.86$ V vs. NHE. Despite the very positive potential, the Co(II) complex was still capable to intercept efficiently the oxidized mediator, leading to a dye regeneration yield of ca. 90 %, which compares relatively well with the regeneration efficiencies (98 %) observed with the I_3^-/I^- electrolyte. This fact coupled to a good passivation of the titania layer ensured by the steric bulk of Y123, led to electron collection efficiencies of the order of 90 % allowing to build cells in which the large photovoltage loss typical of the I_3^-/I^- couple was eliminated. As a result efficiencies exceeding 10 % were found in cells equipped with 4.5 μm scattering overlayer and P-PRODOT (polypropylene-dioxy-thiophene) p-type polymer as a catalytic counter electrode, which served to optimize the cathodic regeneration of Co(III). Finally, by exploiting the principles outlined above, record efficiencies of 12 % under full sunlight were reached by using the bulky porphyrin sensitizers of the type YD2-o-C8 (Fig. 2.13) which block electron recapture. In the presence of co-sensitizing Y123 and $Co(bpy)_3^{3+/2+}$ a panchromatic absorption was achieved with IPCE approaching 90 % in the 400–700 nm range. This, together with the high V_{oc} (~1 V) resulting from the suppression of recombination and by the reduction of the overpotential for dye regeneration is at the basis for the very high efficiency [27]. These performances have been recently surpassed though the collaborative sensitization of two dyes ADEKA-1 and LEG 4, with a silyl and carboxylic anchoring group, respectively, in the presence of $[Co(phen)_3]^{3+/2+}$ (Fig. 2.14). DSSCs optimized with this combination of dyes and electron mediator allowed for a record efficiency of 14.3 % [15].

2.5.2 Cu(I), Ni(III) and Fe (II) Complexes

The search for efficient and robust redox shuttles has been extended to other first row coordination complexes based in Cu(I), Ni(III) and Fe(II) ions. Cu(I)/(II) couples were among the first to be considered, since are usually characterized by slow electron transfer kinetics which could guarantee inefficient electron recapture processes. The electron transfer processes involving Cu(I)/(II) couples are slow due to a large inner sphere reorganizational contribution because the electron transfer is associated to a significant change in the preferred coordination geometry, from tetrahedral Cu(I) to tetragonal Cu(II). However, if the electron transfer is too slow, a low efficiency in dye regeneration is expected and in order to optimize the electron transfer properties of the redox shuttles the coordination sphere of the complex has to be rationally designed. For example, the electron transport systems

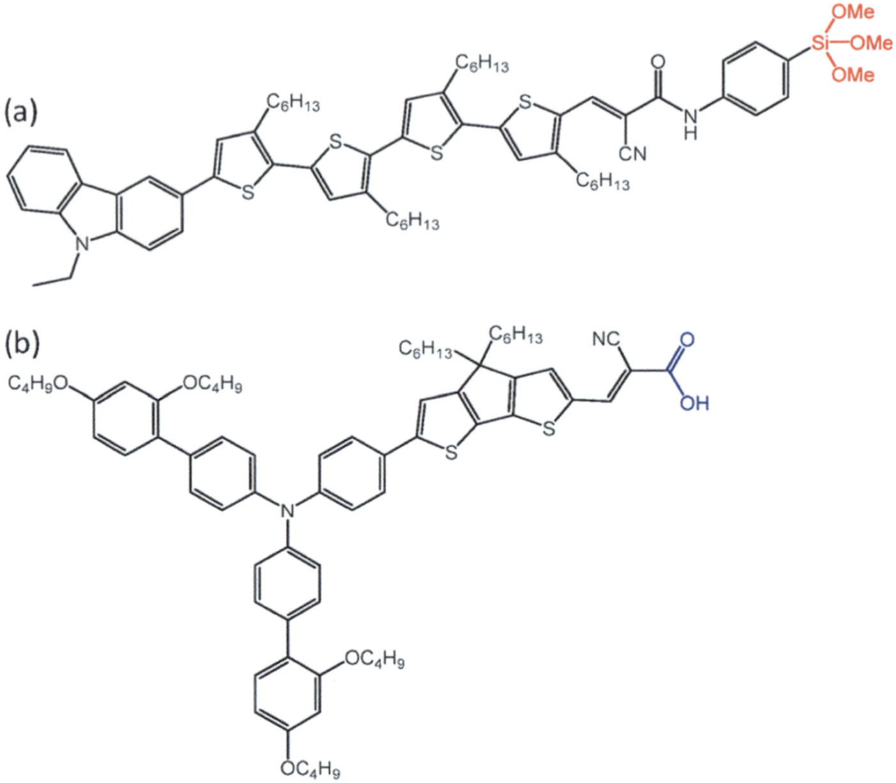

Fig. 2.14 Molecular structures of photosensitizing dyes: (**a**) silyl-anchor dye ADEKA-1 and (**b**) carboxy-anchor dye LEG4

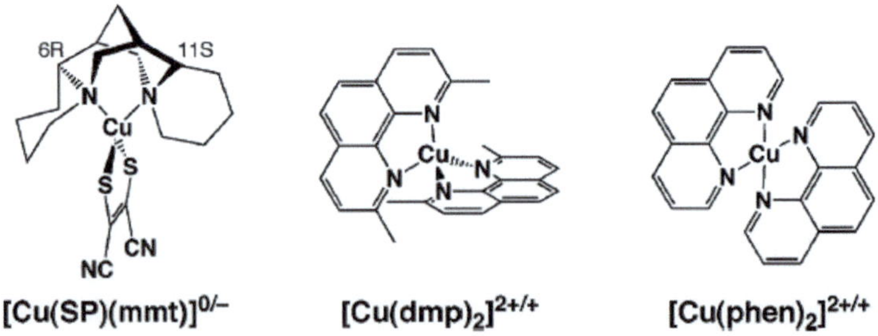

Fig. 2.15 Structure of distorted tetrahedral copper complexes. SP is sparteine N,N′, mmt is maleonitriledithiolato, dmp is 2,9 dimethyl-1,10 phenanthroline, phen is 1,10 phenanthroline

which can be found in nature in blue copper proteins show tetrahedrally distorted geometries intermediate between Cu(I) and Cu(II) geometries.

Blue copper model complexes with a distorted tetrahedral geometry (Fig. 2.15) have been recently employed as efficient electron transfer mediators for DSSC.

The assembly of solar cells based on copper complexes required TiO_2 optimization and multilayered photoanodes with a compact TiO_2 underlayer, necessary to suppress the back recombination from the exposed FTO contact [85]. The photoelectrochemical characterization carried out with a 0.2 M Cu(I)/Cu (II) electrolyte (Cu(II) molar fraction 0.4) 0.5 M Li^+ and 0.2 M Tbpy revealed maximum IPCE ranging from 40 % in the case of Cu(CuSP)(mmt) to 20 % for Cu $(phen)_2{}^{2+}$. The J-V curves obtained under simulated solar irradiation basically followed the same trend observed under monochromatic light, with the best results observed in the case of Cu(CuSP)(mmt) and $Cu(dmp)_2{}^{2+/+}$ which achieved J_{SC} of 4.5 and ~3 mA/cm^2 and V_{oc} of 0.65 and 0.8 V respectively. $Cu(phen)_2{}^{2+}$ was the worst mediator due to the slow electron donation consistent with its small self-exchange rate constant (0.20 $M^{-1}s^{-1}$). The more rigid coordination sphere of Cu (SP)(mmt) and $Cu(dmp)_2{}^{2+}$ allowed for a faster electron transfer and higher self exchange rates (respectively 40 and 100 times higher than $Co(phen)_2{}^{2+}$), beneficial for improving the electron transport within the electrolyte. Thus the use of copper complexes with a distorted tetragonal geometry, in which the structural change between copper (I) and copper (II) complexes is minimized, show promise for developing alternative low cost mediators for photoelectrochemical cells. Interesting results [86] were recently obtained by employing the $Cu(dmp)_2{}^{2+/+}$ shuttle in conjunction with a sterically hindered organic dye C218 (Fig. 2.15) which bears a close structural similarity with the Y123 dye previously described: an aryl-amine donor unit is linked to a rigidly bridged bis-thiophene moiety which is bound to cyano-acrylic linkage group. The alkyl and alcoxy chains act as spacers, constituting an insulating barrier against recombination involving Cu(II) centers, thus the mechanism through which suppression of the dark current is achieved is entirely analogous to what has been described in the cobalt section. The employment of by-layered (transparent+scattering layer) titania substrates allows to obtain very high electron collection efficiencies resulting in a broad IPCEs close to 80 % and in $J > 10$ mA/cm^2. Interestingly, the positive potential of the $Cu(dmp)_2{}^{2+/+}$ couple resulted in $Voc > 0.9$ V, leading to efficiencies comparable to those obtained with the $I_3{}^-/I^-$ couple (Fig. 2.16).

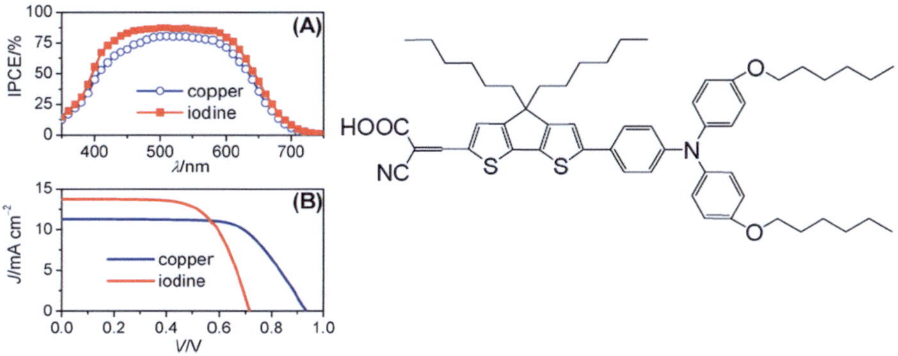

Fig. 2.16 Photoelectrochemical performance (**a**) and (**b**) of bilayered DSSC comprising dye C218 (right) and $[Cu(dmp)_2]^{2+/+}$

Among fast redox couples which ensure a fast reduction of the dye sensitizer, notable examples can be found among Ni(III) and Fe(II) compounds. Ni (IV)/(III) bisdicarbollide complexes are a robust class of lightly colored redox shuttles [87] (Fig. 2.17) which give rise to a fast electron transfer, proving to induce an excellent oxidised reduction which was found to be slightly faster than I^-, as indicated by a half life of ca. 2 μs, yielding to a dye regeneration of about 90 %. The electron recapture by Ni(IV) was however a problem, despite the fact that the Ni(IV) → Ni (III) reduction is a slow process which requires rotation of the dicarbollide cage from a cis → trans conformation. In any case the electron recapture could be quite efficiently suppressed by using a conformal Al_2O_3 coating obtained by atomic layer deposition (ALD). It was found that one ALD cycle, corresponding to the deposition of less than one monolayer, led to the best J-V characteristics, indicating that it was sufficient to passivate surface states causing localization of electron density at the surface. In these conditions N719 sensitized cells gave $J_{SC} = 3.76$ mA/cm^2, $V_{oc} = 650$ mV and an efficiency of 1.5 %. An increased number of ALD cycles led to a better suppression of recombination, causing a complete coating of the titania surface and acting prevailingly by reducing the electron tunnelling probability, but this also implied a decreased efficiency of electron injection, due to the decoupling of the dye from the d band of titania, resulting in a 50 % reduction of the photocurrent.

The electron interception kinetics of Ni(IV) were at least 10^3 times slower than ferrocenium which, originating extremely fast recombination, could not be efficiently employed with conventional dyes in simple untreated titania photoanodes. Passivating treatments [63] aimed at the suppression of electron recapture by ferrocenium cations, based on polysiloxanes, zirconia or alumina overlayers constituted meaningful examples of the possibility of achieving control over the

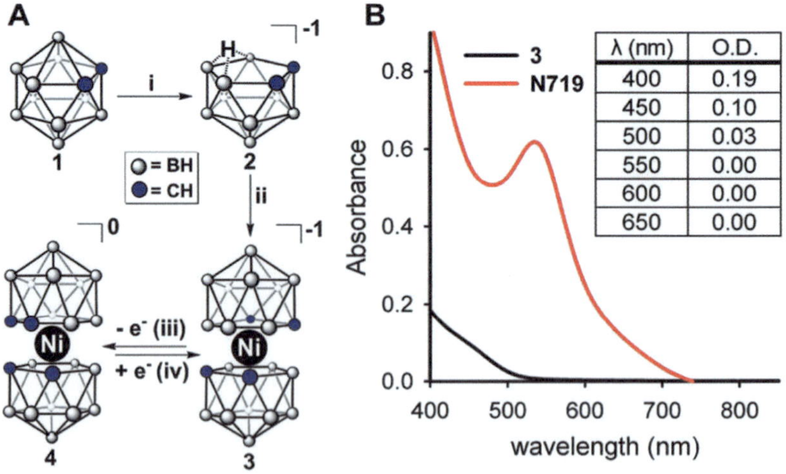

λ (nm)	O.D.
400	0.19
450	0.10
500	0.03
550	0.00
600	0.00
650	0.00

Fig. 2.17 Ni(IV)/(III) biscarbollide complexes and visible absorption spectrum of the shuttles compared to that of N719

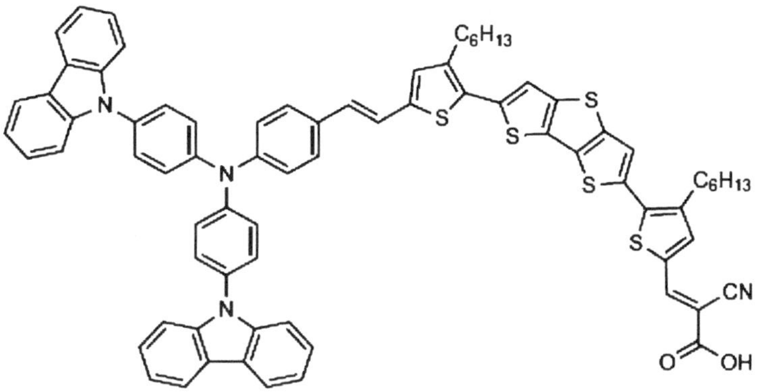

Fig. 2.18 Structure of the Carbz-PAHDTT dye sensitizer

interfacial charge separation dynamics through simple chemical modification of surfaces, but the performances generally remained too low ($J_{SC} < 2$ mA/cm^2) for foreseeing real practical application of fast Fe(II)/(III) couples, until a recent breakthrough [88] which, again, involved the association of ferrocenium/ferrocene couple to a new bulky push-pull dye (Carbz-PAHDTT, Fig. 2.18) which seemed to result in a complete blockage of the back recombination allowing to obtain very high charge collection efficiencies.

The Carbz-PAHDTT is based on a bulky triaryl-amine donor group, functionalised with two carbazole pendants which provide high steric hindrance and improved electron donation characteristics. The π system is based on a chain of four thiophene rings, of which two are bridged by a sulphur atom, and two are functionalised with hexyl chains, introduced to reduce dye aggregation and to contribute to suppress recombination. The long delocalized π system based on four thienil rings enhances both the extinction coefficient of the dye (ca. 3×10^4 M^{-1}cm^{-1} at 500 nm) and the red sensitivity which extends the absorption onset at ca. 650 nm. These features allow to employ a very thin titania layer, instrumental in reducing the probability of electron recapture events during the diffusion of the electrons to the back collector. Indeed this type of dyes was used to sensitize a 2.2 μm thick titania layer modified with a 4.4 μm titania scattering layer composed of 400 nm particles. Under these conditions the Fc^+/Fc led to very good IPCEs (ca. 70 %) in the presence of the co-adsorbed chenodeoxycholic acid, which also helps to reduce dark current phenomena. Under simulated solar illumination $J_{SC} = 12.2$ mA/cm^2 and V_{oc} of 840 mV and FF = 0.73 led to an efficiency of 7.3 %. The improved efficiency over the I_3^-/I^- (~6 %) resulted mainly from the higher V_{oc} obtained by the ferrocenium/ferrocene couple, thanks to its higher formal potential (0.62 V vs. NHE (Normal Hydrogen Electrode) compared to 0.32 V vs NHE for I_3^-/I^- couple) which makes it thermodynamically more suitable for the efficient operation in the DSSC. One significant drawback of the ferrocenium/ferrocene couple is however related to the instability of Fc^+ in the presence of O_2 and tert-butyl pyridine [89, 90].

2.6 Solid State Hole Conductors

In the most efficient DSSC a porous nanocrystalline TiO_2 is in contact with a liquid electrolyte and, as already seen, the key processes take place at the interface between the nanostructured film and the liquid electrolyte. However the electrolyte may degrade over a period of time due to intrinsic chemical instability of the redox species, solvent evaporation or leaking caused by sealing imperfections. The search for suitable solid materials that can replace the liquid electrolyte is therefore an interesting and active area of research. In a solid state DSC the solid hole conducting material captures the holes (the positive charges left on the dye as a result of the photo-induced charge injection in the n-type material) and closes the circuit with the counter electrode. Solid hole conductors include conducting polymers, organic hole conductors and inorganic semiconductors such as CuI and CuSCN [91]. Among these materials the copper (I) compounds have shown the most promising results as hole conducting materials for their application to regenerative photoelectrochemical cells. CuI is a p-type semiconductor with a band-gap of 3.1 eV which can be deposited by spray or dip coating from an acetonitrile solution (typically 3 g of CuI in 100 ml of acetonitrile) onto dye sensitized substrates, using a low temperature deposition technique which avoids the denaturation of the dye monolayer. The casting procedure can be repeated in order to ensure an optimal electrical contact and TiO_2 pore filling by the p-type material. A recent detailed photoemission study [92] showed that the anatase/CuI interface is a type II heterojunction with a substantial space charge which allows for an efficient unimpeded electron transfer from the p-type material to the n-type semiconductor. Indeed in one of the first reports about solid state DSC, Tennakone et al. obtained a global efficiency of about 0.8 % using a natural cyanidine dye, CuI and gold or graphite counter electrodes (either directly evaporated over the hole conductor or simply pressed over it) [91]. A substantial drop in cell performance associated to an increased CuI film resistivity in the presence of a moisture rich environment indicated CuI film degradation. In general CuI based DSC are not stable and, even stored dry in the dark, undergo a rapid decay. One of the reason of such instability seems to reside into the loosening of the electrical contact between the dyed TiO_2 surface and CuI crystallites. CuI deposited from acetonitrile solution produces large crystallites ($\approx$10 μm) that do not penetrate well into the TiO_2 pores and form loose contacts. It has been found [93] that the stability and the response of CuI based DSC can be greatly enhanced by adding small quantities ($\approx 10^{-3}$ M) of 1-ethyl-3-methyl-imidazolium thiocyanate (EMISCN) ionic liquid which acts as a powerful crystal growth inhibitor, probably through strong surface adsorption, leading to a $\times$ 1000 reduction of CuI crystal size. Such effect permits a better permeation of the nanocrystalline TiO_2 matrix and a strengthening of the coupling between the dye and the electron donor. Additionally, the ionic liquid which remains at the CuI grain boundaries after solvent evaporation seem to admit hole conductance, possibly thanks to the hole accepting properties of SCN^-. The photoresponse obtained in the presence of CuI/EMISCN shows an overall efficiency of about 3 % and a maximum monochromatic conversion of about 60 % (Fig. 2.19).

Alternatives to inorganic p-type semiconductors can be found in organic species and conductive polymers. Organic hole conductors like the spiro-compound 2,2′,7,7′-tetrakis(N,N-di-p-methoxyphenyl-amine)9,9′-spirobifluorene (OMeTAD) (Fig. 2.20) have demonstrated some promise for application in DSSC [94].

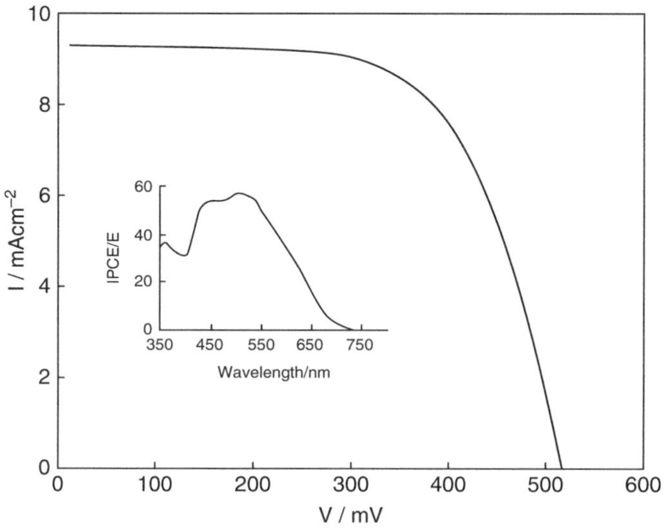

Fig. 2.19 J-V characteristic of the cell made by deposition of CuI from an acetonitrile solution containing 9×10^{-3} M EMISCN. Inset IPCE spectrum (Reproduced with permission from Ref. [93])

Fig. 2.20 Structure of OMeTAD

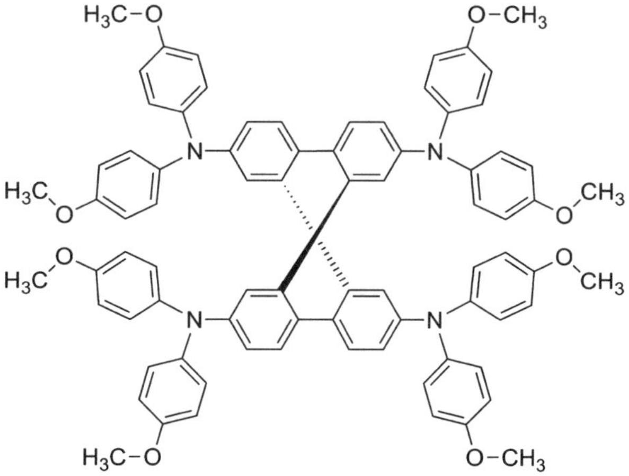

The hole conductor has a spiro-centre (a tetrahedral carbon linking two aromatic moieties) which is introduced in order to improve glass forming properties and prevent crystallization. Crystallization is undesirable since it impairs the formation of a good electrical contact between the TiO_2 surface and the hole transporting material (HTM) which is cast over the dye sensitized surface by spin coating from a chlorobenzene solution. A compact TiO_2 underlayer is required to prevent the short circuiting between HTM and the FTO collector and an evaporated gold layer forms the cathode of the solid state DSC allowing for an intimate contact with the organic hole conductor.

The oxidized dye $Ru(III)(H_2DCB)_2(NCS)_2$ is efficiently regenerated by hole injection in the HTM layer on a nanosecond time scale, as demonstrated by the total disappearance of the Ru(III) bleaching, being compensated by the rise of the oxidized OMeTAD signal ($\lambda_{max} = 530$ nm) within the 6 nanosecond laser excitation pulse. Monochromatic photon to electron conversion as high as 30 % and overall efficiencies of 0.74 % under ca. 0.01 W/cm^2 were obtained by using additives like $N(PhBr)_3SbCl_6$ and $Li((CF_3SO_2)_2N)$:$N(PhBr)_3SbCl_6$ acting as a dopant and introducing free charge carriers in the HTM by amine oxidation; the second additive, a lithium salt, is a source of Li^+ cations which are known to be potential determining ion for TiO_2 and assist electron injection while retarding back recombination. The lithium salt may also compensate for space charge effects: under illumination a net positive space charge is expected to be formed in the HTM which impairs current flow. The salt could screen this field, eliminating space charge control of the photocurrent, leading to J_{SC} of the order of 3 mA/cm^2 under full AM 1.5 conditions.

Conductive polymers based on polythiophenes and polypyrroles could be interesting candidates for replacing the liquid electrolyte in DSSC, due to their low cost, thermal stability and good conductivity [95].

An interesting photovoltaic response was obtained by using the excellent properties of PEDOT as a hole transporting material. PEDOT can be in situ photoelectropolymerized from bis-EDOT monomers by exploiting the oxidizing power of a dye like Z907 [96]. The general mechanism for the photoelectrochemical polymerization is outlined in Scheme 2.1.

The hydrophobic properties of Z907 allow for a good affinity for the scarcely polar polymeric matrix, ensuring a good electrical contact between the p-type material and the dye. The presence of a coadsorbate like deoxycholic acid

1) $TiO_2/Dye + h\nu \longrightarrow TiO_2/Dye^* \longrightarrow TiO_2^-/Dye^+$

2) $TiO_2^-/Dye^+ + Mo \longrightarrow TiO_2^-/Dye + Mo^+ \xrightarrow[nMo]{} \text{oxidized conducting polymer}$

Scheme 2.1 General mechanism for the oxidative photoelectrochemical polymerization of a conductive polymer generated by sensitizer mediated oxidation of the precursor monomers (Mo)

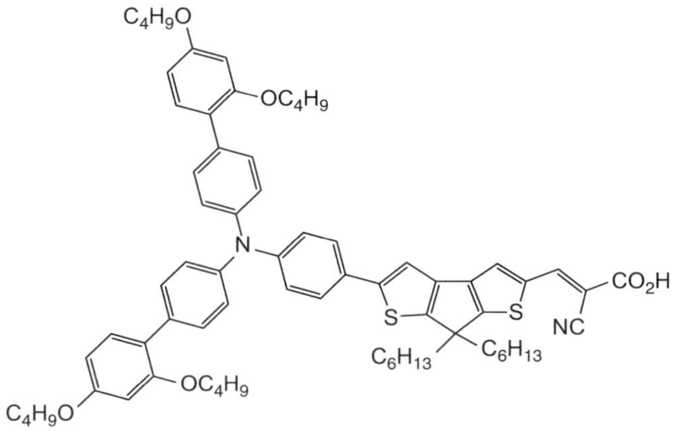

Fig. 2.21 Organic dye LEG4

(DCA), in an optimal 2:1 ratio with respect to the dye, enhanced the global photovoltaic response by reducing dye aggregates. This function seems to be particularly relevant for polymer mediated DSC, since aggregated dye clusters reduce the number of available polymer growth sites and leave less effective space for polymerization initiation and propagation. To optimize the contact between the hole transporting layer and the counter electrode, a PEDOT functionalized FTO was used as the cathode and to achieve optimal performances a drop of BMImTFSI (1-butyl-3methylimidazolium bis(trifluoromethanesulfonyl) amide) containing 0.2 M LiTFSI and 0.2 M Tbpy was cast onto the TiO_2/dye/ PEDOT junction. The ionic liquid may improve the charge transporting capabilities of the heterointerface through screening of space charge effects, lithium cations assist electron injection and percolation through the n-type semiconductor and, as usual, tbpy increases the open circuit photovoltage via suppression of the back recombination. With such a treatment FTO/TiO_2/dye/PEDOT-PEDOT/FTO sandwich cells afforded efficiencies of the order of 2.6 %.

Following the approach reported first by Yanagida, Yang L. et al. fabricated an all-solid state DSC using PEDOT as the hole transporting medium (HTM) in conjunction with the organic dye LEG4 (Fig. 2.21). The solid stated devices displayed 6.0 % and 4.8 % power conversion efficiency with PEDOT deposited from acetonitrile or aqueous micellar phase respectively [97].

The search for low cost semi transparent counter electrode materials which can be easily interfaced with the conductive polymer, avoiding vacuum processing, has recently considered the use of flexible press transferred single-walled carbon nanotube (SWCNT) (Fig. 2.22) [98]. The carbon nanotube film constituted a valuable replacement to the more conventional silver back contact, resulting in devices with 4.8 % efficiencies which were comparable to those recorded with the silver back contact (5.2 %).

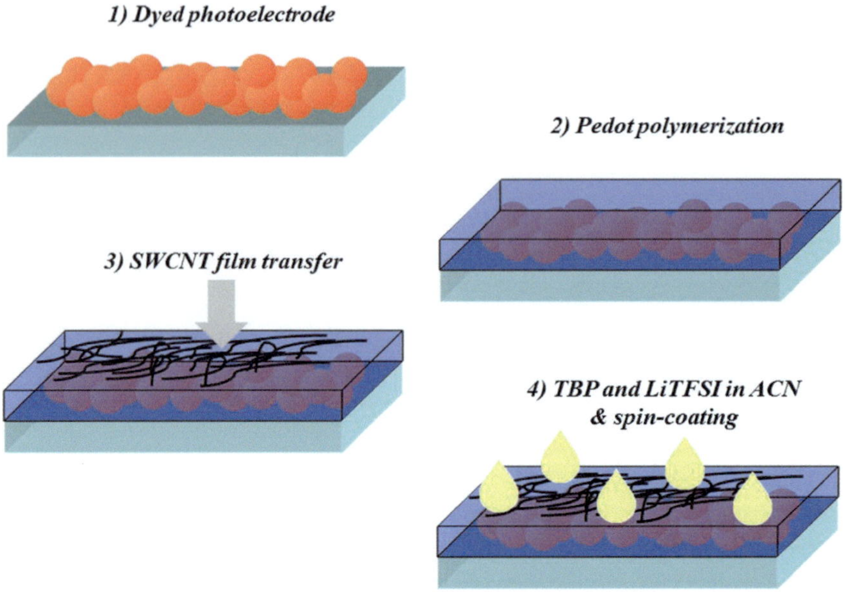

Fig. 2.22 Fabrication by wet methods of solid state DSSCs based on PEDOT as hole transporting layer and on a carbon nanotube composite as integrated counter electrode

2.7 Solid State DSSCs Based on Perovskites

Since the first example reported by T. Miyasaka in 2009 for a liquid based DSC [99], organometal halide perovskites as visible-light sensitizers for photovoltaic cells have attracted particular attention. Some key attributes of these perovskites include ease of fabrication by wet chemical methods, strong solar absorption and low non-radiative carrier recombination rates, reasonably high carrier mobility and the possibility of tuning the electronic properties by forming a series of compounds with controlled stoichiometry. In addition perovskite technology can capitalize the experience developed on over 20 years of research on dye-sensitized and organic photovoltaic cells. One negative aspect of perovskites is the fact that lead has been a major constituent of all highly performing materials, raising toxicity issues during device fabrication, deployment and disposal. Limited stability is an additional issue with organo-halide perovskites, showing a rapid degradation on exposure to moisture, some organic solvents, and ultraviolet radiation.

Methylammonium tin and lead halides with the general formula $(CH_3NH_3)MX_{3-x}Y_x$ (M = Pb or Sn; X,Y = I, Br, Cl) are members of a family of artificial hybrid perovskites with differing dimensionality that were first discovered by Weber and later significantly advanced by Mitzi and co-workers [100, 101]. This class of perovskites has now pushed the limits of the DSSC or hybrid organic-inorganic solar cells technology to photoconversion efficiencies (PCEs) beyond 20 % [102], a benchmark value nobody would have dreamed more than 2 years ago.

2.7.1 Structure

Organic/inorganic perovskites are hybrid layered materials typically with an ABX_3 structure, with A being a large cation, B a smaller metal cation and X an anion from the halide series. They form an octahedral structure of BX_6, which forms a three dimensional structure connected at the corners, as shown in Fig. 2.23.

The cation A fits into the space between four BX_6 octahedra that are connected together through shared corners [103] tuning the distortion of the perovskite lattice (Fig. 2.24).

For the 3-D perovskites the size of the organic A cation is limited by the size of the 3-D hole into which it must fit. For a perfectly packed perovskite structure the geometrically imposed condition for the A, B, and X ions to be in close contact is: $(R_A + R_X) = t\sqrt{2}(R_B + R_X)$, where R_A, R_B, and R_X are the ionic radii for the corresponding ions and the tolerance factor must satisfy $t \approx 1$. Empirically it is found that $0.8 \leq t \leq 0.9$ for most cubic perovskites. Using $t = 1$ and essentially the largest values for R_B and R_X (e.g. $R_{Pb} = 1.19$, $R_I = 2.20$ Å) [104], the limit on R_A is found to be approximately 2.6 Å. Consequently, only small organic cations

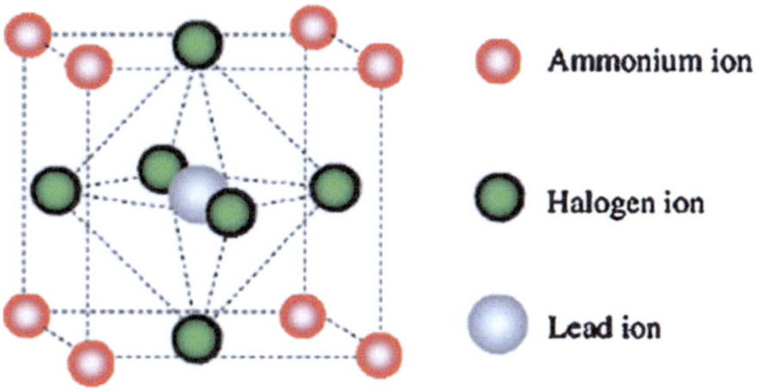

Fig. 2.23 Unit cell of lead hybrid perovskites

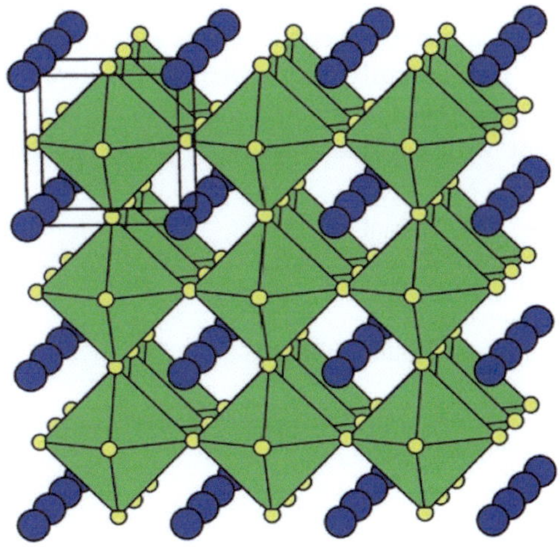

Fig. 2.24 Three-dimensional schematic representation of perovskite structure ABX_3 (Reproduced with permission from Ref. [106])

(i.e. those consisting of three or less C–C or C–N bonds) are expected to enter into the distorted cubic structure and are typically restricted to methylammonium, ethylammonium and formamidinium. In organo halides perovskites used for solar cells metal cations (B) are typically divalent metal ions such as Pb^{2+}, Sn^{2+} and Ge^{2+} while the halide anions (X) are I^-, Cl^- and Br^-.

2.7.2 *Optical and Electrical Properties*

Organic lead halide perovskites are mainly used in photovoltaics as absorbers. Optical absorption measurements indicate that the perovskites with general formula $MAPbX_3$ (where $MA = CH_3NH_3^+$) are direct band-gap semiconductors with a sharp absorption edge [105]. The suitable energy band-gaps, the high absorption coefficient, the large carrier mobilities [106] and diffusion lengths [107, 108], make these material very promising for solar energy applications. Effective diffusion length in $CH_3NH_3PbI_3$ is of the order of 1 μm for both electrons and holes [109], a high value for a semiconductor formed from low-temperature solution processing. The diffusion length of electrons in $CH_3NH_3PbI_{3-x}Cl_x$ was found to be even higher (ca. 1.9 μm) [109]. The performance of perovskite solar cells is strongly affected by the crystallinity of the perovskite layer, however the impact of the perovskite crystal structure on the electronic properties is not well understood. $CH_3NH_3PbI_3$ has a distorted three-dimensional structure that crystallizes in the tetragonal group, while $CH_3NH_3PbBr_3$ has a cubic structure at room temperature (Fig. 2.25), consistent with the differences in the ionic radii of I^- and Br^- (2.2 Å and 1.96 Å respectively) causing less steric repulsion in the case of bromide anions [106, 110].

The band-gap of $CH_3NH_3PbX_3$ can easily be tuned by varying the chemical composition of the hybrid material, resulting in an array of translucent colors useful to optimize the light harvesting in selected spectral regions. Band-gap tuning of $CH_3NH_3PbX_3$ has been achieved via substitution of I^- with Br^- ions, which arises from a strong dependence of electronic energies on the effective exciton mass [111]. Generally, the lighter anions give larger band-gaps, as can be observed by considering the absorption threshold for $CH_3NH_3PbI_3$ and $CH_3NH_3PbBr_3$ consistent with forbidden gaps of 1.55 eV and 2.3 eV, respectively [99, 112, 113].

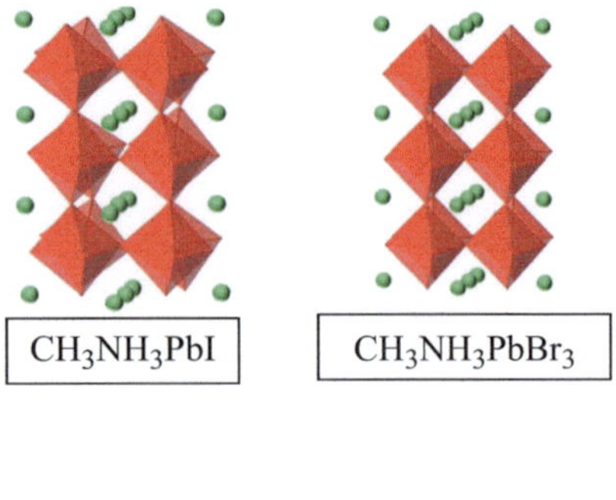

Fig. 2.25 Crystal structures of $CH_3NH_3PbI_3$ and $CH_3NH_3PbBr_3$ (Reproduced with permission from Ref. [115])

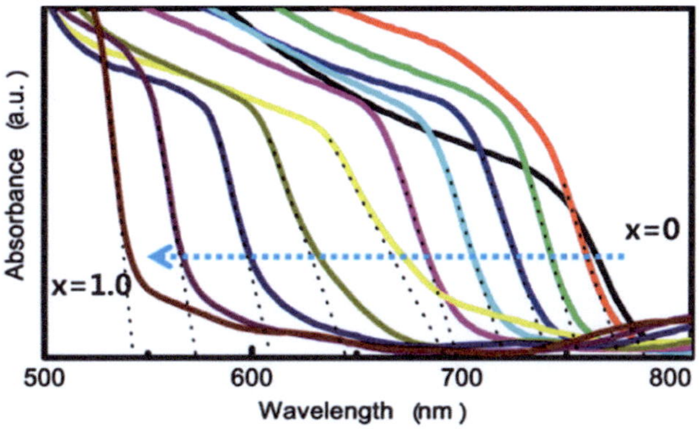

Fig. 2.26 UV-vis absorption spectra of FTO/bl-TiO_2/mp-TiO_2/$MAPb(I_{1-x}Br_x)_3$ (Reproduced with permission from Ref. [115])

Replacement of the heavier halides with the smaller Cl^- (ionic radius of 1.81 Å) [106] gives a band-gap of 3.1 eV in agreement with this rule [114].

The mixed halide $CH_3NH_3PbI_2Cl$ surprisingly display the absorption onset at ~800 nm, similar to what is found for the $CH_3NH_3PbI_3$ and $CH_3NH_3PbI_{3-x}Cl_x$ perovskites, whereas $CH_3NH_3PbI_2Br$ absorbs light below ~700 nm [114]. Thus, through the control of the mixed halide perovskite composition it is possible to tune the final color of the light harvesting medium. For example, the dark brown $CH_3NH_3PbI_3$ (x = 0) can be converted to brown/red for $CH_3NH_3Pb(I_{1-x}Br_x)_3$ and then to yellow for $CH_3NH_3PbBr_3$ (x = 1) with increasing Br content [115] (Fig. 2.26).

The organic cation offers additional possibilities to tune the energy gap but, due to sterical requirement of the cubic or tetragonal cage, the choice of the cation is restricted to few examples satisfying the requisite outlined in Sect. 2.7.1. If the cation is too large, the 3D perovskite structure is unfavourable and lower dimensional layered or confined perovskites will be formed. If the cation is too small, the lattice would be too strained to form [103]. The ethylammonium cation was shown to form a wider bandgap perovskites due to a 2H-type structural rearrangement, since it is too large to maintain the three-dimensional tetragonal ABX_3 lattice [116]. When the ABX_3 structure is maintained, the increase in cation size results in a reduction of the energy gap leading to the extension of the absorption to the infrared [117]. For example, the band-gap of $APbX_3$ decreases from 1.55 to 1.48 eV when the cation in the A site is systematically replaced with cations of increasing size (cesium (Cs) <methylammonium (MA) <formamidinium (FA)), as can be appreciated from Fig. 2.27.

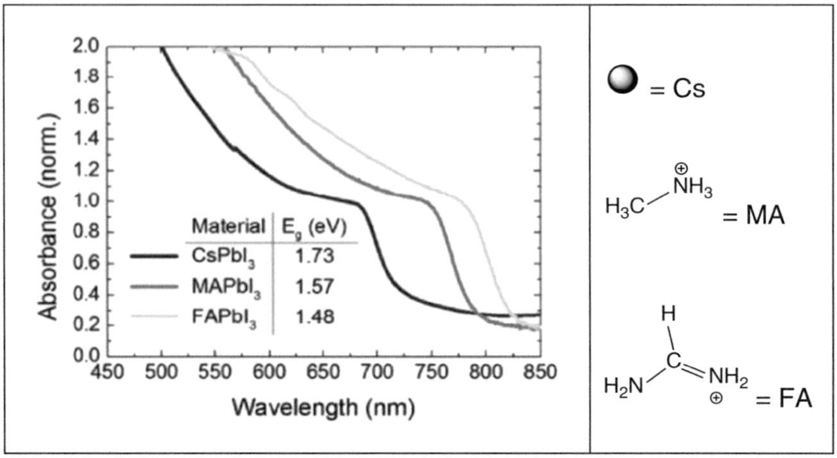

Fig. 2.27 UV-Vis spectra for the $APbI_3$ perovskites formed with different cation A (Reproduced with permission from Ref. [117])

2.7.3 Perovskite Deposition and Cell Engineering

The preparation of hybrid perovskite film generally relies on four main deposition methods, which include spin-coating [118], vacuum vapor deposition [119], two-step deposition technique (TSD) [120], and patterning thin film [121].

One-step spin-coating is the most widely used method to prepare the solar cells and consists in the deposition of a precursor solution (for example PbI_2 and CH_3NH_3I dissolved in DMF) on a nanostructured TiO_2 thin film, followed by a subsequent annealing step at controlled temperature, (typically 100 °C for 45 min) during which solvent evaporates and perovskite crystallizes. The HTM film (typically tri-aryl-amines like Spiro-OMeTAD) is deposited on top of the formed perovskite layer by a further spin coating run. The main drawbacks of this simple method is the uncontrolled precipitation of the perovskite in the pores of the host TiO_2 layer producing large morphological variations and causing a wide spread of photovoltaic performance in the resulting devices, limiting the prospects for practical applications.

An additional experimental difficulty is related to finding a good solvent for dissolving both the inorganic components (typically PbI_2, $PbCl_2$, $PbBr_2$) and the organic cations necessary for the assembly of the perovskite structure.

In order to circumvent these difficulties, vacuum evaporation of the separate inorganic and organic components is considered a viable methodology to grow oriented thin films of layered perovskites with a precise control of the film morphology. Nevertheless, vacuum processing is a more expensive procedure than wet chemical deposition routes. In principle, a good compromise between synthetic simplicity and reproducibility, can be obtained with the so called sequential deposition or two step deposition (TSD), originally developed by Mitzi and coworkers

[103], whereby PbI_2 is firstly deposited onto a nanoporous titanium dioxide film and subsequently transformed into the perovskite by exposing it to a solution of CH_3NH_3I. In the case of the $CH_3NH_3MI_3$ (M = Pb, Sn) perovskites the X-ray diffraction patterns demonstrated that the films prepared by the one step spin-coating method and the TSD display similar structural properties [120].

A comparison between single step and TS deposition routes was carried out by Hagfeldt et al. which also considered the differences in charge recombination, charge transport and energy conversion efficiency in devices based on both ZrO_2 and TiO_2 host layers [122]. SEM analysis revealed that, with both methods, the perovskite can penetrate deeply into the TiO_2 mesoporous layer, achieving a satisfactory permeation of the substrate. Compared with the one-step spin-coating method, the perovskite prepared from TSD has a darker color and stronger absorption indicating that a larger amount of perovskite is formed, manifesting one of the possible advantages of the TSD method (Fig. 2.28) [122].

The mechanism of charge separation in perovskite solar cells is object of intense research, triggered by the fact that also insulating metal oxides like Al_2O_3 and ZrO_2 scaffolds, where injection from the photoexcited perovskite is thermodynamically forbidden, allow to build efficient solar devices, characterized by nearly quantitative photon to electron conversion. In these cases charge transport occurs entirely through the perovskite network reaching the FTO back contact, and the Al_2O_3 and ZrO_2 layers act merely as templates facilitating perovskite deposition and adhesion to the ohmic collector. In general the ZrO_2/perovskite or the Al_2O_3/perovskite heterojunctions allow to generate higher V_{oc} than the equivalent junction based on TiO_2, since the voltage loss resulting from electron injection into the lower lying TiO_2 acceptor levels is avoided [122]. The fact that the electron transport occurs within the $CH_3NH_3PbI_{3-x}Cl_x$ perovskite, received a direct confirmation by photo-induced absorption spectroscopy (PIA) measurements carried out on the Al_2O_3/$CH_3NH_3PbI_{3-x}Cl_x$ layers [123]. PIA measurements also showed that in $CH_3NH_3PbI_{3-x}Cl_x$/spiro-OMeTAD heterointerfaces the photogenerated hole was injected into the HTM layer, following reductive quenching of the perovskite

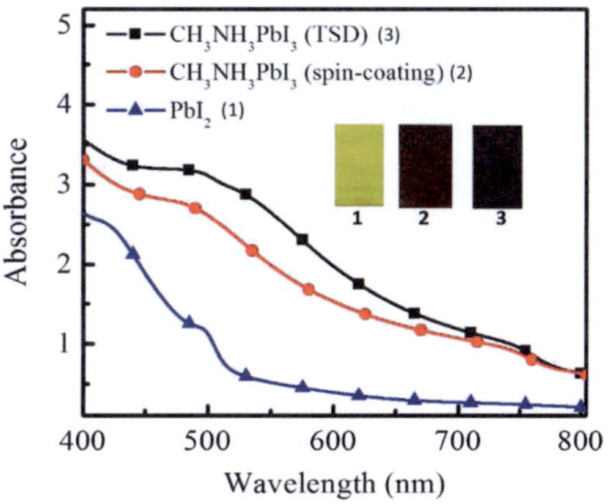

Fig. 2.28 UV-vis spectra and images of spin coated PbI_2 (1) film, and $CH_3NH_3PbI_3$ prepared from spin-coating (2) and TSD method (3) (Reproduced with permission from Ref. [122])

excited state, a mechanism of charge separation which is more reminiscent of p-n junctions rather than of conventional DSSCs. Nevertheless it was also reported, by studying the $TiO_2/CH_3NH_3PbI_3/Au$ architecture generating 5.5 % power conversion efficiency [124], that the perovskite can act directly also as a hole transporting material, indicating that the electrical transport in organic/inorganic metal halides can occur through ambipolar diffusion of both electrons and holes.

The increase in V_{oc} observed by moving from the TiO_2 to the insulating scaffolds like Al_2O_3 is of the order of a few hundred millivolts resulting in a power conversion efficiency close to 11 % under AM1.5 solar illumination. It is however worth to remark that regardless of the chemical nature of the mesoporous layer, the presence of a nanometric compact TiO_2 layer underneath the mesoporous layer is still required to improve the selectivity of the ohmic contact towards charge carriers, allowing a reduction of the local charge recombination by preferentially collecting electrons while blocking holes.

Often, good quality perovskite films prepared by either the single step or the two step deposition techniques allow to achieve power conversion efficiencies of approximately 15 % (measured under standard AM1.5G conditions) [125].

There has been extensive work on investigating how solution-cast materials infiltrate into mesoporous oxides, since a uniform permeation of the metal oxide layer appears to be one of the main requisites to achieve efficient hybrid interfaces. Within an appropriate concentration and solubility range, it has been shown that the solution cast material completely penetrates into pores of the mesoporous metal oxide, forming a wetting layer that uniformly coats the pore walls throughout the thickness of the electrode [126–129]. The degree of pore filling can be controlled by varying the solution concentration: if the concentration of the casting solution is too high, saturation of the pores occurs, and any excess material forms a capping layer on top of the filled mesoporous oxide [127–130].

2.7.4 Future Perspectives of Organo-perovskite Based Solar Cells

In the case of $CH_3NH_3PbI_3$, since the onset of light absorption is ca. 800 nm, the ideal maximum photocurrent is ~27 mA cm^{-2} [131]. Provided 90 % of IPCE is achievable between 400 and 800 nm, 24 mA cm^{-2} of photocurrent can be collected.

Having so far demonstrated the extremely promising photoelectrochemical properties of mixed halide perovskites, for the commercialization of solar devices based on this class of sensitizers one also needs to account for stability and toxicity in addition to high efficiencies. Lead content is, for example, an important drawback for the viability of these cells and there is a drive to replace Pb^{2+} with less toxic elements.

Lead-free solution-processed solid-state photovoltaic devices based on tin, with the general formula $CH_3NH_3SnI_{3-x}Br_x$, have been recently introduced. The featured optical band-gap of 1.3 eV for the $CH_3NH_3SnI_3$ perovskite, that show an absorption onset at 950 nm, is significantly red shifted compared with the benchmark $CH_3NH_3PbI_3$ counterpart (1.55 eV). Bandgap engineering was implemented by chemical substitution in the form of $CH_3NH_3SnI_{3-x}Br_x$ solid solutions, which can be controllably tuned to cover much of the visible spectrum, thus enabling the realization of lead-free solar cells with an initial power conversion efficiency of 5.73 % under simulated full sunlight [132].

Humidity and the interaction between the perovskite and water play an important role in the stability of solar cells based on these materials. In the case of $CH_3NH_3PbI_3$ for example, it was demonstrated that water exposure does not simply cause $CH_3NH_3PbI_3$ to revert to PbI_2. It was shown that, in the dark, water is able to form a hydrate product similar to $(CH_3NH_3)_4PbI_6{\cdot}2H_2O$, and that the degradation processes occurs more rapidly upon illumination, when the $CH_3NH_3^+$ hydrogen-bonding strength is lessened. This causes a decrease in absorption across the visible region of the spectrum and a distinct change in the crystal structure of the material [133]. Strengthening the hydrogen-bonding interaction between the organic cation and metal halide octahedra and/or weakening the hydrogen interaction between the organic cation and H_2O could improve moisture resistance, making this a promising area for further work toward developing commercial perovskite solar cells.

Some authors have demonstrated that an important degradation mechanism in perovskite solar cells is activated by a low shunt resistance occurring during aging, postulating that a direct contact between the metal electrode and the perovskite might result in detrimental electronic shunt pathways, which lowers the power conversion efficiency. Depositing the HTM within an insulating mesoporous "buffer layer" comprised of Al_2O_3 nanoparticles prevents the migration of metal islands from the counter electrode, avoiding electrical shorting while allowing for precise control of the HTM thickness. This enabled an improvement in the solar cell fill factor and enhanced device stability in 350 h durability tests [134].

With a different approach, Snaith's group demonstrated an elegant way to reduce the sensitivity of perovksite solar cells to humidity and thermal stress, by using P3HT/single walled carbon nanotubes (SWCNTs) as organic hole transporting material embedded in an insulating polymer matrix. Despite the presence of an insulating layer, the SWCNT composite was able to establish a good ohmic contact with the evaporated gold counter electrode while acting at the same time, as a barrier against migration of metal atoms of the counter electrode and the ingress of water from the surrounding environment. With this composite structure power conversion efficiencies of up to 15.3 % were accompanied by strong retardation in thermal degradation and by enhanced resistance to ambient moisture [135].

2.8 Solar Fuel Production with Dye Sensitized Photoelectrosynthetic Cells

Despite the many advantages outlined in the previous section, Sun has inherent limitations as energy source. Its radiation is widely distributed but low in intensity (ca. 0.1 W/cm^2 at medium north american and european latitudes, according to AM 1.5 G standards), and large surface areas are required for its collection. To achieve the level of power consumption in the United States of ~3 terawatts (3 TW) at 10 % efficiency, would require ~140,000 km^2 for collection, roughly half the area of Italy. The materials costs with current solar PV technology would be enormous, 60 trillion dollars for 140,000 km^2 based on Si solar cell technology at 400 $\$/m^2$ [136].

There is an even greater drawback to solar energy. It is intermittent, depending on the day/night cycles, seasonal and local weather conditions. Thus in order to serve as a primary energy source, solar has to be coupled with large scale energy storage capabilities. Only the energy density in chemical bonds is adequate for storage at the levels required for our current technology. The energy stored in C-H bonds of hydrocarbon sources has indeed its origin in photosynthetic processes occurred millions of years ago. With an appropriate process, excess solar electrical power could be stored as high-energy chemicals and utilized in existing power plants, or by secondary electricity generation in fuel cells. This leads to the concept of "solar fuels" with targets including: water splitting into hydrogen and oxygen, CO_2 reduction into almost any form of reduced carbon—CO, formaldehyde, methanol, other oxygenates, methane—with liquid hydrocarbons as ultimate goal. Natural gas and liquid fuels are especially appealing since they would utilize our existing energy infrastructure.

The chemical reactions for water splitting into hydrogen and oxygen, and CO_2 reduction into methane, require the multiple transfer of electrons (n) and protons, as reported below:

$$2\,H_2O \rightarrow 2\,H_2 + O_2(n = 4)$$

$$CO_2 + 2\,H_2O \rightarrow CH_4 + 2\,O_2(n = 8)$$

Probably, the first solar fuel available in a large amount will be hydrogen because routes to water splitting are more straightforward than reduction of CO_2. Although the use of hydrogen as primary energy vector may be limited, due to difficult storage and under-developed transportation networks, it could be used for on-site power generation or, reacted with CO_2, will represent a valuable feed stock for CO_2 sequestration and conversion into fuels using well known and consolidated technologies.

At present, the use of biomass strategies (natural photosynthesis) for CO_2 capture and conversion, does not appear efficient enough if consideration is made for the current CO_2 production rate, as conversion efficiencies $\approx$10 % are often considered as a minimum for an economically viable solar fuel device. Thus the search for molecular based high-tech strategies for realizing an efficient artificial photosynthetic device represents a challenging multidisciplinary task which, at present, is still in its infancy.

2.8.1 *Molecular Artificial Photosynthesis in Photoelectrochemical Cells*

As the fundamental knowledge about the functioning of natural photosynthetic structures and processes progresses [137–145], the requirements for artificial photosynthesis become more clear. In fact the key elements for a molecular approach should be tasked with the following functions:

1. Light absorption resulting in the formation of a molecular excited state with appropriate energetics.
2. Electron-transfer quenching of the excited state.
3. Spatially directed electron/proton transfer driven by free energy gradients, to give transiently stored oxidative and reductive equivalents.
4. Catalysis of water oxidation and $H_2O/H^+/CO_2$ reduction, in order to compete successfully with energy wasting charge recombination.

The incorporation and integration, into appropriate configurations, of the elements able to perform functions 1–4 for practical devices, is one of the major challenges in artificial photosynthesis. The key modules are represented, in the case of a photochemical water splitting assembly, in Fig. 2.29. It includes an antenna array for light collection, funnelling energy to a chromophore (C), whose excited state initiates a sequence of electron transfers to chemically linked electron-transfer donors (D) and acceptors (A), in order to activate redox catalysts for water oxidation (Cat_{ox}) or hydrogen evolution (Cat_{red}).

The interfacing of a molecular system with solid electrodes in contact with an electrolytic solution offers the possibility to achieve a better organization and control over the processes as well as a simplified molecular assembly, by separating the overall processes into half-reactions occurring at the separate compartments of a (photo)-electrochemical cell (Fig. 2.30). Such approach is clearly inspired by DSSC, object of extensive discussion in the previous sections and is often referred to as "Dye-Sensitized Photoelectrosynthetic Cell" (DSPEC) configuration.

The essential features in the photoelectrochemical approach are the separation of the charge carriers (electrons and holes) to different phases. The solid phase which collects electrons, can also be controlled, via electrical connections, through the application of an external voltage. The oxidation and reduction intermediates and products are physically separated by an ion-permeable membrane (PEM in Fig. 2.30), resulting in a further decreased probability of recombination by preventing the diffusional encounter of partly reduced and oxidised species in solution. The electronic characteristics of the semiconductor determines the

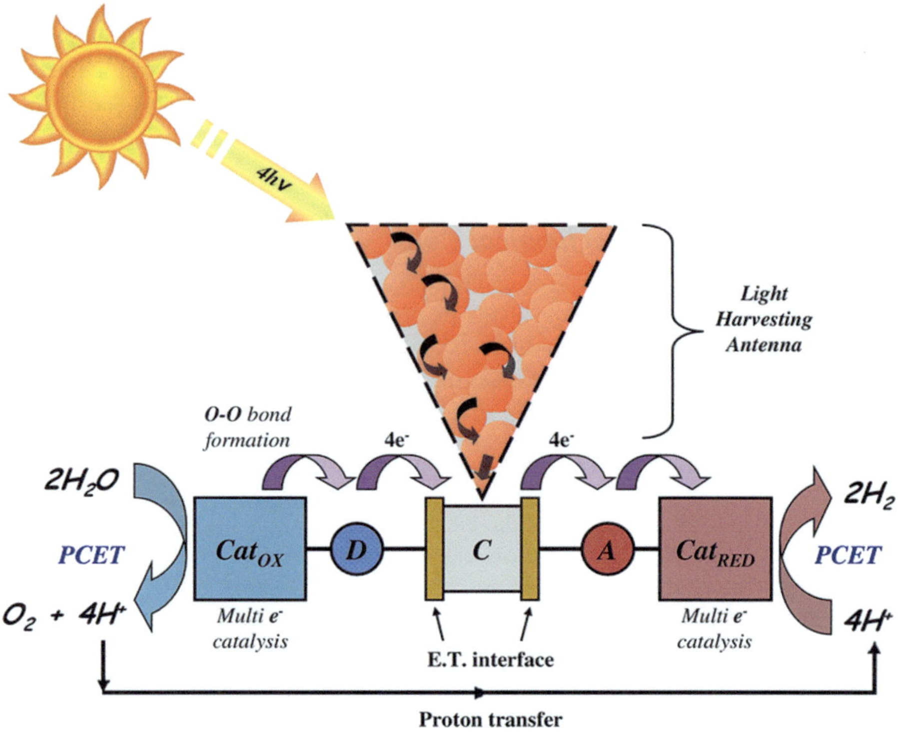

Fig. 2.29 Schematic molecular assembly for water splitting

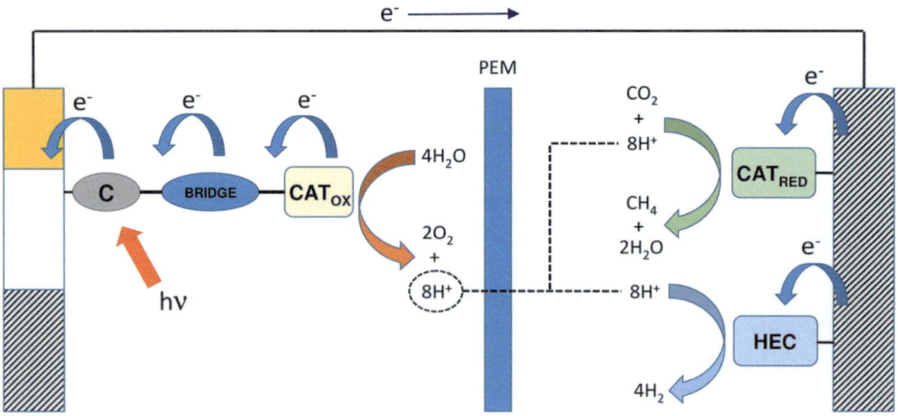

Fig. 2.30 DSPEC for solar fuel production. C is the light absorber injecting electrons into the semiconductor, triggering a sequence of electron transfer events which activate the water oxidation catalyst (CAT_{OX}). At the dark cathode either a water (HEC) or a CO_2 (CAT_{RED}) reduction catalyst drives the production of fuel (H_2 or CH_4 in the scheme). PEM is a proton exchange membrane

direction of the electron flow through the cell. In their most common configuration DSPECs are built with transparent n-type semiconductors like TiO_2, SnO_2, ZnO as photoanodes, and electron conduction results from charge injection into their empty conduction bands, while holes remain transiently confined on the molecular assembly bound at the surface of the material, before hole transfer to the final acceptor (H_2O) in the electrolyte solution occurs. Typically semiconductor thin films consisting of 20–40 nm nanoparticles and of the thickness of 3–5 μm, once covered with a monolayer of dye having $\varepsilon \approx 10^4\ M^{-1}cm^{-1}$, allow for the harvesting of large fractions of light (even >90 % in the dye absorption maximum) making superfluous the use of antenna system as additional light capture strategy and resulting in a further synthetic simplification.

The quasi-Fermi level of the semiconductors used to build the device determine the thermodynamics of the cell and the feasibility of the devised fuel forming process. For example, in order to observe hydrogen evolution without externally applied potential bias, the quasi-Fermi level of the semiconductor (which is in equilibrium with the Fermi level of the dark cathode) must be more negative than the formal potential for hydrogen evolution ($E°(H^+/H_2) = 0$ V vs. NHE). When this condition is not met, the application of an external potential is necessary to drive the reaction of interest. Referring to Fig. 2.31, for hydrogen evolution to occur, since under illumination the Fermi level of the cathode is less negative than that of the H^+/H_2 couple, a positive bias must be applied to the semiconductor electrode as shown in Fig. 2.31b. The application of the voltage will raise the Fermi level of the dark cathode having opposite polarity, activating the charge transfer. This bias, which is usually provided by an external voltage source, should also account for the cathodic overvoltage to sustain the current flow. As a first approximation, the redox levels of the surface bound sensitizing species are not affected by the applied bias, and the photo-produced oxidative equivalents must be capable to oxidize water to oxygen ($E°\ (O_2/H_2O) = 1.23$ V vs. NHE).

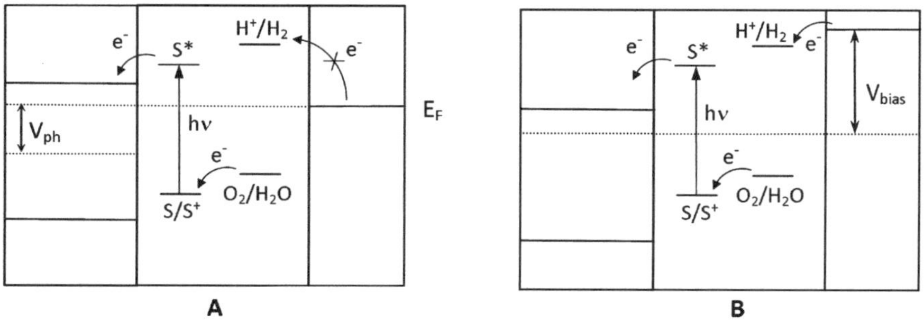

Fig. 2.31 Energy level diagram for a sensitized n-type semiconductor used as a photoanode in DSPEC for water splitting with a single photoactive junction. S/S^+ represents the redox level of the oxidized Dye Sensitizer/Catalyst assembly. Here the quasi-Fermi level (*dotted line*) of the n-type material is approximated with the flat band potential (V_{fb}) of the semiconductor, as commonly accepted

Other DSPEC configurations are possible with photoelectrochemical events driven at a photocathode by hole rather than electron injection. A p-type DSPEC (Fig. 2.32b) operate in an inverse fashion compared to the more common n-type DSPECs (Fig. 2.32a): the excited state of a dye molecule adsorbed on the semiconductor surface injects a hole into the VB of the p-type material (typically NiO). The injected hole diffuses to the back contact of the conducting substrate while the

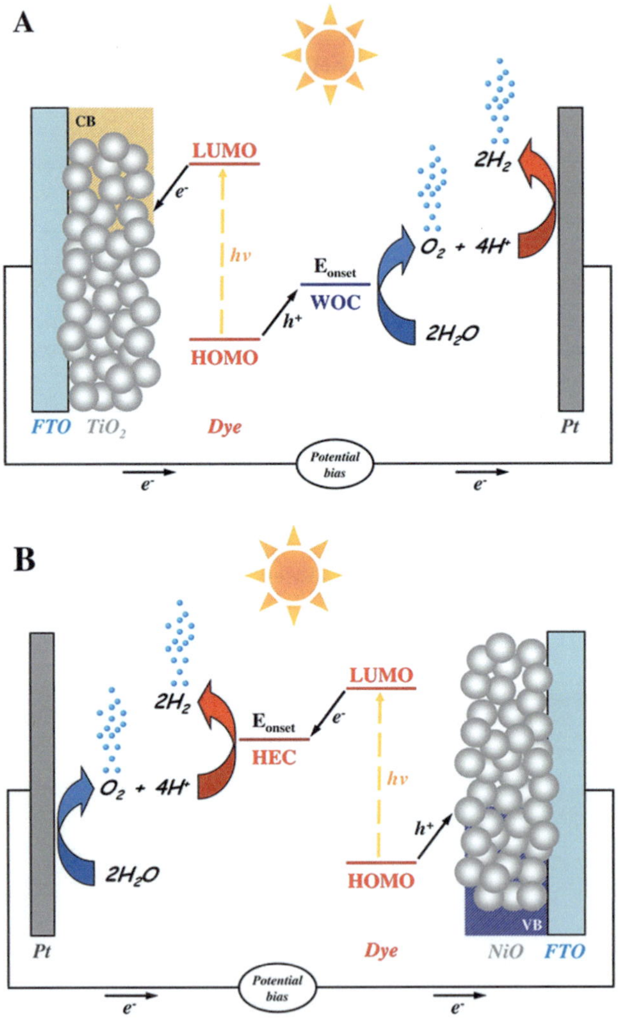

Fig. 2.32 Principles of operation of n-type (**a**) and p-type (**b**) DSPECs for water splitting. TiO_2 and NiO are taken here as common examples of n and p-type semiconductors respectively. Semiconductor electrodes are connected to a catalytic dark counter electrode (platinum wire in the example). In this configuration the application of a potential bias is often necessary to drive hydrogen evolution at the counter electrode of the n-type cell or oxygen evolution at that of the p-type cell (Adapted from Sun et al. [178])

reduced dye is restored to its ground state by delivering an electron to the oxidized component (for example H^+) in the electrolyte, resulting in hydrogen evolution. Clearly, also in this case the coupling of the dye with a catalyst (HEC, Hydrogen Evolution Catalyst, in the case of the p-type water splitting DSPEC) may be necessary to boost the efficiency of the faradic reaction relevant to the production of solar fuel.

When the application of an additional potential is required, the efficiency of the fuel forming solar cell must consider the energy loss due to the need of applying the bias voltage (referred to the counter electrode of the cell) according to Eq. (2.13), where the photocurrent density (J) produced at a given voltage (V_{bias}) under the incident solar power (P) appear. $E°_{fuel}$ is the standard potential for the fuel forming reaction ($E°_{fuel} = 1.23$ V, in the case of water splitting into O_2 and H_2) having a faradic efficiency η, which should ideally approach 100 %. The quantity at the left side of Eq. (2.13) is rigorously called ABPE (Applied Bias Photon to current Efficiency): [146]

$$\text{ABPE} = \frac{\left[J(mA/cm^2) \times \left(E^0{}_{fuel} - V_{bias}\right)(V)\right]\eta}{P(mW/cm^2)} \tag{2.13}$$

In order to avoid the need of the applied bias, a DSPEC cell architecture with separate n-p junctions, which sum their respective photovoltages (Fig. 2.33), can be devised, reminiscent of the concept of photochemical diode [147].

As solar fuel production is a complex matter, where several subsystem must cooperate in order to drive with success the fuel forming reaction, resulting from the minimization of recombination events (Fig. 2.34) [148], the current approach is to try to optimize the separate interfaces, which can be later assembled together in an unassisted photoelectrochemical device. Under this respect, NiO, commonly used as a porous and stable p-type material, has by no means the optimal electronic qualities because it absorbs a considerable amount of visible light and its hole

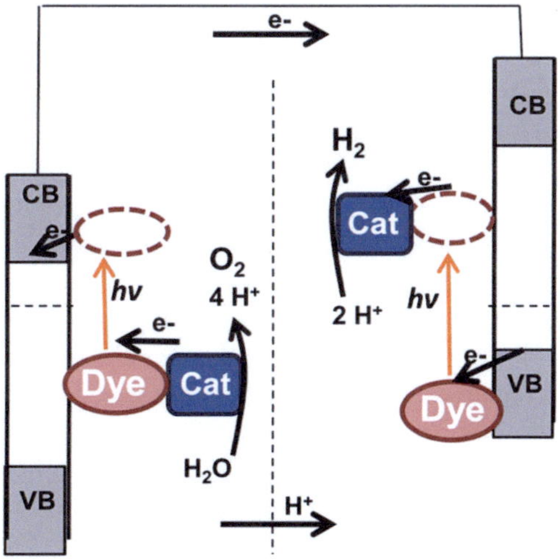

Fig. 2.33 Tandem n-p DSPEC for solar water splitting in the absence of applied bias. At the left interface (n-type), charge injection into n-type results in the activation of the water oxidation catalyst. At the right (p-type) the reductive quenching of the dye (hole transfer to the p-type material) triggers electron transfer to the reduction catalyst. Electrons flow from the n-type to the p-type material (Reproduced with permission from Ref. [147])

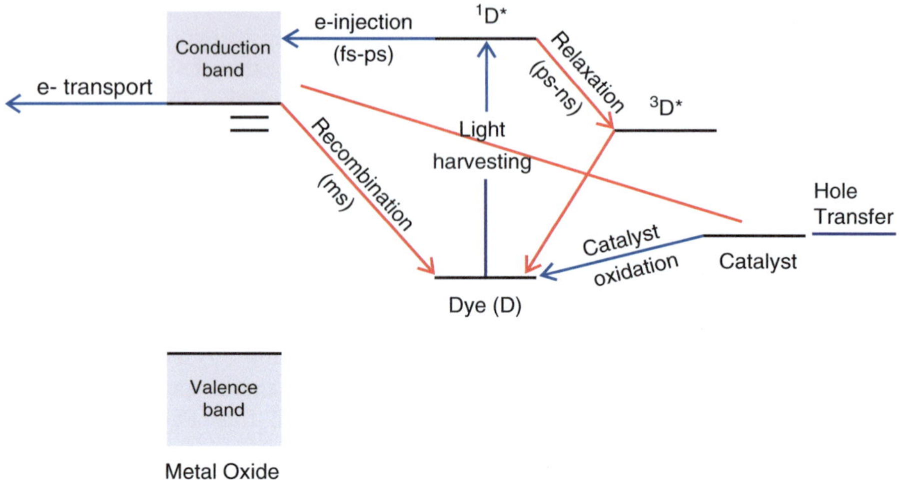

Fig. 2.34 Examples of physical processes relevant for the dye sensitized fuel forming reactions at an n-type interface. *Blue arrows* indicate the desired electron flow, *red arrows* energy wasting deactivation and recombination events (Adapted with permission from Ref. [148])

mobility is relatively low. It is therefore likely to limit the performance of the serially connected n-type interface, for which better materials and a more complete understanding of their interfacial charge separation kinetics are available. The search for alternative more transparent p-type semiconductors, with better electronic properties, porous morphology and with a reactive surface for molecular sensitization is therefore important. Nevertheless the first reports of complete DSPEC cell operating without external bias are beginning to appear in the recent literature (2014–2015) and will be described with a greater detail in the next sections.

2.8.2 Dye Sensitizers for DSPECs: Base Structures and Concepts

The dye is clearly an essential part of DSPECs and should fulfil some general requirements, which are actually more stringent than those outlined for DSSCs: (1) it should have strong visible absorption bands, in order to harvest a large fraction of the solar spectrum; (2) for n-type sensitization, the excited state oxidation potential (E_{ox}*), representing the electron donating capability of the dye at the excited state, should be more negative than the Fermi level of the semiconductor. In the case of p-type sensitization, the excited state reduction potential of the dye (E_{red}*) should be more positive than the Fermi level of the p-type material; (3) in water photo-oxidation processes, the first oxidation potential of the ground state must be positive enough to drive hole transfer to the water oxidation catalyst

(WOC); a similar statement holds for reductive conversion schemes at p-type interfaces, where the first reduction potential of the dye should be negative enough to activate HEC or CRC (CO_2 Reduction Catalyst); (4) it should possess stable anchoring group in hydrolytic and oxidative conditions; (5) chemical and electrochemical stability in operational conditions is a desirable quality for species supposed to resist several redox turnovers without degrading (at present real long term stability with molecular systems is not yet demonstrated).

Ruthenium-polypyridine complexes were the first species to be considered for light-driven water-splitting photoanodes [149–151], because of their broad spectral coverage and high molar absorptivity of Metal to Ligand Charge Transfer (MLCT) transitions ($1–2 \times 10^4\ M^{-1}cm^{-1}$) in the blue-green portion of the visible region. They also show long-lived excited-state lifetimes and good electrochemical stability with acceptable ground and excited state redox potentials.

Spectroscopic and electrochemical properties can be optimized for the intended task by substituting the chromophoric ligands, typically bipyridines or terpyridines, with different electron donating or electron withdrawing functional groups. Electron injection into the most common wide band-gap semiconductors like TiO_2, SnO_2 and ZnO is thermodynamically feasible for most ruthenium-polypyridyl dyes from both the lowest singlet (^{1}MLCT) and triplet (^{3}MLCT) states, but, in order to maximize the charge injection rate, the electronic coupling of the sensitizer with the substrate should be established through the formation of covalent bonds. Carboxylic (–COOH) and phosphonic ($–PO_3H_2$) are the most used anchoring groups for the attachment of molecular species to metal oxides, owing also to their synthetic accessibility. Phosphonic groups were found to bind more strongly to the TiO_2 surface than carboxylates, resulting in greater stability. Desorption studies of ruthenium sensitizers in aqueous solutions at pH 5.7 showed 90 % desorption with a carboxylate anchor, compared to only 30 % desorption for the phosphonate under the same conditions [152], indicating that phosphonic anchors should be preferred to carboxylic groups to enhance the stability of water-splitting cells, despite the less efficient electron injection of the former with respect to the latter, due to the interruption of π conjugation with the organic ligand.

Basic structures of Ru(II) polypryridine sensitizers used for DSPEC photoanodes are reported in Fig. 2.35, where it can be appreciated the predominant presence of $–PO_3H_2$ groups for use in aqueous media.

Organic dyes are a viable alternative to metal complexes in order to avoid the use of rare metals, at least in the light harvesting part of the electrode assembly. Organic species can exhibit π-π* transitions with very high extinction coefficients, even on the order of $10^5\ M^{-1}\ cm^{-1}$. Among these species, porphyrins have been widely studied by the photochemical community as light harvesters in donor-acceptor dyads, in order to study the lifetime and charge recombination processes of the resulting photoinduced charge separated states [153–155]. The singlet ground state (S_0) of porphyrins tend to be more oxidizing with respect to Ru(II) complexes and electron injection into TiO_2 is typically only favorable from the S_2 and S_1 excited states. The lowest triplet excited state tend to be close to the TiO_2 conduction band edge, or even lower (Fig. 2.36). One advantage of porphyrins over tris-bipyridine

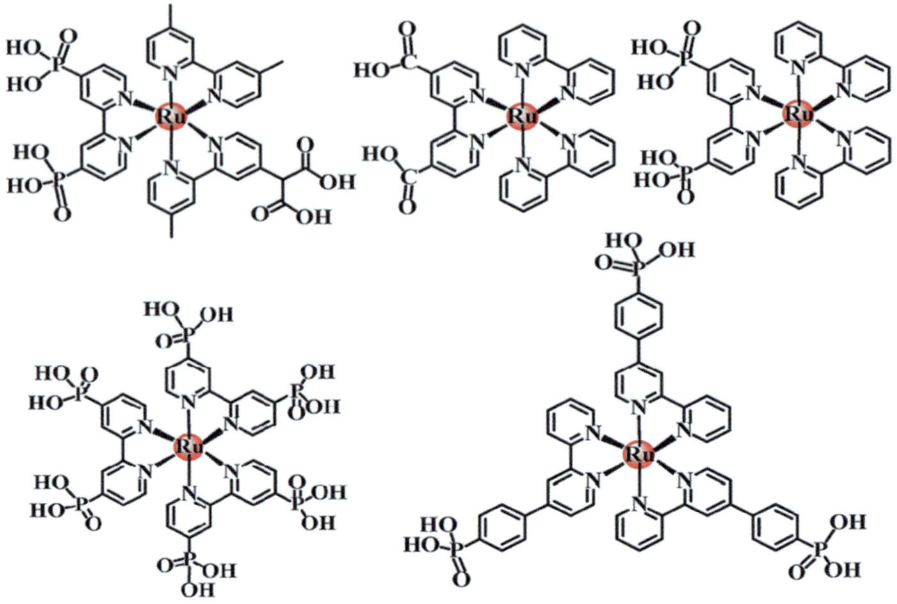

Fig. 2.35 Structures of some MLCT Ru(II) polypyridine complexes as dye sensitizers in DSPEC (Reproduced with permission from from Sun et al. [178])

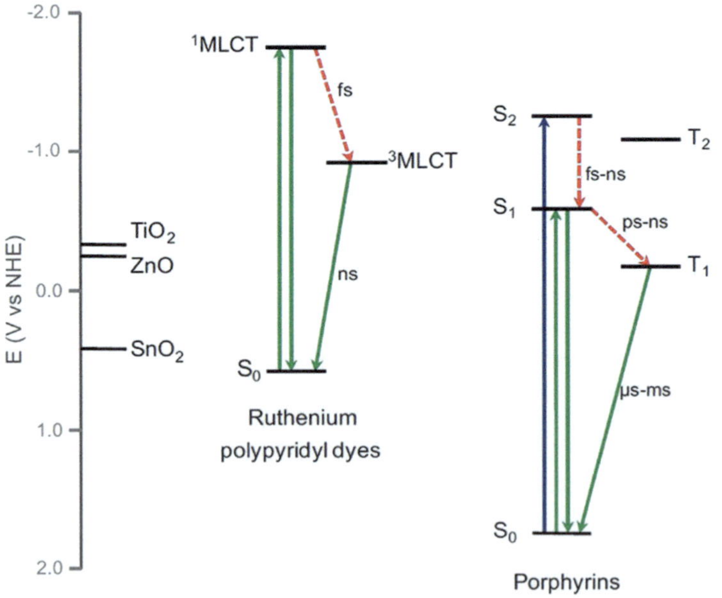

Fig. 2.36 Approximate Jablonsky diagrams showing the first excited state energies and lifetimes of polypyiridine Ru(II) and porphyrin sensitizers compared with the conduction band edge level of common metal oxides used for n-type DSPEC fabrication (Reproduced with permission from Ref. [148])

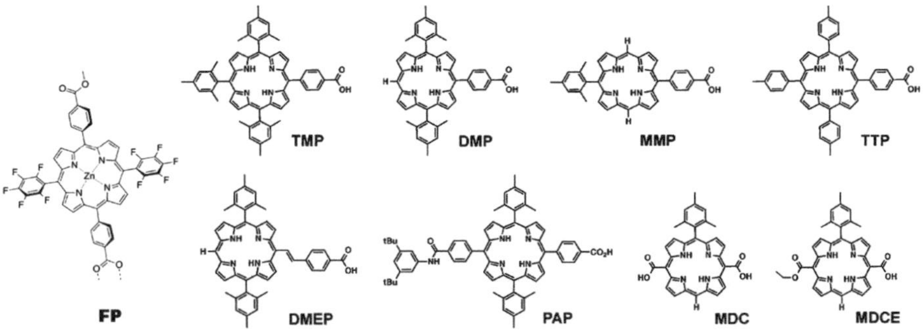

Fig. 2.37 Proposed porphyrin chromophores for n-type DSPEC (Reproduced with permission from Refs. [148, 158])

Table 2.1 Photophysical and electrochemical properties of metal-free porphyrins reported in Fig. 2.35

Porphyrin	E^{0-0} (eV)	E_{pa} (V vs. AgCl)	E_{ox}^{*} (V vs. AgCl)
TMP	1.91	1.05	−0.86
DMP	1.94	1.04	−0.90
MMP	1.96	1.02	−0.94
DMEP	1.87	1.18	−0.67
TTP	1.90	1.18	−0.72
PAP	1.91	1.29	−0.62
MDCE	1.92	1.03	−0.89

E^{0-0} represents the singlet energy gap (S0–S1), E_{pa} is the first oxidation (anodic) potential and E_{ox}^{*} is the excited state oxidation potential obtained from $E_{pa} - E^{0-0}$

ruthenium dyes is that their absorption spectra generally extend further towards the near-IR region, improving the harvesting of the solar spectrum. Considering the minimum thermodynamic requirements for water splitting (1.23 eV) and losses due to cell overpotentials, the minimum absorption onset for realizing water splitting with a 1-photon–1-electron system, as in common single junction cells, would be around 650 nm (1.9 eV).

High potential porphyrins have been proposed as suitable light harvesters for molecular water splitting [156, 157]. Electron-withdrawing pentafluorophenyl groups can be incorporated at the *meso*-carbon of a zinc porphyrin (Fig. 2.37, (FP)) to induce an electron deficiency that leads to a positive shift in the ground state oxidation potential of the complex. Consequently, the holes remaining on a high potential photooxidized porphyrin are thermodynamically capable of driving water oxidation.

Other types of free-base porphyrins considered by Mallouk et al. [158] for n-type DSPEC are reported in Fig. 2.37, with their ground and excited state properties summarized in Table 2.1.

A balance between the ability of performing photoinduced charge transfer to TiO_2 (reflected by large and negative E_{ox}^*) and to promote hole transfer to the WOC (reflected by large and positive E_{pa}) should be searched in order to minimize both excited state deactivation and charge recombination with the oxidized sensitizer.

2.8.3 Molecular Water Oxidation Catalysts: Properties and Mechanism of Action

The catalyst plays a fundamental role in the DSPEC cell, since the complex kinetics, imposed by multi-electron water oxidation, would result in exceedingly high activation barriers for the direct hole transfer from the photo-oxidized dye, immobilized at the semiconductor surface, to water molecules, resulting in predominant recombination events. Up-to-date excellent reviews about molecular water oxidation catalysts exist [159, 160] and the exhaustive and detailed description of a wide class of WOC based on coordination compounds incorporating different metals including first row transition metals like manganese, cobalt and iron, is beyond the scope of this book chapter. For our objective of describing DSPEC at work, only the class of molecular catalysts which were most frequently employed in functioning DSPEC photoanodes will be considered.

As stated before, the water oxidation part of the DSPEC has attracted the large majority of the research work, while, very commonly, hydrogen is simply evolved at a dark catalytic counter electrode under applied bias. This predominant interest about the photoanodic assembly is largely justified by the fact that the water oxidation is considered to be the ubiquitous part in sustainable fuel forming reactions and by the convenience of having a wider arsenal of optimized materials at our disposal as well.

The first homogeneous water-oxidation catalyst was the ruthenium polypyridyl dinuclear complex (blue dimer) reported by Meyer and co-workers in 1982 [161, 162], characterized by the presence of a Ru-O-Ru bridge and of two adjacent water molecules coordinating each Ru center (Fig. 2.38).

The general requirement for WOCs is the production of stable high valent metal-oxo species at low redox potentials, to be able to initiate the multielectron oxidation of water. To enable the rational design of molecular complexes that can fulfill this requirement, extensive research has been directed toward the elucidation of the fundamental steps of H_2O oxidation. Despite the experimental difficulties in determining the reaction mechanisms, the current view proposes two major mechanistic pathways for H_2O oxidation (Fig. 2.39): (i) solvent Water Nucleophilic Attack (WNA) and (ii) interaction of two M-O units (I2M).

In the WNA mechanism, H_2O acts as the nucleophile by attacking the electrophilic high-valent metal – oxo species, which leads to cleavage of the metal – oxo π-bond and the concomitant generation of the crucial O–O bond. This results in the formal two-electron reduction of the metal center to form a metal hydroperoxide

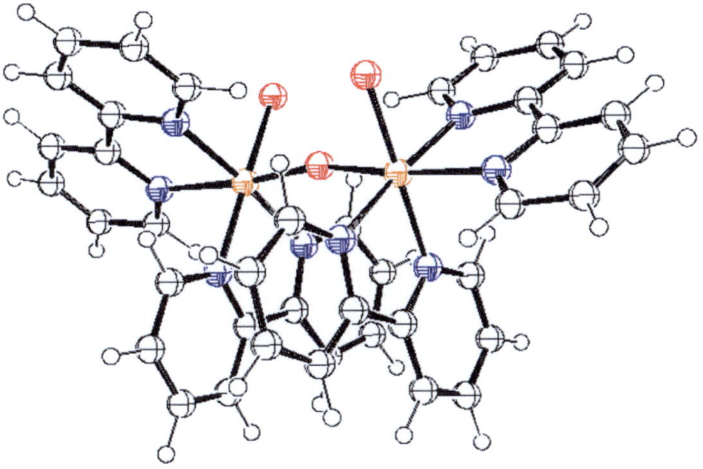

Fig. 2.38 Structure of the blue dimer reported by Meyer and co-workers. Ruthenium atoms are indicated in *orange*, oxygen atoms in *red* and nitrogen atoms in *blue* color. Hydrogens bound to the water molecules are removed for clarity (Reproduced with permission from Ref. [160])

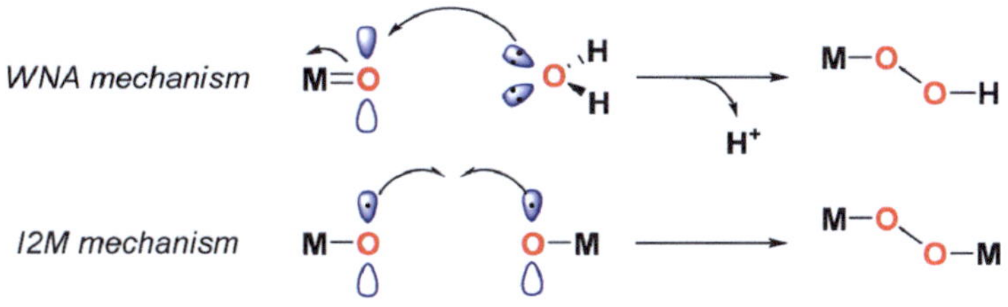

Fig. 2.39 Representation of the two main pathways for oxygen evolution in single site and di-nuclear WOC (Reproduced with permission from Ref. [159])

species (M–OOH), which can subsequently undergo further oxidation to liberate O_2. The other mechanistic pathway (I2M) involves the radical coupling of two metal – oxo species that hold significant radical character, generating a [M–O–O–M] species that may undergo further oxidation to ultimately release O_2. Generally, systems proceeding via the I2M pathway contain flexible ligand scaffolds with large bite angles (>90°) that can promote the O–O bond formation.

The presence of the μ-oxo bridge in the blue dimer promotes a strong electronic coupling between the two metal centers which facilitates the stabilization of the complex at high oxidation states by electronic delocalization. The mechanism proposed by Meyer and coworkers (Fig. 2.40 involved four Proton Coupled Electron Transfer (PCET) steps that give access to a high-valent Ru(V,V) intermediate (intermediate 2, Fig. 2.40) which is subsequently attacked by H_2O, generating a hydroperoxo species (intermediate 4 Fig. 2.40) that is intramolecularly oxidized by the second ruthenium center ultimately resulting in the release of O_2.

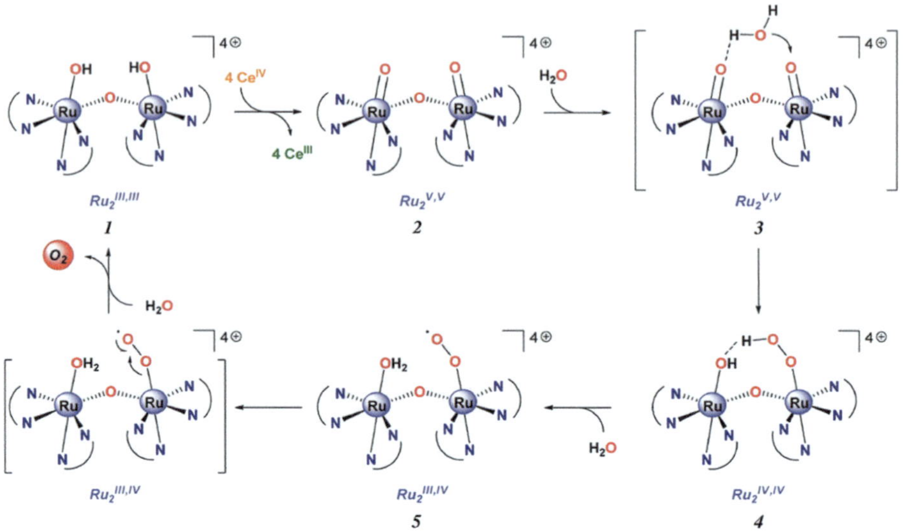

Fig. 2.40 Proposed mechanism of oxygen evolution in μ-oxo bridged blue dimer-type Ru (II) complexes (Reproduced with permission from Ref. [159])

Ruthenium blue dimers were generally characterized by moderate to low water oxidation activities, believed to be related to the instability of the μ-oxo bridge, which upon cleavage results in the breakdown of the dimeric structure into non-active monomeric ruthenium complexes. Research therefore focused on finding more stable organic ligands capable of bringing two ruthenium centers together in close proximity, which would enable more efficient oxygen evolution. The ruthenium complex reported in Fig. 2.41. ($Ru - Hbpp = \left[Ru_2(OH_2)_2(bpp)(tpy)_2\right]^{2+}$ where Hbpp = 2,2′-(1H-pyrazole-3,5-diyl)dipyridine), developed by Llobet et al. [163], was the first dinuclear ruthenium complex lacking a Ru–O–Ru motif capable of oxidizing H_2O to O_2. In this complex, the two ruthenium metal centers are placed in close proximity, oriented in a cis fashion to one another, by the introduction of a rigid pyrazole ring as bridging ligand between the two metal centers. In Ru-Hbpp the oxygen bond formation proceeds solely via the I2M mechanism with oxygen evolution rates which are three times larger than the blue dimer under comparative conditions. This is due to a more favorable ligand environment which pre-orientates the two metal centers and favors the coupling between the two oxo groups formed upon oxidation. The decomposition of the di-nuclear assembly is largely avoided by the elimination of the labile μ-oxo group. Competitive coordination of anions in lieu of the water molecules (anation), known to limit the performance of the blue dimer, is also reduced.

Because of the multimetallic core of the natural Oxygen Evolving Complex (OEC) in photosynthetic organisms, it was long envisioned that artificial molecular WOCs must accommodate multiple metal centers to cope with the accumulation of the four oxidizing equivalents needed for H_2O oxidation. The initial lack of reports

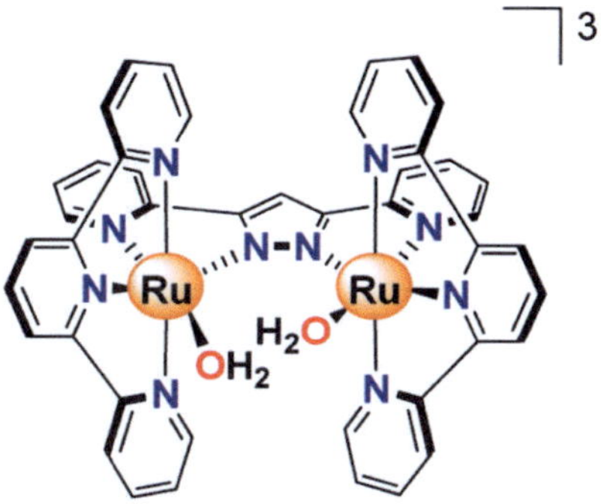

Fig. 2.41 Ru-Hbpp dinuclear oxygen evolution catalyst developed by Llobet et al. [159]

on single-site WOCs led to the creation of a paradigm, which claimed that at least two metal centers were required for H_2O oxidation to occur. However, this early belief has now been disproved, and today there exists a variety of single-site catalysts that can mediate the four-electron oxidation of H_2O. Molecular single-site catalysts offer the possibility of straightforward ligand design, synthesis and characterization. The relative ease with which the ligand environment can be tuned in these molecular systems, both electronically and sterically, makes them amenable for structure – activity relationship studies. More importantly, the incorporation and/or attachment of these catalysts to viable supramolecular assemblies for H_2O photosplitting, resulting from the linkage to a chromophore, is also greatly facilitated.

The first proofs that the four-electron oxidation of H_2O could occur on single-site metal complexes was provided by the group of Thummel in 2005 [164]. The ligand backbone in these ruthenium complexes consisted of a tridentate polypyridyl type ligand, 2,6-di(1,8- naphthyridin-2-yl)pyridine, with uncoordinated naphthyridine nitrogens. The un-coordinated nitrogens interact with the aqua ligand through hydrogen bonding, thus stabilizing the single-site aqua complexes. Solid evidence that H_2O oxidation proceeded through single-site catalysis was provided by Meyer and co-workers in 2008 [165, 166] during the study of complexes obtained by substituting Cl^- with H_2O in the family reported in Fig. 2.42. It was proposed that these single-site ruthenium WOCs catalyzed the oxidation of H_2O through a seven-coordinated ruthenium center, involving a η^2-peroxide intermediate, $[Ru(IV)\text{-}\eta^2\text{-}OO]^{2+}$. Starting from $[Ru(II){-}OH_2]^{2+}$, the addition of three equivalents of Ce(IV) resulted in the formation of the key intermediate $[Ru(V)=O]^{3+}$, which undergoes nucleophilic attack by H_2O molecule to generate a peroxide intermediate $[Ru(III) - OOH]^{2+}$ via the WNA mechanism. Further oxidation of this species through proton-coupled oxidation, formed the $[Ru(IV) - OO]^{2+}$, finally leading to O_2 evolution (Fig. 2.43).

One of the major drawback of WOCs containing neutral and π accepting ligands is the high oxidation potential necessary to reach highly valent Ru (IV) or Ru (V) states, which are difficult to be matched by the photosensitizer in artificial photosynthesis schemes. That is the reason why often strong chemical oxidants like Ce(IV) are used to trigger oxygen evolution in catalytic cycles. A strategy to shift

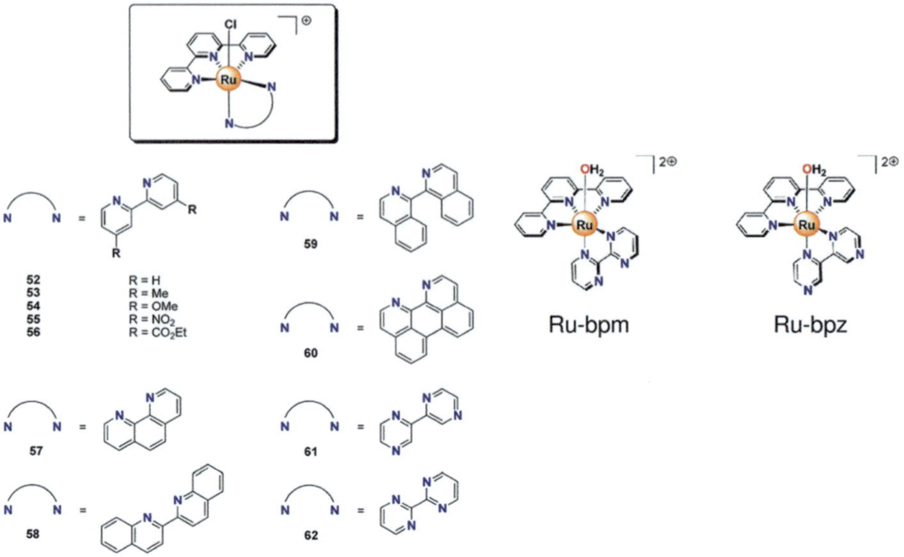

Fig. 2.42 Family of precursor complexes based on different polypyridine ligand and single site water oxidation catalyst based on bipyrimidine (bpm) (Ru-bpm) and bipirazine (bpz) (Ru-bpz) ligands $[Ru(tpy)(bpm)(OH_2)]^{2+}$ and $[Ru(tpy)(bpz)(OH_2)]^{2+}$ investigated by Meyer et al. (Adapted from Refs. [159, 165, 166])

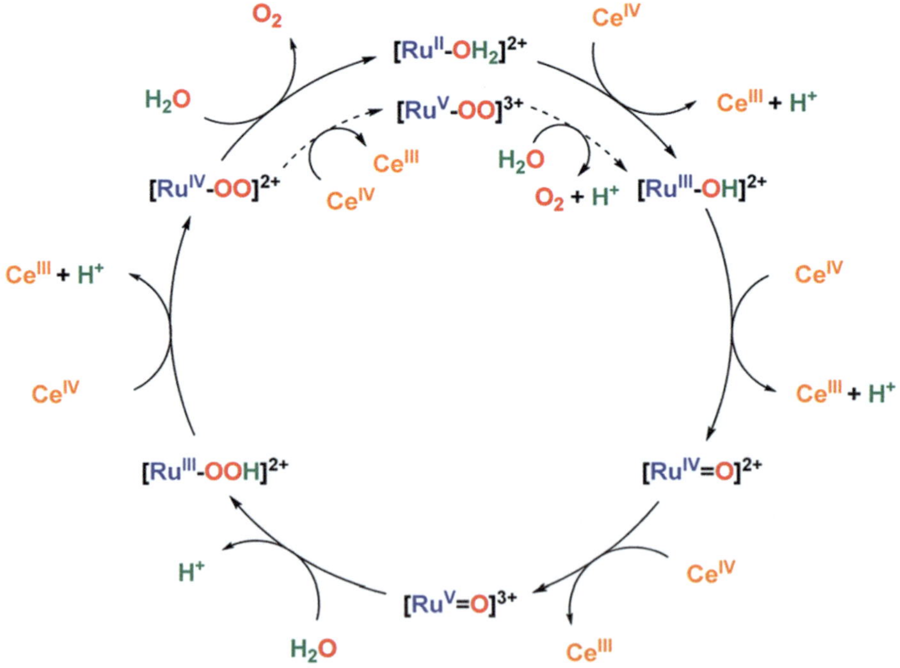

Fig. 2.43 Mechanism of water oxidation by single site Ru catalysts proposed by Meyer et al. for $[Ru(tpy)(bpm)(OH_2)]^{2+}$ and $[Ru(tpy)(bpm)(OH_2)]^{2+}$ type complexes (Reproduced with permission from Ref. [159])

the active redox states of WOCs to less positive potentials is the introduction of ligands with lower or no π-accepting capability and with stronger σ-donation characteristics [167]. The introduction of negatively charged ligands, with stronger donor capabilities has been tested on single site complexes developed by Akermark and Sun, which considered the use of chelating carboxylate functions (H_2bda = 2,2′-bipyridine-6,6′-dicarboxylic acid, H_2pda = 1,10-phenanthroline-2,9-dicarboxylic acid and H_2pdc = 2,6-pyridinedicarboxylic acid, reported in Fig. 2.44) associated to monodentate axial ligands based on functionalized pyridines, isoquinoline, imidazole and di-methyl-sulfoxyde (DMSO), also combined to generate a library of complexes with equivalent or non equivalent axial ligands. Mechanistic studies coupled to X-ray structure determination revealed that oxidation of [Ru(bda)(pic)$_2$] results in a 7-coordinated oxo-intermediate which dimerized resulting in the two [Ru(bda)(pic)$_2$(OH)]$^+$ coupled through a central bridging ([HOHOH]−) suspected to be the active precursor for oxygen evolution, according to an I2M mechanism (Fig. 2.45). This showed also that the presence of an open coordination site for an aqua ligand may not be a strict prerequisite for a ruthenium WOC. The effects of the ligand environment are however subtle and different mechanism are thought to occur in pda and pdc complexes. [Ru(pda)(pic)$_2$] is believed to undergo a mononuclear cycle similar to that reported for the Meyer's single site complexes (Fig. 2.45). Thus, the slightly larger bite angle in Ru-bda may be responsible of the formation of the 7-coordinated intermediate, which is not observed in pda derivatives. In the Ru-pdc complexes the increased electron density at the Ru(II) center, resulting from the use of ligands having relatively low π-accepting capabilities, results in oxidation potentials which are accessible to many polypyriridine Ru(II) sensitizers. The mechanism of action of Ru-pdc complexes is probably the release of a picoline, which is substituted by a water molecule. The oxidation of such pre-catalyst is likely to promote a WNA mechanism with the involvement of highly valent Ru-oxo species.

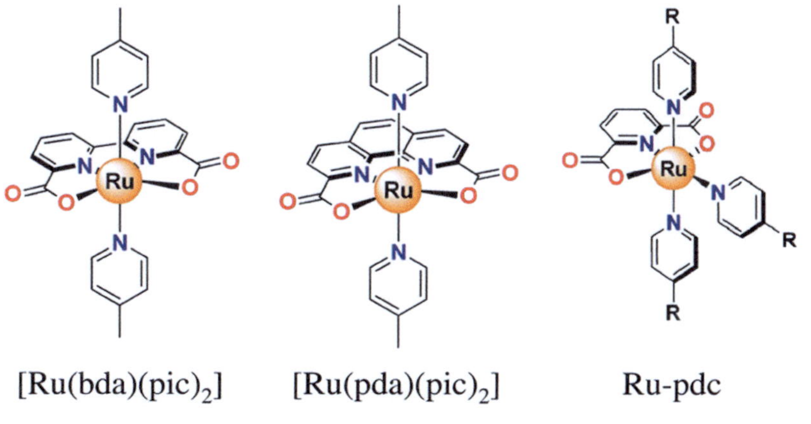

Fig. 2.44 Ligands based on ortho-carboxylic derivatives of (poly)-pyridines as equatorial chelating ligands proposed by Sun et al. (*pic* = picoline. Reproduced with permission from Ref. [159])

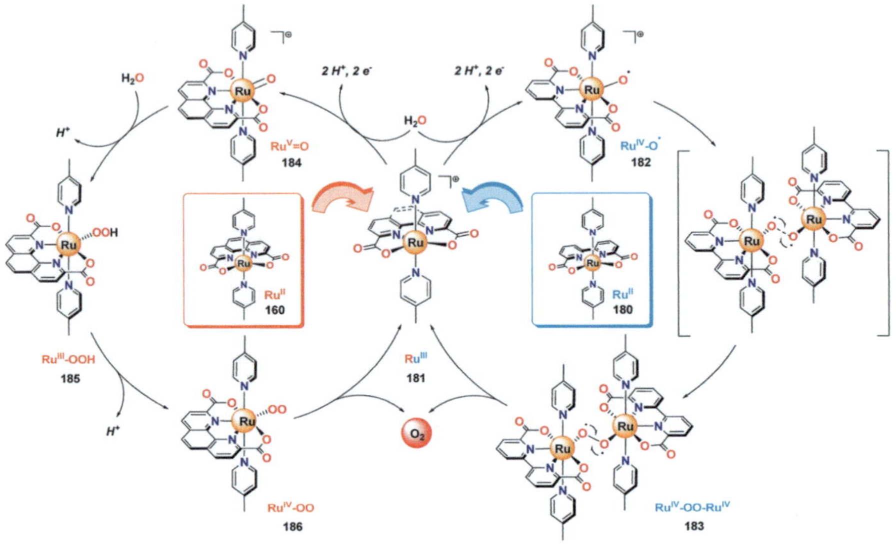

Fig. 2.45 Mechanisms of oxygen evolution in complexes based on equatorial tetra-coordinating polypyridine ligands: pda *left* and bda *right*. [Ru(pda)(pic)$_2$] is believed to follow a mononuclear pathway, according to a MWA mechanism, while [Ru(bda)(pic)$_2$] follows a cooperative mechanism with a 7-coordinated dimeric species (Ru(IV)-OO-Ru(IV)) responsible for oxygen evolution (Reproduced with permission from Ref. [159])

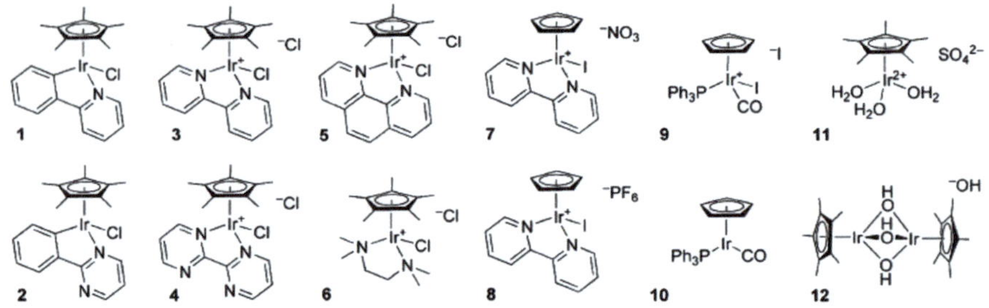

Fig. 2.46 Examples of a recent class of mono-nuclear Ir catalysts: half sandwich iridium pre-catalysts bearing cyclopentadienyl ligands (Reproduced with permission from Ref. [160])

The use of iridium also led to a class of promising molecular catalysts for water oxidation where the pre-requisite seems to be the possibility of having open or labile coordination sites to which water can bind. The mechanism of action, at least for certain classes of Ir catalysts (Figs. 2.46 and 2.47), is proposed to follow the formation of high valent intermediates, undergoing the nucleophilic attack of water (NWA) according to a mononuclear mechanism similar to that already seen for ruthenium single site catalysts. Nevertheless the increased oxygen evolution rate and non linear effects at high Ir catalyst concentration suggest that speciation equilibria and cooperative effects forming dimers or polymers may contribute to the mechanism of oxygen evolution.

Fig. 2.47 Proposed mechanism of action for oxygen evolution in iridium based catalysts (Reproduced with permission from Ref. [160])

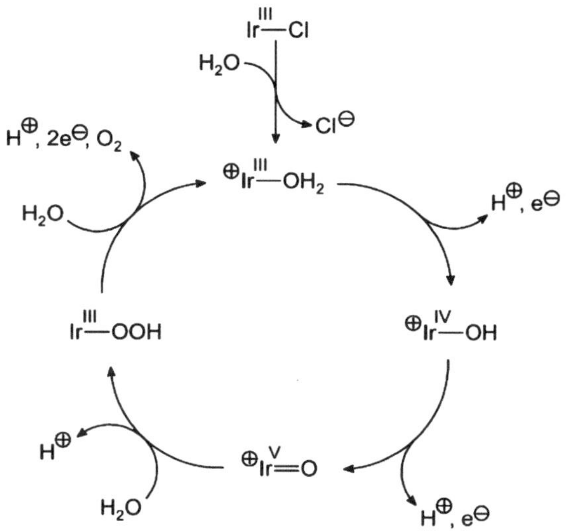

2.8.4 *Molecular Artificial Photosynthesis at Work in Photoelectrochemical Cells*

One of the first approaches for assembling operational DSPEC photoanodes was proposed by Mallouk et al. [149], and involves the anchoring of hydrated IrO_2 nanoparticles, having a long time known activity as WOC, to a phosphonated Ru (II) sensitizer modified with malonate binding groups (Fig. 2.48). The hole transfer to the IrO_2 takes place on a millisecond (2.2 ms) time scale, while charge recombination occurs on a sub-millisecond (0.37 ms) time scale. Thus, the competition between hole transfer and recombination is unfavorable to an effective charge separation. Nevertheless, under a small positive bias ($\geq$330 mV vs. NHE) a stable photoanodic current (few tens of microamperes) was effectively detected, indicating the occurrence of photoinduced water oxidation under steady state conditions. Although the efficiency is low, with a photoanodic current density of less than 30 $\mu A/cm^2$, the device represents a successful proof of concept of water splitting in a n-type DSPEC. Possible efficiency improvements could be related to synthetic modifications of the sensitizer/catalyst assembly, aimed to improve the binding between the photoactive dye and the IrO_2 nanostructures, to slow down recombination and to speed up the hole transfer to Ir(IV), for improving the turnover number (TON).

In another interesting design, by Brimblecomb et al. [151], a manganese cage complex catalyst ($[Mn_4O_4L_6]^+$, or "*cubium*", L = $MeOPh)_2PO_2^-$), able to undergo multiple proton coupled electron transfer, is encapsulated in a nafion membrane which has the role of providing both an high local surface concentration of redox active catalyst and a reasonable coupling with the Ru(II) dye sensitizer bearing carboxylic anchoring functions (Fig. 2.49a). The molecular level solar water

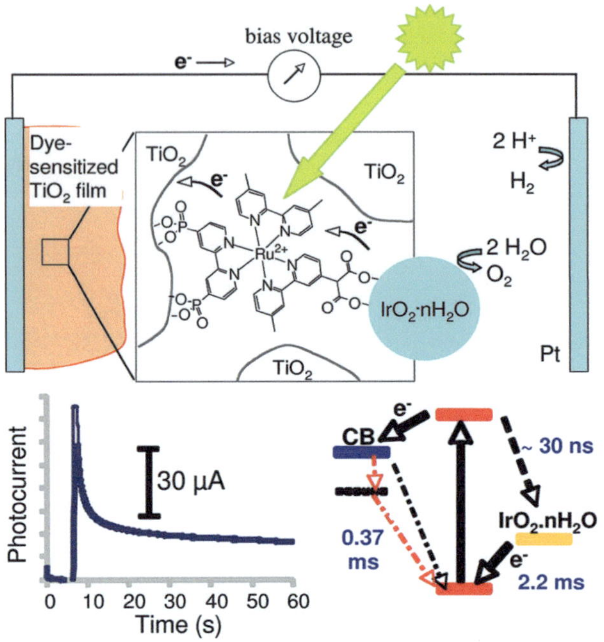

Fig. 2.48 Schematic diagram of a IrO_2 catalyzed DSPEC. Following light excitation and oxidative quenching of the excited state of Ru(II), the hole is transferred to Ir(IV), activating the IrO_2 catalyst toward water oxidation. *Bottom left*: chronoamperogram under constant potential showing a steady state photocurrent. *Bottom right*: energy diagram and time constants for the relevant interfacial electron transfer and recombination (Adapted with permission from Ref. [149])

splitting recalling that occurring in natural photosynthesis was thus beautifully demonstrated, but also in this case the performances were far from a practical application of the device, with IPCEs below 2 % corresponding to a stable photocurrent, with no externally applied bias, of the order of 30–5 $\mu A/cm^2$ depending on the illumination intensity and spectral bandwidth. A subsequent in-depth study of this TiO_2-sensitizer/Nafion-$[Mn_4O_4L_6]^+$ photoanode assembly by using in-situ X-ray absorption spectroscopy and transmission electron microscopy revealed that the tetranuclear Mn cluster is not the real active species for the water oxidation catalysis. It dissociates and oxidatively decomposes leading to the formation of disordered Mn(II/IV)-oxide phase in the form of dispersed nanoparticles which are responsible for the catalytic oxygen evolution [168]. In parallel, Sun and co-workers [150] developed a conceptually similar dye-sensitized photoanode by incorporating the Ru-based WOC $[Ru(bda)(pic)_2]^+$ into a pH-modified Nafion membrane, which was coated onto a $[Ru(II)(4,4'\text{-}(PO_3H_2)_2bpy)_2\ (bpy)]$ sensitized TiO_2 (Fig. 2.49b). Water oxidation by this type of photoanode in a DSPEC was confirmed, even in the absence of applied bias, at pH 7, by the formation of molecular oxygen. Hydrogen could be gas chromatographically detected at the platinum counter electrode of the cell under moderate bias.

A step forward in the construction of DSPEC is the realization of covalently linked WOC-sensitizer assemblies, where the loading ratio between the chromophore and the catalyst is optimized and the intramolecular electron transfer kinetics are facilitated by the close proximity of the two units. Meyer and co-workers have carried out several attempts to immobilize supramolecular assemblies, composed of

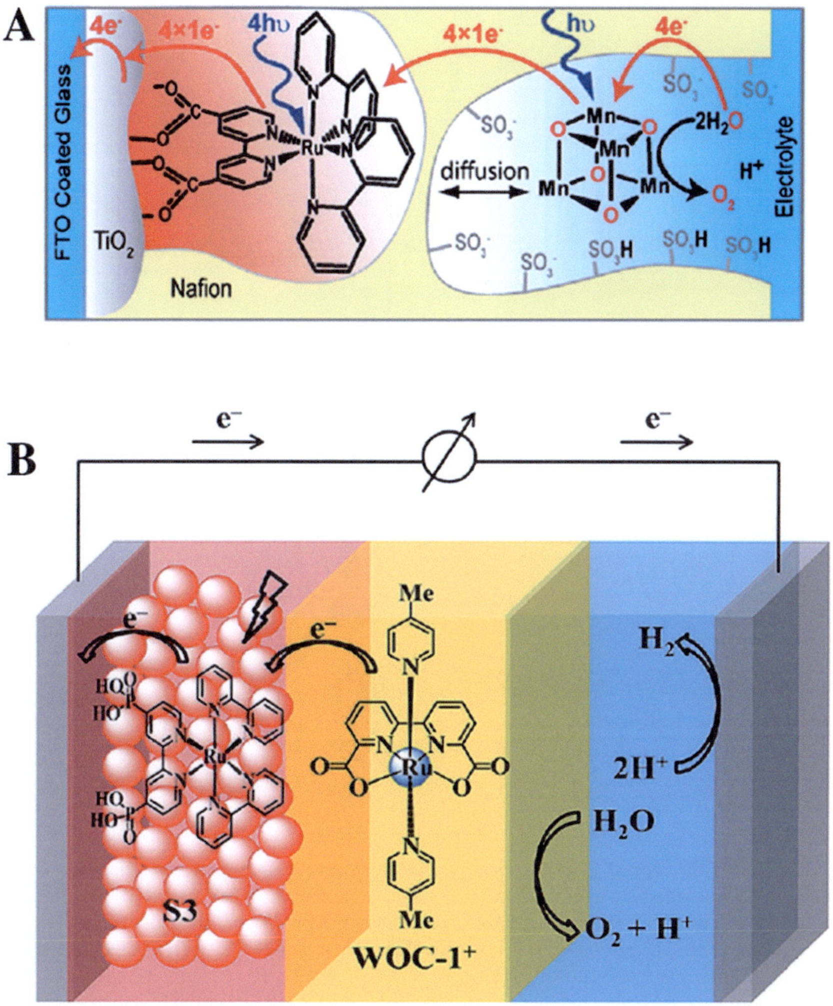

Fig. 2.49 Examples of n-type DSPEC in which the WOC is incorporated and stabilized on the sensitized semiconductor surface by use of a polymeric nafion membrane

Ru polypyridyl sensitizers and Ru based WOCs, onto the surface of nanoporous semiconductive metal oxides. In Fig. 2.50c it is shown the chromophore catalyst assembly $[((PO_3H_2)_2bpy)_2Ru(a)(4\text{-}Mebpy\text{-}4'\text{-}bimpy)Ru(b)(tpy)(OH_2)]^{4+}$ where Ru (a) and Ru(b) identify the ruthenium cores of the sensitizer and of the single catalyst respectively. As illustrated, on oxide semiconductor surfaces, the MLCT excitation of the chromophore results in excited state formation and electron injection into the conduction band of the semiconductor, triggering the sequence of multi-electron

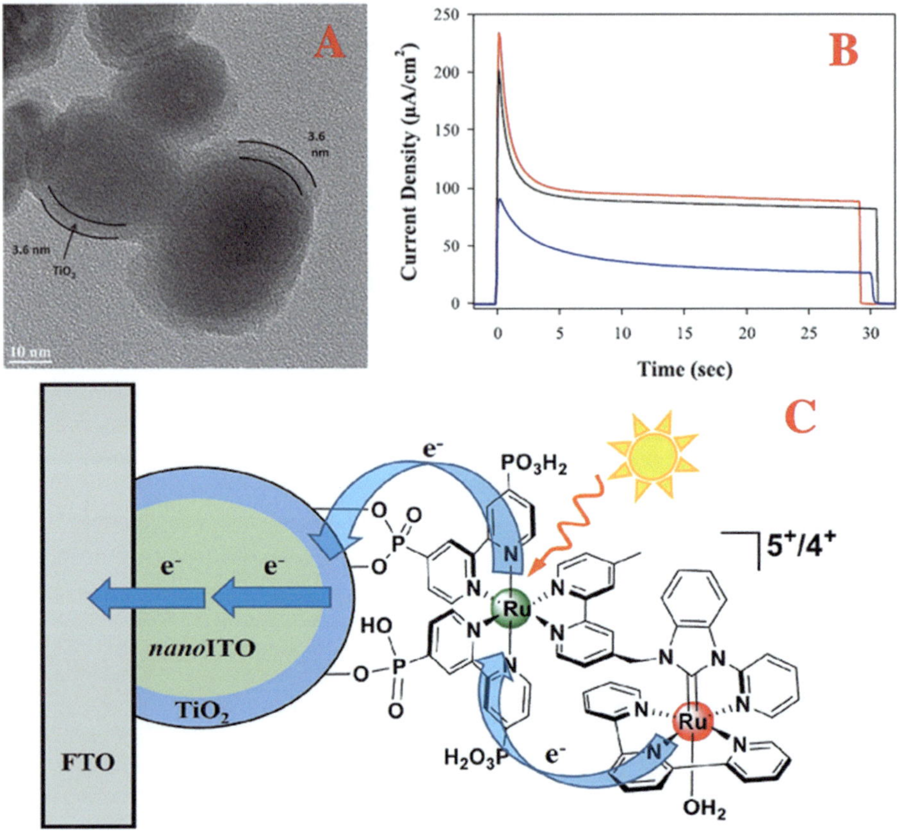

Fig. 2.50 (**a**) TiO_2 capped ITO nanoparticles at the Transmission Electron Microscope (TEM) imaging; (**b**) Photoanodic current generated at constant potential (0.2 V vs. NHE) as a function of the incident 445 nm irradiance: 91 mW/cm^2 (*red*); 58 mW/cm^2 (*black*) and 7.8 mW/cm^2 (*blue*) (**c**) Schematic operational principles of the binuclear sensitizer-WOC supramolecular assembly (Adapted with permission from Ref. [169])

transfers from Ru(a) to Ru(b). To underline the difficulties of artificial photosynthesis, three sequential single-photon-single-electron excitation events are required for a single oxidative activation cycle to reach the key precursor intermediate ((Ru (III)(a)–Ru(IV)(b) = O)$^{5+}$) which undergoes attack by water. Given that, back electron transfer occurs on a timescale of μs to ms and three injection–oxidation cycles are required to activate the catalyst, its reactive form is present in trace amounts under ambient solar illumination and efficiencies for water splitting are negligible. Indeed, photolysis of the Ru(a)-Ru(b) assembly on mesoscopic thin films of TiO_2 results in negligible photocurrents above background. Thus the retardation of recombination events and the acceleration of electron transport across the semiconductor film acquire a paramount importance for achieving success with covalent sensitizer/photocatalyst dyads. The core-shell particles developed by Meyer group were thus intended to minimize recombination losses. ITO (Indium-Tin Oxide) colloidal particles are covered with a thin (3–4 nm) shell of TiO_2, obtained by Atomic Layer Deposition (ALD), which binds the molecular catalyst at its surface [169]. Charge injection results in the fast transit of the TiO_2 shell by electrons, which reach the ITO, having a good conductivity for the fast transport of electrons to the electron collector. A further reduction in the recombination rate may result from the energetically up-hill hopping process from ITO to TiO_2 to the oxidized molecular assembly anchored at the surface. The sensitized photoanode, under monochromatic illumination, was shown to be capable to generate a photocurrent of the order of 2×10^2 μA/cm^2 which dropped to ca. 20 μA/cm^2 over a 30 min illumination period with a monochromatic 445 nm source. During a 30 s short term measurement at pH 4.6 acetate buffer, the highest average stabilized photocurrent density is found to be ca. 100 μA/cm^2 at a monochromatic intensity of 91 mW/cm^2, under a bias of 0.2 V vs. NHE. Based on the light harvesting at the excitation wavelength and on the peak photocurrent, the maximum photon to electron conversion efficiency of this system is 4.4 %.

Along the same lines, SnO_2/TiO_2 core/shell particles modified with the same Ru (a)-Ru(b) assembly [170], object of the previous discussion (Fig. 2.51),

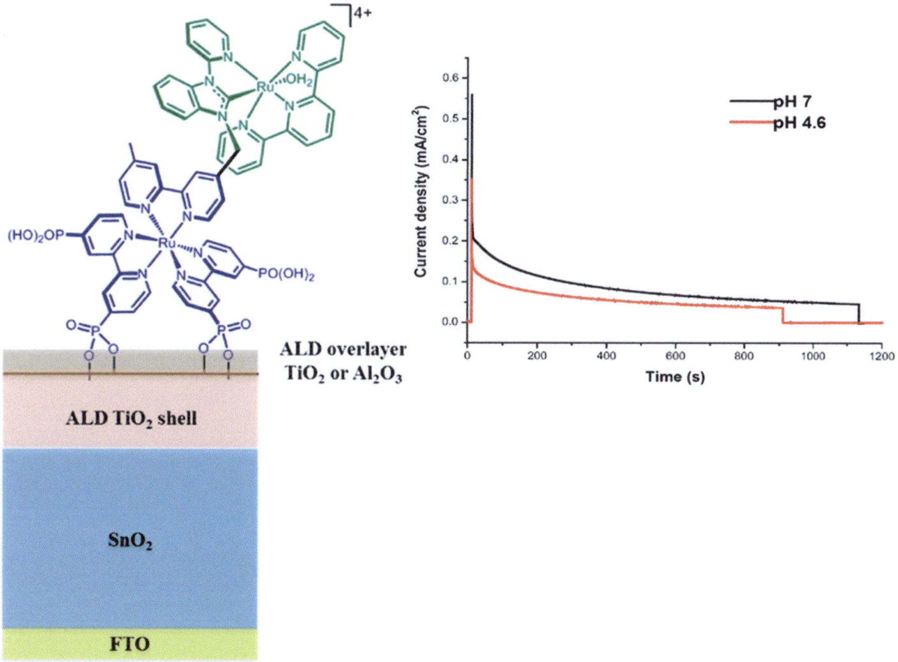

Fig. 2.51 Photosensitizer–WOC dyad adsorbed on core-shell SnO_2-TiO_2 nanoparticles, stabilized with a sub-nanometer Al_2O_3 overlayer. Photoanodic current generated under 0.6 V vs. NHE Potential bias and 445 nm illumination in acetate buffer at pH 4.6 and in phosphate buffer at pH 7. Ionic strength in the two different buffers was adjusted with the addition of $LiClO_4$ and $NaClO_4$ salts (Adapted with permission from Ref. [170])

demonstrated a great promise for enhanced efficiency, which came from the use of SnO_2 having a conduction band more positive than TiO_2 by ca. 0.4 V, resulting in a large internal potential gradient which is instrumental in gaining the suppression of back recombination. An enhanced stability was obtained with an additional sub-nanometric ALD overlayer of Al_2O_3 which was demonstrated [171] to stabilize the molecular adsorbate against hydrolytic phoenomena, by limiting the access of water to the phosphonate-TiO_2 group, buried by the protective alumina. This latter insulating overlayer has to be thin, compared to TiO_2 and SnO_2, in order to allow for efficient charge injection by quantum mechanical tunneling from the excited state of the dye.

Without protective overlayer, at pH 4.6 in acetate buffer, loss of the assembly from the surface by hydrolysis is noticeable after a few minutes. At pH 7 in phosphate buffer, the loss is too rapid for current–time measurements. However, the assembly is stable on the surface under these conditions, with ALD added overlayers of TiO_2 and Al_2O_3, as demonstrated by long term photocurrent measurements on the ALD stabilized photoanode (SnO_2 capped with 6.6 nm of TiO_2 and 0.6 nm of Al_2O_3), showing a significant increase in photocurrent density (up to 0.6 mA/cm^2 peak under 0.6 V vs. NHE and 445 nm illumination) and stability under illumination cycles of the order of 1000 s (Fig. 2.51). The drop in photocurrent is mainly attributed to ligand loss by the chromophore in its oxidized state (Ru(III) (a)).

To date, most approaches explored for the construction of sensitizer–catalyst dyads suffer from complex multi-step synthesis, lack of versatility and often low loading of the semiconductor surface. In order to address this problem, Meyer and co-workers developed a facile "layer-by-layer", self-assembly approach [172], in which the sensitizer and the catalyst were synthesized independently and then bound to the nanoporous semiconductor surface in a stepwise fashion via phosphonate/Zr^{4+} coordination linkages, which served to the realization of models incorporating either a two chromophore (Fig. 2.52) or a chromophore-catalyst assembly. The bi-layer molecular structures could be produced first by immersion of the semiconductor film in the dye solution, followed by treatment with $ZrOCl_2$

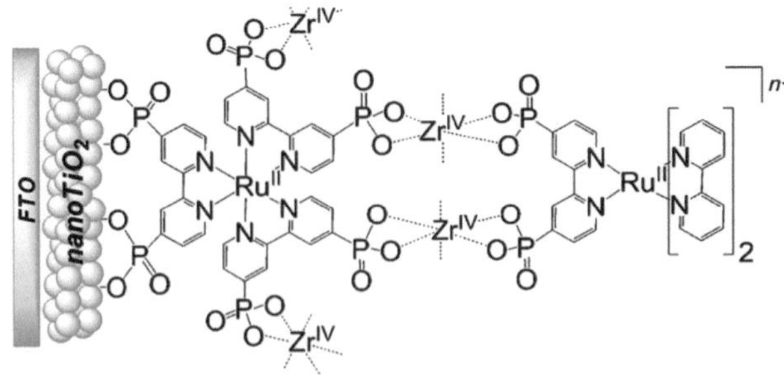

Fig. 2.52 Bi-layered molecular assembly based on Zr(IV) bridges between phosphonic anchors. Herein a two-chromophore assembly is schematized (Reproduced with permission from Ref. [172])

and finally by dipping in a bath containing a last chemical species, which could be either a different dye or a WOC complex. Transient absorption spectroscopy revealed that the assembly was able to undergo inside to outside electron transfer resulting in hole transfer from the dye to the remote unit, either the catalyst or other types of Ru(II) complexes. Although remote injection could in principle occur, resulting in the formation of WOC^+, the short excited state lifetime of the WOC complexes (ca. 10 ns) suggests a very small contribution to the total WOC^+ population by this direct mechanism. Moreover, the recombination events, involving the photoinjected electron and WOC^+ ($e^-(TiO_2)/WOC^+$) are considerably slowed down by the spatial separation between the TiO_2 surface and the external unit.

This concept was exploited by Sun and coworkers which developed a triad where a phosphonated $Ru(bpy)_3$ complex coordinates two WOC units based on the Ru-bda core which binds Zr((IV) through axial pyridines substituted with carboxylic group in the para-position of the ring [173]. This system under an illumination intensity of 300 mW/cm^2 (an incident irradiance equivalent to three suns) in pH 6.4 aqueous buffer under a bias of 0.2 V vs. NHE was able to deliver a stable photocurrent density of the order of 0.5 mA/cm^2 with a maximal IPCE of $\approx$4 % at 450 nm. Alternatively, WOCs can be co-adsorbed onto the surface of semiconductor oxides together with sensitizers to facilitate intermolecular electron transfers. Brudvig and co-workers co-grafted a zinc porphyrin sensitizer together with an iridium-based catalyst onto the nanoporous TiO_2 surface showing that lateral charge hopping can occur between the light harvester and the charge transfer catalyst immobilized on the same surface, generating 30 $\mu A/cm^2$ at the applied bias of 0.3 V vs. Ag/AgCl at 200 mW/cm^2 incident irradiance [156].

The use of organic chromophores like perylenes, co-adsorbed in the presence of metal oxide nanoparticles acting as WOC, was recently explored by the Finke [174] and Bignozzi groups [175] in water splitting electrodes.

Perylenes are simple organic molecules possessing high molar extinction coefficient and good thermal and photochemical stability. Most importantly, they show high oxidation potentials, resulting compatible with the activation of many WOCs reported in literature, including ruthenium, iridium, nickel and iron based catalysts, but are limited by a low excited state oxidation potential. This fact precludes charge injection into many semiconductors and, in general, their use in energy conversion schemes relying on the oxidative quenching of their excited state. The dicationic perylene PBI = [(N,N′-bis(2-(trimethylammonium)ethylene) perylene 3,4,9,10-tetracarboxylic acid bisimide) $(PF_6)_2$] reported in Fig. 2.53 with its frontier molecular orbitals, was demonstrated capable to adsorb and undergo excited state oxidative quenching on SnO_2 and WO_3 n-type semiconductors, resulting in the transfer of photogenerated holes to sacrificial electron donors in solution and to co-deposited IrO_2 nanoparticles (NPs) acting as WOCs (Fig. 2.54)

PBI is characterized by a planar geometry of the aromatic core, while the ethyl-trimethyl-ammonium chains are bent over the π-plane with a dihedral angle close to 90°. The HOMO-LUMO energy gap of 2.48 eV is in good agreement with both the electrochemical gap (2.5 eV; $E_{ox} = 1.7$ V vs. SCE; $E_{red} = -0.8$ V vs. SCE) and with

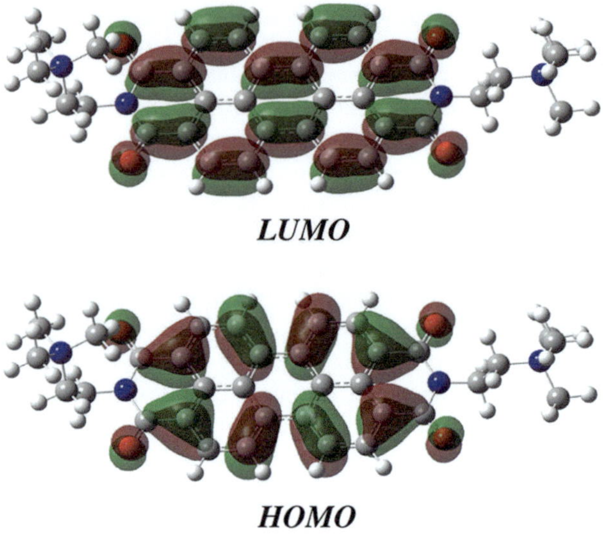

Fig. 2.53 Frontier molecular orbitals of PBI used for water splitting in DSPEC cells based on WO_3 and SnO_2 semiconductors. *HOMO* Highest Occupied Molecular Orbital, *LUMO* Lowest Unoccupied Molecular Orbital (Reproduced with permission from Ref. [175])

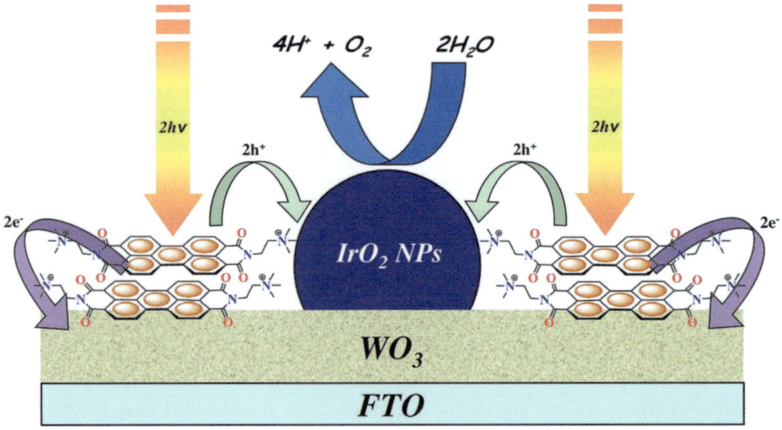

Fig. 2.54 Photoinduced electron injection by PBI loaded on WO_3 by exploiting aggregation forces, triggers water oxidation at co-deposited IrO_2 nanoparticles

the spectroscopic energy gap of 2.36 eV. The resulting excited state oxidation potential (E_{ox}*) is −0.66 V vs. SCE, while the ground state oxidation potential of the order of 1.7 V vs. SCE allows to drive demanding photo-oxidation reaction at the photoanode of DPSECs, as demonstated by the photoanodic current, recorded at sensitized WO_3 electrodes, of the order 70–100 μA/cm^2 at 0.5 V vs. SCE under simulated solar illumination (0.1 W/cm^2 AM 1.5 G + 435 nm cut off), with APCE (Absorbed Photon to Current Conversion Efficiency) of the order of 1 %. Interestingly, WO_3 was found to be the best substrate for obtaining the oxidative quenching of the PBI excited state, with estimated electron injection rates which are from four to eight times higher than those observed on SnO_2 and TiO_2 respectively. Although transient absorption spectroscopy reveled that hole transfer to IrO_2 was one of the kinetic bottlenecks of the system, with the hole transfer rate occurring on the same time scale of the charge recombination (μs), WO_3-PBI type photoanodes may constitute a convenient mean to explore and optimize the interaction of the PBI chromophores, with selected catalytic species, including amorphous metal oxides and molecular species, in order to increase the efficiency of the interfacial charge transfer over competitive carrier recombination, representing, at present, the most serious limitation to molecular level artificial photosynthetic processes.

A complete DSPEC design with sensitized TiO_2 and NiO electrodes was proposed by Sun et al. [176], using the dye/catalyst configuration schematized in Fig. 2.55. The dye sensitizers are based on organic designs of the donor-acceptor type (Fig. 2.55a). Excitation of the molecular dye L0, at the TiO_2 surface, triggers electron transfer from the tryphenyl-amine to cyano acrylic group, coupled with the TiO_2 surface, resulting in electron injection. The hole, confined on the amine donor,

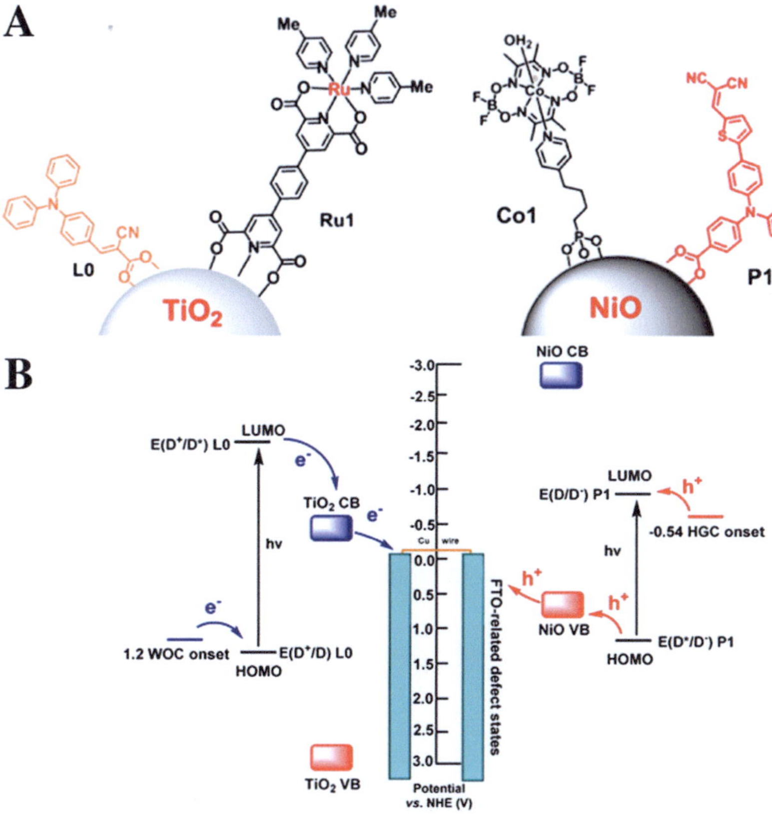

Fig. 2.55 (**a**) Dye sensitizer/catalyst used a sensitized n-type (TiO_2) and p-type (NiO) photoelectrodes. (**b**) Energy levels involved in the charge separation and water splitting reaction at the sensitized electrodes (Adapted with permission from Ref. [176])

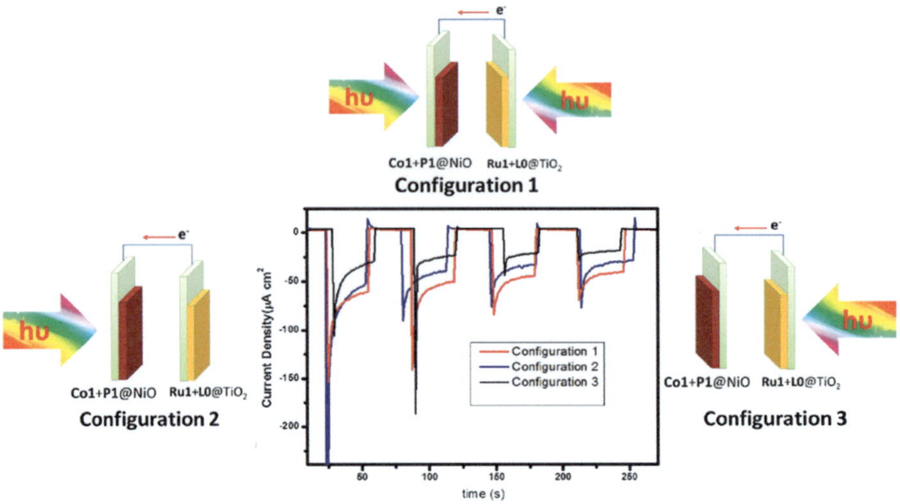

Fig. 2.56 Photocurrent response in phosphate buffer of the tandem configuration in the presence of different illumination geometries. (2) From the p-side; (1) from both sides; (3) from the n side (Reproduced with permission from Ref. [176])

is transferred to a WOC based on the pdc design (Ru1 in Fig. 2.55a), whose oxidation potential is accessible to the oxidized sensitizer. At the p-type surface an "inverted" push pull design [177] results in hole transfer to the amine donor, proximal to the NiO surface, while the electron is transferred to the acceptor arms characterized by the presence of 4 cyano electron withdrawing groups. An HEC based on a cobalt dimetylglyoximate (Co1), co-adsorbed with the inverted P1, allows to drive hydrogen evolution at the photocathode.

This is one of the first examples of DSPEC which can operate without externally applied bias, generating a stabilized photocurrent (in short term measurements) of the order of 70 μA/cm^2 with spikes of the order of 200 μA/cm^2 in neutral phosphate buffer (Fig. 2.56). Since the photocurrent is generated at 0 bias, the ABPE is 0.05 % in the presence of a faradic efficiency of 55 %. Clearly, the best results are achieved when both electrodes are simultaneously illuminated, avoiding shadowing and filtering effects which are particularly important when light strikes first the sensitized n-type photoelectrode.

2.9 Conclusion

The exploitation of molecular systems interfaced with solid state materials represents a unique way toward the realization of energy conversion devices where each subsystem (dye sensitizer, redox mediator or hole transport medium) can be finely tuned through molecular design in order to optimize light absorption and charge separation processes. This approach, moving from fundamental observations dating back to the 1960s of the twentieth century, led, in this new century, to molecular based devices that can achieve efficiencies comparable to that of polycrystalline silicon. These results were achieved through the development of cocktails of dyes allowing for the panchromatic absorption, coupled to electron transfer mediators which are able to generate open circuit voltages exceeding 1 V. A further step towards high efficiency photoelectrochemical devices was realized by incorporating organo halides perovskites in solid state devices, where the electron transport takes place through molecular films of hole transport materials. Although perovkites cannot be considered molecular systems, the fundamental principles of operation of perovskite solar cells recall, under some aspects, and at least for the TiO_2/perovskite heterojunction, those of conventional molecular-sensitized photoelectrochemical cells. At present, in perovskite cells, issues of stability and toxicity (related to the use of lead) are still to be solved. Moreover, the tuning of the electronic and spectral properties of perovskites appears to be limited with respect to traditional molecular sensitization, for which a wide selection of efficient structures is available through synthetic modification.

The last, and most challenging aspects in solar energy conversion through molecular means, involves the production of solar fuels. Although the basic principles are very similar to those met in the description of DSSCs, the energetic constraints that must be satisfied by the dye sensitizers are much more stringent, with ground state oxidation potentials which must exceed those of water by at least 200 mV. The interfacing of the dye with molecular or nanostructured catalyst is another critical aspect, which should minimize the back recombination by maximizing the hole transfer rate to the electrolyte. Long term stability of the organic dyes, in the presence of strongly oxidizing and hydrolytic conditions represent an additional significant challenge. At present the efficiencies obtained with metal oxides sensitized by molecular species are generally lower than those found with other photoelectrochemical devices based on visible absorbing metal oxides, where charge generation follows directly band-gap excitation. Nevertheless, the realization of hybrid interfaces based on molecular chromophores which are able to mimick the fundamentals of photosynthetic process holds a great potential, with plenty of room for optimization, and is, at present days, actively pursued.

References

1. Tan MX, Laibnis PE, Nguyen ST, Kesselman JM, Stanton CE, Lewis NS (1994) Progress in inorganic chemistry, vol 41. Wiley, New York, pp 21–144
2. Gerischer H (1980) Pure Appl Chem 52:2649–2667
3. Hannay NB (1959) Semiconductors. Reinhold Publishing Corporation, New York
4. Becquerel E (1839) Compt Rendus 9:561–567
5. Desilvestro J, Grätzel M, Kavan L, Moser J, Augustynski J (1985) J Am Chem Soc 107:2988–2990
6. O'Regan B, Grätzel M (1991) Nature 353:737–740

7. Gerischer H, Tributsch H (1968) Ber Bunsenges Phys Chem 72:437–445
8. Memming R (1984) Prog Surf Sci 17:7–73
9. Grätzel M (2005) Inorg Chem 44:6841–6851
10. Wenger B, Grätzel M, Moser J-E (2005) J Am Chem Soc 127:12150–12151
11. Moser JE, Grätzel M (1993) Chem Phys 176:493–500
12. Bauer C, Boschloo G, Mukhtar E, Hagfeldt A (2002) J Phys Chem B 106:12693–12704
13. Gardner JM, Giaimuccio JM, Meyer GJ (2008) J Am Chem Soc 130:17252–17253
14. Marton A, Clark CC, Srinivasan R, Freundlich RE, Narducci-Sarjeant, Meyer GJ (2006) Inorg Chem 45:362–369
15. Kakiage K, Aoyama Y, Yano T, Oya K, Fujisawa J, Hanaya M (2015) Chem Commun 51:15894–15897
16. Argazzi R, Bignozzi CA, Heimer TA, Castellano FN, Meyer GJ (1995) J Am Chem Soc 117:11815–11816
17. Wang Q, Moser J-E, Grätzel M (2005) J Phys Chem B 109:14945–14953
18. Fabregat-Santiago F, Bisquert J, Garcia-Belmonte G, Boschloo G, Hagfeldt A (2005) Sol Energy Mater Sol Cells 87:117–131
19. Hardin BE, Snaith HJ, McGehee MD (2012) Nat Photonics 6:162–169
20. Calogero G, Di Marco G, Caramori S, Cazzanti S, Argazzi R, Bignozzi CA (2009) Energy Environ Sci 2:1162–1172
21. Calogero G, Yum J-H, Sinopoli A, Di Marco G, Grätzel M, Nazeeruddin MK (2012) Sol Energy 86:1563–1575
22. Fantacci S, De Angelis F (2011) Coord Chem Rev 255:2704–2726
23. O'Regan BC, Durrant JR (2009) Acc Chem Res 42:1799–1808
24. Imahori H, Umeyama T, Ito S (2009) Acc Chem Res 42:1809–1818
25. Cid J-J, Yum J-H, Jang S-R, Nazeeruddin MK, Martínez-Ferrero E, Palomares E, Ko J, Grätzel M, Torres T (2007) Angew Chem Int Ed 46:8358–8362
26. Nazeeruddin MK, Péchy P, Renouard T, Zakeeruddin SM, Humphry-Baker R, Comte P, Liska P, Cevey L, Costa E, Shklover V, Spiccia L, Deacon GB, Bignozzi CA, Grätzel M (2001) J Am Chem Soc 123:1613–1624
27. Yella A, Lee H-W, Tsao HN, Yi C, Chandiran AK, Nazeeruddin MK, Diau EW-G, Yeh C-Y, Zakeeruddin SM, Grätzel M (2011) Science 334:629–634
28. Odobel F, Le Pleux L, Pellegrin Y, Blart E (2010) Acc Chem Res 43:1063–1071
29. Odobel F, Pellegrin Y, Gibson EA, Hagfeldt A, Smeigh AL, Hammarström L (2012) Coord Chem Rev 256:2414–2423
30. Nazeeruddin MK, Kay A, Rodicio I, Humphry-Baker R, Mueller E, Liska P, Vlachopoulos N, Grätzel M (1993) J Am Chem Soc 115:6382–6390
31. Hagfeldt A, Grätzel M (1995) Chem Rev 95:49–68
32. Argazzi R, Murakami Iha NY, Zabri H, Odobel F, Bignozzi CA (2004) Coord Chem Rev 248:1299–1316
33. Nazeeruddin MK, Zakeeruddin SM, Lagref JJ, Liska P, Comte P, Barolo C, Viscardi G, Schenk K, Grätzel M (2004) Coord Chem Rev 248:1317–1328
34. Polo AS, Itokazu MK, Murakami Iha NY (2004) Coord Chem Rev 248:1343–1361
35. Meyer GJ (2005) Inorg Chem 44:6852–6864
36. Robertson N (2006) Angew Chem Int Ed 45:2338–2345
37. Xie P, Guo F (2007) Curr Org Chem 11:1272–1286
38. Cecchet F, Gioacchini AM, Marcaccio M, Paolucci F, Roffia S, Alebbi M, Bignozzi CA (2002) J Phys Chem B 106:3926–3932
39. Nour-Mohhamadi F, Nguyen SD, Boschloo G, Hagfeldt A, Lund T (2005) J Phys Chem B 109:22413–22419
40. Nguyen HT, Ta HM, Lund T (2007) Sol Energy Mater Sol Cells 91:1934–1942
41. Agresti A, Pescetelli S, Quatela A, Mastroianni S, Brown TM, Reale A, Bignozzi CA, Caramori S, Di Carlo A (2014) RSC Adv 4:12366–12375
42. Harikisun R, Desilvestro H (2011) Sol Energy 85:1179–1188

43. Nazeeruddin MK, Zakeeruddin SM, Humphry-Baker R, Jirousek M, Liska P, Vlachopoulos N, Shklover V, Fischer C-H, Grätzel M (1999) Inorg Chem 38:6298–6305
44. Islam A, Sugihara H, Yanagida M, Hara K, Fujihashi G, Tachibana Y, Katoh R, Murata S, Arakawa H (2002) N J Chem 26:966–968
45. Bomben PG, Robson KCD, Koivisto BD, Berlinguette CP (2012) Coord Chem Rev 256:1438–1450
46. Bessho T, Yoneda E, Yum J-H, Guglielmi M, Tavernelli I, Imai H, Rothlisberger U, Nazeeruddin MK, Grätzel M (2009) J Am Chem Soc 131:5930–5934
47. Kinoshita T, Dy JT, Uchida S, Kubo T, Segawa H (2013) Nat Photonics 7:535–539
48. Hagfeldt A, Boschloo G, Sun L, Kloo L, Pettersson H (2010) Chem Rev 110:6595–6663
49. Ooyama Y, Harima Y (2009) Eur J Org Chem 2009:2903–2934
50. Hara K, Kurashige M, Ito S, Shinpo A, Suga S, Sayama K, Arakawa H (2003) Chem Commun 252–253
51. Kitamura T, Ikeda M, Shigaki K, Inoue T, Anderson NA, Ai X, Lian T, Yanagida S (2004) Chem Mater 16:1806–1812
52. Hara K, Sayama K, Ohga Y, Shinpo A, Suga S, Arakawa H (2001) Chem Commun 569–570
53. Sayama K, Tsukagoshi S, Hara K, Ohga Y, Shinpou A, Abe Y, Suga S, Arakawa H (2002) J Phys Chem B 106:1363–1371
54. Haid S, Marszalek M, Mishra A, Wielopolski M, Teuscher J, Moser J-E, Humphry-Baker R, Zakeeruddin SM, Grätzel M, Bäuerle P (2012) Adv Funct Mater 22:1291–1302
55. Do K, Kim D, Cho N, Paek S, Song K, Ko J (2012) Org Lett 14:222–225
56. Choi H, Baik KC, Kang SO, Ko J, Kang M-S, Nazeeruddin MK, Grätzel M (2008) Angew Chem Int Ed 47:327–330
57. Ito S, Zakeeruddin SM, Humphry-Baker R, Liska P, Charvet R, Comte P, Nazeeruddin MK, Pechy P, Takata M, Miura H, Uchida S, Grätzel M (2006) Adv Mater 18:1202–1205
58. Ito S, Miura H, Uchida S, Takata M, Sumioka K, Liska P, Comte P, Pechy P, Grätzel M (2008) Chem Commun 5194–5196
59. Wu S-L, Lu H-P, Yu H-T, Chuang S-H, Chiu C-L, Lee C-W, Diau EW-G, Yeh C-Y (2010) Energy Environ Sci 3:949–955
60. Lee C-W, Lu H-P, Lan C-M, Huang Y-L, Liang Y-R, Yen W-N, Liu Y-C, Lin Y-S, Diau EW-G, Yeh C-Y (2009) Chem Eur J 15:1403–1412
61. Di Carlo G, Caramori S, Trifilletti V, Giannuzzi R, De Marco L, Pizzotti M, Orbelli Biroli A, Tessore F, Argazzi R, Bignozzi CA (2014) ACS Appl Mater Interf 6:15841–15852
62. Barea EM, Gónzalez-Pedro V, Ripollés-Sanchis T, Wu H-P, Li L-L, Yeh C-Y, Diau EW-G, Bisquert J (2011) J Phys Chem C 115:10898–10902
63. Gregg BA, Pichot F, Ferrere S, Fields CL (2001) J Phys Chem B 105:1422–1429
64. Gregg BA (2004) Coord Chem Rev 248:1215–1224
65. Pichot F, Gregg BA (2000) J Phys Chem B 104:6–10
66. Heimer TA, Heilweil EJ, Bignozzi CA, Meyer GJ (2000) J Phys Chem A 104:4256–4262
67. Montanari I, Nelson J, Durrant JR (2002) J Phys Chem B 106:12203–12210
68. Schlichthörl G, Huang SY, Sprague J, Frank AJ (1997) J Phys Chem B 101:8141–8155
69. Huang SY, Schlichthörl G, Nozik AJ, Grätzel M, Frank AJ (1997) J Phys Chem B 101:2576–2582
70. Okada K, Matsui H, Kawashima T, Ezure T, Tanabe N (2004) J Photochem Photobiol Part A: Chem 164:193–198
71. Hanke KP (1999) Z Phys Chem 212:1
72. Meyer TJ, Taube H (1987) In: Wilkinson G (ed) Comprehensive coordination chemistry: the synthesis, reactions, properties and applications of coordination compounds, vol 1. Pergamon Press, Oxford, p 331
73. Nusbaumer H, Moser J-E, Zakeeruddin SM, Nazeeruddin MK, Grätzel M (2001) J Phys Chem B 105:10461–10464
74. Sapp SA, Elliott CM, Contado C, Caramori S, Bignozzi CA (2002) J Am Chem Soc 124:11215–11222

75. Ghamouss F, Pitson R, Odobel F, Boujtita M, Caramori S, Bignozzi CA (2010) Electrochim Acta 55:6517–6522
76. Bisquert J, Fabregat-Santiago F, Mora-Seró I, Garcia-Belmonte G, Giménez S (2009) J Phys Chem C 113:17278–17290
77. Wang P, Zakeeruddin SM, Comte P, Charvet R, Humphry-Baker R, Grätzel M (2003) J Phys Chem B 107:14336–14341
78. Klein C, Nazeeruddin MK, Di Censo D, Liska P, Grätzel M (2004) Inorg Chem 43:4216–4226
79. Carli S, Casarin L, Caramori S, Boaretto R, Busatto E, Argazzi R, Bignozzi CA (2014) Polyhedron 82:173–180
80. Liberatore M, Burtone L, Brown TM, Reale A, Di Carlo A, Decker F, Caramori S, Bignozzi CA (2009) Appl Phys Lett 94:173113
81. Nelson JJ, Amick TJ, Elliott CM (2008) J Phys Chem C 112:18255–18263
82. Feldt SM, Gibson EA, Gabrielsson E, Sun L, Boschloo G, Hagfeldt A (2010) J Am Chem Soc 132:16714–16724
83. Tsao HN, Yi C, Moehl T, Yum J-H, Zakeeruddin SM, Nazeeruddin MK, Grätzel M (2011) ChemSusChem 4:591–594
84. Yum J-H, Baranoff E, Kessler T, Moehl T, Ahmad S, Bessho T, Marchioro A, Ghadiri E, Moser JE, Yi C, Nazeeruddin MK, Grätzel M (2012) Nat Commun 3:1–8, Article number 631
85. Hattori S, Wada Y, Yanagida S, Fukuzumi S (2005) J Am Chem Soc 127:9648–9654
86. Bai Y, Yu Q, Cai N, Wang Y, Zhang M, Wang P (2011) Chem Commun 47:4376–4378
87. Li TC, Spokoyny AM, She C, Farha OK, Mirkin CA, Marks TJ, Hupp JT (2010) J Am Chem Soc 132:4580–4582
88. Daeneke T, Kwon T-H, Holmes AB, Duffy NW, Bach U, Spiccia L (2011) Nat Chem 3:211–215
89. Zotti G, Schiavon G, Zecchin S, Favretto D (1998) J Electroanal Chem 456:217–221
90. Hurvois JP, Moinet C (2005) J Organomet Chem 690:1829–1839
91. Tennakone K, Kumara GRRA, Kumarasinghe AR, Wijayantha KGU, Sirimanne PM (1995) Semicond Sci Technol 10:1689
92. Kumarasinghe AR, Flavell WR, Thomas AG, Mallick AK, Tsoutsou D, Chatwin C, Rayner S, Kirkham P, Warren S, Patel S, Christian P, O'Brien P, Grätzel M, Hengerer R (2007) J Chem Phys 127:114703
93. Kumara GRA, Konno A, Shiratsuchi K, Tsukahara J, Tennakone K (2002) Chem Mater 14:954–955
94. Bach U, Lupo D, Comte P, Moser JE, Weissortel F, Salbeck J, Spreitzer H, Grätzel M (1998) Nature 395:583–585
95. Murakoshi K, Kogure R, Wada Y, Yanagida S (1998) Sol Energy Mater Sol Cells 55:113–125
96. Kim Y, Sung Y-E, Xia J-B, Lira-Cantu M, Masaki N, Yanagida S (2008) J Photochem Photobiol A Chem 193:77–80
97. Yang L, Zhang J, Shen Y, Park B-W, Bi D, Häggman L, Johansson EMJ, Boschloo G, Hagfeldt A, Vlachopoulos N, Snedden A, Kloo L, Jarboui A, Chams A, Perruchot C, Jouini M (2013) J Phys Chem Lett 4:4026–4031
98. Aitola K, Zhang J, Vlachopoulos N, Halme J, Kaskela A, Nasibulin A, Kauppinen E, Boschloo G, Hagfeldt A (2015) J Solid State Electrochem 19:3139–3144
99. Kojima A, Teshima K, Shirai Y, Miyasaka T (2009) J Am Chem Soc 131:6050–6051
100. Weber ZD (1978) Naturforsch B 33:1443–1445
101. Mitzi DB, Field CA, Harrison WTA, Guloy AM (1994) Nature 369:467–469
102. Green MA, Emery K, Hishikawa Y, Warta W, Dunlop ED (2015) Prog Photovolt Res Appl 23:1–9
103. Mitzi DB (1999) Prog Inorg Chem 48:1–121
104. Shannon RD (1976) Acta Crystallogr Sect A 32:751–767
105. Stoumpos CC, Malliakas CD, Kanatzidis MG (2013) Inorg Chem 52:9019–9038

106. Mitzi DB (2001) J Chem Soc Dalton Trans 1:1–12
107. Xing G, Mathews N, Sun S, Lim SS, Lam YM, Grätzel M, Mhaisalkar S, Sum TC (2013) Science 342:344–347
108. Stranks SD, Eperon GE, Grancini G, Menelaou C, Alcocer MJ, Leijtens T, Herz LM, Petrozza A, Snaith HJ (2013) Science 342:341–344
109. Edri E, Kirmayer S, Henning A, Mukhopadhyay S, Gartsman K, Rosenwak Y, Hodes G, Cahen D (2014) Nano Lett 14:1000–1004
110. Cheng Z, Lin J (2010) CrystEngComm 12:2646–2662
111. Tanaka K, Takahashi T, Ban T, Kondo T, Uchida K, Miura N (2003) Solid State Commun 127:619–623
112. Kim H-S, Lee C-R, Im J-H, Lee K-B, Moehl T, Marchioro A, Moon S-J, Humphry-Baker R, Yum J-H, Moser JE, Grätzel M, Park N-G (2012) Sci Rep 2:1–7
113. Kitazawa N, Watanabe Y, Nakamura Y (2002) J Mater Sci 37:3585–3587
114. Mosconi E, Grätzel M, Amat A, Nazeeruddin MK, De Angelis F (2013) J Phys Chem 117:13902–13913
115. Noh JH, Im SH, Heo JH, Mandal TN, Seok SI (2013) Nano Lett 13:1764–1769
116. Im J-H, Chung J, Kim S-J, Park N-G (2012) Nanoscale Res Lett 7:353
117. Eperon GE, Stranks SD, Menelaou C, Johnston MB, Herz LB, Snaith HJ (2014) Energy Environ Sci 7:982–988
118. Kitazawa N, Enomoto K, Aono M, Watanabe Y (2004) J Mater Sci 39:749–751
119. Era M, Hattori T, Taira T, Tsutsui T (1997) Chem Mater 9:8–10
120. Liang KN, Mitzi DB, Prikas MT (1998) Chem Mater 10:403–411
121. Xia YN, Whitesides GM (1998) Annu Rev Mater Sci 28:153–184
122. Bi D, Häggman L, Boschloo G, Yang L, Johansson EMJ, Hagfeldt A (2013) RSC Adv 3:18762–18766
123. Lee MM, Teuscher J, Miyasaka T, Murakami TN, Snaith HJ (2012) Science 338:643–647
124. Etgar L, Gao P, Xue Z, Peng Q, Chandiran AK, Liu B, Nazeeruddin MK, Grätzel M (2012) J Am Chem Soc 134:17396–17399
125. Burschka J, Pellet N, Moon SJ, Humphry-Baker R, Gao P, Nazeeruddin MK, Grätzel M (2013) Nature 499:316–320
126. Snaith HJ, Humphry-Baker R, Chen P, Cesar I, Zakeeruddin SM, Grätzel M (2008) Nanotechnology 19:424003
127. Kyriazi JM, Ding I-K, Marchioro A, Punzi A, Hardin BE, Burkhard GF, Tétreault N, Grätzel M, Moser J-E, McGehee MD (2011) Adv Energy Mater 1:407–414
128. Abrusci A, Ding I-K, Al-Hashimi M, Segal-Peretz T, McGehee MD, Heeney M, Frey GL, Snaith HJ (2011) Energy Environ Sci 4:3051–3058
129. Ding I-K, Tétreault N, Brillet J, Hardin BE, Smith EH, Rosenthal SJ, Sauvage F, Grätzel M, McGehee MD (2009) Adv Funct Mater 19:2431–2436
130. Docampo P, Hey A, Guldin S, Gunning R, Steiner U, Snaith HJ (2012) Adv Funct Mater 22:5010–5019
131. Snaith HJ (2010) Adv Funct Mater 20:13–19
132. Hao F, Stoumpos CC, Cao DH, Chang RPH, Kanatzidis MG (2014) Nat Photonics 8:489–494
133. Christians JA, Herrera PAM, Kamat PV (2015) J Am Chem Soc 137:1530–1538
134. Guarnera S, Abate A, Zhang W, Foster JM, Richardson G, Petrozza A, Snaith HJ (2015) J Phys Chem Lett 6:432–437
135. Habisreutinger SN, Leijtens T, Eperon GE, Stranks SD, Nicholas RJ, Snaith HJ (2014) Nano Lett 14:5561–5568
136. Song W, Chen Z, Brennaman MK, Concepcion JJ, Patrocinio AOT, Iha NYM, Meyer TJ (2011) Pure Appl Chem 83:749–768
137. Yachandra VK, Sauer K, Klein MP (1996) Chem Rev 96:2927–2950
138. Renger G, Renger T (2008) Photosynth Res 98:53–80
139. Meyer TJ, Huynh MHV, Thorp HH (2007) Angew Chem Int Ed 46:5284–5304
140. McEvoy JP, Brudvig GW (2006) Chem Rev 106:4455–4483

141. Kern J, Renger G (2007) Photosynth Res 94:183–202
142. Dau H, Zaharieva I (2009) Acc Chem Res 42:1861–1870
143. Brudvig GW (2008) Philos Trans R Soc B Biol Sci 363:1211–1219
144. Barber J, Andersson B (1994) Nature 370:31–34
145. Barber J (2006) Biochem Soc Trans 34:619–631
146. Chen Z, Jaramillo TF, Deutsch TG, Kleiman-Shwarsctein A, Forman AJ, Gaillard N, Garland R, Takanabe K, Heske C, Sunkara M, McFarland EW, Domen K, Miller EL, Turner JA, Dinh HN (2010) J Mater Res 25:3–16
147. Hammarström L (2015) Acc Chem Res 48:840–850
148. Young KJ, Martini LA, Milot RL, Snoeberger RC III, Batista VS, Schmuttenmaer CA, Crabtree RH, Brudvig GW (2012) Coord Chem Rev 256:2503–2520
149. Youngblood WJ, Lee S-HA, Kobayashi Y, Hernandez-Pagan EA, Hoertz PG, Moore TA, Moore AL, Gust D, Mallouk TE (2009) J Am Chem Soc 131:926–927
150. Li L, Duan L, Xu Y, Gorlov M, Hagfeldt A, Sun L (2010) Chem Commun 46:7307–7309
151. Brimblecombe R, Koo A, Dismukes GC, Swiegers GF, Spiccia L (2010) J Am Chem Soc 132:2892–2894
152. Gillaizeau-Gauthier I, Odobel F, Alebbi M, Argazzi R, Costa E, Bignozzi CA, Qu P, Meyer GJ (2001) Inorg Chem 40:6073–6079
153. Wasielewski MR (1992) Chem Rev 92:435–461
154. Gould SL, Kodis G, Palacios RE, de la Garza L, Brune A, Gust D, Moore TA, Moore AL (2004) J Phys Chem B 108:10566–10580
155. Liddell PA, Kuciauskas D, Sumida JP, Nash B, Nguyen D, Moore AL, Moore TA, Gust D (1997) J Am Chem Soc 119:1400–1405
156. Moore GF, Blakemore JD, Milot RL, Hull JF, Song H-e, Cai L, Schmuttenmaer CA, Crabtree RH, Brudvig GW (2011) Energy Environ Sci 4:2389–2392
157. Moore GF, Konezny SJ, Song H-e, Milot RL, Blakemore JD, Lee ML, Batista VS, Schmuttenmaer CA, Crabtree RH, Brudvig GW (2012) J Phys Chem C 116:4892–4902
158. Swierk JR, Méndez-Hernández DD, McCool NS, Liddell P, Terazono Y, Pahk I, Tomlin JJ, Oster NV, Moore TA, Moore AL, Gust D, Mallouk TE (2015) Proc Natl Acad Sci 112:1681–1686
159. Kärkäs MD, Verho O, Johnston EV, Åkermark B (2014) Chem Rev 114:11863–12001
160. Blakemore JD, Crabtree RH, Brudvig GW (2015) Chem Rev 115:12974–13005
161. Gersten SW, Samuels GJ, Meyer TJ (1982) J Am Chem Soc 104:4029–4030
162. Gilbert JA, Eggleston DS, Murphy WR, Geselowitz DA, Gersten SW, Hodgson DJ, Meyer TJ (1985) J Am Chem Soc 107:3855–3864
163. Sens C, Romero I, Rodríguez M, Llobet A, Parella T, Benet-Buchholz J (2004) J Am Chem Soc 126:7798–7799
164. Zong R, Thummel RP (2005) J Am Chem Soc 127:12802–12803
165. Concepcion JJ, Jurss JW, Templeton JL, Meyer TJ (2008) J Am Chem Soc 130:16462–16463
166. Concepcion JJ, Jurss JW, Brennaman MK, Hoertz PG, Patrocinio AOT, Murakami Iha NY, Templeton JL, Meyer TJ (2009) Acc Chem Res 42:1954–1965
167. Yoshida M, Masaoka S, Sakai K (2009) Chem Lett 38:702–703
168. Hocking RK, Brimblecombe R, Chang L-Y, Singh A, Cheah MH, Glover C, Casey WH, Spiccia L (2011) Nat Chem 3:461–466
169. Alibabaei L, Brennaman MK, Norris MR, Kalanyan B, Song W, Losego MD, Concepcion JJ, Binstead RA, Parsons GN, Meyer TJ (2013) Proc Natl Acad Sci 110:20008–20013
170. Alibabaei L, Sherman BD, Norris MR, Brennaman MK, Meyer TJ (2015) Proc Natl Acad Sci 112:5899–5902
171. Vannucci AK, Alibabaei L, Losego MD, Concepcion JJ, Kalanyan B, Parsons GN, Meyer TJ (2013) Proc Natl Acad Sci 110:20918–209225
172. Hanson K, Torelli DA, Vannucci AK, Brennaman MK, Luo H, Alibabaei L, Song W, Ashford DL, Norris MR, Glasson CRK, Concepcion JJ, Meyer TJ (2012) Angew Chem Int Ed 51:12782–12785

173. Ding X, Gao Y, Zhang L, Yu Z, Liu J, Sun L (2014) ACS Catal 4:2347–2350
174. Kirner JT, Stracke JJ, Gregg BA, Finke RG (2014) ACS Appl Mater Interf 6:13367–13377
175. Ronconi F, Syrgiannis Z, Bonasera A, Prato M, Argazzi R, Caramori S, Cristino V, Bignozzi CA (2015) J Am Chem Soc 137:4630–4633
176. Li F, Fan K, Xu B, Gabrielsson E, Daniel Q, Li L, Sun L (2015) J Am Chem Soc 137:9153–9159
177. Qin P, Zhu H, Edvinsson T, Boschloo G, Hagfeldt A, Sun L (2008) J Am Chem Soc 130:8570–8571
178. Yu Z, Li F, Sun L (2015) Energy Environ Sci 8:760–775

Chapter 3
Organic Light-Emitting Diodes (OLEDs): Working Principles and Device Technology

Umberto Giovanella, Mariacecilia Pasini, and Chiara Botta

Abstract Organic electronics is a field of material science that has encountered a rapid advance over the last few decades and has now reached the commercial marketplace. Its most relevant example is represented by Organic Light-Emitting Diodes (OLEDs) technology, able to combine the device low energy consumption and low production costs with many additional appealing features, such as large emitting surfaces, transparency and flexibility, color-tunability and color-quality. These unique properties of OLEDs allow to design low cost, large area flexible displays and white lighting sources that can fit to many different environmental requirements, resulting in tremendous benefits in imaging, lighting, automotive, transportation, communication, agriculture and medicine.

This chapter provides an overview on the basic working principles of the devices with the analysis of the different kinds of emission mechanisms and the methods to improve quantum efficiency by optimization of the device architecture. The main classes of materials employed in OLED technology are presented focusing on few representative examples while the challenges to be faced by future research on material and device stability are discussed in view of commercialization applications. Some of the outstanding results recently obtained in white OLEDs (WOLEDs), able to produce a revolution in the next generation lighting industry, are also presented at the end of the chapter.

Acronyms

η_{PL}	Efficiency of photoluminescence
CCT	Colour temperature
CIE	Commission Internationale de l'Eclairage
CRI	Colour rendering index
CT	Charge transfer
DF	Delayed fluorescence

U. Giovanella (✉) • M. Pasini • C. Botta
Istituto per lo Studio delle Macromolecole (ISMAC), CNR, Via E. Bassini 15, 20133 Milan, Italy
e-mail: u.giovanella@ismac.cnr.it; m.pasini@ismac.cnr.it; c.botta@ismac.cnr.it

G. Bergamini, S. Silvi (eds.), *Applied Photochemistry*,
Lecture Notes in Chemistry 92, DOI 10.1007/978-3-319-31671-0_3

EA	Electron affinity
EBL	Electron blocking layers
EIL	Electron injection layers
EL	Electroluminescence
EML	Emitting material layer
EQE	External quantum efficiency
ET	Energy transfer
E_T	Triplet energy levels
ETL	Electron transport layers
FRET	Förster resonance ET
HBL	Hole blocking layers
HIL	Hole injection layers
HOMO	Highest occupied molecular orbital
HTL	Hole transport layers
IC	Internal conversion
ICS	Inter system crossing
IP	Ionization potential
IQE	Internal quantum efficiency of electroluminescence
ITO	Indium tin oxide
J_0	Critical current density
K	Kelvin
L	Luminance
LE	Luminous or current efficiency
LUMO	Lowest unoccupied molecular orbital
*M*W	Molecular weight
OLED	Organic light emitting diode
PE	Power efficiencies
PHOLED	Phosphorescent OLED
PL	Photoluminescence
PLED	Polymer OLED
RISC	Reverse inter system crossing
SMOLED	Small molecule OLED
TADF	Thermally activated delayed fluorescence
T_g	Glass transition temperatures
TTA	Triplet–triplet annihilation
V	Applied voltage
WOLED	White light organic emitting diode
ΔE_{ST}	Singlet-triplet energy splitting
η_C	Fraction of light coupled out of the structure into the viewing direction
ϕ	Work function

3.1 Introduction

The rapid development of optoelectronic device technologies based on organic materials in the last two decades has been motivated both by the need of light and flexible devices and by the increasing interest in technologies with low environmental impact [1–3]. Among the class of organic based devices, organic light-emitting diodes (OLEDs) have introduced a revolution in the display and lighting industry, due to their outstanding merits such as high-luminance efficiency, full-colour/large-area display, wide-viewing angle, lightness and transparency, low power consumption, etc. [2, 3].

The first electroluminescence (EL) of an organic compound was observed in 1965 for an anthracene crystal by applying bias voltages (V) of a few hundred volts [4]. Over 20 years later, Tang and Van Slyke reported the first OLED working at low bias V by using a thin film of evaporated tris(8-hydroxyquinolinato)aluminium (Alq_3) as an emitter [5]. Electroluminescence from conjugated polymers was reported for the first time in 1990 by using a poly(p-phenylene vinylene) (PPV) film sandwiched between two metallic electrodes [6]. Afterwards, the advancements of OLEDs grew very fast and since 2000 their practical applications were demonstrated in small display devices, such as mobile phones and cameras, and flat panel TV screens. Nowadays organic electronics is not only an important academic field but also an established industry, gaining additional market value every year. In fact, thanks to their special properties (surface light source, transparent, flat, flexible), OLEDs open up new applications such as windows which emit light only at night, easy integration into home furniture, for example in kitchens and bathrooms, allowing revolutionary lighting design.

As for other technological fields, OLEDs competitiveness can be further increased by research efforts pointing to solve some problems: the low stability of the blue light emission, the low device lifetimes, the efficiency of light extraction from planar structures, the high cost and the dependence on limited global resources of the best performing noble metals, such as Iridium. Thus, the potential excellent device characteristics of organic semiconductors have not yet been fully consolidated. Researchers are still actively involved in the improvement of OLEDs performances by working on both the materials and the optimization of the device structures [7, 8].

In the next Sect. 3.2, we give an overview of the basic OLEDs working principles and the most important parameters for technological applications; in Sect. 3.3 the typical OLED architecture and manufacturing techniques are presented and the main class of organic semiconducting materials forming the OLEDs layers are discussed; finally, in Sect. 3.4 we focus on WOLEDs. WOLEDs, besides LEDs, are arising as the next generation of indoor lighting sources since conventional systems such as incandescent lamps and compact fluorescent lamps are expected to be replaced by solid state technologies (see Fig. 3.1) [9, 10].

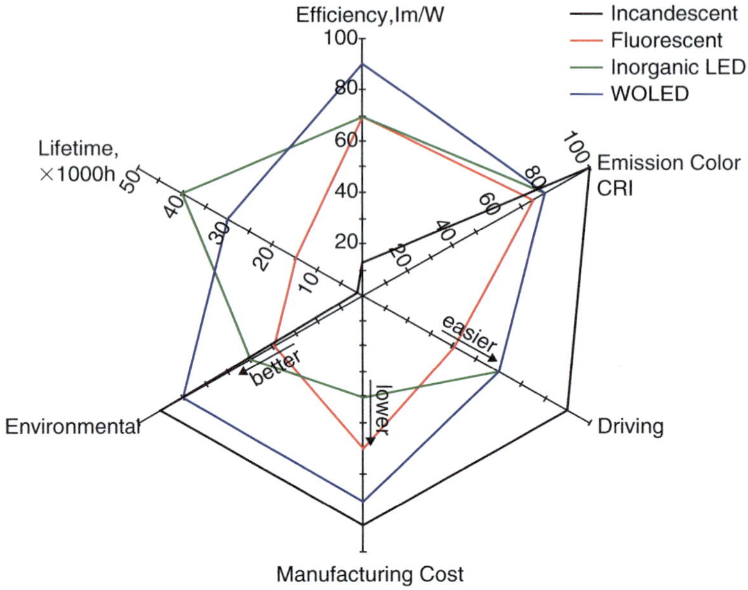

Fig. 3.1 Semiquantitative comparison of the key performance parameters of WOLEDs relative to incandescent lamps, fluorescent tubes, and inorganic white LEDs. Data for incandescent lamps, fluorescent devices, and inorganic LEDs is representative for commercially available light sources. WOLED parameters reflect data from different literature reports (Reprinted from Ref. [10] by permission from John Wiley and Sons, copyright (2010))

3.2 Working Principles

3.2.1 Electronic Properties of Organic Materials

OLEDs active materials, i.e. materials producing light emission, are based on conjugated molecules and polymers or on organometallic compounds (composed by a metal coordinated by organic ligands). The active material must possess both good charge transport and emissive properties. Conjugated compounds are often defined as organic semiconductors since they show electrical and optical properties typical of inorganic semiconductors with the added value of being processable as plastic materials (Sect. 3.3). Their semiconducting properties arise from the delocalized π-electrons that are able to move along the conjugated chain. The overlap of π (bonding) and π^* (antibonding) molecular orbitals forms valence and conduction wavefunctions that provide a continuous system of electron density along the conjugated backbone, supporting mobile charge carriers. The extent of this overlap defines the conjugation length of the chain and determines the Highest Occupied Molecular Orbital - Lowest Unoccupied Molecular Orbital (HOMO-LUMO) energy gap (E_g) of the conjugated material, generally in the range of 1–4 eV (see Fig. 3.2). The E_g can be finely tuned by modification of the chemical

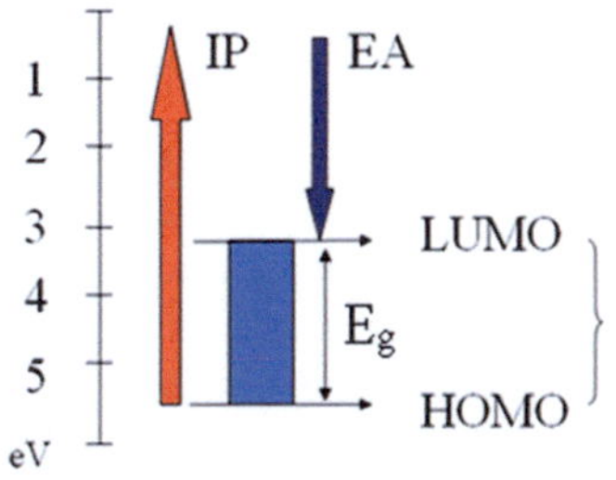

Fig. 3.2 Scheme of the energy position of the LUMO and HOMO levels

structure, allowing optical excitations from ultraviolet to near infrared by extending the conjugation length. In the same way the Ionization Potential (IP) and Electron Affinity (EA) of the conjugated compound can be modulated by insertion of electron-accepting or donating moieties in the chain (Sect. 3.3).

The electronic properties of conjugated polymers and molecules are generally described in terms of molecular orbitals rather than on semiconducting bands. By photon absorption, one electron is promoted from the ground state S_0 to an excited state S_n generating an exciton (bound couple of electron and hole). After photon absorption, radiative recombination of the exciton produces light emission, photoluminescence (PL). PL in organic materials generally takes place after fast (in the ps scale) Internal Conversion (IC) processes from S_n to S_1. Since the spin selection rules do not allow the transition from the triplet state T_1 to the ground state $S_{0,}$ organic materials display fluorescence ($S_1 \rightarrow S_0$), whose lifetimes are typically in the ns scale, and very weak phosphorescence ($T_1 \rightarrow S_0$). In organometallic compounds the spin selection rule is relaxed due to the presence of the heavy metal that increases the Inter System Crossing (ICS) from the S to the T states. Therefore T states recombine radiatively giving rise to phosphorescence with long (100 μs–10 ms) lifetimes due to the metastable nature of the T state [11]. One of the advantages of using organic materials, with respect to inorganic semiconductors, is that by proper modification of the chemical structure the energy position of the emitting S_1 or T_1 states can be changed and the emission colour finely tuned accordingly to the specific application needs.

The efficiency of PL (η_{PL}) is defined as the ratio of the number of emitted photons for absorbed photons. If all the absorbed photons are emitted radiatively, η_{PL} is 1. The value of η_{PL} depends on the nature of the molecular excited states but it is also influenced by the supramolecular organization of the conjugated units. In fact very often molecular aggregation introduces intermolecular interactions that reduces the efficiency of the emission [12]. In a bulk material charge mobility is determined by hopping of the charges from chain to chain in polymeric systems, or from molecule to molecule in molecular materials and is therefore high if strong inter-molecular interactions, i.e. $\pi-\pi$ stacking, are present [13]. A proper control of the inter-molecular interactions in the active material is therefore necessary to keep high the efficiency of emission without reducing the charge transport properties [14, 15].

Besides S and T excitons, other emitting species, as excimers and exciplexes, can originate from inter-molecular interactions when the inter-molecular distances are short (typically lower than 4 Å). The interaction between one molecule in the excited state and another in the ground state produces an excimer state if the two molecules are identical or an exciplex state if the two molecules are different [12]. The emission of these states occurs in different spectral regions with respect to the exciton emission. Therefore the intermolecular interactions can induce relevant spectral changes in the emission properties of a material.

3.2.2 Light Emission Process in OLEDs

The process of light generation in OLEDs can be schematically described by considering four main steps, as depicted in Fig. 3.3. Electrical generation of excited states occurs through three steps: the injection (1) of negative (electron) and positive (hole) charges at the electrodes, their migration (2) into the bulk material till they meet each other, the formation (3) of bound couples of electrons and holes, named excitons. The radiative recombination (4) of the excitons will generate light giving rise to electroluminescence. The efficiency of the emission inside the device, named internal quantum efficiency of EL (IQE or η_{int}) is defined as the number of photons emitted per couple of charges injected. In order to inject a comparable number of the electrons and the holes, the work functions (ϕ) of the two metal electrodes must fit the energy position of the HOMO and LUMO levels so that low potential barriers of the same height for electrons and holes ($\Delta h \approx \Delta e$) are generated (see Fig. 3.3). After injection, the charges should move with a comparable mobility in order to form the excitons in a region far from the metal electrodes, often acting as quenching points.

Indeed a single material generally displays different mobilities for electrons and holes. Therefore, in order to obtain efficient OLEDs, additional layers (called charge regulating layers) are often inserted between the electrodes and the Emitting Material Layer (EML) to optimize the device performances. As shown in Fig. 3.4, Electron or Hole Injection Layers (EIL, HIL) favour the charge injection from the electrodes, while Electron or Hole Transport Layers (ETL, HTL) facilitate the

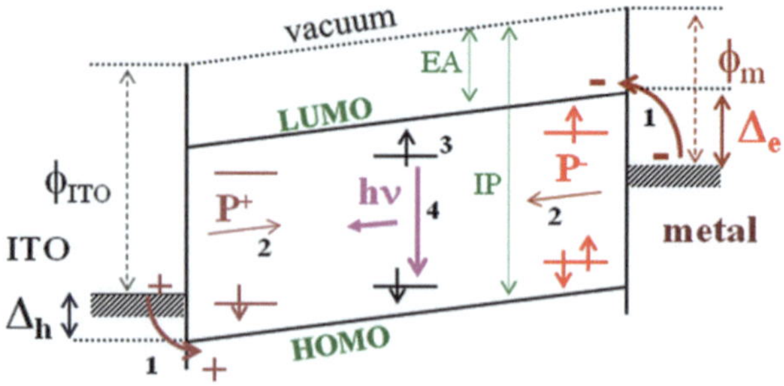

Fig. 3.3 Scheme of the main steps involved in the process of light emission in an OLED: (1) charge injection from the electrodes, (2) charge transport, (3) exciton formation, and (4) exciton radiative recombination

charges transport. Moreover, exciton formation and recombination can be enhanced by insertion of Electron or Hole Blocking Layers (EBL, HBL) that confine the charges within the EML.

The spins of the electrons and holes injected from the electrodes follow the spin statistics. Since the T state is threefold degenerate, exciton formation in OLEDs results in 25 % S (antiparallel spins) and 75 % T excitons (parallel spins). The mechanism of exciton generation in OLEDs therefore favours T excitons with respect to photoexcitation that produces only S excitons. Once the excitons are formed, their radiative recombination occurs exactly with the same mechanism of the PL. An organic material that does not show phosphorescence and has $\eta_{PL} = 1$ will therefore display maximum IQE of 0.25, since the T states will recombine non-radiatively for the spin selection rules (see Fig. 3.5).

The relaxation of the spin selection rules appears therefore to be crucial to reach the theoretical IQE = 1. Incorporation of metals into the organic frameworks provides materials with high ISC where S excitons are converted into T and the phosphorescent $T_1 \rightarrow S_0$ transition is allowed [16], as shown in Fig. 3.5. Nowadays

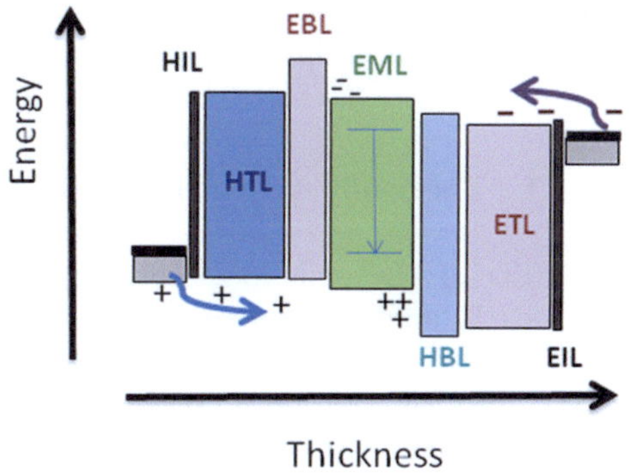

Fig. 3.4 Scheme of the energy levels of the different materials in a multilayer device

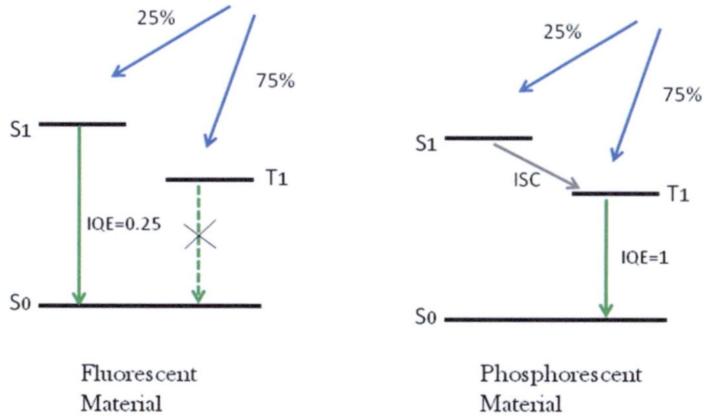

Fig. 3.5 Scheme of exciton generation and recombination by electrical excitation in a fluorescent (*left*) and phosphorescent (*right*) material

IQEs close to this theoretical value have already been reported in red, green, and blue phosphorescent OLEDs (PHOLEDs) by several research groups, using vacuum thermal evaporation processes with Ir(III) and Pt(II) based complexes [17].

The main problem of such phosphorescent materials is the high cost and the dependence on limited global resources of these metals that impose several limitations to their use for the next generation of solid-state displays and light sources. Another problem of the phosphorescent devices is that they often suffer from efficiency roll-off (reduction of efficiency at high densities of current, see later) that limits their practical use. This is mainly caused by the long lifetimes of the T states that increase the probability of quenching due to exciton-charge and exciton-exciton annihilation or exciton dissociation by electric field or at trap sites. In order to solve these problems several strategies have been proposed: i.e. searching phosphorescent complexes with heavy atoms (whose T lifetimes are shorter), reducing charge densities by optimization of the device structure, favouring the conversion of the electrically generated T into S excitons by T–T annihilation (TTA) or by up-conversion of T excitons to S excitons using thermal energy [18]. Among them, Thermally Activated Delayed Fluorescence (TADF) was found to have the most rapid progress in recent investigations [19, 20] and will be briefly discussed in this section.

3.2.2.1 Thermally Activated Delayed Fluorescence (TADF)

In OLEDs, TADF emission occurs through the following steps (Fig. 3.6): (1) triplet excitons T_1 are formed after electron/hole injection in a triplet-to-singlet ratio of 3:1; (2) T_1 triplet excitons are converted to S_1 via a Reverse Inter System Crossing (RISC) process with the aid of thermal energy; (3) the S_1 state recombine quickly after formation, giving Delayed Fluorescence (DF). DF has typical lifetimes of the order of μs, longer with respect to prompt fluorescence (ns) obtained by direct singlet generation but shorter than the typical lifetime of phosphorescence

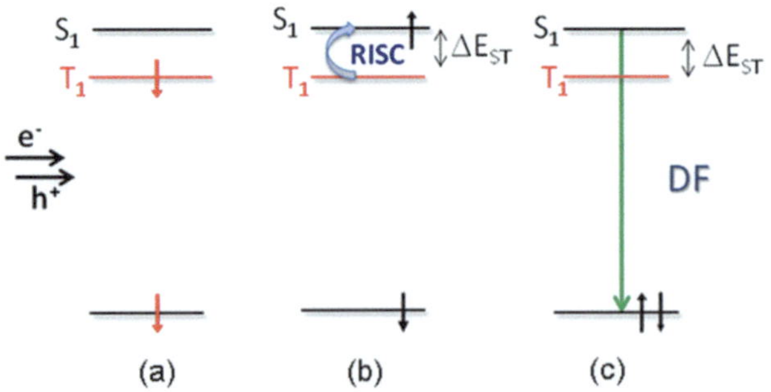

Fig. 3.6 Scheme of the main steps leading to Thermally Activated Delayed Fluorescence (TADF) emission in OLEDs: (1) Triplet generation after injection of a couple of electron and hole, (2) Triplet to Singlet conversion by Reverse Inter System Crossing (RISC), (3) Singlet radiative recombination by Delayed Fluorescence (DF)

(100 μs–10 ms), therefore reducing the efficiency roll-off in OLEDs devices. For efficient $T_1 \rightarrow S_1$ transition stimulated by thermal energy (efficient RISC process) small energy level difference between T_1 and S_1 (singlet-triplet energy splitting, ΔE_{ST}) is essential.

For these reasons in the last few years particular efforts have been devoted to the development of materials displaying TADF activity and a special interest in Cu (I) based complexes for light emitting devices has arisen [21]. In fact cuprous complexes appear to be the ideal alternative to Ir and Pt heavy atom complexes because they are relatively abundant, inexpensive, and non-toxic, and because their lowest triplet excited states have often a small ΔE_{ST}, providing an efficient pathway for RISC leading to TADF emission.

If a heavy metal ion is incorporated into the complex system both efficient phosphorescence ($T_1 \rightarrow S_0$) and DF ($S_1 \rightarrow S_0$) by TADF can potentially occur simultaneously. Due to their efficient triplet emission process, these emitters can harvest all of singlet and triplet excitons [22].

Another class of materials with TADF activity is that of organic molecules with electron donor and acceptor moieties, whose emission occurs from charge transfer (CT) states [3, 23]. ΔE_{ST} is proportional to the exchange integral and decreases as the overlap between the wave functions of the ground and excited states decreases. However, a large overlap between the wave functions of the ground and excited states is necessary in order to have a high efficiency of radiative recombination. Organic molecules containing both electron-donating and electron-accepting substituents may possess HOMO and LUMO spatially separated when the donor and acceptor units are largely twisted. It has been shown by Adachi and co-workers [19] that in such systems the triplet CT states might lay at higher energy with respect to the locally excited T state, providing simple molecular systems with TADF properties.

3.2.2.2 Energy Transfer (ET)

Energy transfer (ET) is an important mechanism involved in the emission process of devices whose EML is composed by two (or more) species, such as the commonly used host-guest systems. After exciton generation on a host (energy donor or D), the excitation can be transferred to a guest (energy acceptor or A) that relaxes radiatively giving light emission. The most efficient ET mechanism is the dipole–dipole coupling-based Förster resonance ET (FRET) that allows to shift the emission to lower energy without dissipation processes as those normally occurring in a two step process as absorption and re-emission. FRET occurs even at long distances, up to 5–10 nm and is fast (about 10 ps). The ET rate k_{DA} scales as $1/R^6$, where R is the intermolecular distance between A and D, as shown in Eq. 3.1 [24]

$$k_{DA} = \frac{1}{\tau_D}\left(\frac{R_o}{R}\right)^6 \tag{3.1}$$

where τ_D is the lifetime of the D in absence of the acceptor and R_o is the Förster radius, defined as the distance at which half of the emission of the D is quenched due to ET to the A. R_o can be estimated from the spectroscopic properties of A and D according to Eq. 3.2

$$R_o{}^6 \approx \frac{k^2 \eta_D}{n^4} \int \varepsilon_A(\lambda) F_D(\lambda) \lambda^4 d\lambda \tag{3.2}$$

where k^2 is an orientation factor that depends on the relative orientations between the A and D transition moments, n is the refractive index of the medium, η_D is the PL emission efficiency of the donor in the absence of the acceptor, F_D is the fluorescence spectrum of D and ε_A is the molar extinction coefficient of A [24]. As shown in Eqs. 3.1 and 3.2, efficient FRET requires that the emission spectrum of D overlaps the absorption spectrum of A and that the acceptor possesses a high molar extinction coefficient.

At D/A proximities under one nanometer, quantum mechanical effects such as orbital overlap and tunneling can result in ET through electron exchange interactions, known as Dexter ET (see Fig. 3.7). Both Förster and Dexter ET processes are involved in the emission mechanism of OLEDs. FRET is especially efficient in fluorescent materials since it is based on dipole-dipole interactions, while the Dexter transfer is generally at the base of ET processes in phosphorescent materials, even though the presence of heavy atoms, relaxing the selection rules, allows FRET processes to take place in phosphorescent materials.

The management of T and S excitons by a proper use of FRET and Dexter ET between emissive materials in multilayer devices allows to reach IQE of 100 %. The complete harvesting of all electrically generated excitons and proper management of the exciton distribution between the fluorophors and the phosphors make it in principle possible to simultaneously achieve white emission and unity IQE [25]. Several different ways to realize the concept have been presented.

As shown in Fig. 3.8, in properly designed multilayer devices for white light emission [26] blue can be obtained from a fluorescent dopant after a fast FRET process from the S excitons generated on the host. The phosphor-doped region is located in the centre of the EML and separated from the exciton formation zones by spacers of undoped host material at distances larger than R_o. Diffusion of S excitons to the phosphor dopants is negligible due to their intrinsically short diffusion

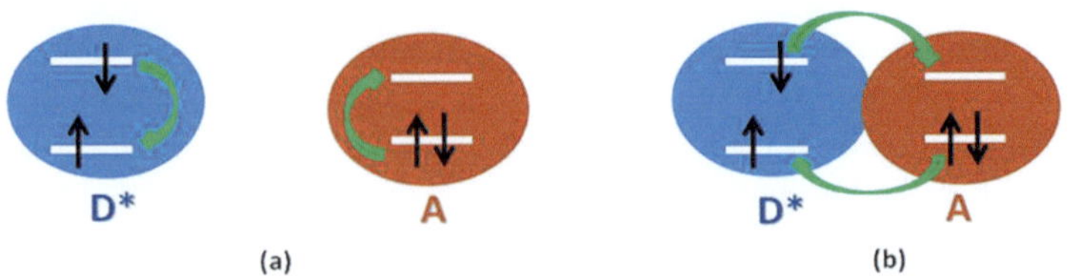

Fig. 3.7 Schematic representation of FRET (**a**) and Dexter (**b**) ET mechanisms from an excited Donor (*D*) to an Acceptor (*A*)

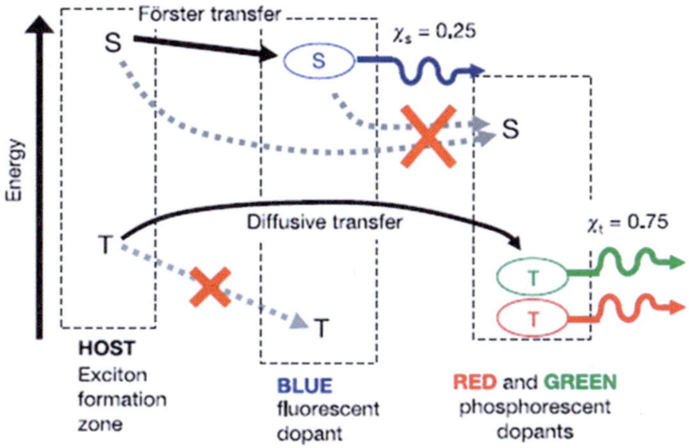

Fig. 3.8 Diagram of the channels for triplet (*T*) and singlet (*S*) formation and transfer directly onto their corresponding emissive dopants (Reprinted by permission from Macmillan Publishers Ltd: Nature, Ref. [26], copyright (2006))

lengths. Differently, the long lifetimes of the T excitons allow them to reach the phosphor-doped regions by diffusive transfer before their natural decay, here they transfer excitation to the lower energy green or red phosphor dopants that give phosphorescence emission.

Another example of proper management of the exciton distribution between the fluorophors and the phosphors is reported in Fig. 3.9. The exciton generation from the injected charges takes place close to the inter-layer region between the blue fluorescent dye and the green phosphorescent dye. There, several ET paths (Förster and Dexter transfer) are possible (solid arrows), which determine the ratio of blue fluorescence to green phosphorescence but do not reduce the quantum efficiency. In fact S excitons generated in the phosphorescent layer will be transferred to the singlet states of the blue fluorescent dye. Triplet excitons will either recombine giving green phosphorescence, or provide orange phosphorescence after diffusive transfer to the orange phosphorescent material [27, 28].

Since most of the emitting dyes suffer from aggregation quenching phenomena, they are generally dispersed (at low doping concentrations) into a host matrix that has the main functions of preventing dye aggregation, transporting the charges and transferring excitons after their generation in the host. Two main mechanisms may occur within a host-guest EML: (i) exciton formation on the host and subsequent energy transfer to the guest dye; (ii) direct formation of the exciton at the dye site after one (or both) charge has been trapped by the guest. One of the main parameters to be considered when a host is chosen for a phosphoresent guest is that energy position of the triplet state (E_T) of the host must be higher than that of the guest. In fact if the triplet level of the host (normally a fluorescent system) is lower or too

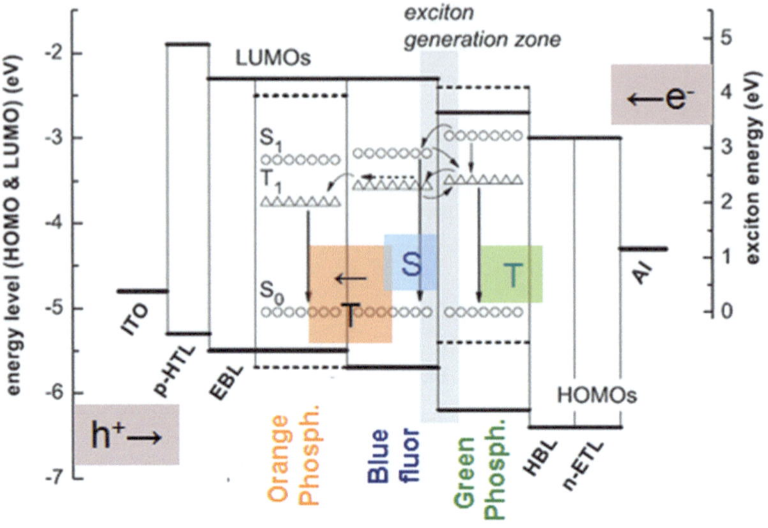

Fig. 3.9 Energy level scheme of a device with separate emission layers together with the exciton energy diagram. *Arrows* indicate possible exciton transfer mechanisms (Adapted from Ref. [27] by permission from John Wiley and Sons, copyright (2009))

close to the triplet level of the phosphorescent dye, the triplet exciton can be transferred back to the host and will not recombine radiatively.

3.2.3 Outcoupling

OLEDs are generally considered as Lambertian emitters characterized by an isotropic emission with equal radiance into any solid angle within the forward viewing hemisphere. The intensity of the emitted light follows the relation $I = I_0 cos\,(\theta)$, where I_0 is measured in the forward direction. Recently this assumption has been questioned [29]. In fact, by introducing outcoupling structures the profiles can be drastically changed.

In the OLED with planar geometry, only a fraction of photons created within the EML is able to escape to air. There are four main light propagation modes in OLEDs determined by the thin-film structure of the device and the respective optical properties (i.e., refractive indices and absorption coefficients) (Fig. 3.10).

The refractive index difference at the organic/substrate interface causes a large fraction of light to be trapped inside the organic EML (organic mode) and inside the substrate (substrate mode). Based on similar considerations, only a fraction of the light that is originally coupled to the substrate can escape the device (outcoupled/air mode). Additionally, the emitting molecules can be directly coupled to surface plasmons of the highly reflective cathode (coupling to metal surface plasmon mode)

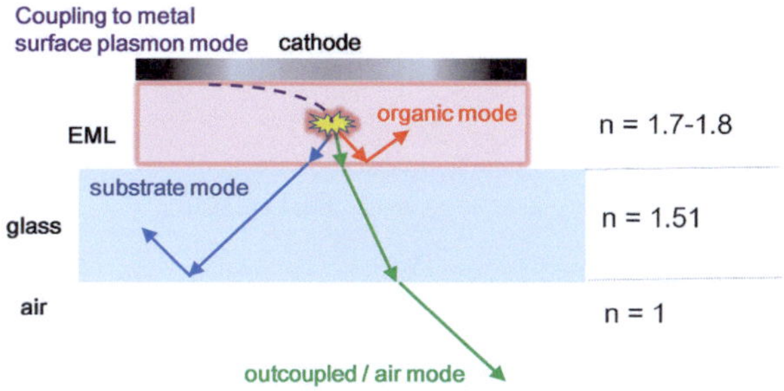

Fig. 3.10 Cross section of an OLED with indication of different light modes

[30]. The fraction of generated light effectively escaping from the substrate (fraction of light coupled out of the structure into the viewing direction, η_C), is $\eta_C \cong \frac{1}{2n_{EML}^2}$ with n_{EML} being the refractive index of the EML [31]. Hence, the light emitted from the substrate surface of the OLED is only about 20 % of the total emitted light from the OLED.

Two exemplifying concepts for improved light outcoupling by applying modifications of the substrate/air interface (e.g. periodic substrates, micro-lens array) will be discussed in Sect. 3.4.

3.2.4 Efficiencies, Luminance

We define the various efficiencies employed in evaluating the fundamental emission properties of OLEDs. As mentioned before, only a fraction of photons created within the EML is able to escape to air, hence, beside the IQE, the external quantum efficiency (EQE or η_{EXT} or η_{EL}) has to be defined. The commonly accepted definition for the EQE is the ratio of the number of photons that are extracted to air per injected electrons [32].

Accurate measurement of the EQE of a device can be achieved by measuring the total light output in all directions in an integrating sphere (Fig. 3.11a), by following the detailed procedure reported by Forrest et al. [32] through the Eq. 3.3:

$$EQE = \frac{\int \lambda I_{detector}(\lambda) d\lambda}{f I_{OLED} \int \lambda \eta_{detector}(\lambda) d\lambda} \tag{3.3}$$

Where I_{OLED} is the current circulating within the device, $I_{detector}$ is the incremental photocurrent generated in the photodetector by the OLED power emitted at a centre wavelength, λ, and $f < 1$ is the fraction of light emitted to that coupled into the detector, $\eta_{detector}$ is the detector quantum efficiency. EQE is typically quoted independently of the emission wavelength. The internal and external efficiencies are related by the equation $EQE(\lambda) = IQE(\lambda)\eta_C(\lambda)$.

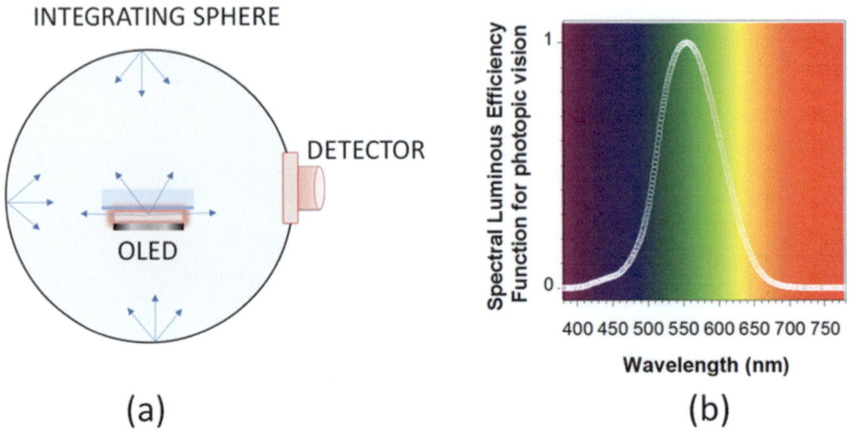

Fig. 3.11 Experimental geometry for measurement of OLEDs efficiency (**a**); $\phi(\lambda)$ photopic curve for human eye sensitivity (**b**)

While EQE is important for material evaluation, artificial light sources performance are usually given in terms of luminous efficacy (LE or η_{LE}) and power efficiency (PE or η_{PE}). These are photometric quantities that take the sensitivity of human eye (photopic curve, Fig. 3.10b) into account.

In an OLED emission with Lambertian profile, the device performance are registered by using a commercial luminance meter that collects light emitted from a fixed diameter spot on the display surface and into a specified (small) solid angle about the surface normal, and calculates a value of the surface luminance indicated as L or B (in candela per square meter or cd/m^2). The recorded L value then allows for a calculation of the LE (measured in candela per ampere, cd/A) at a given current density (J) passing through the device by the Eq. 3.4:

$$LE = L/J \tag{3.4}$$

EQE is related to LE by the expression (3.5) [33]:

$$EQE = \frac{5 \times 10^{-3}}{h\nu(eV)\phi(\lambda)} LE \tag{3.5}$$

PE is defined as luminous flux per unit power of the device (measured in lumen per Watt or lm/W). For flat light sources such as OLEDs emitting into the half plane, luminous intensity is related to luminous flux by 1 cd = π lm, hence (3.6):

$$PE = \pi \frac{LE}{V} \tag{3.6}$$

Typical values of PE for various light sources are listed in Table 3.3. For a summary of all photometric quantities please refer to [32].

3.2.5 *Roll-Off Efficiency and Device Lifetime*

Many applications of OLEDs, present and future, require high-brightness levels, a critical issue to the success of the technology. Unfortunately, in most OLED structures known today, the efficiency of OLEDs tends to decrease with increasing brightness (Fig. 3.12).

This process is called efficiency "roll-off" and can be caused by a range of different non-linear effects, with a rather complex underlying physics [34]. Besides the negative impact on power consumption, this effect also means that the device will be subject to increased electrical stress to achieve a given brightness, which in turn will again reduce the useful device duration (device lifetime).

In efficient devices, the efficiency roll-off can be quantified by the critical current density J_{90}, i.e. the current density at which the EQE drops to 90 % of its maximum value [35]. Thus, devices with strong roll-off will have low critical current densities. Besides, for practical applications, the roll-off efficiency with increasing L could be often relevant as well.

The OLEDs' lifetime is strictly related to its operational current density and usually follows a power-law dependence $\propto J^{-\beta}$, where typically $1.5 < \beta < 3$.

The dominant mechanisms leading to efficiency roll-off at high brightness are different for fluorescent and phosphorescent emitters [34]. While in phosphorescent emitter systems, TTA and charge carrier imbalance appear to be the most important, fluorescent emitters suffer especially from S–T annihilation (STA) and S–polaron annihilation (SPA), but loss of charge carrier balance at high currents can also be an issue in these systems.

High efficiencies are mainly achieved by devices based on phosphorescent emitters, but the EQE rolls off to 90 % of its maximum at currents as low as 1–30 mA/cm^2 (roughly corresponding to a luminance of 1000–10,000 cd/m^2), which is however in the range required for applications in OLED lighting. Fluorescent materials achieve much higher critical current densities in the range of 50–1000 mA/cm^2, but lower EQE. Concerning white light emission, phosphorescent emitters outperform fluorescent approaches in efficiency but suffer from the lowest critical current densities.

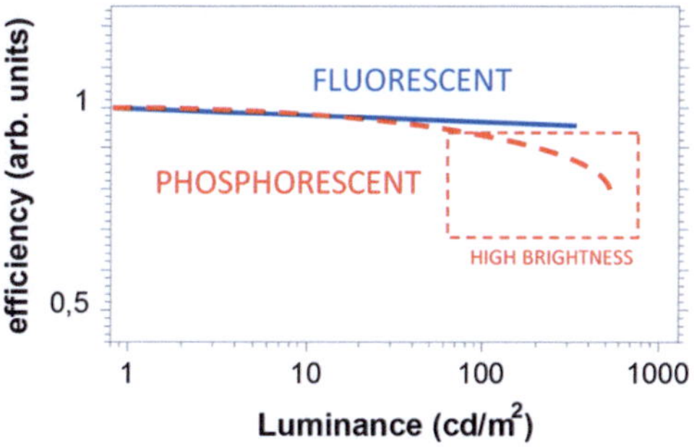

Fig. 3.12 Typical efficiency *vs* L characteristics. In the case of phosphorescence, the efficiency drastically decreases at high brightness as a consequence of quenching processes

3.2.6 CIE, CCT and Colour Rendering Index

The colour of a emitted light is defined by a Commission Internationale de l'Eclairage (CIE) (x, y) coordinate system [36]. The inverted-U-shaped locus boundary (Fig. 3.13) represents monochromatic visible light (450–650 nm). Colours on the periphery of the locus are saturated, while the point at $x = y = 0.33$ represents the white. Primary colours have been defined, normally red (R), green (G), and blue (B) (since human colour vision is usually trichromatic). Every colour, included white, within a triangle defined by three primaries can be created by additive mixing of varying proportions of those primary colours. The CIE 1931 colour space [36] is the most largely used, and defines monochromatic primary colours with wavelengths of 435.8 nm (B), 546.1 nm (G) and 700 nm (R).

Generally, the colour quality of a light source is assessed by two important colorimetric quantities: chromaticity or correlated colour temperature (CCT) and colour rendering index (CRI). The colour temperature of a light source is determined by comparing its chromaticity with that of an ideal black-body radiator: the white colour is not unique but has a distribution along the Planckian locus (Fig. 3.13). Hence the source colour temperature is the temperature in Kelvin (K) at which the heated black-body radiator matches the colour of the light source. Higher Kelvin temperatures (above 5000 K) are called "cool" (green–blue) colours, and lower CCT (2700–3000 K) "warm" (yellow–red) colours. Specific CCT are recommended for different ambient (see Table 3.1).

The human eye is very sensitive to any deviation from the Planckian locus so it is important to design a device with emission as close to the blackbody line as possible. The proper combination of RGB colours, but also of the two complementary colours blue and yellow, allow to tune CCT of the light emission.

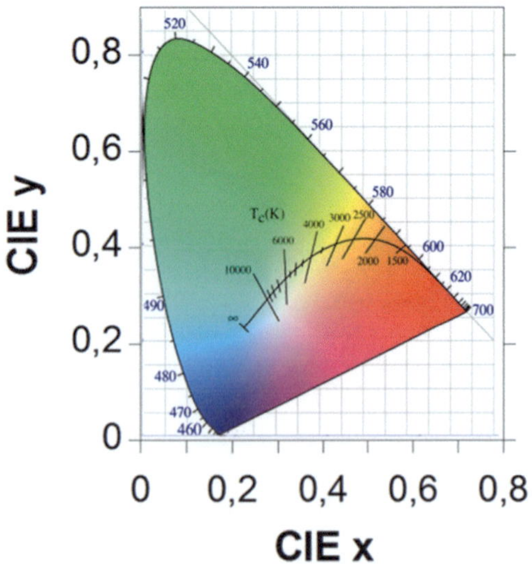

Fig. 3.13 CIE 1931 chromaticity diagram

Table 3.1 Colour temperature (merely characteristic) for common sources

Source	Temperature (K)
Candle flame, sunset, sunrise	1850
Incandescent light bulb	2800–3300
Studio lamp	3200
Moonlight	4100
Tubular fluorescent lamps	5000
Daylight	6500

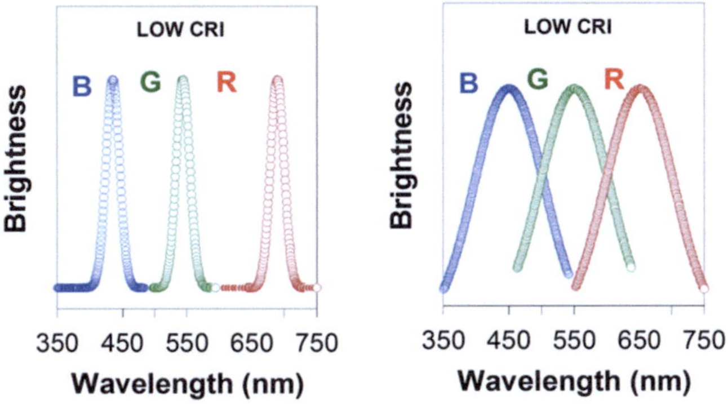

Fig. 3.14 RGB white light with different CRI values

Another important parameter to define the quality of a white light source is the CRI defined as the measure of how true the colours of an object look when illuminated by that light source. It is a dimensionless value ranging from 0 to 100 scale, with 100 being the highest ability for colour rendering.

As example white electroluminescence, achievable either by RGB monochromatic and RGB wideband emissions, evidence a huge difference in CRI values (Fig. 3.14).

3.3 Technology

3.3.1 How to Build Up an OLED

The preparation of the organic thin EML in OLED is one of the crucial point in OLED manufacturing. There are several methods to fabricate the device often resumed in two major categories: thermal vacuum evaporation and wet methods (spin-coating, Doctor blade, ink-jet printing, roll-to-roll, etc.).

Generally, π-conjugated materials are divided into two classes, small molecules and polymers. A polymer is a large molecule (macromolecule) composed of n repeating structural units ($n >> 1$). Small molecules usually have a much lower molecular weight (MW), less than 1000 g/mol, and are much easier to purify. Accordingly, OLEDs fabricated with the two classes of materials are called small molecule OLEDs (SMOLEDs), usually obtained by thermal evaporation method, and polymer OLEDs (PLEDs), prepared by solution techniques.

3.3.1.1 Thermal Vacuum Evaporation

The vacuum deposition process offers a kinetic and thermodynamic control to form homogeneous films with precise morphology and to assemble very complex multilayer devices structures. The high freedom in layer design gives the possibility to optimize charge carrier injection and transport by selecting the more appropriate material and is the main reason for the high efficiencies of evaporated OLEDs. Moreover this technique is solvent free thus eliminating all the risk due to solvent manipulation and storage. A big appeal of this technique is that the vacuum equipment already existing in the semiconducting industry is employed. It is a mature technology and up to now almost all the commercial OLEDs are fabricated in this way. However, due to the requirement of high vacuum regime ($<10^{-6}$ Torr), the fabrication process is more elaborated and expensive than alternative techniques from solution, the device size is limited and the material utilization rate is much less than 10 %. The high waste of expensive organic materials is one of the major reason of the current high cost in OLEDs manufacturing and the problem becomes more severe as large area size panels are demanded.

3.3.1.2 Solution Techniques

Solution techniques are generally cheaper and it is possible to prepare large area display with low material waste. On the other hand, the multilayered optimized OLED architecture responsible of the high performance in vacuum growth devices is harder to achieve due to the solubility of the organic HTL/ETL and the active layer in similar solvents. Moreover, even with the optimization of the process parameters (like concentration of the solution, solvent, rotation speed and temperature) the control of the layers morphology is often difficult.

Today a big effort is devoted to develop multi-layer systems from solution process using materials soluble in orthogonal solvents, so that the sequential layer deposition does not dissolve the previous ones.

3.3.2 Structure of an OLED

The basic OLEDs architecture, consisting of a single film of an electroluminescent organic conjugated material sandwiched between suitable metallic anode and

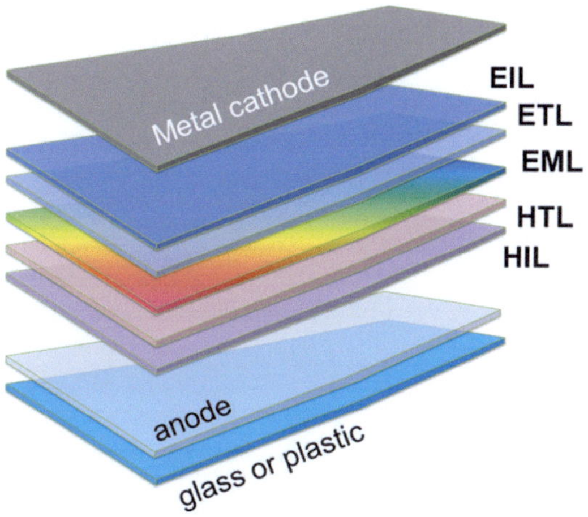

Fig. 3.15 Schematic representation of multilayered OLED

cathode, has evolved towards a multilayered structure (Fig. 3.15). The introduction of one or more layers to improve charge injection, transport, exciton generation and radiative recombination provides a powerful pathway to increase the efficiency of OLEDs dramatically (Sect. 3.2.2).

For each layer with specific functions, such as the HIL/EIL and HTL/ETL, specific materials are developed. In the following, we selected as an example some of the most interesting compounds, discussing their role in the OLEDs architectures as charge regulating and emitting layers.

3.3.2.1 Substrate

The substrate must be transparent to the OLED emitted light, posses high barrier for oxygen and water and sustain the whole device architecture. Generally glass is employed, but also flexible plastic substrates can be used [37].

3.3.2.2 Anode

Anode must be transparent with a work function compatible with HOMO energy of the electroluminescent materials (to give ohmic contact). Indium tin oxide (ITO) is the most common transparent conducting oxide used as anode in organic optoelectronics. It is a ternary non stoichiometric composition of indium, tin and oxygen in varying proportions. Thin films of ITO are most commonly deposited by physical vapour deposition. The ϕ_{ITO} is in the range of 4.5–4.7 eV. Due to the rising cost of ITO, new kind of transparent anode materials have been developed in these years. Among them ZnO and MoO_3 seem to be promising also for photovoltaic technology [38].

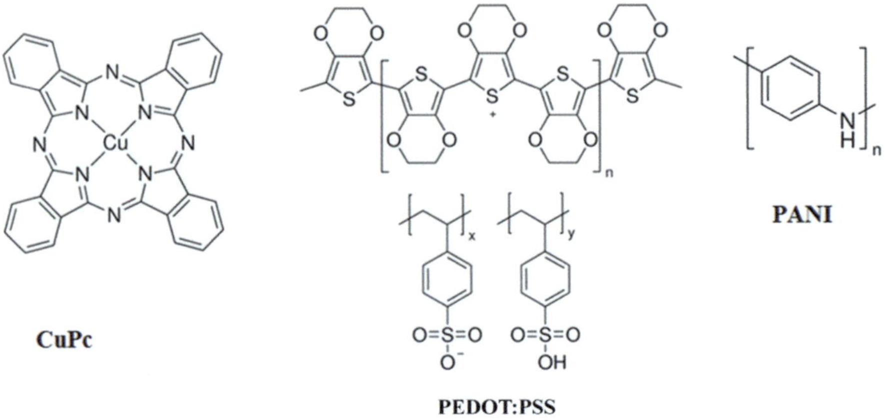

Fig. 3.16 Common materials for HILs

The request for flexible devices lead to the development of flexible anodes for the fabrication of all plastic PLEDs. Polyaniline (PANI) [39] and Poly(3,4-ethylene dioxy-2,4-thiophene)-polystyrene sulfonate (PEDOT:PSS) [40] are water soluble polymers used for this application (Fig. 3.16).

More recently also carbon nanotubes have demonstrated to posses the appropriate characteristics as transparent anode [41].

3.3.2.3 Hole Injection Layer (HIL)

The HIL material is designed to reduce the Δ_h between anode and EML thus enhancing hole injection. The important properties that a HTL material has to fulfil include: appropriate suitable HOMO energy level intermediate between the ϕ_{anode} and HOMO of hole transporting or of the emitting materials; high glass transition temperature (T_g) and good thermal stability; ability to smooth the rough ITO surface; and has to be transparent to permit light outcoupling. Examples of conventional HIL materials are: copper phthalocyanine (CuPc) [42] and PEDOT:PSS [43] (Fig. 3.16). CuPc shows good thermal properties and flat surface in film; it has a HOMO energy level of 4.8 eV suitable to mitigate the potential barrier between anode and EML. However, its absorption in red and blue wavelength regions limits the transparency and the thermal evaporation is imperative due to its low solubility.

The PEDOT:PSS is one of the most widely used HIL polymers for PLEDs and often for SMOLEDs. The aqueous dispersion of PEDOT:PSS forms a uniform, flat, conductive and transparent thin film, enhances the adhesion to the organic layer, and possesses an appropriate HOMO level (5.1 eV). Therefore, PEDOT:PSS reduces the turn-on voltage and prolongs the device lifetime.

3.3.2.4 Hole Transport Layer (HTL)

HTL consists of a material with high hole mobility to efficiently deliver charge carriers towards the active layer. The other important properties of a HTL material are: an appropriate HOMO energy level to ensure a low potential barrier for hole injection from the anode into the emissive layer and a suitable LUMO level to block electron injection from the EML to the HTL (electron blocking role); easily oxidized and fairly stable in the one-electron oxidized form; high T_g to prevent recrystallization during device operation; good adhesion to the anode and able to smooth the anode surface; high solvents resistance in case of solution processed devices and for phosphorescent OLEDs, E_T higher than those of host material of the EML to confine triplet exciton.

Materials containing electron-donating moieties and aromatic amine or carbazole based compounds with various core structures have been used in HTL. Typical examples are N,N′-bis(3-methylphenyl)-N,N′-diphenylbenzidine (TPD) [44] and N,N′-di-[(1-naphthyl)-N,N′-diphenyl]-1,10-biphenyl)-4,40-diamine (NPB) [45] (see Fig. 3.17). They have good hole mobility (~10^{-4} cm^2/Vs), high transparency to visible light and well matched HOMO level (5.4 eV) for hole injection. However, some of these molecules suffer for long-term morphological instability due to their low T_g (65 °C and 95 °C, for TPD and NPB respectively). Several approaches have been used to overcome this problem and successful strategies are the blending of TPD with a polymer like poly(methyl methacrylate) (PMMA) [46] or, alternatively, the use of spiro or star-shaped molecules [47]. Especially, this last strategy increases the T_g and imparts good solubility and good film forming properties to the molecules fruitful for solution processed OLEDs.

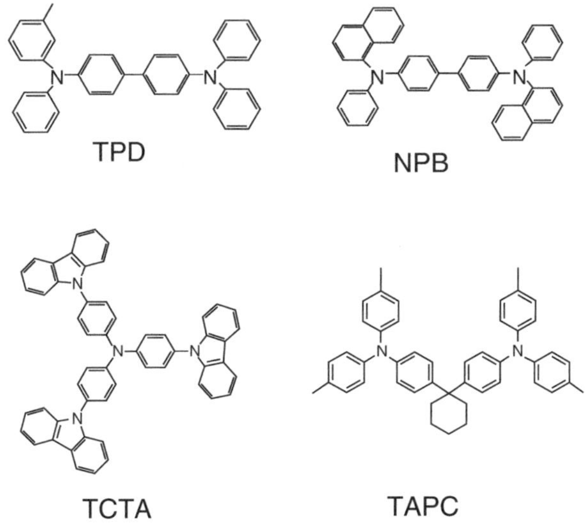

Fig. 3.17 Selected hole transport materials

Unfortunately, TPD and NPB have E_T of 2.34 eV and 2.29 eV respectively, lower than those of green or blue phosphorescent emitters. Therefore, back ET from the emitter's triplet to the HTL materials is observed, which reduces the efficiency of PHOLEDs (Sect. 3.2.2.2). HTL materials with high E_T such as 4,4′,4″-tris (N-carbazolyl)triphenylamine (TCTA) and 1,1-bis[(di-4-tolylamino)-phenyl] cyclohexane (TAPC) (Fig. 3.17) have been preferred in PHOLED devices [48, 49].

Many research groups are still looking for alternative materials to provide thermally and morphologically stable amorphous films. Recently, Lee and his co-workers developed a series of tetraphenylmethyl-based HTL materials, including 4,4′-(diphenyl-methylene)bis(N,N-diphenylaniline) (TCBPA), and by using TCBPA as HTL in a blue PHOLED based on bis((3,5-difluorophenyl)- pyridine) Iridium picolinate, they obtained EQE exceeding 21 % [50].

The most commonly used hole-transporting polymer is poly(vinylcarbazole) (PVK), (see Fig. 3.17) that, after thermal annealing, becomes less soluble thus allowing the fabrication of multilayered solution processed OLEDs [51].

Similarly, thermo- or photo-crosslinkable macromolecular materials for HTLs provide excellent solvent resistance [52, 53].

3.3.2.5 Emissive Layer (EML)

General requirements for OLED emitters are: high η_{PL} in the solid state, good charge carrier mobility (both n and p type), good film forming properties, thermal and oxidative stability, suitable HOMO/LUMO position for easy injection of hole and electron, high T_g and precise CIE coordinates of the emission.

Several classifications are possible in the context of the broad class of active materials for OLEDs. As previously mentioned, one of the most common classification is to distinguish between small molecules for thermal evaporated SMOLEDs and polymeric/high MW materials for solution processed PLEDs.

MW however is only one of the key differences between these two classes of materials. For SMOLED, good thermal stability, high T_g and appropriate temperature of sublimation of the compounds are crucial. In fact thermal isomerization or decomposition during the deposition process may occur [54, 55]. On the other side, excellent solubility and film forming abilities of the constituting materials are imperative for solution processed PLEDs.

In general, a large molecule, such as a polymer, allows for more flexibility in chemical design, but even if the low-cost of solution techniques made PLED appealing, they have not reached expectations.

This classification is nowadays outdated since a new class of material, namely solution processable small molecules or glassy materials has recently emerged.

These materials possess some characteristic typical of the small molecules (well defined chemical structure, precise MW and very high purity), combined to some peculiar properties of polymeric materials (high T_g, good solubility and good film forming properties). The development of OLEDs based on solution processable small molecules would open up to commercialization of printed OLEDs for displays and lighting technology.

For sake of simplicity, in the following we will present different classes of emissive materials according to their emission mechanisms: fluorescence, phosphorescence or TADF (Fig. 3.18).

Fluorescent Materials They consist of a very large class of materials (Fig. 3.18). Among the fluorescent metal complexes those based on Aluminum (Al) are certainly the most important and studied. Alq_3 (Fig. 3.19) is widely used as green ETL in vacuum growth SMOLEDs [5]. Since the film forming ability of Alq_3 by solution process is poor, its derivatives have been developed, mainly by modification of the 8-hydroxyquinoline ligand. As an example, Park et al. synthesized a quinoline derivative, tris(4-tridecyl-8-quinolinolato) aluminium (TDALQ), which is soluble in organic solvents [56]. Alternatively, Chen et al. [57] reported Al(III) 8-hydroxyquinolates with N-ethylanilinesulfonamide substituents, or $Al(Saq)_3$ [tris(5-*N*-ethylanilinesulfonamide-8-quinolato-N1,O8)-aluminum] (Fig. 3.19).

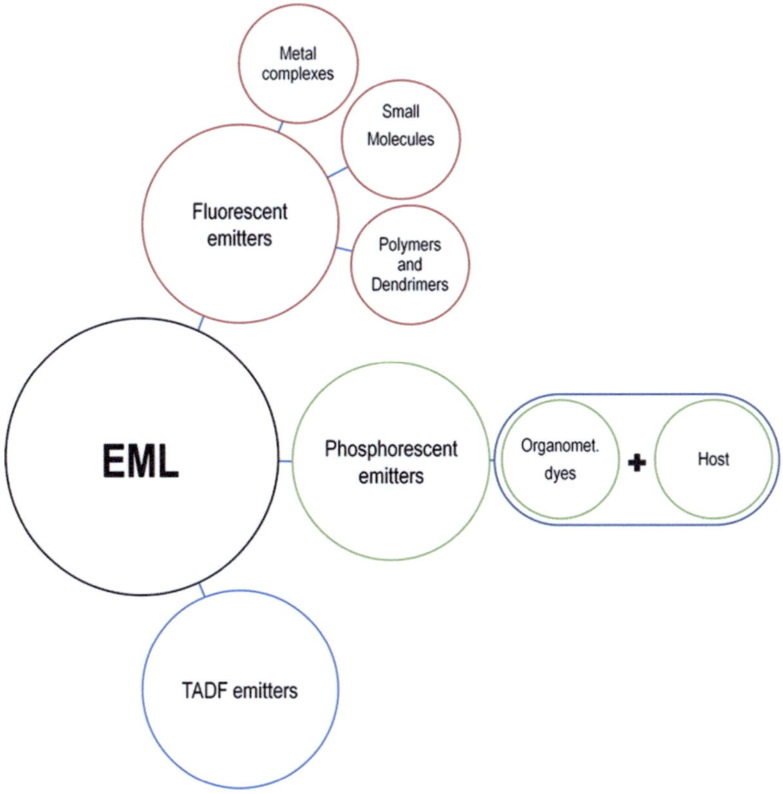

Fig. 3.18 Schematic classification of common classes emitting materials according to their emission mechanisms

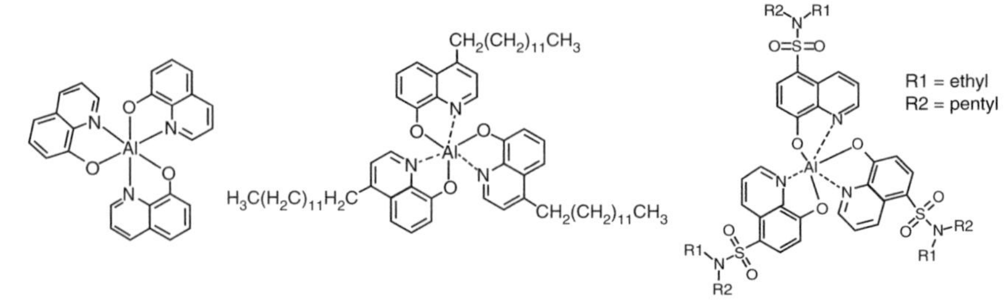

Alq$_3$ **TDAlQ** **Al(Saq)$_3$**

Fig. 3.19 Selected aluminium fluorescent metal complexes

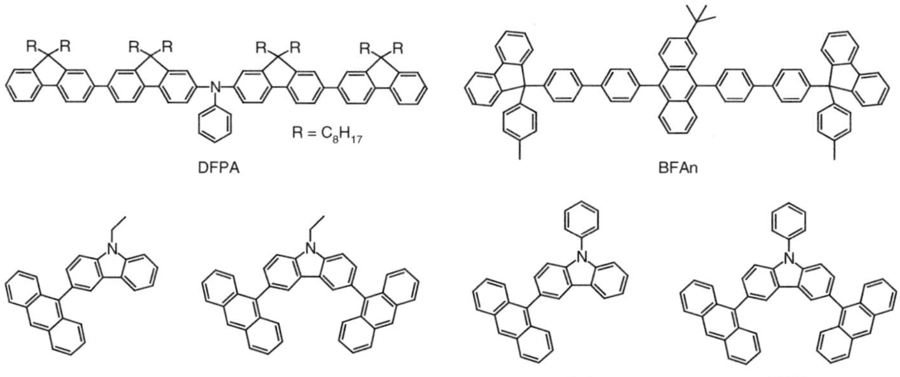

Ac DAc P-Ac P-DAc

Fig. 3.20 Selected blue emitting fluorescent small molecules

Electroluminescent small organic molecules are another important class of materials developed for OLEDs EMLs. A schematic classification of this kind of all organic materials is very difficult. In fact it is possible to use an incredible number of monomers connected in different positions to synthesize materials with an infinite colour tunability, efficiency and lifetime.

Some of the most common monomers are simple aromatic rings like phenylene, thiophene or fused ring as antracene, fluorene, heteroaromatic rings as carbazole, oxidiazole and benzothiadiazole. To increase the solubility and favour good film forming ability, the most common strategies are the introduction of alkyl chains, the use of spiro or star shaped compounds, or the insertion of lateral substituents with increased steric demand [58].

Blue light-emitting small molecules (Fig. 3.20) are the most widely investigated and are mainly based on different combination of fluorene, carbazole and triphenylamine (TPA). Blue fluorescent fluorene containing arylamine oligomer, bis(9,9,9′,9′-tetra-n-octyl-2,2′-difluorenyl-7-yl)- phenylamine (DFPA) has been reported by Kido et al. [59]. Highly efficient deep-blue emitting OLED based on

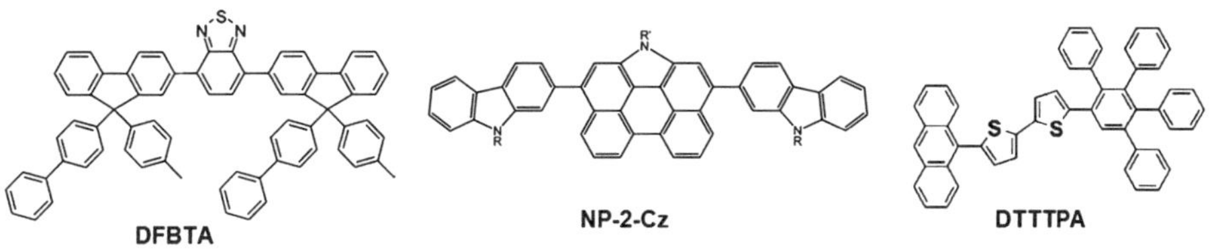

Fig. 3.21 Selected green emitting fluorescent small molecules

the multifunctional fluorescent molecules, comprising covalently bonded carbazole and anthracene moieties as 3-(anthracen-9-yl)-9-ethyl-9H-carbazole) (Ac), 3,6-di (anthracen-9-yl)-9-ethyl-9H-carbazole (DAc), 3-(anthracen-9-yl-)-9-phenyl-9H-carbazole (P-Ac), and 3,6-di(anthracen-9-yl)-9-phenyl-9Hcarbazole(P-DAc) have been developed by Kim et al. [60]. OLEDs based on an anthracene core end-capped with 9-phenyl-9-fluorenyl group 2-tert-butyl-9,10-bis[4′-(9-p-tolyl-fluoren-9-yl) biphenyl-4-yl]anthracene (BFAn) as emitter exhibits an excellent EQE of 5.1 % (5.6 cd/A) with (0.15, 0.12) CIE coordinates. The presence of the sterically congested fluorene groups imparts to BFAn an high thermal stability and high T_g (227 °C) [61].

Green emitting fluorescent small molecules are often based on the combination of fluorene, thiophene and benzothidiazole (BT) moieties. Highly efficient non-doped green OLED incorporating a 9,9-diarylfluorene-terminated 2,1,3-benzothiadiazole green emitter (DFBTA) (Fig. 3.21) exhibits EQE of 3.7 %, LE of 12.9 cd A^{-1} and a maximum L of 168,000 cd/m^2 [62]. OLEDs based on carbazole-substituted N-annelated perylenes, 1-(2-ethylhexyl)-3,10-bis(9-isopentyl-9H-carbazol-2-yl)-1H-phenanthro-[1,10,9,8-cdefg]carbazole (NP-2-Cz), (Fig. 3.21) gives L exceeding 62,000 cd/m^2 and EQE up to 4.2 % [63]. Asymmetrical substituted bithiophene with anthracene and phenyl derivative (DTTTPA) 5-(2′,3′,4′,5′-tetraphenyl) phenyl-5′-(9-anthracenyl)-2,2′-bithiophene (Fig. 3.21) as emitting materials shows a maximum L of 8659 cd/m^2 and LE of 4.9 cd/A [64].

Recently green emitters based on oligoquinoxaline [65] and phenothiazine with EQE up to 5.7 % have been reported [66].

For red emission, fluorescent small molecules with a push-pull planar structure containing electron acceptor and electron donor unit red fluorophore are developed. The red emitting small molecules are more prone to aggregation in the solid state, with an unwanted fluorescence quenching, due to strong intermolecular interactions induced by the extended planarization and for this reason the dispersion in a host matrix (host-guest system) is often mandatory. Nevertheless, by using twisted structures it is possible to avoid the aggregation in neat small molecules films. Chiang et al. reported a electron donor-acceptor substituted spirobifluorene (2-((2-(diphenylamino)-9,9′-spirobi[fluoren]-7-yl)methylene)malononitrile (pTSPDCV, see Fig. 3.22) suitable also for non doped red-emitting EML. In pTSPDCV, the red fluorescence is achieved by incorporating a dicyanovinyl electron acceptor and a diphenylamino electron donor, while the non planar arylamino group and the rigid spiro-annulation prevents the aggregation, and hence the PL quenching, in the solid state [67].

The naphtho[2,3-c][1, 2, 5]thiadiazole (NT) derivatives show both good carrier transporting property and high η_{PL}. Among them, NT with two 4-(2,2-diphenylvinyl)phenyl (4,9-bis-[4-(2,2-diphenyl-vinyl)-phenyl]-naphtho [2,3-c][1, 2, 5]thiadiazole (NTD, see Fig. 3.22) substituent is promising for non-doped red OLEDs. The NTD chromophore pushes the emission bathochromically because of the extended conjugation length, so red emission is obtained even without any strong donor group in the molecular structure. Using this NTD derivative, a non-doped double-layer and pure-red OLED with EQE of 2.5 % is achieved [68, 69]. Yang et al. reported an interesting push-pull structure based on TPA as the electron donor and BT as the electron acceptor, with the vinylene bond as the π-bridge connecting the TPA and BT units, (TPA-BT, 4-(2-(7-(4-(diphenylamino)phenyl)benzo[c][1, 2, 5]thiadiazol-4-yl)vinyl)-N,N-diphenylaniline, see Fig. 3.22) [70].

Red fluorophore based on dithienylbenzothiadiazole and maleimide have been reported [71, 72].

Polymeric materials are even more complex than molecules. In fact in a polymer it is possible to tune the optoelectronic properties by modifying the π backbone, by modifying the lateral substituents and by changing the end capping groups (Fig. 3.23) [73].

The most common examples of homopolymers with π-conjugated backbones are: PPV [74], its derivative Poly[2-methoxy-5-(2-ethylhexyloxy)-1,4-

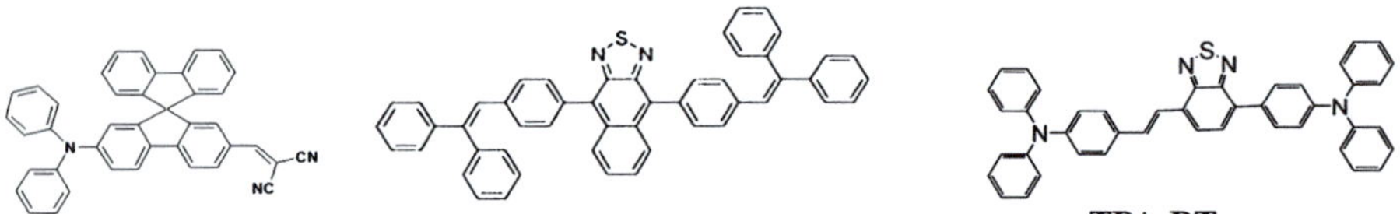

pTSPDCV NTD TPA-BT

Fig. 3.22 Selected red emitting fluorescent small molecules

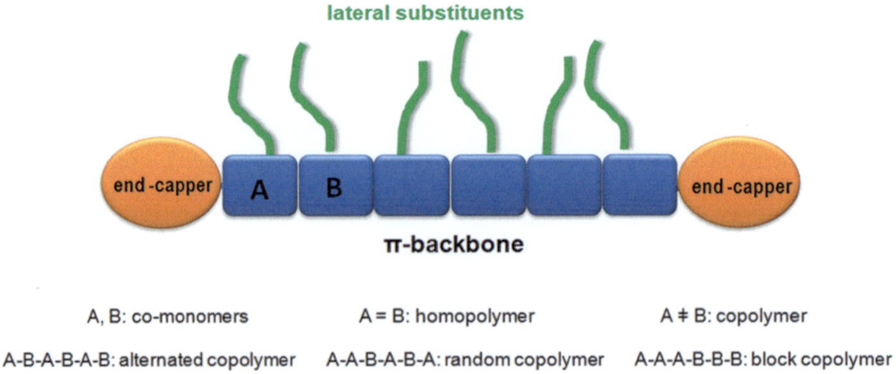

Fig. 3.23 Schematic representation of polymeric modulability via synthetic modifications

phenylenevinylene] (MEHPPV) [75], Polyfluorene (PF) [76–79], Poly-p-phenylene (PP) [80] and Polyalkylthiophene (PAT) [81] (gathered in Fig. 3.24).

The emission colour of polymeric materials can be tuned by modifying the chemical structure of the backbone, by using different co-monomers in alternating or random copolymers [82], or through the insertion of suitable lateral substituents [83].

PF and its derivatives have emerged as the dominant class of polymers for commercial applications. The possibility to tune the emission colour across the entire visible spectrum by appropriate chemical modification, the position of the energy levels suitable to match $\phi_{electrode}$, and the good charge-transport properties, are their main advantages [84]. For this reason, we have chosen some examples of this class of polymers.

By copolymerizing fluorene based monomers with different co-monomers it's possible to develop alternating copolymers with different emission colours. A deep blue emitting material incorporating perfluorinated phenyl group as co-monomer

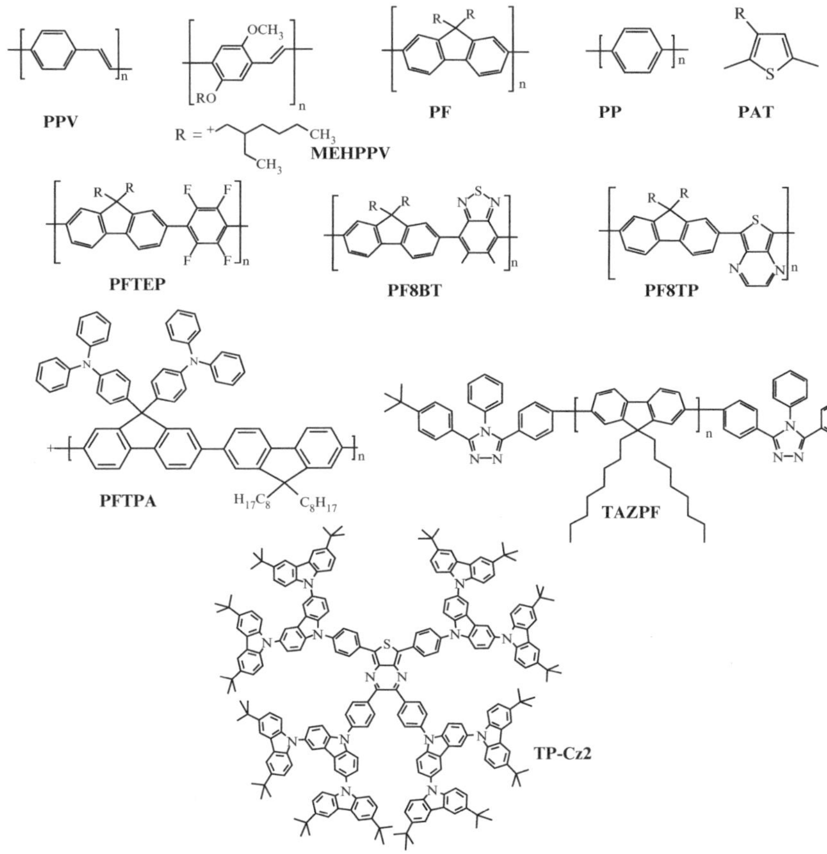

Fig. 3.24 Representative examples of emitting polymers and dendrimers

(PFTFP, Fig. 3.24) allowed the fabrication of OLEDs with a remarkable EQE of 5.03 % and CIE (0.17; 0.06) which fulfils the requirements for both full-colour displays and biomedical applications [85]. The green emitting polymer obtained by using BT as co-monomer (PF8BT, Fig. 3.24) [86] with device's EQE of 7.1 %, is one of the most attractive materials [87]. The red emission is achieved by using thiophene-BT monomers or red emitting thieno[3,4-b]-pyrazine (Fig. 3.24) [88].

Besides these alternating copolymers, the use of random copolymers is another very effective strategy to achieve high η_{PL} and EQE [89].

The alkylic lateral chains introduced to increase the solubility of the system are often functionalized with active groups able to modulate the spectral position of the emission, to increase the stability of the polymer and to improve hole/electron injection, A representative example is the case of TPA as lateral substituent in PF derivative (PFTPA, Fig. 3.24) [90]. Farther, a proper use of end-capping groups modifies the properties of the materials improving charge balancing, as in triazole-end-capped PF (TazPF) (Fig. 3.24) [91] or changing the emission colour [92].

Rod-coil triblock copolymers based on PF rod [93] and rod-rod all-conjugated diblock copolymers (PF-PF8BT) based on PF and PF8BT rods [94] show superior device performance with respect to the PF homopolymer.

In addition to linear polymers, dendrimers are another type of emitting fluorescent materials attractive for solution processable OLEDs [95]. The emissive core of dendritic molecules is surrounded by branched dendrons, which prevents self-aggregation and concentration-quenching in the solid state [96]. Here is reported as example a soluble thieno-[3,4-b]-pyrazines cored dendrimer TPCz2 with carbazole dendrons (Fig. 3.24) [97].

Phosporescent Materials Phosphorescent materials used in OLED technology are based mainly on complexes of transition metals or lanthanides. The first ones emit from a mix of metal and ligand states, display generally broad emissions whose spectral position can be varied by changing the organic ligand. The second ones show emission based on the electronic transition of the metals, therefore displaying sharp lines characteristic of the metal. The most efficient systems are those containing a heavy-metal. These complexes are highly emissive at room temperature and most of them possess radiative lifetimes in the range of microseconds, which are advantageously shorter than the typical lifetimes (ms-s) of pure organic phosphorescent materials. Electrophosphorescent complexes of heavy metal ions, particularly Ir(III) and Platinum(II) have been demonstrated with IQE approaching 100 %. In the following some examples of this class of materials are reported.

The archetypal structural unit of a transition metal phosphorescent complex comprises an aryl heterocycle, frequently a 2-phenylpyridine unit, that is bound to the metal through the heteroatom of the heterocycle and through a metallated carbon atom *ortho* to the interannular bond. The most well-known example is tris [2-phenylpyridinato-C^2,*N*]iridium(III) [$Ir(ppy)_3$] (Fig. 3.26) where three phenylpyridine ligands are bound to Ir(III) and oriented in a facial arrangement, with each pyridyl group *trans* to a phenyl group [98]. This complex emits green light ($\lambda_{max} = 510$ nm) with excellent $\eta_{PL} = 0.97$. By changing the cyclometalated ligands and/or ancillary ligands, the emission colors of these complexes can be tuned over the visible spectrum. In general blue-shift can be achieved by introducing electron-withdrawing substituents into the metallated aryl ring, and red-shift by introducing electron-donating substituents into the metallated aryl ring.

The most studied phosphorescent blue emitters are FIrpic [99] and Ir(III)bis (4,6-difluorophenylpyridinato) tetrakis(1-pyrazolyl)-borate [FIr_6] [100]. However, they struggled to find a commercial applications due to their poor colour purity, short lifetime and weak solubility. The tuning of the emission towards the deep blue region is still challenging and the introduction of electron-withdrawing groups as fluorine, trifluoromethyl, and cyanide, or replacing phenyl rings with electron acceptor substituents [101–103] is the most current viable strategy as in the case of tris[2-(3′,4′,5′,6′-tetrafluorophenyl)pryidinato-$N,C^{2'}$]iridium(III)) [$Ir(F_4ppy)_3$] containing fluorine-substituted and phenylpyridine ligands (Fig. 3.25).

Compared to Ir analogues, there are relatively few reports on Pt complexes with blue phosphorescence. Recently, Pt(II) complexes with tetradentate ligands have shown a high η_{PL} and the compound 3-methyl-1-(3-(9-(pyridin-2-yl)-9H-carba-zol-2-yloxy)phenyl)-1H-imidazolyl platinum(II) [Pt1] (Fig. 3.25) shows a device's EQE of 23.7 % with CIE coordinates of (0.14, 0.15) [104].

$Ir(ppy)_3$-based complexes (Fig. 3.26) are the most common phosphorescent materials for green emitting PHOLED reaching an excellent device's EQE of 21.6 %, PE of 82.0 lm/W [105]. Nonetheless though widely used in vacuum evaporated PHOLEDs, $Ir(ppy)_3$ has a low solubility in common organic solvents (toluene, THF, dichloromethane, chloroform, and chlorobenzene). This low solubility is insufficient for solution deposition of $Ir(ppy)_3$-based EML with the

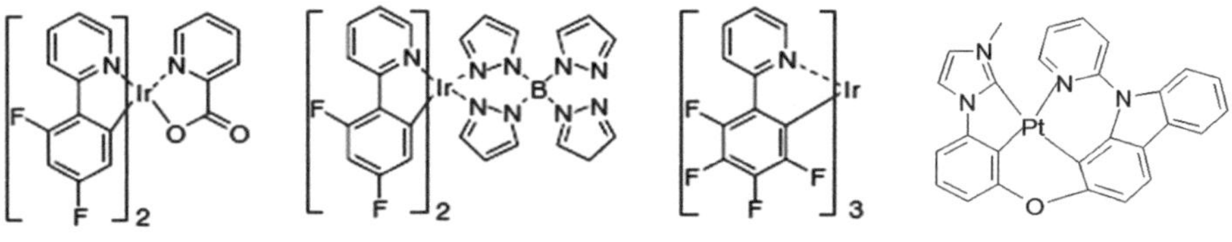

FIrpic **FIr_6** **$Ir(F_4ppy)_3$** **Pt1**

Fig. 3.25 Selected blue phosphorescent dyes

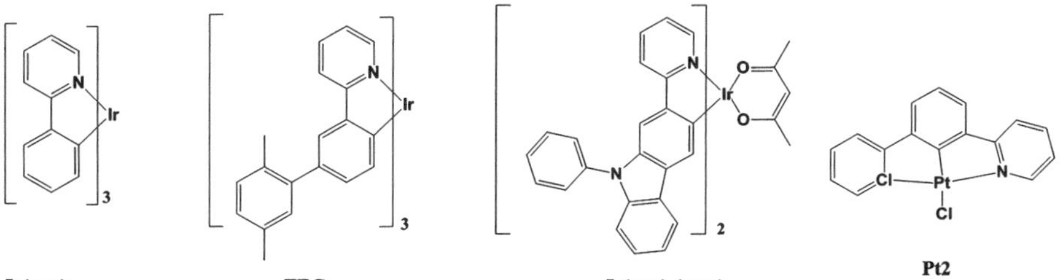

$Ir(ppy)_3$ **TEG** **$Ir(ppy)_2(acac)$** **Pt2**

Fig. 3.26 Selected green phosphorescent dyes

optimum doping levels, but emitters with good solubility in common organic solvents such as tris(2-(3-p-xylyl)phenyl)pyridine iridium(III) (TEG) have been developed [106].

Forrest and Thompson et al. [107] reported a series of cyclometalated Ir(III) complexes, $(C\text{^}N)_2Ir(acac)$ [(acac)=acetylacetonate], and a bis [2-phenylpyridinato-C^2,*N*](acetylacetonato)Ir(III) [$Ir(ppy)_2(acac)$], (Fig. 3.26) based OLED shows a high EQE of 17 % and LE of 62 lm/W [108].

In the class of Pt organometallic dyes with green emission a series comprising cyclometalated Pt(II) complexes by using the tridentate ligand 1,3-di(2-pyridyl) benzene [Pt2] (Fig. 3.26) with device's EQE of 16 % and low roll-off efficiency has been reported [109].

As red emitting dyes, in addition to Ir(III) and Pt(II) complexes, there are many organometallic dyes based on other heavy metal ions as: Os(II), Ru(II), Re(I) [110], Cu(I) [21] and Eu(III) [111].

Cyclometalated Ir(III) complexes represent one of the most studied class of red emitters and some important ligands are: 1-phenylisoquinoline derivatives as in well-known tris(1-phenylisoquinoline)Ir(III) [$Ir(piq)_3$] (Fig. 3.27) [112] with a maximum EQE of 14.8 %; 2-phenylquinoline derivative like tris (2-phenylquinoline)iridium(III) [$Ir(phq)_3$] (Fig. 3.27) [113, 114]; pyridine derivatives as in [Ir-py] (Fig. 3.27) which exhibited, in a vacuum deposited PHOLED, a L_{max} of 51,000 cd/m^2, EQE of 13.7 % and CIE coordinates of (0.60, 0.39) [115].

Cyclometalated Pt(II) complexes have also attracted considerable attention in the context of the red emitters for PHOLED [116, 117]. Highly efficient and stable PHOLEDs by applying [[3,5-bis(1,1-dimethylethyl)-phenylimino]bis [6-(2-pyridinyl-κN)-2,1-phenylene-κC]]platinum(II) [Pt3] (Fig. 3.27) with low power consumption are demonstrated with CIE coordinates (0.66, 0.34), and EQE of 19.5 % [118].

The selection of the host materials is of great importance for the preparation of efficient PHOLEDs both vacuum and solution processed, to avoid aggregation quenching mechanisms discussed in Sect. 3.2.2.2.

A good host material, mainly requires: ET to the phosphorescent dopant/s; E_T higher than that of triplet emitters; bipolar charge transport properties; HOMO and LUMO energy levels suitable for hole injection and electron injection respectively; thermal, chemical and electrochemical stability; high T_g and stable film

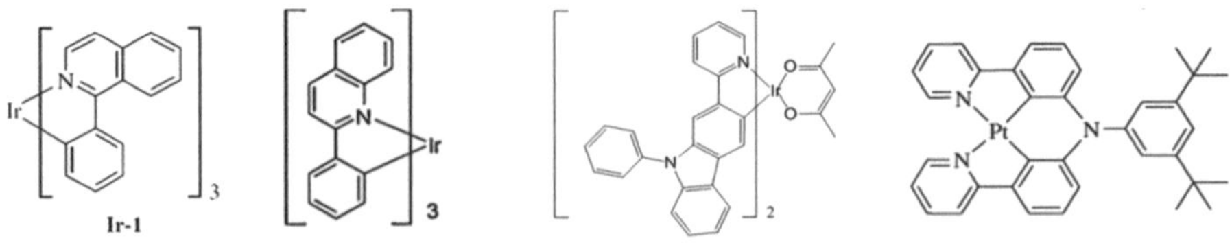

$Ir(piq)_3$ **$Ir(phq)_3$** **Ir-py** **Pt3**

Fig. 3.27 Selected red phosphorescent dyes

morphology; for solution processed EML, both a good solubility in standard organic solvents and miscibility with the dopant/s are mandatory.

Typical host materials can be summarized as molecular host (with hole transporting, electron transporting and bipolar transporting characteristics) and polymeric host.

The use of small molecules host is advantageous in terms of E_T and material purity, but they have the problem of the poor morphological stability due to low T_g.

Carbazole is the most widely used moiety for small molecules hole transporting hosts. Typical carbazole derivatives such as CBP and mCP (Fig. 3.28) [119] with high E_T (2.6 eV and 2.9 eV respectively), are processed both by vacuum thermal evaporation and wet methods. Thought they are affected by poor morphological stability (low T_g around 60 °C) and spiro or star-shaped moieties, as in the case of CBF and TCB, are introduced to increase the T_g above 160 °C [120].

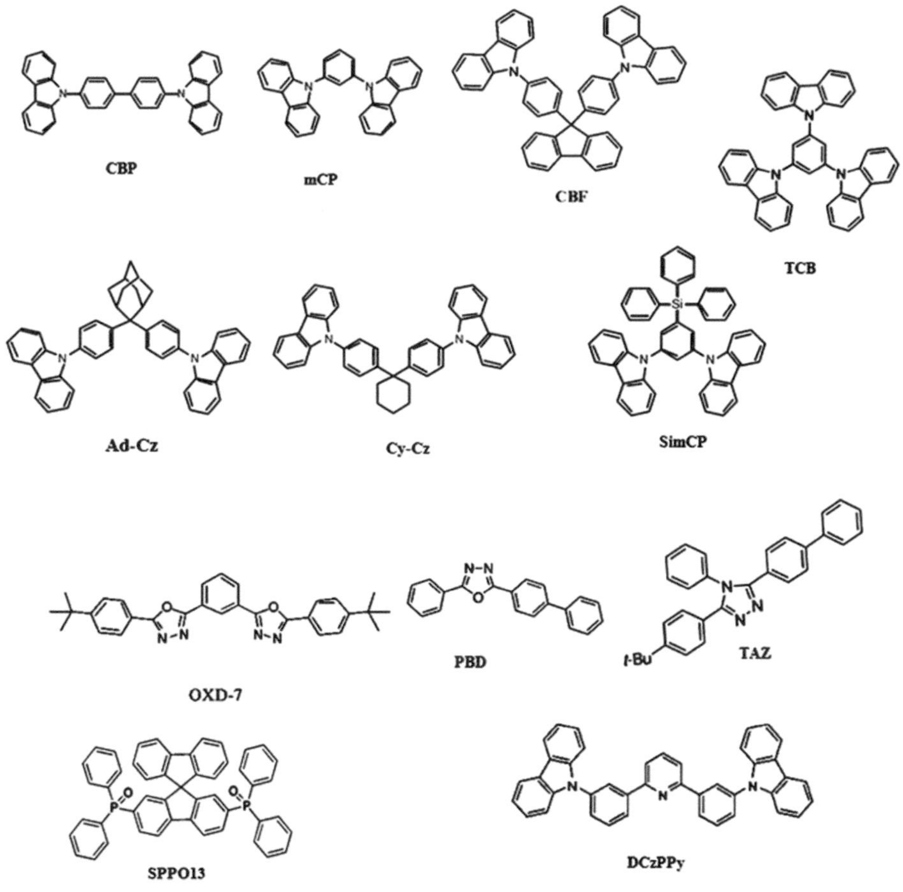

Fig. 3.28 Selection of molecular host with hole transporting characteristics (*top*), electron transporting characteristics (*middle*) and bipolar transporting characteristics (*bottom*)

The insertion of groups able to interrupt the conjugation is a different strategy to obtain soluble host materials with morphological stability and high E_T. For example, the E_T of biphenyl is close to 2.8 eV and to achieve the higher E_T necessary for efficient blue PHOLEDs, more than two phenyl rings directly linked in *para* positions should be avoided.

The interruption of the conjugation could be achieved by inserting cyclohexanyl or adamantyl groups between two phenyl rings of CBP, as in Cy-Cz and Ad-Cz. (Fig. 3.28), and green $Ir(ppy)_3$-doped PHOLEDs with EQE of 11.0 % was reported by Watanabe et al. [121].

The use of Si atoms inserted between conjugated moieties is another approach to interrupt the conjugation. Wang et al. [122] introduced triphenylsilyl moiety into mCP to obtain 3,5-di(*N*-carbazolyl)tetraphenylsilane (SimCP) (Fig. 3.28).

These strategies are also effective in the development of electron transporting hosts. Triazole derivative 3-(4-Biphenylyl)-4-phenyl-5-tert-butylphenyl-1,2,4-triazole (TAZ) (Fig. 3.28) [123], oxidiazole based materials as 2,2′-(1,3-Phenylene)bis[5-(4-tert-butylphenyl)-1,3,4-oxadiazole] (OXD-7) [124] and 2-(4-Biphenyl)-5-(4-tert-butylphenyl)-1,3,4-oxadiazole (PBD) (Fig. 3.28) [125] are popular moieties for electron transporting host [126].

Phosphine oxide (PO) derivatives have been also developed as an electron transporting host in particular for blue-emitting PHOLEDs [127]. In this case the conjugation of the molecules is interrupted by the insertion of P=O moieties and, host materials with high E_T based on fluorene, spirobifluorene (SPPO13) (Fig. 3.28) [128], dibenzothiophene and dibenzofuran are reported.

However hosts with bipolar charge transport properties seem to be very promising to achieve high charge recombination efficiency [129]. Two possible approaches can be pursued to obtain bipolar charge transport: a single engineered molecular host having both electron and hole transporting moieties [130] or a simple blend of an electron and hole transporting materials.

Kido and his co-workers designed bipolar host materials by introducing both an electron acceptor pyridine moiety and an electron donor carbazole component, 2,6-bis(3-(carbazol-9-yl)phenyl)pyridine (DCzPPy) (Fig. 3.28) and the corresponding FIrpic-based device showed an EQE of 24.3 % [131].

The use of mixed host instead of single molecular host has the advantage that charge transport properties can be simply and finely manipulated by changing the ratio between the components. However, the mixture should be carefully chosen because of the compounds' immiscibility or demixing during the time. There are several possible combination [132] and for example a device based on a ternary host TPD:PBD:CBP embedding green phosphorescent Iridium showed an EQE of 22 % [133].

The most used polymeric host for solution processed PHOLED is PVK (Fig. 3.17) due to its good solubility, good film forming ability, high morphological stability (T_g over 200 °C depending on the molecular weight), relatively high E_T of about 2.5–3.0 eV [134] as a result of its low conjugation and good hole transporting properties. However PVK is a poor electron transport material and the unbalanced charge transport properties limit the performance of the devices. The addition of an electron transport component can solve the problem of imbalance of holes and electrons in the EML and reduces the roll-off efficiency. The most widely used bipolar polymer-small molecule host was a blend of PVK and OXD-7. High EQE of 16.0 % and LE of 30.5 cd/A have been demonstrated in Firpic-based devices [135, 136].

Materials for TADF Emission Both metal-organic complexes and organic systems display TADF properties able to strongly increase the efficiency of OLEDs. Among metal-organic complexes those containing d^{10} metals are the best performing. While the most efficient phosphorescent metal-organic complexes (Ir, Pt, Os, Ru) possess large spin-orbit coupling to allow the triplet to singlet ground state transition ($T_1 \rightarrow S_0$), d^{10} metal complexes often have much weaker spin-orbit coupling, therefore displaying lower phosphorescence efficiency, but possess a small ΔE_{ST} and a stable T_1, which are two key features of the TADF phenomenon (see Sect. 3.2.2.1). The most widely investigated d^{10} metal is Cu(I), also thanks to its low cost and toxicity [21]. Other metal-organic complexes with TADF properties are Ag(I), Au(I) and Sn(IV) complexes [20]. Research is now especially focused on increasing the complex rigidity. In fact the distortion of the excited states in these systems increases dramatically non-radiative decay thus reducing the emission efficiency.

Among organic molecules, systems based on electron donor-acceptor moieties with pronounced intra/inter-molecular CT character often possess small values of ΔE_{ST} since HOMO and LUMO are well separated, especially when the electron acceptor and electron donor group are connected by a bridge with high steric hindrance. In the following we report some examples of this latter class.

Typical molecular systems with TADF include spiro-acridine, triazine, spirobifluorene, ephthalonitrile [137–140] and diphenyl sulfone derivates [141]. EQE up to 20 % is reported for OLEDs using TADF materials as the emitters, which are approaching the best performance of OLEDs based on organic phosphorescent materials.

Adachi and co-workers synthesized a TADF emitter [139] with two cyano electron accepting units and two di*p*-tolylamino electron donating units 2′,7′-bis(di-p-tolylamino)-9,9′-spirobifluorene-2,7-dicarbonitrile (spiro-CN)) (Fig. 3.29) exhibiting yellow emission with a η_{PL} of 27 % and high EQE of 4.4 % in the device. Further improvement of the TADF device efficiency up to 19.3 % has been realized

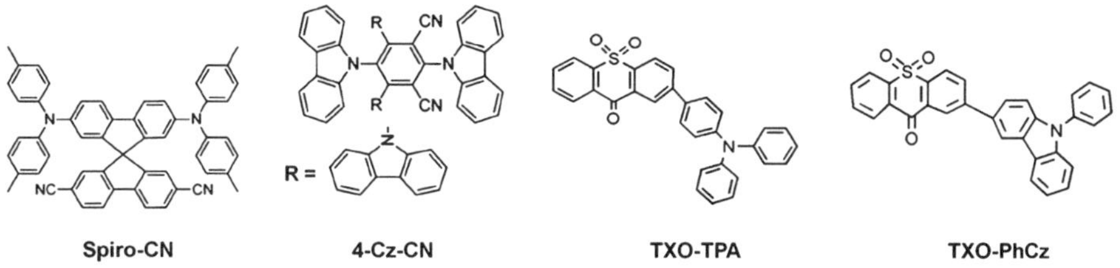

Fig. 3.29 Examples of electron donor/acceptor molecular systems for TADF emission

in 2012 [23] by employing a series of highly efficient TADF emitters based on carbazolyl dicyanobenzene, where carbazole acts as an electron donor and the dicyanobenzene as an electron acceptor, see 4-Cz-CN (Fig. 3.29). Roll-to-roll techniques have been developed for the low-cost fabrication of TADF OLEDs based on 4-Cz-CN and high EQEs (up to 19.1 %) comparable to those achieved by conventional vacuum deposition are reported [144].

Other typical molecular systems with TADF [20] are nitrogen heterocycle-based acceptors as triazine derivatives, oxadiazole and triazole derivatives, heptazine derivatives [142], diphenyl sulfoxide-based acceptors and diphenyl ketone-based acceptors [143]. TADF OLEDs based on diphenyl ketone-based acceptors as thioxanthone (TX), TXO-TPA and TXO-PhCz (Fig. 3.29), are reported. Both of them have a typical electron donor/acceptor structure with 9-H-thioxanthen-9-one-10,10-dioxide (TXO) as an electron acceptor unit and TPA/N-phenylcarbazole (N-PhCz) as an electron donor and exhibit EQE of 18.5 % and 21.5 %, respectively [145].

3.3.2.6 Electron Transport Layer (ETL)

Generally, the hole mobility is much higher (about 1000 times) than the electron mobility in organic semiconducting materials. The development of efficient electron transport materials is therefore crucial to improve the efficiency of OLEDs otherwise reduced by unbalanced charge carriers [146].

The important properties an electron transport material has to posses in order to substantially enhance OLED performance include: suitable HOMO and LUMO levels allowing minimization of the potential barrier for electron injection, reduction of turn-on/operating voltage and effective hole blocking ability; a reversible electrochemical reduction with a sufficiently high reduction potential to improve electron transport; high electron mobility to move the charge recombination zone away from the cathode and improve the exciton generation rate; high T_g and thermal stability; processability to give uniform, pinhole-free, thin films either by vacuum evaporation (low molar mass ETL) or by spin-coating, printing, and related wet techniques (soluble polymeric ETL); stable morphology.

High electron mobility can be expected if a molecule contains an electron-withdrawing group such as oxadiazole, pyridine, triazina, phenantroline, siloli and phosphine oxide [147]. Oxadiazole derivatives are the most widely investigated materials and OLED based on PBD are very common. However the T_g is very low (60 °C) and the uncontrolled crystallization during device operation takes place. The strategy to overcome the problem is the same used for HTL and involves spiro, star-shaped or dendritic molecules.

PO-based materials is a class of promising ETL. The PO functional group comprises an electron donating phosphorus atom double bonded to an electron withdrawing oxygen atom. It is easily connected to a core structure because phosphorus can make five covalent bonds. The high electronegativity of oxygen makes the PO group highly polar and electron withdrawing.

One of the most effective PO-based ETLs is diphenylphosphine oxide-4-(triphenylsilyl)phenyl (TSPO1) (Fig. 3.30) [148] with tetraphenylsilane as the high E_T core and PO as electron withdrawing units. A high EQE of 25.4 % has been obtained in deep blue-emitting PHOLEDs with TSiPO1 as the ETL due to both the triplet exciton blocking and charges confinement. Its utility has also been demonstrated in solution-processed B, R, G and white PHOLEDs [149].

Often moieties of different classes of electron transporting compounds are combined to form multifunctional optimized materials as in the case of those developed by Kido and co-workers using phenylpyridine as lateral substituent and benzene, silicon or boron as the central group. They synthesized the starburst electron transporting and hole blocking materials 1,3-bis(3,5-dipyrid-3-yl-phenyl) benzene (BmPyPB) [131], diphenyl-bis[4-(pyridin-3-yl)phenyl]silane (DPPS) [150], tris-[3-(3-pyridyl)mesityl] borane (3TPYMB) (Fig. 3.30) [151]. By using BmPyPB, Kido and his co-workers obtained PHOLEDs with an EQEs of 26 % [99].

3.3.2.7 Electron Injection Layer (Interfacial Materials) (EIL)

The role of electron injecting material is often played by the electron transporting compounds already described in the previous section due to their suitable LUMO energy level that guarantees an ohmic contact with the electrode [146].

Recently, the interfacial engineering has been identified as an essential approach for maximizing device performance in particular for OLEDs and organic photovoltaics [152]. The development of conjugated polyelectrolyte (CPE), which can be processed from water/alcohol and other polar solvents, have opened up to full solution processable multi-layer organic electronic devices. By using a thin layer of CPE as the cathode interfacial material, the resulting POLED and polymer solar cells exhibit significant enhancements in their device performances due to an interfacial dipole formation that tunes the electron injection potential barrier [153].

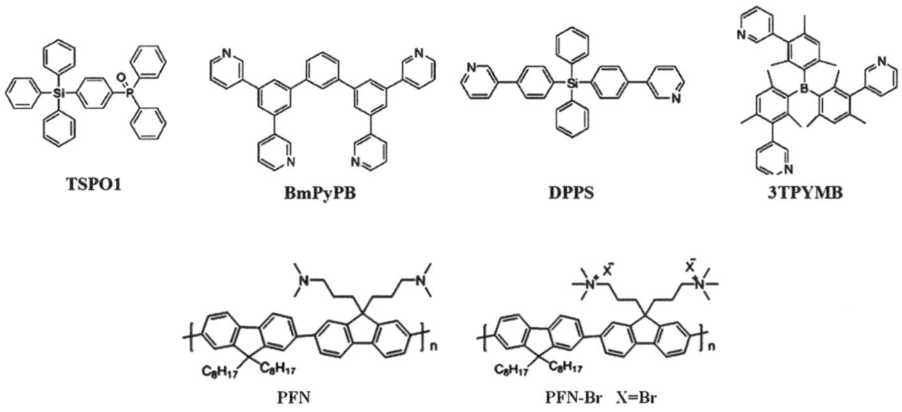

Fig. 3.30 Examples of electron transport small molecules and electron injection polymers

The most common polymers are based on PF backbone with polar or ionic groups at the end of the lateral alkylic chains as the amino-functionalized CPE poly[9,9-bis-(3′-(N,N-dimethylamino)propyl)-2,7-fluorene)-alt-2,7-(9,9-dioctylfluorene)] (PFN), (Fig. 3.30) and its quaternizated derivative poly[(9,9-bis(3′-((N,N-dimethyl)-N-ethylammonium)-propyl)-2,7-fluorene)-alt-2,7-(9,9-dioctylfluorene)] dibromide (PFN-Br) (Fig. 3.30) [154].

3.3.2.8 Cathode

The efficiency of OLEDs is significantly influenced by the performance of the electron-injecting contact (cathode). Lowering the energetic barrier between the contact and the LUMO energy level of the adjacent ETL should enhance the carrier injection, lowering the turn on voltage, thus improving the overall OLEDs performance and reducing energy consumption. This is mostly due to the fact that in many devices holes seem to be the majority carriers and the balanced charge carrier injection is of crucial importance.

In metallic cathodes the electron-injecting contact should approach ohmic behaviour (which has already been achieved for hole injection from ITO into various HTLs/HILs) and metals with very low ϕ such as Barium or Calcium are employed.

However, due to their chemical reactivity, low ϕ metals are highly sensitive against moisture and oxygen which implies firstly a capping with Aluminium or other less reactive metals, and then the necessity for hermetic OLED packaging. In Table 3.2 the ϕ of some typical cathode materials used in OLEDs are reported. The cathode materials are generally deposited by thermo-evaporation in high vacuum, but lot of research is devoted to the development of printable electrodes in view of full wet fabrication processing.

Recently, the use of interfacial layer such as polar polymers (Sect. 3.3.2.7), thanks to the creation of an interface dipole which enhances electron injection, has permitted to simplify OLEDs cathodes by avoiding the use of high reactive materials.

The further request for transparent OLEDs has pushed researcher to develop new high transparent cathodes or fabricate devices with an inverted architecture [155].

Table 3.2 Work functions of typical cathode materials used in OLEDs

Metal	Work function, ϕ (eV)
Cs	2.1
Li	2.9
Ba	2.5
Ca	2.9
Mg	3.66
Al	4.2
Cu	4.6
Ag	4.64
Au	5.1

3.4 White Emitting OLEDs (WOLEDs)

WOLEDs are gaining increasing importance in applications like full-colour display panels, wearable intelligent electronics and eco-friendly interior lighting due to their steadily improving efficiency and other desirable features including superior white colour balance (CCT and CRI), fascinating flexibility, and wide viewing angle. Combining important "eco" features with a very thin, lightweight and bendability, WOLEDs offer significant new lighting design opportunities.

State-of-the-art WOLEDs with highest efficiency reported in the scientific literature has far exceeded beyond fluorescent tube efficiency (90 lm/W) with EQE and PE of 54.6 % and 123.4 lm/W at 1000 cd/m^2 (Fig. 3.31), with an extremely small roll-off in efficiency at high L (106.5 lm/W at 5000 cd/m^2) [30], and highlights the potentiality of WOLEDs in offering significant energy savings and environmental benefits (Table 3.3).

Phosphorescent OLED technology has dominated the last decade record efficiencies, but recently TADF has been proposed as a breakthrough mechanism to boost all-fluorescence-based WOLEDs to higher efficiencies and operational stability. Despite the excellent values, the efficiencies are still far from the theoretical limit of 249 lm/W [156], hence further enhancements in performance of WOLEDs are still possible and urgently required.

Light sources for industrial application need not only to be energy efficient but also to have excellent colour stability and a large CRI. Unfortunately, many of the WOLEDs reported to exhibit impressive EL efficiencies suffer from strong voltage

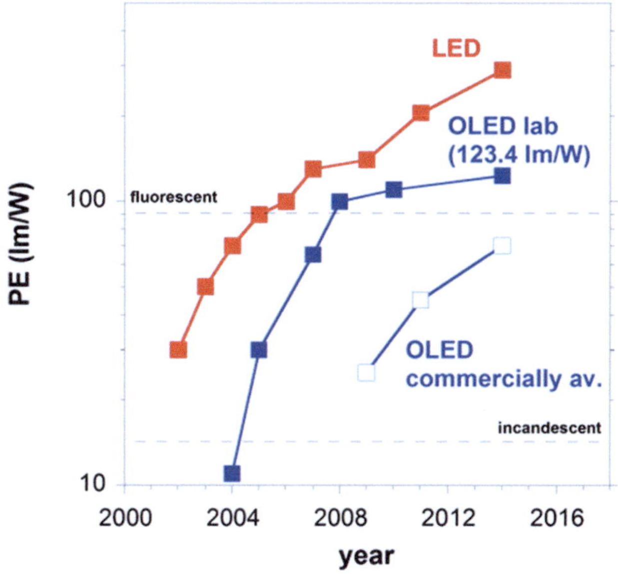

Fig. 3.31 Annual increase of OLEDs' PE with respect to conventional white light sources

Table 3.3 Overview of indoor light sources

Type of light source	PE (lm/W) at lamp level	CRI	Price $/1,000 lm at luminare level	Lifetime (hours)
Standard incandescent	10–15	100	1	1000–2000
Fluorescent (linear, compact)	40–100	>75	2–5	5000–50,000
LED	80–280 at the chip level	>92	10–125	10,000–50,000
WOLED	50–100	>87	300–200	up to 30,000

dependent colour shifts and/or low CRI values, below the requested values for high-quality white light source, thereby limiting their practical use. Conversely, some other WOLEDs with high colour stability have unsatisfactory EL efficiency. The simultaneous optimization of WOLED efficiency, colour stability and CRI remains an important target.

3.4.1 Approaches for WOLEDs

There are a large number of strategies for producing white emission, involving both emitting layer development (Fig. 3.32) and engineering of the architecture (Fig. 3.33). In this section, we will report some examples of single white emitting layer approach (Fig. 3.33a) to manufacture WOLEDs, with a special regard to the solution techniques for the EML deposition (Sect. 3.3.1.2) in view of a more sustainable manufacturing process. For a detailed discussion about engineering of device architecture to achieve white light the reader can refer to [157, 158].

Generation of white light emission in a single active layer is a well-established technology. Currently the most efficient WOLEDs produce white light through the simultaneous emission of multiple chromophores, typically from combination of the three primary colours (R,G and B), but also in some cases of the two complementary ones (blue-greenish and orange-yellow) within a single emissive layer (additive colours, Fig. 3.32).

In a multifluorophore system, the FRET mechanisms must be controlled to achieve contemporary emission from all components (Sect. 3.2.2). This is not straightforward due to aggregation and phase segregation phenomena. Moreover, the use of multiple dopants can lead to several possible operational challenges such as voltage-dependent emission and colour-aging effects due to differing electrical properties or degradation processes of each emissive dopant. Because of those reasons, the rational molecular design of new chromophores (see Sect. 3.3.2.5) led to the development of single materials emitting over the whole visible spectrum.

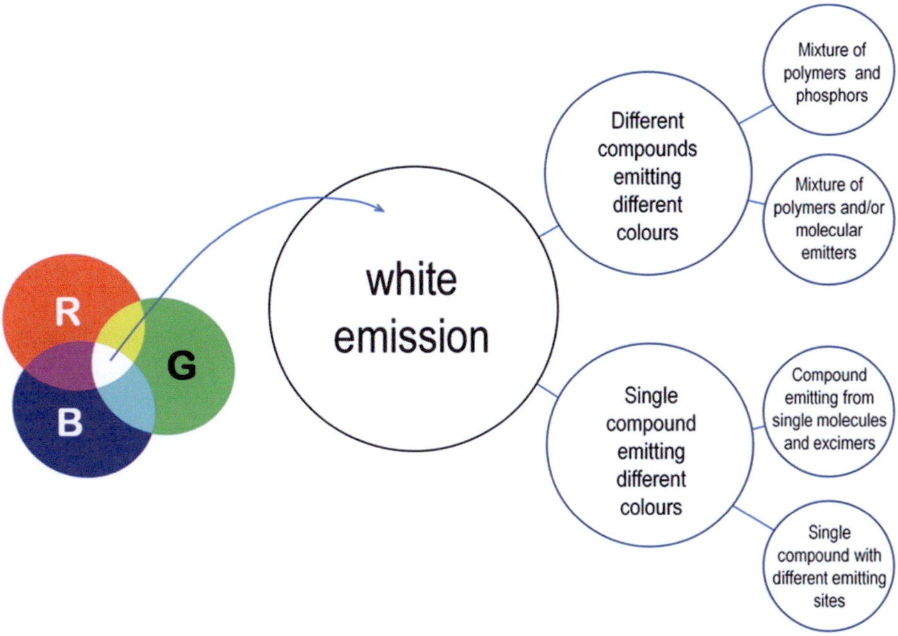

Fig. 3.32 Additive RGB primary colours and main concepts for single-emitting layer white light emission

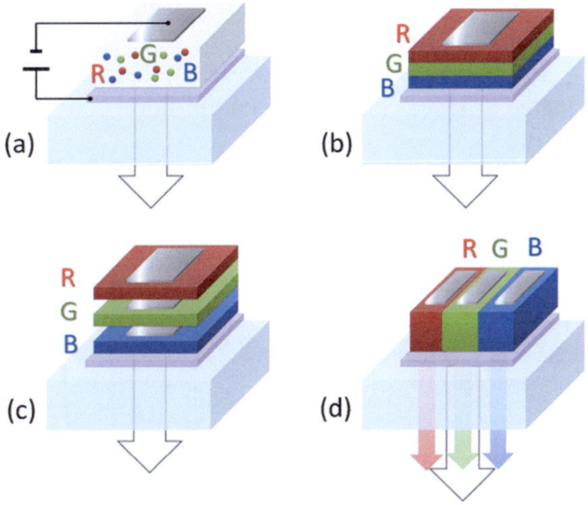

Fig. 3.33 Various approaches to generate white light from OLEDs: (**a**) single RGB multi doped emitting layer; (**b**) single device with RGB staked layers; (**c**) multiple monochromatic staked devices; (**d**) single layer with adjacent monochromatic RGB pixels

In the following paragraphs we will discuss some representative examples of single EML-based diodes (Fig. 3.33a) that fall into two main categories: Mixture of different colour emitting compounds and single multicolour emitting compound. Afterwards, some of the key elements of WOLEDs' research are discussed.

3.4.1.1 Mixture of Different Colour Emitting Compounds

The simplest way to realize WOLEDs uses blends of R, G and B polymers as EML. In this view, Bradley and co-workers [159] reported on the synthesis and characterization of novel R, G, and B light-emitting PF derivatives containing different co-monomers. By optimized R, G and B copolymer blend composition, highly efficient WOLEDs were achieved. The use of polymers which exhibit broader emissions with respect to small molecules, led to high CRI $\geq$ 90 and attractive CCT close to 4700 K. The typical WOLEDs chromatic stability upon current density increase is reported in Fig. 3.34.

A very high efficient solution-processed WOLED can be achieved also by using an appropriate polymer as host for highly efficient phosphorescent dopants (see Sect. 3.3.2.5). A large offer of transition metal organometallic complexes [mainly Ir (III) or Pt(II)] with different emission colours is available [98–109]. In this framework, a phosphor-doped WOLEDs based on a two component system was proposed by Zhang et al.: a dendrimer host and an orange Ir(III) complex [160]. The solution-processed WOLEDs demonstrated a LE, PE, and EQE of 70.6 cd/A, 47.6 lm/W, and 26 %, respectively, at a $L = 100$ cd/m^2. The achievement of such high efficiency is attributed to two main factors. First, the solution-processable carbazole-based dendritic host provides efficient hole injection because of its high-lying HOMO level and possesses a high E_T level (usually hardly achievable in conjugated polymers), which effectively reduces the quenching of triplet excitons on the blue phosphor (i.e. back ET). Second, the highly efficient orange Ir(III) complex is used as the dopant that shows good miscibility with the host matrix. In Fig. 3.35, typical J–L–V characteristics, LE-PE-L characteristics, and EL spectra for WOLEDs are reported (Fig. 3.36).

In contrast to solution processing, thermal evaporation allows for a much higher degree of layer complexity, composition control and thickness accuracy thanks to

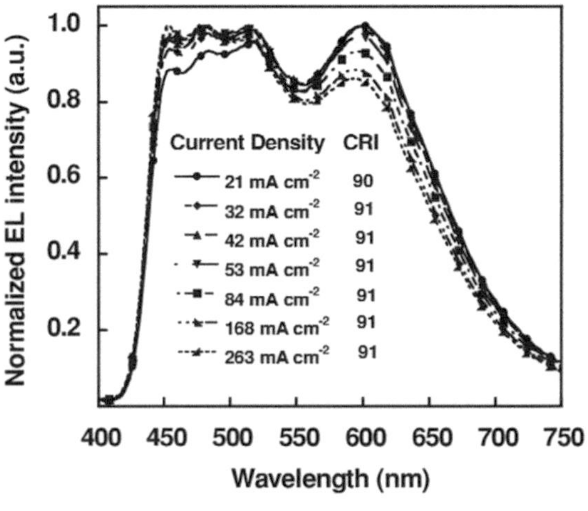

Fig. 3.34 EL spectra for ITO/PEDOT:PSS/RGB-polymers blend/EIL/CsF/Al WOLED driven at different currents (Adapted from Ref. [159] by permission from John Wiley and Sons, copyright (2013))

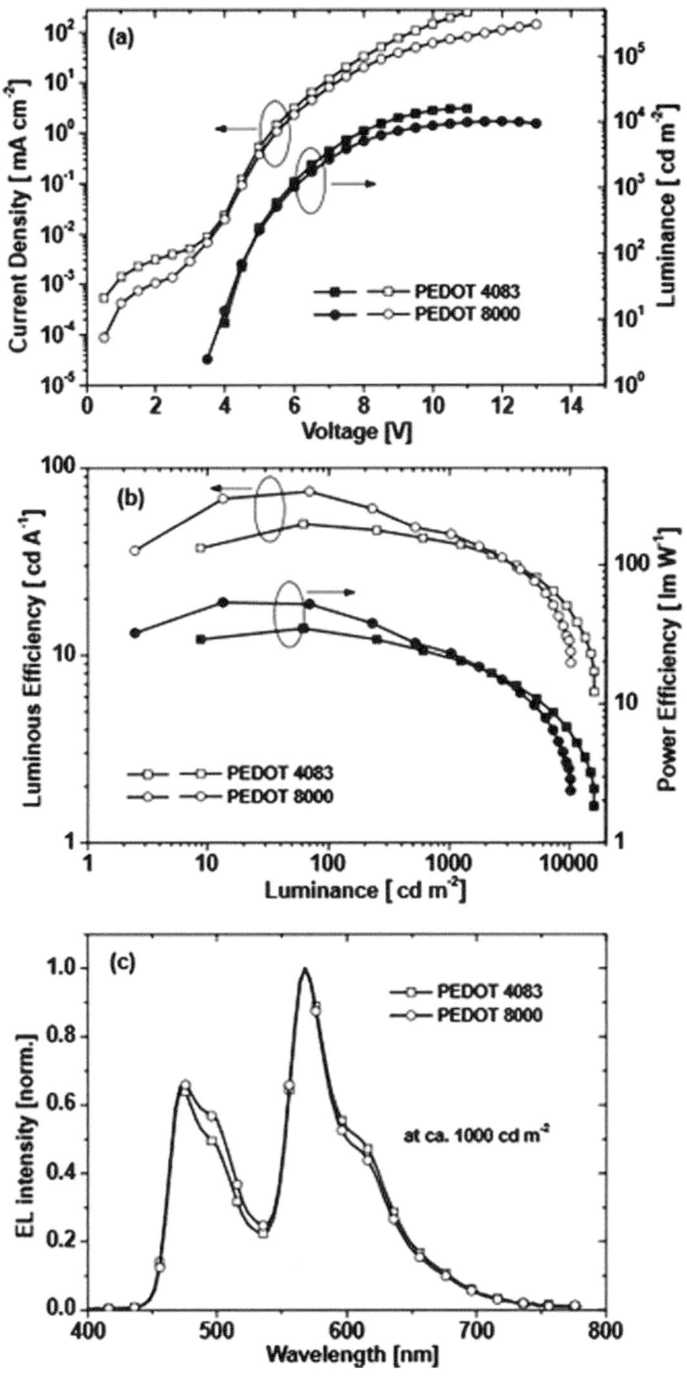

Fig. 3.35 (**a**) J–L–V characteristics, (**b**) LE-PE- L characteristics, and (**c**) EL spectra (at a luminance of 1000 cd/m^2) of the solution-processed WOLEDs based on a dendrimer host and an orange iridium (Adapted from Ref. [160] by permission from John Wiley and Sons, copyright (2012))

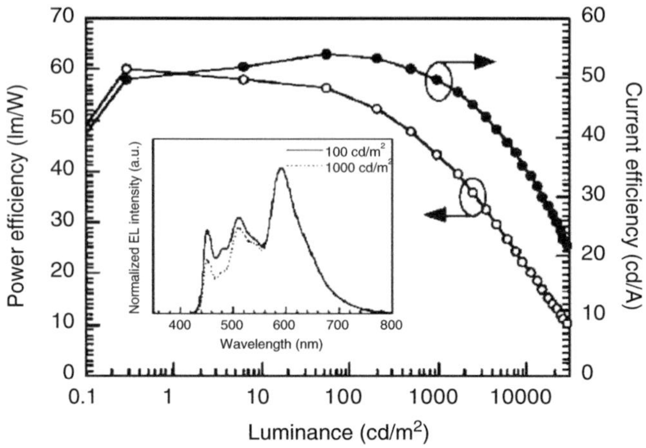

Fig. 3.36 PE-L-CE characteristics and EL spectrum for the WOLED fabricated by Kido and co-workers (Adapted from Ref. [161] by permission from John Wiley and Sons, copyright (2010))

the deep control of the deposition process (Sect. 3.3.1.1). Often the evaporated device consists of many subsequently deposited thin films, which are designed to play specific roles within the device such as hole/electron injection, transport or blocking, exciton blocking and of course emission of light. On the other hand, the preparation by thermal evaporation also allows subnanometer control of the deposited layers, opening more design freedom, which enables better device engineering and optimization. The vacuum growth approach lead to WOLED with the record PE of 44 lm/W at 1,000 cd/m^2, which is the typical brightness for lighting purposes [99].

The work carried out by Kido and co-workers is a representative example of this strategy. They developed WOLED with high performance and CRI acceptable for the illumination light source [161]. The *mer*-tris(*N*-dibenzofuranyl-*N*-'-methylimidazole)Ir(III) [Ir(dbfmi)] embedded in 3,6-bis(diphenylphosphoryl)-9-phenylcarbazole (PO9) was chosen as blue emitter, while Ir(ppy)$_3$ and iridium bis (2-phenylquinoly-N,C$^{2\prime}$)dipivaloylmethane [PQ$_2$Ir(dpm)] in CBP emitted green and orange-red light. WOLED with the elaborated multilayered structure ITO/TAPC/TCTA/PQ$_2$Ir(dpm) 2 wt.% doped CBP/Ir(ppy)$_3$ 6 wt.% doped CBP/Ir (dbfmi) 10 wt.% doped PO9/B3PyPB/LiF/Al were fabricated by a vacuum process. Excellent device performances were recorded: PE of 59.9 lm/W, LE of 49.9 cd/A and EQE of 21.6 % were kept high even at high brightness (PE of 28.7 lm/W, 40.7 cd/A, EQE 16.7 % at 5,000 cd/m^2). The angular dependence of luminous intensity was well fitted by a Lambertian distribution. The CIE coordinates and CRI changes were negligibly small from (0.40, 0.40) and 82.2 at 100 cd/m^2 to (0.43, 0.43) and 80.2 at 1,000 cd/m^2.

3.4.1.2 Single Multicolour Emitting Compounds

Single white-emitting compounds, typically polymers, have the advantages over physical blend systems of no potential phase segregation and excellent spectral stability. Since they generally consist of high MW compounds, the wet processing methods are employed for EML preparation.

With respect to the molecular design, two main concepts have been proposed: the main polymer conjugated backbone embeds all chromophores and contributes to the overall emission (Fig. 3.37a). In the second approach (Fig. 3.37b), the chromophores are grafted as lateral groups in the polymer main chain, which is often not conjugated (hence not contributing to the emission) and they can be seen as isolated molecules dispersed in a host polymer. The individual emission from the incorporated chromophores can be achieved simultaneously by a fine-tuning of their contents to manage the ET and charge trapping mechanisms (3.2.2.2). Therefore, the resulting broad EL spectrum can be tuned to achieve a specific CCT and CRI.

Following the first approach, by using two emitting fluorescent species, Tu et al. [162] reported on an efficient white light emitting polymer by admixing moieties of an orange fluorophore (1,8-naphthalimide derivative) into the blue emitting PFO main polymer (P3-0.5). The emitter concentration was kept low to achieve contemporary emission from the dye and the main chain. Efficiencies of 5.3 cd/A and 2.8 lm/W at 6 V, with CIE (0.26, 0.36), were measured. Later, the same approach was explored by different groups leading to 8.99 cd/A, 5.75 lm/W, and 3.8 % EQE even with improved colour quality [163].

Besides, phosphorescent WOLEDs were extensively studied by Wang's group. They developed a white emitting polymer by grafting a conjugated backbone with blue and yellow phosphors simultaneously [164]. FIrpic and bis[2-(9,9-diethyl-9H-fluoren-2-yl)-1-phenyl-1H-benzimidazolate-κN,κC](acetylacetonato)-iridium(III) [$(fbi)_2Ir(acac)$] were selected as the blue and yellow emitters respectively, because their combination has been applied earlier in the fabrication of highly efficient WOLEDs (Sect. 3.3.2.5). The fluorinated poly(arylene ether phosphine oxide) backbone with a high E_T and appropriate HOMO/LUMO levels, was selected as main backbone. Tuning the contents of FIrpic and $(fbi)_2Ir(acac)$ resulted in the generation of individual blue and yellow emissions to give white EL. A device

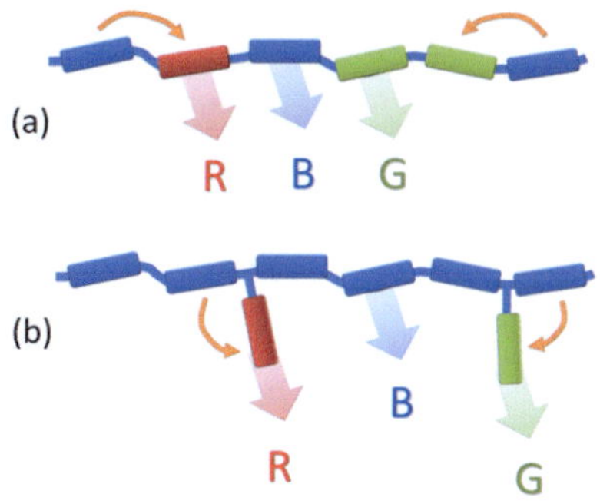

Fig. 3.37 Single white-emitting polymers main concepts; intramolecular ET is highlighted

configuration ITO/PEDOT:PSS/EML/ETL/LiF/Al was used. When a ETL layer (based on 9,9′-spirobis(fluorine)-2,7-diylbis(diphenylphosphine oxide), or SPPO13, Fig. 3.28) was used, an optimized LE of 18.4 cd/A (8.5 lm/W, 7.1 %) was obtained with CIE coordinates of (0.31, 0.43). Even at a brightness of 1000 cd/m^2 (and high J), the LE remains as high as 14.2 cd/A, indicating a weak efficiency roll-off at high J and almost unaffected CIE coordinates (Fig. 3.38).

In response to the call for a physiologically-friendly light, WOLEDs' CCT can be suitably tuned to match to the costumers demands even farther from the Planckian locus (Sect 3.2.6) typical of conventional solid-state lighting. A representative example in this view was reported by Kalinowski and co-workers [165]. Highly efficient, variable-colour OLEDs were realized via mixing of molecular exciton and excimer phosphorescent emissions (see Sect. 3.2.1) from an organic phosphor Pt(N^C^N) complex derivative. In doped TCTA blends the Pt (II) complex has been used as either the low-concentration bluish-green (molecular) phosphorescence emitter or high-concentration red (excimer) phosphorescence emitter. By adjusting the relative amount of B and R emissive species, the colour of the emission was easily and finely tuned. High purity white light emission with

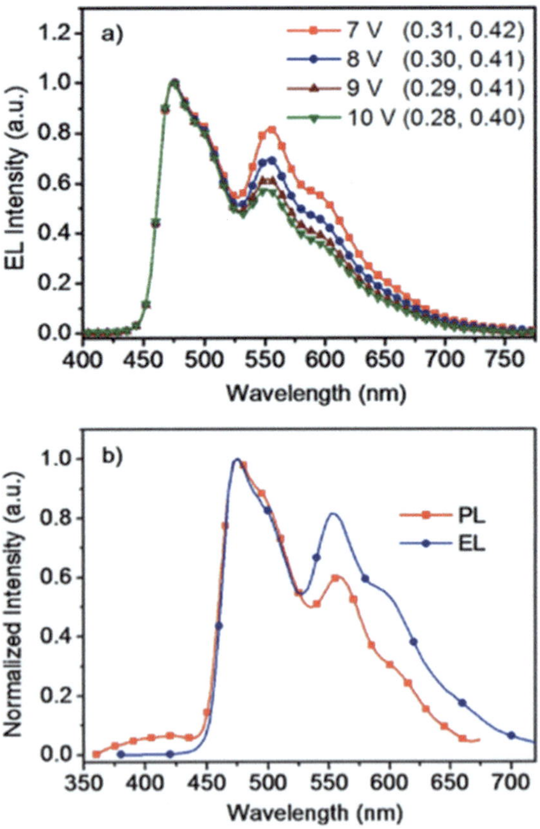

Fig. 3.38 (**a**) EL spectra at different driving voltage and (**b**) EL vs PL spectra for WOLED based on fluorinated poly(arylene ether phosphine oxide) backbone grafted with FIrpic and [(fbi)$_2$Ir(acac) (Reprinted with permission from Ref. [164]. Copyright (2012) American Chemical Society)

CIE coordinates (0.34, 0.35) was recorded, with EQE = 11.5 %, PE of 6.8 lm/W and CRI = 74, though CRI = 81 has been achieved with CIE coordinates (0.42, 0.38).

3.4.2 Key Elements of WOLEDs Research: Some Examples

Key elements of WOLEDs research have to be addressed to achieve such a high performance and maximum power savings. Among them, IQE, outcoupling efficiency operating voltage and device lifetime are discussed in the following.

IQE has already approached ≈100 % thanks to phosphorescence or TADF [166], which allows the harvesting of both singlet and triplet excitons generated by electrical injection (Sect. 3.2.2), and a wide series of phosphors and TADF molecules with various emission colours has been demonstrated (Sect. 3.3.2.5).

All-fluorescence-based WOLEDs using red, green, and blue TADF materials as emissive dopants was reported by Adachi's group [167]. The WOLEDs achieved the best reported EQE of over 17 % with CIE coordinates of (0.30, 0.38). As a representative recent example, a solution-processed multilayer phosphorescent WOLEDs [168] incorporating the commercial G-emitting $Ir(ppy)_3$ (0.2 wt.%) and R-emitting $Ir(phq)_3$ (0.7 wt.%) into a blue host-based EML is presented. The WOLED shows a high power efficiency of 34 lm/W and an EQE of 21 % (with peak at 45 lm/W and 22 %, respectively) at 100 cd/m^2 for a white emission with a CRI of 70 and CIE coordinates of (0.43, 0.43) with no perceived change in the EL spectra under varying J.

Another challenge is the EQE limited by light outcoupling to the ≈20 % in the conventional OLED architecture (Sect. 3.2.3). The development of light extraction technologies is an increasingly popular strategy to enhance the OLEDs' efficiency. Various photonic structures have been proposed to the appropriate interfaces [169], and among them, two examples are discussed: one for external (micro-lens array) and one for internal (periodic nanostructured interface) light extraction technologies.

In the first example, flexible microlens arrays using a mixture of polydimethylsiloxane prepolymer by breath figure method (a low technology content approach and cost-effective materials) [170], allowed a 34 % of EQE enhancement for OLEDs [171]. An increase of light intensity at 30–60°, hence deviating from a theoretical Lambert distribution, was also observed, a result which is certainly attractive for illumination purposes (Fig. 3.39).

In the second example, efficient WOLED architecture was reported by Tang and co-workers [172] that combines deterministic aperiodic nanostructures for broadband quasi-omnidirectional-light extraction (Fig. 3.40). The EQE and PE are raised to 54.6 % and 123.4 lm/W at 1000 cd/m^2 with an extremely small roll-off in efficiency at high L. Moreover, a superior angular colour stability over the visible wavelength range compared to conventional OLEDs is attained.

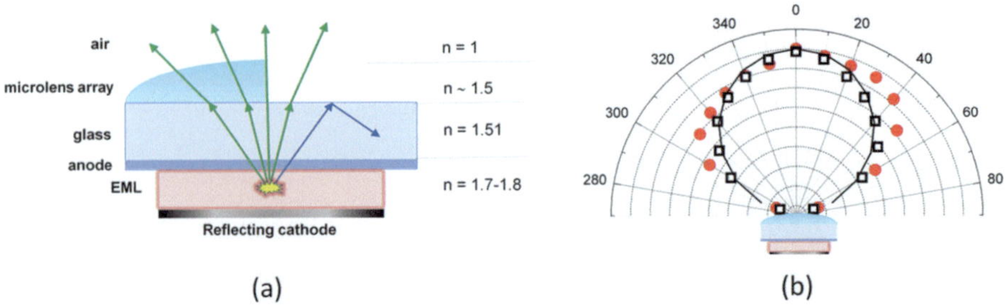

Fig. 3.39 (**a**) Cross section of an OLED with indication of different light pathways and lens effect; (**b**) Normalized angular dependence of EL intensity of a sample device without pattern (*black square*) and with microlens array (*red circle*). Theoretical Lambert distribution (*line*) is also presented. 0 denotes normal to the device

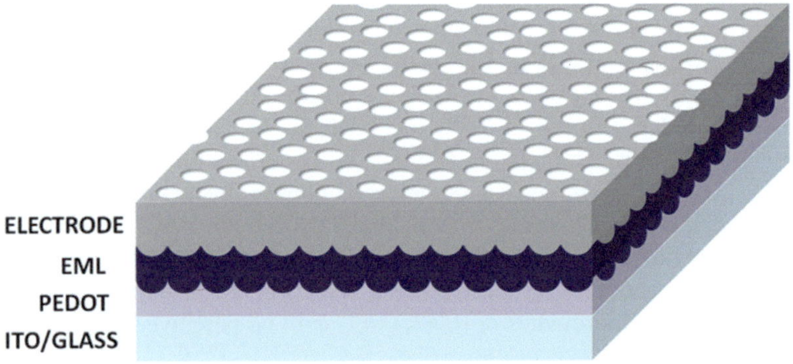

Fig. 3.40 Schematic of the WOLED in which PEDOT:PSS is patterned. The organic layer and metal electrode are subsequently deposited onto the substrate

Strategies to reduce operating voltage have also been adopted with the use of the charge regulating layers (EILs/HILs, ETLs/HTLs and EBLs/HBLs) and electrical doping for multilayer energy level matching (Sect. 3.3.2) which can remarkably reduce the energy loss during charge injection and transport processes. This is easily achievable in multilayer vacuum deposited WOLEDs but the development of cross-linkable layers or soluble in orthogonal solvents with respect to the EML (Sect. 3.3.2.7) has allowed the fabrication of multilayer solution processed devices.

The OLEDs lifetime (Sect. 3.2.5) thanks to rational development of organic semiconductors with improved chemical stability (Sect. 3.3.2.5), have reached values two to three times greater than typical LCDs, with lifetimes exceeding 50,000 h.

Besides the intrinsic emitter stability, the WOLEDs lifetime enhancement can be achieved by a wise control of the recombination zone. In fact, in conventional multilayer OLEDs with an electron transporting host, charge build-up at the HTL/EML heterojunction may limit the device stability. Hence the elimination of the abrupt HTL/EML interface by mixing the two materials would reduce the electric field across the interface and dilute the concentration of any degradation related quenching species [173].

3.5 Summary and Outlook

OLEDs technology offers many advantages, going from the manufacturing costs to the eco-compatibility and low energy consumption. OLEDs can be printed onto a substrate using traditional inkjet technology, which can significantly lower the cost than LCDs and allow for much larger displays. OLED screens provide clearer images, even in bright light, can be viewed from almost any angle up to 160°, produce good brightness and unprecedented contrast and can be operated at low voltages (3–4 V). Unlike LCDs, they can be very thin and flexible since they do not need backlights or chemical shutters to open or close. In OLEDs each pixel is a single light source providing full range of colours with high resolution, able to turn on and off as fast as any light bulb and hence with fast response time. Moreover they are tough enough to use in portable devices such as smartphones, tablets and so on. By improving the efficiency, performance and lifetime of OLED light sources, a relevant reduction of global energy consumption would be obtained. In fact, solid state lighting, which play a significant role in the reduction of energy consumption, might introduce eco-friendly and energy efficient green technology lighting systems.

Despite the many advantages of OLEDs there are still some drawbacks to overcome. The major one is related to device degradation, leading in some cases to OLED lifetimes shorter with respect to other sources. Degradation is caused by several, often concomitant, effects, including molecular crystallization, electrochemical reactions at the electrode/organic interface, migration of ionic species, cathode oxidation and delamination. The low values of lifetimes are mainly due to the fact that atmospheric environment is detrimental for the device performance and encapsulation is never perfectly hermetic.

Currently, more than 80 companies, universities and other non-industrial laboratories worldwide are engaged in order to find out the solutions for the main drawbacks of OLEDs. Moreover, certain technical barriers, such as manufacturing cost, are still challenging in order to globalize these power saving and eco-friendly light sources. The uncertainty of product performance, particularly the required lifetime to justify the investment, can negatively influence decision makers. This market barrier is mainly associated with the high cost of capital required to make large investments in the implementation of OLED technologies. There is a lot of research to be done to develop an economical and reliable mass production process and to find out new stable materials.

What is more fascinating in the development of OLEDs technology is that the ideas are endless and almost anything is possible with the expected advancements in new materials and fabrication technologies. By using OLEDs for solid state lighting, our future lighting will be colourful, prosperous and eco-friendly, with newly designed wall and furniture decorations. This new generation of light in portable and flexible displays, computer displays, mobile phones, cameras and high resolution televisions is already changing our life.

Another rapidly growing field in OLEDs technology is that of medical care, in which the use of flexible and biodegradable devices will open new perspectives [174]. Examples include carriers for controlled drug release and scaffolds for tissue engineering. The ability to integrate fully biodegradable, high-performance electronics and sensors could significantly expand the functional capabilities in medicine. Integrated biochips exploit a multi-disciplinary approach to produce portable point-of-care medical diagnostic systems that uncouple diagnosis from centralized laboratories. These portable devices are cost-effective and have several advantages including broader accessibility to health care worldwide. Intriguingly, a wearable ultraviolet or infrared OLEDs would be used as smart bandage, eventually equipped with sensors and electronics for real time monitoring of the wound environment [85].

References

1. Forrest SR (2004) Nature 428:911
2. Kamtekar KT, Monkman AP, Bryce MR (2010) Adv Mater 22:572
3. Adachi C (2014) Jpn J Appl Phys 53:060101
4. Helfrich W, Schneider WG (1965) Phys Rev Lett 14:229
5. Tang CW, VanSlyke SA (1987) Appl Phys Lett 51:913
6. Burroughes JH, Bradley DDC, Brown AR, Marks RN, Mackay K, Friend RH, Burns PL, Holmes AB (1990) Nature 347:539
7. Wang J, Zhang F, Zhang J, Tang W, Tang A, Peng H, Xu Z, Teng F, Wang Y (2013) J Photochem Photob C Photochem Rev 17:69
8. Kim M, Kyu Jeon S, Hwang S-H, Yeob Lee J (2015) Adv Mater 27:2515
9. Thejokalyani N, Dhoble SJ (2014) Renew Sust Energ Rev 32:448
10. Gather MC, Köhnen A, Meerholz K (2011) Adv Mater 23:233
11. Elliott PIP (2013) Annu Rep Prog Chem Sect A Inorg Chem 109:360
12. Pope M, Swenberg CE (1982) Electronic processes in organic crystals. Clarendon, Oxford
13. Dong H, Fu X, Liu J, Wang Z, Hu W (2013) Adv Mater 25:6158
14. Facchetti A (2011) Chem Mater 23:733
15. Kuik M, Wetzelaer G-JAH, Nicolai HT, Craciun NI, De Leeuw DM, Blom PWM (2014) Adv Mater 26:512
16. Chou P-T, Chi Y (2007) Chem Eur J 13:380
17. Kim SY, Jeong WI, Mayr C, Park YS, Kim KH, Lee JH, Moon CK, Brütting W, Kim JJ (2013) Adv Funct Mater 23:3896
18. Chiang C-J, Kimyonok A, Etherington MK, Griffiths GC, Jankus V, Turksoy F, Monkman AP (2013) Adv Funct Mater 23:739
19. Zhang Q, Li B, Huang S, Nomura H, Tanaka H, Adachi C (2014) Nat Photonics 8:326
20. Tao Y, Yuan K, Chen T, Xu P, Li H, Chen R, Zheng C, Zhang L, Huang W (2014) Adv Mater 26:7931
21. Cariati E, Lucenti E, Botta C, Giovanella U, Marinotto D, Righetto S (2015) Coord Chem Revi 306:566

22. Zhu Z-Q, Fleetham T, Turner E, Li J (2015) Adv Mater 27:2533
23. Uoyama H, Goushi K, Shizu K, Nomura H, Adachi C (2012) Nature 492:234
24. Lakowicz JR (2006) Principles of fluorescence spectroscopy, 3rd edn. Kluwer Academic/ Plenum Publishers, New York
25. Chen J, Zhao F, Ma D (2014) Mater Today 17:175
26. Sun Y, Giebink NC, Kanno H, Ma B, Thompson ME, Forrest SR (2006) Nature 440:908
27. Schwartz G, Reineke S, Conrad Rosenow T, Walzer K (2009) K Leo Adv Funct Mater 19:1319
28. Schwartz G, Reineke S, Conrad Rosenow T, Walzer K (2007) K Leo Adv Mater 19:3672
29. Rosenow TC, Furno M, Reineke S, Olthof S, Lüssem B, Leo K (2010) J Appl Phys 108:113113
30. Reineke S, Lindner F, Schwartz G, Seidler N, Walzer K, Lüssem B, Leo K (2009) Nature 459:234
31. Gu G, Garbuzov DZ, Burrows PE, Vendakesh S, Forrest SR, Thompson ME (1997) Opt Lett 22:396
32. Forrest SR, Bradley DDC, Thompson ME (2003) Adv Mater 15:1043
33. Gong X, Robinson MR, Ostrowski JC, Moses D, Bazan GC, Heeger AJ (2002) Adv Mater 14:581
34. Murawski C, Leo K, Gather MC (2013) Adv Mater 25:6801
35. Baldo MA, Adachi C, Forrest SR (2000) Phys Rev B 62:10967
36. http://www.cie.co.at
37. Gaspar DJ, Polikarpov E (2015) OLED fundamentals: materials, devices, and processing of organic light, CRC Press, Boca Raton (chapter 2, Substrates, A. Bhabdari, D.J. Gaspar)
38. Facchetti A, Marks TJ (2010) "Transparent electronics". From synthesis to applications. Wiley, Chichester
39. Yang Y, Westerweele E, Zhang C, Smith P, Heeger AJ (1995) J Appl Phys 77:694
40. Groenendaal L, Jonas F, Freitag D, Pielartzik H, Reynolds JR (2000) Adv Mater 12(7):48
41. Wu J, Agrawal M, Becerril HA, Bao Z, Liu Z, Chen Y, Peumans P (2010) ACS Nano 4(1):43
42. Kwong RC, Nugent MR, Michalski L, Ngo T, Rajan K, Tung Y-J, Weaver MS, Zhou TX, Hack M, Thompson ME, Forrest SR, Brown JJ (2002) Appl Phys Lett 81:162
43. Lamansky S, Djurovich PI, Abdel-Razzaq F, Garon S, Murphy DL, Thompson ME (2002) J Appl Phys 92:1570
44. VanSlyke SA, Chen CH, Tang CW (1996) Appl Phys Lett 69:2160
45. Adachi C, Tsutsui T, Saito S (1991) Optoelectron Devices Technol 6:25
46. Schein LB (1992) Philos Mag B 65:795
47. Salbeck J, Yu N, Bauer J, Weissortel F, Bestgen H (1997) Synth Met 91:209
48. Sun Y, Forrest SR (2007) Appl Phys Lett 91:236503
49. Eom S-H, Zheng Y, Chopra N, Lee J, So F, Xue J (2008) Appl Phys Lett 93:123309
50. Lee CW, Lee JY (2013) Org Electron 14:370
51. Giovanella U, Betti P, Bolognesi A, Destri S, Melucci M, Pasini M, Porzio W, Botta C (2010) Org Electron 11:2012
52. Zuniga CA, Barlow S, Marder SR (2011) Chem Mater 23:658
53. Huang F, Cheng Y-J, Zhang Y, Liu MS, Jen AK-Y (2008) J Mater Chem 18:4495
54. Scopelliti R, Zuppiroli L, Graetzel M, Nazeeruddin MK (2008) Inorg Chem 47:6575
55. Jou JH, Hsu MF, Wang WB, Chin CL, Chung YC, Chen CT, Shyue JJ, Shen SM, Wu MH, Chang WC, Liu CP, Chen SZ, Chen HY (2009) Chem Mater 21:2565
56. Ryu DW, Kim KS, Choi CK, Park YI, Kang IN, Park JW (2007) Curr Appl Phys 7:396
57. Cheng JA, Chen CH, Liao CH (2004) Chem Mater 16:2862
58. Duan L, Hou L, Lee T-W, Qiao J, Zhang D, Dong G, Wang L, Qiu Y (2010) J Mater Chem 20:6392
59. Li TX, Yamamoto T, Lan HL, Kido J (2004) Polym Adv Technol 15:266
60. Kim SH, Cho I, Sim MK, Park S, Young Park S (2011) J Mater Chem 21:9139
61. Wu C-H, Chien C-H, Hsu F-M, Shih P-I, Shu C-F (2009) J Mater Chem 19:1464

62. Ku S-Y, Chi L-C, Hung W-Y, Yang S-W, Tsai T-C, Wong K-T, Chenc Y-H, Wu C-I (2009) J Mater Chem 19:773
63. Bucinskas A, Volyniuk D, Danyliv Y, Grazulevicius JV, Baryshnikov G, Minaev B, Ivaniuk K, Cherpak V, Stakhira P (2015) RSC Adv 5:78150
64. Zhang H, Xu X, Qiu W, Qi T, Gao X, Liu Y, Lu K, Du C, Yu G, Liu Y (2008) J Phys Chem C 112:34
65. Gifford AP, Zhu Y, Lou Y, Jenekhe SA (2006) Chem Mater 18:20
66. Kulkarni AP, Kong X, Jenekhe SA (2006) Adv Funct Mater 16:1057, J.M. Hancock
67. Chiang CL, Wu MT, Dai DC, Wen YS, Wang JK, Chen CT (2005) Adv Funct Mater 15:231
68. Wei P, Duan L, Zhang DQ, Qiao J, Wang LD, Wang RJ, Dong GF, Qiu Y (2008) J Mater Chem 18:806
69. Qiu Y, Wei P, Zhang DQ, Qiao J, Duan L, Li YK, Gao YD, Wang LD (2006) Adv Mater 18:1607
70. Yang Y, Zhou Y, He Q-G, He C, Yang C-H, Bai F-L, Li Y-F (2009) J Phys Chem B 113:7745
71. Thangthong A, Prachumrak N, Sudyoadsuk T, Namuangruk S, Keawin T, Jungsuttiwonga S, Kungwan N (2015) V Promarak Org El 21:117
72. Yeh H-C, Chan L-H, Wua W-C, Chen C-T (2004) J Mater Chem 14:1293
73. Grimsdale AC, Chan KL, Martin RE, Jokisz PG, Holmes AB (2009) Chem Rev 109:897
74. Burroughes JH, Bradley DDC, Brown AR, Marks RN, Mackay K, Friend RH, Burn PL, Holmes AB (1990) Nature 345:539
75. Braun D, Heeger AJ, Kroemer H (1991) J Electron Mater 20:945
76. Bernius M, Inbasekaran M, O'Brien J, Wu W (2000) Adv Mater 12:1737
77. Bernius M, Inbasekaran M, Woo E, Wu W, Wujkowski L (2000) J Mater Sci Mater Electron 11:111
78. Bernius M, Inbasekaran M, Woo E, Wu W, Wujkowski L (2000) Thin Solid Films 363:55
79. Inbasekaran M, Woo E, Bernius M, Wujkowski L (2000) Synth Met 111–112:397
80. Grem G, Leditzky G, Ullrich B, Leising B (1992) Adv Mater 4:36
81. Roncali J (1992) Chem Rev 92(4):711
82. Bolognesi A, Pasini M (2007) Synthetic methods for semiconducting polymers, Semiconducting polymers. Wiley-VCH, Weinheim
83. Pal B, Yen W-C, Yang J-S, Su W-F (2007) Macromolecules 40(23):8189
84. Chen P, Yang G, Liu T, Li T, Wang M, Huang W (2006) Polym Int 55:473, and references therein
85. Giovanella U, Botta C, Galeotti F, Vercelli B, Battiato S, Pasini M (2013) J Mater Chem C 1 (34):5322
86. Morii K, Ishida M, Takashima T, Shimoda T, Wang Q, Nazeeruddin MK, Grätzel M (2006) Appl Phys Lett 89:183510
87. Kabra D, Lu LP, Song MH, Snaith HJ, Friend RH (2010) Adv Mater 22:3194
88. Wu W-C, Lee W-Y, Chen W-C (2006) Macromol Chem Phys 207:1131
89. Wu W-C, Liu C-L, Chen W-C (2006) Polymer 47:527
90. Pasini M, Giovanella U, Betti P, Bolognesi A, Botta C, Destri S, Porzio W (2009) ChemPhysChem 10:2143
91. Hung M-C, Liao J-L, Chen S-A, Chen S-H, Su A-C (2005) J Am Chem Soc 127:14576
92. Sung H-H, Lin H-C (2004) Macromolecules 37(21):7945
93. Lu S, Liu T, Ke L, Ma D-G, Chua S-J, Huang W (2005) Macromolecules 38:8494
94. Giovanella U, Betti P, Botta C, Destri S, Moreau J, Pasini M, Porzio W, Vercelli B, Bolognesi A (2010) Chem Mater 23(3):810
95. Li J, Liu D (2009) J Mater Chem 19:7584
96. Burn PL, Lo S-C, Samuel IDW (2007) Adv Mater 19:1675
97. Li J, Li Q, Liu D (2011) ACS Appl Mater Interfaces 3:2099
98. King KA, Spellane PJ, Watts RJ (1985) J Am Chem Soc 107:1431
99. Su S-J, Gonmori E, Sasabe H, Kido J (2008) Adv Mater 20:4189
100. Zhen Y, Eom S-H, Chopra N, Lee J, So F, Xue J (2008) Appl Phys Lett 92:223301

101. Xu M, Zhou R, Wang G, Xiao Q, Du W, Che G (2008) Inorg Chim Acta 361:2407
102. Ragni R, Plummer EA, Brunner K, Hofstraat JW, Babudri F, Farinola GM, Naso F, De Cola L (2006) J Mater Chem 16:1161
103. Lee SJ, Park K-M, Yang K, Kang Y (2009) Inorg Chem 48:1030
104. Hang X-C, Fleetham T, Turner E, Brooks J, Li J (2013) Angew Chem Int Ed 52:6753
105. Fukase A, Dao KLT, Kido J (2002) Polym Adv Technol 13:601
106. Rehmann N, Hertel D, Meerholz K, Becker H, Heun S (2007) Appl Phys Lett 91:103507
107. Lamansky S, Djurovich P, Murphy D, Razzaq FA, Lee HE, Adachi C, Burrows PE, Forrest SR, Thompson ME (2001) J Am Chem Soc 123:4304
108. Zhao WQ, Ran GZ, Liu ZW, Bian ZQ, Sun K, Xu WJ, Huang CH, Qin GG (2008) Opt Express 16:5158
109. Cocchi M, Virgili D, Fattori V, Rochester DL, Williams JAG (2007) Adv Funct Mater 17:285
110. Ho C-L, Li H, Wong W-Y (2014) J Organomet Chem 751:261
111. Freund C, Porzio W, Giovanella U, Vignali F, Pasini M, Destri S, Mech A, Di Pietro S, Di Bari L, Mineo P (2011) Inorg Chem 50(12):5417
112. Zhou G-J, Wong W-Y, Yao B, Xie Z, Wang L (2007) Angew Chem Int Ed 46:1149
113. Chang Y-L, Puzzo DP, Wang Z, Helander MG, Qiu J, Castrucci J, Lu Z-H (2012) Phys Status Solidi C 9:2537
114. Wu F-I, Su H-J, Shu C-F, Luo L, Diau W-G, Cheng C-H, Duan J-P, Lee G-H (2005) J Mater Chem 15:1035
115. Ho C-L, Chi L-C, Hung W-Y, Chen W-J, Lin Y-C, Wu H, Mondal E, Zhou G-J, Wong K-T, Wong W-Y (2012) J Mater Chem 22:215
116. Xia ZY, Xiao X, Su JH, Chang CS, Chen CH, Li DL, Tian H (2009) Synth Met 159:1782
117. Hu Z, Wang Y, Shi D, Tan H, Li X, Wang L, Zhu W, Cao Y (2010) Dyes Pigment 86:166
118. Fukagawa H, Shimizu T, Hanashima H, Osada Y, Suzuki M, Fujikake H (2012) Adv Mater 24:5099
119. Jou J-H, Sun M-C, Chou H-H, Li C-H (2005) Appl Phys Lett 87:043508
120. Xiao L, Chen Z, Qu B, Luo J, Kong S, Gong Q, Kido J (2011) Adv Mater 23:926
121. Watanabe K, Kanai D, Tsuzuki T, Takanaga E, Tokito S (2007) J Photopolym Sci Technol 20:39
122. Wu M-F, Yeh S-J, Chen C-T, Murayama H, Tsuboi T, Li W-S, Chao I, Liu S-W, Wang J-K (2007) Adv Funct Mater 17:18879
123. Ye T, Shao S, Chen J, Wang L, Ma D (2011) ACS Appl Mater Interface 3:410
124. Jiang W, Duan L, Qiao J, Zhang D, Dong G, Wang L, Qiu Y (2010) J Mater Chem 20:6131
125. Yang XH, Jaiser F, Klinger S, Neher D (2006) Appl Phys Lett 88:02110
126. Tao Y, Yang C, Qin J (2011) Chem Soc Rev 40:2943
127. Burrows PE, Padmaperuma AB, Sapochak LS, Djurovich P, Thompson ME (2006) Appl Phys Lett 88:183503
128. Jeon SO, Yook KS, Joo CW, Lee JY (2009) Appl Phys Lett 94:013301
129. Yook KS, Lee JY (2014) Adv Mater 26:4218
130. Jeon SO, Lee JY (2012) J Mater Chem 22:4233
131. Su S-J, Sasabe H, Takeda T, Kido J (2008) Chem Mater 20:1691
132. Kim H, Byun Y, Das RR, Choi BK, Ahn PS (2007) Appl Phys Lett 91:093512
133. Cai M, Xiao T, Hellerich E, Chen Y, Shinar R, Shinar J (2011) Adv Mater 23:3590
134. Jankus V, Monkman AP (2011) Adv Funct Mater 21:3350–3356
135. Ahmed E, Earmme T, Jenekhe SA (2011) Adv Funct Mater 21:3889
136. Yang X, Neher D, Hertel D, Däubler TK (2004) Adv Mater 16:161
137. Méhes G, Nomura H, Zhang Q, Nakagawa T, Adachi C (2012) Angew Chem Int Ed 51:11311
138. Lee SY, Yasuda T, Nomura H, Adachi C (2012) Appl Phys Lett 101:093306
139. Nakagawa T, Ku S-Y, Wong K-T, Adachi C (2012) Chem Commun 9580
140. Nakanotani H, Masui K, Nishide J, Shibata T, Adachi C (2013) Sci Rep 3:2127
141. Zhang Q, Li J, Shizu K, Huang S, Hirata S, Miyazaki H, Adachi C (2012) J Am Chem Soc 134:14706

142. Li J, Nakagawa T, MacDonald J, Zhang Q, Nomura H, Miyazaki H, Adachi C (2013) Adv Mater 25:3319
143. Lee SY, Yasuda T, Yang YS, Zhang Q, Adachi C (2014) Angew Chem Int Edn 53:6402
144. Kawano K, Nagayoshi K, Yamaki T, Adachi C (2014) Org Electron 15:1695
145. Wang H, Xie L, Peng Q, Meng L, Wang Y, Yi Y, Wang P (2014) Adv Mater 26:5198
146. Kulkarni AP, Tonzola CJ, Babel A, Jenekhe SA (2004) Chem Mater 16:4556
147. Hughes G, Bryce MR (2005) J Mater Chem 15:94
148. Jeon SO, Jang SE, Son HS, Lee JY (2011) Adv Mater 23:1436
149. Yook KS, Lee JY (2011) Org Electron 12:1293
150. Xiao L, Su S-J, Agata Y, Lan H, Kido J (2009) Adv Mater 21:1271
151. Tanaka D, Agata Y, Takeda T, Watanabe S, Kido J (2007) Jpn J Appl Phys 46:L117
152. Liu B, Bazan GC (2012) Conjugated polyelectrolytes: fundamentals and applications. Wiley-VCH, Weinheim, ISBN: 978-3-527-33143-7
153. Lee BH, Jung IH, Woo HY, Shim H-K, Kim G, Lee K (2014) Adv Funct Mater 24:1100 (and ref therein)
154. Hu Z, Zhang K, Huang F, Cao Y (2015) Chem Commun 51:5572
155. Hofmann S, Thomschke M, Lüssem B, Leo K (2011) Opt Express 19(s6):A1250
156. Tyan Y-S, Photonics J (2011) Energy 1:011009
157. Thejo Kalyani N, Dhoble SJ (2012) Renew Sustain Energy Rev 16:2696
158. Chiba T, Pu Y-J, Kido J (2015) Adv Mater 27:4681
159. Yu L, Liu J, Hu S, He R, Yang W, Wu H, Peng J, Xia R, Bradley DDC (2013) Adv Funct Mater 23:4366
160. Zhang B, Tan G, Lam C-S, Yao B, Ho C-L, Liu L, Xie Z, Wong W-Y, Ding J, Wang L (2012) Adv Mater 24:1873
161. Sasabe H, Takamatsu J, Motoyama T, Watanabe S, Wagenblast G, Langer N, Molt O, Fuchs E, Lennartz C, Kido J (2010) Adv Mater 22:5003
162. Tu GL, Zhou QG, Cheng YX, Wang LX, Ma DG, Jing XB, Wang FS (2004) Appl Phys Lett 85:2172
163. Liu J, Zhou QG, Cheng YX, Geng YH, Wang LX, Ma DG, Jing XB, Wang FS (2006) Adv Funct Mater 16:957
164. Shao S, Ding J, Wang L, Jing X, Wang F (2012) J Am Chem Soc 134:20290
165. Cocchi M, Kalinowski J, Murphy L, Williams JAG, Fattori V (2010) Org Electron 11:388
166. Nishide J, Nakanotani H, Hiraga Y, Adachi C (2014) Appl Phys Lett 104:233304
167. Higuchi T, Nakanotani H, Adachi C (2015) Adv Mater 27:2019
168. Aizawa N, Pu Y-J, Watanabe M, Chiba T, Ideta K, Toyota N, Igarashi M, Suzuri Y, Sasabe H, Kido J (2014) Nat Commun 5:5756
169. Gather MC, Rein S (2015) J Photon Energy 5:057607–1
170. Yabu H, Shimomura M (2005) Langmuir 21:1709
171. Galeotti F, Mróz W, Scavia G, Botta C (2013) Org El 14:212
172. Ou Q-D, Zhou L, Li Y-Q, Shen S, Chen J-D, Li C, Wang Q-K, Lee S-T, Tang J-X (2014) Adv Funct Mater 24:7249
173. Duan L, Zhang D, Wu K, Huang X, Wang L, Qiu Y (2011) Adv Funct Mater 21:3540
174. Ye E, Chee PL, Prasad A, Fang X, Owh C, Jing Jing Yeo V, Jun Loh X (2014) Mater Today 17:194–202

Chapter 4
Light-Emitting Electrochemical Cells

Chia-Yu Cheng and Hai-Ching Su

Abstract Recently, solid-state light-emitting electrochemical cells (LECs) have attracted intense attention due to simple device structure, low operation voltage and high power efficiency. Compatibility with simple solution processes and superior device efficiency are beneficial in display and lighting applications. In this chapter, the working mechanism of LECs is introduced and some previous important works on LECs, such as LECs with various emission colors, topics on device lifetime and turn-on time of LECs and novel device technologies on LEC device structures are reviewed. Finally, conclusions and outlooks for LECs are discussed.

Abbreviations

CCL	Color conversion layer
CIE	Commission Internationale de L'Eclairage
CRI	Color rendering index
ECD	Electrochemical doping
ED	Electrodynamical
EDLs	Electric double layers
EL	Electroluminescence
EQE	External quantum efficiency
FWHM	Full width at half maximum
HOMO	Highest occupied molecular orbital
iTMCs	Ionic transition metal complexes
LECs	Light-emitting electrochemical cells
LUMO	Lowest unoccupied molecular orbital
MLCT	Metal-to-ligand charge transfer
NIR	Near-infrared
OLEDs	Organic light-emitting diodes
PLECs	Polymer light-emitting electrochemical cells

C.-Y. Cheng • H.-C. Su (✉)
Institute of Lighting and Energy Photonics, National Chiao Tung University, Tainan 71150, Taiwan
e-mail: haichingsu@mail.nctu.edu.tw

G. Bergamini, S. Silvi (eds.), *Applied Photochemistry*,
Lecture Notes in Chemistry 92, DOI 10.1007/978-3-319-31671-0_4

PLQY	Photoluminescent quantum yield
SKPM	Scanning Kelvin probe microscopy
WLEC	White light-emitting electrochemical cells

4.1 Introduction

4.1.1 Features of Solid-State Light-Emitting Electrochemical Cells

Organic light-emitting diodes (OLEDs) have received much attention in recent years due to their potential applications in flat-panel and solid-state lighting. However, OLEDs encountered some challenges such as complicated multi-layered evaporation processes and using low-work-function electrodes which normally are highly reactive metals. To overcome such difficulties, solid-state light-emitting electrochemical cells (LECs) were proposed by Pei et al. in 1995 [1].

Compared with conventional OLEDs, solid-state LECs possess several promising advantages, including simple fabrication processes, low-voltage operation, and high power efficiency. LECs are generally composed of only a single emissive layer, which is compatible with large-area solution processes. Furthermore, LECs can conveniently utilize air-stable electrodes such as gold and silver and hence would reduce stringent packaging procedures. The emissive layer of LECs contains an ionic luminescent material, which can be roughly fallen into two categories: fluorescent conjugated light-emitting polymers and phosphorescent ionic transition metal complexes (iTMCs). For the former, the emissive layers of polymer LECs (PLECs) contain ionic salts to provide mobile ions. Ion conducting polymers (e.g., poly(ethylene oxide) (PEO)), are commonly incorporated into the emissive layer to avoid phase separation between nonpolar conjugated light-emitting polymers and polar salts. For the latter, iTMC-based LECs have also been intensively studied in view of their several advantages over PLECs. No ion-conducting material is needed in iTMC-based LECs since iTMCs are intrinsically ionic. In addition, electroluminescence (EL) efficiencies of LECs based on iTMC are generally higher owing to the phosphorescence nature of iTMCs.

4.1.2 Organization of This Chapter

In the following section, we will briefly review the development of LECs. First, operating mechanisms of LECs are introduced. Then we review the advances of LECs, including topics of device color, efficiency, lifetimes and turn-on times. Works on device engineering of LECs including tandem LECs, probing recombination zone position of LECs by employing microcavity effect and novel optical structure to improve light extraction from white LECs are also covered. Finally, conclusions and outlooks are mentioned.

4.2 Operation Mechanism of Light-Emitting Electrochemical Cells

Two theories have been adopted to explain the operational mechanism of LECs so far. One is electrodynamical (ED) model and the other is electrochemical doping (ECD) model. Mobile ions incorporated in the emissive layer of LECs can drift toward electrodes under an applied bias and consequently decrease injection barrier of carriers in both models. Nevertheless, it is noticed that the reasons for decrease in injection barrier for carriers are distinct in these two models.

In the ED model, as shown in Fig. 4.1a, when a bias is applied, anions and cations are moving toward anode and cathode, respectively. When a steady state is achieved, sufficient ions accumulate and cause significant enhancement of electric field near electrodes. Subsequently, electric double layers (EDLs) are formed near electrodes and thus it demonstrates a significant drop of the electric potential at the electrode interfaces. It is the reason that injection barrier for carrier reduces in the ED model. Furthermore, injected carriers drift and recombine to emit light in the field-free region between the EDLs.

In the ECD model, as shown in Fig. 4.1b, however, in order to compensate additional charges from the accumulation of separated ions near the electrodes

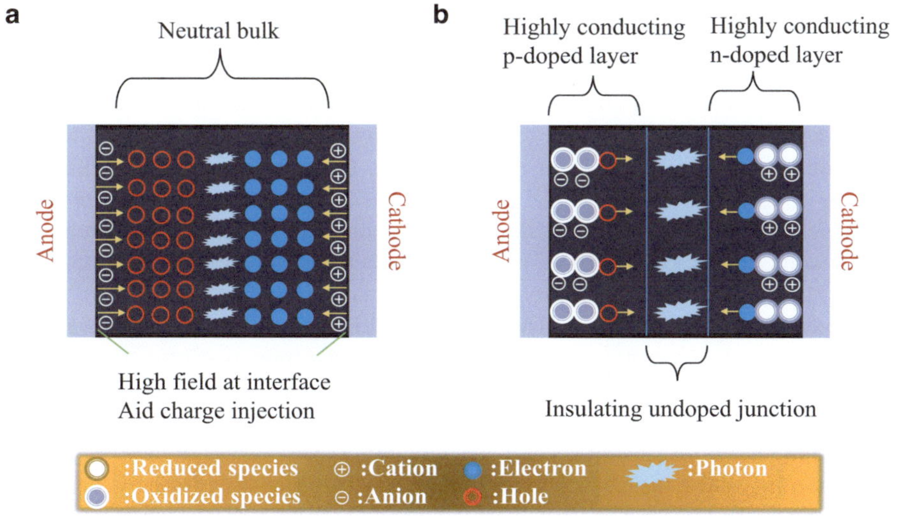

Fig. 4.1 Schematic diagrams for operation mechanisms of (**a**) electrodynamical (ED) model and (**b**) electrochemical doping (ECD) model

under an applied bias, oxidized and reduced molecules are formed at the anode and the cathode, respectively. That is, highly conductive p-doped region is near the anode and n-doped region is near the cathode and consequently the ohmic contacts with the electrodes interface are formed to facilitate injection of both electrons and holes. The doped layers extend into the intrinsic region of the emissive layer over time and a p-i-n junction is formed finally. In the intrinsic region, electric potential drops substantially and light emit due to charge recombination.

Both the ED and the ECD models were supported by experimental data and numerical modeling. To study operation mechanism of LECs, a planar PLEC was used instead because it is difficult to probe the potential profile in a thin active layer of sandwiched PLEC. One of such planar devices was reported by Pei et al. [1]. It showed the existence of a p-i-n junction through the light emission located within a narrow zone in the large interelectrode gap. In addition, formation of doped regions (p- and n-doped) in planar devices was also observed by monitoring the photoluminescence image and the electrostatic potential from planar PLECs using scanning Kelvin probe microscopy (SKPM) [2–5]. These results revealed a sharp potential drop in the center of the device [6, 7]. It was consistent with the ECD model and most PLEC devices followed this model [1–4, 8–10].

However, in the study of PPV-based planar LEC device, Pingree et al. reported that over 90 % of the potential dropped at the polymer/cathode interface in the PPV device and very little electric field appears across the bulk of the polymer film [11]. Similar phenomena which showed strong potential drops at the electrode interface and little potential drop at the center of the device were discovered in planar iTMC-based LECs device [12]. These results cannot be explained well by the ECD model but was consistent with the ED model [12–14].

Experimental results shown above provided evidences to support both models and thus the operational mechanism of planar LECs has been debated. Recently, van Reenen et al. demonstrated that employing two different contacts, that is, non-injection limited and injection limited contacts, in planar LEC obtained a completely contrast potential profile. In non-injection limited ohmic contacts, the LECs followed the ECD model, characterized by the formation of a dynamic p-n junction in the bulk of the device. However, in the injection-limited regime, the device followed the ED model [15]. Analysis on operation mechanism of sandwiched LEC by employing impedance spectroscopy has also been reported. The results supported the ECD model for either PLECs or iTMC-based LECs [16, 17]. Recenty, JL-V measurements were used to clarify the operating model on sandwiched LECs based on iTMCs [18]. At driving voltages above the bandgap of the iTMCs, the separation of ionic charges led to the formation of p- and n-doped regions and the central region is ionically neutral, which corresponds the ECD model. However, at biases below the bandgap of the iTMCs, current transients showed the behavior predicted by the ED model [18].

According to the recent observations, the model responsible for LEC operation depended on the injection regime. That is, the ECD model happens when there is no limited injection, e.g., for ohmic contacts or bias voltage higher than bandgap. In contrast, the ED model dominates when there is only limited carrier injection into the emissive layer.

4.3 Review of Solid-State Light-Emitting Electrochemical Cells

4.3.1 Device Color and Efficiency

We review LEC devices with emission color across the whole visible spectrum, including white LECs. The emission color can be identified by the wavelength of EL spectrum peak and Commission Internationale de L'Eclairage (CIE) coordinates. For white LECs, additional color rendering index (CRI) values should be calculated to evaluate their suitability for lighting applications. To examine device efficiency of LECs, there are three types of efficiencies including external quantum efficiency (EQE, photons per electron), current efficiency (emitted flux in candelas per electrical ampere) and power efficiency (emitted flux in lumens per electrical watt).

4.3.1.1 Infrared LECs

Most ruthenium-based complexes typically emit in the red and orange spectral region owing to low metal-to-ligand charge transfer (MLCT) energies. First infrared-emitting material in LECs originated from Ng et al. who synthesized a series of polyimides based ruthenium(II) complex in 1999 [19]. The devices based on complex **1** and complex **2** (Fig. 4.2) showed a low EQE (0.1 %) and a broad emission with deep-red EL centered at about 650 nm and near infrared centered near 750–800 nm. It is noticed that the full width at half maximum (FWHM) is about 200 nm and thus the good color purity was not achieved [19].

In 2000, Suzuki demonstrated that the EL peak wavelength located about 800 nm from near-infrared (NIR) LECs employing poly(9-vinylcarbazole) (PVK) host doped with a laser dye (LDS821) and an electron transport material 2-(4-biphenylyl)-5-(4-tert-butyl-phenyl)-1,3,4,-oxadizole (PBD). However, still

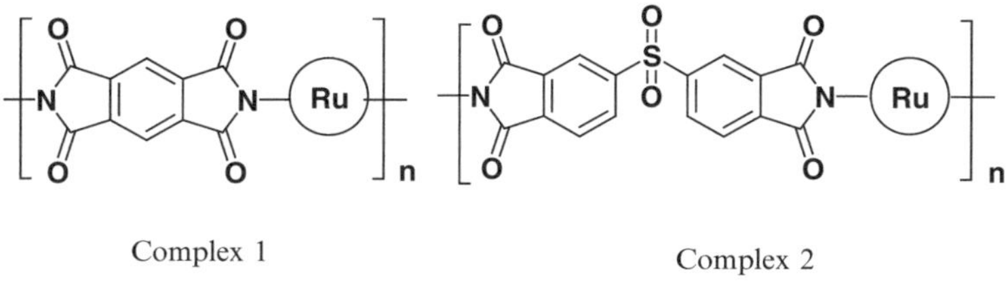

Fig. 4.2 Molecular structures of complexes 1 and 2

low EQE was obtained, only 0.015 % [20]. In 2004, Suzuki also published the NIR LECs using commercial dye (IR1051) and synthesized rare-earth metal complex (complex **3**, Fig. 4.3). These NIR LECs showed the maximums of emission spectrum at 1100 nm and 1500 nm respectively [21]. In 2005, Hosseini et al. reported NIR LECs utilizing a host-guest system, in which an Os complex was used as the guest, to achieve EL spectrum centered at 710 nm and the EQE was 0.75 % [22]. In addition, Bolink et al. obtained the EL peak wavelength of 705 nm from LECs based on complex **4** [23] (Fig. 4.4). Nevertheless, EL spectra of the two NIR LECs mentioned above were not pure because of wide FWHM and some spectrum extending into 550–600 nm. In 2006, Wang et al. synthesized a series of NIR electrochromic and electroluminescent polymers containing the pendant dinuclear ruthenium complexes. The LEC based on complex **5** (Fig. 4.5) emitted NIR EL centered at 790 nm and the efficiency is 5.4×10^{-6} % [24]. In 2008, a series of ruthenium complexes with different ligands were used in LECs with a configuration of ITO/complex (100 nm)/Au (100 nm) [25]. The devices performed the EL from 630 to 1040 nm by tuning the lowest unoccupied molecular orbital (LUMO) and highest occupied molecular orbital (HOMO) levels of these complexes. The device based on complex **6** (Fig. 4.6) emitted NIR EL centered at 880 nm and the EQE of 0.075 % can be achieved [25]. In 2009, Bolink et al. reported the LECs utilizing

Fig. 4.3 Molecular structure of complex 3

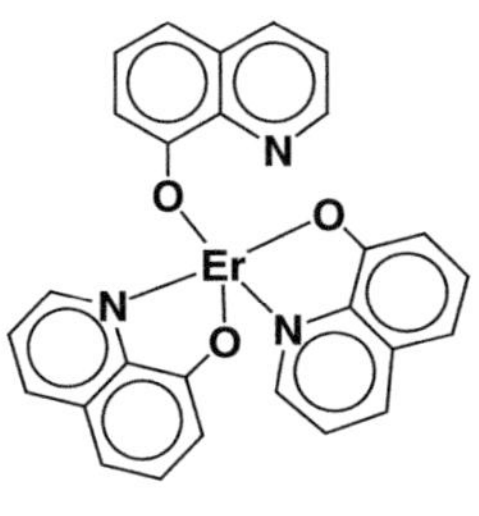

Complex 3

Fig. 4.4 Molecular structure of complex 4

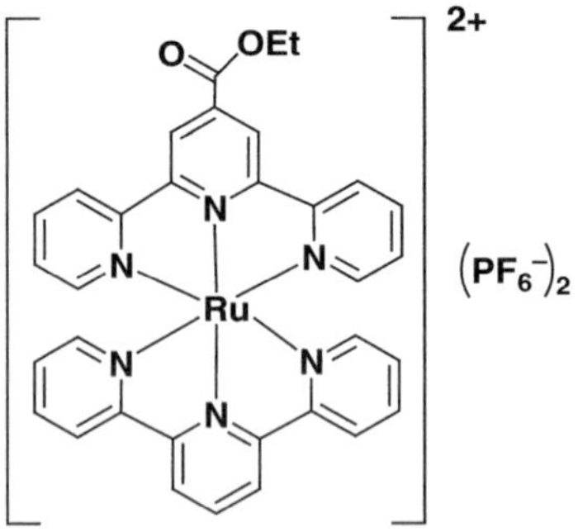

Complex 4

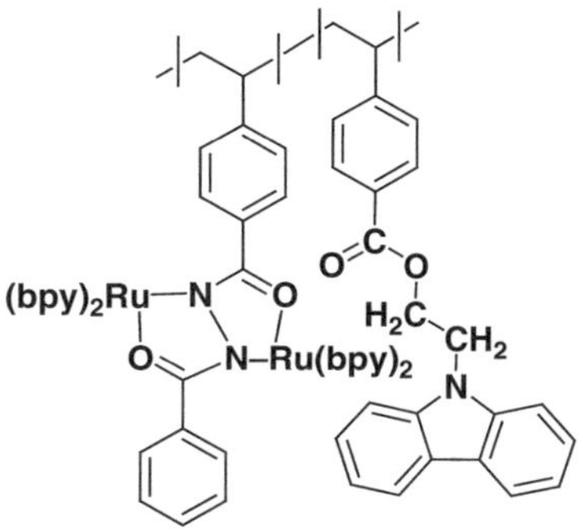

Fig. 4.5 Molecular structure of complex 5

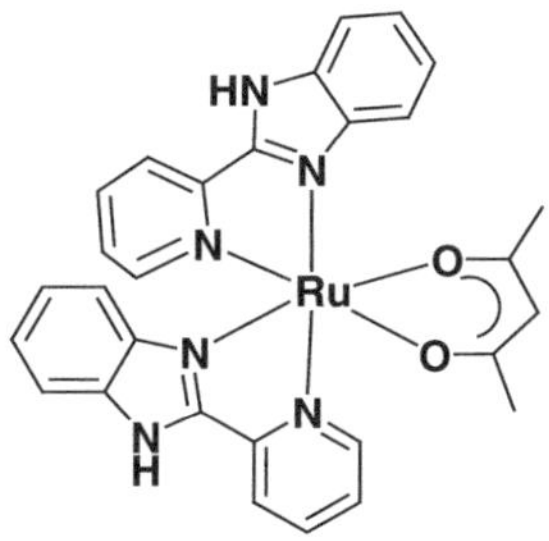

Fig. 4.6 Molecular structure of complex 6

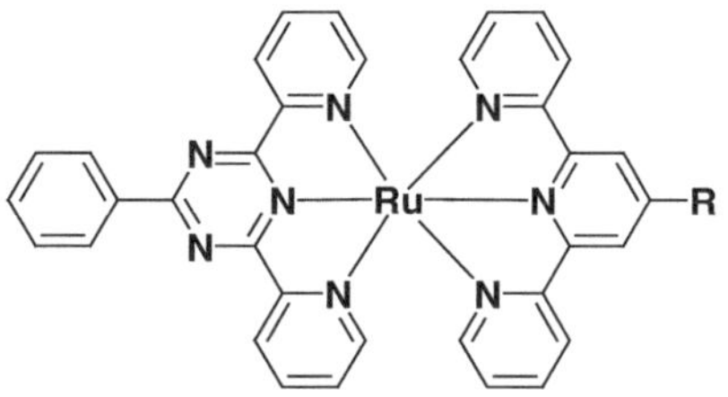

Fig. 4.7 Molecular structures of complexes 7a and 7b

complex **7a** and **7b** (Fig. 4.7) as active components which presented NIR EL at 717 nm and 725 nm, respectively [26].

It is noticed that all NIR LECs mentioned above showed low device efficiencies. In 2011, Su et al. demonstrated the NIR LECs based on phosphorescent sensitized fluorescence [27]. The EL spectra of the NIR LECs doped with commercial NIR dyes 3,3′-diethyl-2,2′-oxathiacarbocyanine iodide (DOTCI) and 3,3′-diethylthiatricarbocyanine iodide (DTTCI) centered at ca. 730 and 810 nm, respectively. Moreover, the DOTCI and DTTCI doped NIR LECs exhibited peak EQE (power efficiency) up to 0.80 % (5.65 mW W^{-1}) and 1.24 % (7.84 mW W^{-1}), respectively [27]. More recently, in 2014, Su et al. reported the NIR LECs employing a tandem device structure and the EQE achieved was up to 2.75 %, which is over tripled enhancement as compare to previously reported NIR LECs based on the same NIR dye [28].

4.3.1.2 Red LECs

In 2005, Tamayo et al. proposed efficient red LECs based on complex **8** (Fig. 4.8) [29]. The peak EQE and peak EL wavelength of the device were 7.4 % and 635 nm, respectively. The CIE coordinates of the EL spectrum was (0.67, 0.32). In 2008, Su et al. proposed red-emitting complex **9** (Fig. 4.9) ($\lambda_{max} = 656$ nm), which was used to fabricate efficient white LECs [30]. In 2009, He et al. also reported red LECs based on complex **10** (Fig. 4.10) emitting at ca. 650 nm (CIE coordinates = (0.66, 0.33), EQE = 2.6 %), which has also been doped to fabricate white LECs [31].

Attaching electron-withdrawing groups onto the ancillary ligands is also effective in achieving red-emitting iTMC. For example, in 2009, Costa et al. utilized complex **11** and polymer complex **12** (Fig. 4.11), which introduced electron-withdrawing in the bipyridine (bpy) ligand to realize the deep-red-emitting LECs [32]. The LEC based on polymer complex **12** showed red EL peaked at 660 nm, which is red-shifted as compared to the EL of devices employing complex **11**

Fig. 4.8 Molecular structure of complex 8

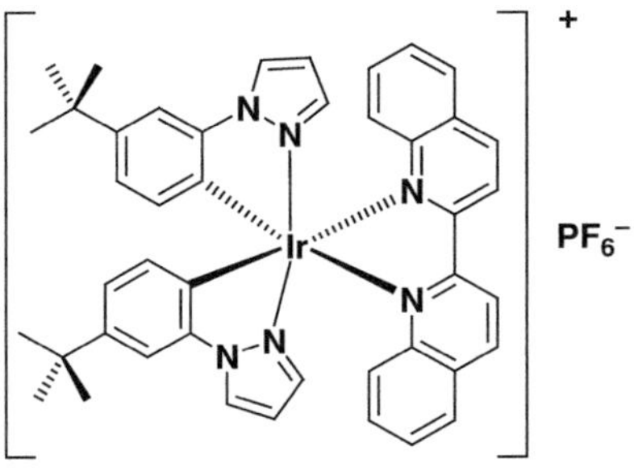

Complex 8

Fig. 4.9 Molecular structure of complex 9

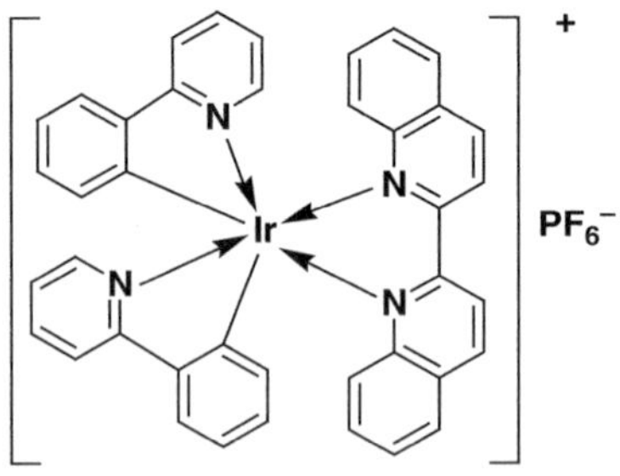

Complex 9

Fig. 4.10 Molecular structure of complex 10

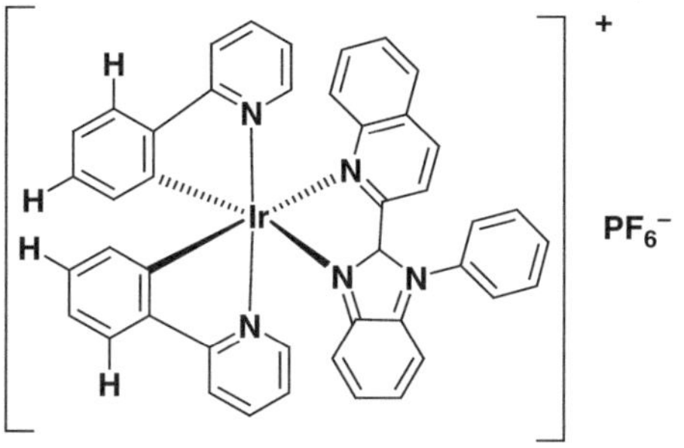

Complex 10

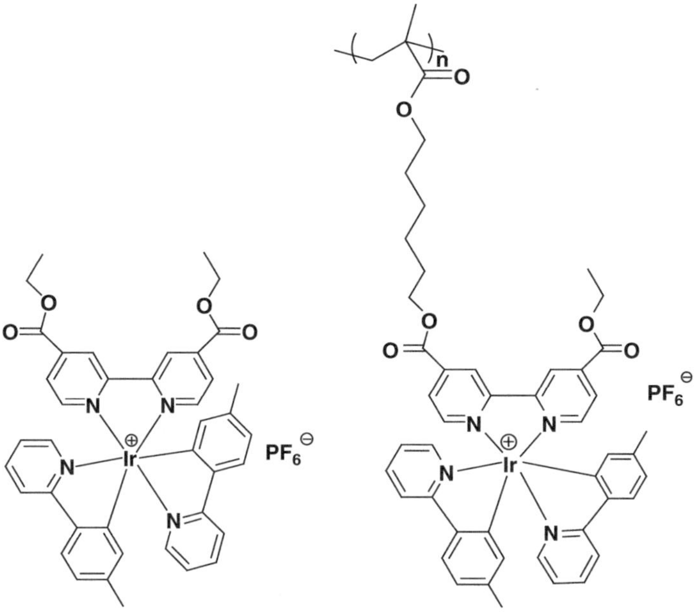

Complex 11

Polymer complex 12

Fig. 4.11 Molecular structures of complex 11 and polymer complex 12

(630 nm). In addition, longer device lifetimes can be obtained in devices based on polymer complex **12**. It resulted from the fact that the spatial distribution of the complexes in the polymer reduced the interaction between adjacent iTMCs and protected the complex against degradation reactions. Similarly, Chen et al. created a red-emitting complex **13** (Fig. 4.12) by adding an electron-withdrawing diphenylamino group to the ancillary ligand in 2011 [33]. The device showed a peak current efficiency of 0.013 cd A^{-1}.

Costa et al. demonstrated deep-red-emitting LECs by anchoring a perylenediimide red fluorescent group to an ionic iridium(III) complex (complex

Fig. 4.12 Molecular structure of complex 13

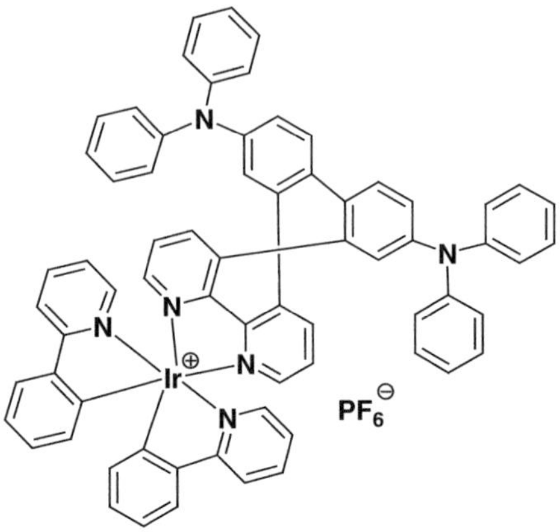

Complex 13

Fig. 4.13 Molecular structure of complex 14

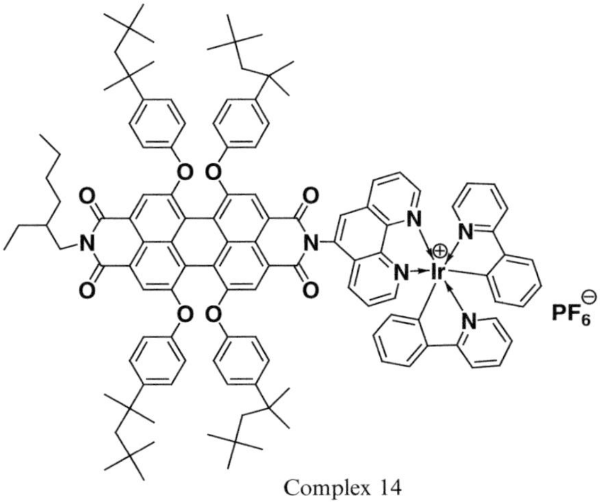

Complex 14

14, Fig. 4.13). The LEC based on complex **14** demonstrated deep-red emission peak at 644 nm with CIE coordinates of (0.654, 0.344) and a peak EQE of 3.27 % [34]. To obtain efficient red-emitting PLECs, Yang et al. proposed phosphorescent LECs based on a blue-emitting polymer doped with a red phosphorescent dye [35]. The EL spectrum of the device exhibited a maximum wavelength at 616 nm. Under 3.1 V, a peak current efficiency of 1.2 cd A^{-1} and a peak power efficiency of 1.0 lm W^{-1} were achieved.

4.3.1.3 Orange, Yellow and Green LECs

Incorporating a fluorescent dye into an iTMC is also a feasible way to obtain variable EL from LECs. Su et al. reported highly efficient phosphor-sensitized solid-state LECs utilizing complex **15** (Fig. 4.14) as the host and a fluorescent cationic dye (R6G) as the guest. Such yellow LECs achieve EQE and power efficiency up to 5.5 % and 21.3 lm W^{-1}, respectively [36]. Costa et al. reported the orange-emitting LECs (570–590 nm) based on complex **16** and **17** (Fig. 4.15) in 2009. The LECs based on complex **16** showed a peak EQE of 5.6 % and a peak power efficiency of 16.3 lm W^{-1} through optimizing the doping concentration of ionic liquid (1:1) in the active layers [37]. Furthermore, to enhance the efficiency of orange LECs, Su et al. proposed iridium complex **18** (Fig. 4.16) with a bulky ancillary ligand and demonstrated the orange LECs with EQE and power efficiency of 7.1 % and 22.6 lm W^{-1}, respectively, at 2.5 V [38]. By incorporating complex **18** (guest) into complex **15** (host), the host–guest LECs showed yellow EL from the guest complex **18**. High EQE (power efficiencies) up to 10.4 % (36.8 lm W^{-1}) was realized due to the significantly suppressed concentration quenching of the guest complex [39].

Fig. 4.14 Molecular structure of complex 15

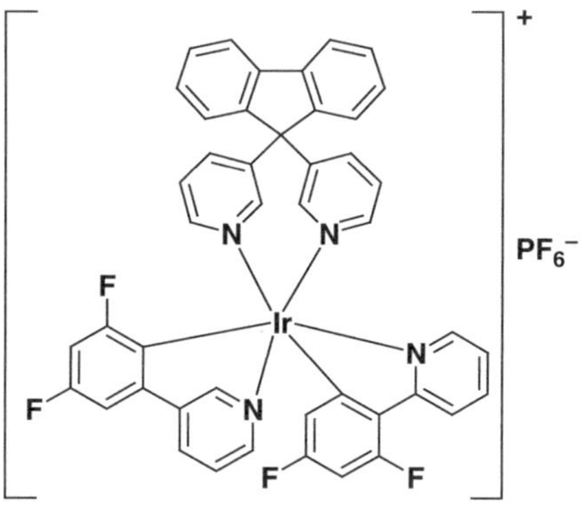

Complex 15

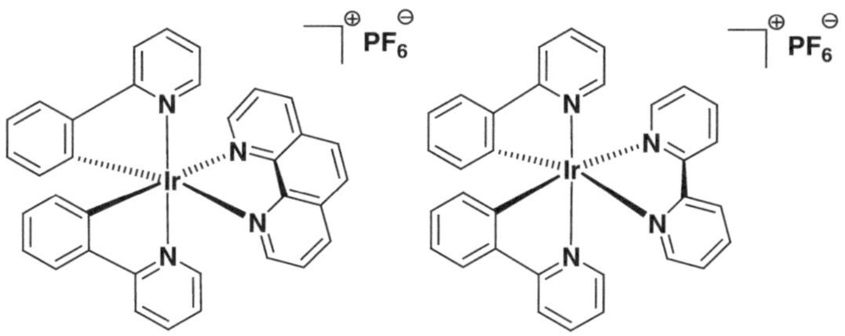

Complex 16 Complex 17

Fig. 4.15 Molecular structures of complexes 16 and 17

Fig. 4.16 Molecular structure of complex 18

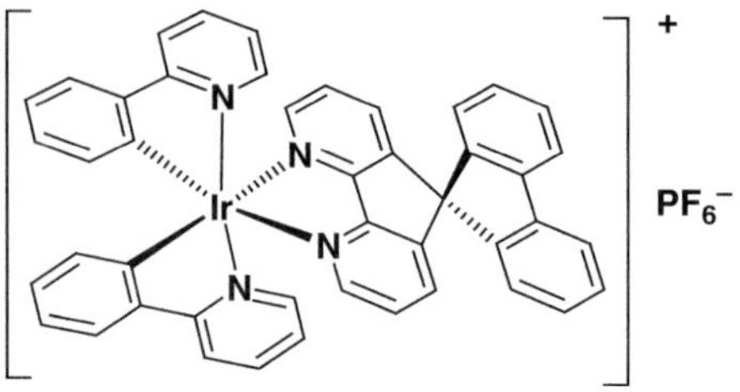

Complex 18

Fig. 4.17 Molecular structure of complex 19

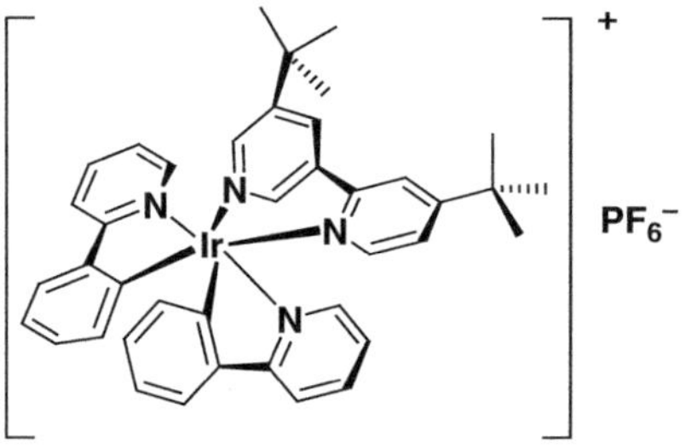

Complex 19

To develop the yellow-emitting iTMCs, attachment of electron-donating groups on the 4,4′-positions of the bpy ligand was employed to increase the HOMO–LUMO gap. Slinker et al. reported LECs based on the complex **19** (Fig. 4.17) in 2004. Devices employing complex **19** emitted yellow EL ($\lambda_{max} = 560$ nm) with EQE of 5 % and power efficiency >10 lm W^{-1} at 3 V [40]. The fluoro substituent on the phenylpyridine (ppy) ligands of complex **20** (Fig. 4.18) led to blue-shifted EL (531 nm) as compared to complex **19** [41]. Similarly, destabilizing the LUMO level of complex **21** (Fig. 4.19) further increased the energy gap. The EL of LECs based on complex **21** showed blue-green emission (500 nm, with EQE of 0.75 % and CIE coordinates (0.198, 0.512)) [42]. To achieve even blueer EL, Bolink et al. have demonstrated a novel cationic iridium-(III) complex by introducing positively charged tributylphosphonium side groups at the periphery of the cyclometalating ligands [43]. The blue-green EL ($\lambda_{max} = 487$ nm) can be obtained by the LECs based on complex **22** (Fig. 4.20) doped with 20 % poly(methyl methacrylate (PMMA).

For PLECs, in 1995, Pei et al. fabricated orange LECs using poly[2-methoxy-5-(2-ethylhexyloxy)-1,4-phenylenevinylene] (MEH-PPV) as the fluorescent emissive material [1]. At the same time, they also fabricated green-emitting PLECs utilizing poly(phenylene vinylene) (PPV) as the semiconducting luminescent polymer. The ion conducting polymer PEO and salt were added into polymers in the devices. When the LECs were fabricated with PPV, PEO and lithium trifluoromethanesulfonate ($LiCF_3SO_3$), the yellow-green EL from the dynamically

Fig. 4.18 Molecular structure of complex 20

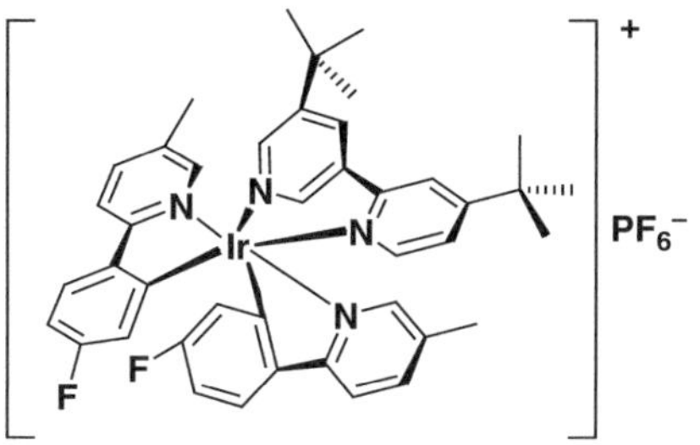

Complex 20

Fig. 4.19 Molecular structure of complex 21

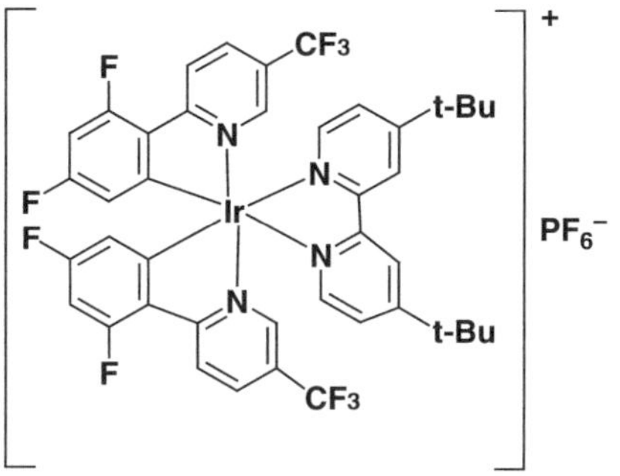

Complex 21

Fig. 4.20 Molecular structure of complex 22

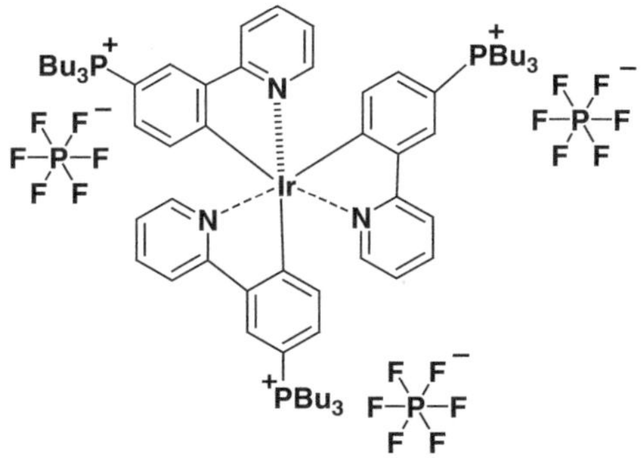

Complex 22

formed p-n junction can be measured. The EQEs of 0.1–0.2 % were obtained in these LECs.

One of the possible ways to achieve switchable emission (orange or green) from bilayer PLECs was proposed by Yang and Pei [44]. Such devices consisted of PPV/MEH-PPV emissive layers. Both layers were blended with ionically conductive PEO and $LiCF_3SO_3$. When a forward bias was applied (ITO wired as anode), the orange EL was measured because the p-n junction was completely formed inside the

MEH-PPV layer. In contrast, the green EL was shown from PPV layer under reverse bias due to the formation of p-n junction therein. Fang et al. demonstrated orange-red emitting LECs, which were composed of a mixture of (MEH-PPV: PEO:KCF_3SO_3 (mass ratio of 1:0.085:0.03)) sandwiched between ITO and Al electrodes [45]. These devices can attain a respectable operational lifetime of ~1000 h at a significant brightness of >100 cd m^{-2} and a high power efficiency of 2 lm W^{-1}.

Yang and Pei proposed poly[9,9-bis(3,6-dioxaheptyl)-fluorene-2,7-diyl] (BDOH-PF) to fabricate blue-green emitting LECs [46]. The polyether-type side groups included in BDOH-PF were incorporated to promote the ionic conductivity necessary for the operation of LECs. High EQE up to 4 % and power efficiency up to 12 lm W^{-1} were achieved under 3.1 V.

Yu et al. reported yellow-emitting PLECs based on a PPV derivative, Super Yellow [47]. These devices showed lifetime ($t_{75\%}$) of 27,000 h at peak brightness of 100 cd/m^2 ($t_{75\%}$ is defined as time for the peak brightness dropped to 75 % of the initial value). Furthermore, the power efficiency up to 15 lm W^{-1} was obtained.

4.3.1.4 Blue and UV LECs

There were generally two ways that could be used to realize blue-emitting LECs. The first one is to attach the functional ligands, such as electron-withdrawing substitutes (–F, or –CF_3) onto the cyclometalating ligands to stabilize the HOMO level of iTMC. The other one is to employ electron-donating substitutes (–$N(CH_3)_2$) onto the ancillary ligands to destabilize the LUMO levels [41, 42, 48, 49]. Using complex **23** (Fig. 4.21) containing 1-phenylpyrazolyl (ppz)-based cyclometalating ligands, Tamayo et al. reported blue-emitting LECs with peak EL centered at 492 nm and CIE coordinates of (0.20, 0.41) [29]. The peak EQE of the blue-emitting LECs is 4.6 % at 3 V. By using diimine ancillary ligands involving electron-donating nitrogen atoms, He et al. obtained two cationic iridium complexes (complex **24** and **25**, Fig. 4.22) which have significantly destabilized LUMO levels owing to 2-(1H-pyrazol-1-yl)pyridine (pzpy) as the ancillary ligands [50]. LECs based on complex **24** exhibited green-blue EL (λ_{max} = 486 nm) and had a relatively high efficiency of 4.3 cd A^{-1} when the ionic liquid 1-butyl-3-

Fig. 4.21 Molecular structure of complex 23

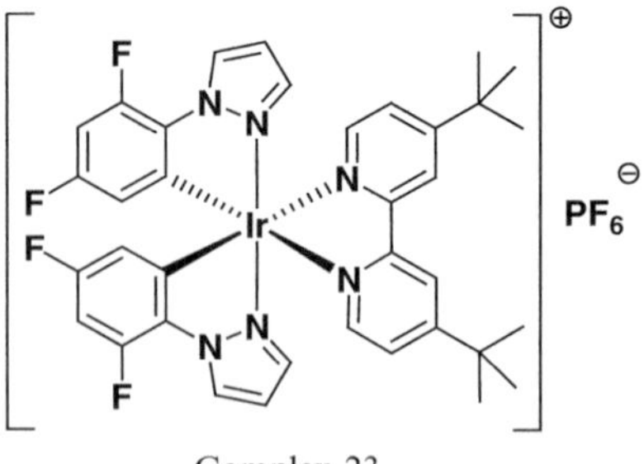

Complex 23

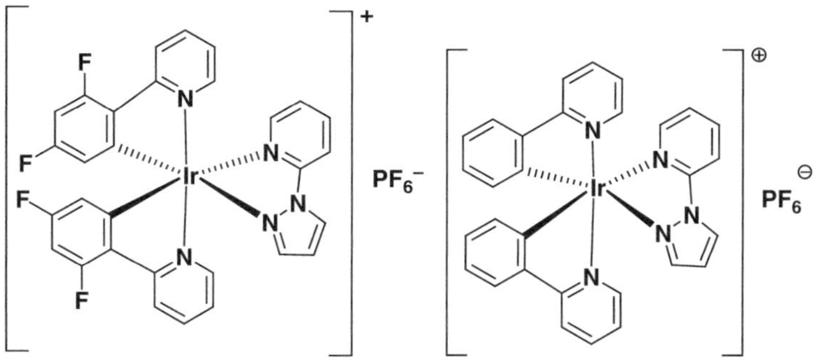

Fig. 4.22 Molecular structures of complexes 24 and 25

methylimidazolium hexafluorophosphate ($BMIMPF_6$) was incorporated into the emissive layer. The CIE coordinates of the EL spectrum of LECs based on complex **24** was (0.27, 0.50). LECs based on complex **25** gave even bluer EL ($\lambda_{max} = 460$ nm) with CIE coordinates of (0.20, 0.28).

In 2010, He et al. proposed a highly efficient blue-green-emitting cationic iridium complex **26** (Fig. 4.23), which contains a bulky side group in the ancillary ligand [51]. The LECs based on complex **26** showed highly efficient blue-green EL, with peak current efficiency, EQE and power efficiency of 18.3 cd A^{-1}, 7.6 % and 18.0 lm W^{-1}, respectively [51]. The blue-emitting iridium complex with the fastest response (short turn-on times of a few minutes at 5 V) was reported by Myldak et al. in the same year [52]. These complexes were based on a series of pyridine-1,2,3-triazol ligand together with the 2-(2′,4′-fluorophenyl)pyridine (dfppy) ligand and showed blue emission in solutions ($\lambda_{max} = 452$ nm). Nevertheless, the LECs using these complexes exhibited blue-green EL peaked at 488 nm and a shoulder was observed at 460 nm. This phenomenon of red-shifts in the EL emission for LECs based on blue-emitting cationic iridium complexes should be attributed to the change of emissive excited state from ^{3}LC π–π^* in dilute solution to ^{3}MLCT in neat films [53].

Yang et al. proposed a new strategy to develop the blue-emitting complex by using carbene-type ligands [54]. They employed methyl- or n-butyl substituted bisimidazolium salt carbine-type ligands as the ancillary ligands (complexes **27a** and **27b**, Fig. 4.24). Owing to the molecular-design strategy based on using high ligand-field carbene ligands, the emission of complexes **27a** and **27b** can be efficiently tuned to a blue or deep-blue region. In thin films, complexes **27a** and **27b** gave blue emission with peaks at 456 nm and shoulder peaks around 484 nm. However, the EL spectra of complexes **27a** and **27b** showed a peak at 488 nm, which was red-shifted as compared to the PL spectra in thin film.

Chen et al. obtained saturated deep-blue EL ($\lambda_{max} = 424$ nm) with CIE coordinates of (0.151, 0.122), an EQE of 1.04 %, and a peak power efficiency of 0.63 lm W^{-1} under 3.4 V from LECs incorporating compound **28** (Fig. 4.25) [55]. It was

Fig. 4.23 Molecular structure of complex 26

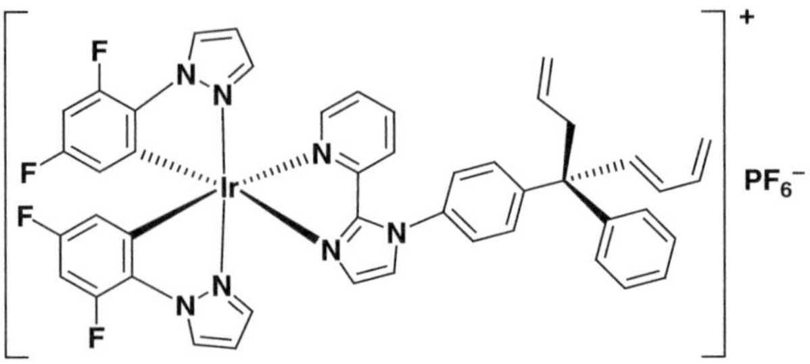

Complex 26

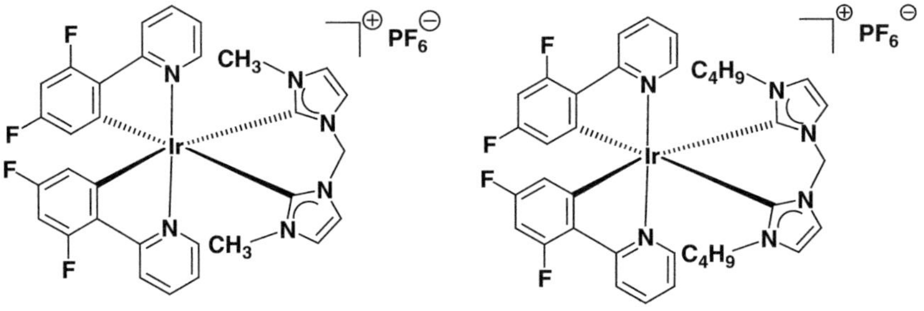

Complex 27a

Complex 27b

Fig. 4.24 Molecular structures of complexes 27a and 27b

Fig. 4.25 Molecular structure of compound 28

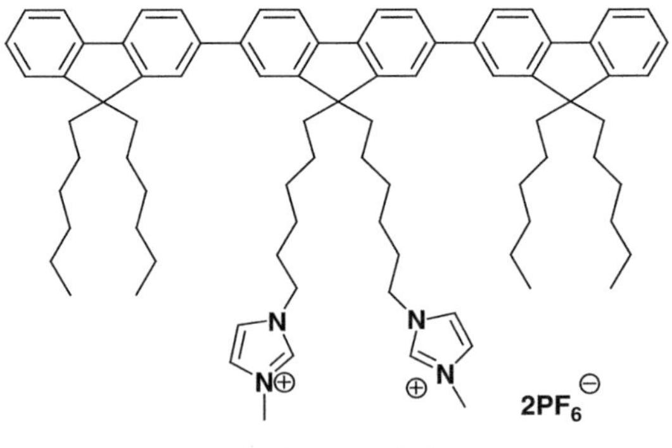

Compound 28

formed by covalent tethering of methylimidazolium moieties as pendent groups of terfluorene. Polymers are also good candidates as the emissive materials for blue LECs. Blue PLECs were reported by Neher et al. in 1996 [56]. The blue-emitting polymer was a rigid-rod polyelectrolyte-sulfonated poly(p-phenylene), which possessed both luminescent and ionic moieties. The LECs based on these polymers exhibited deep-blue EL.

Fig. 4.26 Molecular structure of compound 29

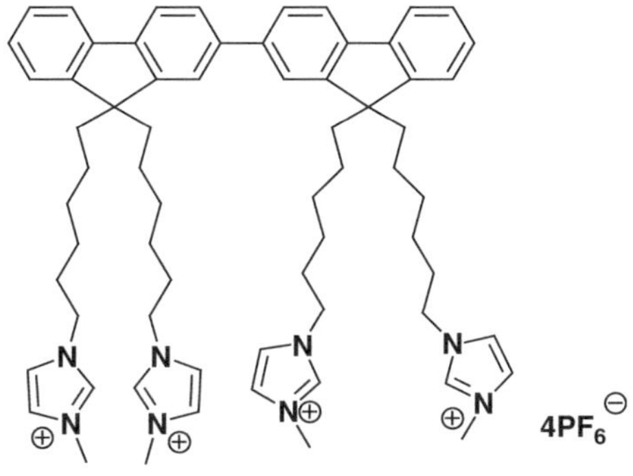

Compund 29

To develop UV-emitting, i.e., λ_{max} <400 nm, Chen et al. have synthesized a UV-emitting ionic bifluorene derivative compound **29** (Fig. 4.26) [57]. It exhibited essential ionic character required for LECs by chemically tethering methylimidazolium moieties as pendent groups to the bifluorene. Devices based on compound **29** exhibited UV emissions at 386 nm, which were the first example of UV LECs ever reported.

4.3.1.5 White LECs

White organic light-emitting devices play an important role in solid-state lighting. The first white LEC (WLEC) was based on phase-separated from the mixture of BDOH-PF and PEO [46]. Blue-green LECs based on a blend of BDOH-PF and $LiCF_3SO_3$ demonstrated the EQE of 4 % and power efficiency of 12 lm W^{-1} under 3.1 V. While PEO was incorporated into the emissive layer, white EL was obtained due to phase separation. Such WLECs exhibited EQE of 2.4 % and brightness of 400 cd m^{-2} at a bias of 4 V [46]. Similarly, in 2010, Sun et al. proposed WLECs employing blend of polymer **30** (Fig. 4.27) and $LiCF_3SO_3$ [58]. Such devices exhibited EL spectrum with FWHM of 160 nm and CIE coordinates of (0.24, 0.31). It is noted that such white emission may result from excimers and it was the first report of WLECs using a single polymer in the emissive layer [58].

Tang et al. reported trichromatic WLEC with good color rendering properties based on blue, green and red-emitting polyspirobifluorene-based copolymers [59]. The ion-transport material trimethylolpropane ethoxylate (TMPE) and the salt $LiCF_3SO_3$ were mixed as the electrolyte in the emissive layer. White emission with CIE of (0.39, 0.43) and a CRI value of 83 were achieved from a B:G:R mass ratio of 100:1:3. The device showed peak current efficiency and power efficiency of 3.1 cd A^{-1} and 1.6 lm W^{-1}, respectively. More notably, the turn-on time to reach 100 cd m^{-2} was relatively fast at ~5 s. Furthermore, device lifetimes up to 17 h can be obtained for WLECs utilizing polyspirobifluorene-based copolymers. This work was the first demonstration of polymer WLECs employing the host-guest strategy.

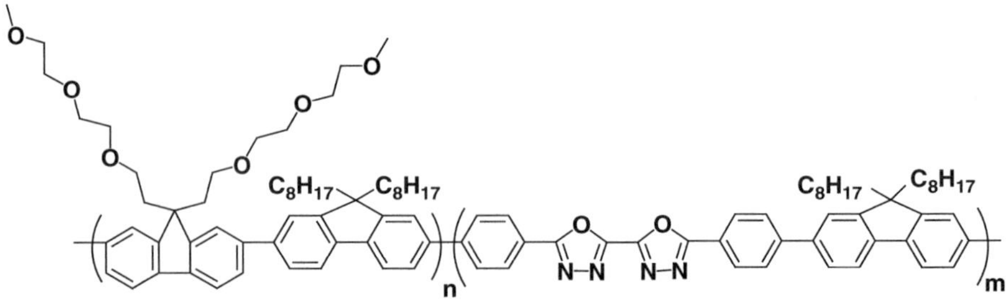

Polymer 30

Fig. 4.27 Molecular structure of polymer 30

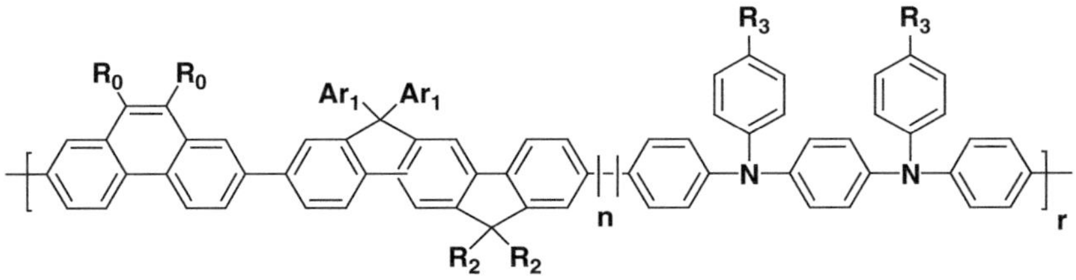

Polymer 31

Fig. 4.28 Molecular structure of polymer 31

Compared to multiple-component emissive layer, single-emitter WLECs possess relatively easier fabrication processes. In 2013, the same group, Tang et al. proposed a multifluorophoric conjugated copolymer (polymer **31**, Fig. 4.28) doped with an electrolyte ($LiCF_3SO_3$-TMPE) to inhibit energy-transfer interactions and thus demonstrate broad-band WLECs [60]. The EL spectrum showed high CRI of 82, CIE coordinates of (0.41, 0.45) and current efficiency of 3.8 cd A^{-1} from such WLECs was measured. Moreover, color-drift problems resulting from phase separation, which commonly occurred in conventional blended multi-emitter devices, can be removed when using single-emitter LEC configuration. Single-component LEC device has also been proposed by Tsai et al. in 2013 [61]. They synthesized a series of polyfluorene (PF) electrolytes bearing Br^-, BF_4^-, or PF_6^- counterions on the side chain. Yellow-emitting 2,1,3-benzoselenadiazole moieties were incorporated into blue-emitting PF main chains (polymer **32**, Fig. 4.29) to produce single-component WLECs. Peak EQE of 0.69 % and peak power efficiency of 1.56 lm W^{-1} can be obtained in WLECs based on polymer **32**. Furthermore, for this single-component multiple-chromophore WLECs, the EL can be varied by adjusting the bias and/or the thickness of the emissive layer.

Compared to white PLECs, WLECs based on iTMCs generally showed higher device efficiencies due to phosphorescent nature of iTMCs. In 2008, the first iTMC-based WLECs was reported by Su et al. [30]. Such WLECs were composed of blue-

Fig. 4.29 Molecular structure of polymer 32

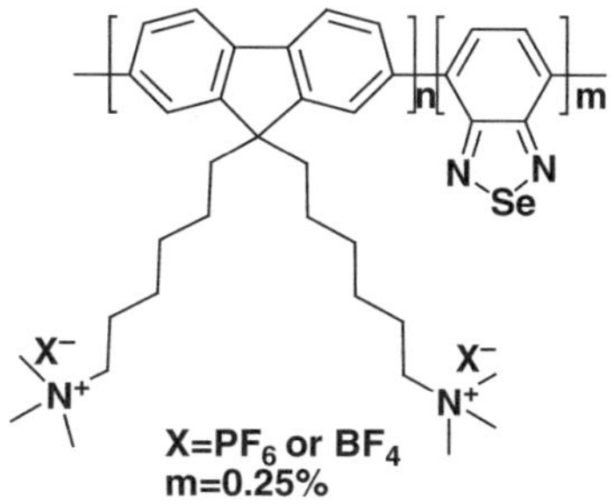

Polymer 32

Fig. 4.30 Molecular structure of complex 33

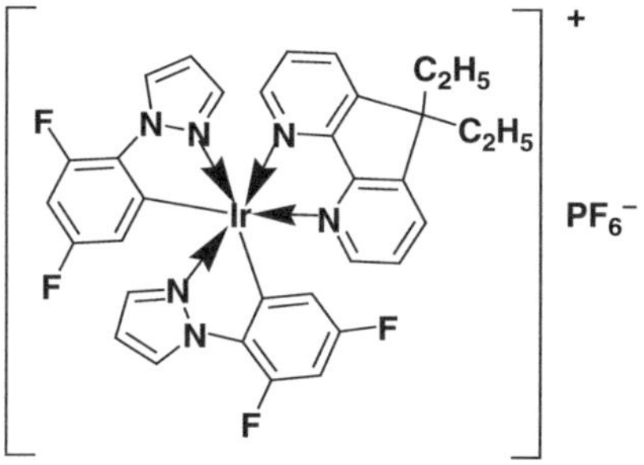

Complex 33

green-emitting complex **33** (Fig. 4.30) doped with a small amount of red-emitting complex **9** (Fig. 4.9). Additional ionic liquid [$BMIM^+(PF_6^-)$] was added to the emissive layer to shorten the device response time [62]. The devices show white EL spectra with CIE coordinates from (0.45, 0.40) to (0.35, 0.39) at 2.9–3.3 V and high CRI value up to 80. Peak EQE of 4 % and peak power efficiency of 7.8 lm W^{-1} can be achieved under 2.9 V.

According to the same conception, in 2009, He et al. reported WLECs by doping the red-emitting complex **10** (Fig. 4.10) into the blue-green-emitting complex **34** (Fig. 4.31) [31]. These devices exhibited warm-white light with CIE coordinates of (0.42, 0.44) and CRI of 81. Under 3.5 V, the peak EQE, current efficiency, and power efficiency were 5.2 %, 11.2 cd A^{-1}, and 10 lm W^{-1}, respectively. To acquire further enhancement of efficiency, in 2010, the same group fabricated WLECs with CIE coordinates of (0.37, 0.41) and CRI value up to 80 by doping the same guest red-emitting complex **10** (Fig. 4.10) into the host blue-green emitting complex **26** (Fig. 4.23) [51]. For complex **26**, superior photoluminescent quantum yields (PLQYs) obtained in neat film was due to the sterically bulky 4-tritylphenyl group as the ancillary ligand, which suppressed the intermolecular interaction and excited state self-quenching. These WLECs showed peak EQE, current efficiency and power efficiency of 5.6 %, 11.4 cd A^{-1} and 11.2 lm W^{-1}, respectively.

In 2011, Su et al. introduced a double-doped strategy to improve device efficiency of WLECs [62]. An orange-emitting guest complex **35** (Fig. 4.32) was incorporated into an emissive layer composed of an efficient blue-green emitting host complex **23** (Fig. 4.21) and a red-emitting guest complex **9** (Fig. 4.9). Complex **35** improved carrier balance of WLECs and thus peak EQE of 7.4 % and peak power efficiency of 15 lm W^{-1} were achieved in these double-doped WLECs. More recently, Chen et al. reported WLECs based on complex **36a** (Fig. 4.33) as the host and complex **36b** (Fig. 4.33) as a guest [63]. Peak EQE of 5 % and current efficiency of 12.4 cd A^{-1} can be achieved in these WLECs.

Fig. 4.31 Molecular structure of complex 34

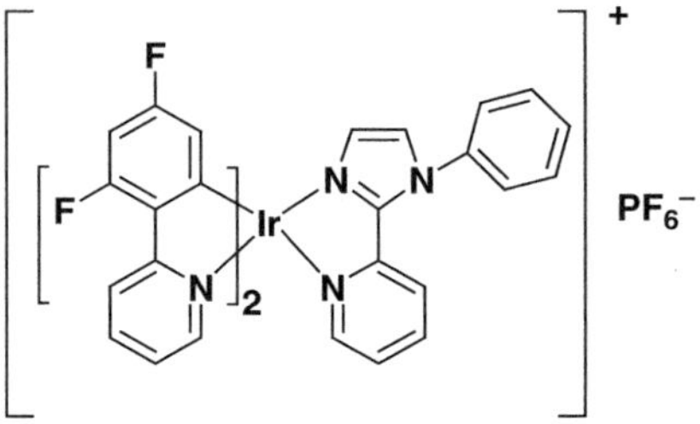

Complex 34

Fig. 4.32 Molecular structure of complex 35

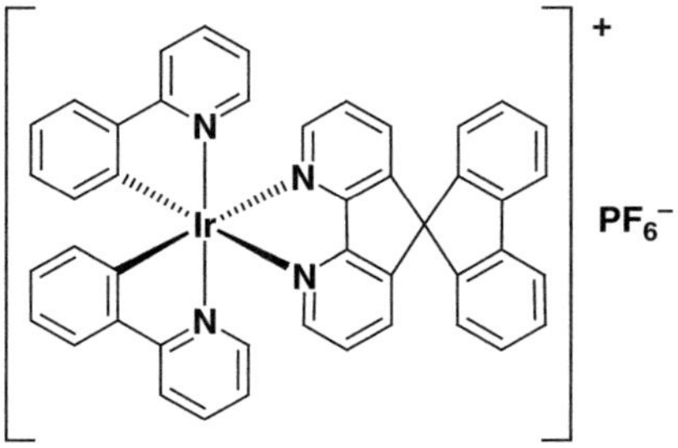

Complex 35

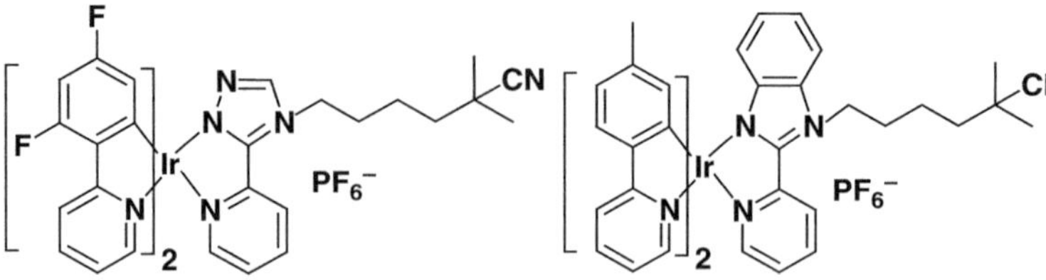

Complex 36a Complex 36b

Fig. 4.33 Molecular structures of complexes 36a and 36b

4.3.2 Device Lifetime

Lifetime of LECs is an important indicator in evaluating durability for display or lighting applications, which is generally defined as the time it takes for the brightness of the device to decay from the maximum to half of the maximum. Some works about improving device lifetime of LECs are reviewed in the following.

For PLECs, short lifetime is due to the existence of chemical degradation reaction and phase separation [64]. Waagberg et al. revealed the spatial position at which device degradation occurs and identify the mechanism behind the chemical degradation reaction [65]. They utilized a planar surface cell with 1 mm interelectrode gap and utilized blend of MEH-PPV, PEO and $LiCF_3SO_3$ as the active material. By exposing a number of pristine MEH-PPV+PEO+$LiCF_3SO_3$ films to various temperatures and photo-excitation intensities, the photo-induced degradation accelerated by increased temperature was observed. Thus, improving thermal heat sinking of the active material, especially in the light-emitting p–n junction region, would be a feasible way to improve lifetime. In addition, operating LECs at higher initial applied voltages and under conditions at which the active material exhibited high ionic conductivity were possible to alleviate the extent of this undesired side-reaction and attained a relatively centered p-n junction [66].

Polymer end groups also affects the lifetime of LECs. Phase separation is generally inevitable in PLECs due to large difference of polarity between the electrolyte and polymer. Hence, implanting polarity groups into polymer main chains is effective to tune the difference of polarity between the electrolyte and polymer and suppress phase separation [67]. Furthermore, lifetime of LECs also depended on types of ionic electrolyte and applied voltages [8]. Lifetime generally degraded as the bias increased.

For iTMC-based LECs, degradation of $[Ru(bpy)_3]^{2+}$-based device was associated to water-assisted ligand-exchange reactions to form photoluminescence quenchers. Possible quenchers were complex $[Ru\text{-}(bpy)_2(H_2O)_2]^{2+}$ and $[Ru\text{-}(bpy)_2(H_2O)_2]_2O^{4+}$. Both Kalyuzhny et al. [68] and Zhao et al. [69] supported the water-induced formation of photoluminescence quenchers by observing reduced lifetime owing to humidity. For LECs based on ionic iridium complex, the reason of degradation also resulted from similar water-assisted ligand-exchange reactions. Hence, the design of robust hydrophobic complexes in Ir-iTMCs was one of the feasible strategies to lengthen device lifetime. For instance, hydrophobic groups were attached on the periphery of the ligands and intramolecular π–π stacking interactions were introduced into the iTMC. Both approaches have been shown to significantly enhance the stability of LECs based iTMCs. Bolink et al. reported efficient and stable LECs based on complex **37** (Fig. 4.34), which showed intramolecular π–π stacking interaction [70]. Such devices showed long lifetime >3,000 h at an average brightness of 200 cd m^{-2} under 3 V.

Fig. 4.34 Molecular structure of complex 37

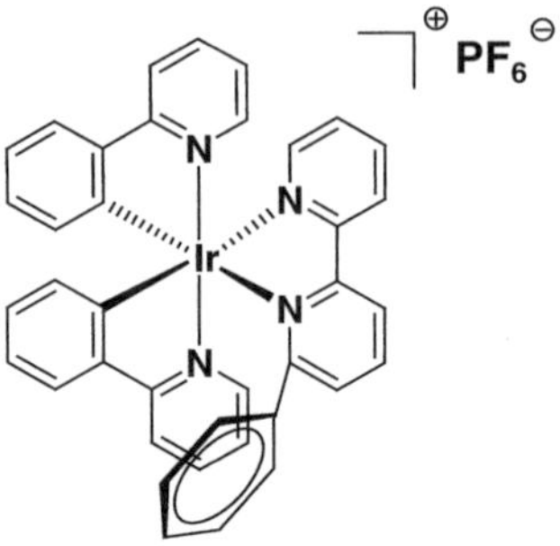

Complex 37

4.3.3 *Device Turn-On Time*

4.3.3.1 Bias Voltage and Counterions

Under an applied constant bias, current density and brightness of LECs increase with time owing to the motion and redistribution of ions in LECs. Turn-on time of LECs, which is generally defined as the time required for LECs to reach maximum brightness under a constant bias, commonly ranges from a few minutes to several hours. For practical applications, turn-on time of LECs should be significantly improved. However, it is inevitable that short turn-on time is usually obtained at the expense of device stability.

Applying higher voltages, either continuously or alternatively, leads to faster turn-on because of faster movement of ions towards the electrodes. For the former, it results in shorter device lifetimes but the latter, e.g., a pulsed-driving scheme, can improve both turn-on time and device stability. Turn-on time of LECs also depends on the thickness of the emissive layer of LECs. Thinner emissive layer reduced the turn-on time, but led to reduced device efficiency due to exciton quenching near the electrodes [71]. Environmental conditions for device operation may influence turn-on time, e.g. faster response was observed in air than in nitrogen atmosphere [72]. It was caused by the residual moisture and/or solvent existing in the emissive layer and thus the ionic mobility increased [22–24]. Compared with larger counterions, e.g. PF_6^-, utilizing complexes with smaller counterions such as BF_4^- or ClO_4^- led to shorter turn-on times since smaller ions increased ionic conductivity of the emissive layer. All approaches mentioned above are beneficial in enhancing ionic conductivities and thus induce faster response. However, device lifetime was deteriorated as well due to a faster rate of irreversible redox processes.

4.3.3.2 Chemical Modifications

Chemical modifications by attaching ionically charged ligands to metal complexes increase the ionic conductivity and subsequently reduce turn-on time of LECs. In

2008, Zysman-Colman et al. reported a series of homoleptic ruthenium and heteroleptic iridium iTMCs containing pendant triethylammonium groups (complex **38a–d** and complex **39a–d**, Fig. 4.35) [73]. Turn-on time of LECs based on complex **38c** and complex **39b** were only 5 s and 150 s under 3 V. Su et al. also demonstrated reduction in turn-on time of single-component Ir-based LECs with tethered imidazolium moieties [74]. The proposed iTMC was fused with two imidazolium groups at the ends of the two alkyl side chains on its ligand (complex **40**, Fig. 4.36). Both complexes possessed similar EL properties and comparable device efficiencies. However, the turn-on time of the LECs based on complex **40** (200 min) was much shorter than that of devices based on the model complex without imidazolium groups (500 min) due to increasing concentrations of mobile counterions (PF_6^-) in the emissive layer. In 2009, Kwon et al. reported the LECs employed complex **41** (Fig. 4.37), which was modified from a neutral iridium complex attached with a pendant Na^+ ion [75]. The LECs based on complex **41** showed a sixfold increase in device lifetime and a threefold shorter turn-on time as compared to the devices based on a model complex with PF_6^- counter anion. These results were attributed to faster delivery of the Na^+ ions to the electrode as compared to PF_6^- ions.

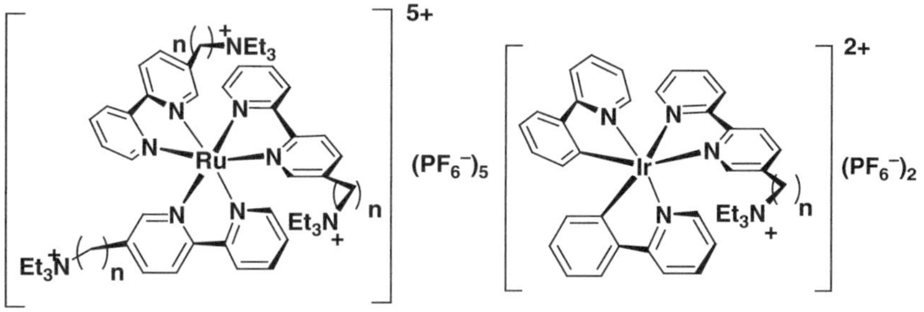

n= 1 (38a), 4 (38b), 6 (38c), 8 (38d)
Complex 38a-d

n= 1 (39a), 4 (39b), 6 (39c), 8 (39d)
Complex 39a-d

Fig. 4.35 Molecular structures of complexes 38a–d and 39a–d

Fig. 4.36 Molecular structure of complex 40

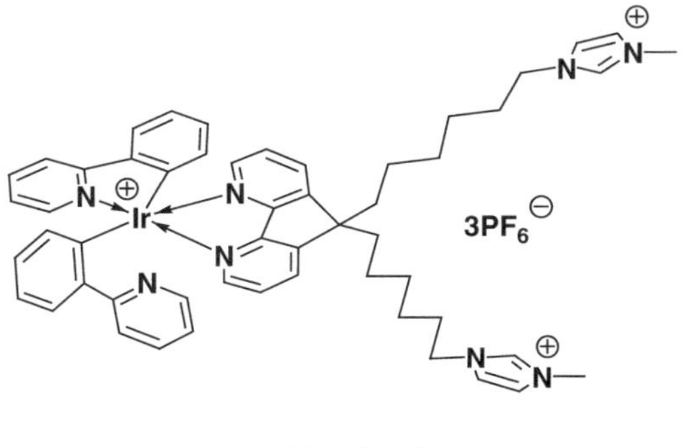

Complex 40

Fig. 4.37 Molecular structure of complex 41

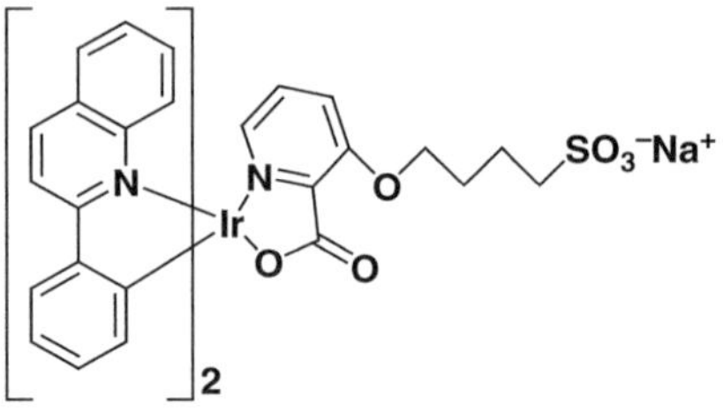

Complex 41

4.3.3.3 Ionic Additives

The ionic liquid $[BMIM]^+[PF_6]^-$ has been commonly incorporated into the emissive layer of LECs to reduce turn-on time. Malliaras et al. demonstrated improved turn-on time of LECs based on an iTMC by introducing the ionic liquid $[BMIM]^+$ $[PF_6]^-$ [76]. Turn-on time was reduced from approximately 5 h to 40 min as a result of an increase in the ionic conductivity upon addition of the ionic liquid. Similar phenomena were also found in LECs utilizing complex **16** (Fig. 4.15) doped with various amounts of $[BMIM]^+[PF_6]^-$. Compared with pristine devices, reduced turn-on time from 70 to 0.7 h was achieved when the molar ratio of Ir-iTMC: $[BMIM]^+$ $[PF_6]^- = 1{:}1$ was used. Furthermore, device current and power efficiency increased by almost one order of magnitude [37]. However, in above examples, improvement in turn-on time was unfortunately accompanied by deteriorated device stability. That is, a trade-off between turn-on time and device stability was inevitable.

Costa et al. blended various ionic liquids into Ir-iTMCs to study the balance between device response and stability of LECs [77]. Three ionic liquids were evaluated: $[EMIM]^+[PF_6]^-$, $[BMIM]^+[PF_6]^-$ and $[HMIM]^+[PF_6]^-$. Both LECs doped with $[EMIM]^+[PF_6]^-$ and $[HMIM]^+[PF_6]^-$ possessed faster device response. Nevertheless, the LECs doped with $[EMIM]^+[PF_6]^-$ showed higher luminance level while the LECs doped with $[HMIM]^+[PF_6]^-$ led to lower luminance. It was shown that it is possible to simultaneously optimize the turn-on time, luminance, and device stability upon the addition of an ionic liquid with a high intrinsic ionic conductivity.

4.3.3.4 Driving Mode

In 2001, Rudman et al. reported pulse operation of LECs dramatically increased device lifetime as compared to constant-voltage operation [78]. Next year, they operated a device under a constant current, i.e. applying a higher initial operating voltage automatically to achieve the set current flow. The initial high voltage rapidly decreased as the injection barriers were reduced by the displacement of the ions in the emissive layer. Hence, the device emits around 25 cd m^{-2} after ~0.5 s after a constant current has been applied [79]. Compared to voltage-driven pulsed operation, which led to slower device response and poorer device stability, current-driven pulsed operation exhibited better device performance. In 2012, Tordera et al. proposed LECs based on current-driven pulsed operation to reach much shorter turn-on times and promising device stability due to slowing down the extension of the doped zones from electrodes to intrinsic layer [80].

4.3.4 Device Engineering

Generally, LECs are composed of a single emissive layer sandwiched between two electrodes. However, to improve device performance, multi-layer configurations of LECs have also been reported. Device thickness is also a non-trivial parameter to optimize device efficiency since microcavity effect plays an important role in output characteristics of LECs. Furthermore, design of optical structure for enhancing light outcoupling efficiency is also an important issue to further improve device efficiency of WLECs.

4.3.4.1 Tandem LECs

Tandem organic light-emitting devices, in which multiple devices are piled vertically, have been shown to achieve 2X enhancement in light output and current efficiency. In 2013, the first tandem LECs were proposed by Lu et al. [81]. Tandem LECs were demonstrated by employing a thin poly(3,4-ethylenedioxythiophene): polystyrene sulfonate (PEDOT:PSS) layer as a connecting layer between two emitting units. Light outputs of the tandem devices were enhanced by ca. 1.8 times when compared to those obtained from the single-layered devices at similar current densities. Peak EQE obtained from the tandem LECs was up to 5.83 %, which is 2.37 times higher than that of the single-layered devices. Furthermore, as compared to the single-layered device, the tandem device can also provide higher device efficiencies and longer lifetimes at relatively higher light output.

In 2014, Su et al. reported a tandem device structure to enhance device efficiencies of phosphorescent sensitized fluorescent NIR LECs [28]. The emissive layers, which were composed of a phosphorescent host and a fluorescent guest to harvest both singlet and triplet excitons of host, were connected vertically via a thin transporting layer, rendering multiplied light outputs. Output EL spectra of the tandem NIR LECs were shown to change as the thickness of emissive layer varied due to altered microcavity effect. By fitting the output EL spectra to the simulated model concerning microcavity effect, the stabilized recombination zones of the thicker tandem devices were estimated to be located away from the doped layers. Therefore, exciton quenching near doped layers mitigated and longer device lifetimes can be achieved in the thicker tandem devices. The peak external quantum efficiencies obtained in these tandem NIR LECs were up to 2.75 %, which is over tripled enhancement as compare to previously reported NIR LECs based on the same NIR dye.

4.3.4.2 Microcavity Effect

Employing microcavity effect to dynamically monitor temporal evolution of recombination zone position of an operating sandwiched LEC was proposed by Su et al. in 2013 [82]. The recombination zone of LECs doped with a low-gap hole trapper has been shown to be closer to the anode as compared to neat-film devices. These results provided direct experimental evidence to confirm that carrier balance of LECs can be modified by adjusting carrier transport. Furthermore, the proposed technique would be a powerful tool for studying device physics of LECs.

Meanwhile, Su et al. subsequently demonstrated improving device efficiencies of WLECs by optimizing the thickness of the emissive layer [83]. They compared EL characteristics of WLECs with various thicknesses of active layers and the corresponding recombination zone positions were estimated by employing microcavity effect, i.e., fitting the measured EL spectrum and the simulated one with properly adjusted recombination zone positions. As compared to thinner devices (<200 nm), the recombination zones in thicker devices were closer to the center of the active layer, resulting in lesser exciton quenching. Hence, a high EQE up to 11 % can be achieved in WLECs with an active-layer thickness of 270 nm. However, further increasing the thickness to 400 nm resulted in deteriorated balance of carrier mobilities, rendering asymmetric recombination zone position in the emissive layer. These results confirmed that tailoring the thickness of the emissive layer to avoid exciton quenching would be a feasible approach to improve device efficiencies of WLECs.

Recently, Su et al. also reported non-doped WLECs by employing the microcavity effect [84]. With a relatively thicker emissive layer (490 nm), red emission can be additionally enhanced by the microcavity effect when the recombination zone moves to appropriate positions. Hence, white EL can be harvested by combining blue emission from the complex and red emission from the microcavity effect. These non-doped WLECs showed EQE and power efficiency up to 5 % and 12 lm W^{-1}, respectively. These results show that efficient white EL can be obtained in simple non-doped LECs.

4.3.4.3 Design of Optical Structure

In a typical bottom-emitting organic light-emitting devices, the majority of the light generated in the emissive layer is trapped in the organic and ITO layers due to significant differences in the refractive indices between these two layers ($n_{org.} \approx 1.7$ and $n_{ITO} \approx 1.9$) and the glass substrate ($n_{glass} \approx 1.5$). The proportion of such waveguided light is up to ca. 40–60 % of the total emission [85]. Another portion of emission is confined in the glass substrate due to total internal reflection at the glass/air interface. Finally, only ca. 20 % of total emission can be out-coupled into the air mode, which can be used eventually. To further enhance device efficiency, extracting the trapped light in waveguided and substrate mode would be a feasible way.

In 2014, Edman et al. demonstrated 60 % improvement in the outcoupled luminance in the forward direction by laminating a light-outcoupling film, featuring a hexagonal array of hemispherical microlenses as the surface structure, onto the front side of the device and a large-area metallic reflector onto the back side [86]. However, such enhancement resulted from extracting light from the glass substrate (substrate mode) only.

In 2015, Su et al. proposed a novel technique employing embedded red color conversion layers (CCLs) to recycle the trapped blue EL in the ITO layer (waveguide mode) and in the glass substrate (substrate mode) into the air mode [87]. High EQE and power efficiency up to 12.5 % and 27 lm W^{-1}, respectively, were obtained in such WLECs. Nevertheless, for the waveguide mode, only the evanescent tail of the blue EL in the red CCL can be recycled. A large amount of waveguided blue EL was still trapped inside the ITO layer.

To further recycle the trapped EL in waveguide mode, Su et al. demonstrated efficient WLECs employing waveguide coupling [88]. Waveguide mode in ITO layer and substrate mode in glass substrate can be extracted simultaneously. Hence, peak EQE and power efficiency of WLECs up to 19 % and 34 lm W^{-1}, respectively, can be achieved. These values were among the highest reported for WLECs and thus confirmed that waveguide coupling would be a feasible way to realize highly efficient WLECs.

4.4 Conclusion and Outlook

Solid-state LECs possess advantages of simple solution processes, low-voltage operation, and high power efficiency, and thus may be good candidates for display and lighting applications. In this chapter, we have introduced the working mechanism of LECs. Then we have provided overview of LECs, which have exhibited respectable efficiency across the visible spectral region, from blue to deep red and even infrared region. Furthermore, highly efficient WLECs are also reviewed. To overcome the drawbacks of LECs, e.g., long turn-on times and moderate lifetimes, several useful techniques have been mentioned. Finally, works on device engineering to improve carrier balance and to increase light outcoupling efficiency are reviewed.

To further improve device efficiency of LECs, further tailoring molecular structures to achieve more balanced electron and hole transport is required. Improving device lifetimes will also be essential for lighting applications. Introducing functional groups with resistance against quencher formation and/or electrochemical degradation in emissive materials will facilitate realization of long-lifetime

LECs for lighting applications. In addition, further increasing light harvesting from trapped waveguide mode and substrate mode would be beneficial in reaching device efficiencies required for practical applications.

References

1. Pei Q, Yu G, Zhang C et al (1995) Science 269:1086
2. Gao J, Dane J (2003) Appl Phys Lett 83:3027
3. Dane J, Tracy C, Gao J (2005) Appl Phys Lett 86:153509
4. Dane J, Gao J (2004) Appl Phys Lett 85:3905
5. Shin JH, Matyba P, Robinson ND et al (2007) Electrochim Acta 52:6456
6. Matyba P, Maturova K, Kemerink M et al (2009) Nat Mater 8:672
7. Rodovsky DB, Reid OG, Pingree LS et al (2010) ACS Nano 4:2673
8. Edman L (2005) Electrochim Acta 50:3878
9. Sun Q, Li Y, Pei Q (2007) IEEE/OSA J Disp Technol 3:211
10. Yang C, Sun Q, Qiao J et al (2003) J Phys Chem B 107:12981
11. Pingree LS, Rodovsky DB, Coffey DC et al (2007) J Am Chem Soc 129:15903
12. Slinker JD, Defranco JA, Jaquith MJ et al (2007) Nat Mater 6:894
13. Pei Q, Heeger AJ (2008) Nat Mater 7:167
14. Hu Y, Gao J (2011) J Am Chem Soc 133:2227
15. Van Reenen S, Matyba P, Dzwilewski A et al (2010) J Am Chem Soc 132:13776
16. Rudmann H, Shimada S, Rubner MF (2003) J Appl Phys 94:115
17. Li Y, Gao J, Yu G et al (1998) Chem Phys Lett 287:83
18. Lenes M, Garcia-Belmonte G, Tordera D et al (2011) Adv Funct Mater 21:1581
19. Ng WY, Gong X, Chan WK (1999) Chem Mater 11:1165
20. Suzuki H (2000) Appl Phys Lett 76:1543
21. Suzuki H (2004) J Photochem Photobiol A Chem 166:155
22. Hosseini AR, Koh CY, Slinker JD et al (2005) Chem Mater 17:6114
23. Bolink HJ, Cappelli L, Coronado E et al (2005) Inorg Chem 44:5966
24. Wang S, Li X, Xun S et al (2006) Macromolecules 39:7502
25. Xun S, Zhang J, Li X et al (2008) Synth Met 158:484
26. Bolink HJ, Coronado E, Costa RD et al (2009) Inorg Chem 48:3907
27. Ho CC, Chen HF, Ho YC et al (2011) Phys Chem Chem Phys 13:17729
28. Lee CL, Cheng CY, Su HC (2014) Org Electron 15:711
29. Tamayo AB, Garon S, Sajoto T et al (2005) Inorg Chem 44:8723
30. Su HC, Chen HF, Fang FC et al (2008) J Am Chem Soc 130:3413
31. He L, Qiao J, Duan L et al (2009) Adv Funct Mater 19:2950
32. Rodríguez-Redondo JL, Costa RD, Ortí E et al (2009) Dalton Trans 9787
33. Chen HF, Wong KT, Liu YH et al (2011) J Mater Chem 21:768
34. Costa RD, Céspedes-Guirao FJ, Ortí E et al (2009) Chem Commun 3886
35. Chen FC, Yang Y, Pei Q (2002) Appl Phys Lett 81:4278
36. Su HC, Lin YH, Chang CH et al (2010) J Mater Chem 20:5521
37. Costa RD, Ortí E, Bolink HJ et al (2009) Adv Funct Mater 19:3456
38. Su HC, Fang FC, Hwu TY et al (2007) Adv Funct Mater 17:1019
39. Su HC, Wu CC, Fang FC et al (2006) Appl Phys Lett 89:261118
40. Slinker JD, Gorodetsky AA, Lowry MS et al (2004) J Am Chem Soc 126:2763
41. Slinker JD, Koh CY, Malliaras GG et al (2005) Appl Phys Lett 86:173506
42. Lowry MS, Goldsmith JI, Slinker JD et al (2005) Chem Mater 17:5712
43. Bolink HJ, Cappelli L, Coronado E et al (2006) Chem Mater 18:2778
44. Yang Y, Pei Q (1996) Appl Phys Lett 68:2708

45. Fang J, Matyba P, Edman L (2009) Adv Funct Mater 19:2671
46. Yang Y, Pei Q (1997) J Appl Phys 81:3294
47. Yu Z, Wang M, Lei G et al (2011) J Phys Chem Lett 2:367
48. Slinker JD, Rivnay J, Moskowitz JS et al (2007) J Mater Chem 17:2976
49. Terki R, Simoneau L-P, Rochefort A (2008) J Phys Chem A 113:534
50. He L, Duan L, Qiao J et al (2008) Adv Funct Mater 18:2123
51. He L, Duan L, Qiao J et al (2010) Chem Mater 22:3535
52. Mydlak M, Bizzarri C, Hartmann D et al (2010) Adv Funct Mater 20:1812
53. Bolink HJ, Cappelli L, Cheylan S et al (2007) J Mater Chem 17:5032
54. Yang CH, Beltran J, Lemaur V et al (2010) Inorg Chem 49:9891
55. Chen HF, Liao CT, Chen TC et al (2011) J Mater Chem 21:4175
56. Cimrová V, Schmidt W, Rulkens R et al (1996) Adv Mater 8:585
57. Chen HF, Liao CT, Kuo MC et al (2012) Org Electron 13:1765
58. Sun M, Zhong C, Li F et al (2010) Macromolecules 43:1714
59. Tang S, Pan J, Buchholz H et al (2011) ACS Appl Mater Interfaces 3:3384
60. Tang S, Pan J, Buchholz HA et al (2013) J Am Chem Soc 135:3647
61. Tsai CS, Yang SH, Liu BC et al (2013) Org Electron 14:488
62. Su HC, Chen HF, Shen YC et al (2011) J Mater Chem 21:9653
63. Chen B, Li Y, Chu Y et al (2013) Org Electron 14:744
64. Shao Y, Bazan GC, Heeger AJ (2007) Adv Mater 19:365
65. Wågberg T, Hania PR, Robinson ND et al (2008) Adv Mater 20:1744
66. Fang J, Matyba P, Robinson ND et al (2008) J Am Chem Soc 130:4562
67. Zhang Y, Gao J (2006) J Appl Phys 100:084501
68. Kalyuzhny G, Buda M, Mcneill J et al (2003) J Am Chem Soc 125:6272
69. Zhao W, Liu CY, Wang Q et al (2005) Chem Mater 17:6403
70. Bolink HJ, Coronado E, Costa RD et al (2008) Adv Mater 20:3910
71. Lee K, Slinker J, Gorodetsky A et al (2003) Phys Chem Chem Phys 5:2706
72. Buda M, Kalyuzhny G, Bard AJ (2002) J Am Chem Soc 124:6090
73. Zysman-Colman E, Slinker JD, Parker JB et al (2007) Chem Mater 20:388
74. Su HC, Chen HF, Wu CC et al (2008) Chem Asian J 3:1922
75. Kwon TH, Oh YH, Shin IS et al (2009) Adv Funct Mater 19:711
76. Parker ST, Slinker JD, Lowry MS et al (2005) Chem Mater 17:3187
77. Costa RD, Pertegás A, Ortí E et al (2010) Chem Mater 22:1288
78. Rudmann H, Rubner M (2001) J Appl Phys 90:4338
79. Rudmann H, Shimada S, Rubner MF (2002) J Am Chem Soc 124:4918
80. Tordera D, Meier S, Lenes M et al (2012) Adv Mater 24:897
81. Lu JS, Kuo JC, Su HC (2013) Org Electron 14:3379
82. Wang TW, Su HC (2013) Org Electron 14:2269
83. Jhang YP, Chen HF, Wu HB et al (2013) Org Electron 14:2424
84. Lin GR, Chen HF, Shih HC et al (2015) Phys Chem Chem Phys 17:6956
85. Chutinan A, Ishihara K, Asano T et al (2005) Org Electron 6:3
86. Kaihovirta N, Larsen C, Edman L (2014) ACS Appl Mater Interfaces 6:2940
87. Lu JS, Chen HF, Kuo JC et al (2015) J Mater Chem C 3:2802
88. Cheng CY, Wang CW, Cheng JR et al (2015) J Mater Chem C 3:5665

Chapter 5
Industrial Photochromism

Andrew D. Towns

Abstract This chapter outlines how industrial photochromic dyes are designed, manufactured and used. It explains the properties that are crucial to their commercial utility and, in doing so, gives a definition of 'industrial photochromism'. Such colorants possess several key molecular features. This Chapter reveals how these structural elements influence photochromism. Additionally, it describes the typical chemistries employed in dye construction and takes a critical look at the limitations of current commercial colorants. Along the way, the directions that the exploitation of existing types might take in order to improve performance receive attention as do the potential future industrial applications to which photochromic dyes may be put.

5.1 Introduction

Global business worth many millions of dollars relies on photochromic colorants. The industrial utility of these dyes depends upon their defining characteristic, which is a reversible colour change that occurs in response to differing ambient light levels [1]. Their most important commercial application is the production of ophthalmic lenses that darken reversibly when exposed to strong sunshine. While industry actively tries to improve upon current light-responsive systems, researchers continue to attempt to develop disruptive new technologies that might capture, grow or kick-start new markets for industrial photochromism. This Chapter aims to give the reader an overview of the phenomenon and the colorants behind it, concentrating on those in commercial use today.

Numerous types of photochromic materials are known to science [2–4]. Collectively they exhibit wide variations in behaviour. Currently, however, commerce exploits only a limited number of these colorant classes for niche industrial applications. The properties demanded of them are narrow but exacting. This Chapter prescribes these traits, providing a definition for industrial photochromism. It outlines the history of its successful exploitation over the past 25 years, describing how industry took advantage of photochromic colorant chemistry.

A.D. Towns (✉)
Current Affiliation: Lambson Ltd., York Road, Clifford House, Wetherby LS227NS, UK

Vivimed Labs Europe Ltd., B3, Leeds Road, Huddersfield HD1 6BU, UK

G. Bergamini, S. Silvi (eds.), *Applied Photochemistry*,
Lecture Notes in Chemistry 92, DOI 10.1007/978-3-319-31671-0_5

Developing commercial products from photochromic colorants presents a different set of challenges to that posed by conventional dyes and pigments. Not only are these light responsive organic compounds less robust and orders of magnitude more expensive, but their dynamic nature also adds extra dimensions of complexity to their application. These additional requirements must be tackled by appropriate molecular design. The structures of industrial photochromic dyes may appear at first glance to be arbitrary and complicated, but each feature serves at least one purpose: this Chapter delineates their influence on properties and performance. It also puts into context important external factors that significantly impact colorant behaviour. The medium into which the dye is incorporated, as well as the additives that accompany it, exert a profound effect. In addition, the means by which photochromic colorants are applied can make the difference between success and failure because they are relatively delicate and sensitive to their environment. Since the manufacture of ophthalmic lenses consumes the bulk of photochromic dyes produced industrially, this Chapter will focus primarily on properties which matter most to this particular technology. However, consideration will also be given to other outlets, providing the reader with a deeper appreciation for the limitations of photochromic dyes as well as which techniques are appropriate for their application.

What follows is therefore not intended to be a comprehensive account of the synthesis or photochemistry of photochromic colorants – there are plenty of resources which go into such detail elsewhere as will be pointed out later. Instead, this Chapter provides the reader with an introduction to the design, manufacture and application of industrial photochromic dyes. It aims to furnish insights into the molecular features encountered in the structures of such colorants and impart rules of thumb that describe their influence on behaviour.

Finally, against the backdrop of numerous successful years of industrial photochromic dye use, much effort continues to be dedicated to bringing new technologies to market based on photochromism. While this endeavour has not yet borne fruit, it could lead, if successful, to the commercialisation of colorant classes at unprecedented industrial scales. Consequently, these applications will receive brief mention along with the dyes that form the focus for what could be the next 'killer app' for photochromism.

Before then, we shall take a look at what is meant by photochromism as exhibited by industrial photochromic dyes.

5.2 Industrial Photochromism

The conventional definition of photochromic behaviour is much broader than the type of light-responsive effect which forms the basis for most industrial activity. Photochromism is usually defined as a reversible colour change that is driven in one or both directions by the action of light [5]. The word "reversible" is of key importance. If light causes an irreversible colour change, then the effect should not be described as photochromism. In certain fields, a permanent one-way switch in colour though exposure to light is said to be "irreversibly photochromic". However, this phrase is incorrect by definition. Just as one would not describe the permanent fading in colour of a dyed fabric as an example of photochromism, an irreversible change in coloration must not be referred to as a photochromic effect. The reversible appearance of colour, disappearance of colour, or alteration in hue are all phenomena validly described as photochromism. This section attempts to narrow down the above behaviour into a definition of industrial photochromic behaviour.

5.2.1 A Definition of Industrial Photochromism

Photochromic phenomena can be split into one of two types as shown in Fig. 5.1. Those materials in which colour changes are driven entirely photochemically in both directions exhibit what is known as "P-type" photochromism. In contrast, illumination switches "T-type" compounds in only one direction – they return to their original state in its absence. The reversion occurs thermally. An easy way to remember this distinction between the two kinds of behaviour is that the label T-type denotes that a Thermal transformation is involved, whereas P-type signals that purely Photochemical changes are concerned.

P-type compounds typically change colour upon irradiation with ultraviolet (UV) or visible light and remain in that state, even when no longer exposed to the illumination. They only return to their original state after light of another range of wavelengths strikes them. True P-type colorants therefore generally behave like switches. They flip between states of different absorption profiles, but only by subjecting them to appropriate wavelengths of light. As shall be mentioned later in this Chapter, the potential for the states of P-type materials to be manipulated in a controlled way has captured the attention of many researchers and technologists for use in a wide variety of different fields [6].

Despite this interest, the most commercially important photochromic compounds have a T-type character. The vast majority of such colorants in industrial use exhibit a specific subset of T-type photochromism: UV radiation, and in some cases short wavelength visible light, triggers the conversion of colourless (or almost colourless) species to coloured forms. Although the exact wavelengths of light that effect the change are dependent on dye structure, they typically lie within the UV-A region (320–400 nm) for commercial photochromic colorants. (Some applications

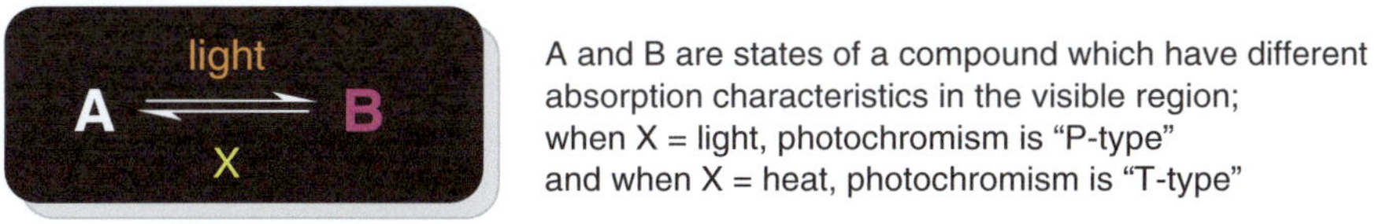

Fig. 5.1 General depiction of P- and T-type photochromism

rely on the ability of particular photochromic molecules to react efficiently also to blue light of wavelengths around 410–420 nm.) Industrial dyes respond well to sunlight since UV-A makes up a significant proportion of the solar radiation incident at the Earth's surface: they become rapidly coloured and deliver pronounced photochromic effects in unfiltered sunshine outdoors. However, their effectiveness decreases markedly behind architectural glazing and automotive windows because these structures attenuate UV light. For example, laminated car windscreens block transmission of UV-A wavelengths very efficiently. On the other side of the coin, light sources that emit only a small proportion of their radiation in this region are, unsurprisingly, not very effective at activating commercial photochromic dyes. Consequently, artificial thermoluminescent illumination that radiates little energy within the near UV, such as tungsten filament light bulbs, is typically incapable of bringing about photochromism.

Industrial T-type colorants revert thermally to their colourless form in the absence of activating UV-A light as fast as design considerations and other performance aspects will allow. Not all dyes used commercially behave in a completely T-type manner. In these instances, certain wavelengths within the visible region also drive the back-reaction from coloured to colourless, conferring an element of P-type character on their photochromism. The presence of these visible wavelengths thus partly counteracts the effect of UV-A radiation by causing photobleaching. Consequently, the photochemical component of the back-reaction must remain inefficient.

Another way of classifying photochromic effects involves the nature of the colour change. The T-type dyes described above are said to exhibit "positive photochromism" [5]. This label means that the longest wavelength absorption band shifts from the UV upon irradiation to the visible region of the spectrum, i.e. moves from shorter to longer wavelengths and usually entails a transition from a colourless state to a coloured one through activation by UV. The opposite behaviour is "negative photochromism". This term does not refer to the absence of any photochromism! It typically applies instead to a coloured material becoming colourless upon irradiation with light, returning to its original coloured state in the dark, i.e. a shift of the longest wavelength band in the visible region to a shorter set of wavelengths in the UV. Examples of the latter type of switching are rarer than those of the former kind. As might be deduced from the preceding paragraph, industrial photochromism is overwhelmingly of the positive type.

The properties of commercially significant photochromic colorants outlined above thus represent only a small subset of the permutations under the general umbrella of possibilities covered by photochromism. The term "heliochromism" [7] accurately captures the specific characteristics of the photochromism typically exhibited by industrial dyes. This label, derived from the Greek word for sun, was coined to define a special type of photochromism where the response is dependent on the intensity of incident unfiltered sunlight. A "heliochromic" colorant is efficiently activated by near UV radiation, exhibits a moderately efficient thermal fade, and is photobleached with low efficiency by white light. While heliochromism is by far the most commercially important variant of photochromism, industry and academia usually apply the more general term "photochromic" when describing the behaviour of colorants and products which falls within the definition of heliochromic.

Industrial photochromism arguably takes the form of an even narrower definition (see Fig. 5.2). Economic considerations come into play in addition to technical criteria. Photochromic effects must meet consumer expectations of longevity and cost without infringing third party intellectual property. In addition to forming the basis for marketable products, industrial photochromism has to be economically viable as well as safe to remain a commercial proposition otherwise the phenomenon just reverts to being photochromism!

The relationship that industrial photochromism has with the other classes of photochromic behaviour is depicted in Fig. 5.3. The hierarchy as shown is a little

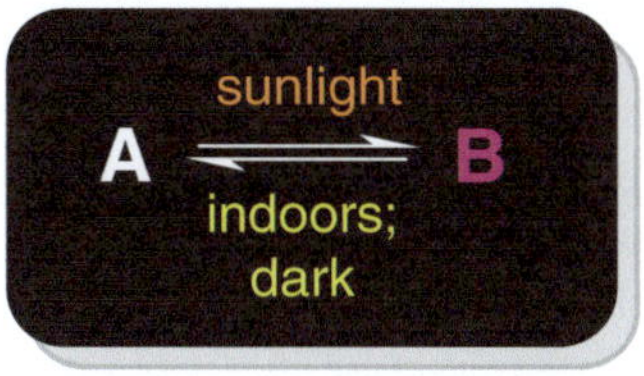

A reversible colour change induced in a dye, which:

(i) rapidly becomes coloured on exposure to a source of strong UV light and quickly becomes colourless when exposure to the source is stopped;

(ii) can be repeated over many cycles during a time period that is acceptable to the expectations of consumers;

(iii) is cost-effective and safe to implement.

Fig. 5.2 A definition of industrial photochromism

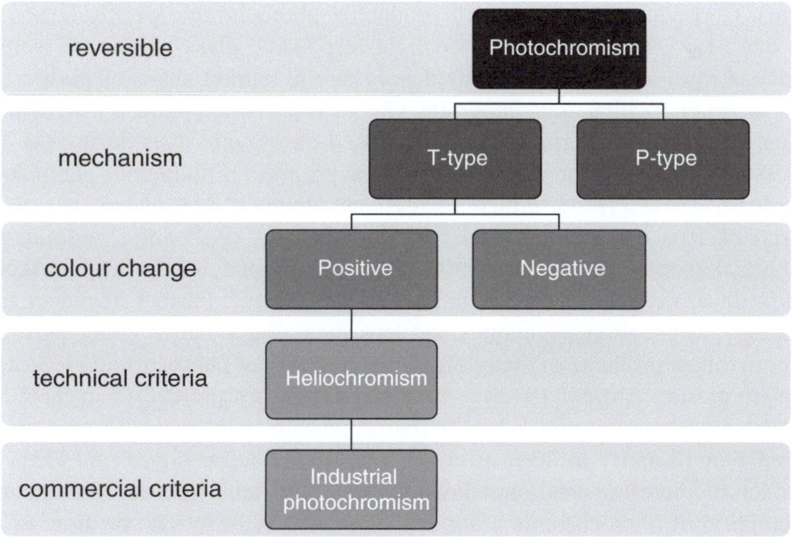

Fig. 5.3 A hierarchy of photochromic behaviour

simplistic in that many industrial dyes are not purely T-type and it refers only to commercialised mainstream applications. The introduction of significant new potential uses for P-type photochromic colorants mentioned in the following section would lead to major revision of Fig. 5.3.

As we shall see in the next section, although heliochromism forms the basis for a commercially successful global enterprise, the genesis behind the dyes upon which it is reliant may be traced back to efforts to develop materials by the military.

5.2.2 History, Current Status and Future Development of Industrial Photochromism

The roots of the industrial exploitation of photochromism lie in the first modern systematic efforts to study organic light-responsive materials in the mid-twentieth century. In its pioneering attempts to make use of photochromic colorants for the visual protection of its personnel, the US military synthesised the initial examples of a class of dye [8] that eventually became one of the most industrially important families. However, the first commercially successful outlet for industrial photochromism marketed to the consumer depended not on organic colorants, but on an inorganic system [9]. Photochromic glass lenses containing silver halide microcrystals [10] have been produced in their millions since the 1960s for prescription spectacles. Since then, polymer materials have largely replaced mineral glass in the fabrication of lenses because the former substances enable the construction of more lightweight and robust optics – but they are not compatible with silver halide systems. Consequently, the goal of producing photochromic plastic lenses grew from one of wishful thinking, when light-responsive glass spectacles were first introduced, into a real and unfulfilled need as the market share of plastic lenses grew. In order to plug this important gap in the market, industry searched for commercially viable alternative systems based on organic dyes during the 1970s and 1980s – it formed the primary driver for progress in industrial photochromic technology. These efforts, which focused not only on dye design, but also the material of lens construction and mode of colorant application, culminated in commercial success during the 1990s in the US, Europe and Japan [9]. Accounts of early developments in the US are available [11, 12]. Further advances to dye technology, application techniques and lens media have led to better performing and more robust products so that globally the majority of photochromic lenses worn today are plastic. Around 15–20 % of corrective lenses prescribed in the US and Australia are photochromic, whereas the proportion is lower in Europe and Asia (around 1 in 10 and 1 in 20 respectively) but nevertheless significant [13]. Their manufacture therefore has spread worldwide. The lens industry dominates the consumption of photochromic colorants, acquiring them by one or more of these means:

(i) Use of photochromic lens monomer mixtures that are marketed by companies with the expertise in dye and polymer chemistry to formulate them. This business model enables lens manufacturers who do not have in-depth knowledge of the intricacies of formulating to cast or coat photochromic lenses themselves. It also gives them freedom to operate where there is patented technology in place.
(ii) Purchase of colorants 'off-the-shelf' from specialised photochromic dye producers (such as Vivimed Labs Europe Ltd). This route allows lens manufacturers flexibility in creating their own formulations, but without having to concern themselves about colorant chemistry.
(iii) Synthesis by external contract manufacture or an in-house facility. Since this option demands substantial investment of resource as well as an in-depth knowledge of dye design, it tends only to be used by larger players within the industry.

The industry associated with the production of photochromic plastic lenses continues to account for a substantial proportion of the research taking place into photochromism. While it has achieved its overall goal of producing neutral-coloured photochromic eyewear based on plastic lenses with performance and robustness levels generally acceptable to the consumer, more remains to be done. The business still strives to improve properties like responsiveness and durability as well as develop new ways of introducing photochromism to lens materials that have proved resistant to it – more will be said about this aspect later. The improvements made in photochromic dye technology for lenses have crossed over into other avenues of use, such as coloration of plastics and surface coatings, from the perspective of colorant choice and performance. Nevertheless, what was said in a previous review [1] holds true: "*... achieving commercially acceptable performance is by no means a trivial matter for those unused to working with photochromic colorants. They tend not to be as robust or as easy to use as conventional dyes and pigments. In addition, they are more expensive, cost being of the order of thousands of dollars per kilogram of photochromic dye as opposed to merely dollars per kilogram for commodity colours.*" In order to create a product with satisfactory economics, behaviour and longevity, a commercial enterprise must take heed of numerous technical factors which are peculiar to industrial photochromism. As we shall see in the next section, various disparate applications have been envisaged for photochromic colorants. Some present technical challenges that remain beyond the reach of current dye technology. Others, however, have met with commercial success.

5.3 Focus on Important Applications of Photochromic Dyes

Significant investment has been sunk into research and development work over the past half century or so that has attempted to make a wide range of aesthetic and functional applications for light-responsive colorants a reality. This section looks at the various potential and actual uses for which photochromic dyes have been advocated.

5.3.1 Types of Application

So far, one main application – sun-responsive lenses – has dominated in terms of commercial success. Other diverse outlets have also been industrialised, but all rely on organic molecules exhibiting T-type photochromism. Each usage depends upon one or more of three aspects of photochromic behaviour [2, 14]:

(a) Reversibility
In the case of sun lenses for spectacles, the importance of reversible adaptation to varying ambient lighting levels of transmittance is self-evident. For these items to constitute useful products, they must automatically modulate the intensity of light received by eyes of the spectacle wearer on a repeated basis for a significant duration. More will be said about the ophthalmic exploitation of photochromism later, but for most uses of photochromic colorants, robust reversibility is crucial. The term is relative as light-responsive systems can be rather delicate – more so than conventional dyes and pigments. Nevertheless, even low-end aesthetic applications depend upon repeatable colour change effects to bring about variations in lighting levels without intervention by other means. Resilient switching is an essential requirement for potentially wide-ranging functional applications, such as adaptive windows and agricultural films, which are intended to spontaneously regulate transmitted light for the comfort and protection of the people and crops inside. These last two types of use are a highly sought after prize, but are beyond the reach of known photochromic colorant systems. Unfortunately, the durability of current materials fall far short of the minimum of a decade of working lifetime required for commercial success. It is unlikely that the types of industrial dye in circulation today can ever be modified to stand up to the levels of prolonged solar exposure which must be endured. Alternative solutions to the problem of furnishing sufficiently long-lasting reversible colour changes posed by glazing and related applications are being intensively researched. One active field attempting to tackle the issue of longevity centres on electrochromic materials, for windows [15] and even spectacle lenses [16].

(b) Sensitivity to radiation
As well as being essential to the above functional uses, this property forms the basis for the creation of arresting visual changes that are purely for aesthetic reasons. These encompass attention-grabbing promotional novelty items, toys and even pieces of jewellery as well as artwork. While there have been demonstrations of striking heliochromic effects in large structures, the aforementioned shortcomings in photostability prevent them becoming a commercial reality. For example, the concept of vehicles coated with T-type

photochromic materials was explored in the 1980s, but durability remains orders of magnitude too low for it to become anything more than a curiosity. Challenges of a different kind hold back the exploitation of photochromic compounds in cosmetics and personal care products. They are regulatory in nature: the business might not justify the cost of proving that use in such applications is safe, or be barred legally or ethically from doing so.

Photochromic colorants have, however, been employed commercially in textile coloration for many years. This usage firmly remains niche for a small market size where poor fastness, particularly photostability, is not critical. The usual technique for application is screen printing whereby a coating laden with photochromic colorant is applied to the surface of a garment and cured [17]. Photochromic detail may also be imparted to fabric by means of - commercially-available mass-coloured thread, such as poly(propylene) melt-spun with photochromic dye, being woven into it. Other conventional textile coloration techniques are not suited to the application of industrial photochromic colorants. For example, conventional commercial T-type dyes are not substantive to natural fibres, while treating them as disperse dyes in the exhaustion dyeing of important hydrophobic polymers such as polyester [18], nylon [19] and acrylic [20] leads to significant residual colour and/or weak photocoloration. This outcome lies with the fact that the dye molecules are relatively delicate and bulky. The conditions required to exhaustion dye these synthetic fibres degrade photochromic colorants. Additionally, the relatively large molecular size of photochromic colorants compared to conventional disperse dyes greatly hinders diffusion into the polymer. As will be discussed later, the nature of the polymers themselves tends to suppress photochromism. Durable, high-contrast photocoloration with acceptable kinetics will require specific development of new coloration processes and a re-think of dye design, moving away from current conventional T-types which are the product of research to find optimal colorants for ophthalmics. Steps have been made in this direction, such as conversion of commercial types to water-soluble acid dyes for silk [21] and wool [22] or fibre-reactive colorants for cotton [23, 24]. The viability of alternative means of delivery of industrial dyes has also been examined, for example, in conjunction with solvents [25–27] or supercritical carbon dioxide [28] as well as sol-gel coating [29] and nanoparticulate composites [30]. However, the above aims require radical advances to be made.

(c) Functionality

Light-sensitive ophthalmic lenses straddle the boundary between aesthetics and functionality: not only must their colours be appealing to consumers, but they must also be pleasant to wear and capable of modulating lighting levels without degrading visual acuity. However, many proposed applications exploit photochromism purely as a means to an end. In these outlets, colour is of little or no importance: the mere existence of a visually observed or machine-readable change is key. One example of this functional use of photochromic colorants is their incorporation into packaging in the form of security markings. Their light-driven property changes make possible the overt or covert authentication or validation of goods [31]. Variations on this approach that have been investigated range from the discrete marking of currency to the hidden labelling of fuel [32]. However, the durability and cost of the dyes hampers the commercialisation of these applications. An interesting niche functional use of photochromism to reach the market relates to angling. The inclusion of a photochromic colorant in fishing line enables anglers to keep track of the cast line whilst hiding its presence from their prey. Water shields the line from UV, so below the surface the line remains colourless and inconspicuous to fish. In contrast, above the surface of the water, solar UV photoactivates the dye in the line, so it becomes coloured and readily visible.

5.3.2 Applications of P-Type Dyes

A large proportion of the research effort into photochromism expended over the past decade concentrated on highly functional applications. Most of them tend to lie within the realm of information technology, optoelectronics and nanotechnology. All of these hi-tech uses require tightly controlled reversible switching of optical properties between states that do not decay [33, 34]. P-type colorants are suited to these sorts of task [35], whereas T-types are not owing to their thermal fading. For example, optically-switchable molecules have been at the heart of efforts to develop next generation memory technologies because they can in theory offer much greater data storage densities than existing systems [36–39]. Many instances of novel materials that can act as light-activated switches have been created by covalently linking P-type dyes to non-photochromic systems so that the fluorescence [40, 41] or electrical [42, 43] properties of the latter can be photo-regulated. Another area that has attracted great interest is full or partial replacement of electronic circuitry with that based on photons, for example in computer processors, leading to step changes in speed and power consumption [44]. Consequently, P-type dyes have been the focus of work to create optical equivalents to electronic components, such as switches and logic gates [45, 46]. Other avenues of investigation include therapeutic uses, such as the photo-control of drug activity, as well as the development of light-responsive constructs intended as components for machinery at the macro [47] and nano scale [48].

The most well studied P-type colorants present a marked contrast to T-type dyes not only in their photochromic properties, but also in their synthesis and the mechanism behind their light responsiveness. Compounds displaying pure P-type behaviour are rare: just two classes are known [49]. The first, diarylethene, is a relative newcomer and continues to receive the greatest attention, although to date, just a small number of such dyes are listed for sale in lab supply catalogues. The second class, made up of fulgides and their derivatives, is older and comprises examples that were commercialised for a time, but remains of minor importance in an industrial context. These two types will now be reviewed briefly along with another chromophore, azobenzene, which is very familiar to industry. Its P-type photochromic tendencies have led to a surge of interest in teaching this old dog new tricks.

5.3.2.1 Fulgides

Since fulgides were first synthesised a century ago, they and their derivatives attracted attention because of their unusual chromic properties [50, 51]. These include piezochromism (colour change with pressure), tribochromism (colour change with mechanical grinding) and photochromism [52]. Interest in their photochromic properties grew during the early 1980s following the finding that some derivatives, such as derivative **1** shown in Fig. 5.4 [53], exhibited purely P-type photochromism [54], making them candidates for the functional applications listed above [55].

Fulgides and their derivatives are capable of spanning the gamut of the visible spectrum, and even into the infra-red, when photoactivated with UV light [52], which leads to a ring-closure reaction. Figure 5.4 illustrates such a transformation for dye **1** which was once commercially available as Aberchrome 540 for use in chemical actinometry [56]. The low spatial requirements for the isomerisation mean that these compounds are photochromic as solids as well as in solution. However, a major drawback of this class is poor fatigue resistance.

Fulgimides (imido analogues of fulgides) were used commercially for producing lenses in Japan during the early 1990s [57]. Fulgides themselves have been the subject of investigation for conventional textile and printing ink applications [58]. While fulgide derivatives are still available from laboratory chemical

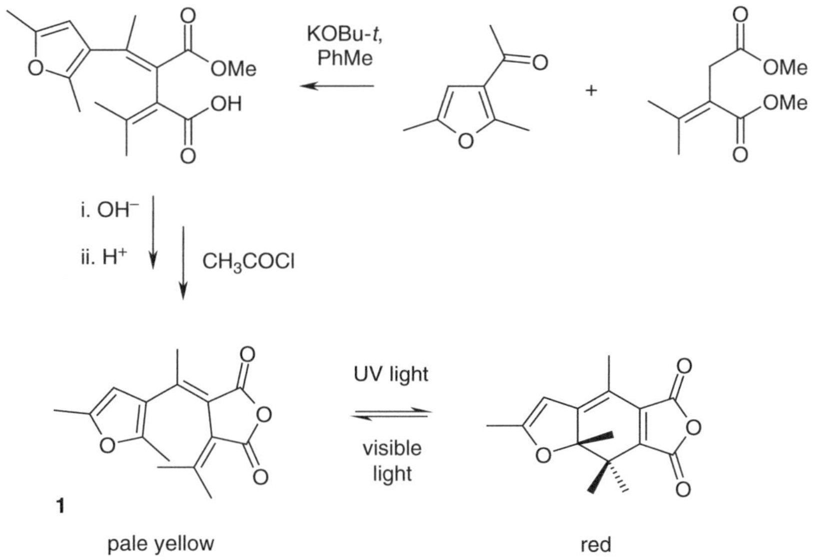

Fig. 5.4 Synthesis and photochromism of a furyl fulgide **1** [52]

suppliers in research quantities, their industrial renaissance awaits the development and commercialisation of an optical switching technology to which they suited.

5.3.2.2 Diarylethenes

This class of compounds remains the most intensely researched group of P-type dyes over the past two decades [54]. Great interest surrounds their potential use as elements in high density memory systems, optical switches for photonic circuitry and light-triggered components in nanomechanical devices as well as for many other applications [59, 60]. The subclass of dithienylhexafluorocyclopentenes (or dithienylethenes, DTE) is arguably the most deeply investigated. An example of the synthesis and photochromism of a generalised symmetrical DTE is shown in Fig. 5.5 [61]. A halothiophene is lithiated, then reacted with perfluorocyclopentene, which is a low boiling (27 °C) liquid, to create a colourless compound **2**. This material undergoes a ring-closure reaction, either in solid form or in solution, upon exposure to UV, generating a thermally stable coloured species. Irradiation of the coloured ring-closed form with particular wavelengths of visible light leads to reversion through ring-opening back to the colourless isomer **2**.

As well as symmetrical DTE, asymmetric compounds can be created by sequential exposure of the cyclopentene to different lithiated heterocycles rather than just two equivalents of the same one [59]. By appropriate design, one can synthesise a range of dyes that spans the visible spectrum. In one elegant proof-of-concept study, three different types of DTE units were linked together to give a 'trimer' that was capable of being switched between yellow, red and blue colours by particular combinations of UV and visible light for potential use as a multiple state memory element [62]. Crystals of DTE undergo pronounced shape changes on irradiation [63, 64], making them of interest as actuators in nano-technology. Despite the significant effort which continues to be expended on diarylethenes, no applications for these colorants have yet become a commercial reality.

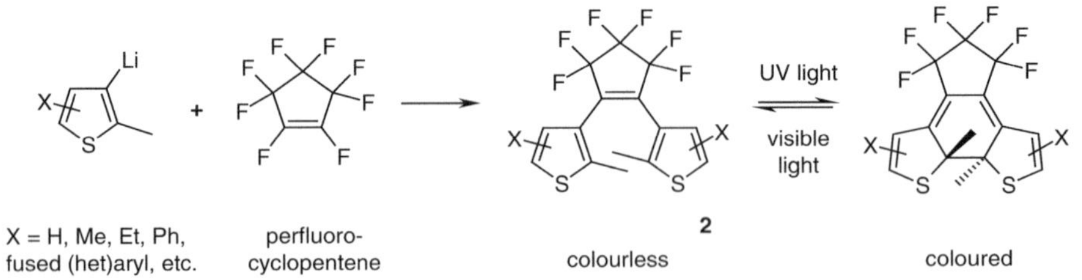

Fig. 5.5 Synthesis and photochromism of a dithienylethene **2**

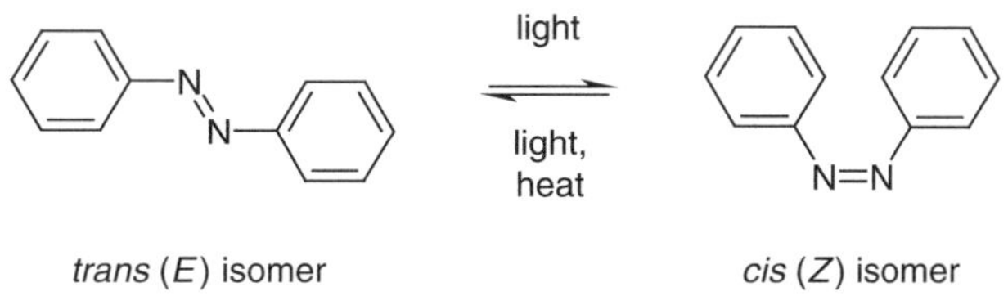

Fig. 5.6 Photochromism of azobenzene

5.3.2.3 Azobenzenes

Although not a pure P-type photochromic material and despite being a very well-known chromophore, azobenzene (and more correctly compounds derived from it) has been receiving much attention recently as a means to create functional materials. Through structural modification, its P-type character can be strengthened, such that the *trans* to *cis* photoisomerisation is effected with one set of wavelengths and the *cis* to *trans* conversion brought about with another range of wavelengths as shown in Fig. 5.6 [65].

As with the other two classes discussed above, azobenzene-based systems have been investigated in connection with optical switching and storage of data. Of particular note is the incorporation of azobenzene units into polymers to produce light responsive materials with novel properties. The geometry changes in azobenzene brought about by photoisomerisation produces deformation: contraction occurs upon irradiation with UV light whereas exposure to visible light leads to expansion. This phenomenon made possible a proof of concept study in which a light-driven motor was demonstrated: the design centred on an azobenzene-containing belt which acted as a photomobile material, transforming light energy into controlled motion that turned the motor [47].

It is remarkable that the azo chromophore, despite being almost as old as the field of synthetic dye chemistry itself, is finding new uses at the cutting edge of material science. However, like the outlets being sought for P-type fulgide derivatives and diarylethenes, its commercial exploitation is likely to be some distance away. However, industry is already exploiting T-type colorants – we shall now look at the ways in which they are being utilised.

5.3.3 *Non-ophthalmic Applications of T-Type Dyes*

To get the best out of commercial T-type dyes, they must be incorporated into the right media using suitable techniques. One cannot simply treat them like conventional dyes and then expect to obtain photochromism of the desired colour with useful kinetics and sufficient resilience. The properties of the medium are hugely important – they can make the difference between obtaining the desired photochromic effect and failing to generate any photochromism at all. For it to perform in the intended manner, care must be taken to ensure that the photochromic colorant ends up in the correct physical state in an appropriate environment without exposing it to conditions that can degrade it during or after application. This section will illustrate this necessity by looking initially at the incorporation of photochromic dyes into polymeric media through mass coloration in which the coloration is mixed and dissolved in molten polymer. It will point out where the limitations of the colorants lie in addition to ways in which these shortcomings can be overcome.

5.3.3.1 Thermoplastics

Injection moulding is an effective means of mass-colouring thermoplastics with industrial T-type dyes. Irrespective of whether the colorant is introduced in the form of neat dye or a masterbatch (i.e. plastic containing a high concentration of dye), the outcome must be a monomolecular dispersion of dye in the polymer because conventional T-type colorants do not furnish observable photochromic effects as solids – they must be in solution. The chemical and physical nature of both colorant and polymer dictate the photochromic properties of the system. The rigidity of the polymer matrix and the free volume it makes available to the dye strongly influence the thermodynamics and kinetics of the photochromism. Typical commercial T-type colorants perform best in polymer matrices, which consist of flexible polymer chains, such as those of polyolefins – these matrices are characterised by high free volume as well as low flexural moduli and glass transition temperatures (T_g). All these factors alter rates and extents of activation and fading because commercial T-type colorants change molecular shape substantially when they switch between colourless and coloured states [66]. Consequently, a host matrix with high rigidity and crystallinity either elicits a poor photochromic response or none at all. A reasonable rule of thumb is that polymers with a flexural modulus of greater than 2GPa, like polycarbonates and styrene–acrylonitriles, should be avoided. Care must also be taken when applying the dye to ensure that it can tolerate the temperature and time required to melt the polymer and mix the components uniformly. Exposure to temperatures above 250 °C causes degradation of conventional commercial photochromic dyes, leading to discoloration and loss of light-responsiveness. Polymers which require high processing temperatures, like nylon, are thus out of the reach of such colorants.

The nature of the polymer influences the lifetime of the photochromic effect. The cumulative dose of incident UV radiation determines the amount of photodegradation that occurs, not the number of cycles the plastic is switched between colourless and coloured states. While dye structure and polymer type are key factors in governing the rate of deterioration with UV exposure, the inclusion of additives can make an enormous difference to the working life of the photochromic system. These materials are typical of those employed in non-photochromic systems but, when employed in the right concentration and combination, can extend lifetimes by an order of magnitude, for example in polyolefins [67, 68]. The following types of additives are often used:

- UV absorbers – such materials can be very effective through the capture of high energy photons, which might otherwise destroy dye; it is crucial that they are selected so that absorption within the window of UV wavelengths which activates the dyes is minimised and responsiveness is not diminished.

- Hindered amine light stabilisers and antioxidants – these compounds scavenge free radicals to prevent them attacking colorants.
- Triplet state quenchers – these additives impede photochemical pathways that do not contribute to photochromism and which can lead to formation of non-photochromic side products, thus preventing dye degradation and often discoloration. They do so by harmlessly dissipating the energy of reactive excited states of the dye arising from absorption of UV, thereby inhibiting photochemical side-reactions.

The usefulness of additives varies from dye to dye: what works well for one colorant–polymer combination is not necessarily beneficial for another [68]. To complicate formulation further, significant interactions between additives may occur. In some cases, additives harm dye performance by adversely affecting kinetics or increasing unactivated system colour. However, many industrial photochromic systems owe their commercial existence to judicious use of stabilising packages of additives.

Through a suitable combination dye, polymer, additives and application technique, pronounced photochromic effects with commercially acceptable lifetimes can be produced. The intensity of the activated forms of industrial dyes enables them to be achieved with relatively low loadings of colorant. These concentrations vary typically within the range of just 0.01–0.30 %w/w depending on the colorant and medium. For example, in the case of poly(ethylene), an inclusion level of only 500 ppm dye is typical. Less is sometimes more in the case of photochromic colorants: application at too high a concentration can diminish or destroy any photochromism. A common mistake made by those who are new to handling photochromic dyes is to assume that the remedy for weak photocoloration is simply to add more dye to the system rather than pay attention to other aspects.

The influence of matrix and difficulties with dye concentration can be avoided by using colorants in forms which enables them to be processed like pigments. One means is microencapsulation in which a solution of the dye in a non-polar solvent is enclosed within a polymer shell of only 1–10 μm in diameter. The resultant microcapsules can be dispersed in a similar manner to a conventional pigment, although higher loadings are needed than in the case of neat colorant. In addition, greater care must be taken during application to ensure that the microcapsules are not ruptured, for example, by shear stress during mixing. Since the dye is present encased in a solvent solution, microencapsulation presents two benefits: the first is that the colour and performance of the dye is independent of the polymeric matrix into which microencapsulate has been dispersed. Thus, even if the matrix is not favourable to photochromism, microencapsulation makes possible a satisfactory photoresponse. Secondly, because it is dissolved in a solvent, the kinetics of the dye will be superior to that typically observed in polymer matrices: not only will activation be more rapid, but thermal fading will occur more quickly too. Microencapulate is available commercially in powder form as well as an aqueous slurry. The subject remains the subject of industrial and academic study [69]. A related approach entails micronising a solid solution of photochromic dye in a polymer and dispersing the resultant photochromic pigment into the medium. One means of preserving the integrity of the dye/polymer system and attaining acceptable photochromic performance is to prepare core-shell particles [70]. The dye is incorporated into a relatively low T_g polymer, such as poly(2-ethylhexyl acrylate) which forms the core. Around it is fashioned a shell of a rigid high T_g polymer, like poly(styrene), conferring the particles with good mechanical properties to enable their application, for example in coatings or thermoplastics by mass coloration.

Other approaches to providing T-type photochromic molecules with localised environments which enable them to exhibit photochromism in media that are hostile to them include doping them into porous materials, such as silicas, zeolites and metal-organic frameworks [71]. The pores give sufficient free volume for the spatial reorganisation needed for interconversion of T-type dyes. An alternative and commercially successful strategy is to attach oligomeric chains to conventional industrial T-type dyes [72]. These chains, which may be primarily silicon- or carbon-based, e.g. polydialkylsiloxane, create a micro-environment for the colorant that enhances its photochromism in polymeric systems which are conventionally used with commercial dyes, and even makes possible colour changes in polymers that are off-limits for oligomer-free colorants [73, 74]. The difference that the oligomeric substituents make to kinetics runs counter to the usual rule of the thumb for conventional dyes that faster thermal fade equates to weaker photoactivated colour: when in a polymeric matrix, the kinetics of a dye substituted with oligomer are more like those of the corresponding chain-free dye dissolved in a solvent, i.e. faster to activate and to thermally fade than in a polymeric matrix, yet without any detriment to intensity of colour when one allows for the difference in molecular mass. These properties are usually mutually exclusive. The technology represents an advance in that it allows those working with photochromic colorants to create more responsive effects without sacrificing depth of photoactivated colour. In the case of ophthalmics, this translates to lenses that are faster to react to changes in light intensity, but which still achieve satisfactorily low levels of transmittance when activated. (One study manipulated the kinetics of three individual dyes by means of adjusting chain length on each to create a set of matched colorants for the production of a grey lens [75].) Another consequence is that successful application to substrates which are too challenging for conventional dyes becomes possible. For example, the production of photochromic effects in rigid thermoset plastics can be achieved by attachment of low T_g polymer chains to naphthopyran dyes [76]. A further benefit is that the oligomer appears to be protective of the chromophore. Polymers whose processing temperatures are too high for conventional dyes, like polyamides, thus become accessible. Conventional industrial photochromic dyes get severely damaged during mass coloration of nylon, leading to unacceptable base colour and ultimately a lack of photochromism. On the other hand, the application to such media of colorants whose structure contains oligomeric chains has been commercially successful. Industrial colorants bearing oligomeric chains are thus also capable of producing commercially acceptable photochromic effects in polymeric media that conventional colorants cannot.

Industry has taken advantage of these techniques for the application of photochromic colorants to polymers to produce a plethora of light-responsive products. For example, a diverse array of items have been fashioned from poly(ethylene) and plasticised poly(vinyl chloride) coloured by means of photochromic masterbatch or microencapsulate, including bottles, hair accessories, mobile phone case and toys. The encapsulation approach has even been claimed for use in lenses [77].

5.3.3.2 Surface Coatings

Photochromic colorants have found their way into inks and varnishes. While products like nail varnish reached the market (in which photochromic dyes are dissolved in organic media), their colour-change properties have been exploited for more functional applications, such as security printing. The simplest means of applying photochromic dyes in a coating form is to dissolve them in a commercially available solvent-based gloss varnish – this approach tends to work well with industrially available dyes owing to their solvent solubility and good photoresponsivity in non-polar solvents. As is the case with other media, photochromism will be suppressed if the concentration of colorant is too high. Steps must be taken to ensure that the colorant does not get damaged during formulation or application and it is provided with the right environment to function once the surface coating has been applied to a substrate. Consequently, because industrial photochromic dyes are not water soluble, alternative approaches must be taken for aqueous systems [78]. As is the case with thermoplastics, one approach is to disperse photochromic microencapsulates into surface coatings in a similar fashion manner to that for conventional pigments. One cannot simply add conventional industrial T-type dyes in neat form since the colorants are water-insoluble, yet must be in solution in order to furnish photochromic effects. Clearly, inclusion levels of colorant are dependent on the form in which the dye is added as well as the purpose of the application. Lithography, flexography and gravure are all printing techniques that produce relatively thin films of ink (2–7 μm thick when dry). Typically therefore, they require more neat colorant (1–3 % w⁄w dye in the ink) than techniques such as conventional screen printing, which produce thicker films (~12 μm when dry) and call for lower loadings (0.5–1 % w/w dye).

Since ophthalmic lenses are by far the most important commercial application of photochromism, the remainder of this review will concentrate on the properties demanded of industrial photochromic dyes for this use. It will describe the characteristics needed for commercial success and how colorants fulfilling these requirements are designed. In doing so, a flavour of the ways in which the molecular features of industrial colorants influence their photochromism will be given together with an outline of the means of their introduction during dye synthesis.

5.3.4 Ophthalmic Applications of T-Type Dyes

This section outlines the demands placed upon industrial T-type colorants by the lens sector and the ways in which they are applied.

5.3.4.1 Key Properties for Ophthalmic Applications

Commercial success demands compliance with numerous stringent requirements. Lenses must possess the following attributes:

(i) Show high transmittance in their unactivated state
This property is often termed low 'residual' or 'base' colour, which demands that the dyes themselves do not absorb appreciably in the visible region of the spectrum before exposure to sunlight.

(ii) Respond rapidly to exposure to sunlight
Lenses must change ('activate') as swiftly as possible from a colourless state to a strongly coloured one in response to an increase in ambient light intensity. Commercial lenses typically approach their peak darkness, i.e. minimum transmittance, after only a minute or two of exposure to strong sunlight at 20–25 °C. In order to do so, the dyes employed must switch from their colourless to coloured forms efficiently, the latter being strongly absorbing, so that only a relatively low concentration of activated colorant produces a pronounced colour change effect.

(iii) Fade quickly in the absence of sunlight
Once light levels drop, lenses should lose their colour rapidly to prevent impairment of vision. The initial fall in intensity tends to approximate first-order decay so a measure of rate of fade that is often used is 'half-life' ($t_{1/2}$), which is the time taken for absorbance to halve following removal of the activating light. Commercial lenses typically exhibit half-lives at 20–25 °C of a minute to a minute and a half. Short half-lives and high observed intensities tend to be mutually exclusive, therefore a balance usually has to be struck between attaining sufficient depth of shade and ensuring bleaching is acceptably quick. Since lens systems based on combinations of industrial dyes, polymers and additives do not always give idealised first order decay in absorbance, other means of capturing behaviour are sometimes employed when comparing kinetics to give a more rounded view of performance, e.g. $t_{3/4}$ – the duration over which absorbance falls to 25 % of its initial value.

(iv) Retain their photochromism
Photochromic dyes are notoriously delicate compounds compared to conventional colorants, especially with regard to photostability. However, it is possible through the inclusion of additives to create lenses that retain sufficiently pronounced photochromic behaviour over a commercially acceptable period, typically around 2 years – this behaviour is usually expressed in terms of minimum transmittance in an activated state and maximum transmittance in an unactivated state, particularly in regard to photoyellowing.

(v) Exhibit relatively little dependence to temperature in their photochromic response.
Given that the reversion of photoactivated commercial colorants back to their colourless forms is primarily driven thermally, higher ambient temperatures promote faster fade-back and thus lower intensities at steady state activation.

For changes in photochromic performance to remain sufficiently small across a range of ambient temperatures in which the spectacles will be worn so that they are not obvious to the wearer, this dependence must ideally be low.

In addition to being capable of satisfying the above criteria, there are other considerations which are vital to successful use of industrial dyes in lens manufacture. Generally T-types must be in solution to exhibit photochromism, i.e. dyes must be sufficiently soluble in the lens medium. There are isolated examples reported of T-type compounds related to families from which commercial derivatives are drawn undergoing colour changes in solid form [79]. However, the stipulation that dyes must be monomolecularly dispersed in solution to generate observable effects is certainly true of industrial colorants: too low a solubility in the medium, be it monomer, polymer or a solvent, will be deleterious to dye performance. Since commercial T-type dyes typically show positive solvatochromism (i.e. their absorption maxima shift to longer wavelengths as the polarity of their environment increases), then one has to allow for the fact that the nature of the lens medium into which the dye is incorporated will not only affect the kinetics of the colorant but also its colour.

5.3.4.2 Methods of Dye Application to Ophthalmic Lenses

The would-be manufacturer of photochromic lenses is faced with a number of choices when deciding how to incorporate dyes. However, these options are restricted by the considerable patent protection that still exists and which claims particular techniques for applying colorants. Additionally, the physical characteristics and optical properties demanded of a lens may dictate its constituent materials, which in turn often narrow down the application methods available. Consequently, in terms of imparting an economic, resilient and high performance effect to a lens, the dye is just one part of the story. The other two important determinants are the lens matrix and the production technique. These factors contribute to the environment which the lens manufacturer provides for the colorant: this is crucial, hence the build-up of intellectual property and the need for different techniques. For a review which provides a useful, albeit dated, insight into photochromic lens manufacturing, see [9]. A brief overview of the main techniques follows:

In mass: Where the lens medium is compatible with the photochromic dyes, then the two can be mixed along with additives, such as stabilisers, in order to give a lens in which the dye is dissolved uniformly throughout. One of the simplest techniques for manufacture in this manner is injection-moulding a mix of thermoplastic and dye. A commonly used current method is dissolution of dye in a monomer or resin system at a concentration of roughly 300–500 ppm which is then thermally or UV-cured into a semi-finished lens that can be ground to the desired prescription. While this methodology is quite heavy in its use of dye and is restricted to polymer types in which the dye exhibits satisfactory photochromism, such as acrylates, the technique is relatively simple technically, especially when the manufacturer buys in a monomer/dye/additive package as a pre-mix. There is considerable scope for expertise in optimising dye performance by blending different monomers and/or achieving a particular level of cure to supply an environment that favours responsive, intense photochromism.

Coating: This manufacturing technique involves applying photochromic dye in a coating to the front or the back face of a lens in solution along with a resin by a technique like spin- or dip-coating. While the coating must contain a relatively high concentration of dye to achieve a satisfactory photochromic effect because the coating is thin, less colorant overall is used compared to in-mass casting. The use of a coating also enables photochromic effects to be achieved with lens materials, like polycarbonates, in which dyes traditionally will not work when applied in-mass. However, since the colorant concentration is high, the dye must possess a low base colour and also good photostability since, unlike in an in-mass lens, there is no reservoir of dye lying deeper into the lens to replace degraded colorant.

Lamination: While technically challenging, this technique offers advantages similar to those for coating. It involves bonding a film containing photochromic dyes to the surface of one half of a lens; the film is then enclosed by being sandwiched between the other half of the lens.

Imbibition: This process is proprietary to Transitions Optical [13]. Dye is applied to the lens surface by spin-coating. The lens is then baked to enable the dye to diffuse from the coating into the polymer matrix. A photochromic layer around 150–200 μm thick is created. The dye remains trapped in the lens surface upon cooling. The spent coating is then washed off.

5.4 Industrially Important T-Type Photochromic Dyes

During its endeavours since the mid-twentieth century to meet the technical and commercial challenges of creating marketable lenses as well as developing other photochromic systems, industry scrutinised many thousands of T-type dyes. In order to identify suitable colorants for this application and others, industrial, governmental and academic laboratories synthesised large numbers of colorants drawn from a variety of photochromic dye classes. However, only a tiny proportion survived to the stage of commercial exploitation. To get that far, these colorants not only had to demonstrate acceptable performance in their targeted end-use – as illustrated for lenses above – but also be practical and economic to make. Consequently, just a select few of the numerous families of T-type dye have been industrialised.

The first modern attempts to develop commercial outlets for organic photochromic materials centred on spiropyrans. This kind of dye still forms the focus of many studies into developing new technologies (for a good recent review, see [80]). However, only two classes of T-type photochromic colorant are currently of industrial importance: spirooxazines and naphthopyrans. This section will therefore describe these two classes in more detail in terms of their general characteristics and preparation.

5.4.1 Spirooxazine Class of Photochromic Dyes

From an industrial perspective, spiropyran dyes were the class of principal interest during the 1950s–1970s. Readily synthesised from accessible starting materials, they furnish intense photocoloration that decays over useful timescales [81]. A particularly well explored set of compounds was the indolino-spirobenzopyrans **3** as shown in Fig. 5.7. However, in the 1970s, isolated reports appeared in the open and patent literature of studies that replaced a methine carbon of the pyran ring with a nitrogen atom to give spirooxazines, for example the indolino-spironaphthoxazines **4**. When industry realised that members of the spirooxazine class were much more photostable than their spiropyran counterparts [8], interest in the latter for use in lens materials waned (although they are still much utilised in academic studies as probes or for the creation of photocontrolled systems [82]). The focus shifted to spirooxazine-based systems, leading in the early 1980s to the launch of the first plastic photochromic lenses into the US market by American Optical [8]. This initial effort failed commercially because the look, performance and lifetime of the product did not fulfil customer expectations [11]. An attempt by PPG Industries using a closely related type of blue oxazine-based dye **5**, which the company started working on in 1983 [11], was also short-lived [12]. However, further developments meant that polymer-based lenses, which employed the

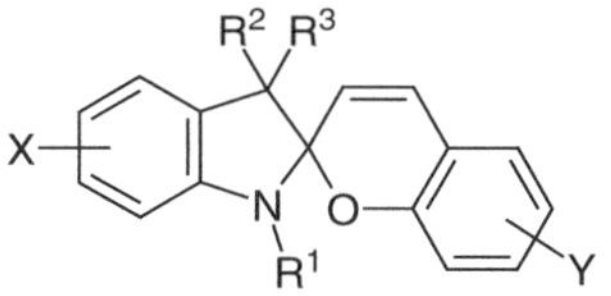

$R^{1\text{-}3}$ = alkyl (usually methyl)
X , Y = H, halogen, nitro, etc.

3

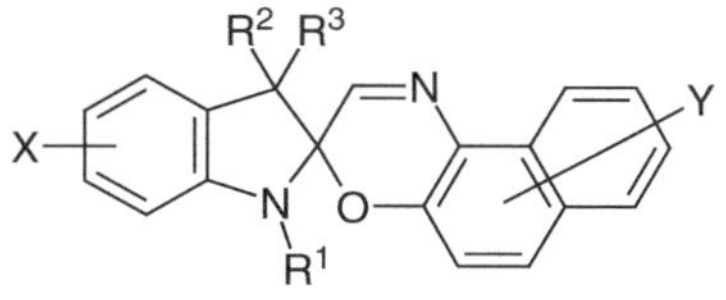

R^1 = alkyl; $R^{2\text{-}3}$ = alkyl (usually both methyl)
X = H, halogen, etc.; Y = H, amino, hetaryl, etc.

4

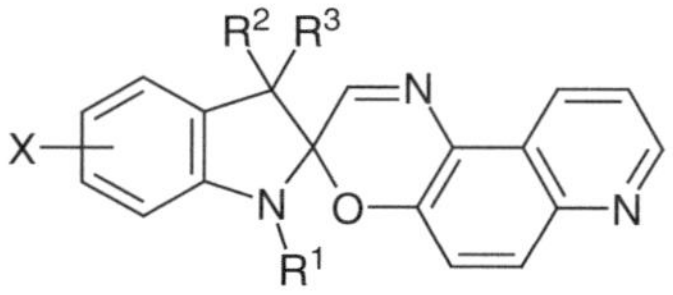

5

Fig. 5.7 General structures of typical industrial spiropyrans and spirooxazines spirooxazine class in their manufacture, did ultimately taste commercial success in the 1990s. At this time, PPG Industries marketed the Transitions® lens, while Rodenstock distributed Colormatic® and Photocolor® lenses produced using this dye class in the US [9]. In Japan, Tokuyama also made use of oxazine colorants [57]. This family of photochromic dye remains of industrial interest, but is now of secondary importance to the naphthopyran class. In 1992, the TransitionsPlus® lens was brought to market with a mix of oxazine and naphthopyran dyes, while the Transitions 3 lens launched in 1996 was formulated with an all-naphthopyran combination [83] since such dyes offer less temperature-dependent behaviour [84].

Given their structural similarity, it should be no surprise that the spiro-pyran and -oxazine classes **3–5** share the same general mechanism of photochromism and that their methods of synthesis are closely related.

Their photochromic properties arise from the ability of the colourless form of the compound to undergo a reversible molecular rearrangement in which the pyran or oxazine ring-opens to generate a coloured photomerocyanine dye in contrast to the mechanisms of the P-type colorants outlined earlier. Figure 5.8 illustrates the transformation for a simple spiropyran. This reaction is sparked by UV irradiation (although in the case of commercial dyes, short wavelength visible light can also be a trigger). Energy is absorbed, leading to a photochemical sequence of events that includes cleavage of the bond between the spiro carbon atom and the pyran/oxazine ring oxygen atom, ultimately leading to the formation of a mixture of coloured photomerocyanine isomers [85]. Many academic studies have investigated the nature of the photochemistry behind this isomerisation; they have been recently summarised [82].

It is important to remember that the equilibrium shown in Fig. 5.8 is dynamic. At any given time, there is constant interconversion between the ring-closed and -opened forms. The balance between these two sides of the equilibrium determines the intensity of the coloured state of the system. When irradiated with UV light, the concentration of dye in its photomerocyanine form rises because of the resultant increase in rate of the ring-opening reaction. In other words, the balance of the equilibrium shifts from the ring-closed side, leading to an intensification of colour. Cessation of the UV irradiation causes the opposite effect. The rate of ring-opening relative to the rate of ring-closure falls, so that the concentration of the colourless

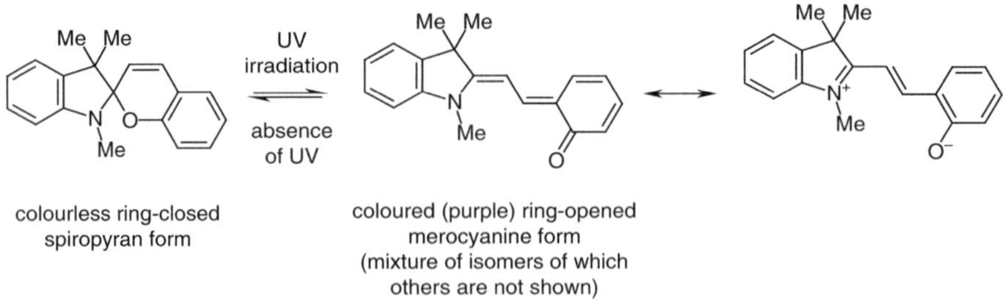

Fig. 5.8 Photochromism of a simple spiropyran

form of the dye rises relative to the photomerocyanine form. Observers see this overall switch in the equilibrium as fading of colour.

The goal of industrial dye design is create colorants that show little tendency to ring-open under low levels of ambient lighting, i.e. the equilibrium is far towards the pyran/oxazine side, so residual colour is minimised and transmission is at its maximum. Conversely, at high ambient lighting levels, the aim is to achieve an equilibrium shifted as far as possible to the photomerocyanine side in order to maximise the intensity of photocoloration. Reality involves compromise. High photoactivated colour strength can be achieved but at the expense of a slower thermal fade and a tendency to exhibit residual colour. Speeding up the rate of fade is one means of lowering base colour but the trade-off is a weakening of photocoloration. As we shall see later in this Chapter, there are ways to modify structure and environment to achieve the better of both worlds.

Because the ring-closure reaction in commercial spirooxazine dyes is essentially entirely thermally driven, the ambient temperature affects the position of the equilibrium with important consequences for commercial products. A rise in temperature speeds up decay of the photomerocyanine dye to the colourless spirooxazine form. At a given intensity of irradiation, more thermal energy therefore leads to the equilibrium shifting away from the coloured ring-opened side in the case of both spiropyrans and spirooxazines. The overall result is thus a perceived reduction in intensity of colour at higher ambient temperature. While more rapid fading back to a colourless state upon removal of UV irradiation accompanies this effect, such temperature dependence is generally undesirable. It can complicate the formulation of dye mixtures and leads to differences in performance that are noticeable to the consumer: pronounced changes in response are perceptible at 5 °C, 20 °C and 35 °C. Consequently, the photochromism of lenses based on spirooxazines is thus less effective in hotter climates: they work better on winter ski-slopes than on summer beaches.

As discussed elsewhere in this Chapter, factors other than temperature influence the photochromism of a given dye, such as the substrate into which it has been dissolved, by affecting the position of the equilibrium between the ring-closed and ring-opened forms as well as the ease of interconversion between them.

A commonly employed synthetic route to non-ionic indolino-spirobenzopyrans **3** involves condensation of Fischer's base (or an analogue) with a salicylaldehyde (see Fig. 5.9), typically in equimolar amounts, in refluxing alcohol. The hot solvent readily dissolves the starting materials while the spiropyrans tend to have limited solubility in cold alcohols, so they can often be isolated following cooling of the reaction mass. However, variations in reaction stoichiometry and solvent choice, e.g. toluene, can bring about improvements in yield and quality. Depending upon the particular materials involved and the intended application for the dye, the product may be sufficiently pure without requiring subsequent clean-up by recrystallization. Reference [81] gives an excellent overview of the important practical considerations involved in spiropyran preparation, including quality of starting materials, reaction conditions and purification. Oxazine analogues are synthesised in an analogous manner as discussed shortly below. Studies aimed at optimising the

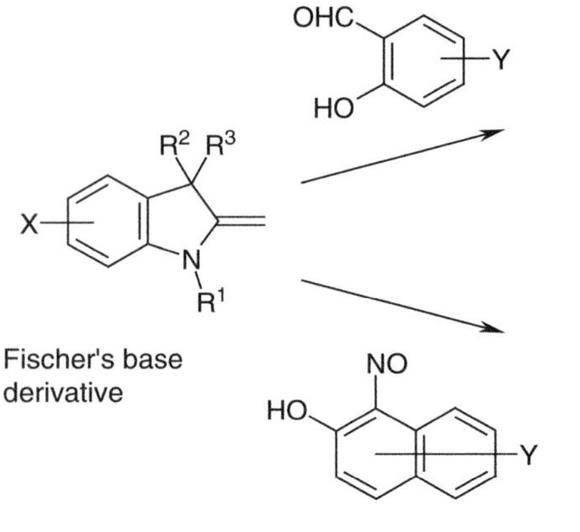

Fig. 5.9 Commercial synthetic routes to indolino-spirobenzopyrans and -spironaphthoxazines

conditions of spirooxazine synthesis do not often appear in the open literature, but a good example is [86].

There are two sub-classes of spirooxazine which are of particular commercial interest. The first type has already been mentioned: naphth[2,1-*b*][1,4]oxazines **4**. These colorants are useful for producing bluish-red, violet and blue photochromic effects. They are generally prepared from a 1-nitroso-2-naphthol derivative in refluxing alcohol. As is the case with spiropyrans, alternative conditions may be beneficial, such as use of toluene as solvent. However, for most purposes, purification of the isolated product is accomplished by recrystallization. An effective means of doing so is to use an alcohol as a non-solvent in combination with an aromatic hydrocarbon, like toluene, an ester or an ether, such as tetrahydrofuran, in which the dyes tend to have good solubility. Column chromatography is sometimes utilised commercially, but only as a last resort because of the relatively high cost of this technique.

Use of a 5-nitroso-6-hydroxyquinoline derivative as the "naphthol" component furnishes indolino-spiropyridobenzoxazines **5**. This type of dye garnered industrial attention in the 1980s for the coloration of plastic lenses [11] and went on to become part of the successful commercialisation of the technology early in the next decade [12].

The second sub-class of spirooxazine is a source of industrially useful greenish-blue colorants: naphth[1,2-*b*][1, 4]oxazines **6**. The synthetic strategy is very similar to their [2,1-*b*] counterparts (see Fig. 5.10), the main difference being a switch to the use of 2-nitroso-1-naphthol synthons [87, 94].

Upon activation with UV light, the parent compounds **4** (R^{1-3} = methyl; X, Y = H) and **6** (R^{1-3} = methyl; X, Ar = H) are relatively fast-fading dyes that produce weak blue photocoloration. Industry has realised the potential of this class to produce intense reddish-blue to turquoise photochromic effects with commercially interesting half-lives by means of judicious selection of substituents. This Chapter covers these design aspects in more detail later.

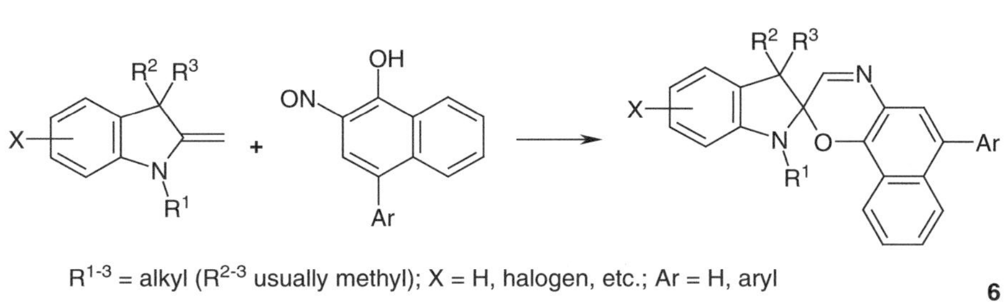

6

Fig. 5.10 Synthetic route to commercial naphth[1,2-b][1, 4]oxazines

Table 5.1 Typical solubilities (g/L) of industrial indolino-spironaphthoxazines in organic solvents

Class	Toluene	Ethyl acetate	THF	Acetone	Methanol
4	60–120	20–80	100–300	10–30	2–5
6	200–300	25–50	300–500	10–30	1–3

Industrial naphthoxazines do not exhibit photochromism in their solid state. While they may show changes in absorption upon irradiation, these are not visible to the naked eye. The colorants tend to be in the form of weakly coloured powders, although certain dyes have sufficiently extended conjugated π-systems in one or both halves of their ring-closed forms to give them an intrinsically yellow physical form. Commercial derivatives **4** and **6** are relatively high melting (typically >150 °C and >100 °C respectively) with some oxazine derivatives possessing more than one melting range as a consequence of polymorphism. Their solubility varies widely (see Table 5.1), but all are essentially water-insoluble. Particularly good solvents are aromatic hydrocarbons, such as toluene and xylene, in contrast to alcohols like methanol or ethanol. Both classes of oxazine dissolve very readily in ethers, while common esters and ketones act as reasonably good solvents, exemplified in Table 5.1 by tetrahydrofuran (THF), ethyl acetate and acetone, respectively. The ranges tabulated are only guidelines, for example, certain [2,1-*b*] oxazines have much lower solubilities in particular solvents. When stored in the dark, sealed at ~20C away from moisture, industrial naphthoxazines are stable for years. They are also non-toxic: LD_{50} values are typically $>2{,}000$ mg/kg (oral, rat).

Indolino-spironaphthoxazines remain a popular option in commerce for the production of photochromic effects. Industry continues to take advantage of their responsive and intense heliochromism for plastics and coatings. However, where higher end performance is required or greater compatibility between components of dye mixtures is desired, particularly in the case of photochromic ophthalmic lens manufacture, then as we shall see in the next section, naphthopyrans have assumed greater industrial significance.

5.4.2 Naphthopyran Class of Photochromic Dyes

Naphthopyrans are the most commercially important class of photochromic molecule. All the major manufacturers of plastic photochromic lenses make use of naphthopyrans. Their chemistry allows ready and cost-effective adjustment of structure, enabling great control of kinetics and colour by appropriate selection and location of many different types of functional group. Different permutations bring into reach colours that span the visible spectrum from yellows through to oranges, reds, purples and blues. In addition, commercial naphthopyrans generally have stability properties that are as good as any other known class. Another reason for their importance is the lower sensitivity of their photochromism towards variations in temperature when compared to spirooxazines. The versatility of the naphthopyran class in providing colorants across the visible spectrum means that formulators can create mixtures from this single class to produce neutral shades. An advantage of doing so is that properties, such as temperature dependence, are likely to be better matched than when working with mixtures of dyes drawn from different families [84].

This class of dyes were initially neglected by industry for many years owing to misinformation and lack of clarity concerning properties published in the academic and patent literature [88]. However, once their potential was recognised as colorants for lenses, these dyes became the subject of intense study over the past three decades [88, 89]. Initially, they were introduced in the early 1990s as hypsochromic components in conjunction with blue oxazine-based colorants to generate neutral shades [90]. As more was learnt about manipulating their colour and kinetics, naphthopyrans displaced oxazines as the bathochromic constituents in such mixtures and thus became the dominant type of industrial photochromic dye.

Of the three possible naphthopyran subclasses, only two are commercially useful: 3*H*-naphtho[2,1-*b*]pyrans **7** and 2*H*-naphtho[1,2-*b*]pyrans **8** (see Fig. 5.11).

Dyes of structure **7** are a source of industrially important yellow, orange and red colorants, whereas colorants based on **8** furnish red, purples and blues. As we shall see later, the latter class also supplies dyes with relatively neutral shades, simplifying the formulation of mixtures to produce brown or grey coloration. Like spiropyrans and -oxazines, the photochromism of naphthopyrans relies on light-induced ring opening. Figure 5.12 illustrates the transformation in the case of a simple naphthopyran. As with the analogous transformation for spirooxazines, a mixture of several photoisomers results. The ground states of the photomerocyanines feature increased charge density on the oxygen atom (from contributions by zwitterionic canonical forms such as that shown in Fig. 5.12). In addition, the charge densities of first excited states involve transfer of electron density to the oxygen. Substitution patterns favouring this transfer, e.g. electron donors on the aryl rings such as the amino group in the [1,2-*b*]naphthopyran shown in Fig. 5.12, therefore lead to bathochromic shifts in absorption. Relationships between colour and constitution of industrial dyes will be covered in a later section.

As is the case with spiropyrans and spirooxazines, the position of the photostationary state equilibrium between ring-opened and -closed forms and the

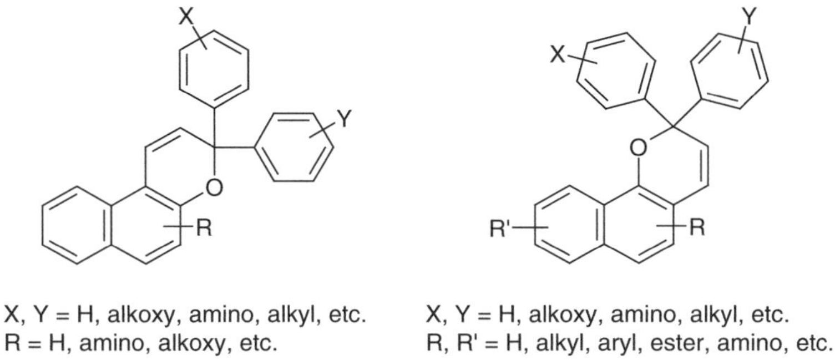

7 **8**

Fig. 5.11 General structures of typical industrial naphthopyrans

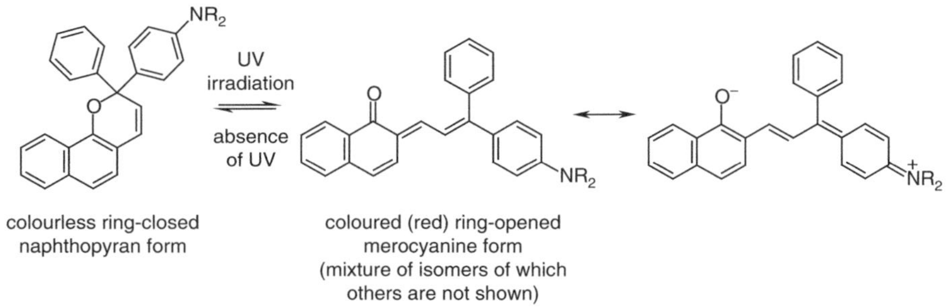

Fig. 5.12 Light-induced photoisomerisation of a simple [1,2-b]naphthopyran

rate of thermal fade can be readily manipulated by modification of dye structure. The flexibility available stems from the synthetic accessibility of naphthopyrans by combination of naphthols (and corresponding analogues) with diarylalkynols as illustrated for [2,1-*b*]naphthopyrans in Fig. 5.13.

While only one of several pathways to such materials, that shown in Fig. 5.13 is a general and important manufacturing route for large-scale production. The medium is often a non-polar solvent, such as toluene or xylene, in the presence of an appropriate acidic substance, for example, an aromatic sulphonic acid or an aluminium oxide. The reaction sometimes benefits from the azeotropic removal of water generated during the reaction (or introduced from the acid). Occasionally, chemical dehydration by means of a reagent such as a trialkyl orthoformate is advantageous. The goal industrially is to isolate a product of sufficient purity directly from the reaction mass since it is the cheapest option. While this is sometimes possible, recrystallisation, or as a last resort column chromatography, of the obtained product is typically needed to obtain the desired product quality. A very good overview of the synthesis (and properties) of naphthopyrans is provided by [91].

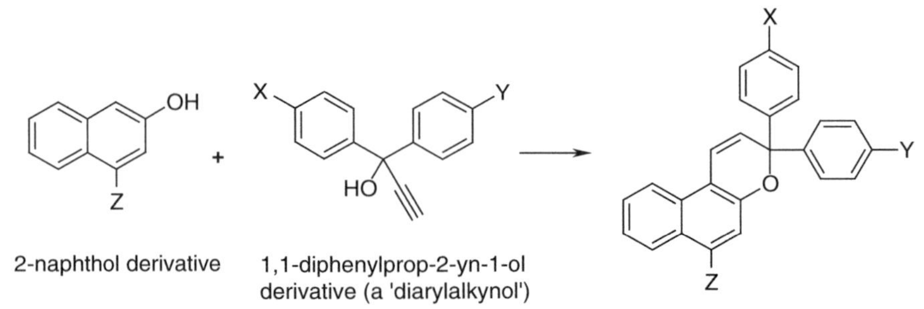

Fig. 5.13 A commercial synthetic route to naphthopyrans using general [2,1-*b*] derivatives as an example

Table 5.2 Typical solubilities (g/L) of industrial naphthopyrans in organic solvents

Class	Toluene	Ethyl acetate	THF	Acetone	Methanol
7	20–100	10–30	10–200	10–30	1–10
8	15–200	10–30	30–400	10–30	1–5

When not provided as solutions in pre-formulated monomer mixtures, industrial naphthopyrans are supplied as powders, typically of m.p. >150 °C, whose appearance is often highly dependent upon purity and method of isolation. With a few exceptions, they have shelf-lives of years when kept in cool, dark, dry conditions. Industrial naphthopyrans are generally classified as non-hazardous for transport and tend to be non-toxic. All commercial dyes are water-insoluble with poor solubility in alcohols. They dissolve far more readily in solvents of moderate to low polarity (see Table 5.2). As is the case with spirooxazines, aromatic hydrocarbons like toluene and xylene function as good solvents whereas aliphatic hydrocarbons like hexane do not. Naphthopyrans tend to have reasonable solubilities in common ethers, esters and ketones, which is reflected in the figures given in Table 5.2 for THF, ethyl acetate and acetone, respectively. However, these ranges are only indications: the solubility of a particular dye–solvent combination might fall outside of them.

5.5 Designing in the Properties: Structural Influences in Industrial Dyes

This section looks in depth at how structural features of industrial dye types influence their photochromic properties. Use of commercially important molecular substitution strategies to manipulate the colour and intensity of the photochromism of indolino-spironaphthoxazines and naphthopyrans receives particular emphasis. These rules of thumb continue to prove helpful in the design of valuable colorants for ophthalmic use, but they are also worth bearing in mind when attempting to employ these types of colorant in other applications. It is often the case that academic studies base an active molecule on a simple pyran or oxazine but report only weak light-responsive behaviour. Researchers should be able to use some of the design rules outlined below to their advantage when constructing novel photochromic materials intended for use as functional dyes in many different spheres of activity.

As mentioned earlier, modifications to the molecular structure of the parent spirooxazines and naphthopyrans successfully led to the creation of industrially useful photochromic dyes with appropriate kinetics and absorption characteristics. As we shall see, the impact of such changes on colour and photochromic behaviour is not usually mutually exclusive. While it is possible to alter structure in ways that minimally affect hue but make a pronounced difference to equilibrium intensity and thermal fade rate, it is often the case that decisions concerning the types of substitution pattern to introduce when targeting a particular set of performance criteria involve compromise.

5.5.1 Substituent Effects in Indolino-Spironaphthoxazines

Industrial indolino-spironaphthoxazine dyes depend upon structural modifications in one or both of the indoline or naphthalene moieties of their parent compounds to slow their thermal fading to useful rates, to improve the efficiency of their interaction with UV light, to reduce fatigue as well as to adjust colour and intensity advantageously.

Commercial [1,2-*b*][1, 4]oxazines **6** tend to be more bathochromic and slower as a class than [2,1-*b*][1, 4]oxazines **4**. For example, dye **9** is a commercial blue dye with an absorption maximum at 600 nm in toluene solution with a half-life of 31 s at 20 °C whereas the industrially useful dye **10** gives turquoise photocoloration in toluene with a longer λ_{max} (618 nm) and $t_{1/2}$ (20 °C) of 120 s. Both dyes feature an *N*-substituent that lengthens their half-lives (and activated intensity) to a suitable duration. We shall now consider this aspect in detail as well as other types of structural modification employed to create dyes of industrial interest.

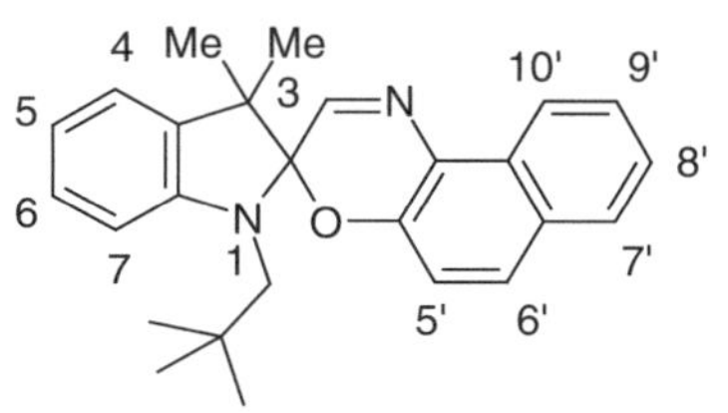

9

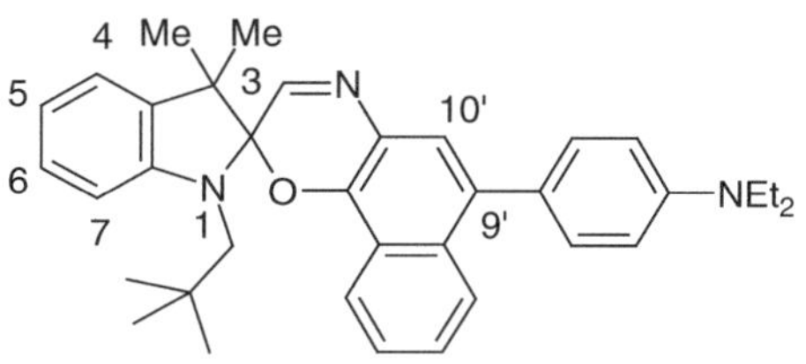

10

Substituents on the indoline fragment exert considerable influence over dye kinetics with varying degrees of effect on colour. Increasing the bulk of the *N*-

substituent is a relatively well-known means of slowing down the fade of indolino-spironaphthoxazines without significantly shifting their absorption maxima. Simple colorants derived from Fischer's base (i.e. **4** where R = Me) tend to give only weak photocoloration owing to rapid rates of thermal reversion by which the coloured photomerocyanine forms revert back to their corresponding colourless ring-closed form. Replacing the *N*-methyl group of a dye with a bulkier alkyl chain of an appropriate kind retards the thermal fade rate of the colorant and increases photostationary state absorbance, while red-shifting its absorption maximum by just a few nm. Dyes modified in this way thus develop a similar colour when photoactivated but are more intensely absorbing. Table 5.3 testifies that the

Table 5.3 Effects of varying substituents at the 1-position of indolino-spironaphthoxazines on absorption maximum and half-life

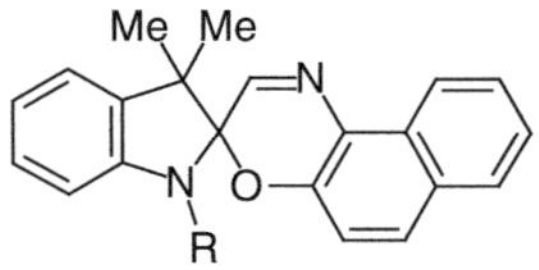

R		Relative intensity[a]	Residual photo-coloration[b]
	Et	0.9	0.9
	*n*-Pr	0.9	0.9
	*n*-Am	0.9	0.9
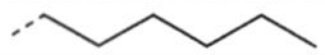	*n*-Hx	0.7	0.9
	*i*-Pr	0.7	0.8
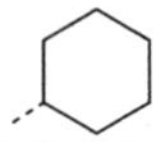	Cy	0.7	0.9
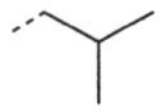	*i*-Bu	1.4	1.2
	Np	1.7	1.3
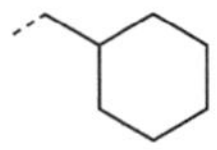	CH_2Cy	1.3	1.2
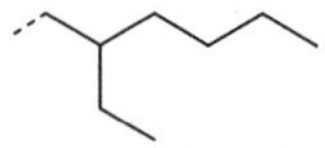	2-EtHx	1.5	1.2
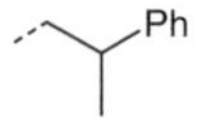	2-PhPr	1.5	1.2
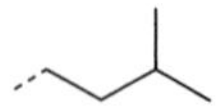	*i*-Am	1.0	1.0

[a]Absorption intensity at photostationary state relative to that for $R = CH_3$
[b]Residual photocoloration 10s into thermal fade relative to that for $R = CH_3$

relationship between chain structure and enhancement of photochromism is not a straightforward one. The presence of branching slows the speed of thermal fade and improves intensity of photocoloration, but only when located at the right point in the alkyl group (see Table 5.3). A branch on the carbon adjacent to the indoline nitrogen (R = *i*-Pr or Cy) actually speeds up rate of thermal fade and weakens the dye. A lack of any branching has a similar effect. However, a branch beta to the nitrogen atom brings about pronounced increases to strength of the photocoloration, for example, R = 2-EtHx compared to R = *n*-Hx. This beneficial effect is lost upon shifting the branching to the gamma carbon (R = *i*-Am). Switching alkyl functions has little effect on fatigue (or activated colour) so tuning properties by appropriate selection of the indolino *N*-substituent is an effective design strategy. Groups such as neopentyl, isobutyl and methylcyclohexyl are introduced readily in an economical manner during synthesis of the requisite Fischer's base intermediates, hence their appearance in industrial dyes like **9** and **10**. The impact of bulky 1-substituents on half-life does not relate to enhancements to the extinction coefficients of the colourless or coloured forms of the dyes since they remain largely unaffected. Nor do they appear to increase the quantum efficiency of photoisomerisation, i.e. the generation of coloured species [92]. Instead, their influence is rooted in the steric crowding they introduce which hinders the ring closure reaction whereby the oxazine C-O bond re-forms, leading to restoration of the spiro geometry. A reduction in the rate of thermal fading, as pointed out earlier, means that for a given intensity of incident UV, the equilibrium concentration of photomerocyanine increases, which results in stronger photocoloration.

Branched *N*-alkyl groups tend to offer another advantage: reduced temperature sensitivity. As discussed earlier, the photochromism of spirooxazine and naphthopyran dyes exhibits an undesirable dependence on temperature. For example, the photochromic performance of conventional N-methyl oxazine-based dyes drops away in hotter environments. However, switching the alkyl function can reduce the rate by which photocoloration weakens as temperature rises. For example, over the temperature range 10–40 °C and for a given intensity of incident light, the photocoloration of the dye in Table 5.3 where R = Np decreases less rapidly than when R = Me.

Strongly inductive electron-donating and -accepting functions may markedly affect colour as well as kinetics when located onto an *N*-alkyl residue in a way that does not insulate their effects from the indoline nitrogen. For example, the ester functionality in **11** produces a hypsochromic shift of 15 nm in absorption maximum to 575 nm compared to the *N*-methyl analogue (toluene solution), presumably through inductive lowering of electron density on the indoline nitrogen.

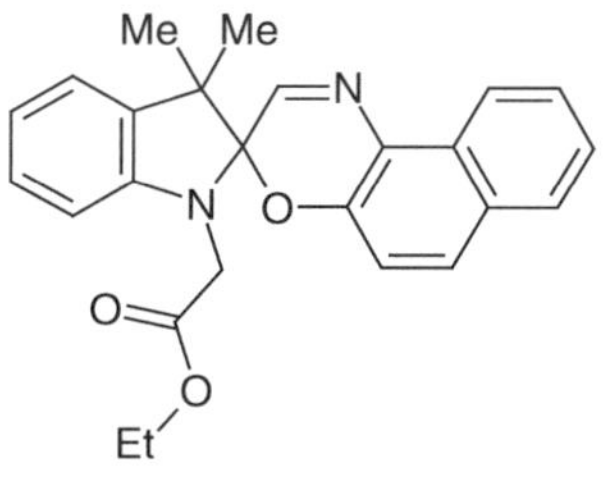

11

Table 5.4 Effects of varying substituents at the 3-position of indolino-spironaphthoxazines on absorption maximum and half-life in toluene solution at 20 °C [94]

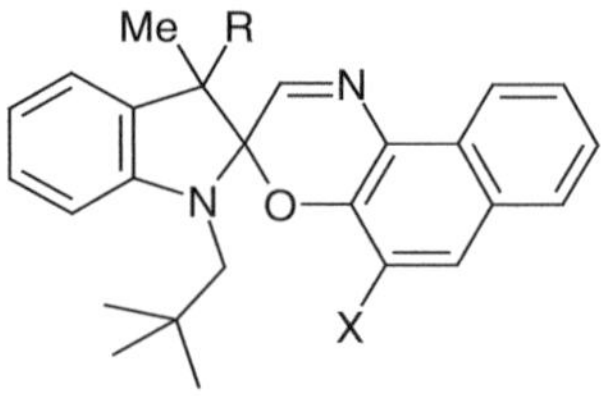

R	X = H		$X = CO_2Me$	
	λ_{max} (nm)	$t_{1/2}$ (s)	λ_{max} (nm)	$t_{1/2}$ (s)
Me	600	31	620	12
Et	600	52	620	20

Another means of tuning dye kinetics without causing a substantial shift in λ_{max} is to make changes to the 3,3-dialkyl motif of the indoline residue. Many commercial oxazine dyes comprise 3,3-dimethyl substitution purely for reasons of synthetic expediency: they are derived from cheap and readily available Fischer's base, or easily-accessible close analogues, all of which possess this substitution pattern. Variations on it, however, have been explored industrially [93] and exploited successfully to slow down thermal fade and thus increase photoactivated intensity. Just a small change can, when coupled with crowding at the 1-position, have a substantial effect on half-life without appreciably altering hue (as illustrated by the data in Table 5.4) to produce strong blue dyes.

Even more pronounced increases in activated strength are possible by switching one of the 3-alkyl functions for an ester group, or both with a lactone ring. The latter change is at the cost of a moderate hypsochromic shift in absorption to give slower-fading reddish-blue dyes.

Replacement of the *gem*-dimethyl fragment with 3,3-spirocyclohexyl substitution is another piece of molecular engineering utilised by industry. The end-result is strongly dependent upon dye structure. The presence of the spiro-aliphatic ring in certain colorants causes substantial reductions in $t_{1/2}$ and losses in photocoloration intensity when compared to those properties for analogues with the 3,3-dimethyl motif. Such dyes are weaker but feature a modest bathochromic shift in absorption maximum. For example, the cyclohexyl [1,2-*b*]oxazine **12** is greener and faster (λ_{max} 626 nm and $t_{1/2}$ 26 s in toluene at 20 °C) than the *gem*-methyl analogue **10** ($\Delta\lambda_{max}$ −6 nm and $\Delta t_{1/2}$ + 94 s) [94]. The same effect is manifest in some blue [2,1-*b*]oxazines of structure **4** (R^1 = Np, R^2R^3 = $-(CH_2)_5-$). However, the impact of introducing a spirocyclohexyl ring may be favourable in other cases: the (now obsolete) industrial colorants **13** (X = N, Y = CH) and **13** (X = CH, Y = N) incorporated this aliphatic system [57].

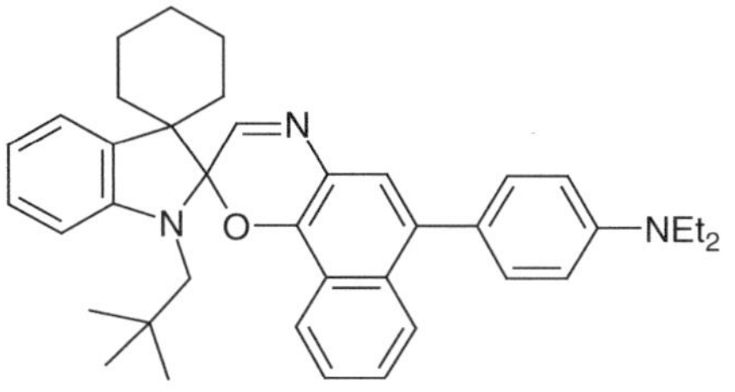

12

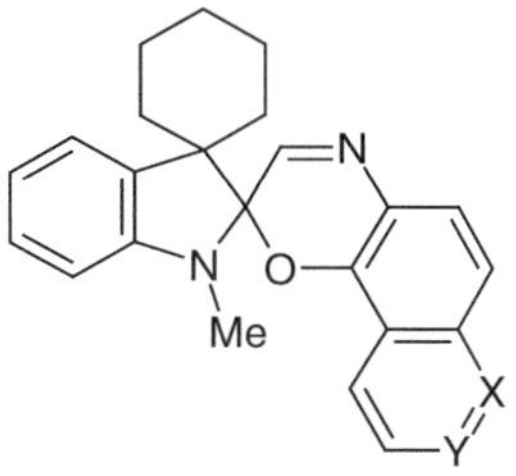

13

Owing to the complex interplay of steric influences that 1- and 3-substituents exert within photomerocyanine species, the impact of modifying 3,3-functionality varies significantly, as noted above. The influence of *N*-substituents was first recognised during the late 1970s [95]. The ability to tune kinetics by means of replacing groups at the 3-position was noted a few years later [96]. Both remain powerful tools with which to manipulate dye kinetics. Nevertheless, academia often overlooks these design strategies as a means of increasing photochromic response. Many studies that attempt to exploit simple indolino-spironaphthoxazines (or more commonly their pyran analogues) achieve limited success because of the weak, fast-fading nature of the dyes involved. Consequently, the focus prematurely moves on to other photochromic systems. Appropriate structural change at the 3-position (and/or at the indolino nitrogen atom) brings about much more intense photocoloration and would potentially yield better photochromic materials with which to undertake investigations and demonstrate proof of concept. The potential enhancements to photochromism that appropriate substitution of the five-membered ring of the indoline moiety can confer are thus well worth taking into consideration when attempting to exploit spirooxazines (or spiropyrans) as light-responsive units in novel applications.

While the aforementioned adjustments in spirooxazines predominantly affect the kinetics of their photochromism rather than their colour, numerous means of manipulating the absorption characteristics of such dyes exist. However, none can do so in isolation from the photochromic properties. Consequently, one must consider the overall impact of making a particular change and perhaps try to mitigate any undesired effects with other modifications, bearing in mind that these may have implications relating to synthesis and economics.

For example, one can fine-tune colour and kinetics by addition of electron-donating and -accepting groups on the 4- to 7-positions of the indoline ring system (see Table 5.5). Donors produce bathochromic shifts, i.e. λ_{max} becomes longer and colour becomes greener, since they facilitate transfer of electron density from the nitrogen heterocycle towards the keto-naphthalene end of the photomerocyanine. The strength, number and position of donors not only affects absorption maximum, but also half-life, as illustrated in Table 5.5 for the [2,1-*b*] class. Similar behaviour is observed for [1,2-*b*]naphthoxazines [94].

Table 5.5 Effect of donor substitution within indoline moiety on spirooxazine properties in toluene solution at 20 °C

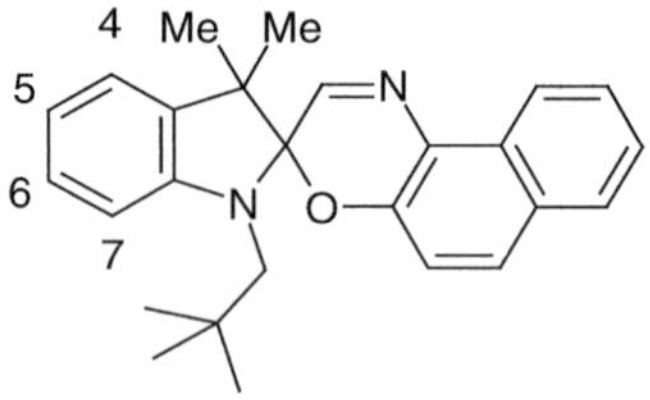

Substitution	λ_{max} (nm)	$t_{1/2}$ (s)
Unsubstituted	600	31
5-methyl	608	21
4,6-dimethyl	605	20
5-methoxy	615	10

Table 5.6 Effects of donor substitution at the 6′-position of spirooxazine properties in toluene solution at 20 °C

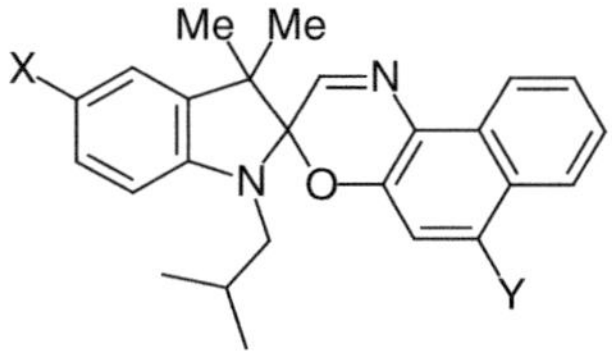

X	Y	λ_{max} (nm)	$t_{1/2}$ (s)
H	H	599	9
H	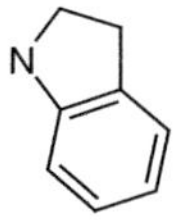	590	25
OMe	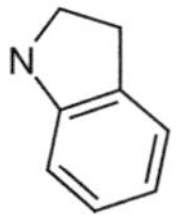	608	9
H	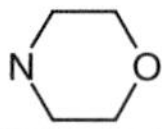	576	33
H	NMe_2	564	32

The introduction of electron donors onto the naphthalene system generally has the opposite effect to their placement on the indolino fragment. As pointed out above, a donor on the indoline ring causes a bathochromic shift: Table 5.6 gives an example of how 5-methoxy substitution produces a bathochromic dye with a faster thermal fade (X = methoxy, Y = indolin-1-yl). In contrast, electron donor functionality on the naphthalene system causes an increase in intensity of photocoloration accompanied by a hypsochromic shift. An important motif of this type is an amine group at the 6′-position, i.e. Y = amino in Table 5.6, because this substitution pattern not only expands the gamut of commercially useful spirooxazines from the blue towards the red, but also confers enhanced photochromism. Industry has exploited it for around a quarter of a century [97]. The size of hypsochromic shift and increase in half-life correspond with a rise in the basicity of the 6′-amino function [98], i.e. indolin-1-yl < *N*-piperidino < *N*-morpholino < dimethylamino (see Table 5.6 for some examples).

Industry has manufactured dyes of this type for over two decades [99], for example **14**, which is available commercially as Reversacol Plum Red [100]. Its 6′-piperidino group transforms a weak fast-fading blue colorant into a much stronger magenta dye (λ_{max} 565 nm and $t_{1/2}$ 23 s in toluene solution at 20 °C). Dyes of this type photoisomerise with greater quantum efficiency to merocyanine forms that intrinsically absorb more intensely [92, 101]. Another reason put forward for better performance is that the 6′-amino function modulates the absorption profile of the photomerocyanines so that they do not compete as effectively for activating UV wavelengths with the ring-closed form, improving photochromic response [97]. While the absorption of their ring-closed forms tends to bleed a little into the visible region, leading to more pronounced residual yellowness (and greater sensitivity to activation by blue light), their augmented photochromism results in deep coloration even at relatively low usage levels. Industry has therefore exploited all of the amino functions shown in Table 5.6 to create colorants that react rapidly to incident UV and short wavelength visible light by developing strong purple and bluish-red photocoloration. Indolinospiro[2,1-*b*][1,4]oxazines containing a 6′-amine function can be prepared in the conventional manner from reaction of a Fischer's base derivative with a 2-nitroso-1-naphthol having the corresponding amino substitution. It is also possible to use a one-pot method to build up the dye from three separate components (Fischer's base, nitrosonaphthol and amine), although yield and quality depend strongly on how the process is executed, e.g. stoichiometry, reaction conditions, order and duration of addition, etc.

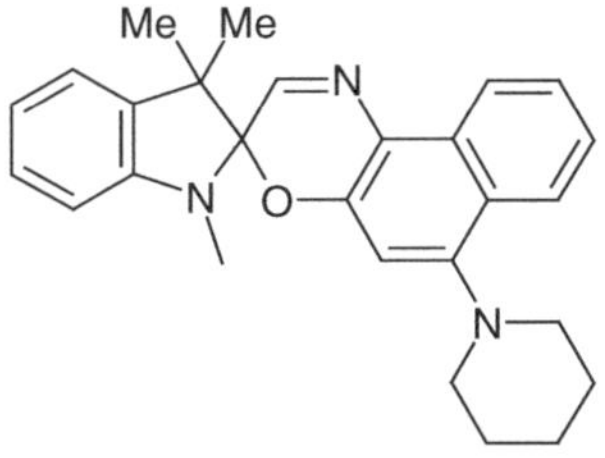

14

Converse rules of thumb hold for electron acceptors: when located on the indoline fragment, they cause hypsochromic shifts in λ_{max}, i.e. a shortening of wavelength of absorption, leading to the photoactivated form becoming redder. In contrast, electron-withdrawing functions placed on the naphthalene system shift absorption maxima to longer wavelengths. As an example, Table 5.4 indicates the impact of locating a (moderately electron-withdrawing) ester group at the 5′-position, i.e. $X = CO_2Me$. In addition to a moderate red-shift in absorption

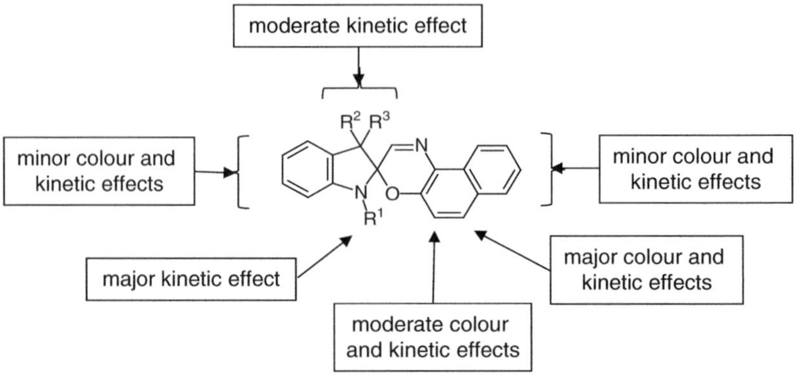

Fig. 5.14 Summary of substituent effects in naphth[2,1-*b*][1,4]oxazines

compared to the corresponding unsubstituted (X = H) analogue, the 5′-ester dyes produce weaker photocoloration as indicated by significantly reduced $t_{1/2}$ values.

Figure 5.14 summarises in a very general way the tools at the disposal of the dye designer to adjust the properties of spirooxazines of the [2,1-*b*]-type. As ever with photochromic colorants, exceptions crop up depending upon the exact nature of the alterations and the structures under modification. For example, increasing the bulk of 3,3-dialkyl fragment on the indoline system leads to pronounced differences in $t_{1/2}$ in certain instances but not others. The rules illustrated apply in an analogous manner in the case of the isomeric [1,2-*b*]-class of oxazines, for example, enlarging the *N*-alkyl function tends to lead an increased half-life whilst making little impact on the position of the absorption maximum of the activated form.

It should also be borne in mind that the long wavelength absorption bands of photomerocyanine species induced from typical commercial spirooxazines are often not clean peaks but exhibit shoulders like those of many other colorants owing to bands arising from a complex array of electronic transitions and the presence of isomeric photoproducts. Consequently, it is dangerous to assume that simple relationships exist between absorption maxima and changes to molecular structure across the spirooxazine class since alterations to band shape may dictate the wavelength of maximal absorption. As has previously been pointed out, environment exerts a substantial influence on dye colour and kinetics. When one also takes in account that appearance is strongly sensitive to subtleties in absorption band shape, it is perhaps no surprise that rules of thumb remain more effective design guides than current computational methods. Consequently, when fine-tuning the performance of photochromic systems based upon spirooxazines, there is no substitute for the preparation of candidate structures for evaluation of their photochromism in terms of colour, kinetics and longevity. The same is true of the naphthopyran class, especially when constructing derivatives that produce complex absorption patterns upon photoactivation in order to generate neutral colours. However, before jumping straight into such intricacies, we will look next at some of the general relationships governing the photochromic properties of the two types of commercial naphthopyran dye and strategies employed in controlling their colour and kinetics.

5.5.2 *Substituent Effects in Naphthopyrans*

Just as is the case with spirooxazines, many ways exist of adjusting the photochromic properties of naphthopyran derivatives through modification of molecular structure. Qualitative rules derived from observations of structure–property trends generally serve as reliable guidelines. They are often helpful in a quantitative sense too. This section illustrates this point by examining some molecular design strategies employed by industry to create naphthopyran colorants with commercially useful performance. While rules associated with specific changes to structure usefully inform decisions on which approaches to take, there are plenty of exceptions where the impact of transformations in structure does not quite match with expectation. Given the many factors that influence the photochromism of organic colorants other than dye structure, nothing can replace synthesising and testing compounds of interest when fine-tuning colour, kinetics and performance.

The parent [1,2-*b*]naphthopyran **15** produces very slow fading red-orange photocoloration (λ_{max} 508 nm and $t_{1/2}$ >400 s in toluene at 20 °C) and is of no commercial significance in its own right. However, appropriate substitution at positions 5–10 and 2–4′ has proven a very powerful means of adjusting colour and kinetics to generate dyes of great industrial importance. The parent [2,1-*b*] naphthopyran **16** is more hypsochromic and much faster fading (λ_{max} 429 nm and $t_{1/2}$ 14 s in toluene at 20 °C) than its isomer **15**. While it is an industrial dye and available commercially as Reversacol Rush Yellow [102], the relatively fast thermal fade leads to somewhat weak photocoloration.

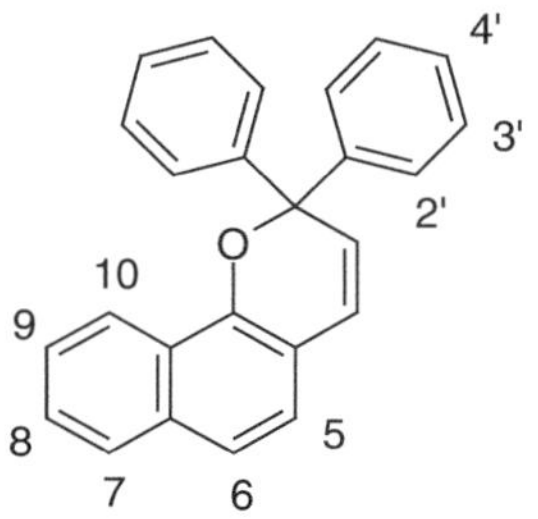

15

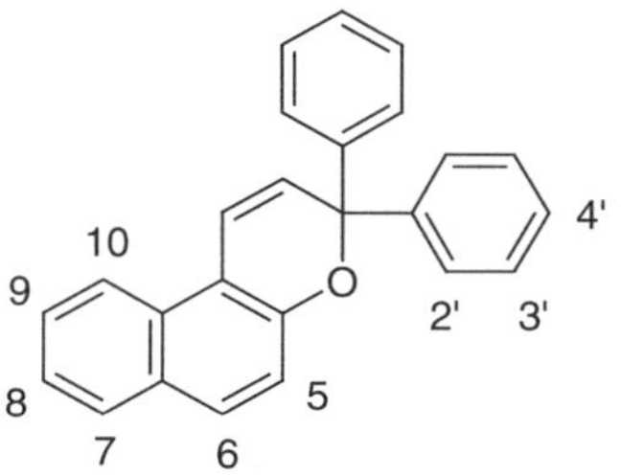

16

However, augmentation of structure **16** (Table 5.7, X = Y = H) enables the achievement of much more. Placing electron-donor groups such as alkoxy and amino functions into positions X and Y (i.e. the 4′- and 4″-positions *para* to pyran ring link) leads to

Table 5.7 Impact of varying 4′,4″-groups on absorption maximum and half-life in toluene (20 °C)

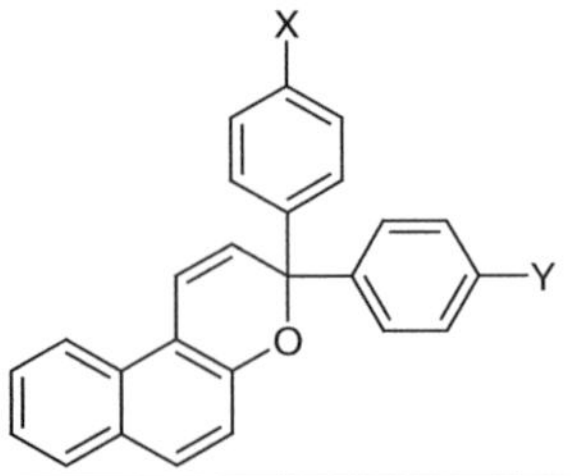

X	Y	λ_{max} (nm)	$t_{1/2}$ (s)
H	H	429	14
OMe	H	460	5
OMe	OMe	472	2
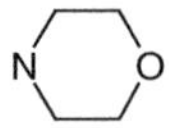	H	501	4
	H	515	3
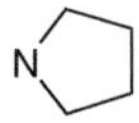	H	542	2
NEt_2	H	544	2
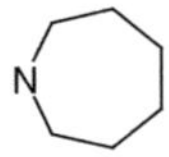	H	549	3
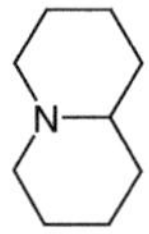	H	581	2
CF_3	H	421	18
CF_3	OMe	446	7
CF_3	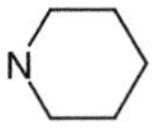	510	4
NEt_2	NEt_2	576	1
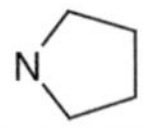	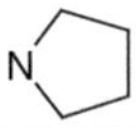	577	1

significant bathochromic shifts in absorption, creating orange and red colorants. Unfortunately, they are too weak to be of much industrial interest. Generally, as overall donor strength at these two positions rises, $t_{1/2}$ falls so strength diminishes. A single methoxy or amino group thus produces marked red shifts in absorption, but at the expense of intensity of photocoloration. These effects grow in size as the electron-donating character of the substituent increases, i.e. H < methoxy < morpholino < piperidino ≈ dimethylamino < diethyamino ≈ pyrrolidino < homopiperidino < julolidino. Addition of a second donor function further enhances the bathochromic shift in λ_{max}, although to a lesser degree than that brought about by introduction of the first group. The same holds true of the reduction in half-life resulting from the presence of the second donor. For example, a single methoxy group (Table 5.7, X = OMe, Y = H) induces a significant red shift of ~30 nm with a substantial shortening in $t_{1/2}$ from 14 s to 5 s; a second methoxy function (X = Y = OMe) gives a fast-fading orange dye as a consequence of a further 12 nm red-shift in λ_{max} and drop in half-life to just a couple of seconds. Placing one amine function into the parent structure (i.e. Table 5.7, X = amino, Y = H) leads to both large bathochromic shifts (~70–120 nm) and decreases in half-life giving fast-fading, weak red dyes. A second amine group of the same type typically pushes absorption maxima another ~20–30 nm longer and further speeds up thermal fade – Table 5.7 shows several examples of such mono- and di-amino pairings. An extreme case is the action of two powerful donors in concert (Table 5.7, X = Y = pyrrolidino or diethylamino). They shift absorption sufficiently to give bluish-purple photocoloration. However, since strong donors also shorten half-life, the dipyrrolidino dye and its diethylamino analogue are only weakly photochromic at ambient temperatures ($t_{1/2} = <1$ s) despite these groups being likely to increase the intensity of absorption of the merocyanine form [103].

Electron acceptors tend to have the opposite effect, i.e. hypsochromic shifts in absorption of photoactivated dye to shorter wavelengths accompanied by an increase in activated intensity and duration of $t_{1/2}$. Table 5.7 exemplifies this relationship with trifluoromethyl groups: placing one of these acceptors onto the parent dye creates a colorant (Table 5.7, X = CF_3, Y = H) that has a slightly slower thermal fade (18 s cf.14 s) and which is modestly blue-shifted (421 nm cf. 429 nm). Given the big influence that it has on absorption and kinetics, substituent choice on the aryl rings of [2,1-*b*]naphthopyrans (as well as [1,2-*b*]naphthopyrans) is therefore a crucial aspect of dye design. The relationship between colour and rate of thermal fade means that the [2,1-*b*]naphthopyran class is restricted to being a source of commercially important yellow, orange and red dyes only. However, many of these dyes rely on additional design features.

One such industrially useful design tactic is the incorporation of a strong electron donor onto the naphthalene moiety in an analogous fashion to spirooxazines. An amino or alkoxy function located at the 6-position boosts the strength of photocoloration, mitigating the kinetic effects of strong donors introduced onto the 3-aryl groups to red-shift absorption (see Table 5.8).

Industry has been utilising this means of creating valued orange and red colorants for more than 20 years [104]. For example, when one compares the figures for 6-morpholino derivatives in Table 5.8 with those in Table 5.7, the addition of the 6-amino group leads to a substantial decrease in rate of thermal fade (accompanied by a large jump in intensity of steady state photocoloration) along with a moderate hypsochromic shift: several commercial naphthopyrans make use of this substitution pattern.

Substitution with an alkoxy group is also effective. For example, swapping the 6-morpholino residue of the derivative in Table 5.8 (X = morpholino) with a methoxy group leads to a minor bathochromic shift (2 nm) and a substantial further increase in half-life from 42 to 92 s. The same switch in the case of the analogue in

Table 5.8 Absorption maxima and half-lives of 6-morpholino derivatives of [2,1-*b*]naphthopyrans in toluene solution at 20 °C

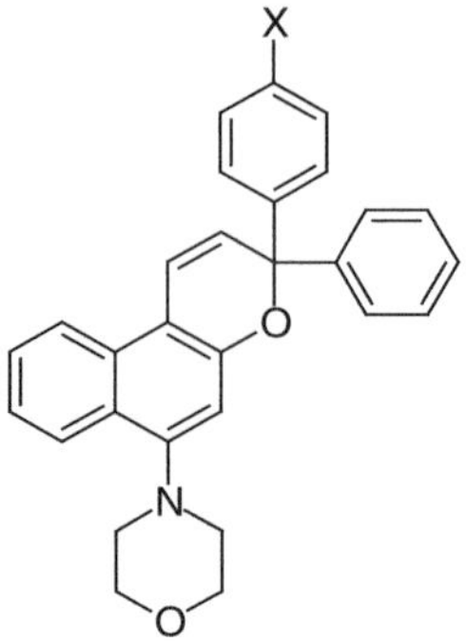

X	λ_{max} (nm)	$t_{1/2}$ (s)
H	418	103
OMe	433	54
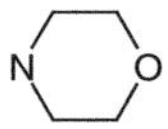	470	42
	488	33
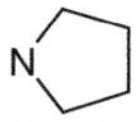	518	29
NEt_2	516	23

Table 5.8 with X = piperidino brings about a minor hypsochromic shift (2 nm) and an increase in $t_{1/2}$ from 33 to 84 s. These derivatives are prepared by reaction of diaryl alkynols as per Fig. 5.13 with 4-morpholino-2-naphthol (Z = morpholino) and 4-methoxy-2-naphthol (Z = methoxy). These two naphthols are readily prepared from 2-naphthol [88].

Placing an electron donor onto the 8-position gives a similar, albeit less pronounced, effect as 6-donor functions. All else being equal, an 8-methoxy group leads to greater bathochromism of photoactivated colour as well as a slower thermal fade and greater photostationary state intensity compared to corresponding substitution at adjacent 7- and 9-positions. An explanation for this observation is that the presence of a donor at the 8-position makes possible an additional zwitterionic canonical form like a 6-donor (see Fig. 5.15). Given that this form involves destruction of the aromaticity of the naphthalene system whereas it is partly preserved in the case of the 6-donor [88] means that the impact of the 8-donor is more limited.

Another well-used approach to enhance photocoloration is the inclusion of a substituent on one of the 3-aryl rings *ortho* to the C3 carbon, i.e. at the 2′-position as shown in Table 5.9. Generally, the imposition of a fluoro function at this location leads to a modest shift in absorption maximum. Far more remarkable is the increase in half-life and intensity of photocoloration. For example, when

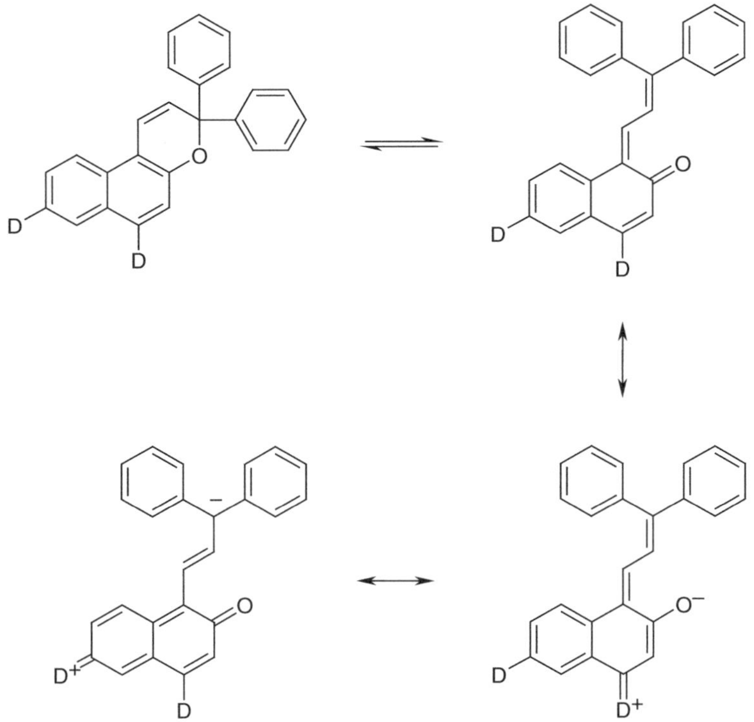

Fig. 5.15 Donor effects at 6- and 8-position of [2,1-*b*]naphthopyrans

Y = 2′-F, the dyes produce stronger photocoloration and are slower to fade (i.e. $t_{1/2}$ is much longer) following activation compared to dyes in which Y = H. This is illustrated in Table 5.9 for the corresponding pairs of dyes where X = H, methoxy, morpholino or piperidino. The more sterically demanding o-methyl and -methoxy groups, i.e. Y = 2′-Me or 2′-OMe, generate even more pronounced effects. While their donor character leads to rather small bathochromic shifts, these substituents greatly lengthen half-life as can be seen from the data shown in Table 5.9. Such patterns are thus of interest for slowing down the rate of fading (and increasing the intensity of photocoloration) of relatively fast and weak [2,1-*b*]naphthopyrans while retaining their hypsochromism. For example, $t_{1/2}$ is increased by two orders of magnitude in dyes where X = alkoxy or X = amino through the introduction of an *ortho* methyl or methoxy group (see Table 5.9). In contrast, their absorption maxima see no more than a 10 nm shift. This strategy has been used commercially for many years, although it is often used in conjunction with other features to adjust kinetics because the change it produces is so marked. The *ortho* substituent is believed to be a source of steric hindrance to ring closure [105], thus increasing the lifetime of the photomerocyanine species, leading to slower thermal fading and increased photostationary intensity. Even more bulkier *ortho*-groups lead to longer half-lives, for example, $t_{1/2}$ 2′-Cl < $t_{1/2}$ 2′-Br < $t_{1/2}$ 2′-I.

Table 5.9 Effect of 2′- and 3′-substitution on absorption maxima and half-lives of [2,1-*b*]naphthopyrans in toluene solution at 20 °C

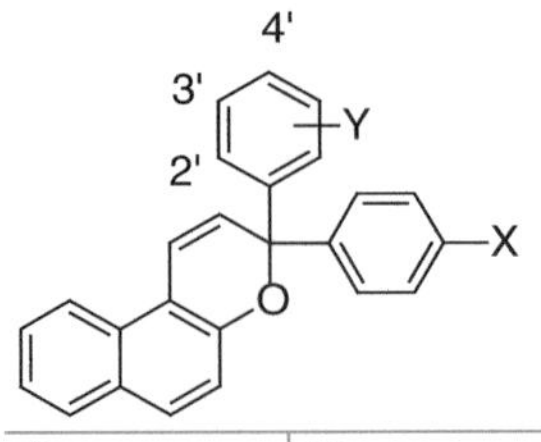

X	Y	λ_{max} (nm)	$t_{1/2}$ (s)
H	–	429	15
	2′-F	421	108
OMe	4′-OMe	472	2
	2′-F-4′-OMe	463	36
	2′-Me-4′-OMe	470	400
	2′,4′-DiOMe	476	320
	3′,4′-DiOMe	482	3
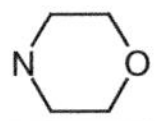	–	501	4
	2′-F	503	87
	2′-OMe	508	1,500
	4′-OMe	505	2
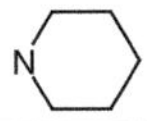	–	515	3
	2′-F	525	75
	2′-OMe	525	880
	3′-OMe	520	6
	4′-OMe	519	2
NMe_2	4′-NMe_2	538	1
	4′-NMe_2-2′-Me	549	99
CF_3	4′-OMe	446	7
	2′,4′-DiOMe	437	600

Notice in the data of Table 5.9 how a 3′-substituent does not bring about a large increase in half-life: a methoxy function in this position confers a bathochromic shift as would be expected of an electron donor, but no large increase in half-life as it does not bring about the hindrance typical of 2′-OMe substitution.

However, a 3′-substituent is capable of usefully producing steric hindrance but for different reasons to 2′-groups. Crowding of a 4′-amino substituent is a tactic sometimes exploited in industrial dye design to adjust colour with relatively little impact on kinetics as illustrated in Table 5.10. Locating a group *ortho* to the pyrrolidino nitrogen (i.e. at the 3′-position) leads to twisting of the amine function out of the plane of the aryl ring, resulting in diminished participation of the amino group in the molecular π-system. Consequently, when X = Me or halogen, significant hypsochromic shifts result because the effectiveness of the amino group as an

Table 5.10 Influence of 3′-substituent on absorption maximum and half-life of 4′-amino functionalised [2,1-*b*]naphthopyran in toluene solution at 20 °C [106]

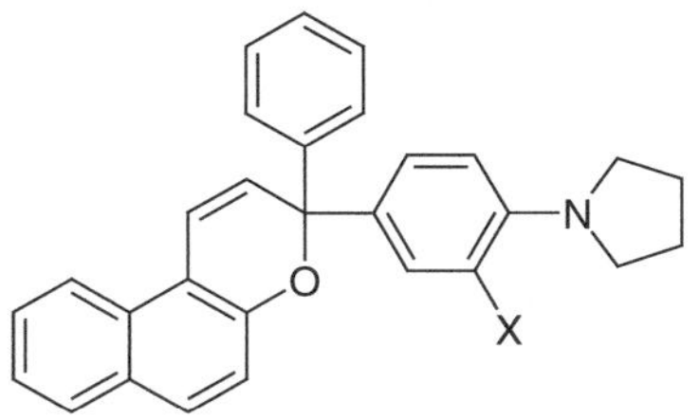

X	λ_{max} (nm)	$t_{1/2}$ (s)
H	538	5
Me	517	3
OMe	557	5
F	529	4
Cl	513	4
Br	493	4
I	494	5

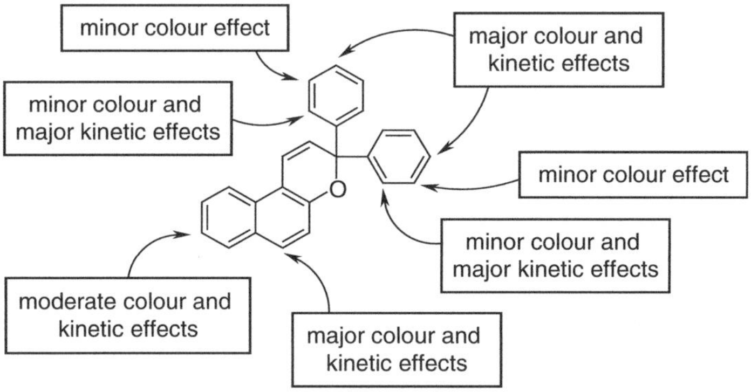

Fig. 5.16 Summary of substituent effects in [2,1-*b*]naphthopyrans

electron donor is reduced through losses in conjugation. A 3′-methoxy function (i.e. Table 5.10, X = OMe) bucks this trend: its capacity as a donor is sufficient to outweigh any steric hindrance caused, producing bathochromism.

A summary of the relationships outlined above regarding the position of a substituent on the [2,1-*b*]naphthopyran system and its potential impact on the colour and kinetics of the colorant is shown in Fig. 5.16.

Some of the relationships given in Fig. 5.16 also hold true for the [1,2-*b*] naphthopyran system. These include substitution on the 2-phenyl rings: for example, introduction of strong donors at the 4′- and 4″-positions leads to bathochromism accompanied by an increase in rate of thermal fade, whereas *ortho* groups, i.e. at 2′- and 2″-positions bring about pronounced lengthening of half-life. As previously mentioned, the parent compound **15** fades too slowly thermally to be of commercial use. Industry overcomes this drawback by the

Table 5.11 Influence of 5- and 6-position on kinetics of [1,2-*b*]naphthopyrans [109]

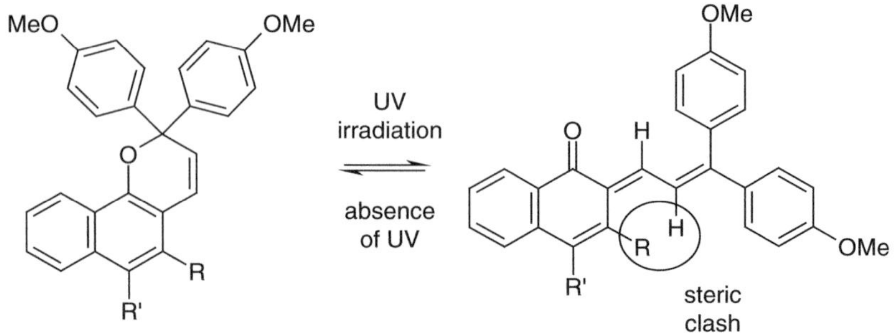

R	R′	$t_{1/2}$ (s)	λ_{max} (nm)
H	H	>1,800	508, 412
Me	H	178	496, 416
Me	Me	65	492, 418
Ph	Ph	20	508, 418
CO_2Me	H	2	492
CO_2Me	Me	4	492, 416
CO_2Me	Ph	7	506, 416

location of functional groups onto the 5- and 6-positions (i.e. Table 5.11, R and R′) to impose steric crowding on the photomerocyanine, reducing stability and prompting a greater tendency for thermal reversion to the ring-closed form of the dye. Replacing the hydrogen at the 5-position with a methyl group (i.e. R = Me, R′ = H) is sufficient to bring the half-life down by an order of magnitude to a length that is approaching commercially utility, yet makes only a small impact on colour. Addition of a methyl group to the structure at the 6-position further shortens half-life to an industrially useful duration while retaining its red colour when activated. The resultant colorant (R = R′ = Me) is available on the market as Reversacol Berry Red [107, 108].

Imposition of even bulkier groups at the 5-position (i.e. R = Ph or CO_2Me) further speeds up thermal fade, giving dyes that are too weak. However, a substitution pattern of 5-Me-6-aryl is commercially valuable. It not only permits the creation of intense bathochromic colorants with suitable half-lives, but also furnishes relatively neutral-coloured dyes with prominent double absorption peaks (see Table 5.12). Donor groups placed into the 9-position and 6-phenyl ring cause bathochromism and increased intensity through longer half-life. In certain cases, two distinct peaks appear, giving duller and more neutral colours. The effect is more pronounced for alkoxy groups compared to alkyl groups, e.g. where X = pyrrolidino and Y = H. Subtle differences in substituent effects can make a substantial impact on colour. For example, when X = Y = piperidino, the gap between the two absorption peak maxima widens to around ~100 nm compared to ~80 nm for dyes with just one amine function, producing a more neutral

Table 5.12 Influence of donor groups in 5-methyl-6-phenylnaphth[1,2-*b*]pyrans in toluene solution at 20 °C

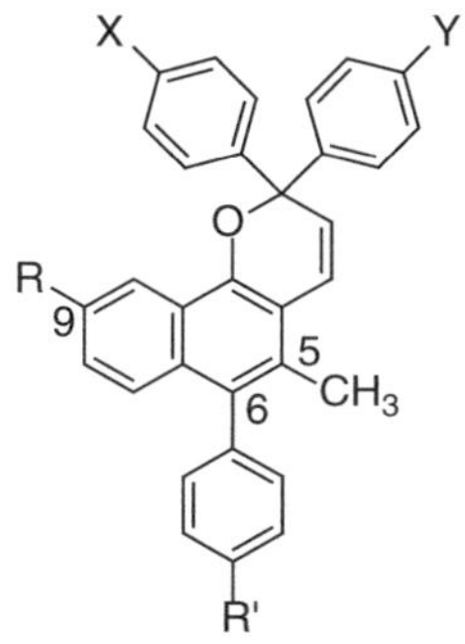

X	Y	R	R′	λ_{max} (nm)	$t_{1/2}$ (s)
OMe	OMe	H	H	498, 419	32
		OMe	OMe	528, 427	53
		Me	Me	506, 422	59
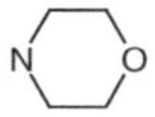	H	H	H	514	63
		OMe	OMe	532, 461	99
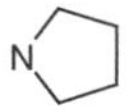	H	H	H	550	42
		OMe	H	559, 488	44
		OMe	OMe	560, 485	52
		Me	Me	549	54
		H	H	551	14
		OMe	H	560, 466	17
		OMe	OMe	560, 465	22
		Me	Me	549, 448	22

appearance. A consequence of the presence of two amino groups is that thermal fade is more rapid, leading to weaker steady-state coloration.

Locating stronger donor functions on the 9-position is a strategy that has proved commercially successful for producing colorants with multiple broad absorption bands, simplifying the formulation of neutral grey lens formulations. Derivatives of general structure **17** have been marketed over the past 15 years featuring 9-amino groups, giving grey [110] and brown [111] photocoloration. The substituent enhances intensity of absorption in the short wavelength region of the visible spectrum when photoactivated. Colour and kinetics can be further adjusted using the rules of thumb already discussed above to fine tune the position of these absorption bands while also trying to keep a reasonable balance of photocoloration intensity and rate of thermal fade. In the case of **17** (X = OMe), brown coloration results as photoactivation gives two peaks (λ_{max} 556 nm and 454 nm in toluene at 20 °C). Increasing donor strength in the aryl rings shifts these bands bathochromically in line with expectation. For example, **17** (X = piperidino) is purplish-grey since the absorption maxima are shifted to 581 nm and 490 nm in toluene. These bands are affected in slightly different ways by the medium in which the dye is applied so colour may be subtly affected by solvatochromism. Nevertheless, this strategy was innovative by enabling formulators to minimise the amount of shading components required to produce a lens of neutral colour.

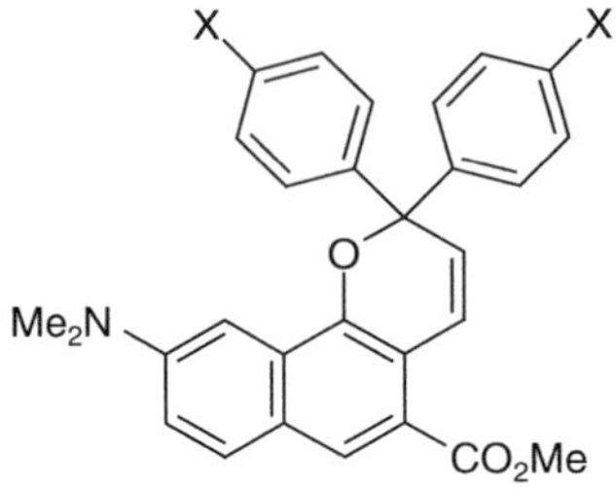

17

Another commercially important class are indeno-fused dyes of general structure **18**, which have been used in lens manufacture for the past 20 years [12, 112]. In the case of these colorants, industry made further use of crowding by fusion of ring systems at the 5- and 6-positions of the [1,2-*b*]naphthopyran system to manipulate kinetics whilst simultaneously tuning absorption to give blue and green colorants with multiple broad absorption bands for neutral shades. Changes at the 13-position enable adjustment of kinetics while the indeno skeleton holds the fused phenyl ring in plane with the rest of the π-system, creating a chromophore of enhanced intensity and bathochromism. Analogues with other fused systems appear in many patents, but only a few have been commercialised, e.g. derivatives with a 6-membered ring **19**. In this field, the existence of a very large proportion of structures in the patent literature owes itself not so much to the supposed technical advantages claimed, but to efforts aimed at skirting round existing patents.

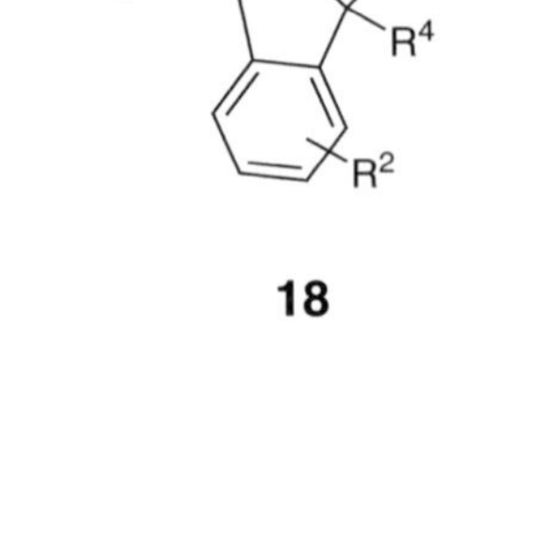

18

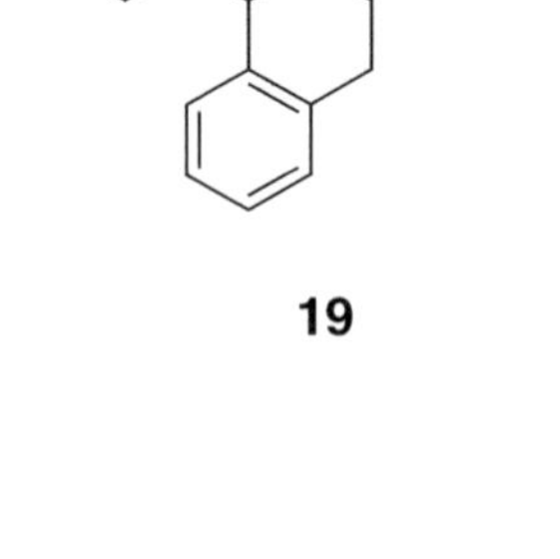

19

Table 5.13 Influence of 13-substituent in indeno-fused [1,2-*b*]naphthopyrans in methacrylate-based polymer at 23 °C

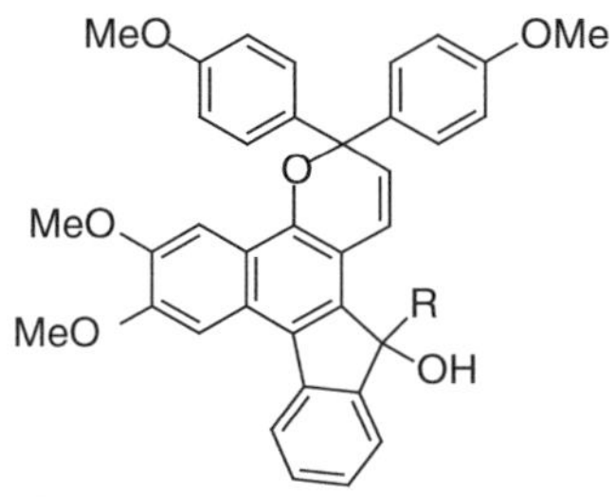

R		$t_{1/2}$ (s)	λ_{max} (nm)
	H	351	460, 570
	Me	230	455, 575
	Et	146	455, 580
	*n*-Pr	115	460, 580
	*n*- Bu	116	455, 580
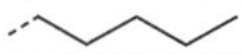	*n*-Am	157	460, 570
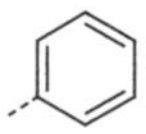	Ph	260	460, 580
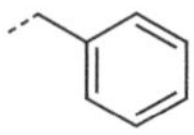	Bzl	258	460, 580
	Vinyl	325	460, 575
	Ethynyl	171	460, 575
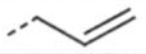	Allyl	186	460, 575

Substituents at the 13-position of **18** strongly influence kinetics with a minor impact on colour (see Table 5.13). Small variations in the position of the two main absorption bands of the dyes in the visible region occur, but significant differences in half-life are apparent. In line with findings from 5- to 5,6-substituted [1,2-*b*] naphthopyrans, locating bulkier groups into the 13-position tends to lead to increased fade-speed. However, the relationship is complex. For example, when R = Me, thermal fading is more rapid than when R = H. Increased crowding results in shorter half-lives as shown in the data for R = Et and R = *n*-Pr. As we saw earlier in the case of *N*-alkyl functions in indolino-spironaphthoxazine dyes, geometry as well as size plays a part. For example, increases in chain length (R = *n*-Bu or *n*-Am) or introduction of aryl groups (R = Ph or Bzl) fail to decrease $t_{1/2}$ further. In addition, there are significant differences between the kinetics of the sets of dyes where R = Et, R = vinyl and R = ethynyl as well as the pair of dyes where R = *n*-Pr and R = allyl. The data demonstrate that once again, when developing industrial colorants, it is difficult to make exact predictions and that, even when one has developed rules of thumb, the adjustment of dye properties remains an exercise in trial and error.

The indeno-fused derivatives can be prepared by several means, two of which are shown in Fig. 5.17 where coloured non-photochromic ketone derivatives are

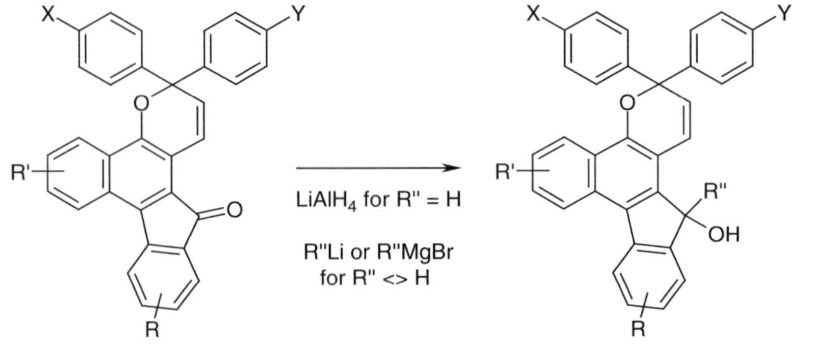

X, Y = H, alkoxy, amino, etc.; R, R' = H, alkyl, alkoxy, etc.; R'' = H, (un)saturated alkyl, aryl

Fig. 5.17 A commercial synthetic route to 13-substituted indeno-fused naphthopyrans

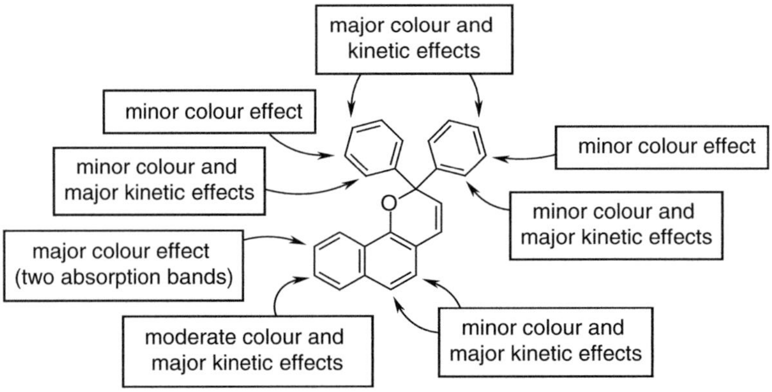

Fig. 5.18 Summary of substituent effects in [1,2-*b*]naphthopyrans

either reduced, or reacted with an organometallic compound, to furnish photochromic products.

This section has given a flavour of the relationships that govern photochromic properties of the [1,2-*b*]naphthopyran system. A summary of them is provided above in Fig. 5.18.

5.6 Conclusions

Heliochromism is a good starting point for the current definition of the most commercially successful kind of industrial photochromism. While P-type photochromic colorants have been the focus of intense efforts to develop new technologies in the areas of optoelectronics, data storage and nano-engineering, the principal commercial use of industrial dyes is to put T-type photochromism to work in ophthalmic lenses. The light-sensitive silver halide-based glasses of half a century ago have largely given way to all-organic polymeric systems. Breakthroughs in colorant chemistry and dye application, whose roots lay in the late 1960s and 1970s, made the switch to more lightweight, more responsive materials possible. These advances included the identification of two broad and versatile classes of colorants with relatively high intrinsic photostability. They came into their own in the 1990s as chemists learnt how to manipulate structure to fine-tune dye colour and kinetics. Formulators and polymer technologists complemented their efforts by finding new ways of enhancing colorant performance through use of additives and providing dyes with right environment in which to function best. Consequently, the industrial usage of photochromism for ophthalmic lenses blossomed, driving the development of more sophisticated dyes. Their structures have been carefully designed to balance the often mutually exclusive demands of colour, speed, strength and stability. This Chapter has attempted to enlighten the reader with some background behind these advances, providing insight into the synthesis, design rules and limitations of these colorants. Even now, over half a century since the first serious modern efforts to develop organic photochromic molecules for real world applications, much remains to be learnt. Considerable progress continues to be made in improving upon the properties of existing T-type molecules for commercial use. By appropriate structural design, one can obtain dyes with multiple absorption bands in the visible region when activated, thereby simplifying the task of the formulator because mixtures of fewer dyes are needed to obtain neutral colours. Other modifications reduce the compromise between speed and strength with, in certain cases, minimal impact on activated hue. Trial and error still plays a part. . .and is likely to continue to do so. Many influences interact in too complex a manner to allow precise predictions of behaviour. Fortunately, there are rules of thumb to help one navigate through this maze. This Chapter outlined some of them.

Numerous applications, either new or old in terms of their inception, await the creation of photochromic entities with the right set of technical properties, economics and, in particular, robustness. Of the many thousands of light-responsive colorants subjected to industrial scrutiny, only a very small proportion have made it through to commercialisation. One could argue that with just a tiny fraction of the design space of photochromic molecules explored, the potential to discover improved variants within existing dye classes, as well as truly novel types of light-responsive colorant, is enormous. Just how much of this space is economically accessible remains to be seen. While the existing industrially significant types of T-type dye will probably retain their dominance in the immediate future, it is conceivable that the rise of another technology reliant upon photochromism might alter the landscape of industrial dye usage radically. If one of the high-tech applications described in this Chapter that make use of P-type dyes became commercially successful, then it could be expected to dwarf the market for T-type colorants. Such a development would redefine industrial photochromism and necessitate a complete re-write of this Chapter. However, efforts in this direction not only need to see off competing technologies that do not utilise photochromic materials, but also surmount substantial technical challenges, which may prevent them seeing the light of day – either literally or metaphorically...

Acknowledgements Thousands of people working during this century and the last within industry, academia and elsewhere in countries across the globe can take a share of the credit for making industrial photochromism what it is. However, the author wishes to acknowledge the contribution to the knowledge, as well as the previously unpublished data, contained within this Chapter of the numerous members of technical staff of Vivimed Labs Europe Ltd (and James Robinson Ltd as was), particularly Dr. D.A. Clarke and Dr. S.M. Partington. In addition, mention must be made of the fruitful collaboration between the organisation with the research groups of Prof. J.D. Hepworth, Dr. B.M. Heron and Dr. C.D. Gabbutt, leading to the creation of insights into the properties of photochromic dyes and its realisation in the form of industrial photochromism.

References

1. Corns SN, Partington SM, Towns AD (2009) Color Technol 125:249–261
2. Brown GH (ed) (1971) Photochromism. Wiley-Interscience, New York
3. Dürr H, Bouas-Laurent H (eds) (1990) Photochromism molecules and systems. Elsevier, Amsterdam
4. Crano JC, Guglielmetti RJ (eds) (1999) Organic photochromic and thermochromic compounds, vol 1–2. Plenum, New York
5. Dürr H, Bouas-Laurent H (2001) Pure Appl Chem 73:639–665
6. Zhang J, Zou Q, Tian H (2013) Adv Mater 25:378–399
7. Heller HG, Oliver SN, Whittal J, Johncock W, Darcy PJ, Trundle C (1986) GB2146327B (The Plessey Company plc)
8. Chu NYC, in Ref. 3, Chapter 10, pp 493–509Please provide year, publisher name, and location for Refs. [8, 10, 14, 50, 55, 61, 81, 99].
9. Crano JC, Kwak WS, Welch CN (1992) In: McArdle CB (ed) Applied photochromic polymer systems. Blackie, Glasgow
10. Hoffman HJ, in Ref. 3, Chapter 22, pp 822–854
11. Bruneni JL (1997) More than meets the eye: the stories behind the development of plastic lenses. PPG Industries, Pittsburgh
12. Van Gemert B (2000) Mol Cryst Liq Cryst 344:57–62
13. Meslin D (2010) Materials and treatments. Essilor Academy Europe, Paris
14. Guglielmetti R, in Ref. 3, Chapter 23, pp 855–878
15. Mortimer RJ, Rosseinsky DR, Monk PMS (eds) (2015) Electrochromic materials and devices. Wiley-VCH, Weinheim
16. Österholm AM, Shen DE, Kerszulis JA, Bulloch RH, Kuepfert M, Dyer AL, Reynolds JR (2015) ACS Appl Mater Interfaces 7:1413–1421
17. Little AF, Christie RM (2011) Color Technol 127:275–281; 126 (2010) 164–170; 126 (2010) 157–163
18. Aldib M, Christie RM (2011) Color Technol 127:282–287
19. Lee S-J, Son Y-A, Suh H-J, Lee D-N, Kim S-H (2006) Dyes Pigment 69:18–21
20. Billah SMR, Christie RM, Shamey R (2008) Color Technol 124:223–228
21. Shah MRB, Christie RM, Morgan KM, Shamey R (2005) Mol Cryst Liq Cryst 431:535–539
22. Billah SMR, Christie RM, Shamey R (2012) Color Technol 128:488–492
23. Son Y-A, Park Y-M, Park S-Y, Shin C-J, Kim S-H (2007) Dyes Pigment 73:76–80

24. Peng L, Guo R, Jiang S, Lan J, He Y, Huang X (2015) Fibers Polym 16:1312–1318
25. Billah SMR, Christie RM, Morgan KM (2008) Color Technol 124:229–233
26. Aldib M, Christie RM (2013) Color Technol 129:131–143
27. Aldib M (2015) Color Technol 131:172–182
28. Coimbra P, Gil MH, Duarte CMM, Heron BM, de Sousa HC (2005) Fluid Phase Equilib 238:120–128
29. Cheng T, Lin T, Brady R, Wang X (2008) Fibers Polym 9:301
30. Feczkó T, Samu K, Wenzel K, Neral B, Voncina B (2012) Color Technol 129:18–23
31. van Renesse RL (1998) In: van Renesse RL (ed) Optical document security, 2nd edn. Artech House, Boston, pp 201–224, Chapter 9
32. McCallien DWJ, Bezer M (2000) (John Hogg Technical Solutions Ltd), British Patent 2344599A
33. Feringa BL, Browne WR (eds) (2011) Molecular switches, 2nd edn. Wiley-VCH, Weinheim
34. Dong H, Zhu H, Meng Q, Gong X, Hu W (2012) Chem Soc Rev 41:1754–1808
35. Irie M, Yokoyama Y, Seki T (eds) (2013) New frontiers in photochromism. Springer Japan, Tokyo
36. Tsujioka T (2005) Mol Cryst Liq Cryst 431:391–395
37. Barachevsky VA, Strokach YP, Puankov YA, Krayushkin MM (2007) J Phys Org Chem 20:1007–1020
38. Barachevsky VA (2008) J Photochem Photobiol A 196:180–189
39. Tsujioka T (2014) J Mater Chem C 2:3589–3596
40. Raymo FM, Tomasulo M (2005) Chem Soc Rev 34:327–336
41. Yildiz I, Deniz E, Raymo FM (2009) Chem Soc Rev 38:1859–1867
42. Tsujioka T, Irie M (2010) J Photochem Photobiol C 11:1–14
43. Orgiu E, Samori P (2014) Adv Mater 26:1827–1845
44. Andréasson J, Pischel U (2013) Israel J Chem 53:236–246
45. Matsuda K, Irie M (2006) Chem Lett 35:1204–1209; Belser P, De Cola L, Hartl F, Adamo V, Bozic B, Chriqui Y, Iyer VM, Jukes RTF, Kühni J, Querol M, Roma S, Salluce N (2006) Adv Func Mater 16:195–208
46. Fang Y, Sun M (2015) Light Sci Appl 4:e294
47. Yamada M, Kondo M, Mamiya J-I, Yu Y, Kinoshita M, Barrett CJ, Ikeda T (2008) Angew Chem Int Ed 47:4986–4988
48. International Symposium on Photochromism 2013 Book of Abstracts, Berlin, Germany; 2013
49. Irie M (2001) In: Feringa BL (ed) Molecular switches, 1st edn. Wiley-VCH, Weinheim, pp 37–62, Chapter 2
50. Fan M-G, Yu L, Zhao W, in Ref. 4, vol 1, Chapter 4, pp 141–206
51. Whittal J, in Ref. 3, Chapter 9, pp 467–492
52. Heller HG, Koh KSV, Köse M, Rowles N (1999) Photochromics by design: the art of molecular tailoring. In: J Griffiths (ed) Colour science'98 conference papers. The University of Leeds, Harrogate
53. Heller HG, Oliver SN (1981) J Chem Soc Perkin Trans I:197
54. Zollinger H (2003) Color chemistry, 3rd edn. VCHA & Wiley-VCH, Zürich
55. Yokoyama Y, Gushiken T, Ubukata T, in Ref. 33, Chapter 3, pp 81–95
56. Kuhn HJ, Braslavsky SE, Schmidt R (2004) Pure Appl Chem 76:2105
57. Kobayakawa T (1999) Third international symposium on photochromism, Fukuoka
58. Bamfield P, Hutchings MG (2010) Chromic phenomena, 2nd edn. The Royal Society of Chemistry, Cambridge
59. Irie M (2000) Chem Rev 100:1685–1716
60. Irie M, Fukaminato T, Matsuda K, Kobatake S (2014) Chem Rev 114:12174–12277
61. Irie M, in Ref. 4, vol 1, Chapter 5, pp 207–222
62. Higashiguchi K, Matsuda K, Tanifuji N, Irie M (2005) J Am Chem Soc 127:8922–8923
63. Morimoto M, Irie M (2005) Chem Commun 3895
64. Irie M (2008) Bull Soc Chem Jpn 81:917

65. Zhao Y, Ikeda T (eds) (2009) Smart light responsive materials: azobenzene-containing polymers and liquid crystals. Wiley, Hoboken
66. Such G, Evans RA, Yee LH, Davis TP (2003) J Macromol Sci C43:547–579
67. Homola TJ (2000) Mol Cryst Liq Cryst 344:63–68
68. Homola TJ (1999) In: ChemiChromics USA'99 conference proceedings. Spring Innovations Ltd., New Orleans
69. Vázquez-Mera N, Roscini C, Hernando J, Ruiz D, Hernando Campos J (2013) Adv Opt Mater 1:631–636
70. Werkman PJ, Rietjens GH, Bernadus TNM, de Vries GC, Klink SI, Hofstraat JW, Kloosterboer JG (2003) (Koninklijke Philips Electronics NV), World Patent Application 03/001555A1
71. For example: Heft A, Pfuch A, Schimanski A, Grünler B (Innovent e.V.), European Patent 2233545 (2012)
72. For example: Partington SM (Vivimed Labs Europe Ltd.), World Patent Application 2010/020770A1 (2010); Evans RA, Skidmore MA, Yee LH, Hanley TL, Lewis DA (Advanced Polymerik Pty Ltd.), US Patent 7807075 (2010)
73. Evans RA, Hanley TL, Skidmore MA, Davis TP, Such GK, Yee LH, Ball GE, Lewis DA (2005) Nat Mater 4:249
74. Evans RA, Such GK (2005) Aust J Chem 58:825
75. Malic N, Campbell JA, Ali AS, York M, D'Souza A, Evans RA (2010) Macromolecular 43:8488–8501
76. Malic N, Campbell JA, Evans RA (2008) Macromolecular 41:1206
77. Roscini C, Vázquez-Mera N, Ruiz-Molina D (2015) (Consejo Superior de Investigaciones Cientificas), US Patent Application 2015/0024126
78. Amon A, Bretler H, Bleikolm A (1993) (SICPA Holding SA), European Patent 0327788
79. For example, Patel DG, Benedict JB, Kopelman RA, Frank NL (2005) Chem Commun 2208–2210
80. Qin M, Huang Y, Li F, Song Y (2015) J Mater Chem C 3:9265–9275
81. Bertelson RC, in Ref. 4, vol 1, Chapter 1, pp 11–83
82. Klajn R (2014) Chem Soc Rev 43:148–184
83. Van Gemert B (1999) Third international symposium on photochromism, Fukuoka
84. Van Gemert B in ChemiChromics'99 conference papers. Spring Innovations, New Orleans
85. Wilkinson F, Worrall DR, Hobley J, Jansen L, Williams SL, Langley AJ, Matousek P (1996) J Chem Soc Faraday Trans 92:1331–1336
86. Pottier E, Sergent M, Phan Tan Luu R, Guglielmetti R (1992) Bull Soc Chim Belg 101:719–739
87. York M, Evans RA (2010) Synth Commun 40:3618–3628
88. Van Gemert B, in Ref. 4, Chapter 3, pp 111–140
89. Hepworth JD, Gabbutt CD, Heron BM (1999) Photochromism of Naphthopyrans. In: J Griffiths (ed, Colour science'98 conference papers. The University of Leeds, Harrogate
90. Crano JC, Flood T, Knowles D, Kumar A, Van Gemert B (1996) Pure Appl Chem 68:1395
91. Hepworth JD, Heron BM (2006) In: Kim S-H (ed) Functional dyes. Elsevier, Amsterdam, pp 85–135, Chapter 3
92. Hobley J, Wilkinson F (1996) J Chem Soc Faraday Trans 92:1323–1330
93. For example:Momota J, Kobayakawa Takashi T (Tokuyama Corp.), Japanese Patent 3227062 (2001); Komura Y, Momota J (Tokuyama Corp.), Japanese Patent 2000-026469 (2000); V Krongauz, A Chif, A Aizikovich and V Tchernoivanov (Yeda Research and Development Co. Ltd.), US Patent 6891038 (2005)
94. Partington SM, Towns AD (2014) Dyes Pigment 104:123–130
95. Hovey RJ, Chu NYC, Piusz PG, Fuchsman H (1982) (American Optical Corp.), US Patent 4342668
96. Melzig M, Martinuzzi G (1984) (Optische Werke G Rodenstock), European Patent 0508219
97. Wilkinson F, Hobley J, Naftaly M (1992) J Chem Soc Faraday Trans 88:1511–1517

98. Rickwood M, Marsden SD, Ormsby ME, Staunton AL, Wood DW, Hepworth JD, Gabbutt CD (1994) Mol Cryst Liq Cryst A 246:17–24
99. Malatesta V, in Ref. 4, vol 2, pp 65–166
100. Parry H, Corns N, Towns A (2003) Speciality Chem Mag 23(8):27
101. Clarke DA (1995) Photochromic dyes. In: ChemiChromics'95 conference papers. Spring Innovations, Manchester
102. Higgins S (2003) Chem Br 39(6):26
103. Gabbutt CD, Hepworth JD, Heron BM, Partington SM, Thomas DA (2001) Dyes Pigment 49:65–74
104. Rickwood M, Smith KE, Gabbutt CD, Hepworth J (1994) (Pilkington Plc), WO 94/22850
105. Gabbutt CD, Gelbrich T, Hepworth JD, Heron BM, Hursthouse MB, Partington SM (2002) Dyes Pigment 54:79–93
106. Gabbutt CD, Heron BM, Instone AC, Horton PN, Hursthouse MB (2005) Tetrahedron 61:463–471
107. Rauzy E, Berro C, Morel S, Herbette G, Lazzeri V, Guglielmetti R (2004) Polym Int 53:455
108. Favaro G, Ortica F, Romani A, Smimmo P (2008) J Photochem Photobiol A 196:190
109. Research Disclosure, 36144 (May 1994) 267
110. Clarke DA, Heron BM, Gabbutt CD, Hepworth JD, Partington SM, Corns SN (2002) (James Robinson Ltd.), European Patent 1117656
111. Clarke DA, Heron BM, Gabbutt CD, Hepworth JD, Partington SM, Corns SN (2011) (Vivimed Labs Europe Ltd.), European Patent 0971914
112. Van Gemert B (1996) (PPG Industries Inc.), World Patent Application 96/14596

Chapter 6
Application of Visible and Solar Light in Organic Synthesis

Davide Ravelli, Stefano Protti, and Maurizio Fagnoni

Abstract This chapter highlights the impressive recent developments (mainly focused in the 2013–2015 period) in the use of visible and solar light to promote valuable organic transformations, virtually inaccessible under thermal conditions. Compact fluorescent lamps (CFL), LEDs and even ambient sunlight were used as the light sources. The photochemical approach was applied, among others, to ring formation, arylations, additions onto C=C bond, alpha- and beta- functionalization of carbonyls, (de)halogenations, oxidations and reductions. The last part comprises some exemplary cases of the synthesis of bioactive compounds.

6.1 Introduction

Apart from physical deactivation, the main phenomenon that one may expect when a molecule absorbs light, is a chemical process. Thus, activation of the substrate (s) and subsequent formation of the product(s) is achieved by promoting one of the reactants to its reactive excited state. Since the birth of photochemistry, the chance to induce a (selective) transformation in the substrate by simply shining light on the reaction mixture has fascinated scientists. Since then, photochemical reactions have been sparsely introduced in synthetic planning, albeit these processes are still underutilized in comparison to their potentiality. The advantages lying in the photochemical approach are apparent. The photon alone is responsible for the activation of the reagent, thus avoiding the use of "brute force" (viz. heating, aggressive chemicals, etc.) otherwise required in conventional thermal reactions. This allows to use mild conditions and, on the other hand, leads to no residue from the chemicals used for activation, and thus to no waste, or to a lesser amount of it (and at any rate, less risky). This simplifies both the work-up and the purification step [1, 2]. Importantly, the high energy of excited states allows deep-seated

D. Ravelli • S. Protti • M. Fagnoni (✉)
PhotoGreen Lab, Department of Chemistry, University of Pavia,
Viale Taramelli 12, 27100 Pavia, Italy
e-mail: fagnoni@unipv.it

G. Bergamini, S. Silvi (eds.), *Applied Photochemistry*,
Lecture Notes in Chemistry 92, DOI 10.1007/978-3-319-31671-0_6

transformations that have no analogy in thermal chemistry, or when they have, this follows a complementary way, as in the case of pericyclic reactions (where opposite pathways are most often followed). Likewise, the generation of (otherwise) hardly accessible intermediates is often a smooth process [3]. In the simplest version of a photochemical process, light is absorbed by the photoactive substrate (A), directly populating the reacting excited state (A*) (Scheme 6.1, *path a*). In many cases, however, another molecule (PC in Scheme 6.1) is present and absorbs light. Thus, irradiation of PC results in the *indirect* activation of the chemical substrate through either a physical (photosensitization, *path b*) or a chemical (photocatalysis, *path c*) step. In the first case, energy transfer between PC* and the non-absorbing reagent is responsible for the promotion of A to a (dark) excited state that then reacts as in *path a*. In the second one, excitation of the photocatalyst causes the chemical activation of the reagent, normally via an electron or an atom transfer process. This generates a deactivated form of the photocatalyst (PC') and a high energy intermediate responsible for the formation of the final product. The photocatalyst may be used in a sub-stoichiometric amount, provided that it is regenerated in one of the ensuing steps of the reaction [4].

Scheme 6.1 Different ways to perform a photochemical reaction

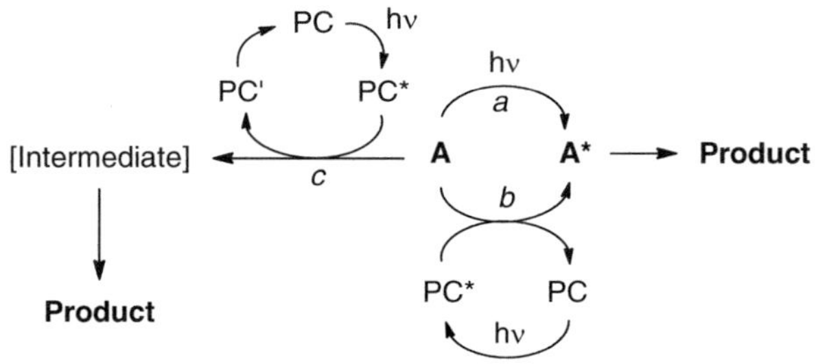

However, the most significant limitation to the use of photochemistry is the large energy expenditure involved in the use of artificial light sources, that often represents the main cost (both economic and environmental) of a photochemical process. In an ideal system, photons should be furnished by a renewable and inexhaustible source of energy, avoiding any cost for generating light. Solar light is the obvious choice, since this is the environmental-friendly energy source par excellence. As a matter of fact, light coming from the sun could easily fulfill our present needs, as the amount of energy reaching our planet in 1 h ($4.3 \cdot 10^{20}$ J) is comparable to that consumed globally in 1 year ($5.598 \cdot 10^{20}$ J, 2012 data) [5]. As for the emission profile, the light produced by the sun is similar to that expected from a 5525 K (5250 °C) blackbody (Fig. 6.1), with a large component (44 %) lying in the visible part of the electromagnetic spectrum and a similar amount (53 %) in the near-infrared part, while there is only a small fraction (3 %) in the UV region [6].

In the last decades, several efforts have been focused on the storage of this enormous amount of energy hitting the surface of the Earth. Accordingly, different

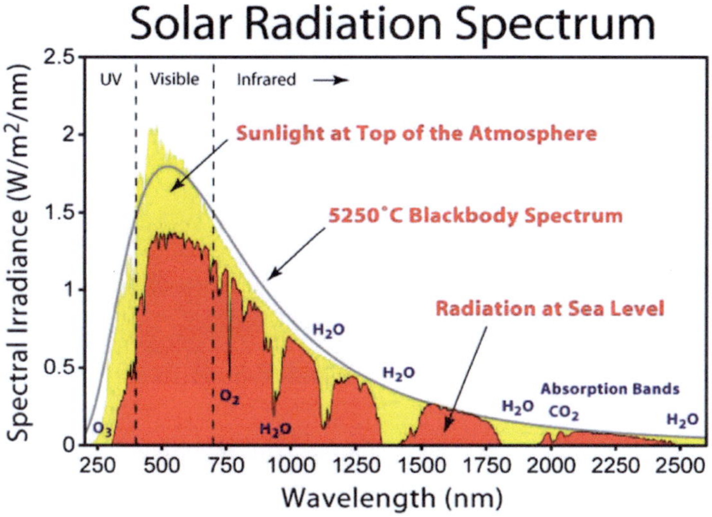

Fig. 6.1 The sun emission spectrum (Reprinted from Wikipedia (Creative Commons License). See: http://www.globalwarmingart.com/wiki/Image:Solar_Spectrum_png)

strategies have been developed. The simplest approach involves solar collectors capable to absorb solar energy as heat and then transfer it to a working fluid (air, water or oil). The collected heat can be directly used to provide domestic heating, or stored in a different form [7–9]. Alternatively, solar energy can be transformed into electric energy, by having recourse to photovoltaic modules [10–12], or into chemical energy, e.g. through systems able to perform water splitting [13–15].

6.1.1 *Solar Photochemistry*

The legend of *Phaeton*, precipitated when he attempted to drive the chariot of his father *Helios*, expresses how well mankind understood the power of solar irradiation and how it was fascinated by the possibility of controlling natural events and harnessing the energy involved. In one of his most famous science fiction novels, called *The Golden Apples of the Sun*, first published in early 1950s [16], Ray Bradbury described the mission of an ammonia-frozen spaceship (the *copa de oro*) to collect a burning fragment of the sun, and bring it to the Earth. The formulation of visionary ideas has its field in novels, but it has a role also in the realm of sciences that, by contrast, have also the duty of providing the required know-how to transform embryonic ambitions into scientific goals. The early pioneers of photochemistry Ciamician and Paternò surmised that solar energy could be harnessed and used for carrying out chemical reactions, or stored as chemical energy. Chemical bonds were their actual *copa de oro*, and while working on the roof of their laboratories (see Giacomo Ciamician and his colleague Paul Silber in Bologna; Fig. 6.2), these scientists devoted themselves to the challenge of stealing *the guarded secret of the plants* [17], that is the ability of using solar energy for the

Fig. 6.2 Giacomo Ciamician and Paul Silber in their laboratory, in Bologna, Italy (*right*, Courtesy of University of Bologna, Italy)

production of chemicals under mild conditions. This represents a wonderful episode that, for some time, brought the photochemistry discipline at the forefront of chemical sciences, and to the prediction of a great future [18, 19]. Thus, in about one decade, photochemical paths alternative to practically all of the organic reactions known at that time were devised (see some examples in Scheme 6.2).

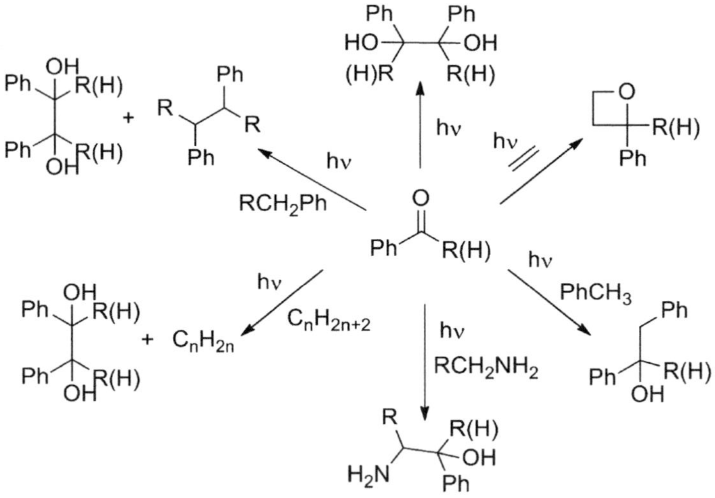

Scheme 6.2 The most widely investigated reactions at the onset of photochemistry

Following an abrupt stop after World War I, research on solar synthesis restarted with the work of A. Schönberg and G. O. Schenck. In particular, the latter assembled in Heidelberg (Germany, see Scheme 6.3) a pilot plant for the synthesis of anthelmintic ascaridol from α-terpinene on a 1 kg day^{-1} scale by using an alcoholic extract of stinging nettles (containing chlorophyll in the role of photosensitizer) and air as the oxidant [20].

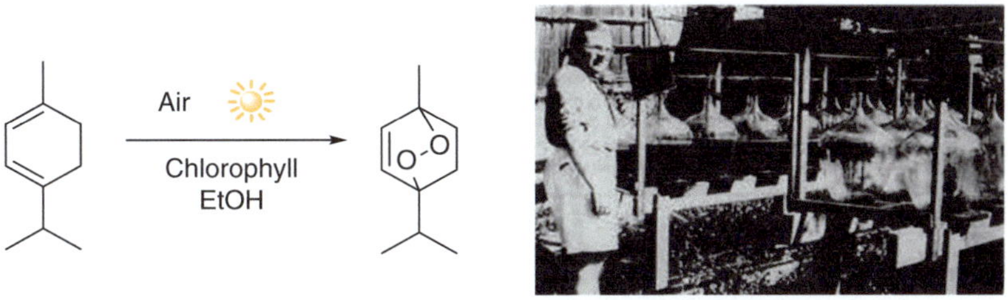

Scheme 6.3 Synthesis of Ascaridole (*left*). Schenck at his ascaridol pilot plant in 1952 (*right*) (Reprinted with permission from Ref. [21])

New technologies introduced at the end of the twentieth century, able to concentrate solar light by means of parabolic trough collectors, allowed for a significant shortening of the reaction time and an easy scale up of the processes [21, 22]. Solar variations of the few photochemical reactions with industrial significance have thus been reported in the literature, including the synthesis of rose oxide (a widely appreciated fragrance) and the Toray process for the production of ε-caprolactam [23–25]. Unfortunately, sun is a discontinuous light source, unevenly distributed on Earth surface, and further depends on weather conditions, imposing severe limitations to the reproducibility of the reactions.

6.1.2 Visible Light Induced Photochemistry

A reasonable approach to overcome the discontinuity of sunlight is the use of artificial visible light sources. In this case, the energy expenditure can be in part compensated for by the low cost and the large availability of the lamps required and by the improved reproducibility and productivity of the process. As hinted above, most organic molecules are colorless and thus rather than direct irradiation (Scheme 6.1, *path a*), visible light is used with organic or inorganic photocatalysts or photosensitizers (Scheme 6.1, *paths b,c*). Early efforts in this field were based on the peculiar photoredox properties of ruthenium based complexes such as tris (bipyridine)ruthenium(II), $[Ru^{II}(bpy)_3]$, used in debromination reactions [26] and in Pschorr cyclizations (see Scheme 6.4) [27].

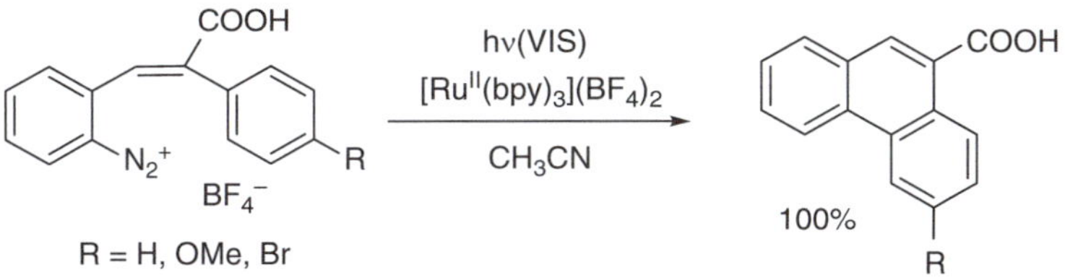

Scheme 6.4 $[Ru^{II}(bpy)_3]$ photocatalyzed Pschorr cyclization

In the last decade, however, visible light photoredox catalysis underwent an impressive advancement and was applied to most of the current synthetic challenges. In these reactions, redox active molecules able to absorb in the visible range, both inorganic (e.g. Ru^{II} and Ir^{III} based complexes) and organic (dyes, such as Eosin Y, Fluorescein and Riboflavin), have been efficiently employed as catalysts [28, 29]. Actually, most of the reactions using solar/visible light are photocatalytic process.

Aim of the Authors of this chapter is to provide the reader with a (representative) picture of the application of solar/visible light-promoted processes in organic synthesis. Since most of the reviews currently available in the literature cover the period until 2012 [6, 28, 29], we mainly focused on papers published in the 2013–2015 period.

The reactions have been classified according to the synthetic target. The scheme of each reaction is presented, along with the corresponding reference. Further pertinent references were added below each scheme (dubbed as *related*) and in some cases other references on processes pertaining to the class considered have been added at the end of each section (dubbed as *other*).

For each of the reactions examined, the most important experimental details have been highlighted (see Table 6.1 below), namely:

- Light sources: *Compact fluorescent lamps* (CFLs) and *Light Emitting Diodes* (LEDs) are the most widely employed light sources. CFLs are energy-saving Hg-containing fluorescent lamps emitting "white" light with a spectral power distribution that depends on the mix of phosphors present inside the bulb, with a rated service life ranging from 6000 to 15,000 h. A lighting efficiency in the 7–10 % range can be achieved in this case. On the other hand, Light Emitting Diodes (LEDs) are devices made of a semiconductor material able to generate an incoherent electroluminescence, fitted with an optics capable to shape and focus the emission. A large variety of (almost) monochromatic LEDs (actually the emission range covers a narrow window, usually around 20 nm) is available, and the kind of LED employed (white, blue, green) has been indicated in the scheme. *Cool-white tubes* are mercury arc fluorescent lamps having rare earth-doped phosphors that produce a bright light convenient for photochemical applications. Most of the reactions showed in this chapter can be carried out under solar exposition, and, when possible, the corresponding results have been reported in the scheme. High-pressure *Xenon arc lamps*, mimicking the solar emission spectrum, have been adopted in some cases.
- Reaction conditions: the generation and/or the ensuing reactivity of excited states (e.g. triplet states) or of intermediates is often incompatible with the presence of oxygen. In this case, degassing of the solution by flushing with an inert gas (Ar, N_2) or by Freeze-Pump-Thaw (FPT) cycles may be required, unless of course an oxidation is carried out. In each scheme, we also indicated when stirring of the solution was specifically required.
- Reactor setup: A key advantage of using visible light is the large availability of materials and light sources to be employed in assembling the photoreactor.

Table 6.1 Symbols used in the present chapter

Light source	Conditions	Temperature	Conditions
CFL: 	Batch: 	rt or a numeric value	Stirring: 
LED: 	Flow: 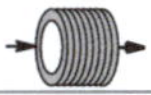		Nitrogen-flushed: N_2
Solar light: 			Argon-flushed: Ar
Xenon Lamp: Xe			Oxygen-flushed: O_2
Cool-White Lamp: CW			Freeze-pump-thaw cycles: FPT
Simulated Solar light: simulated			Air-equilibrated: air

Expensive UV-transparent materials, such as Vycor and quartz (commonly used in photochemistry), can be avoided. Conventional laboratory glassware such as vials, tubes or flasks made from Pyrex can be employed for carrying out batch reactions. However, photochemical flow reactors have been successfully employed as well, and in this case the reaction mixture is circulated in a light-transparent, fluorinated polymer tubing (e.g. polytetrafluoroethylene (PTFE) or fluorinated ethylene propylene (FEP)), by means of a syringe or an HPLC pump [30]. Microreactors, where the reduced optical path of the reaction channels allows for an efficient light absorption by the substrate and a minimization of the residence time, have also been successfully employed [31].

For the reader's convenience, the experimental conditions are indicated by means of symbols (see Table 6.1).

6.2 Photochemical Reactions Carried Out Under Solar/Visible Light

6.2.1 Formation of a Ring

6.2.1.1 Three Membered Ring

Zhang Y, Qian R, Zheng X, Zeng Y, Sun J, Chen Y, Ding A, Guo H (2015) Visible light induced cyclopropanation of dibromomalonates with alkenes via double-SET by photoredox catalysis. Chem Commun 51:54–57 (Scheme 6.5).

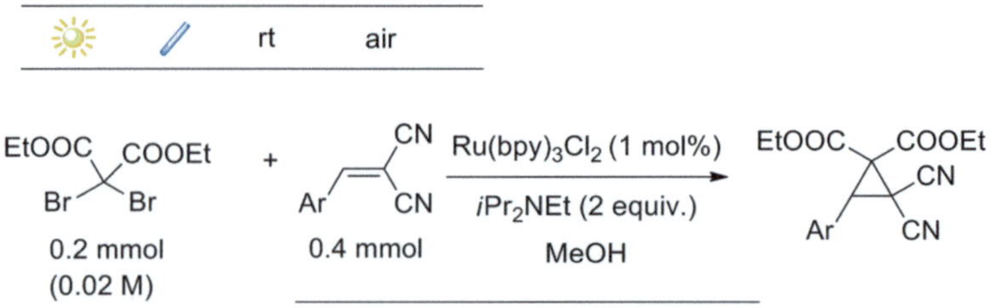

Ar	% yield	time (h)
4-$COOMeC_6H_4$	98	4
C_6H_5	92	5
2-MeC_6H_4	96	2.5
3-$OMeC_6H_4$	85	3

Scheme 6.5 Photocatalyzed cyclopropanation of dibromomalonates with alkenes

6.2.1.2 Four Membered Ring

Shu X-Z, Zhang M, He Y, Frei H, Toste FD (2014) Dual Visible Light Photoredox and Gold-Catalyzed Arylative Ring Expansion. J Am Chem Soc 136:5844–5847 (Scheme 6.6)

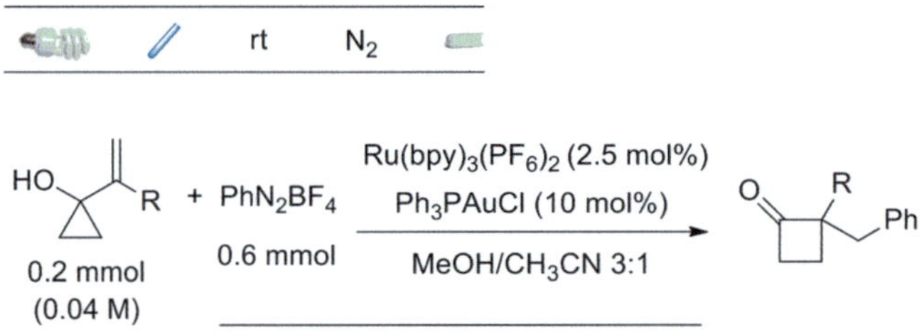

R	% yield	time (h)
Ph	67	15
$PhCH_2$	51	6

Scheme 6.6 Synthesis of cyclobutanones via dual visible light photoredox and gold-catalyzed arylative ring expansion

6.2.1.3 Five Membered Ring

6.2.1.3.1 Cyclopentanes and Cyclopentenes

Woźnica M, Chaoui N, Taabache S, Blechert S (2014) THF: An Efficient Electron Donor in Continuous Flow Radical Cyclization Photocatalyzed by Graphitic Carbon Nitride. Chem Eur J 20:14624–14628 (Scheme 6.7).

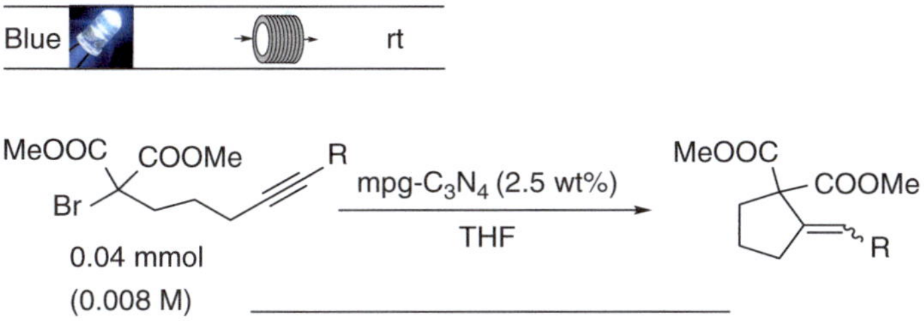

R	% yield	time (min)
Ph	85	5
H	72	5
Me_3Si	85	5

Scheme 6.7 Synthesis of substituted cyclopentanes via metal-free radical cyclization of 2-bromo-1,3-dicarbonyls

Other: Neumann M, Zeitler K (2013) A cooperative hydrogen-bond-promoted organophotoredox catalysis strategy for highly diastereoselective, reductive enone cyclization. Chem Eur J 19:6950–6955.

Nguyen TH, Maity S, Zheng N (2014) Visible light mediated intermolecular [3 + 2] annulation of cyclopropylanilines with alkynes. Beilstein J Org Chem 10:975–980 (Scheme 6.8).

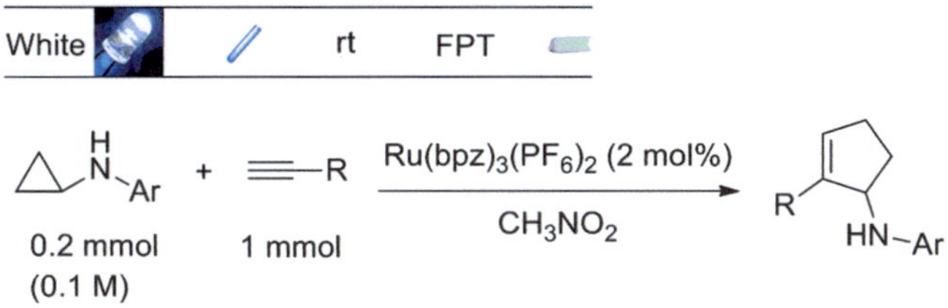

R	Ar	yield	time (h)
Ph	3,5-$Me_2C_6H_3$	62	8
COOMe	4-CNC_6H_4	72	8

Scheme 6.8 Visible light mediated intermolecular [3 + 2] annulation of cyclopropylanilines with alkynes

Other: Gu X, Li X, Qu Y, Yang Q, Li P, Yao Y (2013) Intermolecular Visible-Light Photoredox Atom-Transfer Radical [3 + 2]-Cyclization of 2-(Iodomethyl)cyclopropane-1,1-dicarboxylate with Alkenes and Alkynes. Chem Eur J 19:11878–11882.

6.2.1.3.2 Nitrogen Containing Heterocycles

Xu P, Xie J, Xue Q, Pan C, Cheng Y, Zhu C (2013) Visible-Light-Induced Trifluoromethylation of *N*-Aryl Acrylamides: A Convenient and Effective Method To Synthesize CF_3-Containing Oxindoles Bearing a Quaternary Carbon Center. Chem Eur J 19:14039–14042 (Scheme 6.9).

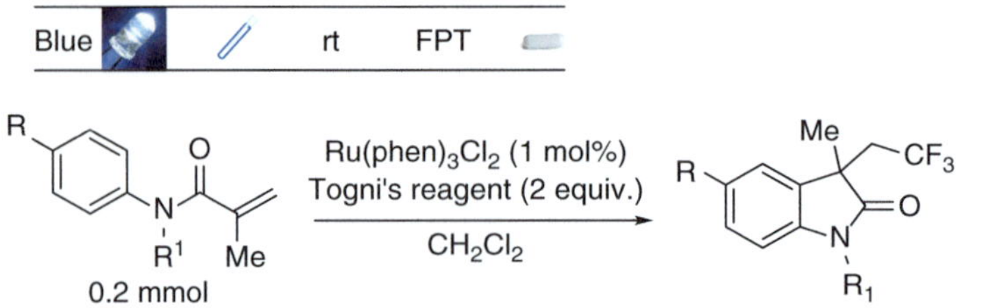

Togni's reagent:

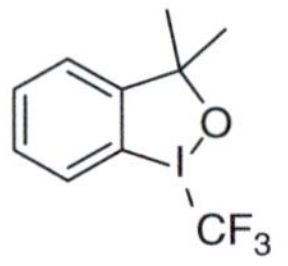

R	R^1	% yield	time (h)
H	Me	82	18
H	Ph	82	18
Br	Me	90	18
MeO	Me	92	18

Scheme 6.9 Visible light-induced synthesis of CF_3-containing oxindoles

<u>*Related*</u>: Fu W, Zhu M, Zou G, Xu C, Wang Z (2014) Visible-Light-Mediated Trifluoroethylation of *N*-Arylacrylamides with Trifluoroethyl Iodide: Synthesis of CF_3-Containing Oxindoles. Synlett 25:2513–2517.

Zhang J-L, Liu Y, Song R-J, Jiang G-F, Li J-H (2014) 1,2-Alkylarylation of Activated Alkenes with Two C-H Bonds by Using Visible-Light Catalysis. Synlett 25:1031–1035.

Tang X-J, Thomoson CS, Dolbier Jr WR (2014) Photoredox-Catalyzed Tandem Radical Cyclization of *N*-Arylacrylamides: General Methods To Construct Fluorinated 3,3-Disubstituted 2-Oxindoles Using Fluoroalkylsulfonyl Chlorides. Org Lett 16:4594–4597.

Yoshioka E, Kohtani S, Jichu T (2015) Direct Photoinduced Electron Transfer from Excited State of Rhodamine B for Carbon-Radical Generation. Synlett 26:265–270 (Scheme 6.10).

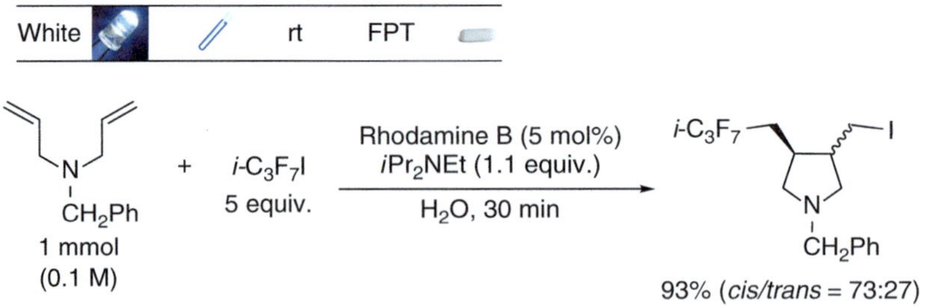

Scheme 6.10 Rhodamine B photocatalyzed synthesis of pyrrolidines

<u>*Related*</u>: Yoshioka E, Kohtani S, Tanaka E, Hata Y, Miyabe H (2015) Carbon radical addition-cyclization reaction induced by ruthenium-photocatalyst under visible light irradiation. Tetrahedron 71:773–781.

Hernandez-Perez AC, Collins SK (2013) A Visible-Light-Mediated Synthesis of Carbazoles. Angew Chem Int Ed 52:12696–12700 (Scheme 6.11).

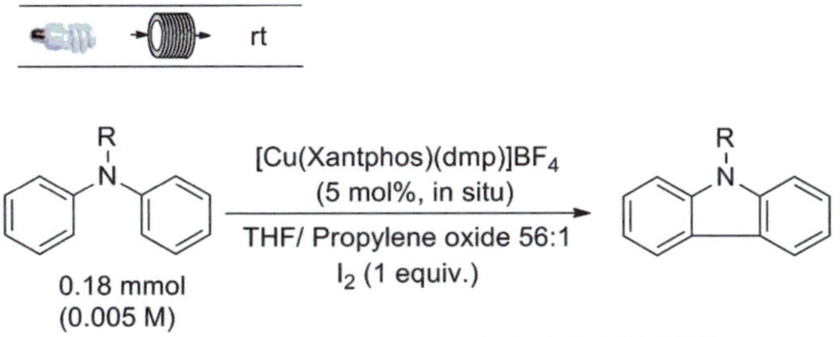

R	% yield	time (h)
Ph	75	10
Me	65	20
$CHMe_2$	65	20

Scheme 6.11 Visible light-mediated synthesis of carbazoles

Cai S, Yang K, Wang DZ (2014) Gold Catalysis Coupled with Visible Light Stimulation: Syntheses of Functionalized Indoles. Org Lett 16:2606–2609 (Scheme 6.12).

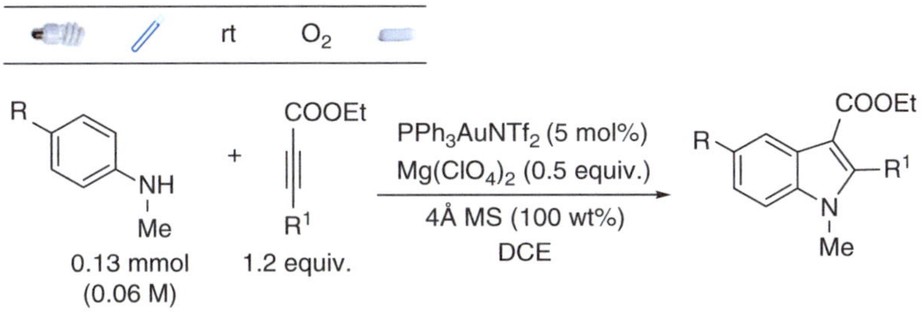

R	R^1	% yield	time (h)
NO_2	Ph	83	52
CN	Ph	79	28
COOMe	4-MeC_6H_4	78	30
COOMe	4-FC_6H_4	82	27

Scheme 6.12 Photochemical synthesis of functionalized indoles

<u>*Related*</u>: Zhang P, Xiao T, Xiong S, Dong X, Zhou L (2014) Synthesis of 3-Acylindoles by Visible-Light Induced Intramolecular Oxidative Cyclization of *o*-Alkynylated *N,N*-Dialkylamines Org Lett 16:3264–3267.

Vila C, Lau J, Rueping M (2014) Visible-light photoredox catalyzed synthesis of pyrroloisoquinolines via organocatalytic oxidation/[3 + 2] cycloaddition/

oxidative aromatization reaction cascade with Rose Bengal. Beilstein J Org Chem 10:1233–1238 (Scheme 6.13).

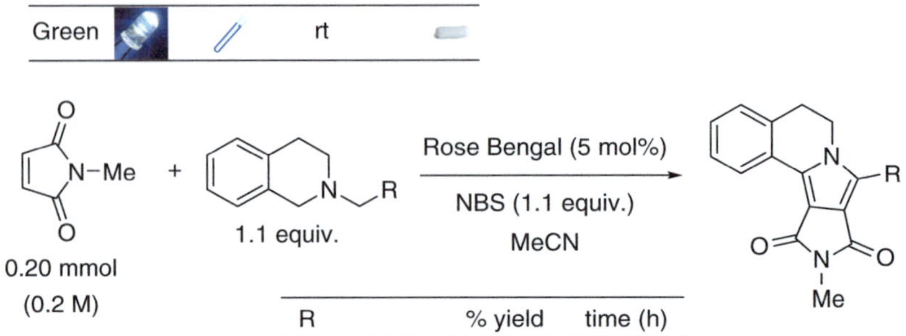

R	% yield	time (h)
COOMe	76	24
CN	42	24
PhCO	56	24

Scheme 6.13 Visible light photoredox catalyzed synthesis of pyrroloisoquinolines

Related: Yadav AK, Yadav LDS (2015) Visible-light-driven electrocyclization of activated allylic amines via azomethine ylide formation. Tetrahedron Lett 56:686–689.

Nakajima K, Kitagawa M, Ashida Y, Miyake Y, Nishibayashi Y (2014) Synthesis of nitrogen heterocycles via α-aminoalkyl radicals generated from α-silyl secondary amines under visible light irradiation. Chem Commun 50:8900–8903.

Xia X-D, J. Wang X-Q, Lu L-Q, Chen J-R, Xiao W-J (2014) Synthesis of 2-Substituted Indoles through Visible Light-Induced Photocatalytic Cyclizations of Styryl Azides. Adv Synth Cat 356:2807–2812 (Scheme 6.14).

Scheme 6.14 Visible light-induced synthesis of 2-substituted indoles

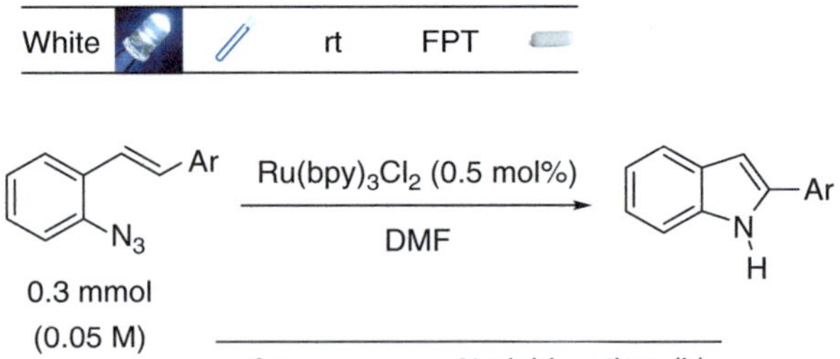

Ar	% yield	time (h)
Ph	96	24
4-tBuC_6H_4	90	24
4-$CF_3C_6H_4$	69	24
2-$NO_2C_6H_4$	46	60

Related: Farney EP, Yoon TP (2014) Visible-Light Sensitization of Vinyl Azides by Transition-Metal Photocatalysis. Angew Chem Int Ed 53:793–797.

Other: Liu Y, Zhang J-L, Song R-J, Li J-H (2014) Visible-Light-Facilitated 5-*exo-trig* Cyclization of 1,6-Dienes with Alkyl Chlorides: Selective Scission of the C(sp^3)-H Bond in Alkyl Chlorides. Eur J Org Chem 1177–1181.

Musacchio AJ, Nguyen LQ, Beard GH, Knowles RR (2014) Catalytic olefin hydroamination with aminium radical cations: a photoredox method for direct C-N bond formation. J Am Chem Soc 136:12217–12220.

Xia X-D, Ren Y-L, Chen J-R, Yu X-L, Lu L-Q, Zou Y-Q, Wan J, Xiao W-J (2015) Phototandem Catalysis: Efficient Synthesis of 3-Ester-3-hydroxy-2-oxindoles by a Visible Light-Induced Cyclization of Diazoamides through an Aerobic Oxidation Sequence. Chem Asian J 10:124–128.

Zoller J, Fabry DC, Ronge MA, Rueping M (2014) Synthesis of Indoles Using Visible Light: Photoredox Catalysis for Palladium-Catalyzed C-H Activation. Angew Chem Int Ed 53:13264–13268.

6.2.1.3.3 Oxygen Containing Heterocycles

Sahoo B, Hopkinson MN, Glorius F (2013) Combining Gold and Photoredox Catalysis: Visible Light-Mediated Oxy- and Aminoarylation of Alkenes. J Am Chem Soc 135:5505–5508 (Scheme 6.15).

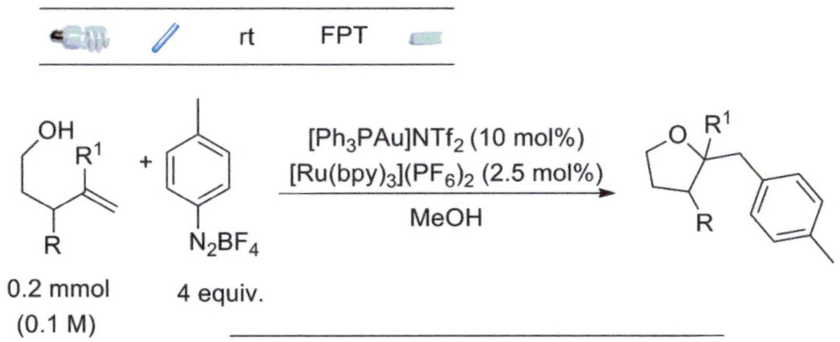

R	R^1	% yield	time (h)
Ph	H	70 (dr 1.6:1)	4-16
H	Me	39	4-16
H	Ph	63	4-16

Scheme 6.15 Visible light-mediated oxyarylation of alkenes

Related: Guo W, Cheng H-G, Chen L-Y, Xuan J, Feng Z-J, Chen J-R, Lu L-Q, Xiao W-J (2014) *De Novo* Synthesis of γ, γ-Disubstituted Butyrolactones through a Visible Light Photocatalytic Arylation-Lactonization Sequence. Adv Synth Catal 356:2787–2793.

Wei X-J, Yang D-T, Wang L, Song T, Wu L-Z, Liu Q (2013) A Novel Intermolecular Synthesis of γ-Lactones via Visible-Light Photoredox Catalysis. Org Lett 15:6054–6057 (Scheme 6.16).

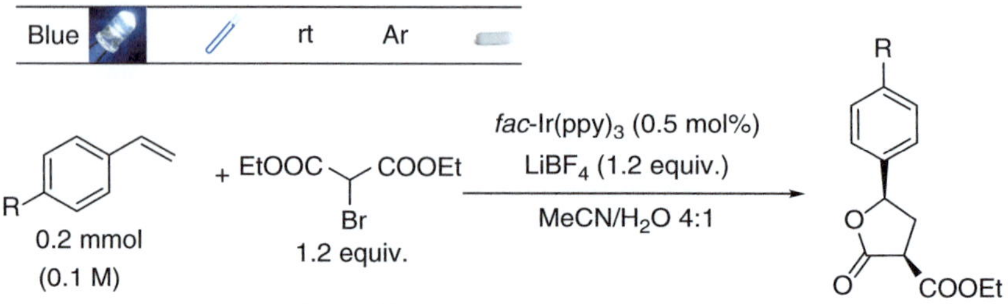

R	% yield	time (h)
H	70 (dr 1:1.5)	24
OMe	72 (dr 1:1)	24

Scheme 6.16 Synthesis of γ-lactones via visible light photoredox catalysis

Related: Lin R, Sun H, Yang C, Yang Y, Zhao X, Xia W (2015) Visible-light-induced bromoetherification of alkenols for the synthesis of β-bromotetrahydrofurans and -tetrahydropyrans. Beilstein J Org Chem 11:31–36.

Yasu Y, Arai Y, Tomita R, Koike T, Akita M (2014) Highly Regio- and Diastereoselective Synthesis of CF_3-Substituted Lactones via Photoredox-Catalyzed Carbolactonization of Alkenoic Acids. Org Lett 16:780–783.

Jiang H, Cheng Y, Zhang Y, Yu S (2013) *De Novo* Synthesis of Polysubstituted Naphthols and Furans Using Photoredox Neutral Coupling of Alkynes with 2-Bromo-1,3-dicarbonyl Compounds. Org Lett 15:4884–4887 (Scheme 6.17).

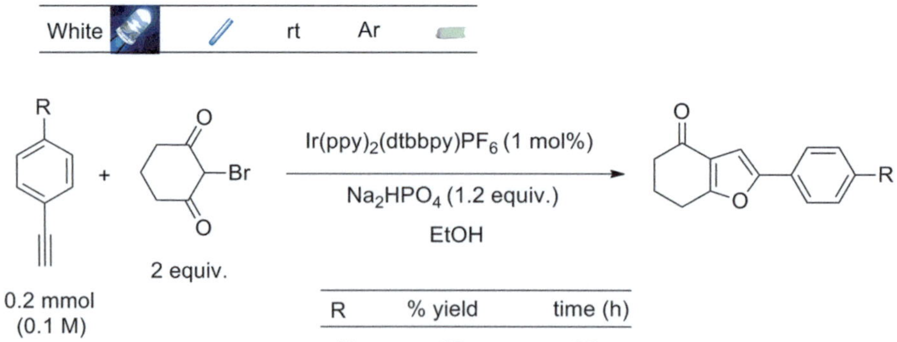

R	% yield	time (h)
H	60	12
OMe	96	12
F	54	12
CF_3	33	12

Scheme 6.17 Synthesis of polysubstituted furans

Blum TR, Zhu Y, Nordeen SA, Yoon TP (2014) Photocatalytic Synthesis of Dihydrobenzofurans by Oxidative [3 + 2] Cycloaddition of Phenols. Angew Chem Int Ed 53:11056–11059 (Scheme 6.18).

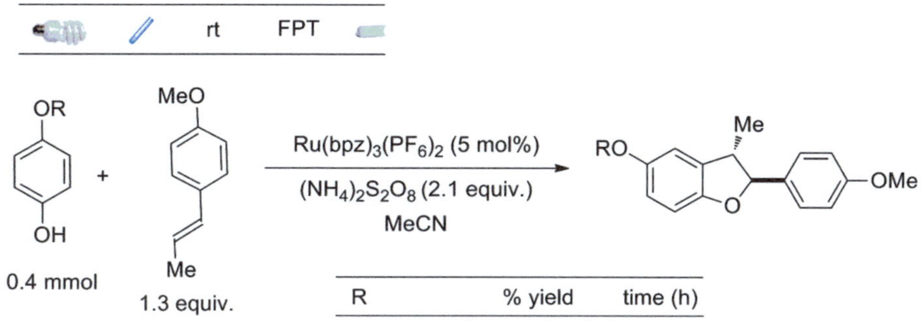

R	% yield	time (h)
Bn	83	30
Allyl	77	30
CH_2CH_2OH	83	30

Scheme 6.18 Synthesis of dihydrobenzofurans by photocatalyzed [3+2] cycloaddition

Other: Gu X, Lu P, Fan W, Li P, Yao Y (2013) Visible light photoredox atom transfer Ueno-Stork reaction. Org Biomol Chem 11:7088–7091.

6.2.1.3.4 More Than One Heteroatom in the Ring

Samanta S, Das S, Biswas P (2013) Photocatalysis by 3,6-Disubstituted-*s*-Tetrazine: Visible-Light Driven Metal-Free Green Synthesis of 2-Substituted Benzimidazole and Benzothiazole. J Org Chem 78:11184–11193 (Scheme 6.19).

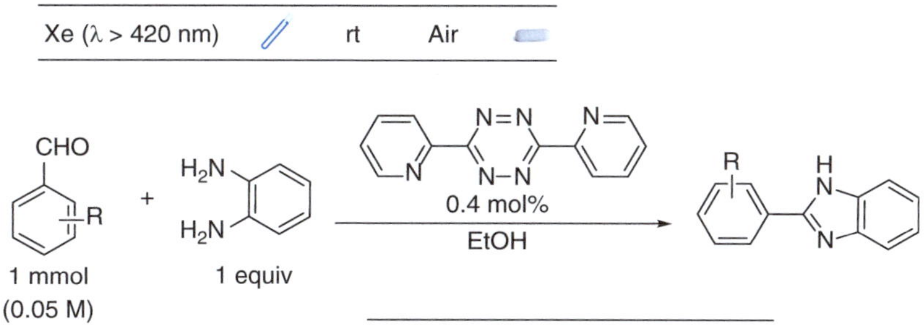

R	% yield	time (h)
H	90	2
2-OH	95	2
5-Cl, 2-OH	90	2.4

Scheme 6.19 Visible light driven synthesis of 2-substituted benzimidazoles

<u>Related</u>: Das S, Samanta S, Maji SK, Samanta PK, Dutta AK, Srivastava DN, Adhikary B, Biswas P (2013) Visible-light-driven synthesis of 2-substituted benzothiazoles using CdS nanosphere as heterogenous recyclable catalyst. Tetrahedron Lett 54:1090–1096.

Park S, Jung J, Cho EJ (2014) Visible-Light-Promoted Synthesis of Benzimidazoles. Eur J Org Chem 4148–4154.

Deng Q-H, Zou Y-Q, Lu L-Q, Tang Z-L, Chen J-R, Xiao W-J (2014) *De Novo* Synthesis of Imidazoles by Visible-Light-Induced Photocatalytic Aerobic Oxidation/[3 + 2] Cycloaddition/Aromatization Cascade. Chem Asian J 9:2432–2435 (Scheme 6.20).

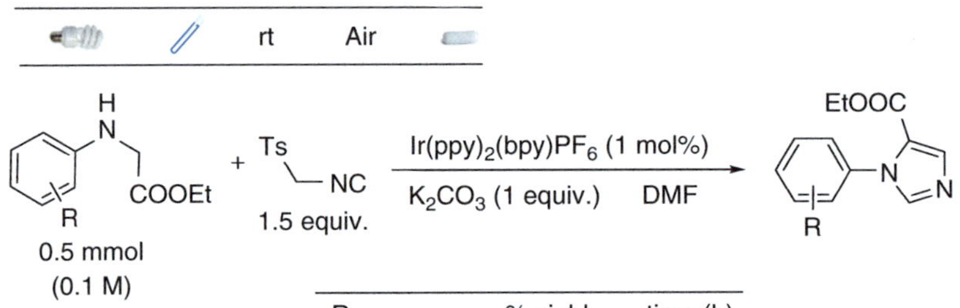

R	% yield	time (h)
4-Me	78	11
H	64	11
4-Br	51	11

Scheme 6.20 Photocatalyzed synthesis of imidazoles

Lu Z, Parrish JD, Yoon TP (2014) [3 + 2] Photooxygenation of aryl cyclopropanes via visible light photocatalysis. Tetrahedron 70:4270–4278 (Scheme 6.21).

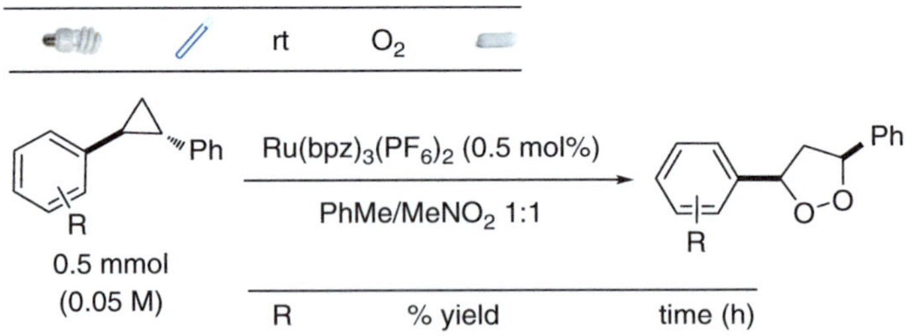

R	% yield	time (h)
4-OMe	99 (syn/anti 5:1)	3
4-OH	78 (syn/anti 10:1)	1
2,4-OMe	99 (syn/anti 10:1)	1

Scheme 6.21 Five membered endoperoxides via [3+2] photooxygenation of aryl cyclopropanes

Hou H, Zhu S, Pan F, Rueping M (2014) Visible-Light Photoredox-Catalyzed Synthesis of Nitrones: Unexpected Rate Acceleration by Water in the Synthesis of Isoxazolidines. Org Lett 16:2872–2875 (Scheme 6.22).

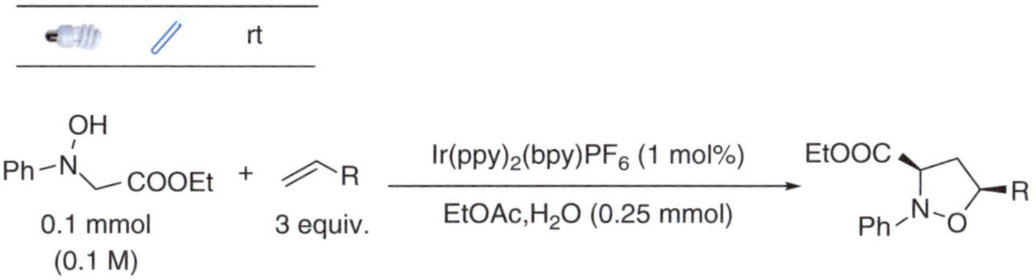

R	% yield	time (h)
OEt	94 (dr 85:15)	12-36
O*i*Bu	74 (dr 85:15)	12-36
Ph	83 (dr 86:14)	12-36

Scheme 6.22 Photocatalyzed preparation of isoxazolidines

Hu X-Q, Chen J-R, Wei Q, Liu F-L, Deng Q-H, Beauchemin AM, Xiao W-J (2014) Photocatalytic Generation of *N*-Centered Hydrazonyl Radicals: A Strategy for Hydroamination of β,γ-Unsaturated Hydrazones. Angew Chem Int Ed 53:12163–12167 (Scheme 6.23).

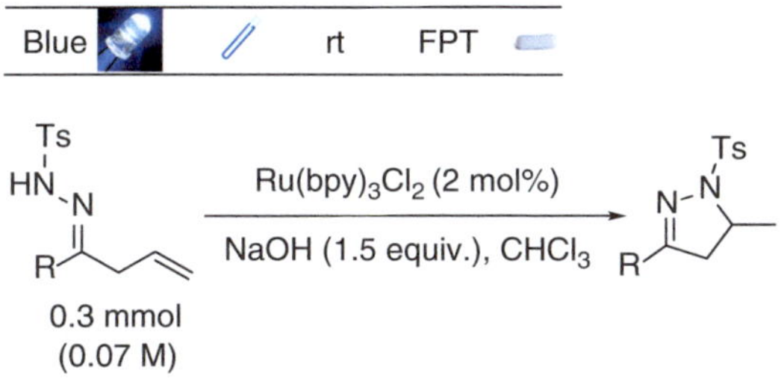

R	% yield	time (h)
Ph	80	12-16
4-ClC$_6$H$_4$	60	12-16
CH$_2$Ph	62	12-16
C$_6$H$_{11}$	76	12-16

Scheme 6.23 Photocatalytic hydroamination of β,γ-unsaturated hydrazones

Morse PD, Nicewicz DA (2015) Divergent regioselectivity in photoredox-catalyzed hydrofunctionalization reactions of unsaturated amides and thioamides. Chem Sci 6:270–274 (Scheme 6.24).

Scheme 6.24 Photoredox-catalyzed hydrofunctionalization of unsaturated amides

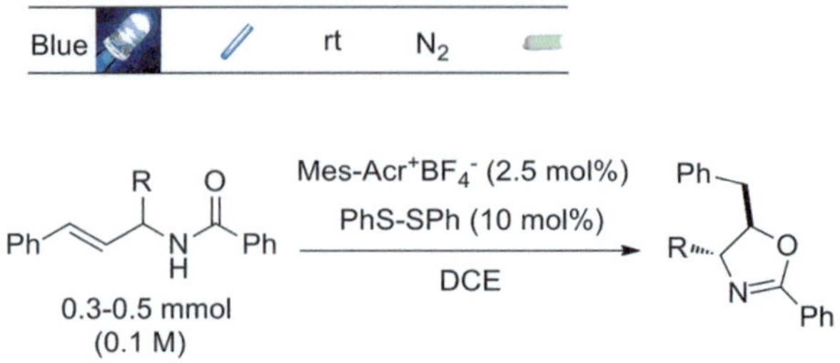

R	% yield	time (h)
H	76	14
Me	64 (dr 2:1)	36
*t*Bu	81 (dr 2:1)	36

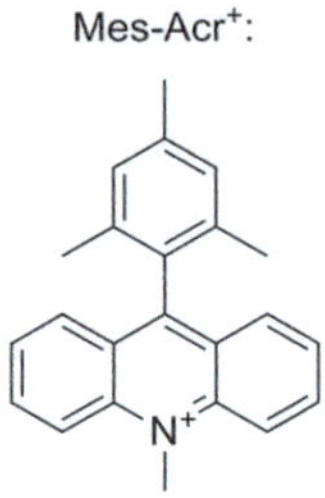

<u>*Other*</u>: Saikh F, Das J, Ghosh S (2013) Synthesis of 3-methyl-4-arylmethylene isoxazole-5(4*H*)-ones by visible light in aqueous ethanol. Tetrahedron Lett 54:4679–4682.

Lin W-C, Yang D-Y (2013) Visible Light Photoredox Catalysis: Synthesis of Indazolo[2,3-*a*]quinolines from 2-(2-Nitrophenyl)-1,2,3,4-tetrahydroquinolines. Org Lett 15:4862–4865.

Yadav AK, Yadav LDS (2014) Visible-light-promoted aerobic oxidative cyclization to access 1,3,4-oxadiazoles from aldehydes and acylhydrazides. Tetrahedron Lett 55:2065–2069.

6.2.1.4 Six Membered Ring

6.2.1.4.1 Carbocycles

Protti S, Artioli GA, Capitani F, Marini C, Dore P, Postorino P, Malavasi L, Fagnoni M (2015) Preparation of (substituted) picenes via solar light induced Mallory photocyclization. RSC Adv 5:27470–27475 (Scheme 6.25).

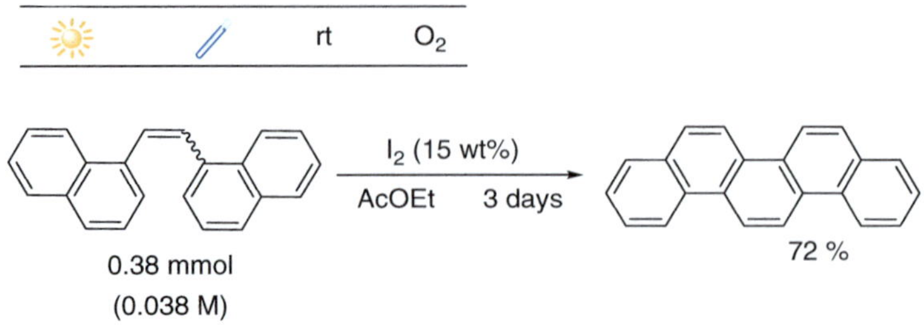

Scheme 6.25 Sunlight-driven synthesis of picene

Rono LJ, Yayla HG, Wang DY, Armstrong MF, Knowles RR (2013) Enantioselective Photoredox Catalysis Enabled by Proton-Coupled Electron Transfer: Development of an Asymmetric Aza-Pinacol Cyclization. J Am Chem Soc 135:17735–17738 (Scheme 6.26).

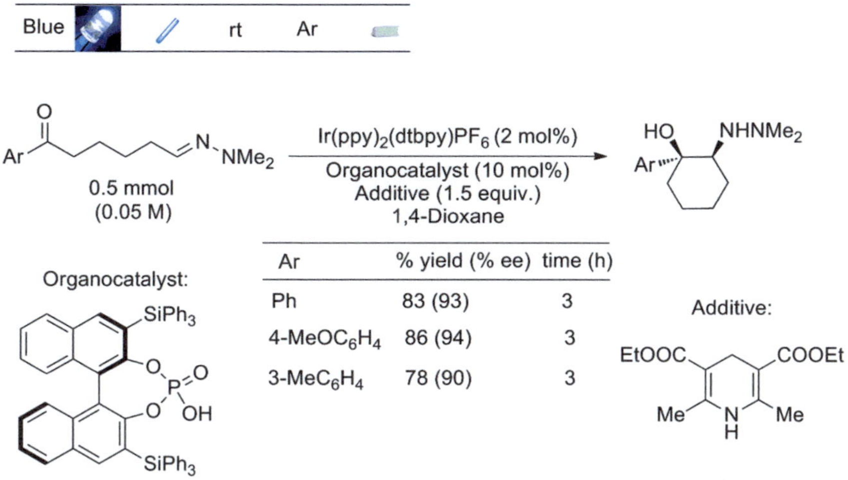

Ar	% yield (% ee)	time (h)
Ph	83 (93)	3
4-MeOC$_6$H$_4$	86 (94)	3
3-MeC$_6$H$_4$	78 (90)	3

Scheme 6.26 Enantioselective photoredox aza-pinacol cyclization

Paria S, Reiser O (2014) Visible Light Photoredox Catalyzed Cascade Cyclizations of α-Bromochalcones or α-Bromocinnamates with Heteroarenes. Adv Synth Catal 356:557–562 (Scheme 6.27).

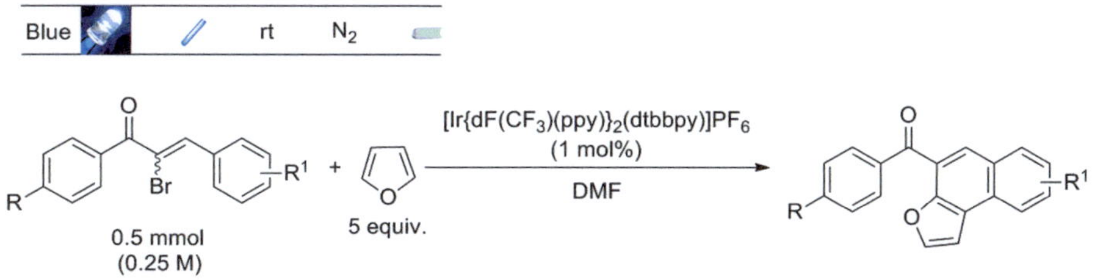

R	R^1	% yield	time (h)
H	H	85	12
Cl	H	91	12
Me	4-Me	83	12

Scheme 6.27 Visible light photoredox catalyzed cascade cyclizations of α-bromocinnamates with heteroarenes

<u>*Related*</u>: Paria S, Kais V, Reiser O (2014) Visible Light-Mediated Coupling of α-Bromochalcones with Alkenes. Adv Synth Catal 356:2853–2858.

6.2.1.4.2 Nitrogen Containing Heterocycles

Dong X, Xu Y, Liu JJ, Hu Y, Xiao T, Zhou L (2013) Visible-Light-Induced Radical Cyclization of Trifluoroacetimidoyl Chlorides with Alkynes: Catalytic Synthesis of 2-Trifluoromethyl Quinolines. Chem Eur J 19:16928–16933 (Scheme 6.28).

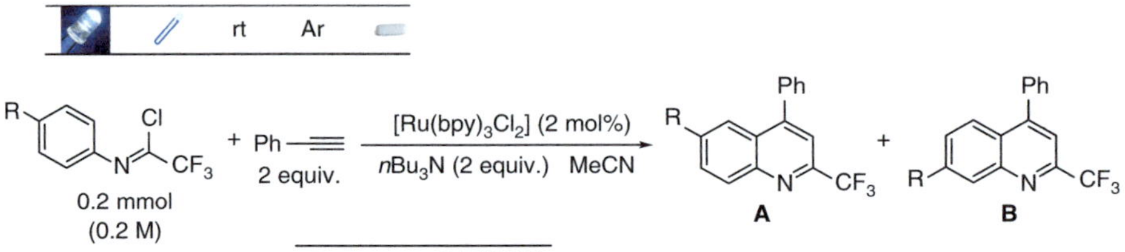

R	yield (**A**/**B**)
Me	79 (7:3)
MeO	69 (4:1)
CN	59 (9:1)

Scheme 6.28 Synthesis of 2-trifluoromethyl quinolines via visible light-induced radical cyclization

Jiang H, Cheng Y, Wang R, Zheng M, Zhang Y, Yu S (2013) Synthesis of 6-Alkylated Phenanthridine Derivatives Using Photoredox Neutral Somophilic Isocyanide Insertion. Angew Chem Int Ed 52:13289–13292 (Scheme 6.29).

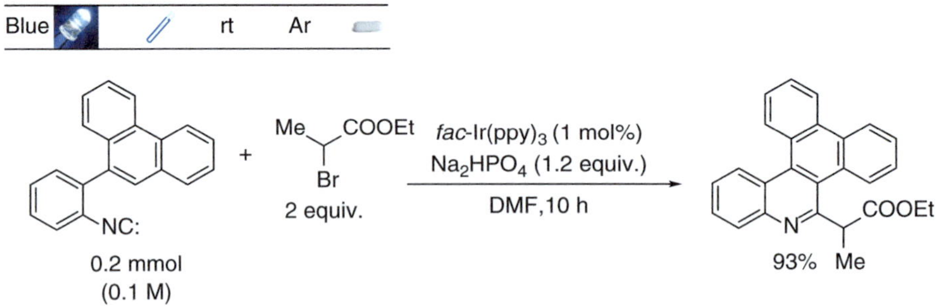

Scheme 6.29 Photoredox catalyzed synthesis of phenanthridine derivatives

Related: Xiao T, Li L, Lin G, Wang Q, Zhang P, Mao Z-W, Zhou L (2014) Synthesis of 6-substituted phenanthridines by metal-free, visible-light induced aerobic oxidative cyclization of 2-isocyanobiphenyls with hydrazines. Green Chem 16:2418–2421.

Jiang H, Cheng Y, Wang R, Zhang Y, Yu S (2014) Synthesis of isoquinolines via visible light-promoted insertion of vinyl isocyanides with diaryliodonium salts. Chem Commun 50:6164–6167.

Cheng Y, Yuan X, Jiang H, Wang R, Ma J, Zhang Y, Yu S (2014) Regiospecific Synthesis of 1-Trifluoromethylisoquinolines Enabled by Photoredox Somophilic Vinyl Isocyanide Insertion. Adv Synth Catal 356:2859–2866.

Gu L, Jin C, Liu J, Ding H, Fan B (2014) Transition-metal-free, visible-light induced cyclization of arylsulfonyl chlorides with 2-isocyanobiphenyls to produce phenanthridines. Chem Commun 50:4643–4645.

Other: An J, Zou Y-Q, Yang Q-Q, Wang Q, Xiao W-J (2013) Visible Light-Induced Aerobic Oxyamidation of Indoles: A Photocatalytic Strategy for the Preparation of Tetrahydro-5*H*-indolo[2,3-*b*]quinolinols. Adv Synth Catal 355:1483–1489.

6.2.1.4.3 Oxygen Containing Heterocycles

Lin R, Sun H, Yang C, Shen W, Xia W (2015) Visible light-induced difunctionalization of electron-enriched styrenes: synthesis of tetrahydrofurans and tetrahydropyrans. Chem Commun 51:399–401 (Scheme 6.30).

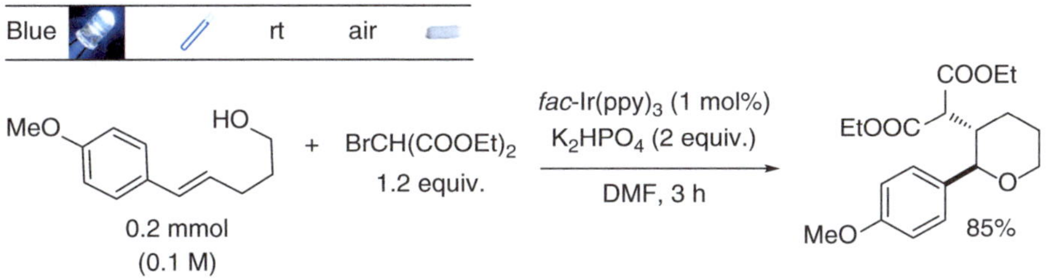

Scheme 6.30 Photocatalyzed synthesis of tetrahydropyrans

6.2.1.4.4 More Than One Heteroatom in the Ring

Revol G, McCallum T, Morin M, Gagosz F, Barriault L (2013) Photoredox Transformations with Dimeric Gold Complexes. Angew Chem Int Ed 52:13342–13345 (Scheme 6.31).

Scheme 6.31 Photoredox intramolecular cyclization of aryl bromides

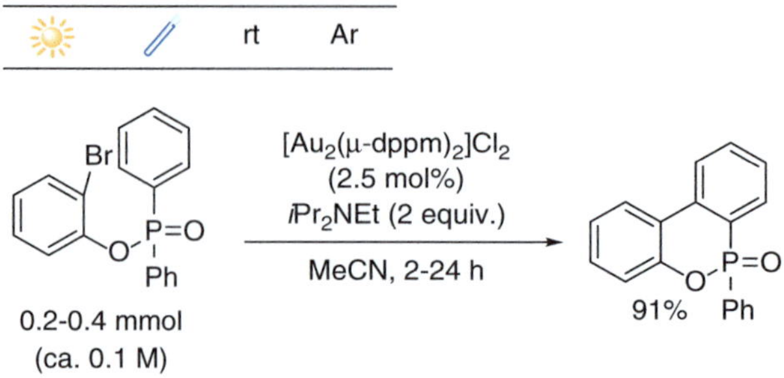

He Z, Bae M, Wu J, Jamison TF (2014) Synthesis of Highly Functionalized Polycyclic Quinoxaline Derivatives Using Visible-Light Photoredox Catalysis. Angew Chem Int Ed 53:14451–14455 (Scheme 6.32).

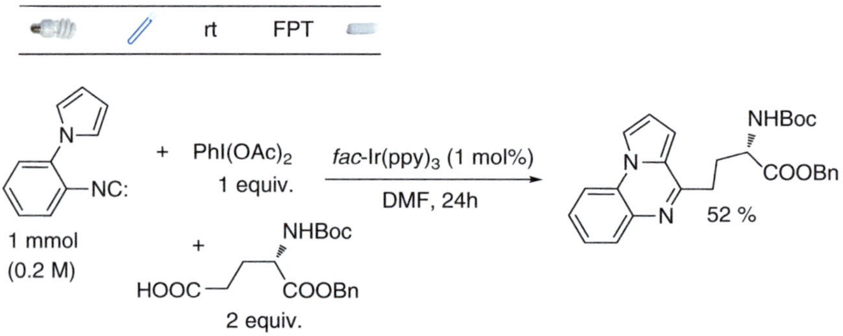

Scheme 6.32 Synthesis of functionalized polycyclic quinoxalines

<u>Other</u>: Sagadevan A, Ragupathi A, Hwang KC (2013) Visible-light-induced, copper(I)-catalysed C-N coupling between *o*-phenylenediamine and terminal alkynes: one-pot synthesis of 3-phenyl-2-hydroxy-quinoxalines. Photochem Photobiol Sci 12:2110–2118.

Guo H, Zhu C, Li J, Xu G, Sun J (2014) Photo-Assisted Multi-Component Reactions (MCR): A New Entry to 2-Pyrimidinethiones. Adv Synth Catal 356:2801–2806.

Deng Q-H, Chen J-R, Wei Q, Zhao Q-Q, Lu L-Q, Xiao W-J (2015) Visible-light-induced photocatalytic oxytrifluoromethylation of *N*-allylamides for the synthesis of CF_3-containing oxazolines and benzoxazines. Chem Commun 51:3537–3540.

6.2.1.5 Two Rings

Alonso R, Bach T (2014) A Chiral Thioxanthone as an Organocatalyst for Enantioselective [2 + 2] Photocycloaddition Reactions Induced by Visible Light. Angew Chem Int Ed 53:4368–4371 (Scheme 6.33).

Scheme 6.33 Enantioselective [2+2] photocycloadditions in the presence of a chiral thioxanthone as organocatalyst

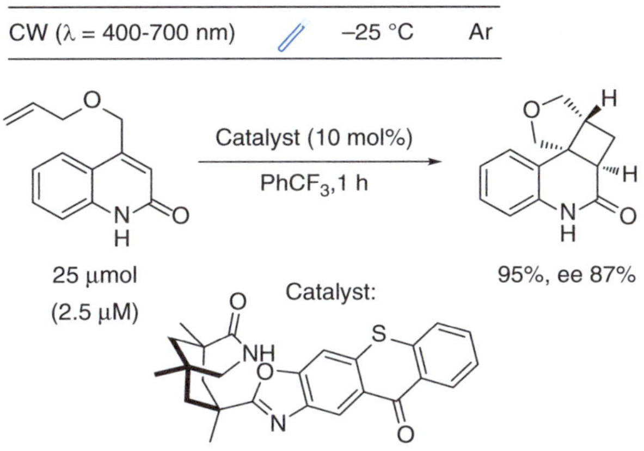

Tang J, Grampp G, Liu Y, Wang B-X, Tao F-F, Wang L-J, Liang X-Z, Xiao H-Q, Shen Y-M (2015) Visible Light Mediated Cyclization of Tertiary Anilines with Maleimides Using Nickel(II) Oxide Surface-Modified Titanium Dioxide Catalyst. J Org Chem 80:2724–2732 (Scheme 6.34).

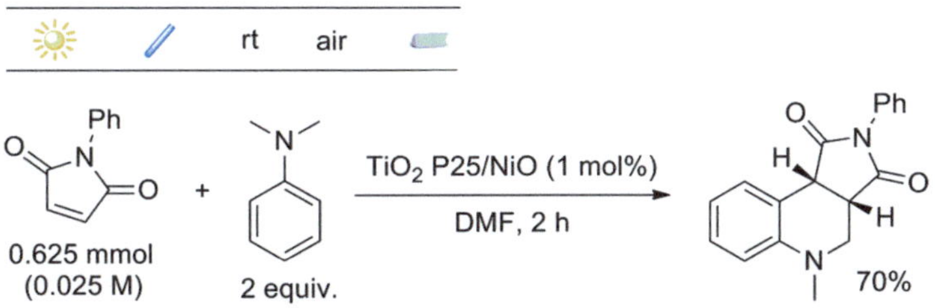

Scheme 6.34 Visible light mediated cyclization of tertiary anilines with maleimides

<u>*Related*</u>: Liang Z, Xu S, Tian W, Zhang R (2015) Eosin Y-catalyzed visible-light-mediated aerobic oxidative cyclization of *N*,*N*-dimethylanilines with maleimides. Beilstein J Org Chem 11:425–430.

Gesmundo NJ, Nicewicz DA (2014) Cyclization-endoperoxidation cascade reactions of dienes mediated by a pyrylium photoredox catalyst. Beilstein J Org Chem 10:1272–1281 (Scheme 6.35).

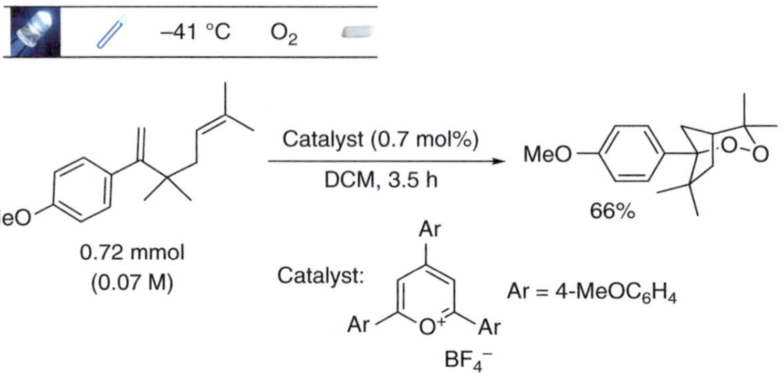

Scheme 6.35 Pyrylium photocatalyzed synthesis of endoperoxides

Other: Hurtley AE, Lu Z, Yoon TP (2014) [2 + 2] Cycloaddition of 1,3-Dienes by Visible Light Photocatalysis. Angew Chem Int Ed 53:8991–8994.

6.2.2 Arylations

6.2.2.1 Formation of Ar-C Bonds

6.2.2.1.1 Ipso-Substitution of a CN Group

McNally A, Prier CK, MacMillan DWC (2011) Discovery of an α-Amino C-H Arylation Reaction Using the Strategy of Accelerated Serendipity. Science 334:1114–1117 (Scheme 6.36).

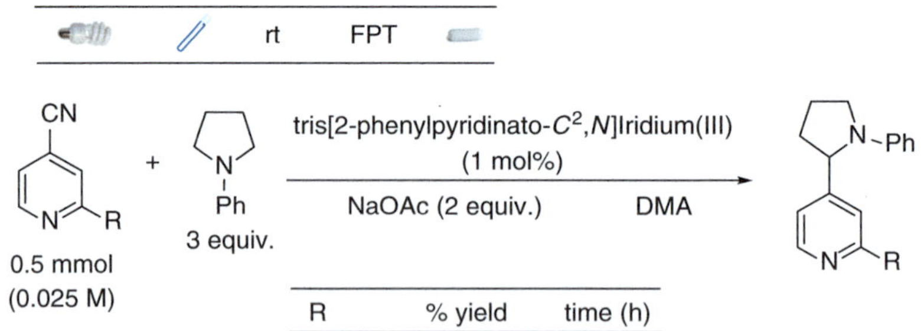

R	% yield	time (h)
H	72	12
Me	87	16

Scheme 6.36 Photocatalyzed α-amino C-H arylations by cyanopyridines

Related: Qvortrup K, Rankic DA, MacMillan DWC (2014) A General Strategy for Organocatalytic Activation of C-H Bonds via Photoredox Catalysis: Direct Arylation of Benzylic Ethers. J Am Chem Soc 136:626–629.

Zuo Z, MacMillan DWC (2014) Decarboxylative Arylation of α-Amino Acids via Photoredox Catalysis: A One-Step Conversion of Biomass to Drug Pharmacophore. J Am Chem Soc 136:5257–5260.

6.2.2.1.2 Ipso-Substitution of a Halogen

Tellis JC, Primer DN, Molander GA (2014) Single-electron transmetalation in organoboron cross-coupling by photoredox/nickel dual catalysis. Science 345:433–436 (Scheme 6.37).

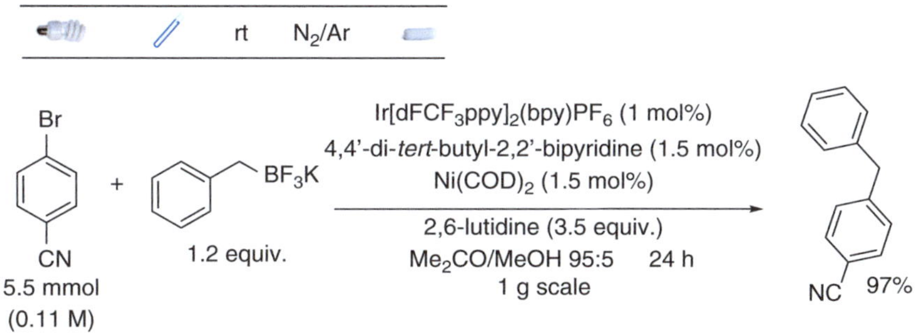

Scheme 6.37 Cross-coupling reaction by photoredox/nickel dual catalysis

Related: Primer DN, Karakaya I, Tellis JC, Molander GA (2015) Single-Electron Transmetalation: An Enabling Technology for Secondary Alkylboron Cross-Coupling. J Am Chem Soc 137:2195–2198.

Zuo Z, Ahneman DT, Chu L, Terrett JA, Doyle AG, MacMillan DWC (2014) Merging photoredox with nickel catalysis: Coupling of α-carboxyl sp^3-carbons with aryl halides. Science 345:437–440 (Scheme 6.38).

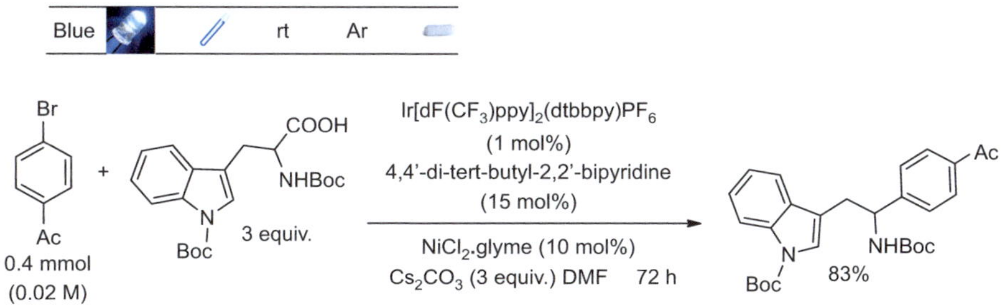

Scheme 6.38 Coupling of sp^3-carbons with aryl halides

Related: Prier CK, MacMillan DWC (2014) Amine α-heteroarylation via photoredox catalysis: a homolytic aromatic substitution pathway. Chem Sci 5:4173–4178.

6.2.2.1.3 Ipso-Substitution of an Onium Group

Guo W, Lu L-Q, Wang Y, Wang Y-N, Chen J-R, Xiao W-J (2015) Metal-Free, Room-Temperature, Radical Alkoxycarbonylation of Aryldiazonium Salts through Visible-Light Photoredox Catalysis. Angew Chem Int Ed 54:2265–2269 (Scheme 6.39).

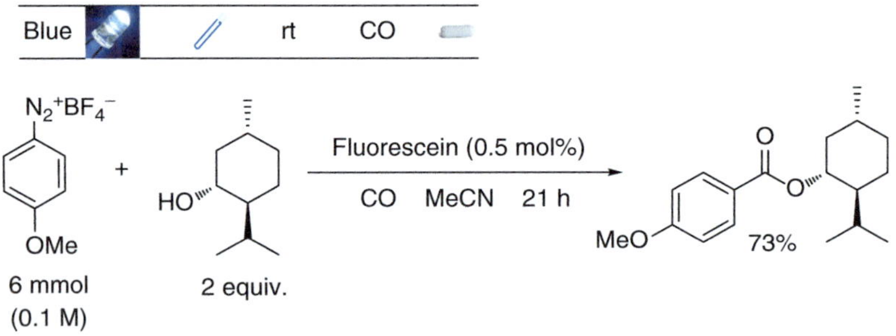

Scheme 6.39 Visible light photoredox catalysed alkoxycarbonylation of aryldiazonium salts

Related: Majek M., von Wangelin AJ (2015) Metal-Free Carbonylations by Photoredox Catalysis. Angew Chem Int Ed 54:2270–2274.

Fumagalli G, Boyd S, Greaney MF (2013) Oxyarylation and Aminoarylation of Styrenes Using Photoredox Catalysis. Org Lett 15:4398–4401 (Scheme 6.40).

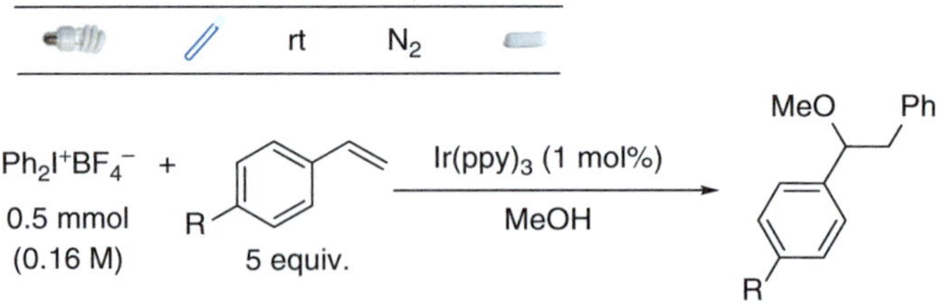

R	% yield	time (h)
H	70	18
Br	57	18
NHAc	42	18

Scheme 6.40 Oxyarylation of styrenes using photoredox catalysis

Related: Hopkinson MN, Sahoo B, Glorius F (2014) Dual Photoredox and Gold Catalysis: Intermolecular Multicomponent Oxyarylation of Alkenes. Adv Synth Catal 356:2794–2800.

Donck S, Baroudi A, Fensterbank L, Goddard J-P, Ollivier C (2013) Visible-Light Photocatalytic Reduction of Sulfonium Salts as a Source of Aryl Radicals. Adv Synth Catal 355:1477–1482 (Scheme 6.41).

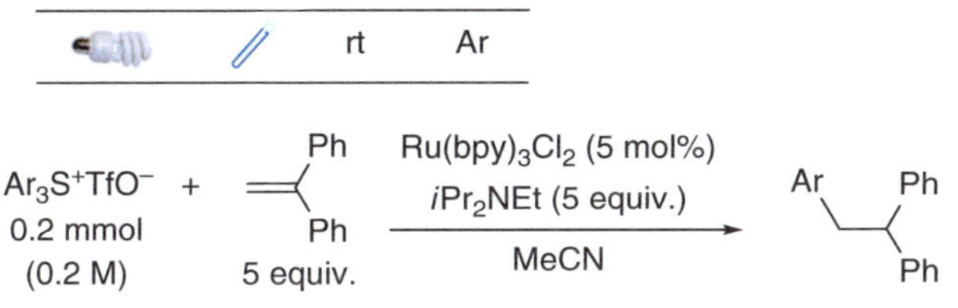

Ar	% yield	time (h)
Ph	70	15
4-MeC_6H_4	70	15
4-FC_6H_4	58	15

Scheme 6.41 Visible light photocatalytic arylation of diphenylethylene

Bu M, Niu TF, Cai C (2015) Visible-light-mediated oxidative arylation of vinylarenes under aerobic conditions. Catal Sci Technol 5:830–834 (Scheme 6.42).

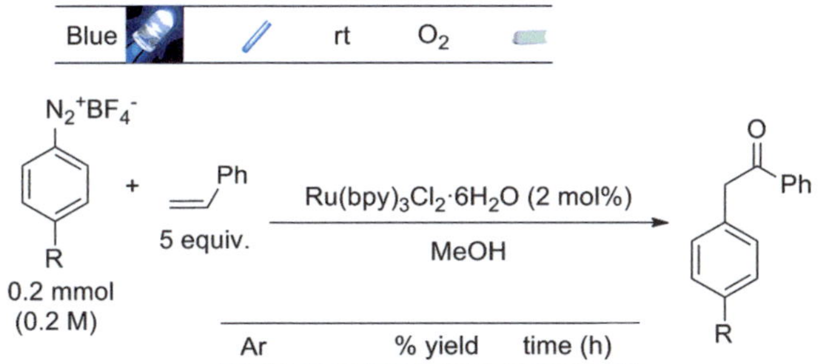

Ar	% yield	time (h)
NO_2	80	10
CN	87	10
OMe	tr	10

Scheme 6.42 Visible-light-mediated synthesis of aromatic ketones

Other: Prasad Hari D, Hering T, König B (2014) The Photoredox-Catalyzed Meerwein Addition Reaction: Intermolecular Amino-Arylation of Alkenes. Angew Chem Int Ed 53:725–728.

Baralle A, Fensterbank L, Goddard J-P, Ollivier C (2013) Aryl Radical Formation by Copper(I) Photocatalyzed Reduction of Diaryliodonium Salts: NMR Evidence for a Cu^{II}/Cu^{I} Mechanism. Chem Eur J 19:10809–10813.

6.2.2.1.4 Ar-H Activation

Ghosh I, Ghosh T, Bardagi JI, König B (2014) Reduction of aryl halides by consecutive visible light-induced electron transfer processes. Science 346:725–728 (Scheme 6.43).

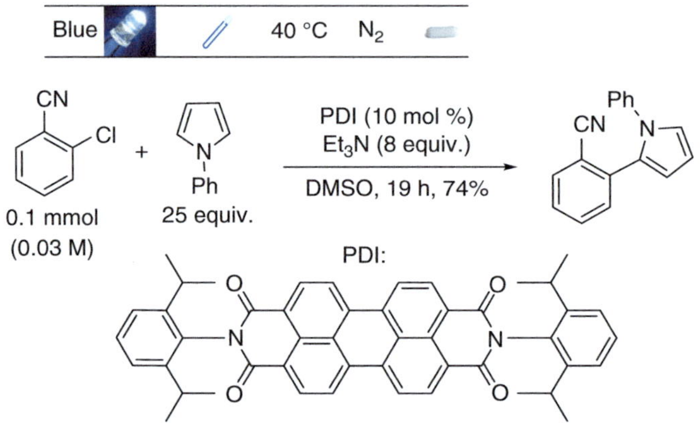

Scheme 6.43 Arylation of pyrroles by consecutive visible light-induced electron transfer processes

Nagib DA, MacMillan DWC (2011) Trifluoromethylation of arenes and heteroarenes by means of photoredox catalysis. Nature 480:224–228 (Scheme 6.44).

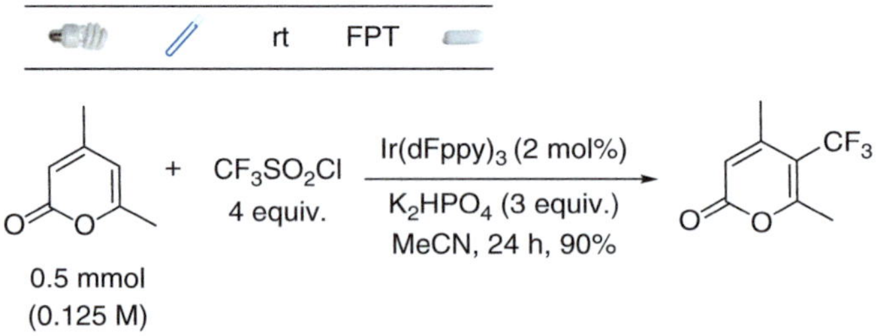

Scheme 6.44 Trifluoromethylation of heterocycles by means of photoredox catalysis

Related: Straathof NJW, Gemoets HPL, Wang X, Schouten JC, Hessel V, Noël T (2014) Rapid Trifluoromethylation and Perfluoroalkylation of Five-Membered Heterocycles by Photoredox Catalysis in Continuous Flow. ChemSusChem 7:1612–1617.

Baar M, Blechert S (2015) Graphitic Carbon Nitride Polymer as a Recyclable Photoredox Catalyst for Fluoroalkylation of Arenes. Chem Eur J 21:526–530.

Nappi M, Bergonzini G, Melchiorre P (2014) Metal-Free Photochemical Aromatic Perfluoroalkylation of α-Cyano Arylacetates. Angew Chem Int Ed 53:4921–4925 (Scheme 6.45).

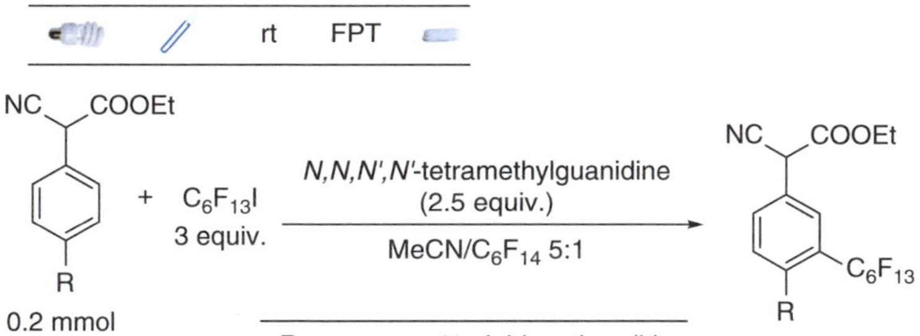

R	% yield	time (h)
MeO	76	8
TMS	86	4
Cl	36	12

Scheme 6.45 Metal-free photochemical aromatic perfluoroalkylation of α-cyano arylacetates

Li X, Gu X, Li Y, Li P (2014) Aerobic Transition-Metal-Free Visible-Light Photoredox Indole C-3 Formylation Reaction. ACS Catal 4:1897–1900 (Scheme 6.46).

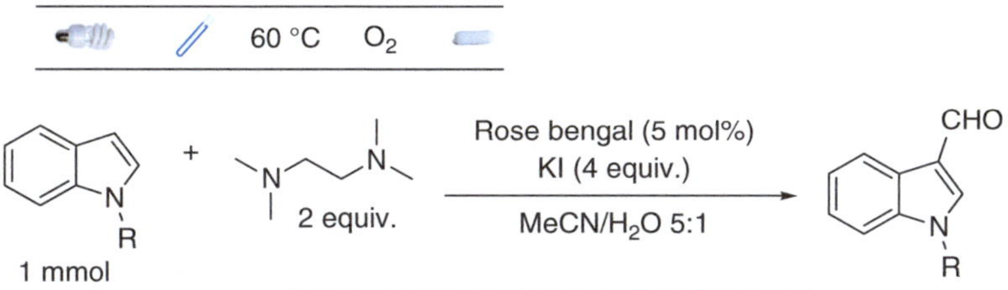

R	% yield	time (h)
Me	70	48
$CH_2CH{=}CH_2$	55	72
Boc	tr	48

Scheme 6.46 Aerobic visible-light photoredox indole C-3 formylation

DiRocco DA, Dykstra K, Krska S, Vachal P, Conway DV, Tudge M (2014) Late-Stage Functionalization of Biologically Active Heterocycles Through Photoredox Catalysis. Angew Chem Int Ed 53:4802–4806 (Scheme 6.47).

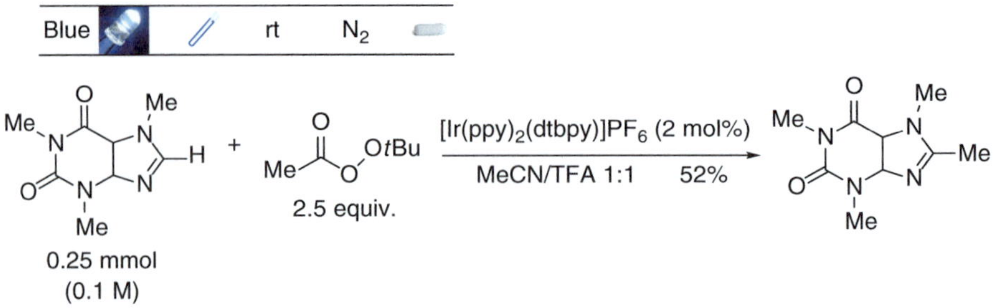

Scheme 6.47 Photoredox catalyzed methylation of biologically active heterocycles

Fabry DC, Zoller J, Raja S, Rueping M (2014) Combining Rhodium and Photoredox Catalysis for C-H Functionalizations of Arenes: Oxidative Heck Reactions with Visible Light. Angew Chem Int Ed 53:10228–10231 (Scheme 6.48).

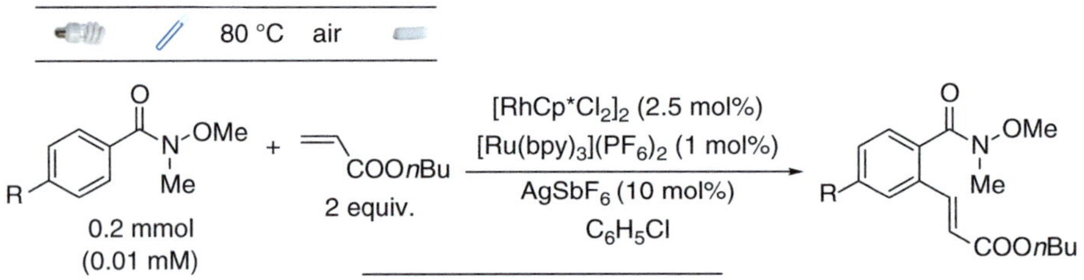

R	% yield	time (h)
H	91	16-24
Cl	84	16-24
*t*Bu	63	16-24

Scheme 6.48 Visible light catalyzed oxidative Heck coupling

Related: Fabry DC, Ronge MA, Zoller J, Rueping M (2015) C-H Functionalization of Phenols Using Combined Ruthenium and Photoredox Catalysis: In Situ Generation of the Oxidant. Angew Chem Int Ed 54:2801–2805.

Other: Carboni A, Dagousset G, Magnier E, Masson G (2014) One pot and selective intermolecular aryl– and heteroaryl-trifluoromethylation of alkenes by photoredox catalysis. Chem Commun 50:14197–14200.

Pitre SP, McTiernan CD, Ismaili H, Scaiano JC (2014) Metal-Free Photocatalytic Radical Trifluoromethylation Utilizing Methylene Blue and Visible Light Irradiation. ACS Catal 4:2530–2535.

Xie J, Yuan X, Abdukader A, Zhu C, Ma J (2014) Visible-Light-Promoted Radical C-H Trifluoromethylation of Free Anilines. Org Lett 16:1768–1771.

Wang L, Wei X-J, Jia W-L, Zhong J-J, Wu L-Z, Liu Q (2014) Visible-Light-Driven Difluoroacetamidation of Unactive Arenes and Heteroarenes by Direct C-H Functionalization at Room Temperature. Org Lett 16:5842–5845.

Jung J, Kim E, You Y, Cho EJ (2014) Visible Light-Induced Aromatic Difluoroalkylation. Adv Synth Catal 356:2741–2748.

Su Y-M, Hou Y, Yin F, Xu Y-M, Li Y, Zheng X, Wang X-S (2014) Visible Light-Mediated C-H Difluoromethylation of Electron-Rich Heteroarenes. Org Lett 16:2958–2961.

Wang L, Wei X-J, Lei W-L, Chen H, Wu L-Z, Liu Q (2014) Direct C-H difluoromethylenephosphonation of arenes and heteroarenes with bromodifluoromethyl phosphonate via visible-light photocatalysis. Chem Commun 50:15916–15919.

6.2.2.2 Formation of Other Ar-X Bonds

Brachet E, Ghosh T, Ghosh I, König B (2015) Visible light C-H amidation of heteroarenes with benzoyl azides. Chem Sci 6:987–992 (Scheme 6.49).

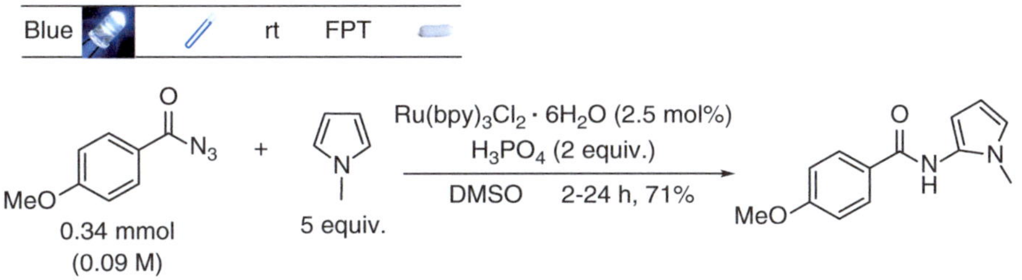

Scheme 6.49 Photocatalyzed C-H amidation of heteroarenes with benzoyl azides

Qin Q, Yu S (2014) Visible-Light-Promoted Redox Neutral C-H Amidation of Heteroarenes with Hydroxylamine Derivatives. Org Lett 16:3504–3507 (Scheme 6.50).

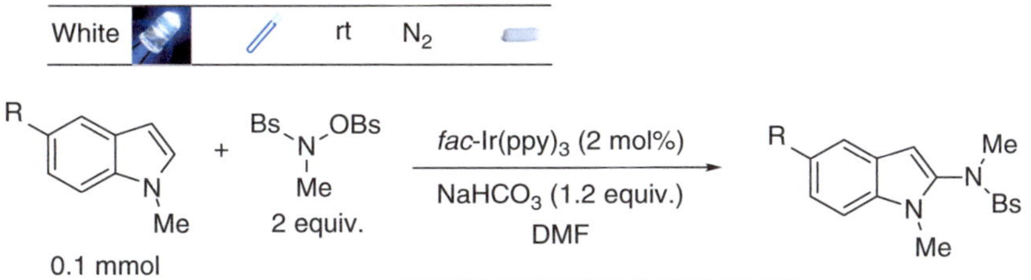

Bs = Benzenesulfonyl

R	% yield	time (h)
H	71	24
Br	47	24
NHBoc	57	24

Scheme 6.50 Visible light catalyzed C-H amidation of heteroarenes with hydroxylamine derivatives

Related: Wang J-D, Liu Y-X, Xue D, Wang C, Xiao J Amination of Benzoxazoles by Visible-Light Photoredox Catalysis. (2014) Synlett 25:2013–2018.

He Y, Wu H, Toste FD (2015) A dual catalytic strategy for carbon-phosphorus cross-coupling via gold and photoredox catalysis. Chem Sci 6:1194–1198 (Scheme 6.51).

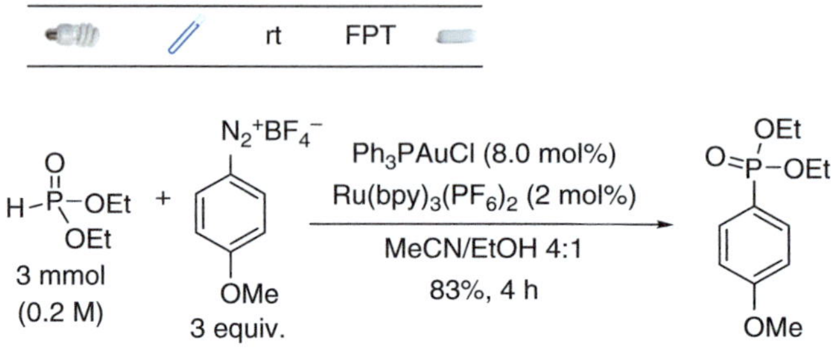

Scheme 6.51 P-arylation of aryldiazonium salts with phosphonates via dual gold/photoredox catalysis

Wang X, Cuny GD, Noël T (2013) A Mild, One-Pot Stadler-Ziegler Synthesis of Arylsulfides Facilitated by Photoredox Catalysis in Batch and Continuous-Flow. Angew Chem Int Ed 52:7860–7864 (Scheme 6.52).

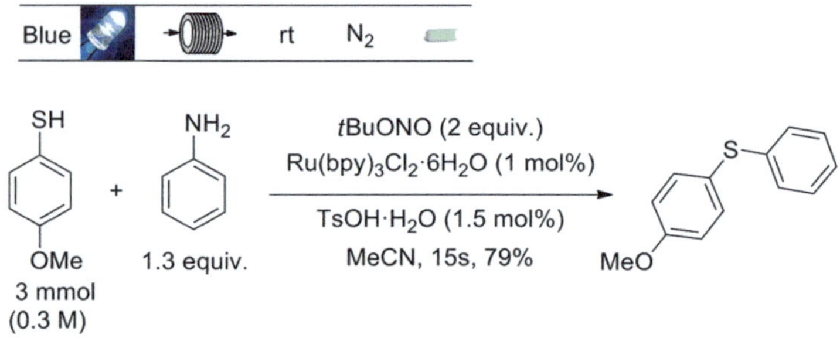

Scheme 6.52 Continuous flow synthesis of aryl sulfides

Related: Kundu D, Ahammed S, Ranu BC (2014) Visible Light Photocatalyzed Direct Conversion of Aryl-/Heteroarylamines to Selenides at Room Temperature. Org Lett 16:1814–1817.

Fan W, Yang Q, Xu F, Li P (2014) A Visible-Light-Promoted Aerobic Metal-Free C-3 Thiocyanation of Indoles. J Org Chem 79:10588–10592 (Scheme 6.53).

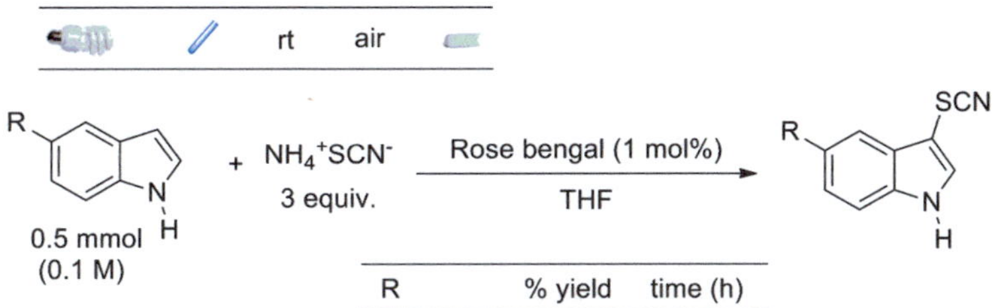

R	% yield	time (h)
H	98	18
Br	76	24
Me	96	18

Scheme 6.53 Visible light-promoted aerobic metal-free C-3 thiocyanation of indoles

Other: Song L, Zhang L, Luo S, Cheng J-P (2014)Visible-Light promoted Catalyst-Free Imidation of Arenes and Heteroarenes. Chem Eur J 20:14231–14234.

Kim H, Kim T, Lee DG, Roh SW, Lee C (2014) Nitrogen-centered radical-mediated C-H imidation of arenes and heteroarenes via visible light induced photocatalysis. Chem Commun 50:9273–9276.

Greulich TW, Daniliuc CG, Studer A (2015) *N*-Aminopyridinium Salts as Precursors for *N*-Centered Radicals - Direct Amidation of Arenes and Heteroarenes. Org Lett 17:254–257.

6.2.3 Additions Onto C=C Bond

6.2.3.1 Addition of H-X

6.2.3.1.1 X = C

Miyazawa K, Koike T, Akita M (2014) Hydroaminomethylation of Olefins with Aminomethyltrifluoroborate by Photoredox Catalysis. Adv Synth Catal 356:2749–2755 (Scheme 6.54).

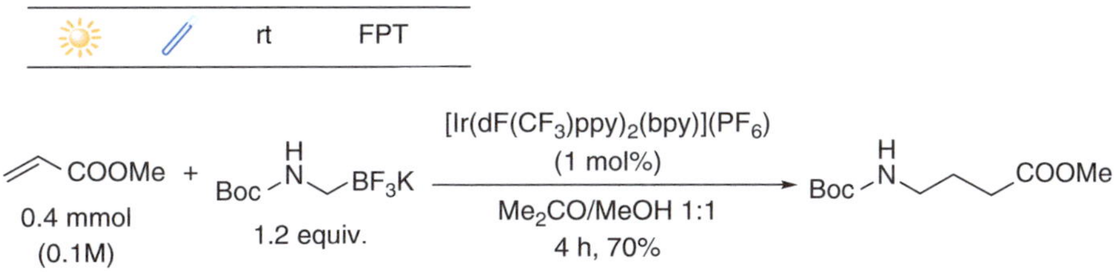

Scheme 6.54 Photoredox catalyzed hydroaminomethylation of olefins

Related: Miyake Y, Nakajima K, Nishibayashi Y (2012) Visible-Light-Mediated Utilization of α-Aminoalkyl Radicals: Addition to Electron-Deficient Alkenes Using Photoredox Catalysts. J Am Chem Soc 134:3338–3341.

Miyake Y, Ashida Y, Nakajima K, Nishibayashi Y (2012) Visible-light-mediated addition of α-aminoalkyl radicals generated from α-silylamines to α,-β-unsaturated carbonyl compounds. Chem Commun 48:6966–6968.

Ruiz Espelt L, McPherson IS, Wiensch EM, Yoon TP (2015) Enantioselective Conjugate Additions of α-Amino Radicals via Cooperative Photoredox and Lewis Acid Catalysis. J Am Chem Soc 137:2452–2455 (Scheme 6.55).

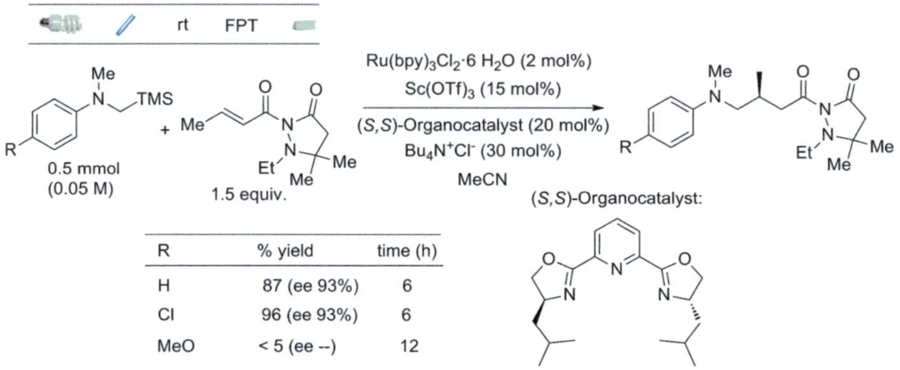

R	% yield	time (h)
H	87 (ee 93%)	6
Cl	96 (ee 93%)	6
MeO	< 5 (ee --)	12

Scheme 6.55 Enantioselective conjugate additions of α-amino radicals onto Michael acceptors

Chinzei T, Miyazawa K, Yasu Y, Koike T, Akita M (2015) Redox-economical radical generation from organoborates and carboxylic acids by organic photoredox catalysis. RSC Adv 5:21297–21300 (Scheme 6.56).

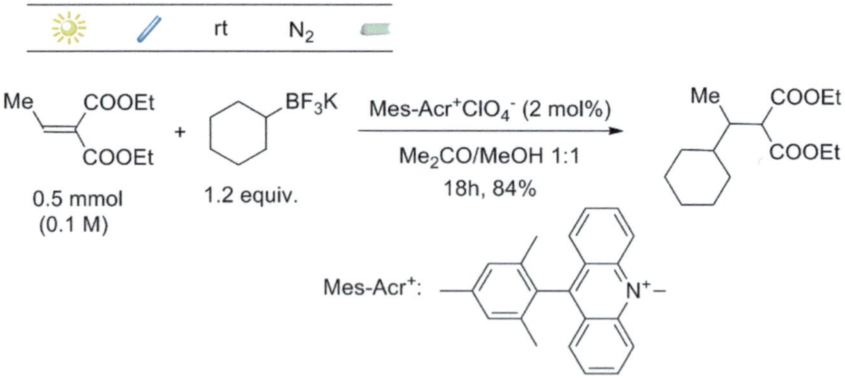

Scheme 6.56 Photocatalyzed radical conjugate addition onto electron-poor olefins

Moteki SA, Usui A, Selvakumar S, Zhang T, Maruoka K (2014) Metal-Free C-H Bond Activation of Branched Aldehydes with Hypervalent Iodine(III) Catalyst under Visible-Light Photolysis: Successful Trapping with Electron-Deficient Olefins. Angew Chem Int Ed 53:11060–11064 (Scheme 6.57).

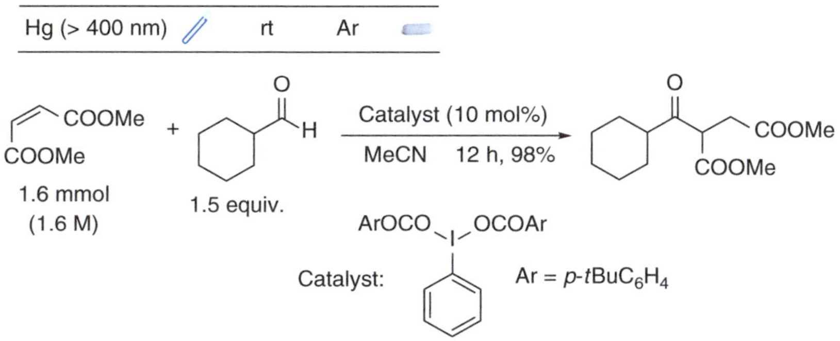

Scheme 6.57 Visible light photochemical hydroacylation processes

Yamada, K, Okada M, Fukuyama T, Ravelli D, Fagnoni M, Ryu I (2015) Photocatalyzed Site-Selective C-H to C-C Conversion of Aliphatic Nitriles. Org Lett 17:1292–1295 (Scheme 6.58).

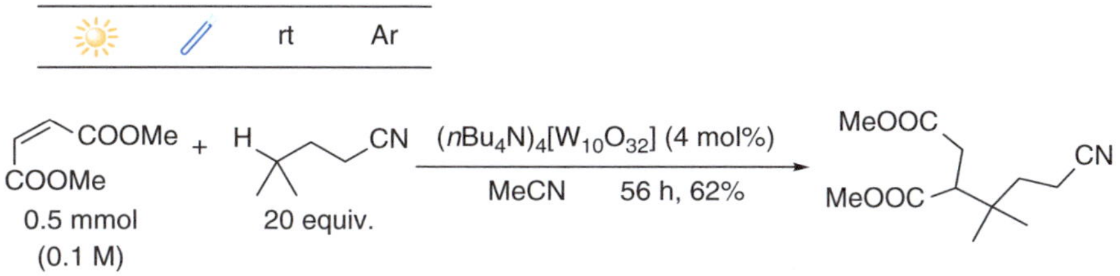

Scheme 6.58 Photocatalyzed hydroalkylation of electron-poor alkenes with aliphatic nitriles

<u>*Related*</u>: Qrareya H, Ravelli D, Fagnoni M, Albini A (2013) Decatungstate Photocatalyzed Benzylation of Alkenes with Alkylaromatics. Adv Synth Catal 355:2891–2899.

<u>*Other*</u>: Lackner GL, Quasdorf KW, Overman LE (2013) Direct Construction of Quaternary Carbons from Tertiary Alcohols via Photoredox-Catalyzed Fragmentation of *tert*-Alkyl *N*-Phthalimidoyl Oxalates. J Am Chem Soc 135:15342–15345.

6.2.3.1.2 X ≠ C

Nguyen TM, Manohar N, Nicewicz DA (2014) *anti*-Markovnikov Hydroamination of Alkenes Catalyzed by a Two-Component Organic Photoredox System: Direct Access to Phenethylamine Derivatives. Angew Chem Int Ed 53:6198–6201 (Scheme 6.59).

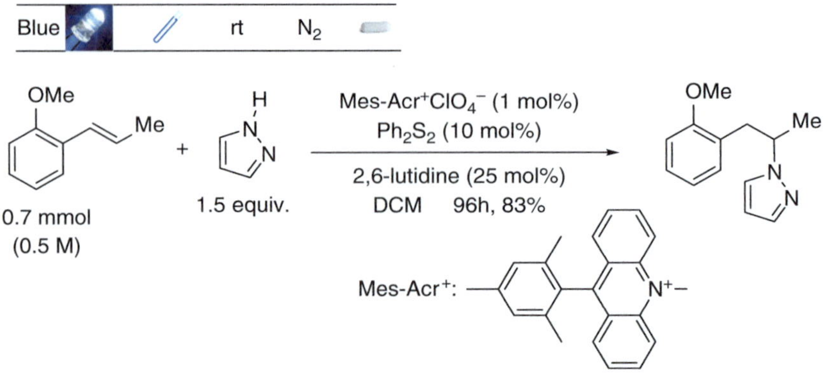

Scheme 6.59 Synthesis of phenethylamines via anti-Markovnikov hydroamination of alkenes

<u>*Related*</u>: Wilger DJ, Grandjean J-M M, Lammert TR, Nicewicz DA (2014) The direct anti-Markovnikov addition of mineral acids to styrenes. Nat Chem 6:720–726.

Yoo W-J, Kobayashi S (2013) Hydrophosphinylation of Unactivated Alkenes with Secondary Phosphine Oxides Under Visible-Light Photocatalysis. Green Chem 15:1844–1848 (Scheme 6.60).

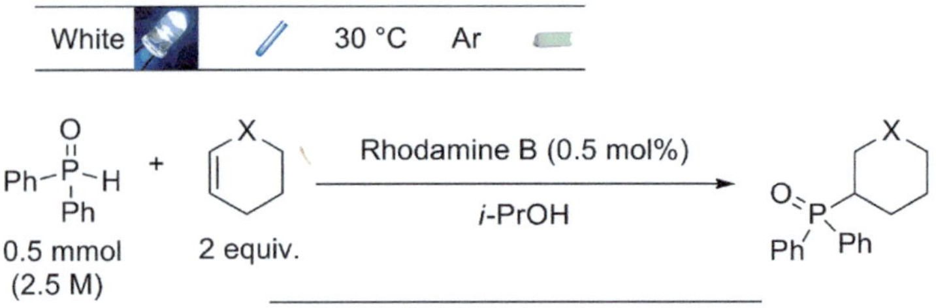

X	% yield	time (h)
CH_2	64	12
O	58	12

Scheme 6.60 Visible light catalyzed hydrophosphinylation of unactivated alkenes

<u>*Other*</u>: Keylor MH, Park JE, Wallentin C-J, Stephenson CRJ (2014) Photocatalytic initiation of thiol-ene reactions: synthesis of thiomorpholin-3-ones. Tetrahedron 70:4264–4269.

Tyson EL, Niemeyer ZL, Yoon TP (2014) Redox Mediators in Visible Light Photocatalysis: Photocatalytic Radical Thiol-Ene Additions. J Org Chem 79:1427–1436.

Bhat VT, Duspara PA, Seo S, Abu Bakar NSB, Greaney MF (2015) Visible light promoted thiol-ene reactions using titanium dioxide. Chem Commun 51:4383–4385.

6.2.3.2 Addition of Y-X

Dagousset G, Carboni A, Magnier E, Masson G (2014) Photoredox-Induced Three-Component Azido– and Aminotrifluoromethylation of alkenes. Org Lett 16:4340–4343 (Scheme 6.61).

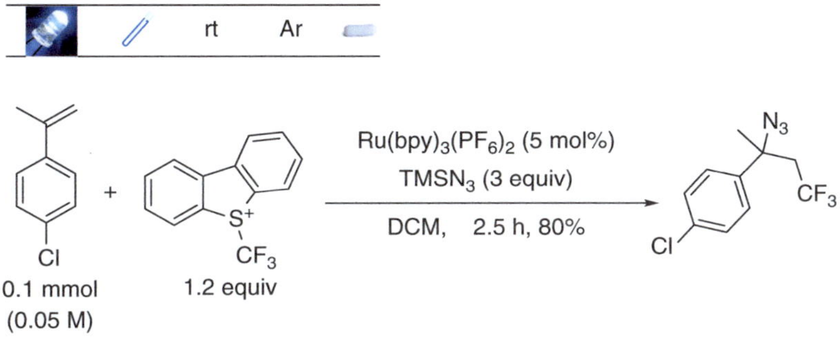

Scheme 6.61 Photoredox catalyzed three component azido- and aminotrifluoromethylation of alkenes

Related: Carboni A, Dagousset G, Magnier E, Masson G (2014) Photoredox-Induced Three-Component Oxy-, Amino-, and Carbotrifluoromethylation of Enecarbamates. Org Lett 16:1240–1243.

Yasu Y, Koike T, Akita M (2012) Three-component Oxytrifluoromethylation of Alkenes: Highly Efficient and Regioselective Difunctionalization of C=C Bonds Mediated by Photoredox Catalysts. Angew Chem Int Ed 51:9567–9571.

Yi H, Zhang X, Qin C, Liao Z, Liu J, Lei A (2014) Visible Light-Induced γ-Alkoxynitrile Synthesis via Three- Component Alkoxycyanomethylation of Alkenes. Adv Synth Catal 356:2873–2877 (Scheme 6.62).

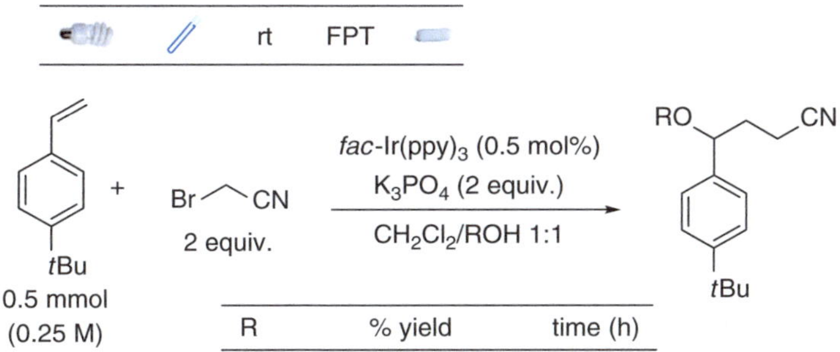

R	% yield	time (h)
Me	90	24
Et	89	24
*i*Pr	69	24

Scheme 6.62 Synthesis of γ-alkoxynitriles via photocatalyzed alkoxycyanomethylation of alkenes

Pirtsch M, Paria S, Matsuno T, Isobe H, Reiser O (2012) [$Cu(dap)_2Cl$] As an Efficient Visible-Light-Driven Photoredox Catalyst in Carbon-Carbon Bond-Forming Reactions. Chem Eur J 18:7336–7340 (Scheme 6.63).

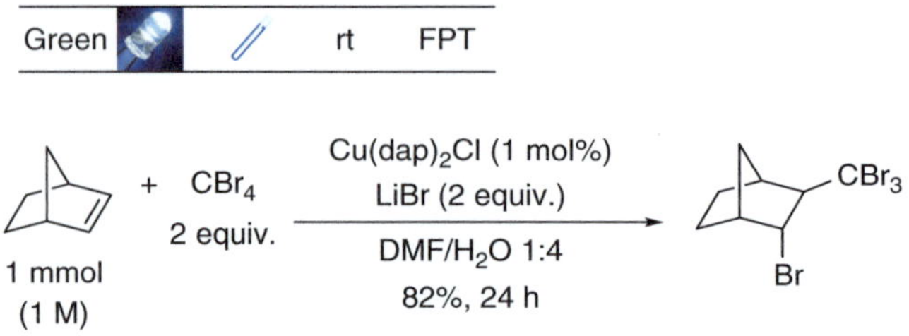

Scheme 6.63 Visible-light-induced copper-catalyzed ATRA reactions

Related: Paria S, Pirtsch M, Kais V, Reiser O (2013) Visible-Light-Induced Intermolecular Atom-Transfer Radical Addition of Benzyl Halides to Olefins: Facile Synthesis of Tetrahydroquinolines. Synthesis 45:2689–2698.

Oh SH, Malpani YR, Ha N, Jung Y-S, Han SB (2014) Vicinal Difunctionalization of Alkenes: Chlorotrifluoromethylation with CF_3SO_2Cl by Photoredox Catalysis. Org Lett 16:1310–1313.

Other: Tang X-J, Dolbier Jr WR (2015) Efficient Cu-catalyzed Atom Transfer Radical Addition Reactions of Fluoroalkylsulfonyl Chlorides with Electron-deficient Alkenes Induced by Visible Light. Angew Chem Int Ed 54:4246–4249.

6.2.3.3 Other Reactions

Jiang H, Cheng Y, Zhang Y, Yu S (2013) Sulfonation and Trifluoromethylation of Enol Acetates with Sulfonyl Chlorides Using Visible-Light Photoredox Catalysis. Eur J Org Chem 5485–5492 (Scheme 6.64).

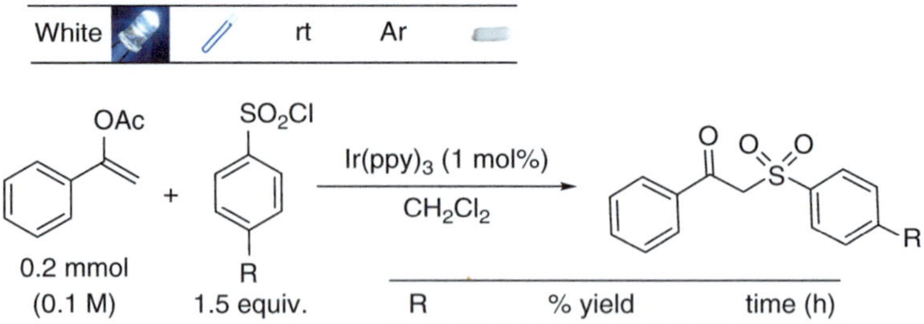

R	% yield	time (h)
Me	84	8-24
NO_2	76	8-24
OCF_3	76	8-24

Scheme 6.64 Sulfonation of enol acetates via visible light photoredox catalysis

Related: Keshari T, Yadav VK, Srivastava VP, Yadav LDS (2014) Visible light organophotoredox catalysis: a general approach to β-keto sulfoxidation of alkenes. Green Chem 16:3986–3992.

Tomita R, Yasu Y, Koike T, Akita M (2014) Combining Photoredox-Catalyzed Trifluoromethylation and Oxidation with DMSO: Facile Synthesis of α-Trifluoromethylated Ketones from Aromatic Alkenes. Angew Chem Int Ed 53:7144–7148.

Hu B, Chen H, Liu Y, Dong W, Ren K, X. Xie, H. Xub, Z. Zhang (2014) Visible light-induced intermolecular radical addition: facile access to γ-ketoesters from alkyl-bromocarboxylates and enamines. Chem Commun 50:13547–13550.

6.2.4 Additions Onto C=N or N=N Bond

Hager D, MacMillan DWC (2014) Activation of C-H Bonds via the Merger of Photoredox and Organocatalysis: A Coupling of Benzylic Ethers with Schiff Bases. J Am Chem Soc 136:16986–16989. (Scheme 6.65).

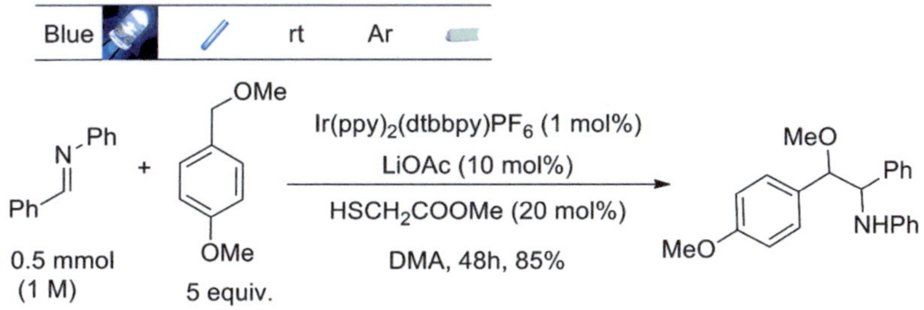

Scheme 6.65 Photocatalyzed coupling of benzylic ethers with Schiff bases

Ryu I, Tani A, Fukuyama T, Ravelli D, Montanaro S, Fagnoni M (2013) Efficient C-H/C-N and C-H/C-CO-N Conversion via Decatungstate-Photoinduced Alkylation of Diisopropyl Azodicarboxylate. Org Lett 15:2554–2557 (Scheme 6.66).

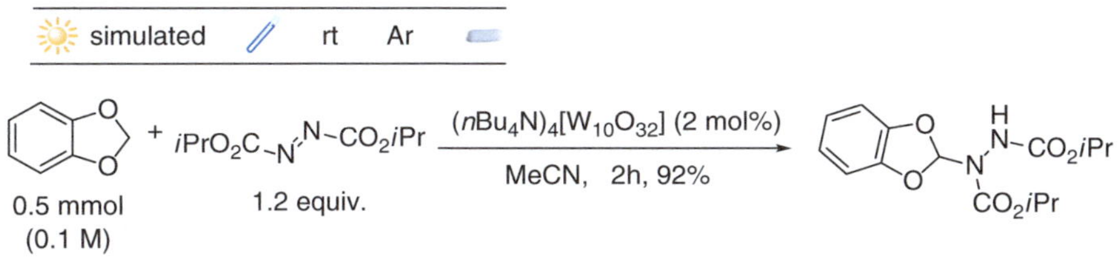

Scheme 6.66 Tetrabutylammonium decatungstate (TBADT) photoinduced functionalization of diisopropyl azodicarboxylate

6.2.5 Synthesis and Reactions of Alkynes

Huang H, Zhang G, Gong L, Zhang S, Chen Y (2014) Visible-Light-Induced Chemoselective Deboronative Alkynylation under Biomolecule-Compatible Conditions. J Am Chem Soc 136:2280–2283. (Scheme 6.67).

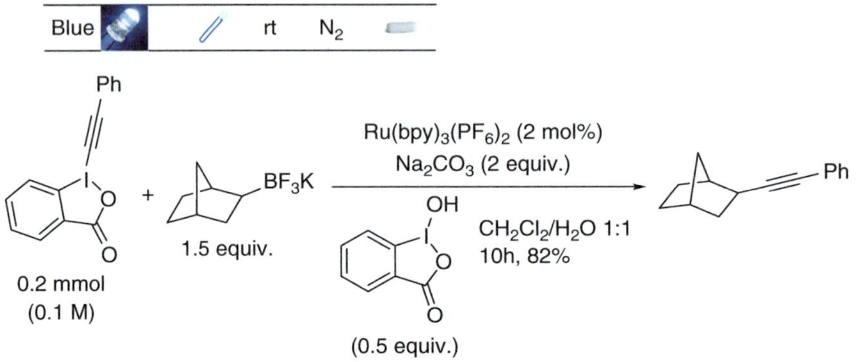

Scheme 6.67 Visible light-induced deboronative alkynylation reactions

Related: Guo S, Tao R, Zhao J (2014) Photoredox catalytic organic reactions promoted with broadband visible light-absorbing Bodipy-iodo-aza-Bodipy triad photocatalyst. RSC Adv 4:36131–36139.

Sagadevan A, Ragupathi A, Lin C-C, Hwu JR, Hwang KC (2015) Visible-light initiated copper(I)-catalysed oxidative C-N coupling of anilines with terminal alkynes: one-step synthesis of α-ketoamides. Green Chem 17:1113–1119 (Scheme 6.68).

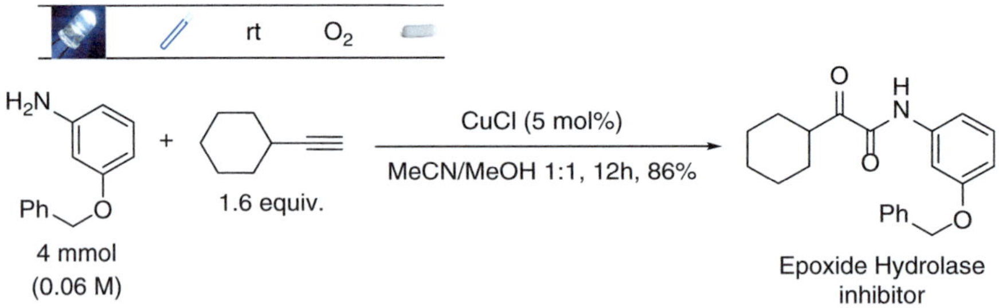

Scheme 6.68 One-step synthesis of α-ketoamides

Gupta S, Do Y, Lee JH, Lee M, Han J, Rhee YH, Park J (2014) Novel Catalyst System for Hydrostannation of Alkynes. Chem Eur J 20:1267–1271 (Scheme 6.69).

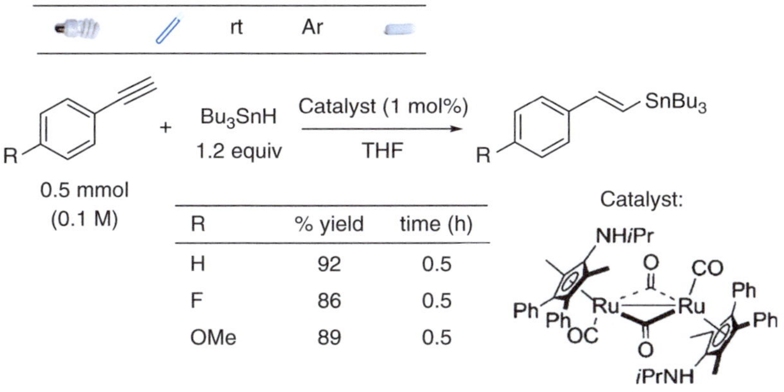

R	% yield	time (h)
H	92	0.5
F	86	0.5
OMe	89	0.5

Scheme 6.69 Photocatalyzed hydrostannation of alkynes

Iqbal N, Jung J, Park S, Cho EJ (2014) Controlled Trifluoromethylation Reactions of Alkynes through Visible- Light Photoredox Catalysis. Angew Chem Int Ed 53:539–542 (Scheme 6.70).

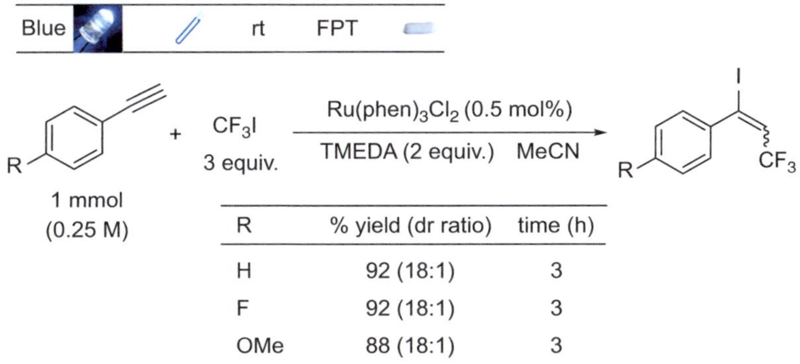

R	% yield (dr ratio)	time (h)
H	92 (18:1)	3
F	92 (18:1)	3
OMe	88 (18:1)	3

Scheme 6.70 Visible light photoredox catalyzed trifluoromethylation of alkynes

Other: Yang J, Zhang J, Qi L, Hu C, Chen Y (2015) Visible-light-induced chemoselective reductive decarboxylative alkynylation under biomolecule-compatible conditions. Chem Commun 51:5275–5278.

6.2.6 Other C-C Bond Formation

6.2.6.1 Allylation

Lang SB, O'Nele KM, Tunge JA (2014) Decarboxylative Allylation of Amino Alkanoic Acids and Esters via Dual Catalysis. J Am Chem Soc 136:13606–13609 (Scheme 6.71).

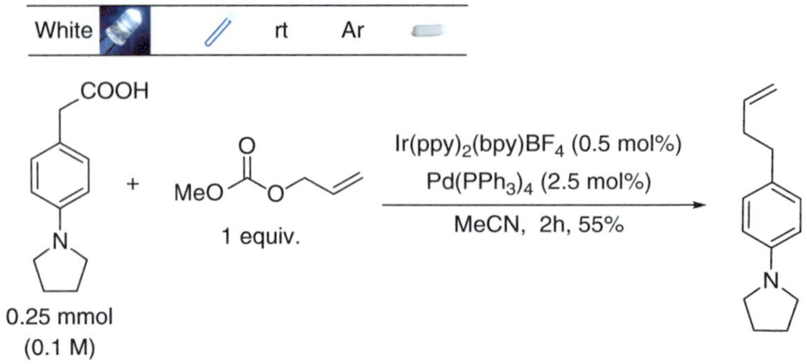

Scheme 6.71 Dual catalytic decarboxylative allylation of amino alkanoic acids

Other: Möhlmann L, Blechert S (2014) Carbon Nitride-Catalyzed Photoredox Sakurai Reactions and Allylborations. Adv Synth Catal 356:2825–2829.

6.2.6.2 Benzylation

Bergonzini G, Schindler CS, Wallentin C-J, Jacobsen EN, Stephenson CRJ (2014) Photoredox activation and anion binding catalysis in the dual catalytic enantioselective synthesis of β-amino esters. Chem Sci 5, 112–116 (Scheme 6.72).

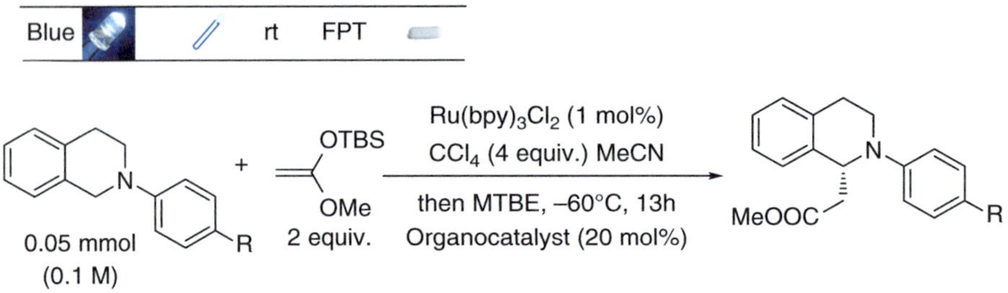

Organocatalyst:

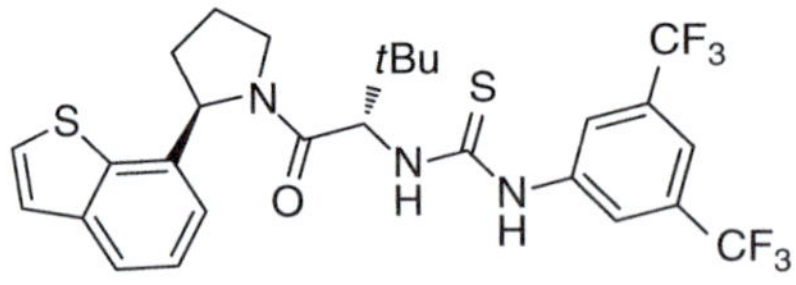

R	% yield (% ee)	time (h)
H	72 (95)	16 (+13)
Br	69 (86)	16 (+13)
OMe	60 (84)	16 (+13)

Scheme 6.72 Enantioselective oxidative C-H functionalization of tetrahydroisoquinolines via combined photoredox and asymmetric anion-binding catalysis

Other: Lerch S, Unkel L-N, Brasholz M (2014). Tandem Organocatalysis and Photocatalysis: An Anthraquinone-Catalyzed Indole-C3-Alkylation/Photooxidation/1,2-Shift Sequence. Angew Chem Int Ed 53:6558–6562.

Li W, Zhu X, Mao H, Tang Z, Cheng Y, Zhu C (2014) Visible-light-induced direct C(sp^3)-H difluromethylation of tetrahydroisoquinolines with the in situ generated difluoroenolates. Chem Commun 50:7521–7523.

Xiao T, Li L, Lin G, Mao Z-w, Zhou L (2014) Metal-Free Visible-Light Induced Cross-Dehydrogenative Coupling of Tertiary Amines with Diazo Compounds. Org Lett 16:4232–4235.

Huang L, Zhao J (2013) Iodo-Bodipys as visible-light-absorbing dual-functional photoredox catalysts for preparation of highly functionalized organic compounds by formation of C-C bonds via reductive and oxidative quenching catalytic mechanisms. RSC Adv 3:23377–23388.

Yamaguchi T, Nobuta T, Tada N, Miura T, Nakayama T, Uno B, Itoh A (2014) Aerobic Photooxidative Carbon-Carbon Bond Formation Between Tertiary Amines and Carbon Nucleophiles Using 2-Chloroanthra-9,10-quinone. Synlett 25:1453–1457.

Zhao Y, Zhang C, Chin KF, Pytela O, Wei G, Liu H, Bures F, Jiang Z (2014) Dicyanopyrazine-derived push-pull chromophores for highly efficient photoredox catalysis. RSC Adv 4:30062–30067.

6.2.6.3 Vinylation

Huang H, Jia K, Chen Y (2015) Hypervalent Iodine Reagents Enable Chemoselective Deboronative/Decarboxylative Alkenylation by Photoredox Catalysis. Angew Chem Int Ed 54:1881–1884. (Scheme 6.73).

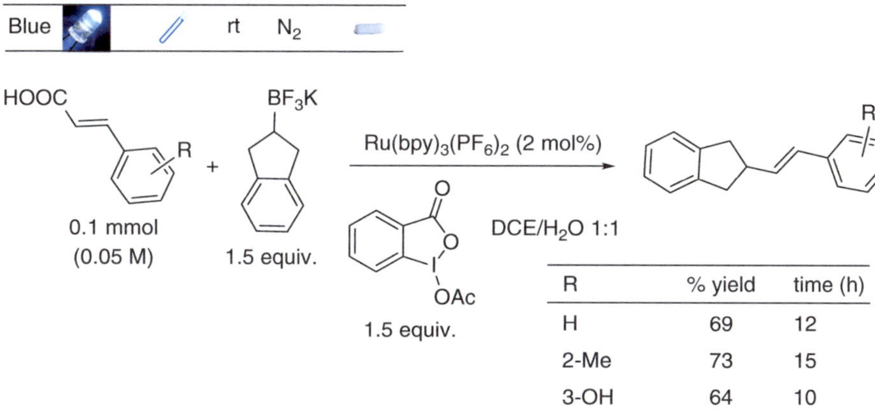

R	% yield	time (h)
H	69	12
2-Me	73	15
3-OH	64	10

Scheme 6.73 Photocatalyzed deboronative/decarboxylative alkenylation

Choi WJ, Choi S, Ohkubo K, Fukuzumi S, Cho EJ, You Y (2015) Mechanisms and applications of cyclometalated Pt(II) complexes in photoredox catalytic trifluoromethylation. Chem Sci 6:1454–1464 (Scheme 6.74).

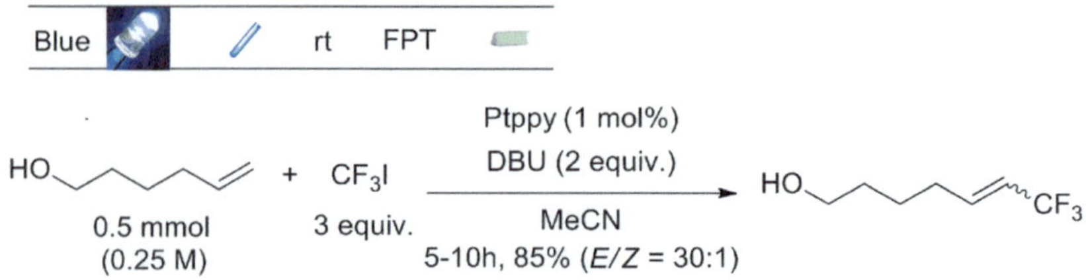

Scheme 6.74 Photoredox catalyzed trifluoromethylations

Related: Tomita R, Yasu Y, Koike T, Akita M (2014) Direct C–H trifluoromethylation of di- and trisubstituted alkenes by photoredox catalysis. Beilstein J Org Chem 10:1099–1106.

Xu P, Abdukader A, Hu K, Cheng Y, Zhu C (2014) Room temperature decarboxylative trifluoromethylation of α,β-unsaturated carboxylic acids by photoredox catalysis. Chem Commun 50:2308–2310.

Noble A, McCarver SJ, MacMillan DWC (2015) Merging Photoredox and Nickel Catalysis: Decarboxylative Cross-Coupling of Carboxylic Acids with Vinyl Halides. J Am Chem Soc 137:624–627 (Scheme 6.75).

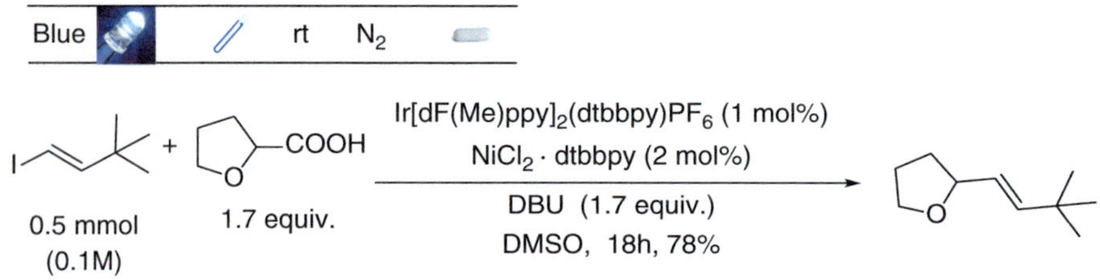

Scheme 6.75 Decarboxylative cross-coupling of carboxylic acids with vinyl halides

Related: Noble A, MacMillan DWC (2014) Photoredox α-Vinylation of α-Amino Acids and *N*-Aryl Amines. J Am Chem Soc 136:11602–11605.

Feng Z-J, Xuan J, Xia X-D, Ding W, Guo W, Chen J-R, Zou Y-Q, Lu L-Q, Xiao W-J (2014) Direct sp^3 C-H acroleination of *N*-aryl-tetrahydroisoquinolines by merging photoredox catalysis with nucleophilic catalysis. Org Biomol Chem 12:2037–2040 (Scheme 6.76).

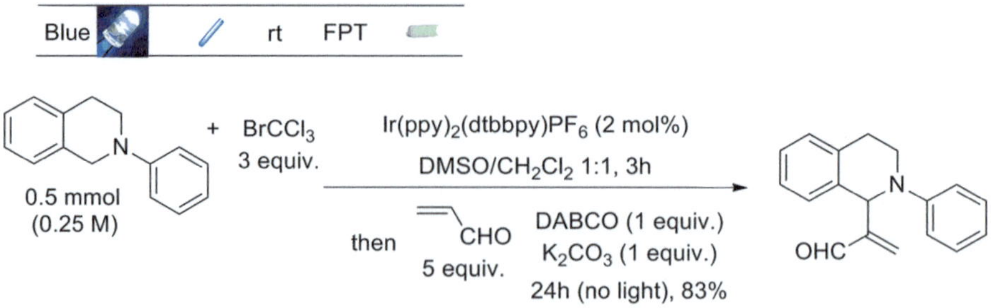

Scheme 6.76 Visible light promoted acroleination of *N*-aryl tetrahydroisoquinolines

<u>Related:</u> Franz JF, Kraus WB, Zeitler K (2015) No photocatalyst required – versatile, visible light mediated transformations with polyhalomethanes. Chem Commun 51:8280–8283.

<u>Other</u>: Yu C, Iqbal N, Park S, Cho EJ (2014) Selective difluoroalkylation of alkenes by using visible light photoredox catalysis. Chem Commun 50:12884–12887.

Zhou S, Zhang D, Sun Y, Li R, Zhang W, Li A (2014) Intermolecular Conjugate Addition of Pyrroloindoline and Furoindoline Radicals to α,β-Unsaturated Enones via Photoredox Catalysis. Adv Synth Catal 356:2867–2872.

Nakajima M, Lefebvre Q, Rueping M (2014) Visible light photoredox-catalysed intermolecular radical addition of α-halo amides to olefins. Chem Commun 50:3619–3622.

6.2.6.4 Cyanation

Srivastava VP, Yadav AK, Yadav LDS (2014) Visible-light-induced cyanation of aza-Baylis-Hillman adducts: a Michael type addition. Tetrahedron Lett 55:1788–1792 (Scheme 6.77).

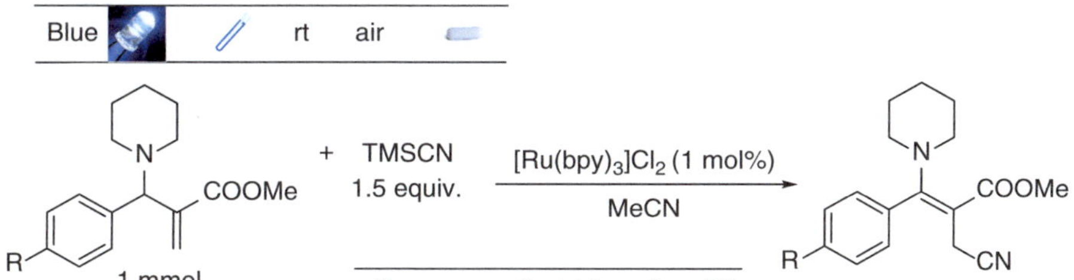

R	% yield	time (h)
H	85	18
OMe	92	14
Br	82	20

Scheme 6.77 Visible light promoted cyanation of aza-Baylis-Hillman adducts

<u>Related</u>: Lin Q-Y, Xu X-H, Qing F-L (2014) Chemo-, Regio-, and Stereoselective Trifluoromethylation of Styrenes via Visible Light-Driven Single-Electron Transfer (SET) and Triplet-Triplet Energy Transfer (TTET) Processes. J Org Chem 79:10434–10446.

Dai X, Cheng D, Guan B, Mao W, Xu X, Li X (2014) Coupling of Tertiary Amines with Acrylate Derivatives via Visible-Light Photoredox Catalysis. J Org Chem 79:7212–7219.

Ushakov DB, Gilmore K, Kopetzki D, McQuade DT, Seeberger PH (2014) Continuous-Flow Oxidative Cyanation of Primary and Secondary Amines Using Singlet Oxygen. Angew Chem Int Ed 53:557–561 (Scheme 6.78).

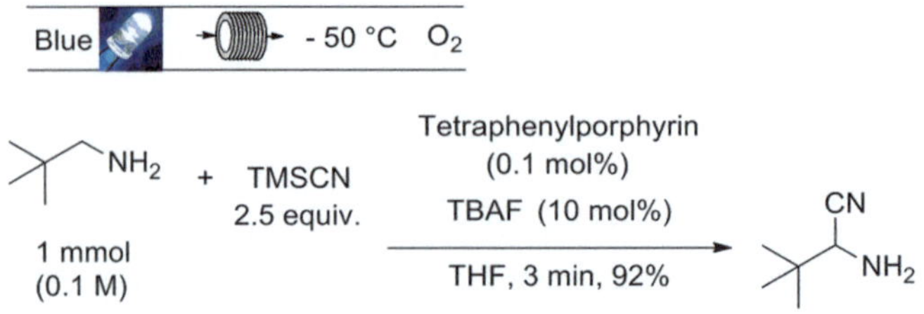

Scheme 6.78 Continuous flow oxidative cyanation of amines via singlet oxygen

Ovoshchnikov DS, Donoeva BG, Golovko VB (2015) Visible-Light-Driven Aerobic Oxidation of Amines to Nitriles over Hydrous Ruthenium Oxide Supported on TiO_2. ACS Catal 5:34–38 (Scheme 6.79).

Scheme 6.79 Visible light-driven aerobic oxidation of amines to nitriles

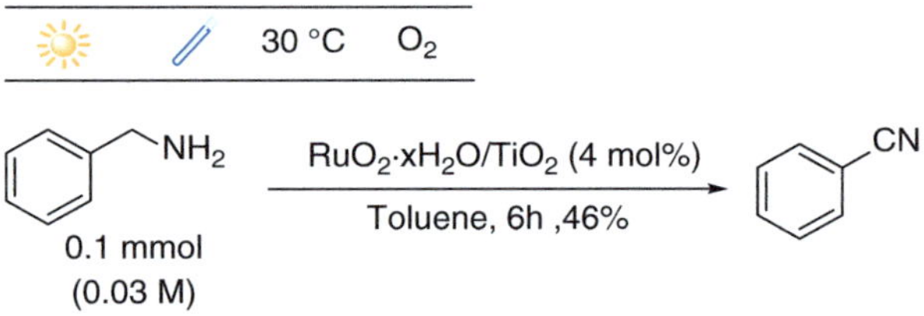

6.2.7 Alpha- and Beta- Functionalizations of Carbonyls

Riente P, Adams AM, Albero J, Palomares E, Pericàs MA (2014) Light-Driven Organocatalysis Using Inexpensive, Nontoxic Bi_2O_3 as the Photocatalyst. Angew Chem Int Ed 53:9613–9616 (Scheme 6.80).

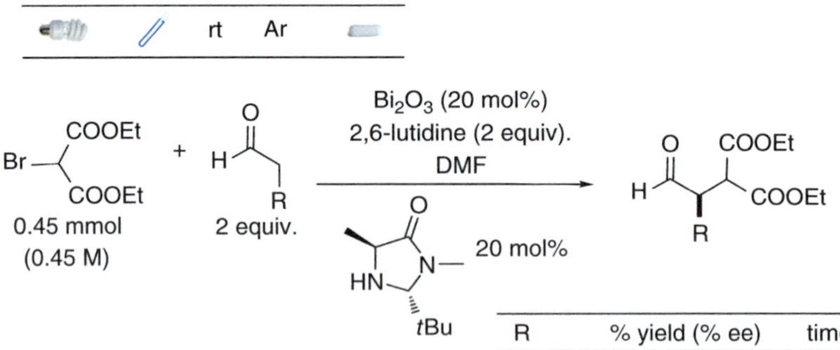

R	% yield (% ee)	time (h)
CH_2Ph	86 (93)	1
Cy	81 (98)	2

Scheme 6.80 Bi_2O_3 photocatalyzed α-alkylation of aldehydes

Cecere G, König CM, Alleva JL, MacMillan DWC (2013) Enantioselective Direct α-Amination of Aldehydes via a Photoredox Mechanism: A Strategy for Asymmetric Amine Fragment Coupling. J Am Chem Soc 135:11521–11524 (Scheme 6.81).

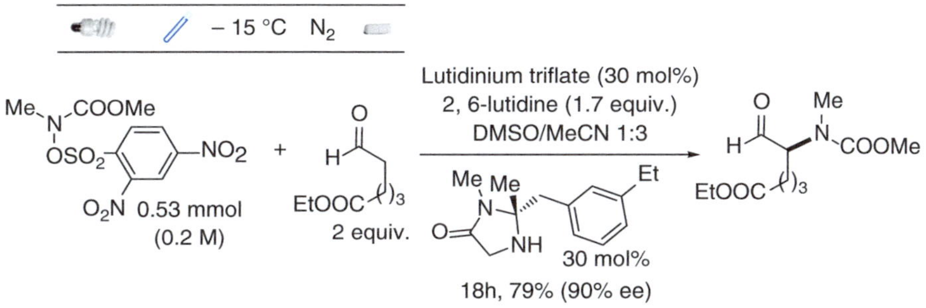

Scheme 6.81 Photoredox catalyzed α-amination of aldehydes

Arceo E, Jurberg ID, Álvarez-Fernández A, Melchiorre P (2013) Photochemical activity of a key donor-acceptor complex can drive stereoselective catalytic α-alkylation of aldehydes. Nat Chem 5:750–756 (Scheme 6.82).

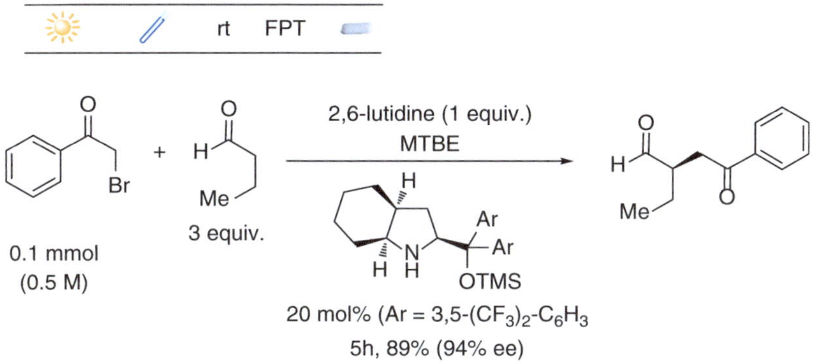

Scheme 6.82 Stereoselective photochemical α-alkylation of aldehydes via donor-acceptor complexes

<u>*Related*</u>: Arceo E, Bahamonde A, Bergonzini G, Melchiorre P (2014) Enantioselective direct α-alkylation of cyclic ketones by means of photo-organocatalysis. Chem Sci 5:2438–2442.

Huo H, Shen X, Wang C, Zhang L, Röse P, Chen L-A, Harms K, Marsch M, Hilt G, Meggers E (2014) Asymmetric photoredox transition-metal catalysis activated by visible light. Nature 515:100–103 (Scheme 6.83).

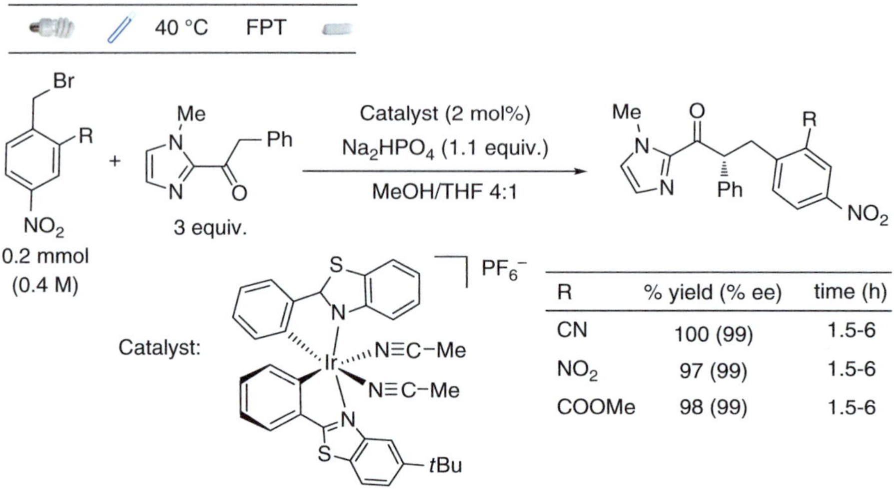

R	% yield (% ee)	time (h)
CN	100 (99)	1.5-6
NO_2	97 (99)	1.5-6
COOMe	98 (99)	1.5-6

Scheme 6.83 Iridium-catalyzed enantioselective photoalkylation of acyl imidazoles

Terrett JA, Clift MD, MacMillan DWC (2014) Direct β-Alkylation of Aldehydes via Photoredox Organocatalysis. J Am Chem Soc 136:6858–6861 (Scheme 6.84).

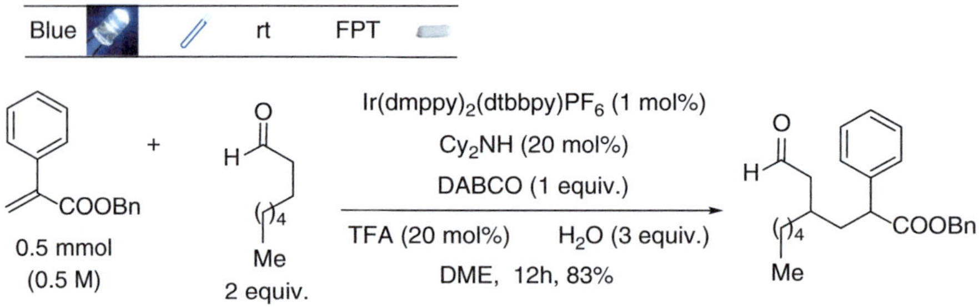

Scheme 6.84 Direct β-alkylation of aldehydes via photoredox organocatalysis

Related: Petronijevic FR, Nappi M, MacMillan DWC (2013) Direct β-Functionalization of Cyclic Ketones with Aryl Ketones via the Merger of Photoredox and Organocatalysis. J Am Chem Soc 135:18323–18326.

Okada M, Fukuyama T, Yamada K, Ryu I, Ravelli D, Fagnoni M (2014) Sunlight photocatalyzed regioselective β-alkylation and acylation of cyclopentanones. Chem Sci 5:2893–2898 (Scheme 6.85).

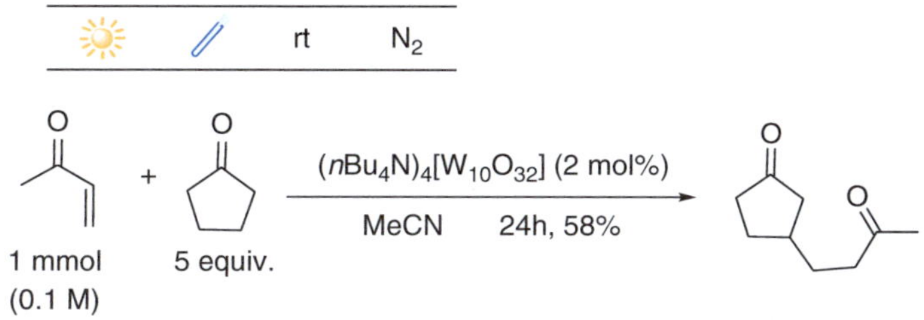

Scheme 6.85 Sunlight photocatalyzed regioselective β-alkylation of cyclopentanones

Other: Zhu Y, Zhang L, Luo S (2014) Asymmetric α-Photoalkylation of β-Ketocarbonyls by Primary Amine Catalysis: Facile Access to Acyclic All-Carbon Quaternary Stereocenters. J Am Chem Soc 136:14642–14645.

Cantillo D, de Frutos O, Rincón JA, Mateos C, Kappe CO (2014) Continuous Flow α-Trifluoromethylation of Ketones by Metal-Free Visible Light Photoredox Catalysis. Org Lett 16:896–899.

Xuan J, Feng Z-J, Chen J-R, Lu L-Q, Xiao W-J (2014) Visible-Light-Induced C-S Bond Activation: Facile Access to 1,4-Diketones from β-Ketosulfones. Chem Eur J 20:3045–3049.

Cai S, Zhao X, Wang X, Liu Q, Li Z, Wang DZ (2012) Visible-Light-Promoted C-C Bond Cleavage: Photocatalytic Generation of Iminium Ions and Amino Radicals. Angew Chem Int Ed 51:8050–8053.

6.2.8 (De)Halogenations

6.2.8.1 Formation of a C-X Bond

Schmidt VA, Quinn RK, Brusoe AT, Alexanian EJ (2014) Site-Selective Aliphatic C-H Bromination Using *N*-Bromoamides and Visible Light. J Am Chem Soc 136:14389–14392 (Scheme 6.86).

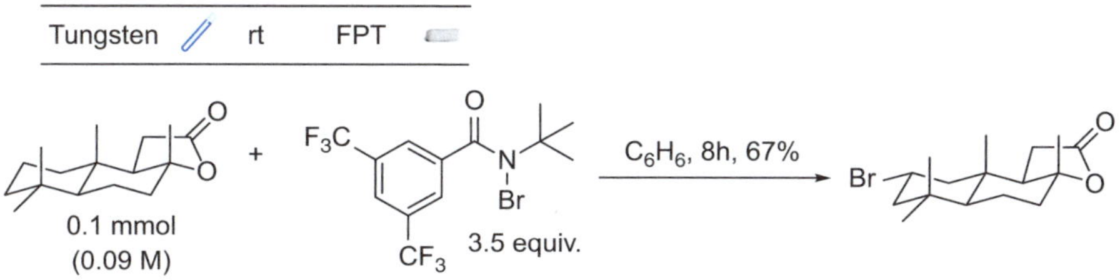

Scheme 6.86 Photochemical site-selective aliphatic C-H bromination reactions

Related: Kee CW, Chan KM, Wong MW, Tan C-H (2014) Selective Bromination of sp^3 C-H Bonds by Organophotoredox Catalysis. Asian J Org Chem 3:536–544.

Cantillo D, de Frutos O, Rincon JA, Mateos C, Kappe CO (2014) A Scalable Procedure for Light-Induced Benzylic Brominations in Continuous Flow. J Org Chem 79:223–229.

Zhao Y, Li Z, Yang C, Lin R, Xia W (2014) Visible-light photoredox catalysis enabled bromination of phenols and alkenes. Beilstein J Org Chem 10:622–627 (Scheme 6.87).

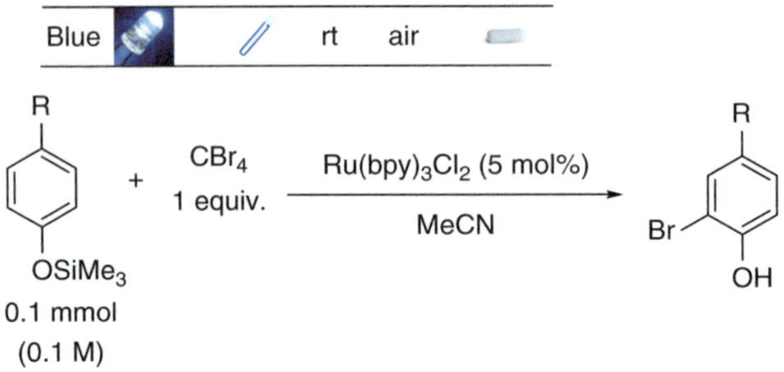

R	% yield
OMe	88
Me	69
Cl	58

Scheme 6.87 Photoredox catalyzed bromination of phenols

Guo L, Yang C, Zheng L, Xia W (2013) Visible light-mediated oxidative quenching reaction to electron-rich epoxides: highly regioselective synthesis of α-bromo (di)ketones and mechanism study. Org Biomol Chem 11: 5787–5792 (Scheme 6.88).

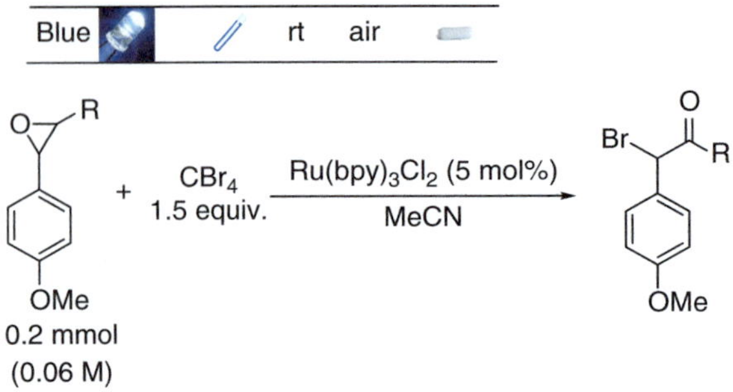

R	% yield	time (h)
Ph	92	5
COOEt	67	9
COPh	78	11

Scheme 6.88 Photocatalyzed synthesis of α-bromoketones

Rueda-Becerril M, Mahé O, Drouin M, Majewski MB, West JG, Wolf MO, Sammis GM, Paquin J-F (2014) Direct C-F Bond Formation Using Photoredox Catalysis. J Am Chem Soc 136:2637–2641 (Scheme 6.89).

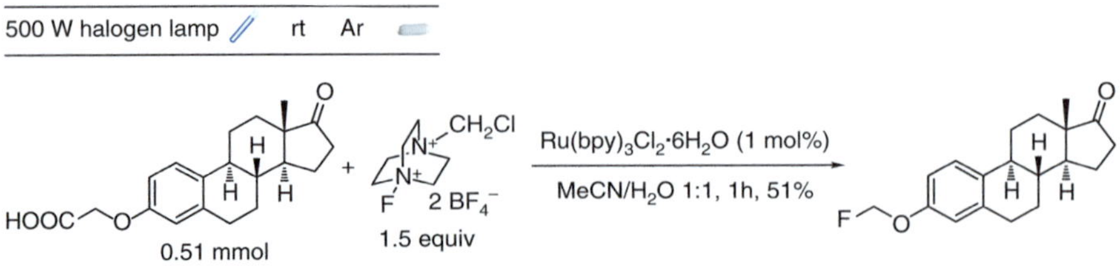

Scheme 6.89 Photoredox catalyzed decarboxylative fluorination

Xia J-B, Zhu C, Chen C (2014) Visible light-promoted metal-free sp^3-C-H fluorination. Chem Commun 50:11701–11704 (Scheme 6.90).

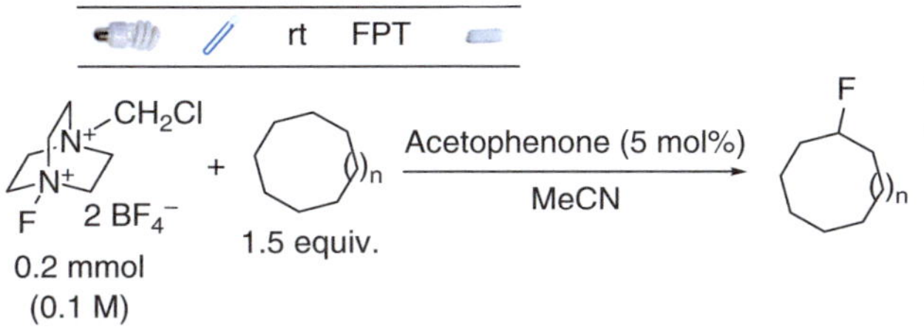

n	% yield	time (h)
1	82 (NMR)	3
3	73	15
5	75	15

Scheme 6.90 Visible light-promoted metal-free sp^3-C-H fluorination

Other: Yuan X, Cheng S, Sru Y, Xue W (2014) Photocatalytic Synthesis of Glycosyl Bromides. Synthesis 46:331–335.

6.2.8.2 Dehalogenations

Nguyen JD, D'Amato EM, Narayanam JMR, Stephenson CRJ (2012) Engaging unactivated alkyl, alkenyl and aryl iodides in visible-light-mediated free radical reactions. Nat Chem 4:854–859 (Scheme 6.91).

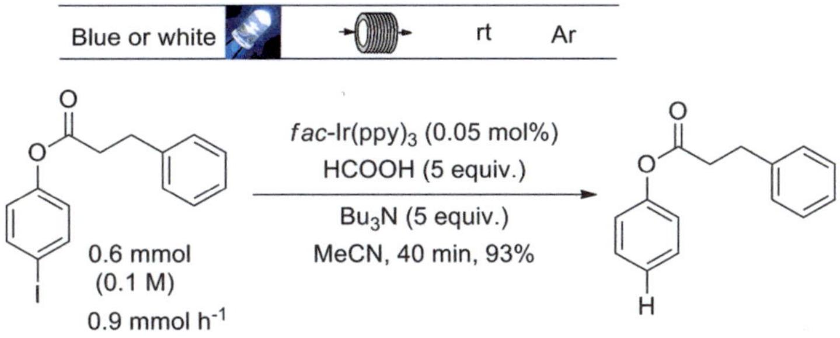

Scheme 6.91 Visible light-mediated dehalogenation of aryl iodides

Senaweera SM, Singh A, Weaver JD (2014) Photocatalytic Hydrodefluorination: Facile Access to Partially Fluorinated Aromatics. J Am Chem Soc 136:3002–3005 (Scheme 6.92).

Scheme 6.92 Site-selective photocatalytic hydrodefluorination processes

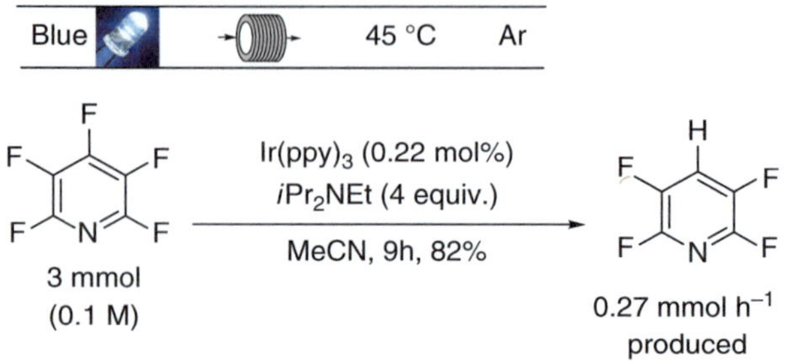

McTiernan CD, Pitre SP, Scaiano JC (2014) Photocatalytic Dehalogenation of Vicinal Dibromo Compounds Utilizing Sexithiophene and Visible-Light Irradiation. ACS Catal 4:4034–4039 (Scheme 6.93).

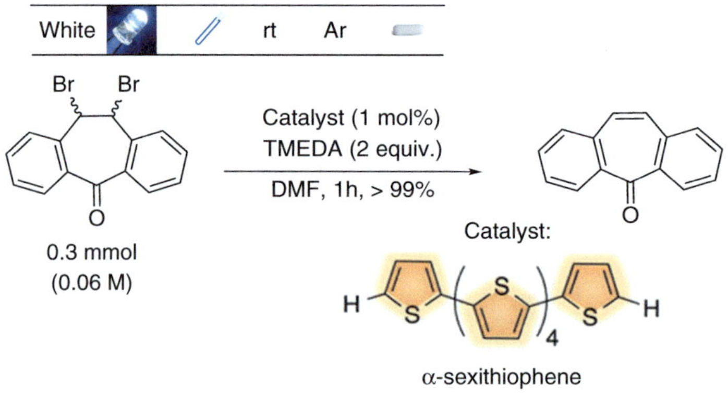

Scheme 6.93 Photocatalytic dehalogenation of vicinal dibromo derivatives

Other: McTiernan CD, Pitre SP, Ismaili H, Scaiano JC (2014) Heterogeneous Light-Mediated Reductive Dehalogenations and Cyclizations Utilizing Platinum Nanoparticles on Titania (PtNP@TiO_2). Adv Synth Catal 356:2819–2824.

6.2.9 *Oxidations and Reductions*

6.2.9.1 Oxidations

Lian M, Li Z, Cai Y, Meng Q, Gao Z (2012) Enantioselective Photooxygenation of β-Keto Esters by Chiral Phase-Transfer Catalysis using Molecular Oxygen. Chem Asian J 7:2019–2023(Scheme 6.94).

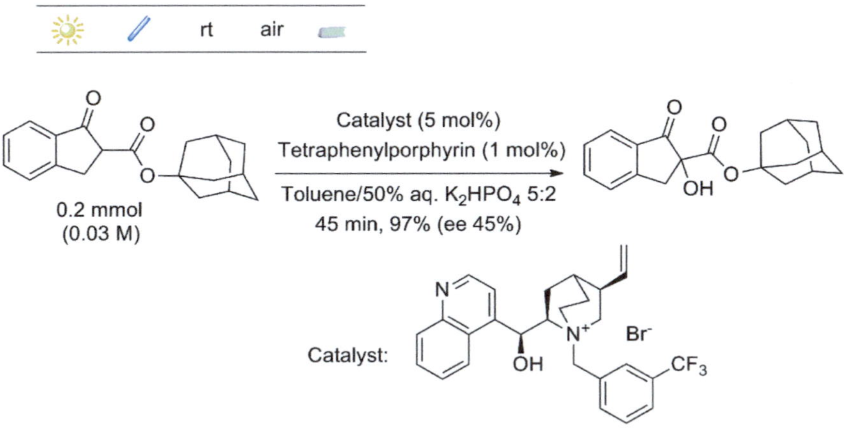

Scheme 6.94 Enantioselective photooxygenation of β-keto esters via chiral phase-transfer catalysis

Fujiya A, Kariya A, Nobuta T, Tada N, Miura T, Itoh A (2014) Photooxidative Cleavage of Aromatic Alkenes into Aldehydes Using Catalytic Iodine and Molecular Oxygen under Visible Light Irradiation. Synlett 25:884–888 (Scheme 6.95).

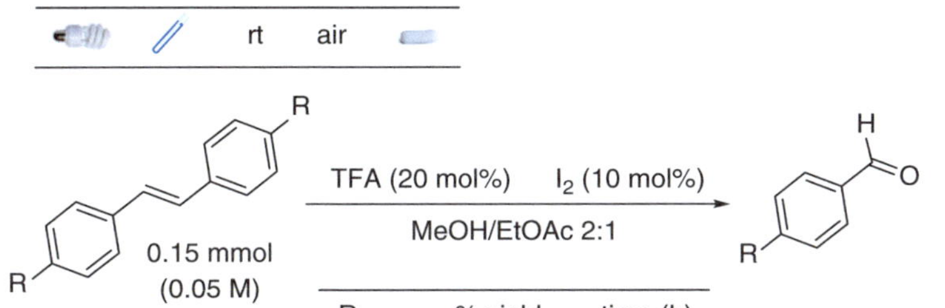

R	% yield	time (h)
Et	79	20
Me	79	20

Scheme 6.95 Iodine catalyzed photooxidative cleavage of aromatic alkenes

Related: Singh AK, Chawla R, Yadav LDS (2015) Eosin Y catalyzed visible light mediated aerobic photo-oxidative cleavage of the C-C double bond of styrenes. Tetrahedron Lett 56:653–656.

Kojima M, Oisaki K, Kanai M (2014) Chemoselective aerobic photo-oxidation of 9*H*-fluorenes for the synthesis of 9-fluorenones. Tetrahedron Lett 55:4736–4738 (Scheme 6.96).

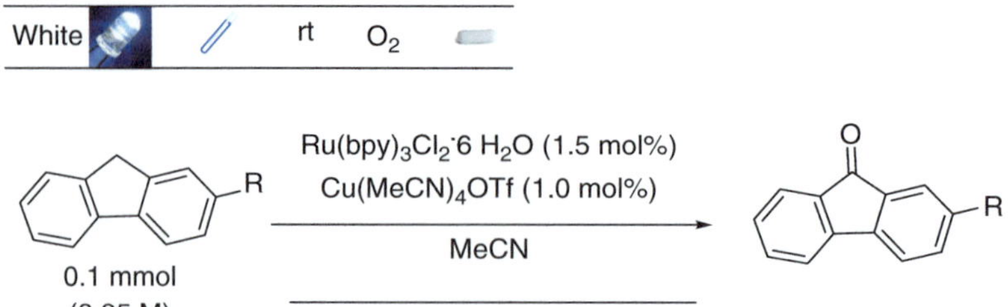

R	% yield	time (h)
H	86	48
CH_3CO	85	48
I	95	96

Scheme 6.96 Synthesis of 9-fluorenones via aerobic photo-oxidation of 9H-fluorenes

<u>*Other*</u>: Ye L, Li Z (2014) $ZnIn_2S_4$: A Photocatalyst for the Selective Aerobic Oxidation of Amines to Imines under Visible Light. ChemCatChem 6:2540–2543.

Huang L, Zhao J, Guo S, Zhang C, Ma J (2013) Bodipy Derivatives as Organic Triplet Photosensitizers for Aerobic Photoorganocatalytic Oxidative Coupling of Amines and Photooxidation of Dihydroxylnaphthalenes. J Org Chem 78:5627–5637.

Yuan B, Chong R, Zhang B, Li J, Liu Y, Li C (2014) Photocatalytic aerobic oxidation of amines to imines on $BiVO_4$ under visible light irradiation. Chem Commun 50:15593–15596.

Xiao Q, Liu Z, Bo A, Zavahir S, Sarina S, Bottle S, Riches JD, Zhu H (2015) Catalytic Transformation of Aliphatic Alcohols to Corresponding Esters in O_2 under Neutral Conditions Using Visible-Light Irradiation. J Am Chem Soc 137:1956–1966.

6.2.9.2 Reductions

Ghosh T, Slanina T, König B (2015) Visible light photocatalytic reduction of aldehydes by Rh(III)-H: a detailed mechanistic study. Chem Sci 6:2027–2034 (Scheme 6.97).

Scheme 6.97 Visible light photocatalytic reduction of aldehydes

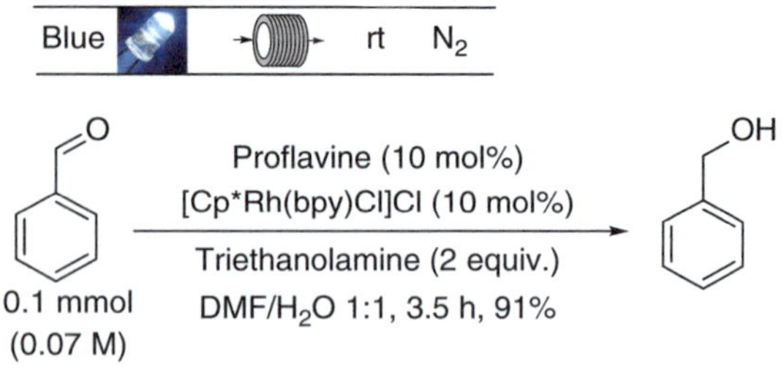

Rackl D, Kais V, Kreitmeier P, Reiser O (2014) Visible light photoredox-catalyzed deoxygenation of alcohols. Beilstein J Org Chem 10:2157–2165 (Scheme 6.98).

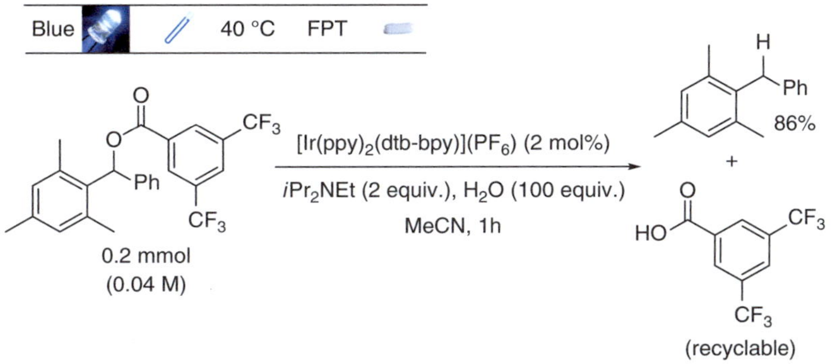

Scheme 6.98 Photoredox-catalyzed deoxygenations

<u>Related</u>: Chenneberg L, Baralle A, Daniel M, Fensterbank L, Goddard J-P, Ollivier C (2014) Visible Light Photocatalytic Reduction of *O*-Thiocarbamates: Development of a Tin-Free Barton-McCombie Deoxygenation Reaction. Adv Synth Catal 356:2756–2762.

<u>Other</u>: Zand Z, Kazemi F, Hosseini S (2014) Development of chemoselective photoreduction of nitro compounds under solar light and blue LED irradiation. Tetrahedron Lett 55:338–341.

Eskandari P, Kazemi F, Zand Z (2014) Photocatalytic reduction of aromatic nitro compounds using CdS nanostructure under blue LED irradiation. J Photochem Photobiol A: Chem 274:7–12.

6.2.10 Synthesis of Bioactive Compounds

Xuan J, Xia X-D, Zeng T-T, Feng Z-J, Chen J-R, Lu L-Q, Xiao W-J (2014) Visible-Light-Induced Formal [3 + 2] Cycloaddition for Pyrrole Synthesis under Metal-Free Conditions. Angew Chem Int Ed 53:5653–5656 (Scheme 6.99).

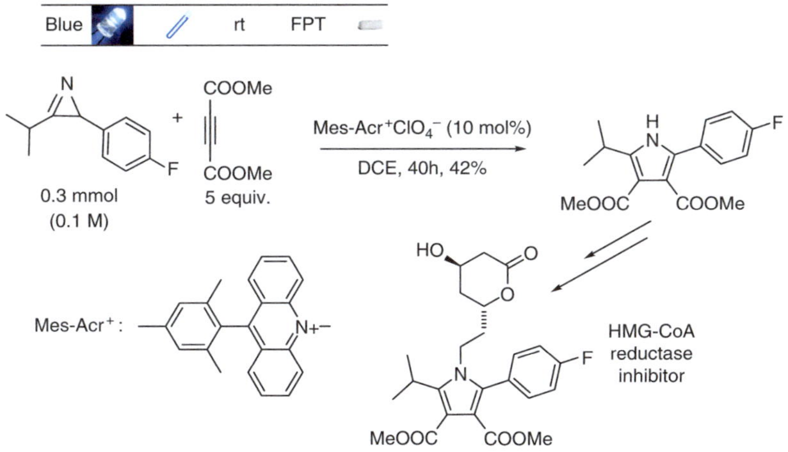

Scheme 6.99 Visible light-induced formal [3+2] cycloadditions for pyrroles synthesis

Yoshioka S, Nagatomo M, Inoue M (2015) Application of Two Direct C(sp[3])-H Functionalizations for Total Synthesis of (+)-Lactacystin. Org Lett 17:90–93 (Scheme 6.100).

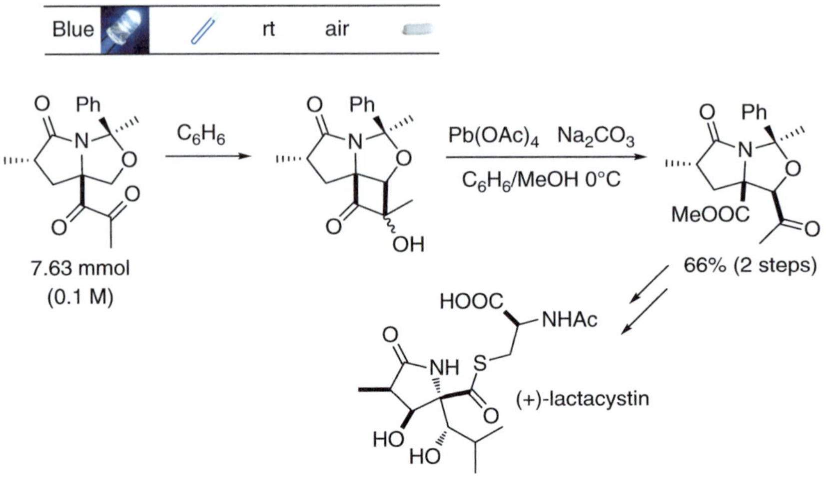

Scheme 6.100 Total synthesis of (+)-lactacystin

Conner ES, Crocker KE, Fernando RG, Fronczek FR, Stanley GG, Ragains JR (2013) Visible-Light-Promoted Selenofunctionalization of Alkenes. Org Lett 15:5558–5561 (Scheme 6.101).

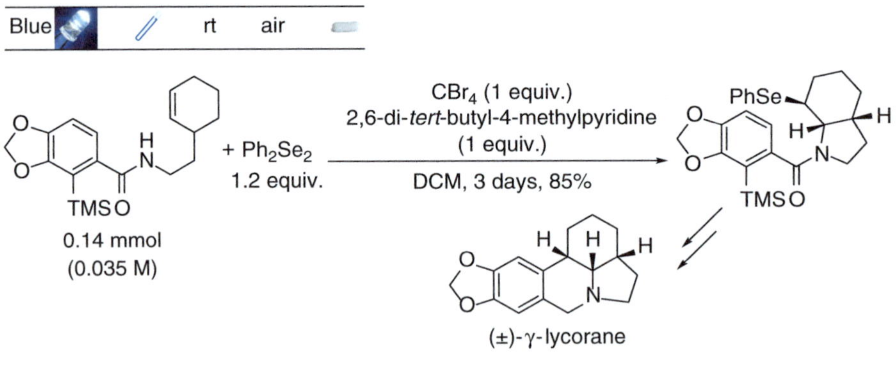

Scheme 6.101 Visible light-promoted selenofunctionalization of alkenes in the synthesis of (±)-γ-lycorane

Zhang J, Chen J, Zhang X, Lei X (2014) Total Syntheses of Menisporphine and Daurioxoisoporphine C Enabled by Photoredox-Catalyzed Direct C-H Arylation of Isoquinoline with Aryldiazonium Salt. J Org Chem 79:10682–10688 (Scheme 6.102).

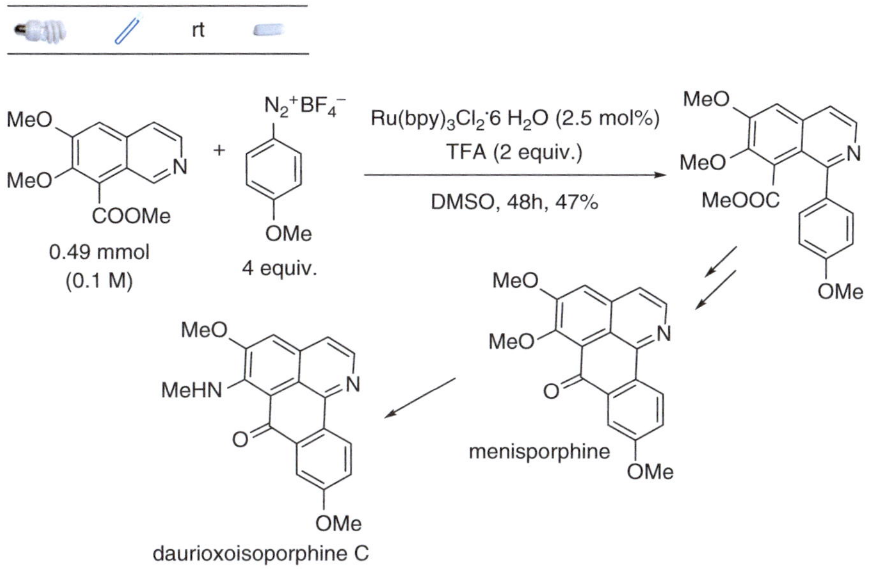

Scheme 6.102 Total syntheses of menisporphine and daurioxoisoporphine C

Related: Xue D, Jia Z-H, Zhao C-J, Zhang Y-Y, Wang C, Xiao J (2014) Direct Arylation of *N*-Heteroarenes with Aryldiazonium Salts by Photoredox Catalysis in Water. Chem Eur J 20:2960–2965.

Maity P, Kundu D, Ranu BC (2015) Visible-Light-Photocatalyzed Metal-Free C-H Heteroarylation of Heteroarenes at Room Temperature: A Sustainable Synthesis of Biheteroaryls. Eur J Org Chem 1727–1734.

Tobisu M, Furukawa T, Chatani N (2013) Visible Light-mediated Direct Arylation of Arenes and Heteroarenes Using Diaryliodonium Salts in the Presence and Absence of a Photocatalyst. Chem Lett 42:1203–1205.

Douglas JJ, Cole KP, Stephenson CRJ (2014) Photoredox Catalysis in a Complex Pharmaceutical Setting: Toward the Preparation of JAK2 Inhibitor LY2784544. J Org Chem 79:11631–11643 (Scheme 6.103).

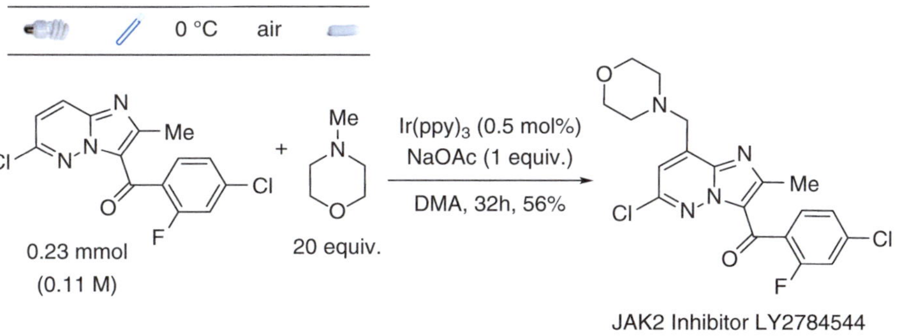

Scheme 6.103 Preparation of JAK2 inhibitor LY2784544

Chu L, Ohta C, Zuo Z, MacMillan DWC (2014) Carboxylic Acids as A Traceless Activation Group for Conjugate Additions: A Three-Step Synthesis of (±)-Pregabalin. J Am Chem Soc 136:10886–10889 (Scheme 6.104).

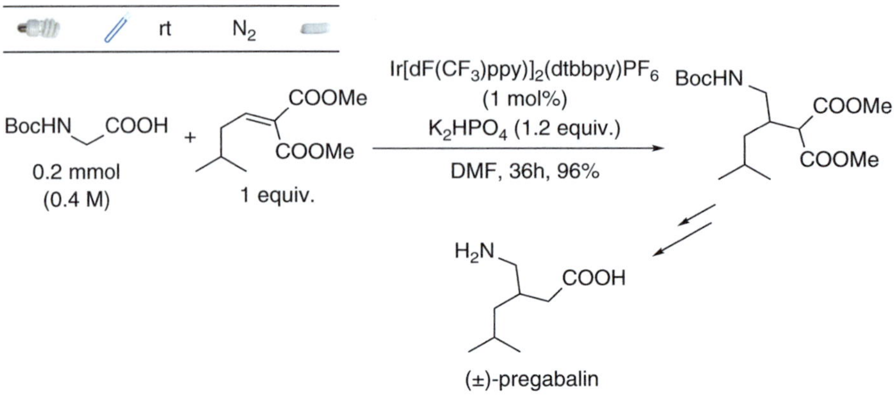

Scheme 6.104 Three-step synthesis of (±)-pregabalin

Xuan J, Zeng T-T, Feng Z-J, Deng Q-H, Chen J-R, Lu L-Q, Xiao W-J, Alper H (2015) Redox-Neutral α-Allylation of Amines by Combining Palladium Catalysis and Visible-Light Photoredox Catalysis. Angew Chem Int Ed 54:1625–1628 (Scheme 6.105).

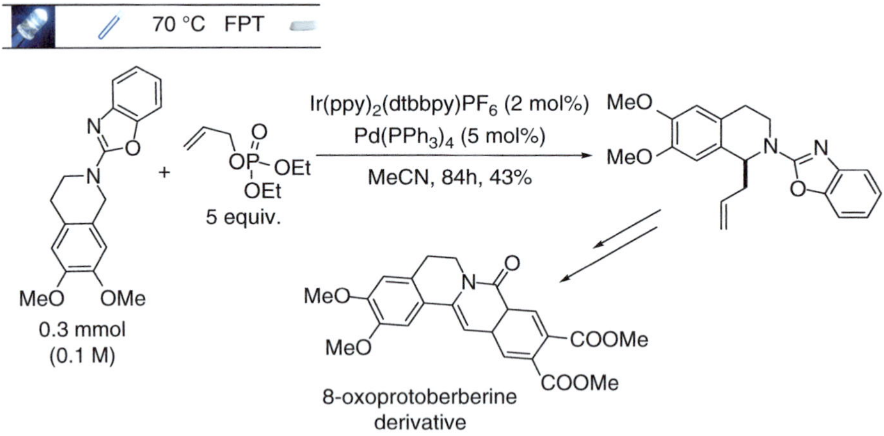

Scheme 6.105 Redox-neutral α-allylation of amines in the synthesis of 8-oxoprotoberberines

Barham JP, John MP, Murphy JA (2014) One-pot functionalisation of *N*-substituted tetrahydroisoquinolines by photooxidation and tunable organometallic trapping of iminium intermediates. Beilstein J Org Chem 10:2981–2988 (Scheme 6.106).

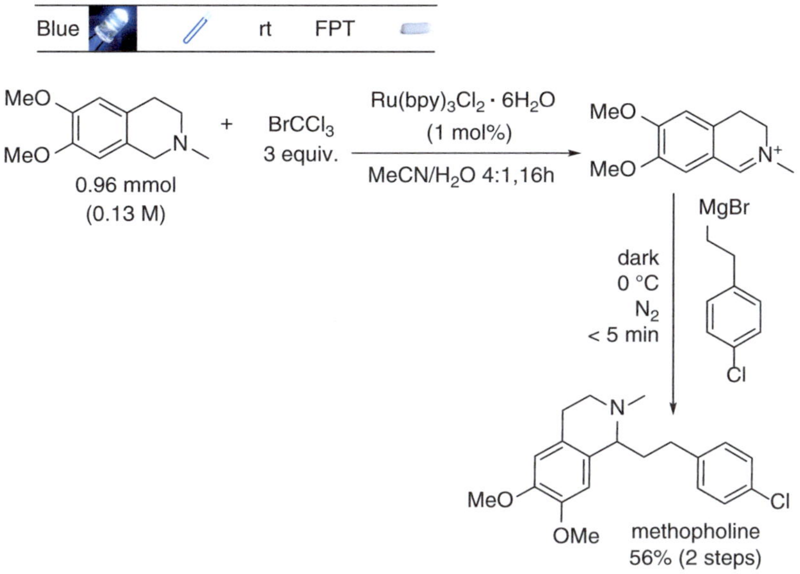

Scheme 6.106 Total synthesis of methopholine

Beatty JW, Stephenson CRJ (2014) Synthesis of (-)-Pseudotabersonine, (-)-Pseudovincadifformine, and (+)-Coronaridine Enabled by Photoredox Catalysis in Flow. J Am Chem Soc 136:10270–10273 (Scheme 6.107).

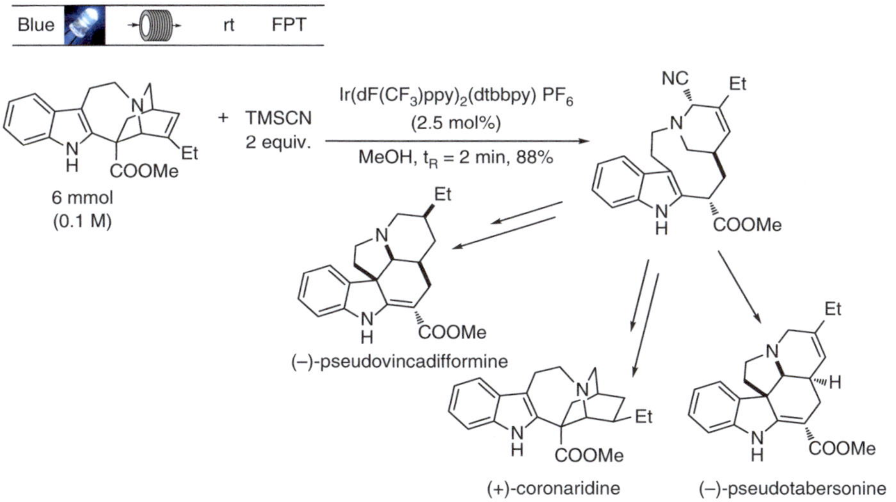

Scheme 6.107 Continuous flow synthesis of (−)-pseudotabersonine, (−)-pseudovincadifformine and (+)-coronaridine

6.3 Conclusions

As apparent from the number of references recently appeared in the literature, visible/solar light-promoted synthetic procedures have been main actors of a real Gold Rush to develop credible alternative to standard thermal protocols. The turning-point that allowed the huge development described in this Chapter has been realizing that some colored derivatives, which have long been known for their photophysical properties and for applications to artificial photosynthesis (e.g. transition-metal polypyridyl complexes, organic dyes, etc.) can be conveniently used also for synthetic purposes. The synthetic targets that can be currently accessed through this approach cover a very wide range of structures, including the construction of different-size rings, the formation of many classes of chemical bonds (e.g. C-C, C-N, C-Halogen, etc.), as well as oxidation and reduction processes. Different classes of substrates (both aliphatic and aromatic) can be subjected to this methodology. Most of the reported procedures exploit the typical reactivity of radical and radical ion intermediates, in turn generated from the starting substrates by having recourse to a convenient photocatalyst/photosensitizer. Indeed, in some instances, this approach enables for the use of unactivated precursors, allowing for the direct activation of X-H (e.g. Ar-H or C-H) bonds, thus skipping the need for introducing suitable reacting moieties. At this regard, the recovery/reuse of the employed photocatalyst, in order to minimize the economic cost of the process is still a challenge, and moving from homogeneous to heterogeneous catalysis (at present, only sparsely reported) is desirable. The devised protocols are mostly characterized by mild conditions, in strict agreement with the general principles of Green Chemistry, though the use of auxiliaries (such as sacrificial electron donors/acceptors) represents a significant contribution in terms of waste generation and care to this point must be paid.

The employment of a soft light source enables the use of easily available and cheap glassware, with no need of dedicated (and more expensive) quartz equipment, required when UV light is used. In terms of reaction conditions, a control of the atmosphere is often required. Indeed, bubbling with an inert (N_2, Ar, etc) or a reactive (e.g. O_2) gas does not represent a major concern. Some of the reported examples requires the use of Freeze-Pump-Thaw (FPT) cycles, that makes the resulting protocols not accessible to all organic chemistry laboratories and thus less appealing under the synthetic profile. Furthermore, sparse examples dealing with the use of (continuous) flow conditions recently appeared in the literature. This technological advancement can help in removing some of the most known problems associated with photochemical processes. Thus, by having recourse to flow conditions, it is possible to scale the synthesis of the desired targets up to the multigram scale and to obtain a further improvement (e.g. in terms of reaction time) with respect to batch conditions.

All these aspects have contributed to raise the interest of an increasing number of laboratories towards visible/solar light-promoted processes and, more in general, to photochemistry. Notably, the last trends involve the use of these processes in the (multi-step) synthesis of bioactive derivatives (see Sect. 6.2.10).

References

1. Fagnoni M, Albini A (2008) The greenest reagent in organic synthesis: light. In: Tundo P, Esposito V (eds) Green chemical reactions. NATO science for peace and security series. Springer, Dordrect/London, pp 173–189
2. Hoffmann N (2012) Photochemical reactions of aromatic compounds and the concept of the photon as a traceless reagent. Photochem Photobiol Sci 11:1613–1643
3. Albini A, Fagnoni M (2013) Photochemically-generated intermediates in synthesis. Wiley, Hoboken
4. Ravelli D, Protti S, Albini A (2015) Energy and molecules from photochemical/photocatalytic reactions. An overview. Molecules 20:1527–1542
5. Key World Energy Statistics (2014) https://www.iea.org/publications/freepublications/publication/key-world-energy-statistics-2014.html, pp 6, 24, 28
6. Schultz DM, Yoon TP (2014) Solar synthesis: prospects in visible light photocatalysis. Science 343:1239176-1-1239176–8
7. Kalogirou SA (2004) See for reviews: solar thermal collectors and applications. Prog Energy Combust 30:231–295
8. Tian Y, Zhao CY (2013) A review of solar collectors and thermal energy storage in solar thermal applications. Appl Energy 104:538–553
9. Lennartson A, Roffey A, Moth-Poulsen K (2015) Designing photoswitches for molecular solar thermal energy storage. Tetrahedron Lett 56:1457–1465
10. Parida B, Iniyan S, Goic R (2011) A review of solar photovoltaic technologies. Renew Sustain Energy Rev 15:1625–1636
11. Razykov TM, Ferekides CS, Morel D, Stefanakos E, Ullal HS, Upadhyay HM (2011) Solar photovoltaic electricity: current status and future prospects. Sol Energy 85:1580–1608
12. El Chaar L, Lamont LA, El Zein N (2011) Review of photovoltaic technologies. Renew Sustain Energy Rev 15:2165–2175
13. Bard AJ, Fox MA (1995) Artificial photosynthesis: solar splitting of water to hydrogen and oxygen. Acc Chem Res 28:141–145
14. Walter MG, Warren EL, McKone JR, Boettcher SW, Mi Q, Santori EA, Lewis NS (2010) Solar water splitting cells. Chem Rev 110:6446–6473
15. Kudo A, Miseki Y (2009) Heterogeneous photocatalyst materials for water splitting. Chem Soc Rev 38:253–278
16. Bradbury R (1953) The golden apples of the sun. Doubleday & Company, Inc, Garden City, New York
17. Ciamician G (1912) Photochemistry of the future. Science 36:385
18. Albini A, Dichiarante V (2009) The belle époque of photochemistry. Photochem Photobiol Sci 11:248–254
19. Fagnoni M, Albini A (2004) Green chemistry and photochemistry were born at the same time. Green Chem 6:1–6
20. Schenck GO, Ziegler K (1945) The synthesis of ascaridole. Naturwissenschaften 32:157
21. Esser P, Pohlmann B, Scharf H-D (1994) The photochemical synthesis of fine chemicals with sunlight. Angew Chem Int Ed 33:2009–2023
22. Oelgemöller M, Jung C, Mattay J (2007) Green Photochemistry: production of fine chemicals with sunlight. Pure Appl Chem 79:1939–1947
23. For a review of the advancements in the field of solar photochemistry until 2010 see: Protti S, Fagnoni M (2009) The sunny side of chemistry: green synthesis by solar light. Photochem Photobiol Sci 8:1499–1516
24. Mumtaz S, Sattler C, Oelgemöller M (2015) Solar photochemical manufacturing of fine chemicals: historical background, modern solar technologies, recent applications and future challenges. In: Letcher TM, Scott JL, Patterson DA (eds) Chemical processes for a sustainable future. The Royal Society of Chemistry, Cambridge

25. Spasiano D, Marotta R, Malato S, Fernandez-Ibañez P, Di Somma I (2015) Solar photocatalysis: materials, reactors, some commercial, and pre-industrialized applications. A comprehensive approach. Appl Catal B-Environ 170:90–123
26. Maidan R, Goren Z, Becker JY, Willner I (1984) Application of multielectron charge relays in chemical and photochemical debromination processes. The role of induced disproportionation of N, N′-dioctyl-4,4′-bipyridinium radical cation in two-phase systems. J Am Chem Soc 106:6217–6222
27. Cano-Yelo H, Deronzier A (1984) Photocatalysis of the Pschorr reaction by tris-(2,2′-bipyridyl)ruthenium(II) in the phenanthrene series. J Chem Soc Perkin Trans 2:1093–1098
28. Ravelli D, Protti S, Fagnoni M, Albini A (2013) Visible light photocatalysis. A green choice? Curr Org Chem 17:2366–2373
29. Schultz DM, Yoon TP (2014) Solar synthesis: prospects in visible light photocatalysis. Science 343:1239176
30. Knowles JP, Elliott LD, Booker-Milburn KI (2012) Flow photochemistry: old light through new windows. Beilstein J Org Chem 8:2025–2052
31. Oelgemöller M (2012) Highlights of photochemical reactions in microflow reactors. Chem Eng Technol 35:1144–1152

Chapter 7
Photochemical Reactions in Sunlit Surface Waters

Davide Vione

Abstract Photochemical processes are important pathways for the transformation of biologically refractory organic compounds, including harmful pollutants, in surface waters. They include the direct photolysis of sunlight-absorbing molecules, the transformation photosensitised by the triplet states of chromophoric dissolved organic matter (CDOM), and the reaction with photochemically generated radical transients. Differently from the direct photolysis, the other processes (often indicated as indirect photochemistry) can also induce the phototransformation of compounds that do not absorb sunlight. The excited triplet states of CDOM, ^{3}CDOM*, play a very important role in surface-water photoprocesses, both directly and as sources of singlet oxygen ($^1O_2$). The most important reactive radical species in surface waters are the hydroxyl radical $^\bullet$OH, the carbonate radical $CO_3^{-\bullet}$, and various peroxy radicals that can be produced upon degradation of dissolved organic matter (DOM), either chromophoric or not. Further radical species such as $^\bullet NO_2$, $Cl_2^{-\bullet}$ and $Br_2^{-\bullet}$ can be involved in the generation of harmful degradation intermediates such as aromatic nitro, chloro, and bromoderivatives.

7.1 Introduction

The persistence in surface water bodies of dissolved organic compounds, including both natural organic molecules and man-made xenobiotics and pollutants, strongly depends on their transformation kinetics due to abiotic and biological processes. Transformation by micro-organisms can be very important for readily biodegradable molecules, including most notably the nutrients. However, even in the presence of biodegradable compounds, biological processes may produce biorefractory intermediates that undergo further biodegradation with difficulty [1]. In these cases, phototransformation can play an important role for the processing of these

D. Vione (✉)
Dipartimento di Chimica, Università di Torino, Via Pietro Giuria 5, 10125 Torino, Italy
e-mail: davide.vione@unito.it; http://www.chimicadellambiente.unito.it; http://naturali.campusnet.unito.it/cgi.bin/docenti.pl/Show?_id=vione

G. Bergamini, S. Silvi (eds.), *Applied Photochemistry*,
Lecture Notes in Chemistry 92, DOI 10.1007/978-3-319-31671-0_7

dissolved molecules. It may also happen that biorefractory organic matter becomes bioavailable after some degree of abiotic processing, with the consequence that the combination of abiotic and biological degradation can lead to the complete mineralisation [2]. Furthermore, phototransformation is very important for the removal of a number of emerging pollutants, which hardly undergo biodegradation in either surface waters or wastewater treatment plants [3, 4].

The abiotic transformation processes include hydrolysis, oxidation mediated by dissolved oxidising species or by metal oxides, such as Fe(III) and Mn(III,IV) (hydr)oxides, and light-induced reactions [5]. Hydrolysis may produce bond cleavage, implying for instance the loss of a lateral functional chain. Hydrolytic reactions are often acid or base-catalysed, but at the ~neutral pH values typical of surface waters the effects of catalysis may not be very marked [6]. Among the oxidising species dissolved in surface waters, there are a number of reactive transients produced upon sunlight irradiation of photoactive compounds [7]. In the dark, Fe and Mn (hydr)oxides can be involved in charge-transfer processes with compounds that are able to form surface complexes (e.g. salicylate and oxalate). The oxidation of the organic molecule would often occur at the expense of the photoreductive dissolution of the oxide, with formation of water-soluble Fe^{2+} and Mn^{2+} [8, 9].

In some cases the removal of a compound from surface waters involves, instead of chemical transformation, the transfer from the dissolved phase to suspended solids or sediment, or the volatilisation from the surface water layer [10]. In all these cases the molecule is simply moved to a different phase, where it can be involved into transformation processes that can be very different from those in water (e.g. atmospheric reactions) [11]. In some cases the original molecule or its transformation intermediates can return back to the water body, as a consequence of dissolution from sediment or of wet and dry deposition from the atmosphere [12].

Many organic pollutants such as polycyclic aromatic hydrocarbons, some pesticides, pharmaceuticals and their transformation intermediates are refractory to biological degradation. In such cases the abiotic transformation processes represent major removal pathways from surface waters. Within the abiotic transformation reactions of xenobiotics, those induced by sunlight are receiving increasing attention because of their importance in the removal of the parent molecules and the possible production of harmful secondary pollutants [13, 14]. The present chapter will be dedicated to the description of photochemical reactions in sunlit surface waters. They include the direct photolysis of the parent compound upon absorption of sunlight, the transformation sensitised by photoexcited CDOM, and in particular by its humic and fulvic components, and the reaction with photogenerated, reactive transients that often have the ability of oxidising the organic substrates [15]. It is the case of $^{\bullet}OH$, $CO_3^{-\bullet}$ and $^1O_2$, while other radical transients such as $^{\bullet}NO_2$, $Cl_2^{-\bullet}$ and $Br_2^{-\bullet}$ may produce secondary pollutants [16].

Direct photolysis, although very important in defining the lifetime of many photolabile compounds in surface waters, will seldom lead to complete depollution. It produces instead a number of transformation intermediates that in some cases may be even more harmful than the parent molecule [17, 18]. The complete removal of the xenobiotic species, e.g. by mineralisation, requires the reaction with oxidising transients such as $^{\bullet}OH$, or microbial processing. In many cases the mineralisation is much slower than the primary step of phototransformation of the parent molecule [19].

Finally, while important advances have been carried out recently in the modelling of the photochemical fate of pollutants in surface waters [20, 21], consideration of the further transformation and phototransformation pathways till complete mineralisation is lagging much behind.

7.2 Direct Photolysis Processes

The direct photolysis of a molecule is the consequence of photon absorption. The absorbed spectral photon flux density by a given compound at the wavelength λ, $p_a(\lambda)$, can be expressed in [einstein L^{-1} s^{-1} nm^{-1}], where 1 einstein = 1 mol of photons [22]. The quantity $p_a(\lambda)$ is related to the absorbance $A_\lambda = \varepsilon_\lambda b\ c$, where ε_λ is the molar absorption coefficient (decadic), b the optical path length of the solution, and c the concentration of the compound. The quantities b and c are usually expressed in cm and in mol L^{-1}, respectively, in which case the unit of ε_λ is [L mol^{-1} cm^{-1}], and A is dimensionless. Be $i_o(\lambda)$ the spectral photon flux density of incident radiation in solution, expressed in [einstein L^{-1} s^{-1} nm^{-1}]. It is $p_a(\lambda) = i_o(\lambda)\left(1 - 10^{-\varepsilon_\lambda \cdot b \cdot c}\right)$. In the environment the incident radiation is not monochromatic, and it is interesting to know the absorbed photon flux P_a in a given wavelength interval, say $\lambda_1 \le \lambda \le \lambda_2$. The following expression holds for P_a [einstein L^{-1} s^{-1}]:

$$P_a = \int_{\lambda_1}^{\lambda_2} p_a(\lambda) d\lambda = \int_{\lambda_1}^{\lambda_2} i_o(\lambda) \cdot \left(1 - 10^{-\varepsilon_\lambda \cdot b \cdot c}\right) d\lambda \tag{7.1}$$

Furthermore, in a surface water body the intensity of the incident radiation is not constant over the whole water column. Absorption and scattering phenomena will reduce the radiation intensity as the depth of the water column increases. As a consequence, the spectral photon flux density of incident radiation at the depth x will be a function of both depth and wavelength, as $i_o(\lambda,x)$. Under the simplified assumption that the scattering of radiation is negligible compared to absorption, the Lambert-Beer relationship applies as $i_o(\lambda,\ x) = i_o(\lambda) 10^{-\alpha(\lambda) \cdot x}$, where $i_o(\lambda)$ is the spectral photon flux density of incident radiation on top of the water column and $\alpha(\lambda)$ is the attenuation coefficient [21]. The coefficient $\alpha(\lambda)$ depends on the absorption spectrum of water and is largely accounted for by CDOM absorption [23]. The measure unit of $\alpha(\lambda)$ is cm^{-1} if x is in cm. In the most general case, for $\lambda_1 \le \lambda \le \lambda_2$ and $0 \le x \le d$, P_a is given by Eq. 7.2, where 2.3 = ln 10:

$$P_a = 2.3 \cdot \int_{\lambda_1}^{\lambda_2} \int_0^d i_o(\lambda) \cdot 10^{-\alpha(\lambda) \cdot x} \cdot \varepsilon_\lambda \cdot c \, dx \, d\lambda \tag{7.2}$$

Figure 7.1 reports the spectral fluence rate density of sunlight [24] at different values of the water column depth, for a given absorption spectrum $\alpha(\lambda)$ of water. It is apparent the rapid decrease of the fluence rate density within the water column, in particular at the shorter wavelengths.

The incident spectral photon flux density at a given depth d is a key factor in defining the rate R of direct photolysis, which is directly proportional to absorbed photon flux P_a. However, not all of the absorbed photons are able to induce photolysis to the same extent. This fact is reflected into the photolysis quantum yield Φ, which measures the probability that the absorption of a photon actually induces photolysis. It is $0 \leq \Phi \leq 1$ and, in the most general case, the quantum yield of photolysis is not constant with wavelength [25].

Figure 7.2 shows the case of nitrite, which has an absorption band at around 280 nm with photolysis quantum yield $\Phi = 0.068$, and one at 365 nm with $\Phi = 0.025$ [26]. Nitrite yields $^{\bullet}OH + {}^{\bullet}NO$ upon absorption of UV radiation [27]:

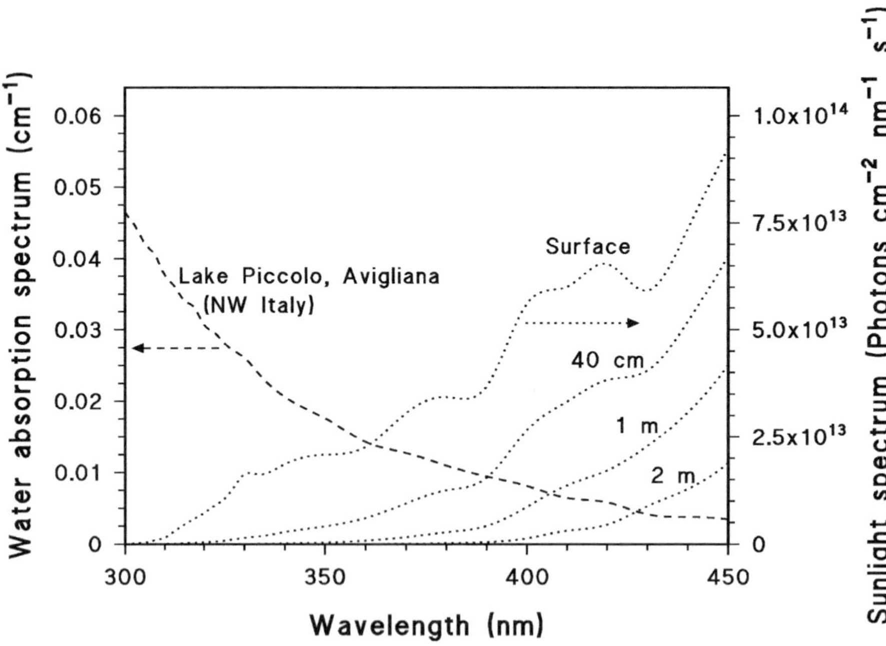

Fig. 7.1 Absorption spectrum $\alpha(\lambda)$ of water from Lake Piccolo in Avigliana (NW Italy), and spectral fluence rate density of sunlight at different depths of the water column, calculated according to the Lambert-Beer approximation. Note that the spectral fluence rate density is the spectral photon flux density times the Avogadro's constant

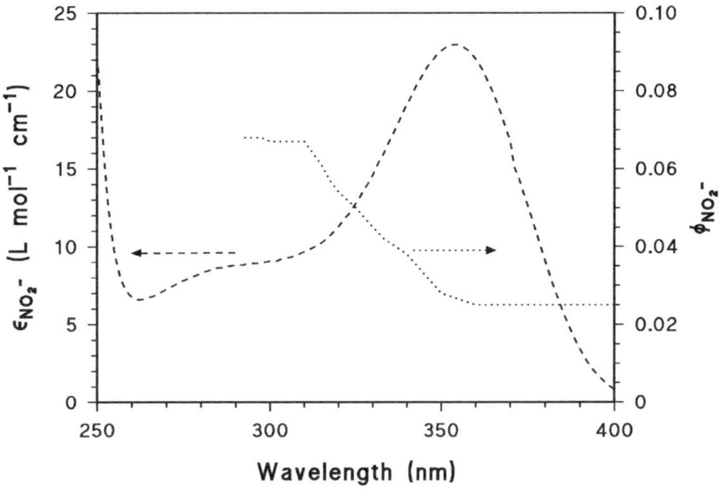

Fig. 7.2 Absorption spectrum and photolysis quantum yield Φ of nitrite in the wavelength interval 250–400 nm. Note the two absorption bands of nitrite at 280 and 365 nm, with different Φ

$$NO_2^- + h\nu + H^+ \rightarrow {}^\bullet OH + {}^\bullet NO \tag{7.3}$$

For irradiation at the wavelength λ the rate of photolysis is $r(\lambda) = \Phi(\lambda)\ p_a(\lambda)$, and the rate R of photolysis for $\lambda_1 \leq \lambda \leq \lambda_2$ and $0 \leq x \leq d$ is [25]:

$$R = 2.3 \cdot \int_{\lambda_1}^{\lambda_2} \int_0^d \Phi(\lambda) \cdot i_o(\lambda) \cdot 10^{-\alpha(\lambda)\cdot x} \cdot \varepsilon_\lambda \cdot c\ dx\ d\lambda \tag{7.4}$$

Direct photolysis has received much attention concerning the degradation of xenobiotic compounds of high environmental concern, such as polycyclic aromatic hydrocarbons (PAHs), haloaromatics (including some pesticides and their metabolites) and, more recently, pharmaceuticals [13, 14].

About PAHs, in the case of naphthalene [28] the direct photolysis proceeds through photoionisation/deprotonation with the net loss of a H atom, followed by either oxidation to naphthoquinone, or ring-opening with formation of monoaromatic carboxylic acids and aldehydes (Fig. 7.3).

Quite interestingly the quinone derivatives are more photochemically active than the parent PAHs, and could undergo more extensive photoprocessing. For instance, the direct photolysis of anthracene [29] in aerated aqueous solution yields 9,10-anthraquinone that is able to absorb a larger fraction of sunlight compared to anthracene [30], and undergoes photo-oxidation as a consequence (Fig. 7.4).

The photoreactivity of anthracene could be strongly substrate-dependent. Indeed, the direct photolysis of anthracene on silica [31] proceeds via dimerisation in addition to the oxidation to anthraquinone (Fig. 7.5). Furthermore, the

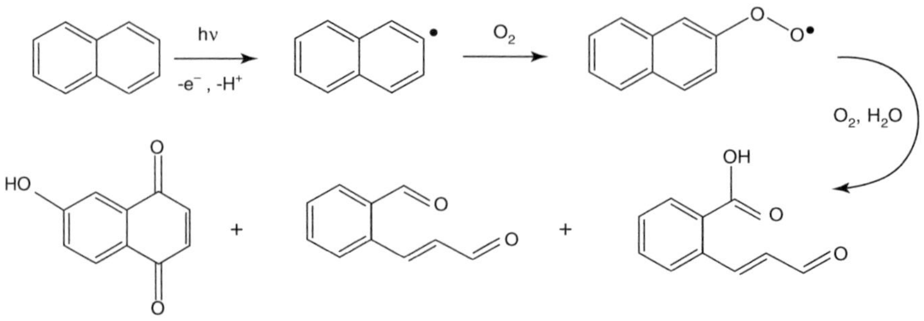

Fig. 7.3 Pathways of the direct photolysis of naphthalene in aqueous solution

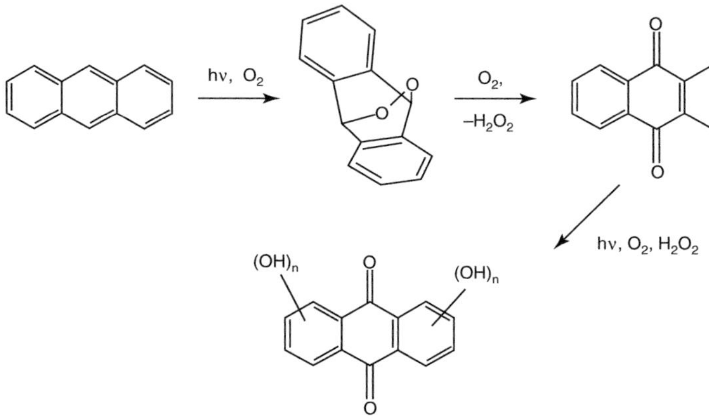

Fig. 7.4 Pathways of the direct photolysis of anthracene in aqueous solution

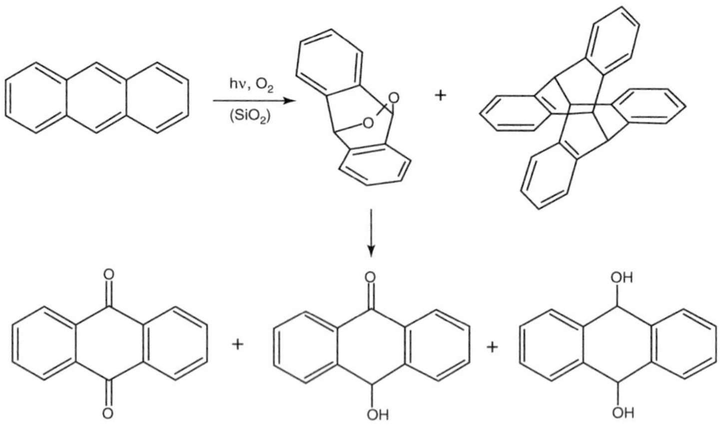

Fig. 7.5 Pathways of the direct photolysis of anthracene on the surface of silica

hydroxylated derivatives are likely to arise on reduction of irradiated anthraquinone [32]. The surface of silica might significantly enhance photodimerisation processes compared to the homogeneous aqueous solution [33], and SiO_2 might be an interesting model for inorganic colloids in surface waters [34].

Chlorophenols are a class of chlorinated aromatic compounds of considerable environmental concern because they can be released as by-products of various industrial activities [35]. They can also be formed as secondary pollutants upon environmental transformation of various pesticides, mainly the chlorophenoxy-acetic and propionic acids [36, 37], and of the antimicrobial agent triclosan [38, 39]. An important issue in the direct photolysis pathways of chlorophenols is the difference between the *ortho*- and the *para*-substituted ones. In both cases the absorption of radiation leads to the first excited singlet state, which is then transformed into the first excited triplet state by inter-system crossing (ISC) [13]. The main difference is that, in the case of *ortho*-chlorophenols [40], the first excited singlet state is sufficiently long-lived to allow chemical reactivity in alternative to ISC, resulting into ring contraction and loss of HCl to form a cyclopentadienyl carboxyaldehyde (Fig. 7.6). The ring-contraction process would be particularly significant for the phenolate anions [41]. In contrast, the first excited triplet state would mainly react by dechlorination [42], either reductive (with the

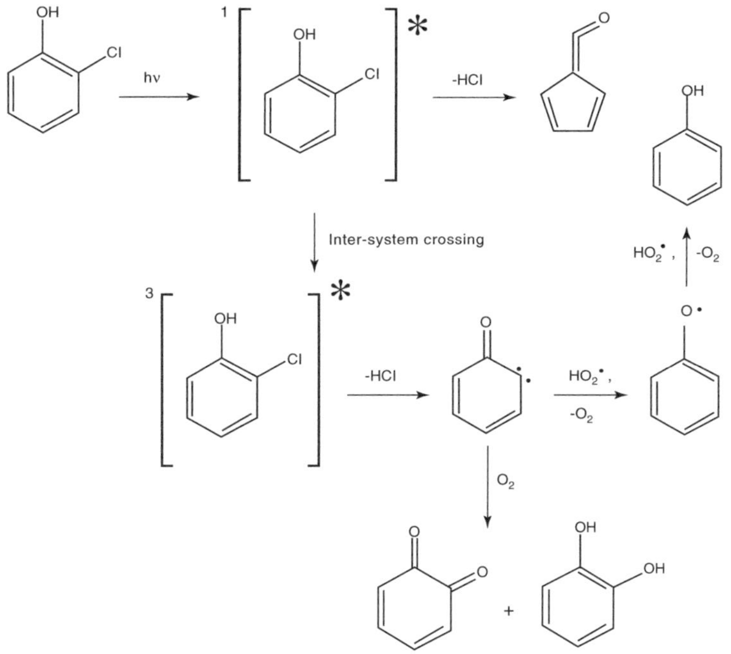

Fig. 7.6 Processes involved in the direct photolysis of 2-chlorophenol in aqueous solution

participation of $HO_2^{\bullet}$) to give the corresponding phenol, or involving oxygen with the final formation of dihydroxyphenols and quinones (Figs. 7.6 and 7.7).

Many xenobiotic compounds of environmental concern undergo different photolysis processes in their protonated or deprotonated form, such as 2-methyl-4-chlorophenoxyacetic acid (MCPA). The photolysis pathways are, therefore, strongly dependent on pH. In the case of MCPA the protonated form undergoes molecular rearrangement, while the deprotonated one follows a dechlorination-hydroxylation pathway (Fig. 7.8). Interestingly, due to significant reactions between MCPA excited states and dissolved organic compounds, the direct photolysis quantum yield of MCPA is also linked to the water content of DOM [43].

Dichlorprop, 2-(2,4-dichlorophenoxy)propionic acid, an herbicide that is extensively used in flooded rice farming, is an interesting compound because it acts as a precursor of various chlorinated phenols in the environment, such as 4-chlorocatechol [44] by direct photolysis (Fig. 7.9) as well as 2,4-dichlorophenol and 4-chloro-2-methylphenol upon hydrolysis in aqueous solution [45]. Another important xenobiotic that can undergo direct photolysis in surface waters is the antimicrobial agent triclosan (Fig. 7.10), which is particularly interesting because its photocyclisation produces a dichlorodibenzodioxin [46]. Interestingly, recent evidence suggests that dioxin formation from triclosan can also take place upon reaction with ^{3}CDOM* [47].

The case of triclosan is a good example of a photolysis process that yields an intermediate that is more harmful than the parent compound. This finding is even

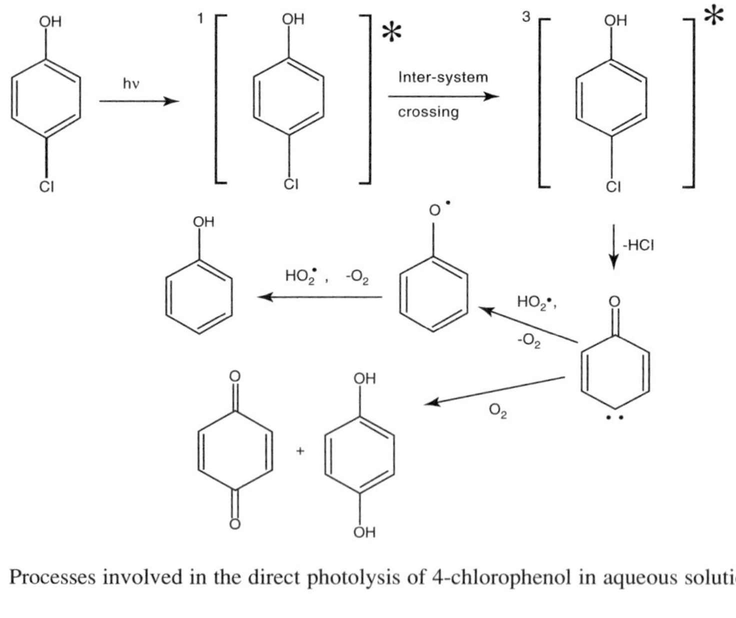

Fig. 7.7 Processes involved in the direct photolysis of 4-chlorophenol in aqueous solution

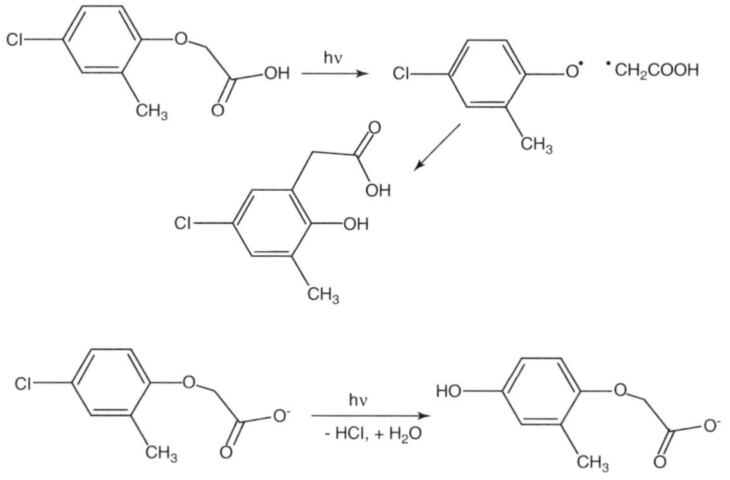

Fig. 7.8 Processes involved in the direct photolysis of MCPA in aqueous solution (both neutral and anionic forms)

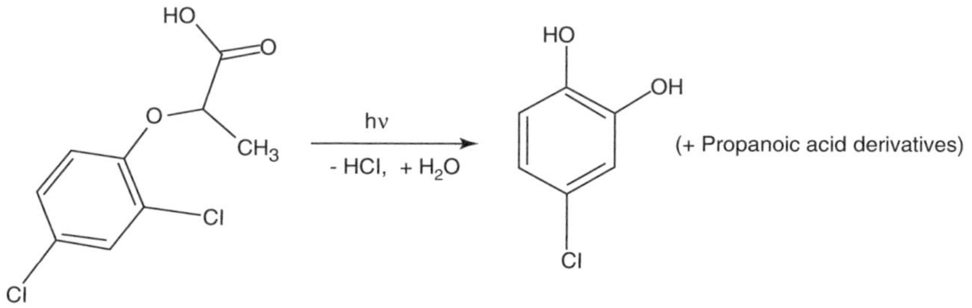

Fig. 7.9 Processes involved in the direct photolysis of dichlorprop in aqueous solution

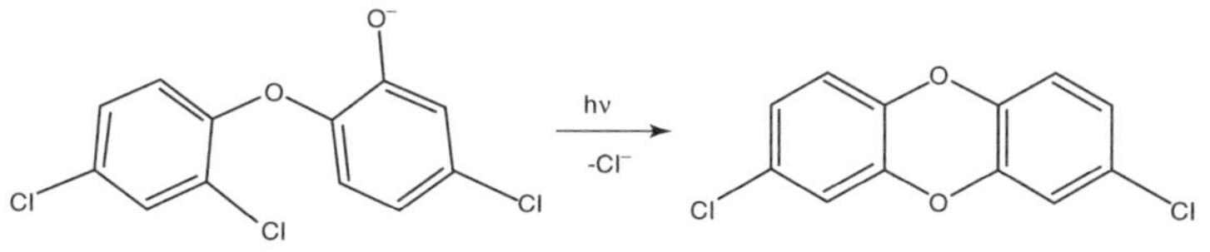

Fig. 7.10 Photocyclisation of triclosan in aqueous solution

more significant because direct photolysis is most likely the main sink of triclosan in surface waters [46, 47]. Indeed the photodegradation of a pollutant is not always beneficial to the environment, and the environmental and health impact of the transformation intermediates is to be considered as well [19, 48]. Another interesting example is the direct photolysis of the anti-epileptic drug carbamazepine [17, 18, 49, 50] that yields, among the other intermediates, the mutagenic acridine.

In summary, direct photolysis can be an important process in the degradation of sunlight-absorbing compounds in surface waters, depending on the irradiation intensity (which is maximum in shallow and clear water bodies), the extent of sunlight absorption by the molecule under consideration, and the photolysis quantum yield. The disappearance of the initial molecule is not necessarily the end of the story, however, because transformation intermediates with different properties, and sometimes even more harmful than the parent compound, can be formed due to direct photolysis [51, 52].

7.3 Transformation Photosensitised by Chromophoric Dissolved Organic Matter (CDOM)

Natural dissolved organic matter in surface waters consists of both autochthonous (aquagenic) material, mainly made up of aliphatic chains that do not absorb radiation (polysaccharides, complex carbohydrates, peptides, proteins), and of allochthonous compounds that derive from soil erosion such as fulvic and humic substances [1]. The latter contain a significant percentage of aromatic groups that absorb sunlight, at such an extent that DOM is the most important radiation absorber in surface waters [23]. A very interesting issue is that sunlight-absorbing DOM (CDOM) is made up of a significant amount of quinonoid substances and aromatic carbonyls, which are known photosensitisers [20]. Therefore, the excitation of CDOM by sunlight can cause the degradation of other dissolved molecules, which do not need to absorb sunlight themselves. The photosensitising ability of CDOM largely depends on the reactions that can be induced by its excited triplet states, ^{3}CDOM* [53].

As shown in Fig. 7.11, the absorption of radiation by a sensitiser causes the transition from the ground state (which for organic molecules is usually a singlet one, S_o) to a vibrationally excited state of an excited singlet state, S_n (where $n = 1, 2\ldots$). It usually follows vibrational deactivation to the ground vibrational state of S_1 (also in the cases in which, initially, $n > 1$), at which point various alternatives are possible. A very common one is the thermal loss of energy, e.g. by collision with the solvent, to reach back the ground state S_o. Some molecules lose energy by radiation, emitting fluorescence photons. Rigid systems, such as the condensed rings of PAHs, are more likely to undergo fluorescence emission than the flexible structures. An alternative is the ISC, in particular when the ground vibrational state of S_1 has the same or similar energy as an excited vibrational state of the first triplet state (T_1) [25].

Vibrational relaxation will bring the sensitiser from an excited to the ground vibrational state of T_1. The subsequent loss of energy to reach S_o can follow various pathways. A possibility is the thermal energy loss, by a combination of vibration and collision. Solid systems (and some liquid ones) may also emit phosphorescence radiation. An interesting alternative in the present context is the chemical reactivity.

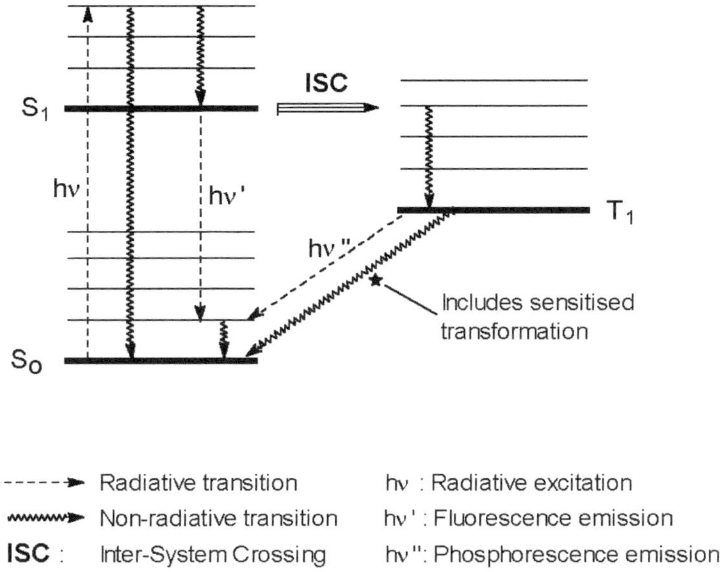

Fig. 7.11 Schematic of the processes that follow the absorption of radiation by a sensitiser

The excited triplet states are sufficiently long-lived to allow transfer of energy, electrons or atoms (usually H ones) to or from other molecules.

Be S the sensitiser and M a generic molecule. In the following reactions the superscripts ^{1} and ^{3} indicate singlet and triplet states, respectively, and * represents surplus energy [25].

$$^1\mathrm{S} + h\nu \rightarrow {}^1\mathrm{S}^* \xrightarrow{\mathrm{ISC}} {}^3\mathrm{S}^* \tag{7.5}$$

$$^3\mathrm{S}^* + \mathrm{M} \rightarrow {}^1\mathrm{S} + \mathrm{M}^* \tag{7.6}$$

$$\mathrm{M}^* \rightarrow \mathrm{Products} \tag{7.7}$$

$$^3\mathrm{S}^* + \mathrm{M} \rightarrow \mathrm{S}^{-\bullet} + \mathrm{M}^{+\bullet} \tag{7.8}$$

$$^3\mathrm{S}^* + \mathrm{M} \rightarrow (\mathrm{S} + \mathrm{H})^\bullet + (\mathrm{M} - \mathrm{H})^\bullet \tag{7.9}$$

The sensitiser would usually undergo reduction when reacting with M. The reduced sensitiser ($\mathrm{S}^{-\bullet}$, $(\mathrm{S} + \mathrm{H})^\bullet$) could be further transformed or be recycled back to S by dissolved oxygen (reactions 7.10 and 7.11):

$$\mathrm{S}^{-\bullet} + \mathrm{O}_2 \rightarrow \mathrm{S} + \mathrm{O}_2^{-\bullet} \tag{7.10}$$

$$(\mathrm{S} + \mathrm{H})^\bullet + \mathrm{O}_2 \rightarrow \mathrm{S} + \mathrm{HO}_2^\bullet \tag{7.11}$$

When reactions (7.10 and 7.11) take place, limited or no transformation of S is observed in the process. If S is maintained and M is transformed, the transformation of M photosensitised by S is a photocatalytic process where S is the photocatalyst [32, 54]. This does not always happen, however. Different reactions than (7.10 and 7.11) could for instance take place, leading to a net transformation of S [55]. As an alternative, photoexcited S could react with ground-state S [32], the latter behaving as M in reactions (7.5–7.9). In such a case, even if $S^{-\bullet}$ or $(S + H)^{\bullet}$ is recycled back to S by oxygen, the transformation of S would proceed via the oxidised $S^{+\bullet}$ or $(S–H)^{\bullet}$.

Reactions (7.5–7.9) suggest that photoexcited CDOM can induce the transformation of dissolved organic compounds. The processes involving definite sensitiser molecules are known to a considerably better extent than the reactions induced by CDOM as a whole. In particular the transformation reactions of phenols, sensitised by quinones or aromatic carbonyls have been subject to many studies. Figure 7.12 reports the processes taking place in the presence of phenol and of 3,4-dimethoxybenzaldehyde under irradiation. The photoexcited triplet state of 3,4-dimethoxybenzaldehyde is able to oxidise phenol to the phenoxyl radical, causing its degradation. In contrast, the reduced 3,4-dimethoxybenzaldehyde is re-oxidised by oxygen and no net transformation of the sensitiser is observed in this case. Differently from this case, however, 2- and 3-methoxybenzaldehyde undergo transformation in the presence of phenol under irradiation [54].

Another interesting group of sensitisers is represented by the quinonoid compounds, among which the anthraquinones are photochemically very active species. Due to its water solubility, anthraquinone-2-sulphonate (AQ2S) is the most studied molecule of its class. Radiation-excited AQ2S (which gives $^3AQ2S^*$ with fairly elevated ISC yield) is able to oxidise both ground-state AQ2S and phenol, producing hydroxylated AQ2S in the former case (however, AQ2S hydroxyderivatives are also formed via evolution of $^3AQ2S^*$) and phenol dimers

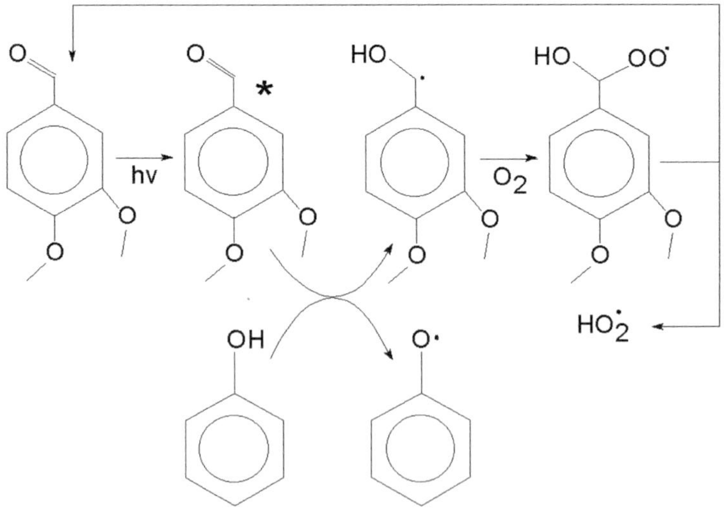

Fig. 7.12 Processes that occur in the presence of phenol and 3,4-dimethoxybenzaldehyde under irradiation (the asterisk denotes the excited triplet state of 3,4-dimethoxybenzaldehyde)

via the phenoxyl radical in the second. Reduced AQ2S is recycled back to AQ2S by oxygen, but the net transformation of AQ2S is assured by further reactions that involve $^3AQ2S^*$ and/or the oxidised AQ2S intermediates [32, 56].

By use of phenol derivatives as probe molecules it has been possible to evaluate as 8×10^{-15} M the steady-state concentration of the $^3CDOM^*$ in the surface layer of the Greifensee lake, Switzerland [57]. 2,4,6-Trimethylphenol (TMP) is particularly reactive toward $^3CDOM^*$ [58] and the relevant bimolecular reaction rate constant ($k_{TMP} = 1.6 \times 10^9\ M^{-1}\ s^{-1}$) has been measured adopting a competitive model [59], in which the scavenging of $^3CDOM^*$ by TMP is able to reduce the production of $^1O_2$ from the reaction between $^3CDOM^*$ and ground-state oxygen. The lower generation of $^1O_2$ significantly reduces the transformation of furfuryl alcohol into pyranone, a very sensitive reaction for the detection of singlet oxygen (see Fig. 7.13). If $[^3CDOM^*] = 8 \times 10^{-15}$ M, the time required for $^3CDOM^*$ to transform 50 % TMP would be around 15 h (note that the half-life time of TMP for reaction with $^3CDOM^*$, in seconds, is given by $t_{½} = \ln 2\ (k_{TMP}\ [^3CDOM^*])^{-1}$).

The transformation processes photosensitised by CDOM have been demonstrated to play a substantial role in the degradation of important classes of pollutants such as phenols, phenylurea herbicides, and sulfonamide antibiotics [20]. In the case of the herbicides isoproturon and diuron, the modelling of $^3CDOM^*$-assisted phototransformation has been able to predict with precision their vertical profiles in the Greifensee lake (Switzerland), thereby confirming the important role of CDOM in the relevant phototransformation [60].

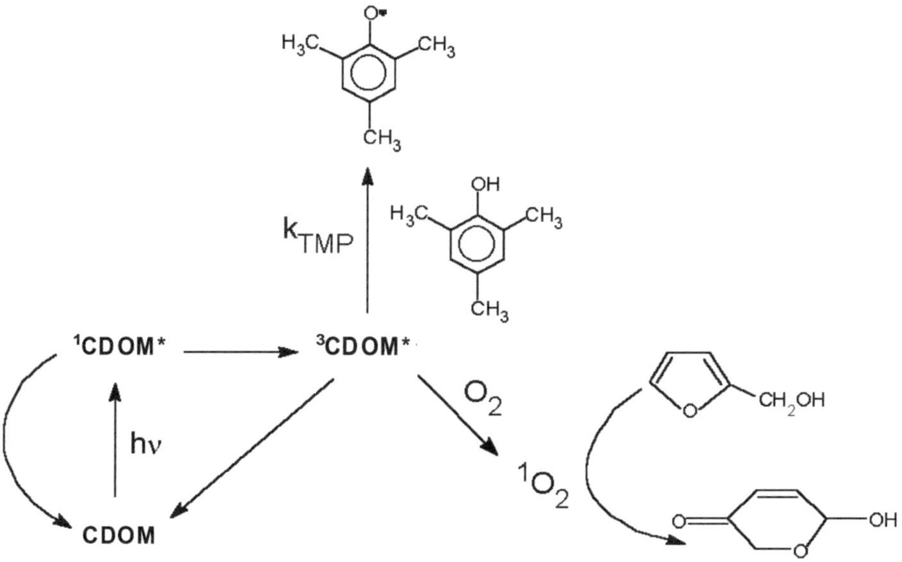

Fig. 7.13 Competitive kinetic model used to measure the rate constant k_{TMP} between $^3CDOM^*$ and 2,4,6-trimethylphenol

Photoexcited $^3CDOM^*$ also induces the oxidation of As(III) to As(V) [61], and of Sb(III) to Sb(V) [62]. Interestingly, some pollutants undergo both direct photolysis upon absorption of sunlight, and transformation photosensitised by $^3CDOM^*$. In this case both CDOM and the substrate compete for radiation absorption, and the effect of CDOM on the phototransformation kinetics can be variable depending on the substrate [53]. If photodegradation in laboratory systems is enhanced in the presence of CDOM, it is a strong indication of an important role played by indirect photolysis processes (among which $^3CDOM^*$ might be significant). If, on the contrary, CDOM inhibits photodegradation, this is often interpreted as an index that $^3CDOM^*$ photoreactions are not important compared to the direct photolysis of the substrate. However, the situation is often more complex in the natural environment because CDOM, as the main sunlight absorber in surface waters, can preserve an active photochemistry even at depths where the direct photolysis is no longer important. For instance, in the case of triclosan, the fact that CDOM inhibits photodegradation in laboratory systems (where direct photolysis predominates due to the shallow water column) is not in contrast with the fact that $^3CDOM^*$-sensitised transformation may be the main photoreaction in deep water [47]. Indeed, it is very possible that $^3CDOM^*$-induced processes may be important for the transformation of a larger number of compounds than commonly believed [63, 64].

Anyway, it is much more likely for $^3CDOM^*$ to enhance degradation of substrates that undergo direct photolysis at a limited extent. Among the classes of compounds that show faster degradation in the presence of CDOM under irradiation there are sulphur-containing molecules [65–67], phenylurea herbicides [60], Mirex [68], carbaryl [69], and some cyanobacterial toxins [70]. In most of these cases it is very interesting to assess the role of $^3CDOM^*$ versus other reactive transients produced by irradiated CDOM (such as $^•OH$ and $^1O_2$).

7.4 Reactions Induced by the Hydroxyl Radical, $^•OH$

The hydroxyl radical is one of the most reactive transients that are formed in natural waters. The high reactivity implies that this species can be involved into the degradation of refractory pollutants, some of which (including most notably alkanes, chloroalkanes such as butyl chloride, benzene and toluene) would be almost exclusively degraded by $^•OH$. Quite interestingly, the cited molecules can also be used as probes to quantify $^•OH$ in surface water samples because they would undergo limited direct photolysis or side reactions with $^3CDOM^*$ and $^1O_2$ [71]. However, the high reactivity of $^•OH$ is also a drawback for its overall significance as an oxidant in surface waters. The very vast majority of $^•OH$ radicals, which are formed upon irradiation of photoactive precursors, are in fact scavenged by DOM and carbonate/bicarbonate, so that only a small fraction is available for the degradation of xenobiotics [72].

The main photochemical $^•OH$ sources in surface waters are nitrate, nitrite, and CDOM. The former two species produce $^•OH$ upon photolysis [27], while the production of $^{\bullet}OH$ by irradiated CDOM is a more complex phenomenon [73–75]. First of all, there are at least a H_2O_2-dependent and a H_2O_2-independent pathways. The latter might involve oxidation of water by $^3CDOM^*$, which could be supported by some findings concerning model sensitisers [76] but it is not at all a general feature of the excited triplet states [77]. As an alternative, hydroxylated and peroxyl groups in photoprocessed CDOM could undergo photolysis to produce $^{\bullet}OH$. The H_2O_2-dependent pathway of CDOM is better known, and a possible explanation for it could be the occurrence of photo-Fenton reactions in the presence of Fe traces. They could involve the photolysis of complexes between Fe(III) and natural organic compounds (reactions 7.16 and 7.17, where L is an organic ligand of Fe) [78, 79]. Fe(III) could also produce $^{\bullet}OH$ directly, upon photolysis of its hydroxocomplexes (mainly $FeOH^{2+}$). However, $FeOH^{2+}$ is present at significant levels only under acidic conditions (typically at $pH < 5$), which are little representative of surface waters [80]. A major exception could be acidic mine drainage water, where Fe(III) photolysis could be a major source of $^{\bullet}OH$ [81]. Hydroxyl groups on the surface of Fe(III) oxide colloids ($\equiv Fe^{III}$-OH) would also be able to yield $^{\bullet}OH$ upon photolysis, but several studies suggest that the efficiency of such an heterogeneous process is very low [49, 80, 82]. Accordingly, if Fe plays a significant role in the generation of $^{\bullet}OH$ radicals under non-acidic conditions, it would mainly be through its possible interaction with DOM (reactions 7.16 and 7.17).

$$NO_3^- + h\nu + H^+ \rightarrow {}^{\bullet}OH + {}^{\bullet}NO_2 \quad (7.12)$$

$$NO_2^- + h\nu + H^+ \rightarrow {}^{\bullet}OH + {}^{\bullet}NO \quad (7.13)$$

$$CDOM + h\nu \rightarrow {}^1CDOM^* - (ISC) \rightarrow {}^3CDOM^* \quad (7.14)$$

$${}^3CDOM^* + H_2O \rightarrow [CDOM + H]^{\bullet} + {}^{\bullet}OH \quad (7.15)$$

$$Fe^{III} - L + h\nu \rightarrow Fe^{2+} + L^{+\bullet} \quad (7.16)$$

$$Fe^{2+} + H_2O_2 \rightarrow Fe^{3+} + {}^{\bullet}OH + OH^- \quad (7.17)$$

$$FeOH^{2+} + h\nu \rightarrow Fe^{2+} + {}^{\bullet}OH \quad (7.18)$$

As far as the relative role of nitrate, nitrite and CDOM as $^{\bullet}OH$ sources is concerned, on average it is $CDOM > NO_2^- > NO_3^-$ [83]. Such a statement is reinforced by the fact that the maximum absorption of sunlight by nitrate lies in the UVB region (316–320 nm depending on the season), where the absorption by other water components is maximum, while that of nitrite is located in the UVA (360–370 nm). CDOM can produce $^{\bullet}OH$ radicals even upon absorption of visible radiation (although with lower quantum yields compared to UVB and UVA [84]) and, because the visible is less attenuated by absorption phenomena than for the UV, CDOM photochemistry is still operational in deeper waters compared to nitrite and nitrate [85, 86].

The main scavengers of $^{\bullet}OH$ radicals in surface freshwaters are, in the order, DOM, inorganic carbon (carbonate and bicarbonate), and nitrite [72, 86]. In contrast, bromide is the main •OH scavenger in seawater [4]. From the budget of the •OH formation and scavenging processes in natural waters one can derive the steady-state [•OH] that, for the surface water layer, usually ranges between 10^{-16} and 10^{-15} M [4, 20]. In contrast, the •OH concentration can be considerably lower in the less illuminated lower depths of a water body [72]. The [•OH] values can now be easily modelled as a function of water chemistry and depth [20], and Fig. 7.14 reports for instance the comparison between the measured and modelled [•OH] in a surface freshwater sample irradiated in the laboratory. Although the significance of the comparison is decreased by the limited water column depth that is allowed by laboratory set-ups, the agreement between model and experiments is more than satisfactory [87].

In addition to [•OH], the other important factor that defines the •OH-related lifetime of an organic compound is its second-order reaction rate constant with •OH ($k_{\bullet OH}$). Among the different transients present in surface waters, •OH affords the highest second-order rate constants with organic compounds [71]. One of the reasons of such elevated reactivity is that, while the vast majority of the other radical species (e.g. $CO_3^{-\bullet}$) are mainly involved into electron-capture reactions and less into additional processes, •OH can abstract electrons, H atoms, or be added to double bonds and aromatic rings. This wide range of possibilities allows •OH to by-pass many kinetic bottlenecks, simply by reacting in an alternative fashion with a molecule that can be particularly refractory to, e.g., one-electron oxidation [88].

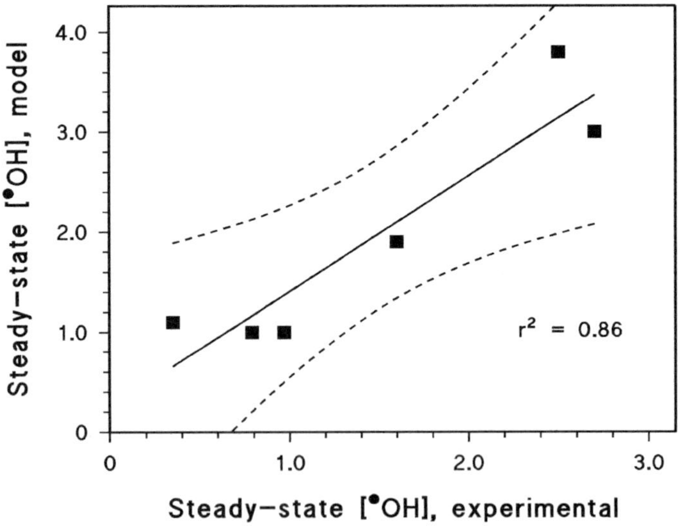

Fig. 7.14 Correlation between the steady-state [•OH] determined experimentally upon irradiation of surface water samples under 22 W m^{-2} simulated sunlight UV irradiance, and the modelled [•OH] concentration. [•OH] units are 10^{-16} M in both cases

The combination of the steady-state [$^{\bullet}$OH] with $k_{\bullet OH}$ defines the half-life $t_{1/2}$ of a given compound, due to reaction with $^{\bullet}$OH. It is $t_{1/2} = \ln\ 2(k_{\bullet OH}[^{\bullet}OH])^{-1}$, but the value of [$^{\bullet}$OH] is highly dependent on the irradiation intensity that shows diurnal and seasonal variations. It has been shown that a constant 22 W m^{-2} sunlight irradiance in the UV for 9 h would deliver the same amount of UV energy as a whole sunny mid-July day at mid latitude [83]. By considering this equivalence, it is possible to correlate the steady-state [$^{\bullet}$OH], corresponding to 22 W m^{-2} sunlight UV as described by the models, with the lifetime of a given compound expressed in mid-July, mid-latitude days. It is $t_{1/2} \approx 2 \cdot 10^{-5} \cdot (k_{\bullet OH}[^{\bullet}OH])^{-1}$, and Fig. 7.15 reports the plot of $t_{1/2}$ vs. [$^{\bullet}$OH] for different values of $k_{\bullet OH}$.

To make an example, 2-methylphenol (o-cresol), with $k_{\bullet OH} = 1.1 \times 10^{10}$ M^{-1} s^{-1} [88], under conditions that give 3×10^{-16} M $^{\bullet}$OH under 22 W m^{-2} sunlight UV, would have $t_{1/2} \approx 6$ days. This is higher than the already-reported 15-h lifetime for the reaction between 2,4,6-trimethylphenol and ^{3}CDOM*, which suggests that transformation photosensitised by CDOM would be a more important sink than $^{\bullet}$OH for electron-rich phenols. Toluene, with $k_{\bullet OH} = 3 \times 10^9$ M^{-1} s^{-1} [81], under the conditions mentioned above would have $t_{1/2} \approx 22$ days for reaction with $^{\bullet}$OH. This is higher than for o-cresol, but toluene is not expected to undergo other important photochemical reactions in surface waters [89].

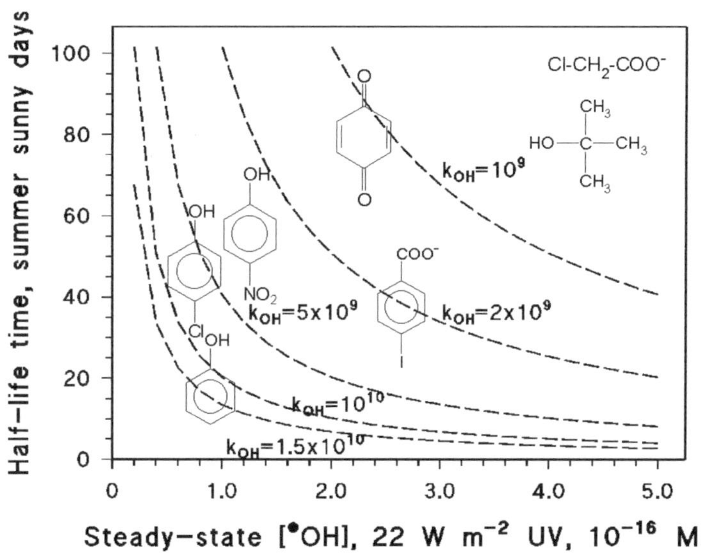

Fig. 7.15 Half-life time (in mid-July, mid latitude sunny days) of a given compound as a function of the steady-state [$^{\bullet}$OH] (valid for 22 W m^{-2} sunlight UV irradiance) and of its second-order rate constant $k_{\bullet OH}$ (M^{-1} s^{-1} units). Some structures are shown as an example of the different $k_{\bullet OH}$ values

7.5 Reactions Induced by the Carbonate Radical, $CO_3^{-\bullet}$

The carbonate radical anion is a fairly reactive transient with relatively elevated reduction potential ($E^0 = 1.59$ V), although its oxidising capability cannot be compared to the hydroxyl radical ($E^0 = 2.59$ V) [90]. The radical $CO_3^{-\bullet}$ is formed in surface waters upon reaction between $^\bullet OH$ and CO_3^{2-} or HCO_3^- [91]. An additional pathway for the generation of $CO_3^{-\bullet}$ is the reaction between carbonate and $^3CDOM^*$, but its weight could at most be 10 % of the $^\bullet OH$-mediated one [92].

$$^\bullet OH + CO_3^{2-} \rightarrow OH^- + CO_3^{-\bullet} \qquad [k_{19} = 3.9 \times 10^8\ M^{-1}\ s^{-1}] \tag{7.19}$$

$$^\bullet OH + HCO_3^- \rightarrow H_2O + CO_3^{-\bullet} \qquad [k_{20} = 8.5 \times 10^6\ M^{-1}\ s^{-1}] \tag{7.20}$$

$$^3CDOM^* + CO_3^{2-} \rightarrow CDOM^{-\bullet} + CO_3^{-\bullet} \qquad [k_{21} = 1 \times 10^5\ M^{-1}\ s^{-1}] \tag{7.21}$$

The main sink of the carbonate radical in surface waters is represented by its reaction with DOM. Literature sources do not agree for the rate constant of such a reaction (40 or 280 L (mg C)$^{-1}$ s^{-1}) [91, 92], but a value of 100 L (mg C)$^{-1}$ s^{-1} could be not very far from reality. By neglecting for simplicity reaction (7.21), and by applying the steady-state approximation to $[CO_3^{-\bullet}]$, one obtains the following result:

$$[CO_3^{-\bullet}] = [^\bullet OH] \cdot \frac{8.5 \cdot 10^6 \cdot [HCO_3^-] + 3.9 \cdot 10^8 \cdot [CO_3^{2-}]}{100 \cdot DOC} \tag{7.22}$$

where the DOC (dissolved organic carbon, units of mg C L^{-1}) is a measure of DOM. The main problem with Eq. 7.22 is that the concentration values of carbonate and bicarbonate are not always determined together in surface water samples, and the value of the inorganic carbon (IC) is more often available. If one neglects H_2CO_3, it is possible to use the approximation that

$$IC\ [mg\ C\ L^{-1}] \approx 1.2 \cdot 10^4 \{[HCO_3^-] + [CO_3^{2-}]\}$$

The pH of the solution allows the separate values of $[HCO_3^-]$ and $[CO_3^{2-}]$ to be derived. In the hypothesis that $[HCO_3^-] = 100\ [CO_3^{2-}]$ (pH ~8, not unreasonable for surface waters [80]) one gets the following expression for $[CO_3^{-\bullet}]$:

$$[CO_3^{-\bullet}] \approx 10 \cdot [^\bullet OH] \cdot \frac{IC}{DOC} \tag{7.23}$$

Figure 7.16 shows the modelled trend of $[CO_3^{-\bullet}]$ as a function of nitrate and DOC, with IC = 10 mg C L^{-1}. It is apparent that $[CO_3^{-\bullet}]$ increases with increasing nitrate (that is an $^\bullet OH$ source and increases $[^\bullet OH]$), while it decreases with DOC because organic matter is the major sink of $CO_3^{-\bullet}$. The role of DOM to decrease $[CO_3^{-\bullet}]$ would not be substantially modified when taking reaction (7.21) into account,

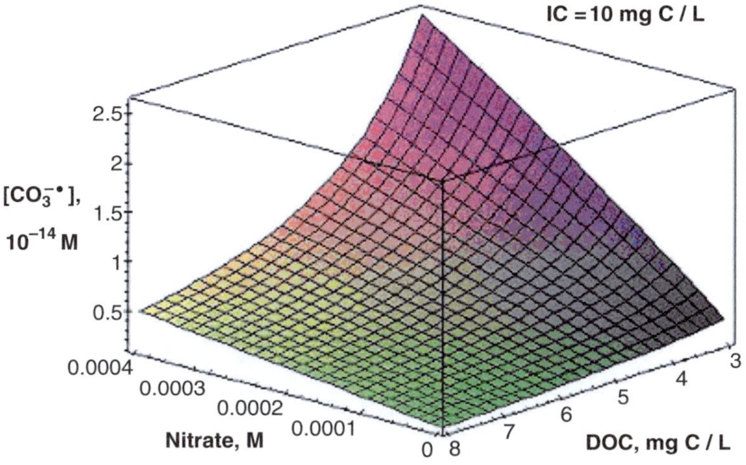

Fig. 7.16 Modelled $[CO_3^{-\bullet}]$ in the surface water layer, as a function of nitrate and DOC, in the presence of IC = 10 mg C L^{-1}. Sunlight UV irradiance: 22 W m^{-2}

because the production of $CO_3^{-\bullet}$ by ^{3}CDOM* would just be a secondary process [92].

Figure 7.17 shows the trend of $[CO_3^{-\bullet}]$ vs. IC and DOC, in the presence of 10^{-5} M nitrate. The concentration of the carbonate radical increases with increasing IC, which is reasonable because carbonate and bicarbonate are the immediate precursors of $CO_3^{-\bullet}$. Moreover, $[CO_3^{-\bullet}]$ decreases with DOC because of the role of DOM as sink. Both Figs. 7.16 and 7.17 indicate that $[CO_3^{-\bullet}]$ would reach concentration values up to around 10^{-14} M in the surface water layer, in agreement with laboratory irradiation data of natural samples [93]. This means that $CO_3^{-\bullet}$ would be one-two orders of magnitude more concentrated than $^{\bullet}$OH, which in some cases could compensate for its lower reactivity. Interestingly, the carbonate radical is mainly formed from $^{\bullet}$OH through a pathway (reactions 7.19 and 7.20 with carbonate and bicarbonate) that on average would account for some 10 % of hydroxyl scavenging in surface waters [86, 87]. The formation rate of $CO_3^{-\bullet}$ would thus be lower than that of $^{\bullet}$OH and, as a consequence, the fact that $[CO_3^{-\bullet}] > [^{\bullet}OH]$ could only be accounted for by slower $CO_3^{-\bullet}$ scavenging. In fact, the second-order reaction rate constant between $CO_3^{-\bullet}$ and DOM is over two orders of magnitude lower compared to $^{\bullet}$OH and DOM [86, 91, 92].

The rate constants of the reactions that involve $CO_3^{-\bullet}$ and organic compounds are very variable. The upper limit is around 10^9 M^{-1} s^{-1} [94], which is easier to be reached by phenolates, anilines and some sulphur-containing molecules. Considering that bimolecular rate constants in aqueous solution have a diffusion-controlled upper limit of around 2×10^{10} M^{-1} s^{-1}, which is also the upper limit for the reactions that involve $^{\bullet}$OH [88], and considering that $[CO_3^{-\bullet}] \sim 10–100\ [^{\bullet}OH]$, the reaction with $CO_3^{-\bullet}$ can be important for compounds that have a $CO_3^{-\bullet}$ bimolecular reaction rate constant around 10^9 M^{-1} s^{-1}.

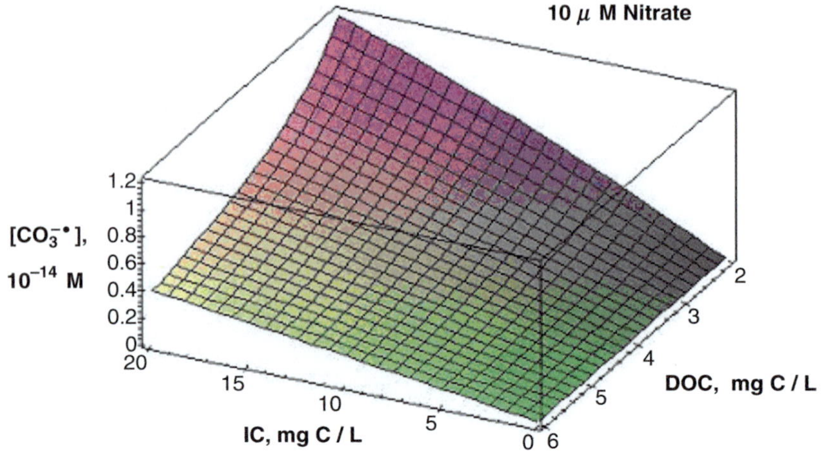

Fig. 7.17 Modelled $[CO_3^{-\bullet}]$ in the surface water layer, as a function of IC and DOC, in the presence of $[NO_3^-] = 10^{-5}$ M. Sunlight UV irradiance: 22 W m^{-2}

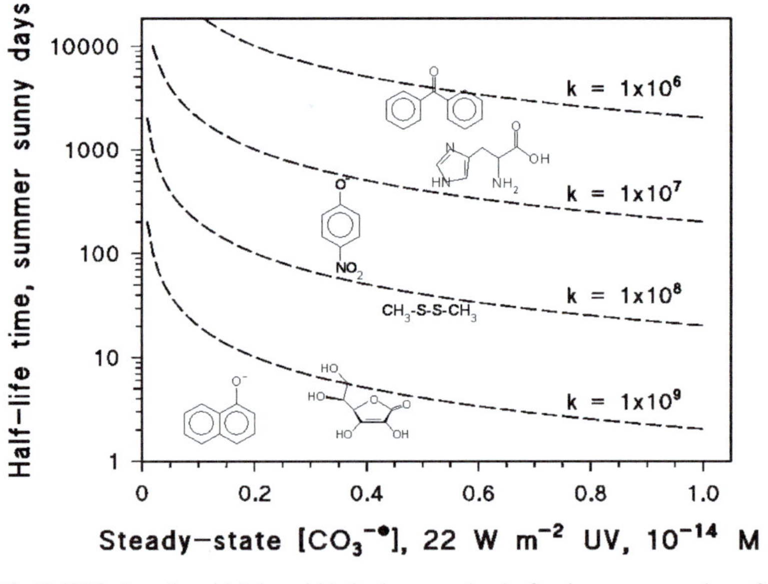

Fig. 7.18 Half-life time (in mid-July, mid latitude sunny days) of a given compound as a function of the steady-state $[CO_3^{-\bullet}]$ (valid for 22 W m^{-2} sunlight UV irradiance) and of its second-order rate constant $k_{CO_3^{-\bullet}}$ (M^{-1} s^{-1} units). Some structures are shown as an example of the different $k_{CO_3^{-\bullet}}$ values. Note the logarithmic scale of the Y-axis

Figure 7.18 shows the half-life times (in mid-July, mid-latitude summer sunny days) of organic compounds, as a function of the steady-state $[CO_3^{-\bullet}]$ and of the second-order rate constants for reaction with $CO_3^{-\bullet}$. Some examples of molecular structures are also shown on the Figure, in connection with the relevant values of the rate constants. The half-life times are calculated as $t_{½} = 2 \cdot 10^{-5} \cdot (k_{CO_3^{-\bullet}}\,[CO_3^{-\bullet}])^{-1}$. A considerable variability of $k_{CO_3^{-\bullet}}$, in the range of 10^6-10^9 M^{-1} s^{-1}, and of the half-life times as a consequence, is apparent in the figure. This variability is generally much higher than that associated with $k_{\bullet OH}$ and, when $k_{CO_3^{-\bullet}}$ is low, $CO_3^{-\bullet}$ plays a negligible role in the degradation of organic compounds.

7.6 Reactions Induced by Singlet Oxygen and Organic Radicals

Singlet oxygen can be formed by activation of ground-state triplet oxygen by $^3CDOM^*$ [53]. It can reach steady-state concentration values in the range $10^{-14} - 10^{-13}$ M in sunlit surface waters, and its formation rate is directly proportional to the amount of radiation absorbed by CDOM. The main removal pathway of $^1O_2$ is the physical quenching upon collision with the water molecules [71].

$$^3CDOM^* + O_2 \rightarrow CDOM + {}^1O_2 \tag{7.24}$$

$$^1O_2 + H_2O \rightarrow O_2 + H_2O + \text{Energy} \tag{7.25}$$

Many molecules show limited reactivity toward $^1O_2$. However, $^1O_2$ can play a significant role in the degradation of compounds such as chlorophenolates and aromatic amino acids [4, 45, 95]. Furfuryl alcohol is another molecule that reacts fast with $^1O_2$ $\left(k_{FFA,{}^1O_2} = 1.2 \times 10^8\ M^{-1}\ s^{-1}\right)$ [96], and its half-life time due to singlet oxygen (in mid-July, mid-latitude sunny days) would be $t_{1/2} \sim 2 \cdot 10^{-5} \cdot \left(k_{FFA,{}^1O_2}[{}^1O_2]\right)^{-1}$. With $[{}^1O_2] = 5 \times 10^{-14}$ M one obtains $t_{½} \approx 3$ days. Interestingly, $^1O_2$ shows a micro-heterogeneous distribution within CDOM and it reaches elevates steady-state levels in the hydrophobic CDOM cores. In this confined environment, it could play an important role toward the degradation of some classes of hydrophobic pollutants [97, 98].

The transformation of DOM under irradiation, induced by photolysis or by reaction with radical species such as $^{\bullet}OH$ and $CO_3^{-\bullet}$, or even the (very limited) scavenging of $^3CDOM^*$ by DOM components can produce various radical species, some of which yield peroxy radicals upon further reaction with oxygen. The radicals thus formed ($R^{\bullet}$, $ROO^{\bullet}$, $RO^{\bullet}$, $R\text{-}H^{+\bullet}$) could be involved in oxidation reactions [99] that are added to the transformation processes mediated by irradiated CDOM. The exact importance of the described radical species in DOM photochemistry is still unclear [53].

$$R - H + h\nu \rightarrow R^{\bullet} + H^{\bullet} \tag{7.26}$$

$$R - H + {}^{\bullet}OH \rightarrow R^{\bullet} + H_2O \tag{7.27}$$

$$R - H + h\nu \rightarrow R - H^{+\bullet} + e^- \tag{7.28}$$

$$R^{\bullet} + O_2 \rightarrow ROO^{\bullet} \quad (7.29)$$

$$2\ ROO^{\bullet} \rightarrow ROO - OOR \quad (7.30)$$

$$ROO - OOR \rightarrow 2\ RO^{\bullet} + O_2 \quad (7.31)$$

7.7 Reactions Induced by Nitrogen Dioxide, $^{\bullet}NO_2$

Nitrogen dioxide is a nitrating agent that can be formed in surface waters upon nitrate photolysis and nitrite photooxidation [27]:

$$NO_3^- + h\nu + H^+ \rightarrow {}^{\bullet}OH + {}^{\bullet}NO_2 \quad (7.32)$$

$$NO_2^- + h\nu + H^+ \rightarrow {}^{\bullet}OH + {}^{\bullet}NO \quad (7.33)$$

$$NO_2^- + {}^{\bullet}OH \rightarrow {}^{\bullet}NO_2 + OH^- \qquad [k_{34} = 1 \times 10^{10}\ M^{-1}\ s^{-1}] \quad (7.34)$$

The discovery of the involvement of nitrogen dioxide in aromatic photonitration processes induced by the irradiation of nitrate and nitrite ions dates back to the mid 1980s [100], but the actual occurrence in the environment of photonitration by $^{\bullet}NO_2$ has been demonstrated only more recently. Photonitration of 2,4-dichlorophenol [36], 4-chloro-2-methylphenol [37] and 4-chlorophenol [101] is likely to take place in the rice fields of the Rhône river delta, where the primary compounds are formed upon transformation of the herbicides dichlorprop and MCPA. In all the cases the expected time trend of the nitroderivatives, under the hypothesis that $^{\bullet}NO_2$ is the nitrating agent, is compatible with the available field data.

It is possible to model the steady-state [$^{\bullet}NO_2$] in surface waters under the hypothesis that reactions (7.32) and (7.34) are the main sources, and hydrolysis (reactions (7.35 and 7.36)) is the main sink [102].

$$2^{\bullet}NO_2 \rightleftarrows N_2O_4 \qquad [k_{35} = 4.5 \times 10^8\ M^{-1}\ s^{-1};\ k_{-35} = 6.9 \times 10^3\ s^{-1}] \quad (7.35)$$

$$N_2O_4 + H_2O \rightarrow NO_3^- + NO_2^- + 2\,H^+ \qquad [k_{36} = 1 \times 10^3\ s^{-1}] \quad (7.36)$$

Additional sources and sinks of $^{\bullet}NO_2$ are possible in surface waters. The oxidation of nitrite by irradiated Fe(III) (hydr)oxides is a very significant pathway leading to aromatic nitration under laboratory conditions [103], but the assessment of its environmental importance is made problematic by the very complex speciation of Fe(III) in surface waters. An important fraction of the total Fe(III) is in fact present in the form of complexes with organic matter [104], the (photo)reactivity of which is poorly known. Indeed, if the average ability of the Fe(III) species to photooxidise nitrite to $^{\bullet}NO_2$ were comparable to that of hematite, Fe(III) could be a major source of $^{\bullet}NO_2$ in surface waters containing over 1 mg Fe L^{-1} [102]. However, it is completely unknown to what extent can hematite be considered representative of the photoreactivity of Fe(III) species toward nitrite.

Reaction with DOM, and in particular with its phenolic moieties, could be a significant sink of nitrogen dioxide. However, at the measured [$^{\bullet}NO_2$] and DOC levels and given the expected rate constants for the reaction between $^{\bullet}NO_2$ and the phenolic compounds, DOM would be a secondary sink compared to hydrolysis. For DOM to be the main sink, it should be almost completely made up of phenolic moieties [83, 102].

From reactions (7.32,7.34–7.36) it is possible to set up an approximate model for the assessment of the steady-state [$^{\bullet}NO_2$] in surface waters, valid, as the previous ones, for 22 W m^{-2} sunlight UV irradiance. From the cited reactions, and applying the steady-state approximation to [$^{\bullet}NO_2$], one gets equation (37) [83]:

$$\left[{}^{\bullet}NO_2\right] = \sqrt{88.8 \cdot \left[{}^{\bullet}OH\right] \cdot \left[NO_2^-\right] + 1.49 \cdot 10^{-15} \cdot \left[NO_3^-\right]} \qquad (7.37)$$

The steady-state [$^{\bullet}OH$] can be modelled based on the known sources and sinks. In many cases, unfortunately, the concentration values of nitrite are not available. In these cases one can assume, with reasonable approximation, [NO_3^-] $\approx$ 200 [NO_2^-] [102] to obtain Eq. 7.38:

$$\left[{}^{\bullet}NO_2\right] = \sqrt{\left[NO_3^-\right] \cdot \left(0.444 \cdot \left[{}^{\bullet}OH\right] + 1.49 \cdot 10^{-15}\right)} \qquad (7.38)$$

Figure 7.19 shows the trend of [$^{\bullet}NO_2$] vs. nitrate and DOC. It is apparent that [$^{\bullet}NO_2$] increases with increasing nitrate (and nitrite as a consequence, which is correlated with nitrate under the adopted approximation) and decreases with DOC. DOM in the adopted model has mainly the effect of scavenging [$^{\bullet}OH$].

The relatively elevated steady-state concentration values that $^{\bullet}NO_2$ can reach in surface waters are compensated for by its low reactivity. Nevertheless, nitrogen

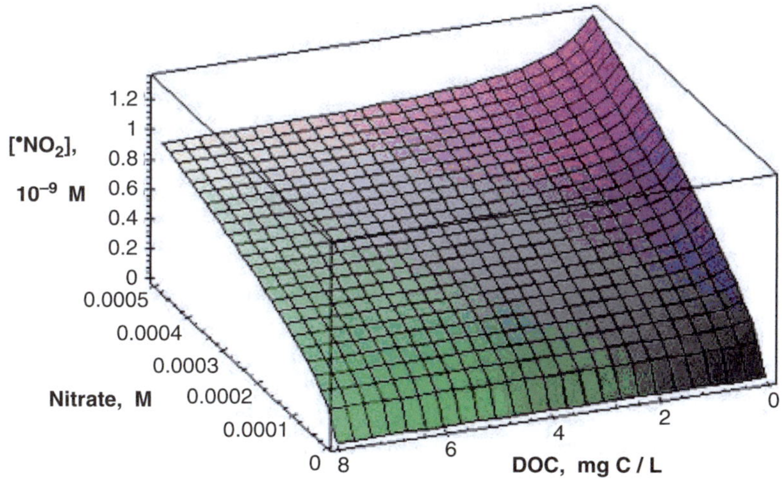

Fig. 7.19 Trend of [$^{\bullet}NO_2$] as a function of nitrate and DOC. Sunlight UV irradiance: 22 W m^{-2}

dioxide can still be a significant sink for some classes of compounds such as phenols and species containing amino groups [102]. However, transformation processes induced by nitrogen dioxide are very likely to generate secondary pollutants such as nitrophenols and nitrosamines [105], with possibly higher environmental impact than the parent species. For instance the nitration of 2,4-dichlorophenol into 2,4-dichloro-6-nitrophenol transforms a cytotoxic molecule, involved in the oxidative stress of cellular tissues, into a less toxic but potentially mutagenic compound [106].

7.8 Reactions Induced by the Dihalogen Radicals $Cl_2^{-\bullet}$ and $Br_2^{-\bullet}$

The radicals $Cl_2^{-\bullet}$ and $Br_2^{-\bullet}$ can be generated in surface waters by the oxidation of the chloride and bromide ions. Chloride can be oxidised to $Cl^{\bullet}/Cl_2^{-\bullet}$ by $^{\bullet}OH$ in acidic solution, but it is practically non-reactive toward $^{\bullet}OH$ under neutral conditions. The reason is that the reaction between Cl^- and $^{\bullet}OH$ is a reversible one, giving back the reactants unless $ClOH^{-\bullet}$ is transformed into $Cl^{\bullet}$ under acidic conditions [88].

$$Cl^- + {}^{\bullet}OH \rightleftarrows ClOH^{-\bullet} \tag{7.39}$$

$$ClOH^{-\bullet} + H^+ \rightleftarrows Cl^{\bullet} + H_2O \tag{7.40}$$

$$Cl^{\bullet} + Cl^- \rightleftarrows Cl_2^{-\bullet} \tag{7.41}$$

The oxidation of chloride by $^{\bullet}OH$ at acidic pH might have some importance in atmospheric waters [107], but it is little likely to affect significantly the chemistry of most surface waters that have pH values in the neutral to basic range [86]. More likely sources of $Cl_2^{-\bullet}$ in surface waters are the oxidation of chloride by semiconductor oxides, such as the Fe(III) (hydr)oxides [49], the photolysis of the complex $FeCl^{2+}$ [108], and the oxidation of chloride by $^3CDOM^*$ [109].

It has been shown that the oxidation of chloride to $Cl_2^{-\bullet}$ by Fe(III) oxide colloids in simulated estuarine water under irradiation could be an important process in the photodegradation of the anti-epileptic drug carbamazepine [49]. Furthermore, the radical $Cl_2^{-\bullet}$ is an oxidising and chlorinating agent. Phenol chlorination by $Cl_2^{-\bullet}$ has been studied in detail [107] and it involves a primary reaction step where phenol is oxidised to the phenoxyl radical, followed by the reaction between phenoxyl and $Cl_2^{-\bullet}$ to give 2- and 4-chlorophenol (Fig. 7.20).

The phenol chlorination yield is relatively low, between 1 and 5 %, and the main products of the reaction between phenol and $Cl_2^{-\bullet}$ are phenoxyphenols, dihydroxybiphenyls, and 1,4-benzoquinone [107]. Still, the oxidation of chloride to $Cl_2^{-\bullet}$ is a potentially important pathway for the formation of chloroaromatic compounds in irradiated seawater [110, 111].

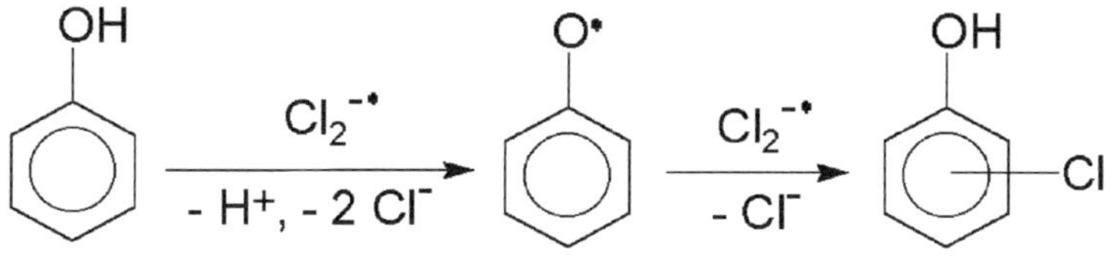

Fig. 7.20 Pathway of phenol chlorination by $Cl_2^{-\bullet}$

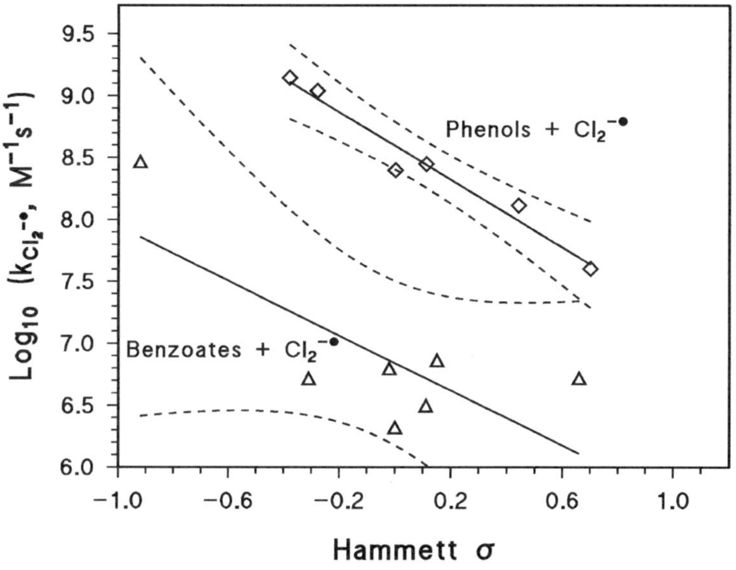

Fig. 7.21 Correlation between the decimal logarithms of the second-order rate constants $k_{Cl_2^{-\bullet}}$ and the Hammett σ for substituted phenols and benzoates, for reaction with the radical $Cl_2^{-\bullet}$

It is often possible to correlate the second-order rate constants of the reactions between an oxidising agent and whole families of aromatic compounds, with the Hammett σ value given by the different substituent groups on the aromatic ring. One usually obtains a line with a negative slope, because the electron-withdrawing substituents (positive σ) reduce the reactivity of the molecule toward one-electron oxidation. Figure 7.21 shows the correlation between the second-order rate constants of the reactions involving phenols and benzoates with $Cl_2^{-\bullet}$, and the Hammett σ [94]. The good linearity and the negative slopes confirm that the radical $Cl_2^{-\bullet}$ mainly reacts by one-electron abstraction in the first step of the processes in which it is involved. Chlorination could occur in the second step.

The radical $Br_2^{-\bullet}$ can be formed upon oxidation of bromide by $^{\bullet}OH$ even in neutral solution [88]. Alternative processes can be the photolysis of the complex $FeBr^{2+}$ and the oxidation of bromide by Fe(III) (hydr)oxides under irradiation [112].

$$Br^- + {}^{\bullet}OH \rightarrow Br^{\bullet} + OH^- \quad [k_{42} = 1.1 \times 10^{10}\ M^{-1}\ s^{-1}] \tag{7.42}$$

$$Br^{\bullet} + Br^- \rightleftarrows Br_2^{-\bullet} \tag{7.43}$$

$$FeBr^{2+} + h\nu \rightarrow Fe^{2+} + Br^{\bullet} \tag{7.44}$$

Reaction (7.42), involving $^{\bullet}OH$, is potentially the main formation process of $Br^{\bullet}/Br_2^{-\bullet}$ in surface waters. In keeping with the importance of reaction (7.42), the bromide ion is the main scavenger of $^{\bullet}OH$ radicals in seawater [113]. Important scavenging processes by Br^- could also occur in deltas and estuaries, the water of which can be enriched by bromide from seawater. Moreover, bromide can also be found at elevated concentration in some inland brines that are usually associated to temporary water bodies, such as ephemeral rivers and lakes [114].

The bromide ion can also increase the quantum yield of photolysis of some photoactive species, and most notably that of nitrate, because of a solvent-cage effect in the scavenging of photoformed $^{\bullet}OH$ radicals [115, 116]. Nitrate photolysis yields $^{\bullet}OH$ + $^{\bullet}NO_2$, but the two radical fragments are very likely to recombine before they can exit the surrounding cage of the solvent molecules [117, 118]. However, the bromide ion at relatively high concentration is able to react with $^{\bullet}OH$ *within the cage* and prevent recombination between $^{\bullet}OH$ and $^{\bullet}NO_2$ (see Fig. 7.22). Therefore, the overall generation rate of $^{\bullet}OH + Br_2^{-\bullet}$ in the presence of bromide is higher than the generation rate of $^{\bullet}OH$ alone in its absence. This fact, and the reactivity of $Br_2^{-\bullet}$ arising from the process, can have important consequences on the degradation of the compounds that undergo sufficiently fast reaction with $Br_2^{-\bullet}$, such as phenolates, anilines, and sulphur-containing molecules. For instance, the bromide ion is able to enhance the degradation of dimethyl sulphide and benzophenone-4 upon irradiation of nitrate [113, 116].

The radical $Br_2^{-\bullet}$ is a brominating as well as an oxidising agent with $E^0 = 1.66$ V [90]. Differently from chlorination by $Cl_2^{-\bullet}$, bromination by $Br_2^{-\bullet}$ is very effective and it can be quantitative in the case of phenol [112]. The bromination pathway proceeds in a similar way as the already described chlorination: oxidation of phenol to the phenoxyl radical by $Br_2^{-\bullet}$, followed by reaction between phenoxyl and another $Br_2^{-\bullet}$ to give 2- and 4-bromophenol.

Figure 7.23 shows the correlation between the rate constants for the reaction of phenolates with $Br_2^{-\bullet}$, and the corresponding Hammett σ [94]. The negative slope suggests the involvement of $Br_2^{-\bullet}$ in one-electron oxidation processes. Moreover, the relatively elevated values of $k_{Br_2^{-\bullet}}$ indicate that the radical could be a significant sink of some classes of organic compounds.

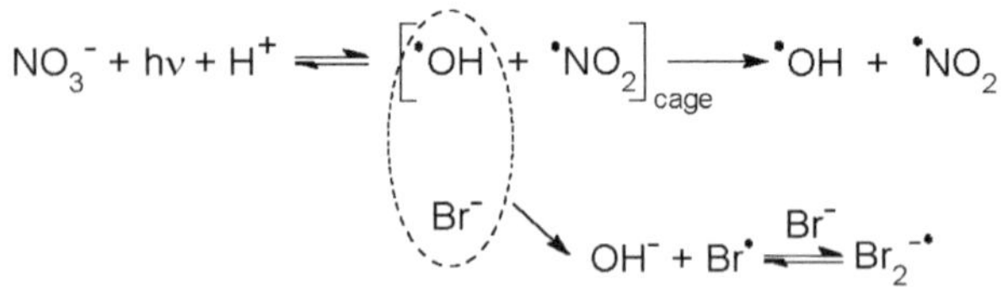

Fig. 7.22 Solvent-cage effect by bromide on the photolysis of nitrate

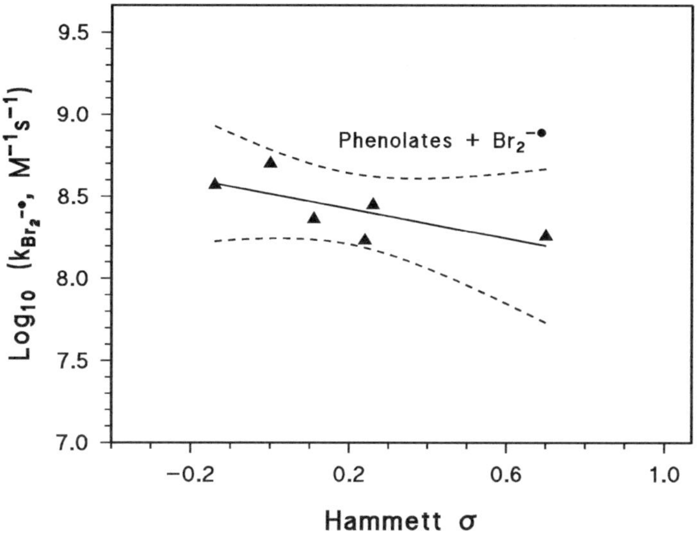

Fig. 7.23 Correlation between the decimal logarithms of the second-order rate constants $k_{Br_2^{-\bullet}}$ and the Hammett σ of substituted phenolates, for reaction with the radical $Br_2^{-\bullet}$

It is, unfortunately, not yet possible to directly measure $Cl_2^{-\bullet}$ and $Br_2^{-\bullet}$ in surface waters, However, photochemical modelling has allowed a tentative assessment of $[Cl_2^{-\bullet}]$ and $[Br_2^{-\bullet}]$, which would both be in the 10^{-14}–10^{-12} M range in seawater.

7.9 Reactions Induced by Fe(III) Complexes

Fe(III) can be involved in photo-Fenton processes, which under definite circumstances can be important sources of $^{\bullet}OH$. Moreover, Fe(III) (hydr)oxide colloids under irradiation can produce $^{\bullet}NO_2$ and $Cl_2^{-\bullet}$ radicals upon oxidation of the nitrite and chloride ions. A very important class of photochemical reactions in surface waters involves the complexes between Fe(III) and a very wide variety of organic ligands. This subject will be treated briefly in the present paragraph, but for a more extended account the reader is referred to a review on the subject [119].

Fe(III) forms photochemically active complexes with a series of carboxylates, and the relevant reactions have been the object of many studies. The photochemistry of ferrioxalate has been studied in depth because of its elevated quantum yield of photolysis and its use as a chemical actinometer. Photolysis proceeds via a ligand-to-metal charge transfer (LMCT) that produces Fe(II) and an oxidised oxalate radical [120]. The latter can undergo mineralisation via decarboxylation, and in fact the photochemistry of Fe(III)-carboxylate complexes can be a significant pathway for the mineralisation of the ligands.

$$Fe^{III}(C_2O_4)_3^{\ 3-} + h\nu \rightarrow Fe^{II}(C_2O_4)_2^{\ 2-} + C_2O_4^{\ -\bullet} \tag{7.45}$$

$$C_2O_4^{\ -\bullet} \rightarrow CO_2^{\ -\bullet} + CO_2 \tag{7.46}$$

$$CO_2^{\ -\bullet} + O_2 \rightarrow CO_2 + O_2^{\ -\bullet} \tag{7.47}$$

LMCT processes are also involved in the degradation of the ligands EDTA and NTA that form complexes with Fe(III). The photochemical transformation of EDTA as complex with Fe(III) is potentially important in surface-water environments because EDTA is refractory to microbial degradation and, for instance, its abatement in wastewater treatment plants is very low [121].

The photoinduced LMCT also involves complexes on the surface of Fe(III) oxides, and most notably those between the surface Fe(III) ions and oxalate [8]. The surface Fe(III)-oxalate complex is able to absorb radiation, which is followed by LMCT with the formation of a transient complex between Fe^{II} and oxidised oxalate ($C_2O_4^{\ -\bullet}$). The complex quickly decomposes with release of $C_2O_4^{\ -\bullet}$ (soon transformed into $CO_2^{\ -\bullet}$ + CO_2) into the solution. A further step is the release of Fe^{2+}. Similar processes of photoreductive dissolution of Fe(III) oxides have been observed in the presence of hydroxycarboxylic acids, most notably the tartaric and gluconic ones [122]. The importance of the cited photochemical processes of Fe(III) photoreduction, where Fe^{2+} is generated upon irradiation of both Fe(III) complexes and Fe(III) (hydr)oxides, is confirmed by field data according to which the concentration of Fe(II) is closely correlated to the sunlight intensity. Such a correlation has been observed under both a monthly and a diurnal scale, in both lake and seawater [123–125]. Indeed, radiation induces the reduction of Fe(III) to Fe(II), and the latter is in turn re-oxidised to Fe(III) in the dark.

An additional pathway that could contribute to the photoreduction of Fe(III) to Fe(II) involves the photoactive components of CDOM (in reaction 7.49, S is an oxidisable substrate) [119]:

$$CDOM + h\nu \rightarrow {}^3CDOM^* \tag{7.48}$$

$${}^3CDOM^* + S \rightarrow CDOM^{-\bullet} + S^{+\bullet} \tag{7.49}$$

$$Fe(III) + CDOM^{-\bullet} \rightarrow Fe(II) + CDOM \tag{7.50}$$

$$CDOM^{-\bullet} + O_2 \rightarrow CDOM + O_2^{\ -\bullet} \tag{7.51}$$

$$Fe(III) + O_2^{\ -\bullet} \rightarrow Fe(II) + O_2 \tag{7.52}$$

In some cases the complexation of Fe(III) by organic ligands enhances the degradation of organic pollutants. For instance, oxalate enhances the degradation of atrazine by irradiated Fe(III). The ligand might have two opposite effects: (i) enhancement of Fe(III) photolysis with generation of Fe(II) (reaction 7.53), which can produce $^{\bullet}OH$ radicals in the Fenton reaction (7.56) with H_2O_2. (ii) Scavenging of $^{\bullet}OH$ radicals (reaction 7.57). Oxalate can enhance degradation if its contribution to the production of $^{\bullet}OH$ is higher than the scavenging [120].

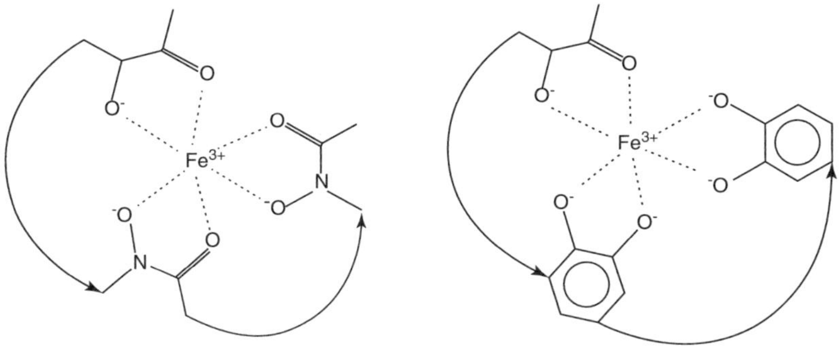

Fig. 7.24 Fe(III)-binding groups in siderophores

$$Fe^{III}(C_2O_4)_3{}^{3-} + h\nu + O_2 \rightarrow Fe^{II}(C_2O_4)_2{}^{2-} + 2\ CO_2 + O_2{}^{-\bullet} \quad (7.53)$$

$$O_2{}^{-\bullet} + H^+ \rightleftarrows HO_2{}^{\bullet} \quad (7.54)$$

$$HO_2{}^{\bullet} + O_2{}^{-\bullet} + H^+ \rightarrow H_2O_2 + O_2 \quad (7.55)$$

$$Fe(II) + H_2O_2 \rightarrow Fe(III) + OH^- + {}^{\bullet}OH \quad (7.56)$$

$$C_2O_4{}^{2-} + {}^{\bullet}OH \rightarrow C_2O_4{}^{-\bullet} + OH^- \quad (7.57)$$

Among the organic ligands of Fe(III), siderophores emitted by bacteria can play a very important role in keeping Fe(III) in its dissolved form. They contain different groups that can bind iron (e.g. hydroxycarboxylates and catecholates, see Fig. 7.24) and in some cases can undergo photolysis with production of Fe(II) [126, 127]. These photoprocesses have an important impact on the biogeochemistry of Fe and on its bioavailability, because Fe (II) is much more available to micro-organisms than Fe(III).

The bioavailability of Fe species is usually lowest for Fe(III) (hydr)oxides and highest for dissolved Fe^{2+}. Fe(III) complexes have an intermediate bioavailability. Obviously the photodissolution of Fe(III) (hydr)oxides and the photolysis reactions involving Fe(III) complexes play an important role to increase the bioavailability of iron [128].

Finally, Fe is known to play a significant role in the photobleaching and photomineralisation of (C)DOM in surface waters. They are the decrease of, respectively, the UV absorbance and the DOC of surface water samples because of irradiation [23, 129].

7.10 Conclusions

Photochemical processes can play an important role in the transformation of biorefractory organic compounds, and most notably of organic pollutants, in surface waters. They include direct photolysis, phototransformation sensitised by

CDOM, and reactions with transient species generated by irradiation of dissolved photoactive compounds. The main photoactive compounds in surface waters are CDOM, nitrate, nitrite and Fe(III), but some reactive transients can arise from the interaction of those species with dissolved anions such as carbonate, bicarbonate, nitrite, chloride and bromide. Interestingly, nitrite is both a photosensitiser and a sink of reactive species in surface waters.

Among the reactive transients, in addition to the excited triplet states of CDOM (^{3}CDOM*), there are $^{\bullet}$OH, $CO_3^{-\bullet}$, $^1O_2$, DOM-derived radicals, $^{\bullet}NO_2$, $Cl_2^{-\bullet}$, and $Br_2^{-\bullet}$. They are involved to a variable extent in the degradation of organic compounds, depending both on the conditions of a particular water body and on the structure of the relevant substrate. The radicals $^{\bullet}NO_2$ and $Br_2^{-\bullet}$ are likely to play a more important role in the generation of secondary pollutants (such as toxic and environmentally persistent aromatic nitro- and bromoderivatives) than in the degradation of the parent molecules. In contrast, ^{3}CDOM*, $^{\bullet}$OH and $CO_3^{-\bullet}$ as oxidising agents would be more often involved in depollution processes connected with the degradation of the primary compounds. For instance Table 7.1 reports the reaction rate constants of the phenylurea herbicide diuron with ^{3}CDOM*, $^{\bullet}$OH and $CO_3^{-\bullet}$, some reasonable values for the steady-state concentrations of the relevant reactive species (valid for 22 W m^{-2} sunlight UV intensity), and the lifetime for the reaction with each species. The lifetime (in mid-July, mid-latitude sunny days) was calculated as $t_{1/2} = 2 \cdot 10^{-5} \cdot (k_i[i])^{-1}$, where k_i is the second-order rate constant and [i] the steady-state concentration of the relevant transient.

It is apparent in the case of diuron that $^{\bullet}$OH would prevail as photoactive transient over ^{3}CDOM* and $CO_3^{-\bullet}$, but the situation could be different for other compounds that are more reactive with ^{3}CDOM* or $CO_3^{-\bullet}$. For instance, in the case of isoproturon, with $k_{^3CDOM^*} = 8 \times 10^8$ M^{-1} s^{-1} [130] and $k_{CO_3^{-\bullet}} = 3 \times 10^7$ M^{-1} s^{-1} [92], the lifetime for reaction with ^{3}CDOM* and $CO_3^{-\bullet}$ would be 3 and 68 days, respectively. Assuming $k_{^{\bullet}OH} \approx 10^{10}$ M^{-1} s^{-1} one could obtain a lifetime of around 20 days for reaction with $^{\bullet}$OH, which would make ^{3}CDOM* the main transient involved in isoproturon degradation. Lifetimes for the reaction with 10^{-14} M $CO_3^{-\bullet}$ would be lower than 10 days for $k_{CO_3^{-\bullet}} > 2 \times 10^8$ M^{-1} s^{-1}, a conditions that is met in the case of anilines and phenoxides, for instance. In the case of the reaction with 8×10^{-15} M ^{3}CDOM*, to obtain $t_{1/2} < 10$ days one would require $k_{^3CDOM^*} > 2.5 \times 10^8$ M^{-1} s^{-1}. This condition is met by some phenylurea herbicides and, more markedly, some electron-rich phenols. In the case of 10^{-16} M $^{\bullet}$OH it would be required $k_{^{\bullet}OH} > 2 \times 10^{10}$ M^{-1} s^{-1} to have $t_{1/2} < 10$ days. This would be a rare event because 2×10^{10} M^{-1} s^{-1} represents the diffusion-control limit for bimolecular rate constants in aqueous solution

Table 7.1 Reaction kinetics of diuron with ^{3}CDOM*, $^{\bullet}$OH and $CO_3^{-\bullet}$. Note that [i] is valid for 22 W m^{-2} sunlight UV intensity. Note that diuron can also undergo direct photolysis

	k_i, M^{-1} s^{-1}	[i], M	$t_{1/2}$, days
^{3}CDOM*	9×10^6 [130]	8×10^{-15}	280
$^{\bullet}$OH	5×10^9 [88]	1×10^{-16}	40.5
$CO_3^{-\bullet}$	8×10^6 [92]	1×10^{-14}	250

[88]. Considering that many compounds have $k_{\cdot OH} > 10^9\ M^{-1}\ s^{-1}$ [88], the reported data could allow a preliminary assessment of the relative role of $^3CDOM^*$, $^{\bullet}OH$ and $CO_3^{-\bullet}$ in the degradation of organic compounds in surface waters under typical conditions.

The relative importance of direct photolysis and of indirect photodegradation is very variable depending on the molecule under consideration and on the composition of water. Direct photolysis is not operational with molecules that do not absorb sunlight, in which case indirect photodegradation has to prevail, but the situation is more complicated for those molecules that absorb sunlight and are reactive toward $^3CDOM^*$, $^{\bullet}OH$, $CO_3^{-\bullet}$ and other transients. The intensity of sunlight absorption and the photolysis quantum yield play a very significant role in defining the rate of the direct photolysis of a given compound. If all the relevant parameters (absorption spectra, quantum yields, second-order reaction rate constants) are known, more precise predictions can be achieved by means of a recently developed photochemical model and software (APEX: Aqueous Photochemistry of Environmentally-occurring Xenobiotics) that determines pollutant photodegradation kinetics as a function of water chemistry and depth [20].

References

1. Oliveira JL, Boroski M, Azevedo JCR, Nozaki J (2006) Acta Hydrochim Hydrobiol 34:608–617
2. Brinkmann T, Hörsch P, Sartorius D, Frimmel FH (2003) Environ Sci Technol 32:3004–3010
3. Fenner K, Canonica S, Wackett LP, Elsner M (2013) Science 341:752–758
4. Vione D, Minella M, Maurino V, Minero C (2014) Chem Eur J 20:10590–10606
5. Kuivila KM, Jennings BE (2007) Int J Environ Anal Chem 87:897–911
6. Comoretto L, Arfib B, Chiron S (2007) Sci Total Environ 380:124–132
7. Fenner K, Canonica S, Escher BI, Gasser L, Spycher S, Tuelp HC (2006) Chimia 60:683–690
8. Siffert C, Sulzberger B (1991) Langmuir 7:1627–1634
9. Xyla AG, Sulzberger B, Luther GW, Hering JG, Van Cappellen P, Stumm W (1992) Langmuir 8:95–103
10. Acharid A, Sadiki M, Elmanfe G, Derkaoui N, Olier R, Privat M (2006) Langmuir 22:8790–8799
11. Muller B, Heal MR (2001) Chemosphere 45:309–314
12. Bucheli T, Müller S, Heberle S, Schwarzenbach R (1998) Environ Sci Technol 32:3457–3464
13. Grabner G, Richard C (2005) Environmental photochemistry part II. In: Boule P, Bahnemann DW, Robertson PKJ (eds) The handbook of environmental chemistry, vol 2M. Springer, Berlin, pp 161–192
14. Pagni RM, Dabestani R (2005) Environmental photochemistry part II. In: Boule P, Bahnemann DW, Robertson PKJ (eds) The handbook of environmental chemistry, vol 2M. Springer, Berlin, pp 193–219
15. Czaplicka M (2006) J Hazard Mater 134:45–59
16. Vione D, Maurino V, Minero C, Pelizzetti E, Harrison MAJ, Olariu RI, Arsene C (2006) Chem Soc Rev 35:441–453
17. Lam MW, Mabury SA (2005) Aquat Sci 67:177–188
18. Vogna D, Marotta R, Andreozzi R, Napolitano A, D'Ischia M (2004) Chemosphere 54:497–505
19. Ruggeri G, Ghigo G, Maurino V, Minero C, Vione D (2013) Water Res 47:6109–6121

20. Bodrato M, Vione D (2014) Environ Sci Process Impacts 16:732–740
21. Vione D, Maddigapu PR, De Laurentiis E, Minella M, Pazzi M, Maurino V, Minero C, Kouras S, Richard C (2011) Water Res 45:6725–6736
22. Braslavsky SE (2007) Pure Appl Chem 79:293–465
23. Brinkmann T, Sartorius D, Frimmel FH (2003) Aquat Sci 65:415–424
24. Frank R, Klöpffer W (1988) Chemosphere 17:985–994
25. Calvert JG, Pitts JN (1966) Photochemistry. Wiley, New York
26. Alif A, Boule PJ (1991) Photochem Photobiol A Chem 59:357–367
27. Mack J, Bolton JRJ (1999) Photochem Photobiol A Chem 128:1–13
28. Endicott JF, Feraudi G, Barber JRJ (1975) Phys Chem 79:630–643
29. Mallakin A, Dixon DG, Greenberg BM (2000) Chemosphere 40:1435–1441
30. Maurino V, Bedini A, Borghesi D, Vione D, Minero C (2011) Phys Chem Chem Phys 13:11213–11221
31. Dabestani R, Higgin J, Stephenson DM, Ivanov IN, Sigman MEJ (2000) Phys Chem B 104:10235–10241
32. Maurino V, Borghesi D, Vione D, Minero C (2008) Photochem Photobiol Sci 7:321–327
33. Sigman ME, Barbas JT, Corbett S, Chen Y, Ivanov IN, Dabestani RJ (2001) Photochem Photobiol A Chem 138:269–274
34. Brunke M, Gonser T (1997) Freshw Biol 37:1–33
35. Czaplicka M (2004) Sci Total Environ 322:21–39
36. Chiron S, Minero C, Vione D (2007) Environ Sci Technol 41:3127–3133
37. Chiron S, Comoretto L, Rinaldi E, Maurino V, Minero C, Vione D (2009) Chemosphere 74:599–604
38. Zhang HC, Huang CH (2003) Environ Sci Technol 37:2421–2430
39. Latch DE, Pacher JL, Stender BL, VanOverbeke J, Arnold WA, McNeill K (2005) Environ Toxicol Chem 24:517–525
40. Guyon C, Boule P, Lemaire J (1984) New J Chem 8:685–692
41. Vione D, Minero C, Housari F, Chiron S (2007) Chemosphere 69:1548–1554
42. Bonnichon F, Richard C, Grabner G (2001) Chem Commun 73–74
43. Vione D, Khanra S, Das R, Minero C, Maurino V, Brigante M, Mailhot G (2010) Water Res 44:6053–6062
44. Meunier L, Gauvin E, Boule P (2002) Pest Manag Sci 58:845–852
45. Maddigapu PR, Minella M, Vione D, Maurino V, Minero C (2011) Environ Sci Technol 45:209–214
46. Latch DE, Packer JL, Arnold WA, McNeil KJ (2003) Photochem Photobiol A Chem 158:63–66
47. Bianco A, Fabbri D, Minella M, Brigante M, Mailhot G, Maurino V, Minero C, Vione D (2015) Water Res 72:271–280
48. Brigante M, DellaGreca M, Previtera L, Rubino M, Temussi F (2005) Environ Chem Lett 2:195–198
49. Chiron S, Minero C, Vione D (2006) Environ Sci Technol 40:5977–5983
50. De Laurentiis E, Chiron S, Kouras-Hadef S, Richard C, Minella M, Maurino V, Minero C, Vione D (2012) Environ Sci Technol 46:8164–8173
51. DellaGreca M, Iesce MR, Previtera L, Rubino M, Temussi F (2004) Environ Chem Lett 2:155–158
52. Cermola M, DellaGreca M, Iesce MR, Previtera L, Rubino M, Temussi F, Brigante M (2005) Environ Chem Lett 3:43–47
53. Richard C, Canonica S (2005) Environmental photochemistry part II. In: Boule P, Bahnemann DW, Robertson PKJ (eds) The handbook of environmental chemistry, vol 2M. Springer, Berlin, pp 299–323
54. Anastasio C, Faust BC, Rao CJ (1997) Environ Sci Technol 31:218–232
55. Alegrìa AE, Ferrer A, Santiago G, Sepúlveda E, Flores WJ (1999) Photochem Photobiol A Chem 127:57–65

56. Bedini A, De Laurentiis E, Sur B, Maurino V, Minero C, Brigante M, Mailhot G, Vione D (2012) Photochem Photobiol Sci 11:1445–1453
57. Canonica S, Jans U, Stemmler K, Hoigne' J (1995) Environ Sci Technol 29:1822–1831
58. Canonica S, Freiburghaus M (2001) Environ Sci Technol 35:690–695
59. Halladja S, Ter Halle A, Aguer JP, Boulkamh A, Richard C (2007) Environ Sci Technol 41:6066–6073
60. Gerecke AC, Canonica S, Muller SR, Scharer M, Schwarzenbach RP (2001) Environ Sci Technol 35:3915–3923
61. Buschmann J, Canonica S, Lindauer U, Hug SJ, Sigg L (2005) Environ Sci Technol 39:9541–9546
62. Buschmann J, Canonica S, Sigg L (2005) Environ Sci Technol 39:5335–5341
63. Vione D, Caringella R, De Laurentiis E, Pazzi M, Minero C (2013) Sci Total Environ 463–464:243–251
64. Avetta P, Marchetti G, Minella M, Pazzi M, De Laurentiis E, Maurino V, Minero C, Vione D (2014) Sci Total Environ 500–501:351–360
65. Hustert K, Moza PN, Kettrup A (1999) Chemosphere 38:3423–3429
66. Krieger MS, Yoder RN, Gibson AJ (2002) Agric Food Chem 50:3710–3717
67. Sakkas VA, Lambropoulou DA, Albanis TAJ (2002) Photochem Photobiol A Chem 147:135–141
68. Burns SE, Hassett JP, Rossi MV (1996) Environ Sci Technol 30:2934–2941
69. Miller PL, Chin YPJ (2002) Agric Food Chem 50:6758–6765
70. Welker M, Steinberg C (2000) Environ Sci Technol 34:3415–3419
71. Hoigné J (1990) In: Stumm W (ed) Aquatic chemical kinetics. Wiley, New York, pp 43–70
72. Brezonik PL, Fulkerson-Brekken J (1998) Environ Sci Technol 32:3004–3010
73. Page SA, Arnold W, McNeill K (2011) Environ Sci Technol 45:2818–2825
74. Lee E, Glover CM, Rosario-Ortiz FL (2013) Environ Sci Technol 47:12073–12080
75. Vermilyea AW, Voelker BM (2009) Environ Sci Technol 43:6927–6933
76. Sur B, Rolle M, Minero C, Maurino V, Vione D, Brigante M, Mailhot G (2011) Photochem Photobiol Sci 10:1817–1824
77. Maddigapu PR, Bedini A, Minero C, Maurino V, Vione D, Brigante M, Mailhot G, Sarakha M (2010) Photochem Photobiol Sci 9:323–330
78. Southworth BA, Voelker BM (2003) Environ Sci Technol 37:1130–1136
79. White EM, Vaughan PP, Zepp RG (2003) Aquat Sci 65:402–414
80. Mazellier P, Mailhot G, Bolte M (1997) New J Chem 21:389–397
81. Allen JM, Lucas S, Allen SK (1996) Environ Toxicol Chem 15:107–113
82. King DW, Aldrich RA, Charneeki SE (1993) Mar Chem 44:105–120
83. Minero C, Chiron S, Falletti G, Maurino V, Pelizzetti E, Ajassa R, Carlotti ME, Vione D (2007) Aquat Sci 69:71–85
84. Marchisio A, Minella M, Maurino V, Minero C, Vione D (2015) Water Res 73:145–156
85. Vione D, Minero C, Maurino V, Pelizzetti E (2007) Ann Chim (Rome) 97:699–711
86. Vione D, Falletti G, Maurino V, Minero C, Pelizzetti E, Malandrino M, Ajassa R, Olariu RI, Arsene C (2006) Environ Sci Technol 40:3775–3781
87. Minero C, Lauri V, Maurino V, Pelizzetti E, Vione D (2007) Ann Chim (Rome) 97:685–698
88. Buxton GV, Greenstock CL, Helman WP, Ross ABJ (1988) Phys Chem Ref Data 17:513–886
89. Hatipoglu A, Vione D, Yalçin Y, Minero C, Çinar ZJ (2010) Photochem Photobiol A Chem 215:59–68
90. Wardman PJ (1989) Phys Chem Ref Data 17:1027–1717
91. Larson RA, Zepp RG (1988) Environ Toxicol Chem 7:265–274
92. Canonica S, Kohn T, Mac M, Real FJ, Wirz J, Von Gunten U (2005) Environ Sci Technol 39:9182–9188
93. Huang JP, Mabury SA (2000) Environ Toxicol Chem 19:2181–2188
94. Neta P, Huie RE, Ross ABJ (1988) Phys Chem Ref Data 17:1027–1228
95. Boreen AL, Edhlund BL, Cotner JB, McNeill K (2008) Environ Sci Technol 42:5492–5498

96. Haag WR, Hoigné J, Gassmann E, Braun AM (1984) Chemosphere 13:631–640
97. Latch DE, McNeill K (2006) Science 311:1743–1747
98. Appiani E, McNeill K (2015) Environ Sci Technol 49:3514–3522
99. Mill T, Hendry DG, Richardson H (1980) Science 297:886–887
100. Vione D, Maurino V, Minero C, Pelizzetti E (2005) Environmental photochemistry part II. In: Boule P, Bahnemann DW, Robertson PKJ (eds) The handbook of environmental chemistry, vol 2M. Springer, Berlin, pp 221–253
101. Maddigapu PR, Vione D, Ravizzoli B, Minero C, Maurino V, Comoretto L, Chiron S (2010) Environ Sci Pollut Res 17:1063–1069
102. Minero C, Maurino V, Pelizzetti E, Vione D (2007) Environ Sci Pollut Res 14:241–243
103. Vione D, Maurino V, Minero C, Pelizzetti E (2002) Environ Sci Technol 36:669–676
104. Cullen JT, Bergquist BA, Moffett JW (2006) Mar Chem 98:295–303
105. McGregor KG, Anastasio C (2001) Atmos Environ 35:1091–1104
106. Heng ZC, Ong T, Nath J (1996) Mutat Res 368:149–155
107. Vione D, Maurino V, Minero C, Calza P, Pelizzetti E (2005) Environ Sci Technol 39:5066–5075
108. Khanra S, Minero C, Maurino V, Pelizzetti E, Dutta BK, Vione D (2008) Environ Chem Lett 6:29–34
109. Brigante M, Minella M, Mailhot G, Maurino V, Minero C, Vione D (2014) Chemosphere 95:464–469
110. Calza P, Vione D, Novelli A, Pelizzetti E, Minero C (2012) Sci Total Environ 439:67–75
111. Calza P, Vione D, Minero C (2014) Sci Total Environ 493:411–418
112. Vione D, Maurino V, Cucu Man S, Khanra S, Arsene C, Olariu RI, Minero C (2008) ChemSusChem 1:197–204
113. Bouillon RC, Miller WL (2005) Environ Sci Technol 39:9471–9477
114. Leroy SAG, Marret F, Giralt S, Bulatov SA (2006) Quart Intern 150:52–70
115. Das R, Dutta BK, Maurino V, Vione D, Minero C (2009) Environ Chem Lett 7:337–342
116. De Laurentiis E, Minella M, Sarakha M, Marrese A, Minero C, Mailhot G, Brigante M, Vione D (2013) Water Res 47:5943–5953
117. Nissenson P, Dabdub D, Das R, Maurino V, Minero C, Vione D (2010) Atmos Environ 44:4859–4866
118. Vione D, Sur B, Dutta BK, Maurino V, Minero CJ (2011) Photochem Photobiol A Chem 224:68–70
119. Waite TD (2005) Environmental photochemistry part II. In: Boule P, Bahnemann DW, Robertson PKJ (eds) The handbook of environmental chemistry, vol 2M. Springer, Berlin, pp 255–298
120. Balmer ME, Sulzberger B (1999) Environ Sci Technol 33:2418–2424
121. Kari FG, Hilger S, Canonica S (1995) Environ Sci Technol 29:1008–1017
122. Kuma K, Nakabayashi S, Matsunaga K (1995) Water Res 29:1559–1569
123. McKnight DM, Kimball BA, Bencala KE (1988) Science 240:637–640
124. Emmenegger L, Schonenberger R, Sigg L, Sulzberger B (2001) Limnol Oceanogr 46:49–61
125. Waite TD, Szymczak R, Espey QI, Furnas MJ (1995) Mar Chem 50:79–91
126. Barbeau K, Rue EL, Bruland KW, Butler A (2001) Nature 413:409–413
127. Barbeau K, Zhang G, Live DH, Butler AJ (2002) Am Chem Soc 124:378–379
128. Barbeau K, Rue EL, Trick CG, Bruland KW, Butler A (2003) Limnol Oceanogr 48:1069–1078
129. Xiao YH, Raike A, Hartikainen H, Vahatalo AV (2015) Sci Total Environ 536:914–923
130. Canonica S, Hellrung B, Müller P, Wirz J (2006) Environ Sci Technol 40:6636–6641

Chapter 8
Photodynamic Therapy

Barbara Krammer and Thomas Verwanger

Abstract Photodynamic therapy (PDT) and fluorescence diagnosis (FD) are effective modalities for treatment or detection of tumors and various other diseases. Both are based on the same principle: accumulation of a photosensitizer (PS) in target cells (or tissue) and irradiation with visible light. Subsequently, the photo-activated PS either returns to the ground state by fluorescence (used for detection and diagnosis, FD) or crosses to its triplet state and reacts chemically with surrounding molecules. When molecular oxygen is present, reactive oxygen species (ROS) including singlet oxygen are generated, which oxidize, amongst others, proteins and lipids affecting the target and leading dose-dependently to its destruction (PDT). Since most PSs accumulate selectively in tumor tissue, PDT and FD are especially suited for tumor therapy or diagnosis, respectively. Although different kinds of tumors can be treated due to the unspecific action of ROS, PDT is mainly restricted to flat tumors located at inner or outer surfaces of the body as light accessibility has to be guaranteed. Various PSs such as porphyrins are approved or currently tested in clinical trials.

After administration to a patient, the PS is transported to the target and localizes in specific cell compartments. Following PS irradiation cells react in different ways to the oxidizing ROS. Low numbers of ROS can be quenched by antioxidants present in the cell and the damage be repaired. Even stimulation of proliferation or of immune reactions was observed. However, ROS levels above a specific threshold induce cell death mainly via apoptosis and necrosis.

B. Krammer (✉) • T. Verwanger
Department of Molecular Biology, University of Salzburg, Hellbrunnerstrasse 34, Salzburg, Austria
e-mail: barbara.krammer@sbg.ac.at; thomas.verwanger@sbg.ac.at

G. Bergamini, S. Silvi (eds.), *Applied Photochemistry*,
Lecture Notes in Chemistry 92, DOI 10.1007/978-3-319-31671-0_8

8.1 Principle and Application of Photodynamic Processes

Photodynamic therapy (PDT) and photodynamic (PDD) or fluorescence diagnosis (FD) offer efficient and clinically approved therapeutic and diagnostic modalities for detection of and intervention in (or ideally cure from) malignant and some non-malignant diseases [1]. The success especially in destruction of tumors, preferentially located at external and internal surfaces of the body, e.g. skin, head and neck, lung, bladder, brain, gastrointestinal tract, cervix uteri and prostate, in small vessels removal and in diseases like microbial infections is based on the same principle: a *per se* harmless and light-sensitive molecule (photosensitizer, PS) gets activated by visible light with a suitable wavelength; by the subsequent deactivation of the molecules it can either emit light (used for detection, diagnosis or fluorescence guided resection) or react with other molecules, leading to a chain reaction (used for therapy). In this case the photo-activated PS generates mainly reactive oxygen species (ROS) in the presence of molecular oxygen, which induces different stimulating or damaging processes in biological matter [23, 24, 36]. The dependence on oxygen limits the photochemical reactions in tissue under hypoxic conditions, which is often the case in tumors. Beside killing of tumor and blood vessel cells by different kinds of cell death, stimulation of a wide range of immune reactions was also reported [17, 20], comprising inflammatory processes, attack of immune cells on the primary tumor and long-lasting systemic immunity against the tumor [72]. Specific low dose protocols of PDT may even be used as PDT vaccine [30, 46] in order to prevent tumor growth by anti-cancer immunity [47, 72], but can also counteract tumor destruction by stimulation of cell proliferation [26].

8.1.1 Selectivity

Since those photochemical processes generate unspecific effects in any biological matter, which would lead to an equal impact on cancerous and healthy tissues, or which would not specifically target e.g. vessels or bacteria, respectively, a selective uptake of the PS to tumor tissue, to small vessels or bacteria is necessary and actually found in most cases. Different factors like tumor-specific cellular uptake (e.g. via increased low-density lipoprotein receptors), facilitated tumor tissue penetration mechanisms (due to loss of E-cadherin; leakiness of tumor vasculature) [37], longer retention in tumor tissue, accumulation of tumor-associated macrophages, which exert phagocytosis and monomerization of aggregated PSs [35, 39], or PS properties as the degree of lipophilicity account for this. Coupling of PSs to tumor-specific antibodies or other carriers can further enhance the selectivity of the PS (s. below) [67, 80].

Furthermore, if possible, also the irradiation is performed selectively in the case of therapy, directed only to the target and sparing the non-target sites. Thus photo-induced damage is restricted to the tumor and surrounding tissue [3, 64].

In case of fluorescence diagnosis, the irradiation area has to be as extensive as possible to increase the probability for the detection and demarcation of malignant tissue. However, false positive results are often obtained when inflammatory milieu and an increased number of macrophages (like tumor-associated macrophages) are present, as macrophages are able to accumulate a multiple of photosensitizing molecules compared to non-macrophagic cells [22].

8.1.2 Clinical Application

When applied for therapy, the photodynamic approach is free of major side effects, relevant mutagenicity and resistance development as it appears to be the disadvantage of traditional radiation- and chemotherapy, or of the more selective targeted therapies. Although PDT is mainly restricted to outer and inner surfaces, clinical trials for prostate and breast cancer are ongoing.

Patients receive the PS either systemically by i.v. injection, instillation or topical administration. For the latter one, the access to tissue may be impeded e.g. by hornification of the skin. This requires procedures to remove the barrier e.g. by scraping the upper cell layers.

Preceding therapeutical intervention, fluorescence diagnosis is often carried out to validate biopsy appraisal.

Following delivery, the PS penetrates the tissue directly (topical administration or instillation) or is firstly transported in the blood stream to the target site (intravenous administration). After selective accumulation e.g. in the tumor tissue and, to a minor degree or retarded in normal tissue, the PS will be exported from the cells and normally excreted by metabolic ways via liver, kidney and bladder. Many PSs show a delay in leaving the target tissue compared to the normal one. This retention is used for photoactivation of the PS, since the accumulation ratio of target: normal tissue is at its maximum. This ratio may reach high values up to 1:60 as for glioblastoma [56]. The time span between PS administration and irradiation varies between a few minutes (elimination of dental bacteria) and a few days (systemic administration of porphyrins), depending on PS properties and delivery modes.

Irradiation is performed by different light sources such as (diode) lasers, lamps (e.g. halogen) with filters or LED devices. The light is transmitted to tissue either directly (e.g. for external surfaces) or by means of fiber optics. A homogenous light distribution in the cavities is achieved by different applicators, appropriate for the respective organ.

After PDT the treatment area or the patients in case of systemic PS application, respectively, have to remain in subdued light for a certain period of time to avoid exposure to bright sunlight. However, to circumvent increased skin sensitivity and to minimize other potential adverse side effects, an accurate dosimetry of both, drug concentration and light application is required [61, 75].

In addition, elimination by photodegradation and physiological clearance of the PS should be as fast as possible to minimize the risk of general photosensitivity. However, for some indications (for example multiple photodynamic treatments) the retention of PS molecules in the target tissue might be desired to avoid re-administration of the drug [3, 64, 77].

PDT can be repeated several times, since it is not mutagenic, or combined with other tumor therapies. The fluorescence guided resection supports the surgical removal of the tumor by precise discrimination between malignant and non-malignant tissue.

The efficiency of PDT depends on the type and concentration of the PS, its uptake to and clearance from tissue and cells, intracellular (re)localization during the incubation time, irradiation parameters (fluence, fluence rate, wavelength and fractionation), oxygen availability and susceptibility of the target including anti-oxidant defense mechanisms.

Subsequently we shall give an overview on photosensitizers, photophysical and photochemical processes involved in irradiation of PSs, and on cellular effects.

8.2 Photosensitizers

Photosensitizers are characterized by extended ring systems of mostly tetrapyrrolic structure, which allow for activation by visible light.

The first clinically applied photosensitizers were hematoporphyrins, which had sub-optimal properties. Subsequently many more PS were designed and synthesized or detected, derived from the basic molecules of chlorin, benzoporphyrin, pheophorbide, purpurin, texaphyrin, porphycene and phthalocyanine; further PSs are natural substances such as hypericin, a constituent of the St. John's wort, or curcumin from turmeric (Table 8.1) and lately artificial PSs such as boron-dipyrromethene (BODIPY) dyes.

A special case is the endogenously generated protoporphyrin IX (PpIX). This molecule is normally formed in most cells in the heme synthesis pathway from the precursor 5-aminolevulinic acid (5-ALA). PpIX is the last synthesis step before heme and its concentration is strongly dependent on the activity of the enzymes converting the respective precursor to the next molecule. Especially critical and limiting is the generation of heme by inserting Fe^{2+} in PpIX by ferrochelatase. When the prodrug 5-ALA is given externally in excess, PpIX is formed and accumulated, since the conversion to heme is executed on a normal physiological rate. Thus, PpIX is available for some hours as a very effective PS. As ferrochelatase is reduced in tumors, the conversion of PpIX to heme is slower, causing a tumor-selective effect.

Many of the PSs are already approved for different applications in different countries and clinically used; some of them are currently being tested in preclinical

Table 8.1 List of photosensitizers with clinical relevance, basic molecule structure and significant characteristics

Class	Basic structure	Photosensitizer	Trade name	Distributed by	Indications	Approval	Absorption maximum (nm)	Extinction coefficient ($M^{-1}\,cm^{-1}$)
Hemato-porphyrin	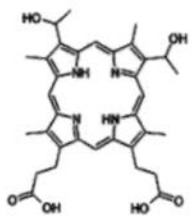	HpD (partially purified), porfimer podium	Photofrin	Axcan Pharma Inc.	Endobronchial non-small-cell lung cancer, esophagial cancers, bladder cancer, gastric and cervical cancer	Canada (1993)	630	3,000
						Japan (1994)		
			Photogem	Moscow Research Oncological Institute		US (1995)		
						Now approved in over 40 countries		
Protoporphyrin prodrug		5-Aminolevulinic acid (5-ALA)	Levulan	DUSA Pharmaceuticals Inc.	Actinic keratosis, basal-cell carcinoma, head and neck and gynaecological tumors; diagnosis of brain, head and neck and bladder tumors	US (1999)	635	<10,000
						Europe (2001)		
		5-ALA-hexylester/H-ALA	Hexvix	PhotoCure ASA	Diagnosis of bladder tumors	US (2010)	375–400	
			Cysview	Photocure Inc.				
		5-ALA-methylester/M-ALA	Metvix	Galderma S.A.	Actinic keratosis, Bowen's disease and basal cell carcinoma	US (2004)	635	
						EU (2001)		
						New Zealand (2002)		
						Australia (2003)		
		5-ALA-benzylester	Benzvix	PhotoCure ASA	Gastrointestinal cancer	Not approved	635	
Proto-porphyrin	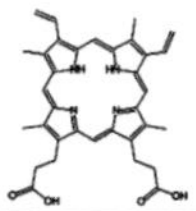	Boronated proto-porphyrin	BOPP	–	Brain tumors	Not approved	630	

(continued)

Table 8.1 (continued)

Class	Basic structure	Photosensitizer	Trade name	Distributed by	Indications	Approval	Absorption maximum (nm)	Extinction coefficient (M^{-1} cm^{-1})
Benzo-porphyrin	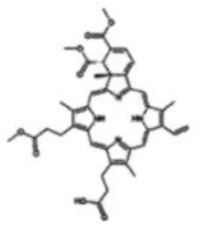	Benzoporphyrin derivative monoacid ring A (BDP-MA)/ Verteporfin	Visudyne	Valeant Pharmaceuticals International, Inc.	Treatment for wet age-related macular degeneration (AMD), pathologic myopia, histoplasmosis	US (2000) EU (2000) Canada (2000)	689–693	35,000
Chlorins	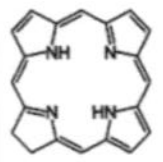	Meta-tetrahydro-phenylchlorin (m-THPC)/Temoporfin	Foscan	Biolitec Pharma Ltd.	Head and neck, prostate and pancreatic tumors	EU (2001)	652	30,000
		Mono-L-aspartyl chlorin e6 (Talaporfin, NPe6, LS11 or MACE)	Aptocine Laserphyrin	Light Sciences Meiji Seika Pharma Co., Ltd	Lung cancer and solid tumors from diverse origins	Japan (2003)	664	45,000
Phthalocyanines	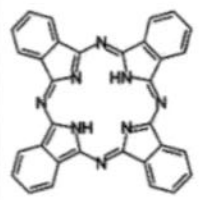	Sulfonated aluminium phthalocyanines ($AlPcS_4$)	Photosens	General Physics Institute Moscow	AMD, various cancers	Russia (2001)	675	110,000
		Silicon phthalocyanines	Pc4	Case Western Reserve University	Actinic keratosis, Bowen's disease, T-cell non-Hodgkin lymphoma and skin cancers	Clinical trials (phase I)/USA	670	40,000
		Zinc phthalocyanine (ZnPc)	CGP55847	Ciba-Geigy Ltd.	Actinic keratosis, Bowen's disease, skin cancer, squamous cell carcinoma of upper aerodigestive tract	Clinical trials (phase I)/USA	675	243,000

Texafrins	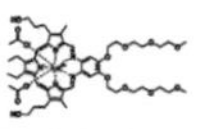	Motexafin lutetium/Lutexaphrin	Antrin	Pharmacylics Inc.	Prostate cancer and photoangioplasty	Clinical trials (phase II)/ USA	732	42,000
Pheophorbide-a	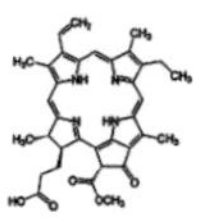	HPPH (2-[1-hexyloxyethyl]-2-devinyl pyro-pheophorbide-a)	Photochlor	Roswell Park Cancer Institute	Early esophageal cancers, non-small cell lung cancer	Clinical trials (phase II)	665	47,500
		Palladium-bacterio-pheophorbide	Tookad (WST09)	The Weisman Institute of Science	Recurrent prostate cancer	Clinical trials (phase III)/ Europe	763	>100,000
Purpurins	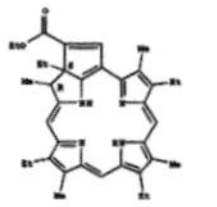	Tin etiopurpurin ($SnET_2$)/ Purlytin/Rostaporfin	Photrex	Miravant Medical Technologies Inc.	Cutaneous metastatic breast cancer, basal-cell carcinoma, Kaposi's sarcoma, and prostate cancer	Clinical trials (phase II)/ USA	660	28,000
Porphycenes	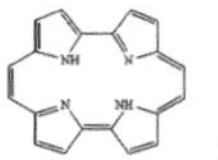	9-Acetoxy-2,7,12,17-tetrakis-(-β-methoxyethyl)-porphycene	ATMPn	Glaxo Wellcome and Cytopharm	Psoriasis and non-melanoma skin cancer	Germany (1997)	610–650	50,000
Anthraquinones	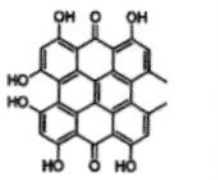	4,5,7,4′,5′,7′-Hexahydroxy-2,2-′-dimethylnaphthodianthrone/ Hypericin	Vidon (PVP-Hypericin)	Sanochemia Pharmazeutika AG	Diagnosis of bladder tumors	Clinical trials (phase II)/EU	590	<52,000
					Treatment of squamous cell carcinoma and basal cell carcinoma			
Curcuminoides	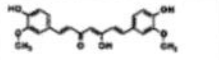	(1E,6E)-1,7-Bis(4-hydroxy-3-methoxyphenyl)-1,6-heptadiene-3,5-dione	Curcumin	–	Oral Disinfection	Clinical trials (phase I)/USA	425	55,000

(continued)

Table 8.1 (continued)

Class	Basic structure	Photosensitizer	Trade name	Distributed by	Indications	Approval	Absorption maximum (nm)	Extinction coefficient (M^{-1} cm^{-1})
Phenothiazines	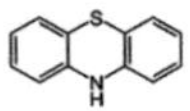	3,7-bis (Dimethylamino)-phenothiazin-5-ium chloride	Methylene blue	–	Basal cell carcinoma, Kaposi's sarcoma, chronic periodontitis	Clinical trials (phase I)/USA	666	82,000
		(7-amino-8-methyl-phenothiazin-3-ylidene)-dimethyl-ammonium	Toluidine blue	–	Chronic periodontitis	Clinical trials (phase II)/ USA	596, 630	<51,000
Xanthenes	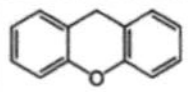	4,5,6,7-Tetrachloro-3′,6-′-dihydroxy-2′,4′,5′,7′-tetraiodo-3H- spiro[isobenzofuran-1,9-′-xanthen]-3-one	Rose Bengal	Provectus Pharmaceuticals	Melanoma, breast cancer, eczema, psoriasis	Clinical trials/ USA	549	100,000
		4,5-Dibromo-rhodamine methyl ester	TH9409	Kiadis Pharma	Graft-versus-host disease, allogeneic stem cell transplantation	Clinical trials/ USA	514	100,000
Cyanines	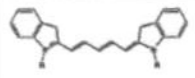	Sodium 3-[(2Z)-2-[(E)-4-(1,3-Dibutyl-2,4,6-trioxo-5-hexahydropyrimidinylidene)but-2-enylidene]-1,3-benzoxazol-3-yl]propane-1-sulfonate	Mero-cyanine 540	–	Leukemia, neuroblastoma	–	556	110,000

studies. Basic structures, clinical indications, state of approval and photodynamically relevant features of the different PSs are displayed in Table 8.1 [1, 49].

An ideal PS should have at least the following properties:

- good solubility and stability,
- purity of known chemical composition,
- efficient and selective accumulation in target tissue,
- (re)localization in those organelles, which are supposed to be responsible for a desired cell death mode, but not in the nucleus,
- low dark toxicity and mutagenicity,
- strong absorption with a high extinction coefficient in the wavelength region between 650 and 850 nm allowing light penetration in tissue as far as possible,
- high intersystem crossing (ISC) efficiency and high singlet oxygen quantum yield,
- photochemical reactions also under hypoxia,
- rapid clearance from healthy tissue and from the body after PDT,
- lack of major side effects.

In all photodynamic treatments, the solubility of the PS plays a prominent role. Especially when a systemic application of the PS is indicated, its transportation in the blood without precipitation or aggregate formation should be ensured, but also effective penetration of the lipid layer of the cell membrane. PS molecules should not form aggregates as this can reduce their capability to absorb and to emit light; in addition a reduction of the lifetime and the quantum yield for ISC is possible [59].

Concerning the photophysical properties of a PS, the wavelength for its photoactivation in the visible spectrum and the quantum efficiency for the triplet state are most important for therapeutic use. The maximal absorption of the PS should ideally match the optical window between 650 and 850 nm where tissue penetration of the light is at least a few mm and the energy level of the triplet state after ISC (s. below) permits singlet oxygen production: the upper wavelength limit for a PS efficiently producing $^1O_2$ is reported to be 850 nm (s. below) [39]. ISC with a high quantum yield and a long-lasting triplet state are the basis for effective generation of cytotoxic products [59]. For diagnostic use of a PS including fluorescence-guided resection and control of the treatment success by comparison of the fluorescence intensities before and after PDT [14], a photoactivated PS should also show a detectable level of fluorescence [80, 83].

Since an ideal PS is not yet available, current research is focused on the improvement of PS properties. One of these advances is the facilitated transport of the PS to tumor tissue by using carriers, another one tumor selective targeting e.g. by coupling the PS to tumor specific antibodies or to LDLs [67], and finally the increased uptake to tumor cells by utilizing effective uptake routes such as endocytosis.

As carriers for PS serve e.g. peptides [34], polymers such as PVP, PEG and polyphosphazenes [26], nanoparticles [19], fullerenes [55] and dendrimers [42].

The technology of photochemical internalization (PCI) is used to transport macromolecules (e.g. toxins such as chemotherapeutics) associated with a PS via endocytosis into the cytoplasm. The endocytic vesicles protect the molecules against lysosomal degradation. When the PS, which is incorporated in the membrane of the vesicles, is photoactivated, it disrupts the vesicles and liberates the load of macromolecules with high precision within the cell. If used in cancer therapy, this technology helps to reduce toxin concentration [73].

8.3 Irradiation with Visible Light

Before a PS – accumulated in target tissue and cells – can be photoactivated, the visible light has often to penetrate tissue along the path to the target, thereby interacting with surrounding molecules. The dosimetry of light application is difficult as the light passes through different media with changing optical properties.

8.3.1 Interaction of Light with Tissue

For light propagation in tissues, the processes of refraction, reflection, absorption and scattering play a role. Besides scattering, absorption of photons in tissue is most relevant in PDT for the attenuation of light intensity with increasing penetration depth [63]. This is not only due to a high portion of water in soft tissue, but also to chromophores. The most important ones are oxy- (HbO_2) and deoxyhemoglobin, melanin and cytochromes. On the basis of the absorption spectra of these molecules a reduction of the absorption allowing an "optical window" for PDT in tissue can be determined between 650 and 1200 nm (Fig. 8.1, [4, 16, 64]). This range is further restricted, as PSs in the triplet state have not enough energy to efficiently produce singlet oxygen when excited by $\lambda > 850$ nm.

That for PDT of solid tumors the actual penetration depth is highly important, shows the clinical use of Photofrin: only when it is photoactivated at its last absorption peak of 630 nm, the light penetrates about 3–5 mm to the tissue [81]. The four peaks at shorter wavelengths are in fact higher, but light with reduced penetration depth is absorbed, which is only applicable in FD.

Applying PS with absorption peaks at wavelengths higher than 700 nm should enable treatment of thicker tumors, since the penetration depth could be doubled [76, 84]. As mentioned above, the upper limit for the excitation wavelength is the lower limit of energy, which allows singlet oxygen production.

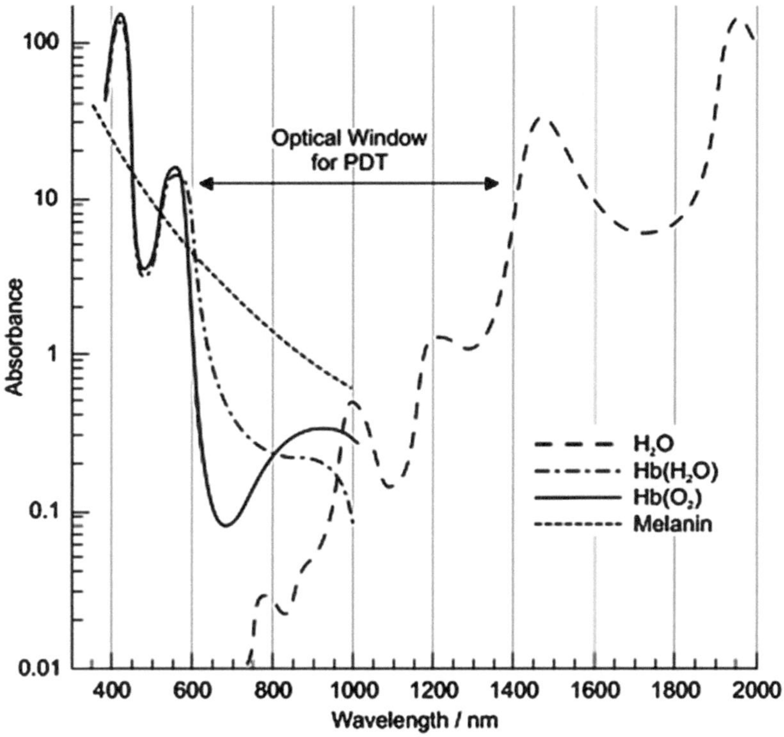

Fig. 8.1 Absorption of light by tissue chromophores and water. Reduction of the absorbance between 650 and 1200 nm allows an ‘optical window’ for PDT (With kind permission from Springer Science+Business Media [63])

8.3.2 Interaction of Light with PS

Those photons, which are not absorbed by the overlaying tissue, finally meet the PS in the target and activate it.

A photosensitizer molecule in its ground state S_0, which absorbs a light quantum with the appropriate energy, is activated to a higher singlet state S_x (with $x \geq 1$). Any excited stated S_x is further divided in so-called vibrational levels with increasing energy. As any excited state is energetically less favorable than the ground state, the PS returns to S_0 as fast as possible. This deactivation of a PS in S_x can occur by several mechanisms (see Fig. 8.2, [64, 79]).

An electron in a high vibrational level of an excited state (e.g. S_1) rapidly falls to the energetically lowest level of that state. This may occur by vibrational relaxation (VR) under conversion of the energy to heat. Molecular relaxation to S_0 may also happen by heat production or emission of a photon within 10^{-11}–10^{-8} s (=fluorescence). In this case light of the PS starts from the lowest vibrational level of S_1 so that the emitted quanta have a lower energy (and a longer wavelength)

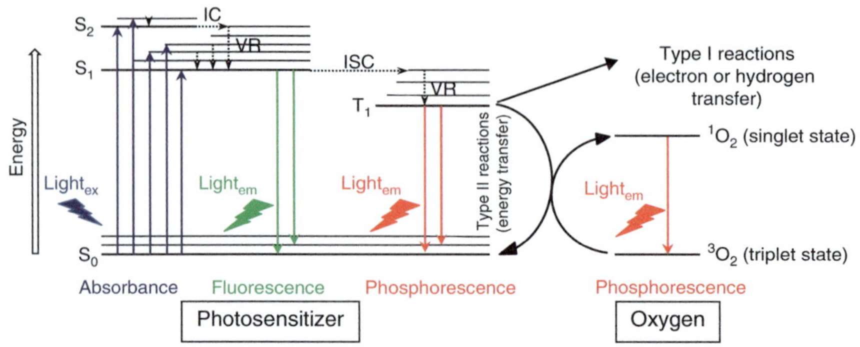

Fig. 8.2 Photosensitization processes illustrated by a Jablonski diagram. *IC* internal conversion, *VR* vibrational relaxation, *ISC* intersystem crossing

than those used for excitation of the molecule [64, 79]. This red-shift is used in fluorescence diagnosis to discriminate between exciting and emitted light.

From the S_1 state the PS molecule can switch with a high probability to the energetically lower level of the triplet state T_1 by ISC (Fig. 8.2). Starting at the T_1 energetic level a relaxation to S_0 by phosphorescence (i.e. emission of light in the triplet state) is possible. Since triplet states are generally characterized by a lifetime in the range of milliseconds, photochemical reactions may be started as alternative to phosphorescence [59].

8.3.3 Photochemical Processes in PDT

The photochemical processes of a PS in the T_1 state, in the presence of surrounding molecules, are operated via two pathways, type-I and type-II photochemical reactions (Fig. 8.2). The type-I photochemical reaction is based on the transfer of electrons (or protons) to oxygen or other adjacent molecules to form radical anions or cations, such as superoxide anions [28, 59]. These radicals can react further with molecular oxygen to produce different ROS. In the case of superoxide anion, this is the production of hydrogen peroxide (H_2O_2), which can easily pass biological membranes and propagate interactions in a wider area. Thus, hydrogen peroxide plays a major role in producing cellular damage [9, 74].

Hydrogen peroxide can also react with superoxide anions in a *Haber Weiss* reaction to form hydroxyl radicals. Those are very active in oxidizing the molecules in a cell and feature the property of rapid diffusion across membranes, similar to hydrogen peroxide. For this reason also the damage of hydroxyl radicals is not restricted to one cellular compartment [9, 32]. If metal ions such as iron or copper are present, hydroxyl radicals can also be produced from hydrogen peroxide by the *Fenton* reaction [32, 33].

The type-II photochemical reaction is the term for the transfer of energy of the PS to molecular oxygen, by which its elementary form, the triplet state ($^3O_2$), is changed to the very reactive singlet oxygen ($^1O_2$) [28]. Almost all PSs employed in PDT show high quantum yields for singlet oxygen formation [39].

Singlet oxygen as an uncharged molecule can diffuse across the cytoplasm and cellular membranes. The lifetime of singlet oxygen in pure water with 3×10^{-6} s [39, 74] is reduced in the cytoplasm by more than one order of magnitude to a maximal value of 1×10^{-7} s due to the interaction with reacting molecules [27, 44, 51, 52, 57]. The largest radius of singlet oxygen interaction was calculated to be 30 nm [51].

Each of these ROS leads to oxidation of substrates at the site of irradiation and within the maximal diffusion range of the single reactive products. As a consequence, the intracellular localization of the PS determines to a large extent the site of cellular damage generated by PDT.

Dependent mainly on the type of photosensitizer and the oxygen concentration, both photochemical reactions may occur in parallel [28, 59]. However, most PSs used in PDT react to irradiation via the type-II way [38, 78, 82]. The ROS level is increasing with the fluence, which can be directly correlated to the number of dead cells [41, 43].

Thus the extinction coefficient of the PS for absorption and the quantum yield for ISC, of the production of singlet oxygen and ROS are of great relevance for PDT efficiency and have therefore to be taken into account for a correct dosimetry [16, 58, 64].

8.3.4 Photodegradation and Photoproduct Formation

Direct interaction of irradiation-induced ROS and singlet oxygen with PS molecules may lead to photoproduct formation or even degradation of the PS [12, 13, 54]. Due to the close proximity of PS and ROS, these processes occur with a high probability.

While in photodegradation the PS molecule is destroyed and broken in smaller fragments, which are inable to absorb in the visible range of the spectrum and have therefore lost their function in PDT ("bleaching") [12, 52, 54], it is subject to minor modifications of the structure in photoproduct formation. The modified PS molecule could still be active in absorption and emission in the visible range, and still be useful for PDT, even if the spectral properties have changed. This is e.g. the case, when porphyrins such as PpIX form as photoproducts typically chlorines, which show a red shift of fluorescence, being even more effective PSs than the initial molecules [5, 18, 25, 45, 54].

Under special conditions, irradiation of the PS may also result in photorelocalization, where the intracellular localization of a PS molecule is changed [6, 7, 40, 69] by an alteration in the mobility of the PS [7, 53]. For example, photorelocalization from lysosomes to the nucleus was reported for some hydrophilic sensitizers [7, 8, 50, 53, 69].

8.4 Effects of ROS on Cells

The kind and severity of damage by ROS is dependent on the ROS concentration. A cellular response to low ROS concentrations may even lead to a more rapid proliferation in some cases, which is likely to be based on cell cycle changes and stimulation of proliferation pathways. Such activation of e.g. $p38^{MAPK}$ and JNK survival pathways [2] and of the transcription factor NF-kappa B [21] were observed.

Stimulatory effects become significant for fluorescence diagnosis by the used low fluences, or for therapy when the light intensity decreases in deeper tissue layers. Stimulation of proliferation, as found for the PSs hypericin and aluminum phthalocyanine [10, 85], is detrimental in tumor treatment, but could be beneficial e.g. in assisting wound healing. Sublethal PDT is also able to modify gene expression patterns or signalling pathways [70, 71].

Up to the beginning of lethal doses, formation of ROS is counteracted by scavengers and the action of ROS by the cell's ability to activate different repair and rescue processes in order to adapt to and to minimize the damage. A first line of defense is the immediate quenching of protein oxidation and lipid peroxidation by existing antioxidants (e.g. glutathione); the second one is the later activation of immediate early genes, which initiate *de novo* gene expression needed for repair.

A rescue and survival mechanism is the removal and recycling of "waste" by the cell. Misfolded proteins or damaged organelles are incorporated, digested and fragmented by lysosomes in autophagic processes. Especially when the PS is accumulated in the endoplasmatic reticulum, mitochondria and lysosomes, autophagy is induced by the photochemically generated ROS [68].

It has to be considered that damaged tumor cells might get support by survival signals even from the microenvironment such as the extracellular matrix. Beside this, malignant cells often have – in contrast to non-malignant cells – improved repair and survival strategies due to specific mutations. Therefore the tumor selective accumulation of the PS is an essential feature to increase the damage to such an extent that it exceeds the repair capacity.

Exceeding an individual threshold PDT dose, the cell is not able to repair the photodynamically induced damage anymore or to adapt to new parameters and dies. Thereby the cell death mode shifts in general with increasing doses from apoptosis to necrosis [62]. When the apoptosis pathway is blocked [15, 68], a special form of autophagic cell death can be switched on [31].

Not only the PDT dose, but also physiological parameters such as the availability of ATP [11] and the cellular targets of the ROS attack determine the cell death mode. Like this the intactness of the plasma membrane may decide between apoptosis or necrosis. E.g. when a PS accumulated in the plasma membrane is photoactivated, the subsequent loss of membrane integrity is highly probable to induce necrosis.

In order to assist tumor elimination, also the tumor microenvironment should be irreversibly damaged. As a first step the effect of PDT on factors in the extracellular matrix, i.e. matrix metalloproteases, adhesion molecules, growth factors and immunological mediators was examined [60]. Immunomodulatory factors such as damage-associated molecular patterns (DAMPs) from dying cells [29] contribute to cell killing by PDT and may even cause systemic immunity [65, 66]. Whole cell lysates, which have been treated by PDT, can be utilized prophylactically as cancer vaccines or have the potential to supplement primary tumor therapies [1, 30, 46]. The impact of immune stimulation on the therapy is currently under investigation.

Some PSs as Visudyne® (treatment of macular degeneration) or Tookad® target the vasculature or, more specifically, the vessel-lining endothelial cells. In this case it comes to a vascular shut-down and hypoxia [48]. Excessive hypoxia in tumor-supplying blood vessels may in turn necrotize tumors [72]; like that the PS could act as anti-vascular tumor agent.

8.5 Advantages and Problems of PDT

PDT and FD have many advantages over other tumor therapies. Some of them are:

- FD is highly useful in supplementing other diagnostic or therapeutic approaches.
- PDT is very effective in selected tumors with special properties (e.g. basal cell carcinoma).
- It is non-mutagenic and shows no major side effects.
- Due to the non-selective and general effect of ROS no resistance is developed.
- PDT can be also applied for several non-malignant diseases such as local microbial infections or unwanted growth of small blood vessels.
- It can be combined with other tumor therapies or strategies (e.g. antivascular approach)
- Due to a mixture of different cell death modes, it induces immune reactions and stimulates the immune system.

PDT has still some problems, although some of them could be resolved or alleviated. In PDT-research the following subjects are currently investigated:

- Further improvement of the PSs towards ideal properties (s. above).
- Solubility of PS, transport to the target: is partly solved by conjugation with solubility enhancers and carriers.
- Photoactivation by visible light is limited by the penetration depth; only flat targets can be treated. Two-photon activation is one approach to overcome this restriction [1]. Treatment is up to now mainly limited to tumors easily accessible

to visible light. However, interstitial PDT for prostate and breast cancer is in clinical trials.

- For clinical use, PDT has to show an advantage over established conventional approaches [14], which is not always the case with respect to efficiency. However, if the relation efficiency vs. side effects would be taken into account, PDT would be superior in many cases.
- Optimization of the treatment protocols [77] for the respective patient.

In conclusion it can be stated that PDT and FD offer effective approaches for tumor therapy and diagnosis, if their limitations are considered. A huge benefit for the patients is – beside lack of major side effects – the stimulation of immune reactions, which could be used for systemic intervention or prophylactic vaccination. In these cases, PDT would not only be restricted to local tumor eradication.

References

1. Agostinis P, Berg K, Cengel KA, Foster TH, Girotti AW, Gollnick SO, Hahn SM, Hamblin MR, Juzeniene A, Kessel D, Korbelik M, Moan J, Mroz P, Nowis D, Piette J, Wilson BC, Golab J (2011) Photodynamic therapy of cancer: an update. CA Cancer J Clin 61(4):250–281
2. Agostinis P, Vantieghem A, Merlevede W, de Witte PA (2002) Hypericin in cancer treatment: more light on the way. Int J Biochem Cell Biol 34(3):221–241
3. Allison RR, Downie GH, Cuenca R, Hu X-H, Childs CJ, Sibata CH (2004) Photosensitizers in clinical PDT. Photodiagn Photodyn Ther 1(1):27–42
4. Anderson RR, Parrish JA (1981) The optics of human skin. J Invest Dermatol 77(1):13–19
5. Bagdonas S, Ma LW, Iani V, Rotomskis R, Juzenas P, Moan J (2000) Phototransformations of 5-aminolevulinic acid-induced protoporphyrin IX in vitro: a spectroscopic study. Photochem Photobiol 72(2):186–192
6. Ball DJ, Mayhew S, Wood SR, Griffiths J, Vernon DI, Brown SB (1999) A comparative study of the cellular uptake and photodynamic efficacy of three novel zinc phthalocyanines of differing charge. Photochem Photobiol 69(3):390–396
7. Berg K, Madslien K, Bommer JC, Oftebro R, Winkelman JW, Moan J (1991) Light induced relocalization of sulfonated meso-tetraphenylporphines in NHIK 3025 cells and effects of dose fractionation. Photochem Photobiol 53(2):203–210
8. Berg K, Western A, Bommer JC, Moan J (1990) Intracellular localization of sulfonated meso-tetraphenylporphines in a human carcinoma cell line. Photochem Photobiol 52(3):481–487
9. Bergamini CM, Gambetti S, Dondi A, Cervellati C (2004) Oxygen, reactive oxygen species and tissue damage. Curr Pharm Des 10(14):1611–1626
10. Berlanda J, Kiesslich T, Engelhardt V, Krammer B, Plaetzer K (2010) Comparative in vitro study on the characteristics of different photosensitizers employed in PDT. J Photochem Photobiol B 100(3):173–180
11. Berlanda J, Kiesslich T, Oberdanner CB, Obermair FJ, Krammer B, Plaetzer K (2006) Characterization of apoptosis induced by photodynamic treatment with hypericin in A431 human epidermoid carcinoma cells. J Environ Pathol Toxicol Oncol 25(1–2):173–188
12. Bonnett R, Martinez G (2001) Photobleaching of sensitizers used in photodynamic therapy. Tetrahedron 57:9513–9574
13. Bonnett R, Martinez G (2002) Photobleaching of compounds of the 5,10,15,20-Tetrakis (m-hydroxyphenyl)porphyrin Series (m-THPP, m-THPC, and m-THPBC). Org Lett 4 (12):2013–2016

14. Brown SB, Brown EA, Walker I (2004) The present and future role of photodynamic therapy in cancer treatment. Lancet Oncol 5(8):497–508
15. Buytaert E, Dewaele M, Agostinis P (2007) Molecular effectors of multiple cell death pathways initiated by photodynamic therapy. Biochim Biophys Acta 1776(1):86–107
16. Castano AP, Demidova TN, Hamblin MR (2004) Mechanisms in photodynamic therapy: part one – photosensitizers, photochemistry and cellular localization. Photodiagn Photodyn Ther 1 (4):279–293
17. Castano AP, Mroz P, Hamblin MR (2006) Photodynamic therapy and anti-tumour immunity. Nat Rev Cancer 6(7):535–545
18. Charlesworth P, Truscott TG (1993) The use of 5-aminolevulinic acid (ALA) in photodynamic therapy (PDT). J Photochem Photobiol 18(1):99–100
19. Chatterjee DK, Fong LS, Zhang Y (2008) Nanoparticles in photodynamic therapy: an emerging paradigm. Adv Drug Deliv Rev 60(15):1627–1637
20. Chen WR, Huang Z, Korbelik M, Nordquist RE, Liu H (2006) Photoimmunotherapy for cancer treatment. J Environ Pathol Toxicol Oncol 25(1–2):281–291
21. Coupienne I, Bontems S, Dewaele M, Rubio N, Habraken Y, Fulda S, Agostinis P, Piette J (2011) NF-kappaB inhibition improves the sensitivity of human glioblastoma cells to 5-aminolevulinic acid-based photodynamic therapy. Biochem Pharmacol 81(5):606–616
22. Demidova TN, Hamblin MR (2004) Macrophage-targeted photodynamic therapy. Int J Immunopathol Pharmacol 17(2):117–126
23. Dolmans DE, Fukumura D, Jain RK (2003) Photodynamic therapy for cancer. Nat Rev Cancer 3(5):380–387
24. Dougherty TJ, Gomer CJ, Henderson BW, Jori G, Kessel D, Korbelik M, Moan J, Peng Q (1998) Photodynamic therapy. J Natl Cancer Inst 90(12):889–905
25. Ericson MB, Grapengiesser S, Gudmundson F, Wennberg AM, Larko O, Moan J, Rosen A (2003) A spectroscopic study of the photobleaching of protoporphyrin IX in solution. Lasers Med Sci 18(1):56–62
26. Feinweber D, Verwanger T, Bruggemann O, Teasdale I, Krammer B (2014) Applicability of new degradable hypericin-polymer-conjugates as photosensitizers: principal mode of action demonstrated by in vitro models. Photochem Photobiol Sci 13(11):1607–1620
27. Firey PA, Rodgers MA (1987) Photo-properties of a silicon naphthalocyanine: a potential photosensitizer for photodynamic therapy. Photochem Photobiol 45(4):535–538
28. Foote CS (1991) Definition of type I and type II photosensitized oxidation. Photochem Photobiol 54(5):659
29. Garg AD, Dudek AM, Agostinis P (2013) Cancer immunogenicity, danger signals, and DAMPs: what, when, and how? Biofactors 39(4):355–367
30. Gollnick SO, Vaughan L, Henderson BW (2002) Generation of effective antitumor vaccines using photodynamic therapy. Cancer Res 62(6):1604–1608
31. Gozuacik D, Kimchi A (2004) Autophagy as a cell death and tumor suppressor mechanism. Oncogene 23(16):2891–2906
32. Halliwell B (1984) Oxygen radicals: a commonsense look at their nature and medical importance. Med Biol 62(2):71–77
33. Halliwell B, Gutteridge JM (1984) Role of iron in oxygen radical reactions. Methods Enzymol 105:47–56
34. Hamblin MR (2008) Covalent photosensitizer conjugates part 2: peptides, polymers and small molecules for targeted photodynamic therapy. In: Hamblin MR, Mroz P (eds) Advances in photodynamic therapy: basic, translational and clinical. Artech House, Norwood
35. Hamblin MR, Newman EL (1994) On the mechanism of the tumour-localising effect in photodynamic therapy. J Photochem Photobiol 23(1):3–8
36. Henderson BW, Dougherty TJ (1992) How does photodynamic therapy work? Photochem Photobiol 55(1):145–157

37. Huygens A, Crnolatac I, Develter J, Van Cleynenbreugel B, Van der Kwast T, de Witte PA (2008) Differential accumulation of hypericin in spheroids composed of T-24 transitional cell carcinoma cells expressing different levels of E-cadherin. J Urol 179(5):2014–2019
38. Ito T (1978) Cellular and subcellular mechanisms of photodynamic action: the 1O2 hypothesis as a driving force in recent research. Photochem Photobiol 28(4–5):493–508
39. Juzeniene A, Nielsen KP, Moan J (2006) Biophysical aspects of photodynamic therapy. J Environ Pathol Toxicol Oncol 25(1–2):7–28
40. Kessel D (2002) Relocalization of cationic porphyrins during photodynamic therapy. Photochem Photobiol Sci 1(11):837–840
41. Kiesslich T, Plaetzer K, Oberdanner CB, Berlanda J, Obermair FJ, Krammer B (2005) Differential effects of glucose deprivation on the cellular sensitivity towards photodynamic treatment-based production of reactive oxygen species and apoptosis-induction. FEBS Lett 579(1):185–190
42. Klajnert B, Rozanek M, Bryszewska M (2012) Dendrimers in photodynamic therapy. Curr Med Chem 19(29):4903–4912
43. Koch S (2015) In vitro tests of a new and improved hypericin-conjugate for application in photodynamic therapy and diagnosis. University of Salzburg, Salzburg
44. Kochevar IE (2004) Singlet oxygen signaling: from intimate to global. Sci STKE 2004(221): pe7
45. Konig K, Schneckenburger H, Ruck A, Steiner R (1993) In vivo photoproduct formation during PDT with ALA-induced endogenous porphyrins. J Photochem Photobiol 18 (2–3):287–290
46. Korbelik M (2011) Cancer vaccines generated by photodynamic therapy. Photochem Photobiol Sci 10(5):664–669
47. Korbelik M, Merchant S (2012) Photodynamic therapy-generated cancer vaccine elicits acute phase and hormonal response in treated mice. Cancer Immunol Immunother 61(9):1387–1394
48. Krammer B (2001) Vascular effects of photodynamic therapy. Anticancer Res 21 (6B):4271–4277
49. Lucky SS, Soo KC, Zhang Y (2015) Nanoparticles in photodynamic therapy. Chem Rev 115 (4):1990–2042
50. Moan J (1986) Effect of bleaching of porphyrin sensitizers during photodynamic therapy. Cancer Lett 33(1):45–53
51. Moan J (1990) On the difusion length of singlet oxygen in cells and tissues. J Photochem Photobiol 6:343–344
52. Moan J, Berg K (1991) The photodegradation of porphyrins in cells can be used to estimate the lifetime of singlet oxygen. Photochem Photobiol 53(4):549–553
53. Moan J, Berg K, Anholt H, Madslien K (1994) Sulfonated aluminium phthalocyanines as sensitizers for photochemotherapy. Effects of small light doses on localization, dye fluorescence and photosensitivity in V79 cells. Int J Cancer 58(6):865–870
54. Moan J, Streckyte G, Bagdonas S, Bech O, Berg K (1997) Photobleaching of protoporphyrin IX in cells incubated with 5-aminolevulinic acid. Int J Cancer 70(1):90–97
55. Mroz P, Tegos GP, Gali H, Wharton T, Sarna T, Hamblin MR (2007) Photodynamic therapy with fullerenes. Photochem Photobiol Sci 6(11):1139–1149
56. Muller P, Wilson BC (2014) Photodynamic therapy. In: Berstein M, Berger MS (eds) Neuro-oncology: the essentials, 3rd edn. Thieme, New York
57. Niedre M, Patterson MS, Wilson BC (2002) Direct near-infrared luminescence detection of singlet oxygen generated by photodynamic therapy in cells in vitro and tissues in vivo. Photochem Photobiol 75(4):382–391
58. Niemz MH (2004) Laser-tissue interactions. Fundamentals and applications. Springer, Berlin
59. Ochsner M (1997) Photophysical and photobiological processes in the photodynamic therapy of tumours. J Photochem Photobiol 39(1):1–18
60. Pazos MC, Nader HB (2007) Effect of photodynamic therapy on the extracellular matrix and associated components. Braz J Med Biol Res 40(8):1025–1035

61. Pervaiz S, Olivo M (2006) Art and science of photodynamic therapy. Clin Exp Pharmacol Physiol 33(5–6):551–556
62. Plaetzer K, Kiesslich T, Verwanger T, Krammer B (2003) The modes of cell death induced by PDT: an overview. Med Laser Appl 18(1):7–19
63. Plaetzer K, Krammer B, Berlanda J, Berr F, Kiesslich T (2009) Photophysics and photochemistry of photodynamic therapy: fundamental aspects. Lasers Med Sci 24(2):259–268
64. Prasad PN (2003) Introduction to biophotonics. Wiley, Hoboken
65. Preise D, Oren R, Glinert I, Kalchenko V, Jung S, Scherz A, Salomon Y (2009) Systemic antitumor protection by vascular-targeted photodynamic therapy involves cellular and humoral immunity. Cancer Immunol Immunother 58(1):71–84
66. Preise D, Scherz A, Salomon Y (2011) Antitumor immunity promoted by vascular occluding therapy: lessons from vascular-targeted photodynamic therapy (VTP). Photochem Photobiol Sci 10(5):681–688
67. Reddi E (1997) Role of delivery vehicles for photosensitizers in the photodynamic therapy of tumours. J Photochem Photobiol 37(3):189–195
68. Reiners JJJ, Agostinis P, Berg K, Oleinick NL, Kessel D (2010) Assessing autophagy in the context of photodynamic therapy. Autophagy 6(1):7–18
69. Ruck A, Beck G, Bachor R, Akgun N, Gschwend MH, Steiner R (1996) Dynamic fluorescence changes during photodynamic therapy in vivo and in vitro of hydrophilic A1(III) phthalocyanine tetrasulphonate and lipophilic Zn(II) phthalocyanine administered in liposomes. J Photochem Photobiol 36(2):127–133
70. Ruhdorfer S, Sanovic R, Sander V, Krammer B, Verwanger T (2007) Gene expression profiling of the human carcinoma cell line A-431 after 5-aminolevulinic acid-based photodynamic treatment. Int J Oncol 30(5):1253–1262
71. Sanovic R, Krammer B, Grumboeck S, Verwanger T (2009) Time-resolved gene expression profiling of human squamous cell carcinoma cells during the apoptosis process induced by photodynamic treatment with hypericin. Int J Oncol 35(4):921–939
72. Sanovic R, Verwanger T, Hartl A, Krammer B (2011) Low dose hypericin-PDT induces complete tumor regression in BALB/c mice bearing CT26 colon carcinoma. Photodiagnosis Photodyn Ther 8(4):291–296
73. Selbo PK, Hogset A, Prasmickaite L, Berg K (2002) Photochemical internalisation: a novel drug delivery system. Tumour Biol 23(2):103–112
74. Sharman WM, Allen CM, van Lier JE (2000) Role of activated oxygen species in photodynamic therapy. Methods Enzymol 319:376–400
75. Sibata CH, Colussi VC, Oleinick NL, Kinsella TJ (2000) Photodynamic therapy: a new concept in medical treatment. Braz J Med Biol Res 33(8):869–880
76. Svaasand LO (1984) Optical dosimetry for direct and interstitial photoradiation therapy of malignant tumors. Prog Clin Biol Res 170:91–114
77. Triesscheijn M, Baas P, Schellens JH, Stewart FA (2006) Photodynamic therapy in oncology. Oncologist 11(9):1034–1044
78. Valenzeno DP (1987) Photomodification of biological membranes with emphasis on singlet oxygen mechanisms. Photochem Photobiol 46(1):147–160
79. Valeur B (2001) Molecular fluorescence: principles and applications. Wiley-VCH, Weinheim
80. van Dongen GA, Visser GW, Vrouenraets MB (2004) Photosensitizer-antibody conjugates for detection and therapy of cancer. Adv Drug Deliv Rev 56(1):31–52
81. Wang HW, Zhu TC, Putt ME, Solonenko M, Metz J, Dimofte A, Miles J, Fraker DL, Glatstein E, Hahn SM, Yodh AG (2005) Broadband reflectance measurements of light penetration, blood oxygenation, hemoglobin concentration, and drug concentration in human intraperitoneal tissues before and after photodynamic therapy. J Biomed Opt 10(1):14004
82. Weishaupt KR, Gomer CJ, Dougherty TJ (1976) Identification of singlet oxygen as the cytotoxic agent in photoinactivation of a murine tumor. Cancer Res 36(7 PT 1):2326–2329

83. Wiegell SR, Stender IM, Na R, Wulf HC (2003) Pain associated with photodynamic therapy using 5-aminolevulinic acid or 5-aminolevulinic acid methylester on tape-stripped normal skin. Arch Dermatol 139(9):1173–1177
84. Wilson BC, Jeeves WP, Lowe DM (1985) In vivo and post mortem measurements of the attenuation spectra of light in mammalian tissues. Photochem Photobiol 42(2):153–162
85. Zancanela DC, Primo FL, Rosa AL, Ciancaglini P, Tedesco AC (2011) The effect of photosensitizer drugs and light stimulation on osteoblast growth. Photomed Laser Surg 29 (10):699–705

Chapter 9
Polymer Nanoparticles for Cancer Photodynamic Therapy Combined with Nitric Oxide Photorelease and Chemotherapy

Fabiana Quaglia and Salvatore Sortino

Abstract Combination cancer therapies are emerging treatment modalities, which aim at exploiting additive/synergistic effects due to the simultaneous generation of multiple active species in the same region of space, with the final goal to maximize the therapeutic action and to minimize side effects. Photodynamic therapy (PDT) is a minimally invasive and clinically approved procedure for cancer treatment and its coupling with either conventional chemotherapeutics or unconventional anticancer agents like nitric oxide (NO) may open intriguing horizons towards new and still underexplored multimodal therapies. Engineered polymer nanoparticles (NPs) offer great potential in this respect since they permit to locate a large number of distinct components for therapy at target site without precluding their individual potential and to encourage the release of therapeutics in the same region of space resulting in an enhancement of the therapeutic outcome. We present here an overview of the most recent advances in polymeric NPs devised for bimodal cancer treatments based on the combination of PDT with either light-triggered NO release or chemotherapy, highlighting the logical design and their potential applications in cancer research.

F. Quaglia (✉)
Laboratory of Drug Delivery, Department of Pharmacy, University of Napoli Federico II, 80131 Naples, Italy
e-mail: quaglia@unina.it

S. Sortino
Laboratory of Photochemistry, Department of Drug Sciences, University of Catania, 95125 Catania, Italy
e-mail: ssortino@unict.it

G. Bergamini, S. Silvi (eds.), *Applied Photochemistry*,
Lecture Notes in Chemistry 92, DOI 10.1007/978-3-319-31671-0_9

9.1 Introduction

Although the knowledge of molecular, cellular, and physiological mechanisms involved in the initiation and progression of cancer has been significant refined, the benefit of cancer treatments remains mainly confined to an increased survival of patients while ensuring a sufficiently good quality of life [1]. Chemotherapy results in severe side effects due to drug accumulation at non-target sites and the necessity of high drug doses to eradicate tumor, sometimes requiring interruption of the treatment and shift to different drug options. Furthermore, only temporary response to chemotherapeutics is achieved and resistance mechanisms are induced when tumor activates alternative pathways inducing tumor survival [2]. Combination chemotherapy is becoming increasingly relevant to attain long-term prognosis and better quality of life in cancer patients. The basic concept in this strategy is to attack tumor on different sides by acting on a single oncogenic pathway through different mechanisms or across parallel pathways without amplification of side effects [3]. Nevertheless, it is not clear if chemo-combination will be able to surmount genetic mutation and resistance mechanisms, once considered the high mutation propensity of several cancer types. Therefore, different treatment modalities are at forefront of cancer research providing a potential solution to addressing the tumor heterogeneity and drug resistance issues.

Treatment modalities based on physical stimuli offer a number of unique advantages compared with conventional chemotherapy, although their clinical applications are still mostly at early stages. Many physical stimulus-responsive therapies not only have the ability to directly kill cancer cells but may enhance the outcome of other treatment modalities to achieve desired synergistic effects via different mechanisms aimed at modulating the tumor microenvironment [4]. In this respect, photodynamic therapy (PDT) is emerging as alternative option for cancer therapy to selectively target neoplastic lesions [5, 6]. PDT consists in the administration (local or systemic) of a photosensitizer (PS) which accumulates in different tissue/cells and, under application of light with a specific wavelength and in the presence of molecular oxygen, produces highly reactive oxygen species (ROS), mainly singlet oxygen ($^1O_2$) finally killing cancer cells and promoting tumor regression. Selectivity is achieved partly by the accumulation of the PS in the malignant cells/tissue, and partly by restricting the application of the incident light to the tumor area. PDT induces many cellular and molecular signaling pathway events, inducing cell death through apoptosis, necrosis, and autophagy, damaging the tumor-associated vasculature, and mediating a strong inflammatory reaction and induction of immune response [6].

Combination of PDT with other cytotoxic agents can result in additive/synergic effects, opening new horizons towards more effective and less invasive cancer treatments. PDT treatment associated with the simultaneous release of nitric oxide (NO) by appropriate NO donors, offers intriguing prospects towards bimodal treatments all based on "non-conventional" chemotherapeutics. Although still confined to the research environment NO holds very promising features in cancer research. In fact, beyond to play multiple roles in the bioregulation of a broad array of physiological processes NO has also proven to be an effective anticancer agent sharing common features with $^1O_2$. Besides, combination of PDT with conventional chemotherapy has been regarded as a potential tool to optimize treatment outcome while limiting toxicity concerns [3].

Nanotechnology offers a valuable tool to target bioactive molecules to solid tumors with the goal to improve response to conventional pharmacological therapies, to alleviate anticancer drug toxicity as well as to overcome multidrug resistance (MDR) [7–14]. Amid nanocarriers, polymer-based nanoparticles (NPs) are at limelight in the burgeoning field of nanomedicine due to the advantage to manipulate their properties by selecting polymer type and mode of carrier assembly [15]. In fact, the advances in polymer chemistry makes it possible to produce an almost infinite number of sophisticated structures which can be engineered in light of strictly defined biological rules. Thus, not only those features affecting distribution of drug dose in the body and interaction with target cells can be controlled, but also spatio-temporal release of the delivered drug can be finely tuned. In analogy to chemotherapeutics, delivery of PS through NPs properly designed to target tumors can be highly beneficial to overcome the main shortcomings of current PDT such as low PS solubility, tendency to aggregation, and specificity [16]. NPs represent the ideal way to deliver multi-modal therapies in the body in a time- and spatial-controlled manner, thus achieving tuned pharmacokinetics and targeting cell/sub-cellular elements of tumor.

This contribution focuses on multifunctional polymer NPs for bimodal cancer treatments based on the combination of PDT with either light-triggered NO release or chemotherapy and exhibiting fluorescence properties, indispensable feature for cellular tracking (Fig. 9.1). Firstly, we shall provide a general overview of polymer

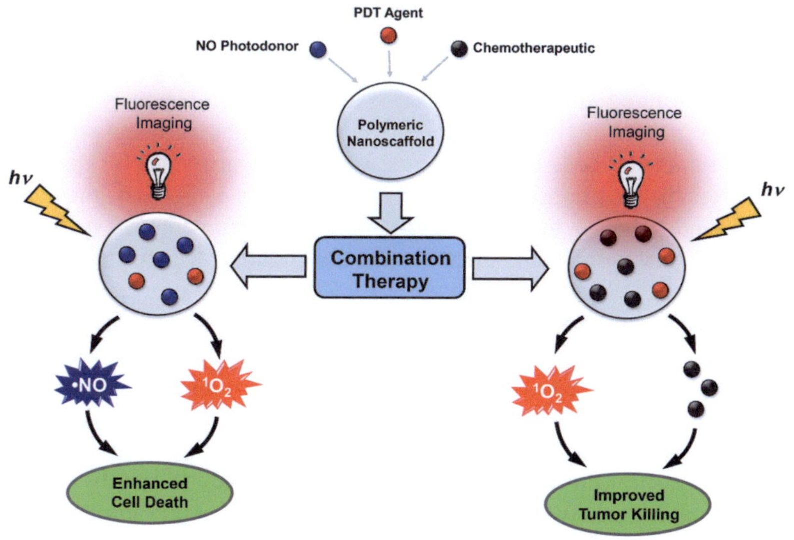

Fig. 9.1 Engineered polymer NPs for PDT combined with NO photorelease (*left*) or chemotherapy (*right*)

NPs for cancer therapy, including some general design guidelines and fabrication methodologies. After a short description of basic concepts of PDT, the peculiar role of specifically designed polymer NPs to overcome PDT shortcomings will be illustrated. Finally, we describe the most recent advances in the application of engineered polymer NPs for the above bimodal therapeutic treatments, highlighting the logical design and their potential in cancer research.

9.2 Polymer Nanoparticles for Cancer Therapy

9.2.1 Generalities

Delivery of chemotherapeutics through NPs has been focused on intravenous route since it permits to reach tumors located in practically all the districts in the body [15, 17]. By exploiting the presence of dysfunctional endothelium of tumor capillary wall and the absence of effective lymphatic drainage in solid tumors, NPs able to long circulate can partly extravasate from the blood through the so-called Enhanced Permeability and Retention (EPR) effect [18, 19] and reach solid tumor interstitium [20–22]. Only an amount around 10 % of the injected dose benefits of EPR effect while the remnant is accumulated in spleen, liver and bone marrow. Drug cargo can be after released systemically or inside tumors and this is found to allow a significant decrease of side effects. Current research efforts are focused on finding strategies to increase tumor to healthy tissue ratio and to improve tumor response.

For long time, NP decoration with ligands that specifically recognize peculiar elements on the membrane of tumor cells has been considered the most favourable option to increase accumulation in cancer tissue, an approach known as active targeting [23, 24]. Despite the enormous attention of the scientific community, this approach has resulted in limited advantages and have been seriously questioned in the last few years [25–27]. Recently, targeting to peculiar elements present on the surface of endothelial cells of tumor blood vessels (integrin receptor) has emerged as a more reliable tool to promote tumor targeting. NPs can be designed also with exquisite responsiveness to tumor environment (pH, temperature) or to external stimuli (light, magnetic field, ultrasound, temperature), which can in principle trigger drug release selectively at tumor level [28–35].

Due to multifaceted nature of cancer, it is nowadays believed that a single drug or an even stand-alone therapeutic strategy may not be enough to treat a specific tumor. Thus, the simultaneous administration of two or multiple therapeutic agents with non-overlapping toxicities in a combination therapy is becoming increasingly pursued to attain long-term prognosis and to attenuate side effects exerted by single drugs [36]. The basic principle of combination therapies is that by modulating different signalling pathways involved in the progression of tumor, synergistic response, maximized therapeutic effect and overcome of drug resistance can be attained [37]. As a logical consequence, administration of multiple drugs in a single nanocarrier is emerging to benefit of targeted delivery to tumors and of releasing multiple agents with precise spatio–temporal control. Furthermore, NPs can be designed to encorporate drugs that have different physical-chemical properties such as hydrophilicity and hydrophobicity. To this end, NPs with core-shell architectures have great potential in delivering therapeutic combinations since they allow to easily change the ratio between transported drugs which can be crucial to attain desired therapeutic effect [38].

When developing NPs for cancer therapy, a rational design should be planned taking into account specific needs dictated by (i) the disease features (tissue, stage, vascularization extent, presence of metastases); (ii) the strategy selected to accumulate the highest dose fraction at tumor level (pharmacokinetics) and to deliver the drug in space and time (intracellular/tumor interstitium, sustained or pulsed release); (iii) the physical-chemical properties of the drug (solubility profile, stability); (iv) the achievement of a product with satisfactory shelf-life (preferably a solid). This is certainly the most critical step in NP development where a multidisciplinary approach at interface between pharmaceutical technology, biology and medicine should be carried out.

9.2.2 Biologically Driven Design Guidelines

The journey of NPs from the site of administration to the site of action is driven by the interactions with the body environment that affect their accumulation in the tumor area. Nevertheless, at cellular level, further critical steps that can limit the final pharmacological outcome of a nanocarrier are the crossing of the extracellular matrix (ECM) and the extent of cellular uptake. All the aforementioned effects are correlated with NP features, including surface charge, size, shape and rigidity [10, 18, 39].

Concerning the accumulation at tumor level, long-circulating intravenous NPs have greater probability to reach the tumor area through EPR effect. To this purpose, NP should be biomimetic to prevent non-specific adsorption of serum proteins/complement/antibodies (opsonization) that predisposes foreign particles to Mononuclear Phagocyte System (MPS) recognition and rapid clearance in liver, spleen and bone marrow. Biomimetic properties is a common feature that passively, actively and physically targeted NPs share and is generally achieved by providing NPs with a hydrophilic polymer shell able to limit opsonization [22]. NP charge strongly contributes to tumor accumulation in that vascular lumen and a great portion of serum proteins carry a negative charge. Positive NPs thereby bind to them giving short circulation time and reduced tumor accumulation [18]. Beside charge, NP size and shape contribute all to MPS escape. NP surface can be decorated with motifs recognizing some receptors overexpressed on the surface of endothelial cells in defective tumor vasculature (i.e. integrin receptor), thus promoting, at least in theory, NP accumulation at tumor level through receptor-mediated intracellular transport [25].

Matrix tissue neighbouring tumor cells, generated by coagulation-derived matrix gel (fibrin gel or stromal tissue) represents another barrier NPs need to cross. Extracellular matrix itself seems not represent an evident obstacle to NPs passage although it has been recently shown that, cationic NPs are intriguing systems to promote deep penetration in tumor tissues [40, 41].

Another key aspect to consider is the ability of NP to enter inside cancer cells, which allows a discrimination between NPs that release intracellularly their drug cargo and NPs that remains entangled in extracellular matrix (ECM) forming an extracellular drug depot. On this matter, the presence of a hydrophilic coating – of benefit to escape MPS recognition- decreases the rate and extent of NP uptake inside cancer cells [42]. Biomimetic coatings that expose motifs recognized by receptors overexpressed on cancer cells can encourage NP uptake through receptor-mediated endocytosis and allow improving both drug selectivity and efficacy. Different receptors have been identified and the corresponding ligands (folate, biotin, transferrin, albumin, hyaluronic acid) employed to decorate NP surface. Nevertheless, it has been recently demonstrated that targeted NPs can paradoxically lose targeting ability in a biological environment due to interaction with other proteins [43] or confine their activity to perivascular regions of a tumor (binding site barrier) [44]. Exocytosis of NPs from cells is another aspect deserving adequate consideration [45].

An added sophistication to selective delivery of NPs can be brought about by utilizing certain cues inherently characteristic of the tumor microenvironment or by applying certain stimuli to this region from outside the body [28]. Stimuli-sensitive NPs based on tailor-made materials can indeed be designed to deliver drug payload sharply and "on demand" by undergoing structural modifications under internal or external stimuli of chemical, biochemical and physical origin. Internal stimuli typical of solid tumors include mainly pH, temperature and reductive conditions [32, 35, 46, 47]. Also an array of tumor-associated enzymes, either extracellular or intracellular, can be used as biochemical trigger of drug release to control spatial distribution of the delivered cargo [47].

9.2.3 Fabrication

A wide array of currently-available materials and their possible combinations can be employed to fabricate NPs spanning from simple biomimetic systems for passive targeting to more sophisticated structures integrating specificity and multifunctionality [48]. Biodegradability and effective elimination of the polymer is a key prerequisite that should be taken into account when developing injectable NPs especially when considering novel materials.

The main biomaterials employed in injectable biodegradable nanoparticles are natural (polysaccharide, proteins, polyaminoacids) or synthetic (polyesters and PEGylated polyesters, poloxamers, acrylic polymers, polyammines). Each of them presents peculiar solubility features which, coupled with drug physico-chemical profile as well as specific needing for surface engineering and drug release rate, drive toward specific fabrication methods to attain core/shell NPs [49].

In their simplest architecture, passively, actively and physically targeted NPs can be formed by: (i) nanostructuring (self-assemblying) of amphiphilic all-in-one copolymers; (ii) physical adsorption of a shell-forming material on preformed polymeric NPs acting as a core template.

In the latter case, polymers that are insoluble in water form the core template of NPs, which is then coated by one or more than one layer of hydrophilic materials/phospholipids to impart tailored properties to the shells (biomimetic and targeted shells). Hydrophilic polymers can be used also as core template after crosslinking with either low molecular weight molecules or an interacting hydrophilic polymer with opposite charge, thus providing nanogels.

Amid shell-forming hydrophilic polymers giving long-circulating NPs, polyethylene glycol (PEG) is undoubtedly the most common material employed to modify hydrophobic NPs, although several others have been proposed [50, 51].

From a therapeutic standpoint, timing of drug release is important not only to drive dosing (number of administrations, frequency) but also useful to optimize therapeutic outcome. For example, sustained extracellular release can be expected to amplify cell response to some chemotherapeutics and extend activity to hypoxic zones of certain tumors resembling a metronomic therapy (subactive doses for longer time frames) [52]. On the other hand, stimuli-responsive polymers, referred to as "environmentally-sensitive", "smart" or "intelligent" polymers are a huge class of chemically diverse structures that respond sharply to small changes in physical or chemical conditions with relatively large phase or property changes mainly used in the cancer field to trigger drug release at tumor level [47]. In all the cases, drug amount released from NPs should be reasonably low in the blood circulation and regulated at tumor level to attain optimal therapeutic response. Time control of the delivered dose can be finely tuned by allocating different drugs in the core or the shell, that is especially important in drug-nucleic acid combination therapies [53].

In general, NPs are prepared by bottom-up approaches, primarily using monomers that are properly cross-linked or preformed polymers that are then nanostructured by techniques such as emulsification/solvent evaporation, interfacial deposition after solvent displacement, dialysis or salting-out [54–56]. It is worth noting that properties of NPs are strictly dictated by the production method, which is especially critical when specific targeting elements have to be exposed on the surface [15, 57].

The unique nano-scale structure of NPs provides significant increases in surface area to volume ratio which results in high propensity to aggregation during manufacturing, storage and shipping. This aspect remains a very challenging issue during pharmaceutical product development. The freeze-drying process is the most diffused method to handle and stabilize NPs, avoiding undesirable changes upon storage. However, freeze-drying also induce aggregation, which mainly affect dispersability in a pharmaceutical vehicle for intravenous administration. Sugars such as trehalose, glucose, sucrose, fructose and sorbitol have been proposed as cryoprotectants to minimize NP collapse upon freeze-drying, although papers focusing on this aspect are very few [58].

9.3 NanoPDT in Cancer

9.3.1 Basic Concepts of PDT

PDT is a well-established therapeutic modality for treating malignant lesions, including cancer, in humans [6, 59]. PDT implies the combined use of three main ingredients: a non-toxic PS, low energy light in the visible or near infrared region, and molecular oxygen (Fig. 9.2). The PS, administered by the intravenous, intraperitoneal, or topical route, accumulates preferentially in tumour cells versus normal cells by passive targeting, with a structure-dependent intracellular distribution [60]. Illumination of the tumor area with an appropriate dose of light of appropriate wavelength excites the PS, which reaches the lowest-energy excited triplet state through intersystem crossing (ISC). Due to its long lifetime, the excited triplet state of the PS can be quenched by molecular oxygen via an energy and/or electron transfer mechanism, producing the highly reactive $^1O_2$, (Type II reaction) and other ROS (Type I reaction) [61–63]. However, the former is nowadays accepted to be the foremost mediator of cytotoxic reactions in PDT treatments [64, 65]. $^1O_2$ is produced photocatalitically exclusively in the illuminated area. This feature allows it to be generated in concentration much larger than those of the PS used. Due to its great oxidizing power, $^1O_2$ is able to cause irreversible damage to lipids, proteins and nucleic acids in the cellular environment, inducing selective destruction of target tissues and cells through a number of mechanisms (Fig. 9.2) [61–66]. Moreover $^1O_2$ does not suffer MDR and, in view of its short lifetime, *ca.* 4 μs in water, it confines its region of action inside the cells within 10–20 nm, minimizing systemic toxicity common to many conventional chemotherapeutics.

Porphyrins, phthalocyanines and, recently, derivatives of 4,4-difluoro-4-bora-3a,4a-diaza-*s*-indacene (BODIPY) [67–69] play a central role as PSs in PDT by virtue of their high ISC quantum yields, very long triplet lifetimes, and strong absorption in the visible region, which are all key parameters for an effective photoproduction of $^1O_2$.

Thanks to their intrinsic fluorescence properties, porphyrinoid-based PSs can be easily visualized in a cellular environment through fluorescence techniques, providing a great advantage in view of image-guided cancer phototherapy. This has recently led to the general term “photosensitizer fluorescence detection” in reference to all applications in which a PS is also used to generate fluorescence contrast

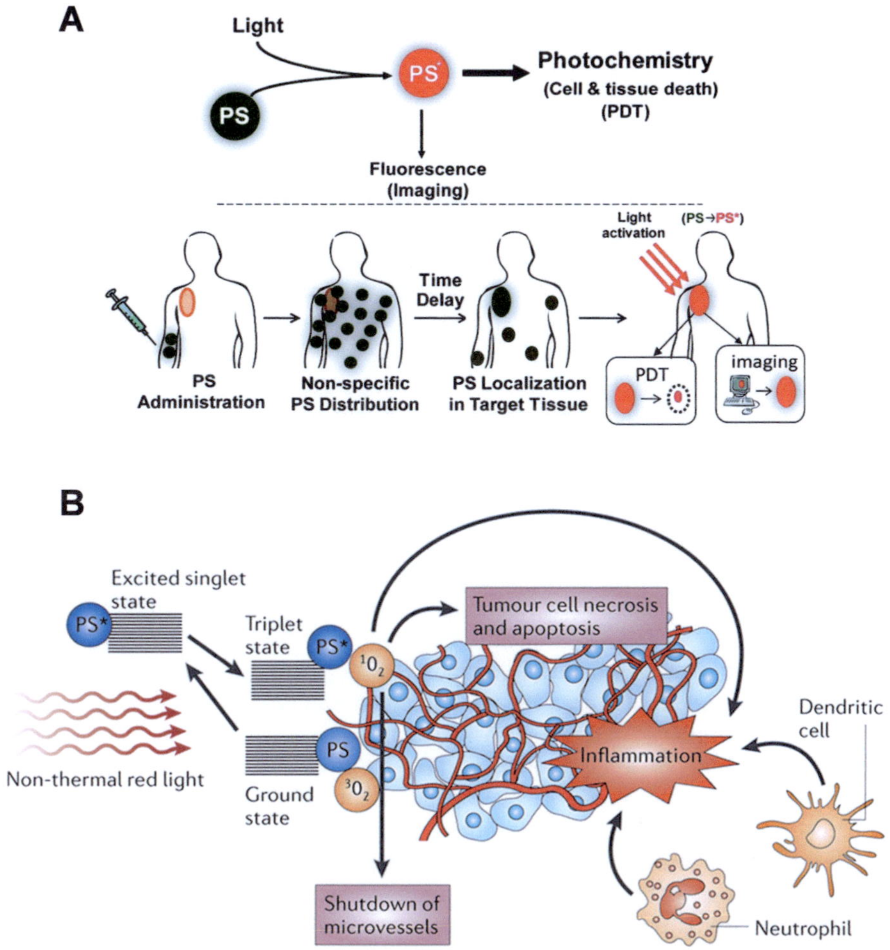

Fig. 9.2 (**a**) Steps of an intravenous photodynamic treatment. (**b**) In tumor PDT, PS absorbs light and an electron moves to the first short-lived excited singlet state. This is followed by ISC, in which the excited electron changes its spin and produces a longer-lived triplet state. The PS triplet transfers energy to ground state triplet oxygen, which produces reactive $^1O_2$ that can (i) directly kill tumor cells by induction of necrosis and/or apoptosis; (ii) cause destruction of tumor vasculature and (iii) produce an acute inflammatory response attracting leukocytes (Adapted with permission from Ref. [64, 65])

[65]. The cellular internalization of the PS and the capacity to effectively photogenerate $^1O_2$ nearby are two crucial prerequisites in PDT. This issue has inspired an intense research activity devoted to integrate the PSs in bio-compatible NPs which merge delivery characteristics and preservation of the photodynamic activity of the PS [70].

9.3.2 Overcoming PDT Shortcomings with Polymer Nanoparticles

The efficacy of a PDT treatment depends on multiple factors related to PS physical-chemical features and its behaviour in the body as well as the illumination conditions. Nanotechnology offers a great opportunity in advancing PDT based on the concept that a PS packaged in a nanoscale-carrier results in controlled pharmacokinetics driven by nanocarrier features rather than PS properties. Thus, the treatment ability to target and kill cancer cells of diseased tissue/organ while affecting as few healthy cells as possible is enhanced [71, 72].

After intravenous injection, accumulation of free PS in the target tissue depends on post-injection time [73]. At time shorter than half-life, PSs predominantly stay in the vascular compartment of the tumor while at longer time, they can accumulate in extravascular sites due to interstitial diffusion. Therefore, drug-light interval may play a crucial role for the therapeutic outcome. Delivery through nanocarriers drives PS in a specific location by proper control of nanocarrier features, potentially altering its intrinsic pharmacokinetics behaviour as well as cellular/subcellular distribution. This strategy can be highly beneficial to improve PS selectivity for solid tumors and to decrease photosensitation in healthy tissues, which remain the main drawbacks of a PDT treatment [74].

Aggregation is another determining factors which can cause a loss of PS efficacy in vivo. In fact, several PS are prone to aggregation in physiological conditions via the strong attractive interactions between π-systems of the polyaromatic macrocycles and, as a consequence, $^1O_2$ is produced with very low yields [75, 76]. Furthermore, most of second generation PSs exhibit poor solubility in aqueous media, complicating intravenous delivery into the bloodstream. The low extinction coefficients of PSs often requires the administration of relatively large amounts of drug to obtain a satisfactory therapeutic response thus demanding for specific vehicles (Chremophor®, propylenglycol), which can lead to unpredictable biodistribution profiles, allergy, hypersensitivity and toxicity [77]. Thus, a carrier specifically engineered for nanoPDT should provide an environment where the PS can be administered in vivo in therapeutic doses in a monomeric form and can retain its photochemical properties without loss or alteration of photoactivity.

9.3.3 Specific Design Requirements of Polymer Nanoparticles for Cancer PDT

The design of NPs for PDT of cancer should primarily follow the biologically-driven guidelines suggested for conventional chemotherapeutics. Additional aspects need to be considered because of the peculiar features of PS and their mode of activation. Parameters such as NP size (scattering effect), oxygen permeability of the matrix, and light absorption of NPs play a pivotal role in controlling quantum yield of $^1O_2$ production [78].

PS are integrated in NPs through physical entrapment or covalent linkage to polymer/monomer precursors and allocated either in the core or the shell in analogy to conventional chemotherapeutics (see Sect. 2.3). The extent of association depends, among others, on the physical-chemical properties of the polymer and the PS (solubility, reciprocal interaction), affinity of PS for the realising medium (aqueous solubility, protein interaction), polymer degradability, and stability of the NPs toward disassembly.

Despite the fact that it is very convenient to adsorb PS onto the shell of preformed NPs, premature PS leaching in the bloodstream before the nanocarrier reaches the target can occur. For this reason encorporation of PS in the NP core or covalent binding of a PS to monomers that are then polymerized or self-assembled in NPs, as well as chemical post-modification of preformed NPs are potentially beneficial. Independently of PS location and loading strategy, aggregation of PS needs to be controlled in order to circumvent loss of PDT efficiency. In fact, unlike chemotherapeutic drugs where higher loading inside NPs nicely correlates with increased cytotoxicity, achievement of high PS loading can be detrimental. If the PS aggregates extensively also within NPs, PS can self-quench and produce less reactive oxygen species upon irradiation overall reducing photodynamic effect.

Release of PS from NPs is not considered as crucial as for other drugs to achieve a therapeutic effect since molecular oxygen can penetrate polymer matrix, get converted into ROS and $^1O_2$, which can then diffuse out of the carrier to cause photodynamic damage. Thus, PS location in the core or the shell, can be useful in principle to finely tune the rate of $^1O_2$ release, which is especially important in combination therapies where a ordered sequence of biological activities needs to be realized.

9.4 Combining PDT and Nitric Oxide Photorelease

9.4.1 Why to Combine PDT with Nitric Oxide Release

After being viewed for many years nothing but a mere environmental pollutant, NO is nowadays one of the most studied molecules in the fascinating realm of the biomedical sciences [79, 80]. Beside to play a multifaceted role in the bioregulation of vital functions such as neurotransmission, hormone secretion and vasodilatation in living bodies, this ephemeral inorganic free radical, with half-life in tissues of ca. 5 s, has proven to be an excellent anticancer, antimicrobial and antioxidant agent [79]. Therefore, NO represents an intriguing therapeutic agent with exciting prospects in tackling important diseases, expecially cancer [81]. The role of NO in tumour biology is however, quite intricate, as this agent can act as a tumour progressor or suppressor depending on its concentration. This results in opposing beneficial and deleterious events. For example, μM range concentrations of NO result in antitumor effects, while pM-nM NO concentrations promote cytoprotective effects [81–83]. This dichotomy has made the development of new strategies and methods for generating NO in a controlled fashion a *hot topic* in the field of nanomedicine [84–88].

The combination of NO with PDT represents a very appealing strategy in view of multimodal therapeutic systems. In fact, analogously to $^1O_2$, NO exhibits small size, absence of charge and good lipophilicity, and capability to attack different types of biological substrate (i.e., lipids, proteins, and DNA), representing a multitarget cytotoxic agent. Furthermore, NO does not suffer MDR and, due to its short lifetime, confines its region of action to short distance from the production site inside the cells, <200 μm, reducing systemic toxicity issues common to many conventional anticancer drugs. Finally, another crucial point to be highlighted is that since NO release is independent of O_2 availability, NO-based anticancer therapy can successfully complement PDT at the onset of hypoxic conditions, typical for some tumours, where PDT may fail.

9.4.2 Light-Triggered NO Release

A number of molecular and macromolecular scaffolds and nanomaterials able to deliver NO with some control have been developed over the last years [85–88]. Among these, those based on photoactivable NO precursors are definitively the most appealing [89–94]. Light represents in fact a powerful tool to control, with superb accuracy, concentration, delivery site and dosage of NO which, as stated above, is a very critical parameter for a positive therapeutic outcome [95]. In addition, light triggering provides fast reaction rates and offers the great advantage of not affecting the physiological values of parameters such as temperature, pH and ionic strength.

The working principle of a NO photochemical precursor bases on the concept of "photocaging." This term refers, in general, to the momentary inactivation of a biologically active molecule by its covalent incorporation within the molecular skeleton of a photoresponsive chromogenic centre (the photocage) [96]. The photosensitive moiety exploits the absorbed excitation light to break a chemical bond, releasing the "caged" molecule, i.e. NO, in its active form (Fig. 9.3). This strategy permits to confine the site of action of NO at the irradiated area with an exquisite spatial precision and to control its dosage with great accuracy by tuning the light intensity and/or duration as needed. These unique features have made the NO photoreleasing compounds a powerful therapeutic arsenal [90–95] for a wide range of therapeutic applications, including cancer. It is important to note that NO photouncaging is accompanied by the formation of a concomitant molecular fragment, the by-product, which deserves specific attention. Indeed, physiological properties and toxicity of this product need to be carefully evaluated for bio-applications of the NO photodonor. Moreover, such a by-product should not

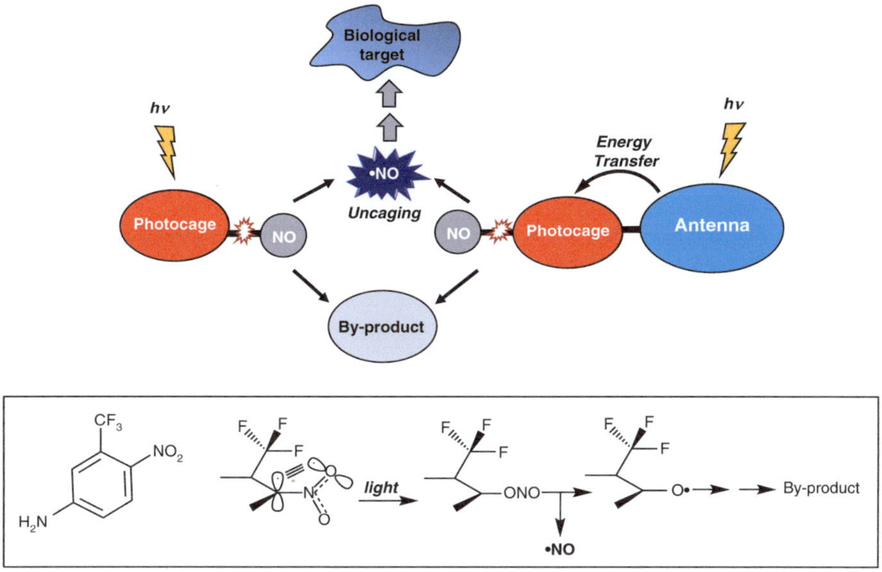

Fig. 9.3 (*Top*) The photouncaging process: light excitation of the photocage triggers cleavage of the covalent bond with the caged and unactive NO molecule, releasing it in its active form; alternatively, photoexcitation of an antenna chromophore encourages the NO uncaging upon intramolecular energy transfer to the photocage. (*Bottom*) Molecular structure of a nitroaniline-based photocage and sketch of the mechanism for its NO photorelease

absorb excitation light in the same region of the NO photoprecursor in order to avoid both inner filter effects and undesired photochemical reactions which can affect the performances in term of NO photorelease. A viable strategy for effective photochemical activation in the case of NO photocages with low molar absorbivity involves the use of antenna chromophores covalently linked to the photocage through a molecular spacer. These chromophores are appropriately chosen in order to act as suitable one- or two- photon light-harvesting centres, which encourage the NO release through effective energy transfer to the NO photocage [91, 94].

A major class of NO photocages are NO complexes of transition metals combined with appropriate chromophores [90, 93, 94]. Despite their excellent absorption properties in the visible region, their elaborate structures make them more difficult to manipulate than organic compounds, and carry some drawbacks due to the intrinsic toxicity of the metal centre and/or the by-products photogenerated. In the last years simple derivatives of nitroaniline bearing a CF_3 substituent in the *ortho* position with respect to the nitro group, have proven to be organic photocages suitable for biological applications [97, 98] and their use in combination with PSs for PDT has been reported [99]. The mechanism leading to the NO photorelease for this NO photocage is similar to that observed for other nitrobenzene derivatives substituted at the *ortho* positions [100, 101] and is recalled in Fig. 9.3. Briefly, due to the presence of the CF_3 substituent the nitro group is placed almost perpendicularly to the aromatic plane. This *out of plane* geometry makes the p orbital of the oxygen atom having a constructive overlap with the adjacent p orbital of the aromatic ring in the ground state. Such a twisted conformation is crucial in triggering the NO photorelease which takes place through a nitro to nitrite photorearrangement followed by the rupture of the O-NO bond leading to the concomitant generation of NO and a low reactive phenoxyl radical.

Furthermore, the simple structure of these NO photoprecursors makes them suitable for derivatization via simple synthetic procedures.

In view of image-guided NO-based phototherapies, the visualization of the NO photodonor in a cellular environment through fluorescence techniques represents an indispensable requisite. This permits to produce a highly localized "burst" of NO precisely at the desired sites. Differently to photosensitizers used in PDT, which are intrinsically fluorescent, the combination of fluorescent imaging and phototherapeutic capacity in the same structure is less common in the case of NO photodonors [99]. Of course, in the case of NO photocages lacking of emissive properties, fluorescent character can be imposed by their covalent combination with suitable flurogenic centres. In this case, inter-chromophoric interactions (i.e. energy or electron transfer) between the fluorophoric unit and the NO photocage need to be intentionally avoided with a careful choice of chromophoric groups by taking into account the energy of the lowest excited states of the components as well as their redox potential. Under these conditions, the fluorophotic centre and the NO photocage can act in tandem even under condition of co-absorption of the excitation light [99].

9.4.3 Polymer Nanoparticles for PDT and Nitric Oxide Photorelease

Polymer NPs represent fascinating scaffolds for the achievements of photoactivable nanoconstructs with distinct photofunctionalities [102]. A facile strategy to achieve polymeric nanoconstructs for bimodal photoactivated therapy based on the simultaneous generation of $^1O_2$ and NO consists in the supramolecular co-encapsulation of PSs and NO photoprecursors in the confined space of the polymeric nanocarrier. The noncovalent approach requires minimal, if any, synthetic efforts permitting the relative amounts of the co-encapsulated photoactive guests to be easily regulated with adjustments in the composition of the mixture of self-assembling components. In most instances, the photoresponsive components are sufficiently hydrophobic to enter spontaneously the interior of the nanosized containers even in a single step. Alternatively, hydrophilic photoresponsive guests can be entangled at the surface of the polymer NPs by Coulombic interactions.

In contrast to polymeric nanoassembly where multiple chromomophoric guests are appropriately designed to interact each other to exchange energy, electrons or protons upon light stimuli [102], the fabrication of polymeric NPs for bimodal phototherapy requires the inter-chromophoric interactions to be avoided intentionally with a careful choice of the chromogenic groups. As illustrated in the following examples, this strategic approach will allow the distinct photoresponsive components to be operated in parallel under light inputs within the same nanoconstruct, permitting the simultaneous photodelivery of $^1O_2$ and NO. Note that, in contrast to non-photoresponsive compounds, the preservation of the photobehavior of independent photoactive components in a confined region of space is not a foregone result. In most cases, the response to light of the single photoactive units located in a so close proximity can be considerably influenced, in both nature and efficiency, by the occurrence of competitive, undesired photoprocesses (i.e., photoinduced energy and/or electron transfer, non-radiative deactivation, etc.), which do not allow the individual components to work in parallel after light absorption.

Another important point to be addressed in the fabrication of such bimodal phototherapeutic systems regards the relative amounts of cytotoxic species photogenerated. It has to be considered that while generation of $^1O_2$ is based on photophysical processes which, in principle, do not consume the PS, photogeneration of NO implies a neat photochemical reaction with consequent degradation of the NO photodonor. As a result, in order to achieve an effective bimodal action, the multichromophoric nanocostructs should be designed taking into account an appropriate balance between the potential reservoir of NO available, the absorption coefficient of the NO photocage and its photodegradation quantum yield in comparison with the photophysical features of the PS for PDT.

The combination of porphyrin centers with nitroaniline-derivative NO photocages is very suited in view of dual modal phototherapeutic applications. In fact, these two chromogenic centers "ignore" each other upon light excitation after their incorporation in restricted environments such as micelles [103], cyclodextrin-based nanoparticles [104] and host-guest complexes [105], preserving well their photochemical behavior. These findings represented a solid basis for the fabrication of both charged and neutral polymeric nanoparticles capable to exhibit simultaneous photorelease of $^1O_2$ and NO and fluorescence properties.

Cationic core-shell polymeric NPs composed of a hydrophobic inner core of polyacrylate surrounded by a hydrophilic shell of quaternary ammonium salts represent suitable scaffolds for the entangling of photoactive components exploiting electrostatic interactions [106]. In this case, the commercial tetrasodium-meso-tetra(4-sufonatophenyl)porphyrine **1**, a well-known $^1O_2$ photosensitizer showing considerable fluorescence in the red region, and the tailored component **2**, in which the a nitroaniline derivative NO photocage was deliberately functionalized with a sulfonate termination, were used to encourage electrostatic binding with the oppositely charged NPs surface (Fig. 9.4).

Bichromophoric NPs, ca. 280 nm in diameter, were obtained by self-assembling procedures allowing to regulate accurately the relative amounts of the chromogenic centres allocated at the NPs surface. This simple procedure permits to obtain a satisfactory absorption of both the PS and the NO photononor around 400 nm. According to the presence of the negatively charged guest components in the positive shell of the NPs, the co-loaded NPs exhibited a larger hydrodynamic

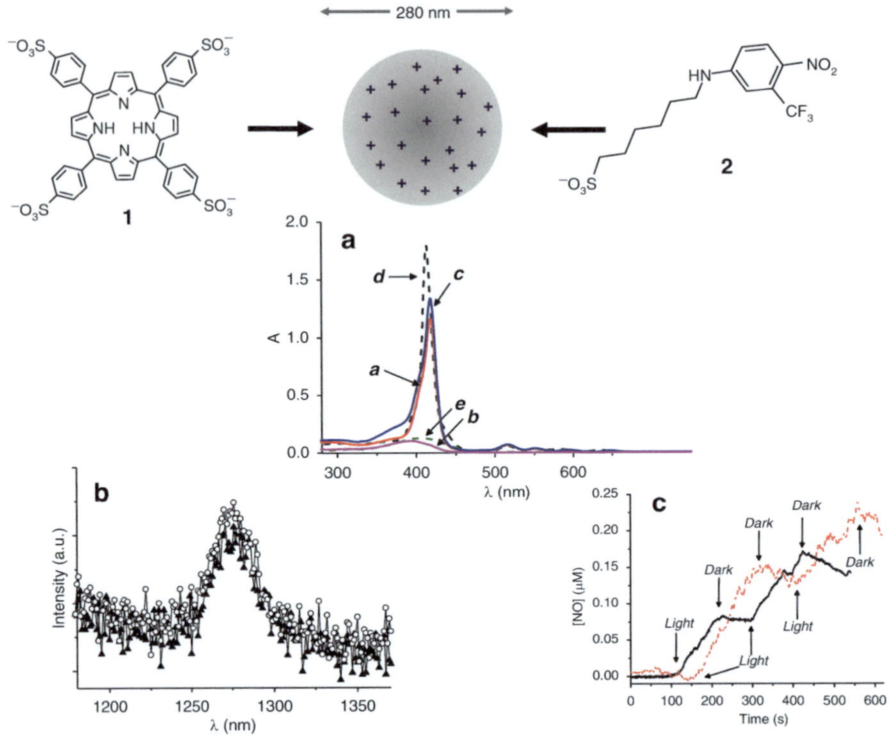

Fig. 9.4 Cationic core-shell NPs co-entangle the anionic components **1** and **2** in the shell, leading to bichromophoric NPs able to simultaneously release NO and $^1O_2$ under Vis excitation. (**a**) Absorption spectra of the cationic NPs loaded with **1** (***a***), **2** (***b***) and **1** + **2** (***c***). Absorption spectra of **1** (***d***) and **2** (***e***) in the absence of NPs. (**b**) Singlet oxygen luminescence detected in D_2O solutions of core-shell NPs loaded with **1** (▲) and **1** + **2** (○) upon 532 nm light excitation. (**c**) NO release profile observed upon irradiation with $\lambda_{exc} = 400$ nm of the NPs **1** loaded with **2** (*dotted*) and **2** + **1** (*solid*) [**NPs**] = 1 mg mL^{-1}; [**1**] = 3.5 μM; [**2**] = 13 μM (Adapted with permission from Ref. [106])

diameter and a reduced zeta-potential value with respect to the unloaded NPs. The spectrum of the NPs co-loaded with **2** and **3** (***c*** in Fig. 9.4a) reveals the absorption of both chromophores. In fact, the Soret band of the porphyrin is accompanied by an additional contribution around 400 nm due to the NO photoreleaser. These spectral characteristics are in excellent agreement with the profile obtained by summing the spectra of the NPs loaded with **2** or **3** separately (***a*** and ***b*** in Fig. 9.4a), indicating the absence of relevant interactions between the individual chromophores, in the ground state. Interestingly, the emission properties of **2** were not affected by the presence of **3** offering a great advantage for cell imaging of the NPs. The lack of communication of the two chromophores in the excited state is also confirmed by their capability to effectively photogenerate the cytotoxic $^1O_2$ and NO with efficiencies comparable to those observed for the mono-chromophoric NPs, supporting the validity of the logical design (Fig. 9.4b, c). The co-loaded NPs transport their

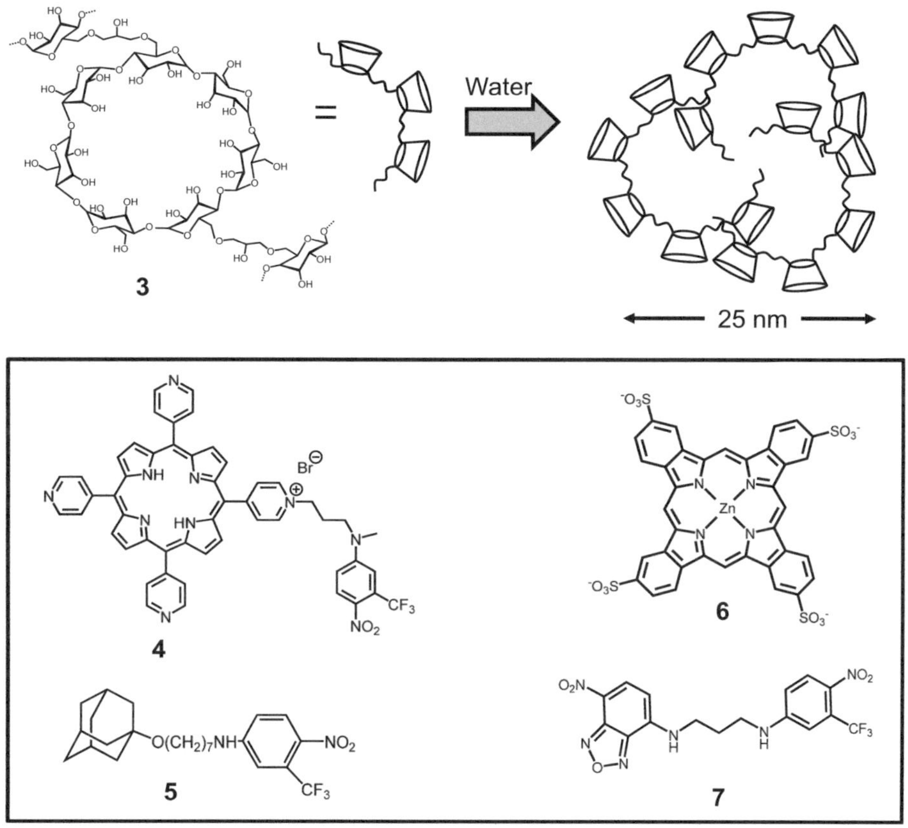

Fig. 9.5 Schematic of the NPs formed by polymer **3** and molecular structures of the photoactive guests that can be encapsulated therein

cargo within melanoma cancer cells without suffering displacement despite the non-covalent binding, were well tolerated in the dark and exhibited strongly amplified cell mortality under visible light excitation most likely due to the combined action of $^1O_2$ and NO [106]. These NPs offer in principle the possibility to exploit the inner hydrophobic core for the further encapsulation of additional components like, for example, chemotherapeutic agents, opening interesting prospects in view of three modal therapeutic applications.

In alternative to charged core-shell polymeric NPs, cyclodextrin (CD) branched polymers represent an interesting class of polymers leading to neutral NPs which offer the possibility of guest interactions with diverse binding sites, i.e. within the 3D macromolecular network and the CD cavities, overcoming solubility and aggregation drawbacks displayed by many guests in aqueous medium [107]. Polymer **3** (Fig. 9.5) consists of β-CD units interconnected by epichlorohydrin spacers to form glyceryl cross-linked β-CD polymer. This polymer shows excellent biocompatibility in vivo [108] and is highly soluble in water where it spontaneously assembles under the form of neutral NPs *ca.* 25 nm in diameter [109]. These NPs are able to entrap a variety of therapeutics and diagnostic agents [110–114] with association constants and payloads larger than that observed for unmodified β-CD, due to the presence of different hydrophobic nanodomains.

The photochemical independence of the porphyrin centre and the nitroaniline derivative motivated the design of the molecular hybrid **4** (Fig. 9.5) in which a pyridyl porphyrin and the NO photodonor are covalently connected through an alkyl spacer [115]. Due to its covalent nature, this molecular conjugate offers the advantage of ensuring that the photodelivery of the cytotoxic species occur exactly in the "very same region of space" of the cell component, with a much more precise control of the delivery site. A main problem of this molecular hybrid is the very low solubility in water medium where it forms large aggregates that are non-responsive to light due to effective self-quenching processes. However, the solubility of the molecular conjugate increases dramatically in the presence of NPs of polymer **3** where the hybrid dissolves mainly as a monomeric species. The resulting polymeric NPs show satisfactory fluorescence, allowing to map the conjugate in melanoma cancer cells and simultaneously release $^1O_2$ and NO under visible light stimuli inducing a good level of amplified cell mortality by bimodal photoaction. Despite the advantage illustrated above, these types of conjugates suffer, in general, a main limitation due to the 1:1 molar ratio of the two photoprecursors of the cytotoxic species. In fact, $^1O_2$, photocatalytically generated in large amounts, could in principle mask the biological effects of NO to a significant extent, this conversely being produced through a photodecomposition pathway, in smaller quantities. Based on these considerations, the use of $^1O_2$ and NO photodonors possessing absorption in distinct spectral region of the visible range is highly desirable. This design rationale should increase the performance of the resulting molecular hybrid, allowing fine-tuning of the cytotoxic species by means of the appropriate selection of the light excitation wavelength and modulation of its intensity. The supramolecular approach offers additional advantages at this regards since it allows regulating also the relative amount of the photoprecursors in favour of the NO photoreleaser, which consumes upon irradiation.

Bichromophoric NPs with excellent bimodal phototherapeutic activity and imaging performances have been obtained by the noncovalent co-encapsulation of the NO photodonor **5** and the zinc phthalocyanine **6** within NPs of polymer **3** [116]. The phthalocyanine chromophore is an excellent $^1O_2$ photosensitizer with absorption falling beyond the absorption of the nitroaniline NO photodonor. Besides, it offers the advantage to exploits two-photon fluorescence for imaging due its high two-photon cross section [117]. However, the porphyrinoid PS is completely aggregated and non-photoresponsive in water medium. In this case, the polymer NPs disrupt significantly the aggregation of **6**, sequestering this chromophore mainly under its photoactive monomeric form. Furthermore, the large amount of CD cavities of the branched polymer (ca. 70 % in weight) tightly accommodate the NO photodonor **5** which was ad-hoc modified with an adamantane appendage, a perfect guest for the β-CD host. The appropriate regulation of the co-entrapment of the two guests in the very same host, generates

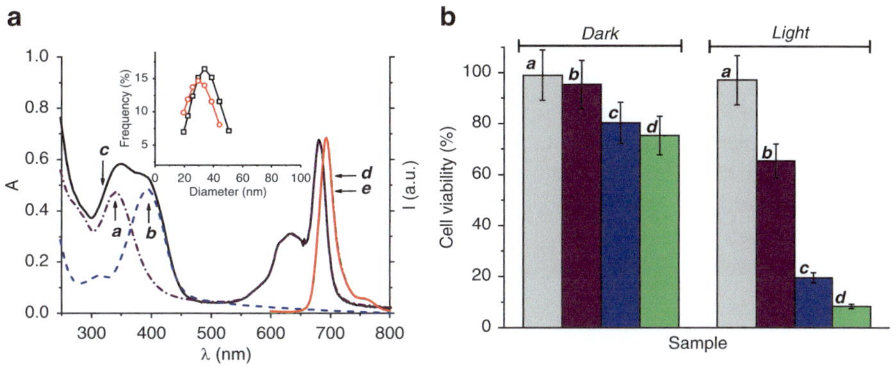

Fig. 9.6 (**a**) Absorption spectra of aqueous dispersions of **3** in the presence of **6** (***a***), **5** (***b***), and **5** + **6** (***c***). Fluorescence spectra of aqueous dispersions of **3** in the presence of **6** (***d***) and **5** + **6** (***e***), recorded at $\lambda_{exc} = 575$. The inset shows the hydrodynamic diameter for aqueous dispersion of **3** in the absence (○) and in the presence (□) of **5** + **6** [**3**] = 11 μM (7.75 mM in β-CD); [**6**] = 15 μM; [**5**] = 40 μM. (Phosphate buffer 10 mM, pH 7.4, 25 °C). (**b**) Dark and photoinduced mortality of A431 human squamous carcinoma cells incubated with aqueous dispersion of **3** in the absence (***a***) and in the presence of **6** (***b***), **5** (***c***) and **5** + **6** (***d***) [**3**] = 11 μM (7.75 mM in β-CD); [**6**] = 15 μM; [**5**] = 40 μM (Adapted with permission from Ref. [116])

photoresponsive NPs, *ca.* 30 nm in diameter, having comparable absorption in well distinct blue and red region of the visible range (Fig. 9.6a) and capable to photogenerate concomitantly $^1O_2$ and NO resulting in an excellent level of cancer cell photomortality (Fig. 9.6b). Moreover, as anticipated above these NPs revealed very suited for fluorescence imaging by two-photon fluorescence microscopy which revealed that the macromolecular host not only delivers the phthalocyanine within the cytoplasm but also in human skin as exemplified in ex vivo experiments [116].

The lack of fluorescence of the NO photodonor does not allow its tracking in cells although the bimodal photomortality provides a good indication of its cell internalization. A convenient way to make this NO photoreleaser fluorescent consists in its covalent joining with a fluorescent label devised in such a way of not altering the photochemical properties. To this end, the nitroaniline photocage has been joined to the 4-amino-7-nitrobenzofurazan moiety, a well-known fluorophore emitting in the green region and extensively used in biology [118], through an alkyl spacer to obtain the molecular conjugate **7**.

This fluorophoric unit was deliberately chosen for the NO photoreleasing properties and the emissive features to be conserved in the molecular conjugate. The rationale behind the design of this molecular hybrid was that the emissive singlet state of the fluorophore lies below the singlet state of the NO photodonor moiety and thus, cannot be quenched by photoinduced singlet-singlet energy transfer. In addition, the reactive short-lived triplet state leading to the NO photorelease is located below the triplet state of the fluorophoric unit making a quenching by triplet-triplet energy transfer thermodynamically forbidden. According to this logical design, light excitation at 480 nm results in the typical green emission of the fluorogenic unit with satisfactory quantum yield, whereas excitation at 400 nm leads to NO release from the nitroaniline chromophore [119]. This fluorescent NO photodonor is not water soluble but can be co-entrapped together with the phthalocyanine **6** within the same NPs of the CD-based polymer **3** [119]. The whole system has been designed in a way that the excited states responsible for the emission and the photogeneration of the cytotoxic species of each compound cannot be quenched by the other and *vice versa*. In fact, as shown in Fig. 9.7, the

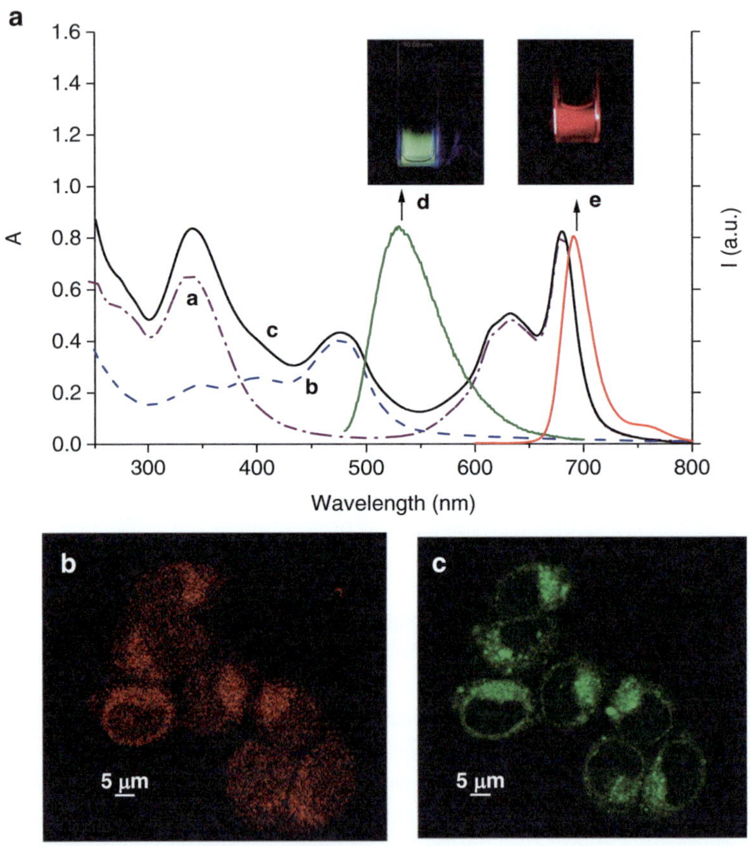

Fig. 9.7 (**a**) Absorption spectra of aqueous dispersions of the polymer **3** in the presence of **6** (*a*), **7** (*b*), and **6** + **7** (*c*). Fluorescence spectra and actual images of aqueous dispersion of **3** in the presence of **6** + **7**, recorded at $\lambda_{exc} = 470$ nm (*d*) and 575 (*e*). (**b**) Confocal images of melanoma cells incubated 4 h with NPs of **3** loading both **6** and **7** and obtained with excitation at 640 and 457 nm collecting fluorescence in the range 660–730 (**b**) and 500–550 (**c**) (Adapted with permission from Ref. [119])

green emission of the NO falls in the hole of absorption of the red emitter PS, making the spectral overlap necessary for photoinduced energy transfer negligible.

Furthermore, the supramolecular approach permits an easy tuning of the relative concentration of the two chromophores in the NPs to obtain comparable absorption in the Vis region. Note that the concomitant presence of both components in the same polymer NPs lead to the average hydrodynamic diameters of the NPs changing from 30 to about 50 nm but does not induce any aggregation phenomena. This nanoplatform shows the convergence of "four-in-one" photoresponsive functionalities, two imaging and two therapeutic simultaneous functions. In fact, illumination with visible light generates parallel green and red fluorescence signals that allows localizing the photoactive agents in melanoma cancer cells. Confocal microscopy analysis provides insights that the NPs do not suffer significant displacement, internalizing in cells with all its cargo. $^1O_2$ and NO are produced with excellent efficiency, producing amplified level of cell mortality due to their simultaneous photoproduction in the same region of space.

9.5 Combining PDT and Chemotherapy

9.5.1 Why to Combine PDT with Chemotherapeutics

Cytotoxic drugs can act in concert with PS as recently evidenced by clinical trials [120] which confirms PDT/chemotherapy as a very promising combined strategy achieving potentiated therapeutic outcome and low side-effects while overcoming drug resistance. Combination of PDT with chemotherapy in a single nanoconstruct can offer an unprecedented opportunity to devise a better scheme for precise and controlled delivery of multiple therapeutic agents in the same area of the body and at a predefined extra/intracellular level. In PDT, any activity of PDT-sensitizing agents will be confined to the illuminated area thus inducing non-systemic potentiated toxicity of the combinations. This should be of special importance in elderly or debilitated patients who tolerate poorly very intensive therapeutic regimens. Moreover, considering its unique $^1O_2$-dependent cytotoxic effects, PDT can be safely combined with other antitumor treatments without the risk of inducing cross-resistance [121, 122]. Despite this potential, few studies on combinations of PDT with standard antitumor regimens have been published to date [123]. Also in this case, for the photodynamic activity of the PS to be preserved in the presence of the chemotherapeutic, the photo- and chemotherapeutic agents must be selected with care in order to avoid any mutual communication upon light excitation. The knowledge of the energy values of the lowest excited states of the components as well as their redox potential may help in predicting the possible occurrence of photoinduced energy/electron transfer between the PS and the chemotherapeutics, which would limit the PDT and chemotherapeutic effects.

9.5.2 Polymer Nanoparticles for PDT and Chemotherapy

In one of the first example reporting combination of PDT and chemotherapy, a chlorin-core star-shaped block copolymer (CSBC) of poly(lactic acid) and PEG (PLA-PEG) was self-assembled to form micelles, which act as nanosized photosensitizing agents and further encapsulate hydrophobic drugs (Fig. 9.8) [124]. The MWs of the CSBC copolymer was around 55 KDa ($Mw/Mn = 1.2$) and chlorin content was approximately 1.23 wt.%. CSBC was loaded with 7-ethyl-10-hydroxy-Camptothecin (SN-38), a drug virtually insoluble in all pharmaceutically acceptable solvents. Smooth, spherical micelles with a size spanning from approximately 115 to 170 nm in diameter depending on drug-to polymer ratio were formed. The cytotoxicity of SN-38-CSBC was enhanced in HT-29 human

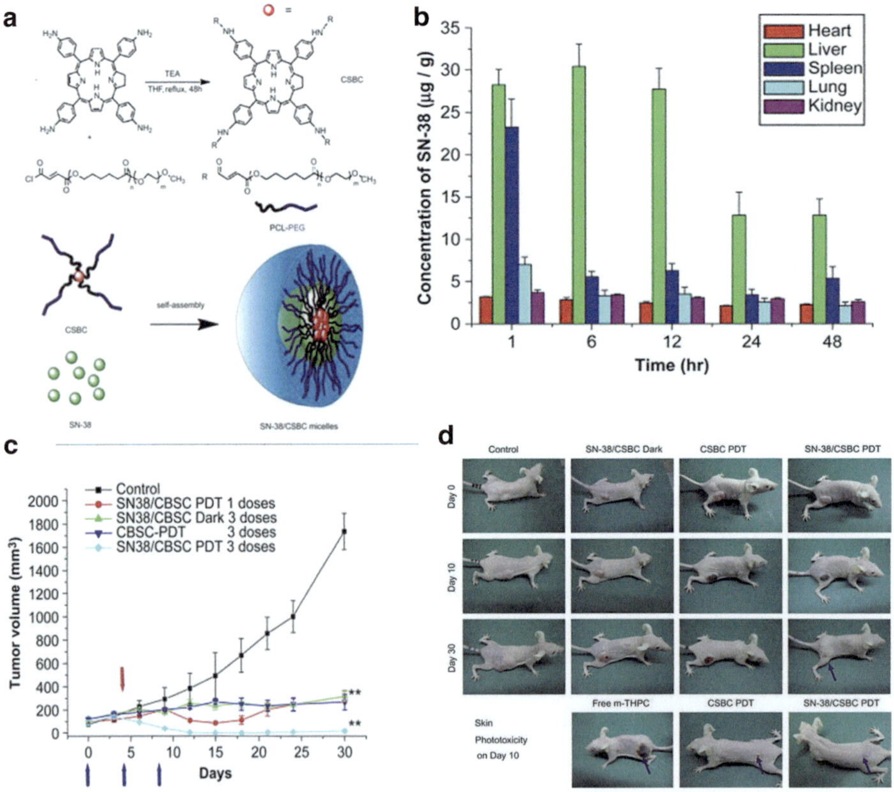

Fig. 9.8 (**a**) Chemical structure of chlorin-core star block copolymers (CSBC) and schematic drawing of self-assembled SN-38/CSBC micelles. (**b**) Biodistribution of SN-38 (10 mg/kg) in mice bearing HT-29 solid tumors. (**c**) Antitumor efficacy of free CPT-11 or SN-38/CBSC micelles in an HT-29 human colon cancer xenograft model.*$P < 0.05$ and **$P < 0.01$, as compared with control. (**d**) The effects of combination therapy on PDT-mediated phototoxicity in nude mice bearing HT-29 human colon cancer xenografts (Adapted with permission from Ref. [124])

colorectal adenocarcinoma cells after irradiation as compared to dark conditions. Furthermore, SN-38/CSBC micelles showed a prolonged plasma residence time as compared with free CPT-11 (a prodrug of SN38) permitting increased tumor accumulation (Fig. 9.8b). The combined effects of SN-38/CSBC micelles with PDT were evaluated in an HT-29 human colon cancer xenograft model. SN-38/CSBC-mediated PDT synergistically inhibited tumor growth (Fig. 9.8c, d), resulting in up to 60 % complete regression of well-established tumors after three treatments. These treatments also decreased the microvessel density (MVD) and cell proliferation within the subcutaneous tumor while attenuating phototoxicity (Fig. 9.8d).

In a following study NPs formulated using an anionic surfactant aerosol-OT, and a naturally occurring polysaccharide polymer, sodium alginate, and carrying a combination of doxorubicin and the PS methylene blue were developed [125]. Methylene blue delivered alone in aerosol OT-alginate NPs enhanced its anticancer photodynamic efficacy in vitro due to increased ROS production and favorable alteration in the subcellular distribution from endolysosome (free methylene blue) to nucleus (in case of loaded NPs) [126]. In a mouse drug-resistant mammary adenocarcinoma tumor model (JC cells) NP-mediated combination therapy resulted in a significant induction of both apoptosis and necrosis. Further, nanoparticle-mediated combination therapy resulted in significantly elevated reactive oxygen species (ROS) production compared to single drug treatment. Tumor levels of doxorubicin and methylene blue were increased for combined NPs compared to the free drug treatments, while accumulation of doxorubicin in the heart was decreased, which could help decrease doxorubicin-associated cardiac toxicity. Increased vasculature damage for combined NPs as compared to free drug combination was also found.

Delivery of drug combination through biodegradable polymers can be highly suited to sustain release rate of the anticancer drug thus mimicking a metronomic therapy (continuous or frequent treatment with low doses of anticancer drugs) and to avoid premature leakage of drug cargo from NPs in the blood before they reach tumor. With this idea in mind, PEGylated nanoparticles for the passive targeting of a conventional anticancer drug docetaxel (DTX) and the second generation photosensitizer zinc-phthalocyanine (ZnPc) were designed (Fig. 9.9) [127]. Amphiphilic biodegradable block copolymers based on poly(ε-caprolactone) and PEG with diblock (PCL-*b*-PEG) and triblock (PEG-PCL-PEG) architectures were assembled in core-shell NPs and loaded with both DTX and ZnPc employing the melting/sonication method (MeSo) (Fig. 9.9a). This technique exploits low melting temperature of the copolymer and consists in the self-assembly of molten copolymer in water assisted by sonication. Resulting NPs are characterized by a PEGylation extent higher than conventional nanoprecipitation technique [128].

NPs with hydrodynamic diameters spanning in the range 60–100 nm, low polydispersity indexes and negative zeta potential were obtained. It was found that the entrapment efficiency was related to the amount of the most abundant drug, showing that proper ZnPc/DTX ratio could allow to load both drugs with high efficiency. Steady-state and time-resolved emission fluorescence

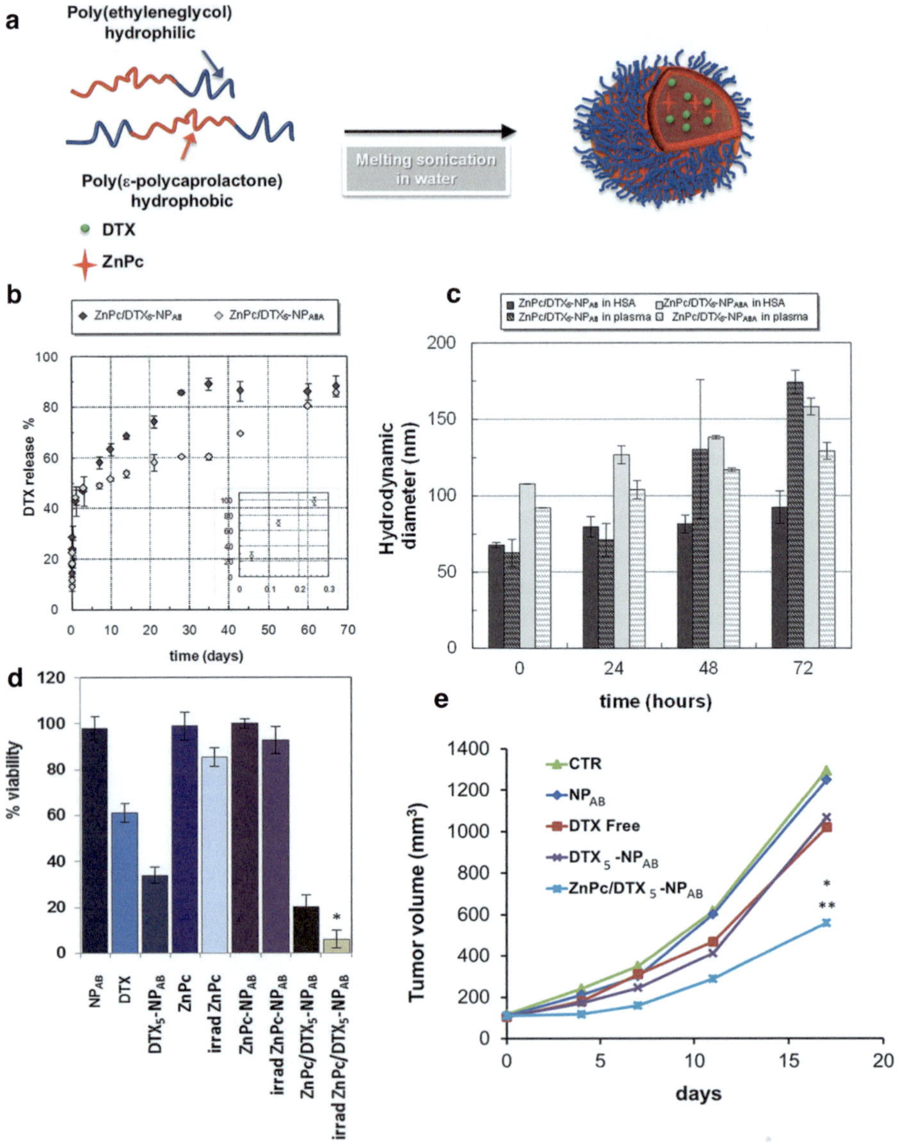

Fig. 9.9 (**a**) Composition of ZnPc/DTX loaded PEGylated NPs. (**b**) Release profile of DTX from ZnPc/DTX-loaded NPs in 10 mM phosphate buffer at pH 7.4 and 37 °C. (**c**) Stability of NPs in the presence of 2 % w/v Human Serum Albumin and human plasma. (**d**) Viability of HeLa cells after 72 photodynamic treatment ("*irrad*" refers to irradiated samples). $^*p < 0.0003$. (**e**) Tumor growth inhibition in female athymic Nu/Nu nude mice bearing subcutaneous A375 tumor. DTX dose was 3.5 mg/Kg whereas ZnPc dose was 0.2 mg/Kg. 24 h post-injection mice were exposed to 15 min (149 J) of 172 mW/cm^2 light. $^*P < 0.05$ DTX_5NP_{AB} vs. ZnPc/DTX_5-NP_{AB}; $^{**}P < 0.01$ CTR vs ZnPc/DTX_5-NP_{AB}; $^{**}P < 0.01$ NP_{AB} vs ZnPc/DTX_5-NP_{AB} (Adapted with permission from Ref. [127])

measurements pointed out the embedding of monomeric ZnPc in the NPs, excluding the presence of ZnPc supramolecular oligomers. The release of DTX was biphasic whereas ZnPc remained mainly associated with NPs (Fig. 9.9b). For PDT this is not considered a drawback since $^1O_2$ generation can occur from PS-loaded NPs rather than free PS. Indeed, when ZnPc-loaded combined NPs were irradiated at 610 nm, $^1O_2$ generation was observed. Combined NPs demonstrated excellent stability also in the presence of human serum albumin (HSA) and in human plasma (Fig. 9.9c) and no toxicity toward red blood cells was found. Since both copolymers presented very similarly colloidal properties, NPs based on the most simple PCL-*b*-PEG were investigated for biological properties. NP cytotoxicity was evaluated in irradiated HeLa cells. After 72 h, viability of cells treated with ZnPc/DTX-loaded NPs strongly decreased as compared to NPs loaded only with DTX, thus showing a combined effect of both DTX and ZnPc (Fig. 9.9d). Superior antitumor activity of ZnPc/DTX-loaded NPs as compared to DTX-loaded NPs was confirmed in an animal model of orthotopic amelanotic melanoma, thus pointing to the application of PEG-b-PCL NPs as carrier for light-activated drugs and in the combined chemo-photodynamic therapy of cancer (Fig. 9.9e). Similar synergic effects were evidenced for PEG-PDLLA nanovesicles (around 176 nm) loaded with doxorubicin (DOX) and hematoporphyrin monomethyl ether (HMME) in HepG2 human hepatocellular carcinoma cells [129].

A key aspect when developing combined NPs is the achievement of the correct PS/chemotherapeutic ratio to attain the desired efficacy. In this respect, coating NPs with different polymer layers is a non-covalent approach useful to entrap different molecules in both the core and the shell as well as to impart specific features to NP surface. On this bases, overwhelming interest has been devoted to engineered layer-by-layer NPs for cancer therapy [130]. Consecutive deposition of ionized hydrophilic polymers with opposite charge onto a hydrophobic nanotemplate results in ultrathin multiple layers of polymer chains. By controlling type and number of alternating layers surrounding the core, multifunctional core-shell nanostructures can be obtained where multiple drugs, magnetic or luminescent layers can be formed. In this context, a simplified version of LbL NPs called double coated NPs (dcNPs) has been proposed for the combined PDT and chemotherapy of cancer cells overexpressing CD44 receptor (Fig. 9.10) [131, 132]. Negatively-charged DTX-loaded NPs of PLGA were sequentially decorated through electrostatic interactions with a polycationic shell of polyethylenimine entangling negatively-charged tetrasodium-meso-tetra(4-sufonatophenyl)porphyrine ($TPPS_4$) and a final layer of hyaluronan (HA), a CD44 ligand (Fig. 9.10a). Polymer layer adsorption was demonstrated by inversion of zeta potential and increase of NP size (Fig. 9.10b). dcNPs bears $TPPS_4$ completely aggregated and photochemically unactive at their surface and are readily internalized in MDA-MB231 breast cancer cells overexpressing CD44 receptor (Fig. 9.10c). The concerted delivery of DTX and TPPS4 resulted in a higher uptake of the hydrophilic PS and tremendous improvement of single drug activity (Figs. 9.10d, e). This feature is of relevance in view of in vivo application since dcNPs are expected to be nonfluorescent and non-photoactive in non-target cells, thereby strongly reducing phototoxicity of

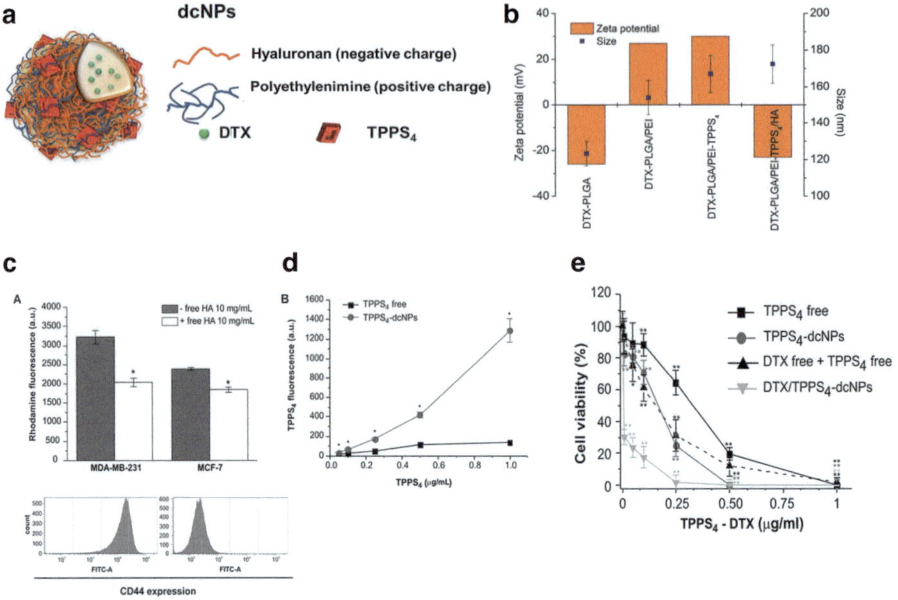

Fig. 9.10 (**a**) Schematic representation of the structure of DTX/$TPPS_4$-dcNPs. (**b**) Evolution of zeta potential and size during NP layering procedure. (**c**) Uptake of RHO-dcNPs and TPPS4-dcNP (50 μg/mL) in MDA-MB-231 (high CD44 expression) and MCF-7 (low CD44 expression) cells after 2 h incubation at 37 °C in medium with or without 10 mg/mL free HA. $^{*}p < 0.001$, with respect to the sample not incubated with free HA (Student's t test). (**d**) Concentration-dependent uptake of free $TPPS_4$ and $TPPS_4$-dcNPs in MDA-MB-231 cells after 24 h incubation at 37 °C. $^{*}p < 0.001$, with respect to $TPPS_4$ free (Student's t test). (**e**) Viability of MDA-MB-231 irradiated with 8 J/cm^2 of blue light after 24 h of exposure to the drugs. Viability was measured 24 h post-irradiation. $^{*}p < 0.01$, $^{**}p < 0.001$, with respect to control (Student's t test) (Adapted with permission from Ref. [131])

carried PS. Nevertheless, taking advantage of targeting to CD44 receptor, dcNPs can localize in tumor tissue, where the PDT component is disassembled from the dcNPs surface and becomes highly fluorescent and phototoxic.

9.6 Concluding Remarks

Biocompatible polymer NPs represent fascinating containers for potentiating PDT effects through combination with NO photoreleasers and chemotherapeutics. NP fabrication needs to be driven by biologically-oriented design guidelines to optimize the co-accommodation of the multifunctional cargo at tumor level as well as to tune the delivery process. In turn, all distinct components with therapeutic action should be ad-hoc selected and assembled in the single NP platform in order to avoid any reciprocal communication upon light excitation. This rational combination ensures the PDT agent, the NO photoreleaser and the chemotherapeutic drug to preserve their individual properties and to operate together in the same region of space, resulting in enhanced therapeutic performances through additive/synergistic effects otherwise impossible with the separate components. In part because of its interdisciplinary nature, the development of NPs for multimodal treatments where PDT is combined with NO and anticancer drugs is its infancy. Many issues still need to be debated and more research is still required before these NPs translate into clinics. However, it appears clear that the present strategies may pave the way towards more effective, minimally invasive and underexplored multimodal anticancer therapies where light plays definitely a dominant role. We really hope this contribution will stimulate chemists, biologists, material scientists and clinicians to proceed in this direction.

Acknowledgements We thank AIRC (Project IG-12834 and IG-15764) and the MIUR (PRIN 2010-2011, 2010C4R8M8 2010H834LS) for financial support.

References

1. Burrell RA, McGranahan N, Bartek J, Swanton C (2013) Nature 501:338–345
2. Turner NC, Reis-Filho JS (2012) Lancet Oncol 13:e178–e185
3. Kemp JA, Shim MS, Heo CY, Kwon YJ (2015) Adv Drug Del Rev. Epub ahead of print
4. Chen Q, Ke H, Dai Z, Liu Z (2015) Biomaterials 73:214–230
5. Dolmans DEJG, Fukumura D, Jain RK (2003) Nat Rev Cancer 3:380–387
6. Agostinis P, Berg K, Cengel KA, Foster TH, Girotti AW, Gollnick SO, Hahn SM, Hamblin MR, Juzeniene A, Kessel D, Korbelik M, Moan J, Mroz P, Nowis D, Piette J, Wilson BC, Golab J (2011) CA Cancer J Clin 61:250–281
7. Jain KK (2010) BMC Med 8:83
8. Jain RK, Stylianopoulos T (2010) Nat Rev Clin Oncol 7:653–664
9. Wang AZ, Langer R, Farokhzad OC (2012) Annu Rev Med 63:185–198
10. Duncan R, Gaspar R (2011) Mol Pharm 8:2101–2141
11. Schroeder A, Heller DA, Winslow MM, Dahlman JE, Pratt GW, Langer R, Jacks T, Anderson DG (2012) Nat Rev Cancer 12:39–50
12. Markman JL, Rekechenetskiy A, Holler E, Ljubimova JY (2013) Adv Drug Deliv Rev 65:1866–1879
13. Palakurthi S, Yellepeddi VK, Vangara KK (2012) Expert Opin Drug Del 9:287–301
14. Iyer AK, Singh A, Ganta S, Amiji MM (2013) Adv Drug Del Rev 65:1784–1802
15. Kamaly N, Xiao Z, Valencia PM, Radovic-Moreno AF, Farokhzad OC (2012) Chem Soc Rev 41:2971–3010
16. Avci P, Erdem SS, Hamblin MR (2014) J Biomed Nanotechnol 10:1937–1952
17. Webster DM, Sundaram P, Byrne ME (2013) Eur J Pharm Biopharm 84:1–20
18. Fang J, Nakamura H, Maeda H (2011) Adv Drug Del Rev 63:136–151
19. Maeda H (2012) J Control Release 164:138–144
20. Moghimi SM, Hunter AC, Andresen TL (2012) Annu Rev Pharmacol Toxicol 52:481–503
21. Bertrand N, Leroux JC (2012) J Control Release 161:152–163
22. Alexis F, Pridgen E, Molnar LK, Farokhzad OC (2008) Mol Pharm 5:505–515
23. Huynh NT, Roger E, Lautram N, Benoit JP, Passirani C (2010) Nanomedicine (Lond) 5:1415–1433

24. Bertrand N, Wu J, Xu X, Kamaly N, Farokhzad OC (2013) Adv Drug Deliv Rev 66:2–25
25. Kwon IK, Lee SC, Han B, Park K (2012) J Control Release 164:108–114
26. Florence AT (2012) J Control Release 164:115–124
27. Ruenraroengsak P, Cook JM, Florence AT (2010) J Control Release 141:265–276
28. Torchilin V (2009) Eur J Pharm Biopharm 71:431–444
29. Mura S, Nicolas J, Couvreur P (2013) Nat Mater 12:991–1003
30. Kim CS, Duncan B, Creran B, Rotello VM (2013) Nano Today 8:439–447
31. Gao W, Chan JM, Farokhzad OC (2010) Mol Pharm 7:1913–1920
32. Tian L, Bae YH (2012) Colloids Surf B-Biointerfaces 99:116–126
33. Du JZ, Mao CQ, Yuan YY, Yang XZ, Wang J (2013) Biotechnol Adv 32:78–803
34. Nowag S, Haag R (2014) Angew Chem Int Ed Engl 53:49–51
35. Abulateefeh SR, Spain SG, Aylott JW, Chan WC, Garnett MC, Alexander C (2011) Macromol Biosci 11:1722–1734
36. Bozic I, Reiter JG, Allen B, Antal T, Chatterjee K, Shah P, Moon YS, Yaqubie A, Kelly N, Le DT, Lipson EJ, Chapman PB, Diaz LA, Vogelstein B, Nowak MA (2013) eLife Sci 2:e00747
37. Hu CMJ, Zhang L (2012) Biochem Pharmacol 83:1104–1111
38. Zheng M, Yue C, Ma Y, Gong P, Zhao P, Zheng C, Sheng Z, Zhang P, Wang Z, Cai L (2013) ACS Nano 7:2056–2067
39. d'Angelo I, Conte C, Miro A, Quaglia F, Ungaro F (2013) Expert Opin Drug Del 11:283–297
40. Stylianopoulos T, Soteriou K, Fukumura D, Jain RK (2013) Ann Biomed Eng 41:68–77
41. Yim H, Park SJ, Bae YH, Na K (2013) Biomaterials 34:7674–7682
42. Gref R, Domb A, Quellec P, Blunk T, Muller RH, Verbavatz JM, Langer R (2012) Adv Drug Del Rev 64:316–326
43. Monopoli MP, Aberg C, Salvati A, Dawson KA (2012) Nat Nanotechnol 7:779–786
44. Lee H, Fonge H, Hoang B, Reilly RM, Allen C (2010) Mol Pharm 7:1195–1208
45. Sakhtianchi R, Minchin RF, Lee KB, Alkilany AM, Serpooshan V, Mahmoudi M (2013) Adv Colloid Interfac 201–202:18–29
46. Mukherjee S, Ghosh RN, Maxfield FR (1997) Physiol Rev 77:759–803
47. Fleige E, Quadir MA, Haag R (2012) Adv Drug Del Rev 64:866–884
48. Tammam SN, Azzazy HME, Lamprecht A (2015) J Biomed Nanotechnol 11:555–577
49. Conte C, Maiolino S, Silva Pellosi D, Miro A, Ungaro F, Quaglia F (2015) Top Curr Chem. in press
50. Amoozgar Z, Yeo Y (2012) Wiley Interdiscip Rev Nanomed Nanobiotechnol 4:219–233
51. Knop K, Hoogenboom R, Fischer D, Schubert US (2010) Angew Chem Int Ed 49:6288–6308
52. Pasquier E, Kavallaris M, Andre N (2010) Nat Rev Clin Oncol 7:455–465
53. Li J, Wang Y, Zhu Y, Oupicky D (2013) J Control Release 172:589–600
54. Muthu MS, Wilson B (2012) Nanomedicine 7:307–309
55. Saraf S (2009) Expert Opin Drug Del 6:187–196
56. Anton N, Benoit JP, Saulnier P (2008) J Control Release 128:185–199
57. Valencia PM, Hanewich-Hollatz MH, Gao WW, Karim F, Langer R, Karnik R, Farokhzad OC (2011) Biomaterials 32:6226–6233
58. Fonte PF, Soares S, Sousa F, Sousa FF, Costa AF, Seabra VF, Reis S FAU – Sarmento B, Sarmento B (2014) Biomacromolecules 15:3753–3765
59. Wainmwright M (2009) Photosensitizers in Biomedicine. Wiley-Blackwell
60. Wilson BC, Patterson MS (2008) Phys Med Biol 53:R61–R109
61. Castano AP, Demidova TN, Hamblin MR (2004) Photodiagn Photodyn Ther 1:279–293
62. Castano AP, Demidova TN, Hamblin MR (2005) Photodiagn Photodyn Ther 2:91–106
63. Castano AP, Demidova TN, Hamblin MR (2005) Photodiagn Photodyn Ther 2:1–23
64. Castano AP, Mroz P, Hamblin MR (2006) Nat Rev Cancer 6:535–545
65. Celli JP, Spring BQ, Rizvi I, Evans CL, Samkoe KS, Verma S, Pogue BW, Hasan T (2010) Chem Rev 110:2795–2838
66. Nowis D, Makowski M, Stoklosa T, Legat M, Issat T, Golab J (2005) Acta Biochim Pol 52:339–352

67. Pandey R, Zhen G (2015) The porphyrin handbook. Academic, San Diego
68. Awuah SG, You Y (2012) RSC Adv 2:11169–11183
69. Kamkaew A, Lim SH, Lee HB, Kiew LV, Chung LY, Burgess K (2013) Chem Soc Rev 42:77–88
70. Wang SZ, Gao RM, Zhou FM, Selke M (2004) J Mater Chem 14:487–493
71. Chatterjee DK, Fong LS, Zhang Y (2008) Adv Drug Deliv Rev 60:1627–1637
72. Davis ME, Chen ZG, Shin DM (2008) Nat Rev Drug Discov 7:771–782
73. Allison RR, Bagnato VS, Sibata CH (2010) Future Oncol 6:929–940
74. Allison RR, Downie GH, Cuenca R, Hu XH, Childs CJ, Sibata CH (2004) Photodiagn Photodyn 1:27–42
75. Aggarwal LPF, Borissevitch IE (2006) Spectrochim Acta A 63:227–233
76. Gabrielli D, Belisle E, Severino D, Kowaltowski AJ, Baptista MS (2004) Photochem Photobiol 79:227–232
77. Gelderblom H, Verweij J, Nooter K, Sparreboom A (2001) Eur J Cancer 37:1590–1598
78. Tang W, Xu H, Kopelman R, Philbert MA (2005) Photochem Photobio 81:242–249
79. Ignarro LJ (2010) Nitric oxide: Biology and pathobiology. Elsevier
80. Ignarro LJ (2009) Arch Pharm Res 32:1099–1101
81. Fukumura D, Kashiwagi S, Jain RK (2006) Nat Rev Cancer 6:521–534
82. Hickok JR, Thomas DD (2010) Curr Pharm Des 16:381–391
83. Chang CF, Diers AR, Hogg N (2015) Free Radic Biol Med 79:324–336
84. (2005) Nitric oxide donors for pharmaceutical and biological applications. Wiley-VCH Verlag GmbH & Co. KGaA, Weinheim
85. Riccio DA, Schoenfisch MH (2012) Chem Soc Rev 41:3731–3741
86. Jen MC, Serrano MC, van Lith R, Ameer GA (2012) Adv Funct Mater 22:239–260
87. Seabra AB, Duran N (2010) J Mater Chem 20:1624–1637
88. Kim J, Saravanakumar G, Choi HW, Park D, Kim WJ (2014) J Mater Chem B 2:341–356
89. Fraix A, Marino N, Sortino S (2015) Top Curr Chem. in press
90. Ostrowski AD, Lin BF, Tirrell MV, Ford PC (2012) Mol Pharm 9:2950–2955
91. Ford PC (2008) Acc Chem Res 41:190–200
92. Sortino S (2010) Chem Soc Rev 39:2903–2913
93. Fry NL, Mascharak PK (2011) Acc Chem Res 44:289–298
94. Ford PC (2013) Nitric Oxide Biol Ch 34:56–64
95. Sortino S (2012) J Mater Chem 22:301–318
96. Klan P, Solomek T, Bochet CG, Blanc A, Givens R, Rubina M, Popik V, Kostikov A, Wirz J (2013) Chem Rev 113:119–191
97. Caruso EB, Petralia S, Conoci S, Giuffrida S, Sortino S (2007) J Am Chem Soc 129:480–481
98. Conoci S, Petralia S, Sortino S (2006) EP2051935A1/US20090191284
99. Fraix A, Sortino S (2015) Chem-Asian J 10:1116–1125
100. Suzuki T, Nagae O, Kato Y, Nakagawa H, Fukuhara K, Miyata N (2005) J Am Chem Soc 127:11720–11726
101. Kitamura K, Ieda N, Hishikawa K, Suzuki T, Miyata N, Fukuhara K, Nakagawa H (2014) Bioorg Med Chem Lett 24:5660–5662
102. Swaminathan S, Garcia-Amoros J, Fraix A, Kandoth N, Sortino S, Raymo FM (2014) Chem Soc Rev 43:4167–4178
103. Caruso EB, Cicciarella E, Sortino S (2007) Chem Commun 47:5028–5030
104. Kandoth N, Vittorino E, Sciortino MT, Parisi T, Colao I, Mazzaglia A, Sortino S (2012) Chem-Eur J 18:1684–1690
105. Fraix A, Goncalves AR, Cardile V, Graziano ACE, Theodossiou TA, Yannakopoulou K, Sortino S (2013) Chem-Asian J 8:2634–2641
106. Fraix A, Manet I, Ballestri M, Guerrini A, Dambruoso P, Sotgiu G, Varchi G, Camerin M, Coppellotti O, Sortino S (2015) J Mater Chem B 3:3001–3010
107. Anand R, Manoli F, Manet I, Daoud-Mahammed S, Agostoni V, Gref R, Monti S (2012) Photochem Photobio Sci 11:1285–1292

108. Daoud-Mahammed S, Grossiord JL, Bergua T, Amiel C, Couvreur P, Gref R (2008) J Biomed Mater Res A 86A:736–748
109. Othman M, Bouchemal K, Couvreur P, Desmaele D, Morvan E, Pouget T, Gref R (2011) J Biomed Mater Res A 354:517–527
110. Renard E, Deratani A, Volet G, Sebille B (1997) Eur Polym J 33:49–57
111. Gref R, Amiel C, Molinard K, Daoud-Mahammed S, Sebille B, Gillet B, Beloeil JC, Ringard C, Rosilio V, Poupaert J, Couvreur P (2006) J Control Release 111:316–324
112. Battistini E, Gianolio E, Gref R, Couvreur P, Fuzerova S, Othman M, Aime S, Badet B, Durand P (2008) Chem Eur J 14:4551–4561
113. Kirejev V, Kandoth N, Gref R, Ericson MB, Sortino S (2014) J Mater Chem B 2:1190–1195
114. Deniz E, Kandoth N, Fraix A, Cardile V, Graziano ACE, Lo Furno D, Gref R, Raymo FM, Sortino S (2012) Chem Eur J 18:15782–15787
115. Fraix A, Guglielmo S, Cardile V, Graziano ACE, Gref R, Rolando B, Fruttero R, Gasco A, Sortino S (2014) RSC Adv 4:44827–44836
116. Kandoth N, Kirejev V, Monti S, Gref R, Ericson MB, Sortino S (2014) Biomacromolecules 15:1768–1776
117. Mir Y, van Lier JE, Allard JF, Morris D, Houde D (2009) Photochem Photobio Sci 8:391–395
118. Uchiyama S, Santa T, Imai K (1999) J Chem Soc Perk T 2 11:2525–2532
119. Fraix A, Kandoth N, Manet I, Cardile V, Graziano ACE, Gref R, Sortino S (2013) Chem Commun 49:4459–4461
120. Allison RR (2014) Future Oncol 10:123–142
121. Kessel D, Erickson C (1992) Photochem Photobio 55:397–399
122. Kessel D (1992) J Photoch Photobio B 12:203–204
123. Zuluaga MF, Lange N (2008) Cur Med Chem 15:1655–1673
124. Peng CL, Lai PS, Lin FH, Yueh-Hsiu WS, Shieh MJ (2009) Biomaterials 30:3614–3625
125. Khdair A, Di C, Patil Y, Ma L, Dou QP, Shekhar MPV, Panyam J (2010) J Control Release 141:137–144
126. Khdair A, Gerard B, Handa H, Mao G, Shekhar MPV, Panyam J (2008) Mol Pharm 5:795–807
127. Conte C, Ungaro F, Maglio G, Tirino P, Siracusano G, Sciortino MT, Leone N, Palma G, Barbieri A, Arra C, Mazzaglia A, Quaglia F (2013) J Control Release 167:40–52
128. Quaglia F, Ostacolo L, De Rosa G, La Rotonda MI, Ammendola M, Nese G, Maglio G, Palumbo R, Vauthier C (2006) Int J Pharm 324:56–66
129. Xiang GH, Hong GB, Wang Y, Cheng D, Zhou JX, Shuai XT (2013) Int J Nanomed 8:4613–4622
130. Venkataraman S, Hedrick JL, Ong ZY, Yang C, Ee PL, Hammond PT, Yang YY (2011) Adv Drug Deliv Rev 63:1228–1246
131. Maiolino S, Moret F, Conte C, Fraix A, Tirino P, Ungaro F, Sortino S, Reddi E, Quaglia F (2015) Nanoscale 7:5643–5653
132. Maiolino S, Russo A, Pagliara V, Conte C, Ungaro F, Russo G, Quaglia F (2015) J Nanobiotech 13:29

Chapter 10
Chemiluminescence in Biomedicine

Mara Mirasoli, Massimo Guardigli, and Aldo Roda

Abstract Chemiluminescence, a luminescence phenomenon in which photons emission is obtained as a result of a chemical reaction, represents a versatile analytical tool, providing high sensitivity and specificity in a wide range of applications in biotechnology, pharmacology, molecular biology, biochemistry and environmental analysis. Owing to its peculiar characteristics, chemiluminescence ensures high detectability, due to the virtual total absence of the background signal, wide dynamic range, which allows analytes to be measured over a broad range of concentrations, and amenability to miniaturization, as no photoexcitation source and wavelength selection systems (e.g., filters) are required.

Chemiluminescence has important applications in analytical chemistry and biomedicine. Owing to its specificity and rapidity, chemiluminescence has been coupled with flow-injection analysis and post-column HPLC and capillary electrophoresis detection, with applications mostly in the pharmaceutical field, but also in food, environmental and biomedical analyses. Furthermore, several bioassays and biosensors based on chemiluminescence have been developed for quantitative and multiplex analysis of molecules (such as hormones, drugs, pollutants), proteins, and nucleic acids of biomedical interest. These methods, which take advantage of chemiluminescence high detectability and simple analytical format, often exploit a variety of miniaturized analytical formats, such as microfluidics, microarrays, lab-on-chip, and paper-based analytical devices.

After a brief introduction on the main chemiluminescence systems that find analytical applications and instrumentation for chemiluminescence detection, this chapter presents the main recent and up-to-date applications in the field and discusses the approaches, challenges, achievements, and open issues.

Luminescence is defined as the emission of electromagnetic radiation (in the ultraviolet, visible or infrared spectrum regions) by atoms or molecules, as a result of the transition from an electronically excited state to a lower energy level, which is usually the ground state.

M. Mirasoli (✉) • M. Guardigli • A. Roda
Department of Chemistry "Giacomo Ciamician", Alma Mater Studiorum – University of Bologna, via Selmi 2, 40126 Bologna, Italy
e-mail: mara.mirasoli@unibo.it

G. Bergamini, S. Silvi (eds.), *Applied Photochemistry*,
Lecture Notes in Chemistry 92, DOI 10.1007/978-3-319-31671-0_10

Luminescence phenomena can be distinguished according to the mechanism of production of the emitting excited state. In photoluminescence, which includes among others fluorescence and phosphorescence, the excited state is obtained upon absorption of an electromagnetic radiation. In chemical luminescence photons emission is obtained as a result of a chemical reaction. Chemical luminescence phenomena can be further distinguished in *bioluminescence* (produced by enzymes or photoproteins present in living organisms or isolated from them), *chemiluminescence* (produced by a chemical reaction triggered by the mixing of the reagents or catalyzed by an enzyme), and *electrochemiluminescence* (in which the reagent species are generated by a redox reaction triggered by the application of an electric potential).

Historically, photoluminescence, in particular fluorescence, represented the most commonly exploited luminescence technique in (bio)analytical applications. Detection techniques based on fluorescence are still today extremely widespread in bioanalytical chemistry, but, even though being characterized by high detectability, they may present limitations. In particular, autofluorescence of the sample components or impurities and scattering of the excitation light may be relevant in complex matrices such as biological samples. In addition, for peculiar applications such as fluorescence microscopy, photobleaching phenomena (i.e., the degradation of the fluorophore upon irradiation) make it difficult to obtain reliable quantitative information. Also, the implementation of fluorescence detection into compact and portable analytical devices (e.g., lab-on-chip devices) is difficult because of the requirement of a photoexcitation source, as well as wavelength selection systems (e.g., filters) both in excitation and emission pathways.

On the other hand, chemical luminescence detection techniques ensure high detectability, as they are characterized by a high signal-to-noise ratio, due to the virtually total absence of the background signal. Indeed, the emission of photons derives from a chemical process involving only the analyte and, differently from fluorescence, there is no non-specific emission from sample matrix components. The background signal of chemical luminescence measurements, essentially due to instrumental noise, is extremely low if detectors optimized for the measurement of low light levels are employed. The absence of an excitation source and wavelength selection filters determines other advantages, such as the increased ability to perform measurements on light-sensitive samples and a greater ease of implementation in miniaturized analytical tools. Another important advantage of chemiluminescence measurements is their wide dynamic range, which allows analytes to be measured over a broad range of concentrations. In addition, the steady-state light emission intensity is usually reached in few seconds or minutes, thus rendering chemiluminescence techniques very rapid [1–3].

Chemical luminescence detection principles have important applications in analytical chemistry, as they are particularly suitable for quantitative analysis and for the determination of analytes present in very low concentrations. Indeed, in addition to their high specificity and high signal-to-noise ratio, chemical luminescence signals are characterized by an intensity of the emitted light which is proportional to the analyte concentration in a wide range. This condition, which occurs in the case of a reaction at a constant rate (zero order reaction), allows the development of quantitative analytical methods characterized by high dynamic ranges and very low detection limits. Another advantage is that such phenomena, in particular bioluminescence and chemiluminescence, can be exploited to perform measurements in biological systems even for long time periods without disrupting or damaging them. This is not always possible, for example, with fluorescence techniques, in particular when exciting radiation at short wavelengths (UV or blue light) and/or high intensity are employed.

Chemical luminescence thus represents a versatile analytical tool, providing high sensitivity and specificity in a wide range of applications in biotechnology, pharmacology, molecular biology, biochemistry and environmental analysis [4–6].

10.1 Introduction

The discussion of the underlying mechanisms of the processes of chemical luminescence requires the analysis of the diagram of the energy levels of a molecule (Fig. 10.1).

In chemical luminescence phenomena, molecules exploit the energy released by a chemical reaction (usually an oxidation reaction) to be promoted from their fundamental electronic state to an excited state. An organic molecule is generally in the ground state S_0, a singlet state (i.e., a state in which all electron's spin are paired), and the chemical reaction leads to the formation of an excited singlet state S_1. If the reaction produces higher energy singlet excited states (S_2, etc.), they normally decay at the S_1 state before any other process can occur. However, the energy released by a chemical reaction is often not enough to populate higher energy singlet excited states.

Subsequently, the excited molecule releases any excess of vibrational energy through collisions with the solvent molecules (vibrational relaxation) and leads to the lowest vibrational level of the S_1 state. The lifetime of the S_1 excited singlet state is however very short, typically of the order of nanoseconds, and the molecule returns to the S_0 ground state following two possible pathways: (a) by dispersing the excess energy in the form of heat (*internal conversion*), and (b) by emitting a radiation, which leads to the *chemical luminescence* phenomenon.

Another possible deactivation processes of the excited singlet state is the *intersystem crossing*, i.e., the non-radiative deactivation to the excited triplet state T_1 (an electronic excited state with lower energy than S_1 and characterized by the presence of two orbitally unpaired electrons that possess parallel spins). As in the case of the singlet state, the transition to the triplet state is immediately followed by vibrational relaxation within this state. The decay of the T_1 triplet excited state to the S_0 ground state mainly occurs through non-radiative processes caused by collisions with the solvent molecules.

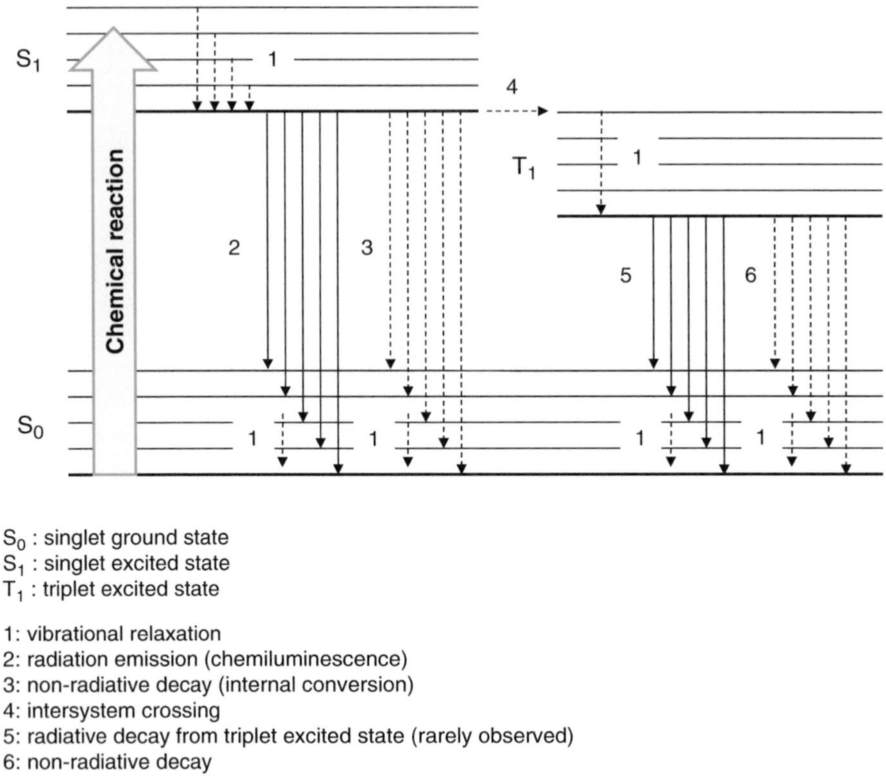

Fig. 10.1 Scheme reporting the energy levels of a molecule and the main processes involved in chemical luminescence emission phenomena

10.2 Chemiluminescence Reactions

Many hundreds of chemical reactions that produce light in the visible range, and which are related to the phenomenon of chemiluminescence, have been described, which can range between extremely bright and brilliant chemiluminescence emission, visible to the naked eye, down to those reactions that are classified as ultra-weak chemiluminescence.

Chemiluminescence is a luminescence phenomenon that is particularly interesting in analytical chemistry, especially for the analysis of complex samples. In fact, chemiluminescent reactions are very specific and, as they do not require an excitation source, they are not subject to those interference phenomena typical of fluorescence measurements, such as background emission due to sample matrix components and diffusion of the excitation radiation. However, the low number of exploitable chemiluminescent reactions limits the practical applications of this phenomenon in analytical field. In addition, chemical interferences can occur, as the presence of radical scavengers or enzyme inhibitors in the sample matrix, which can reduce light output and thus lead to non-specific effects.

In general, chemiluminescent reactions require the presence of one or more substrates that react generating a molecule in the electronically excited state (in case of an oxidation reaction, an electron acceptor, such as oxygen, is necessary). Generally, a catalyst is also present, such as an enzyme or a metal ion, which reduces the activation energy thus leading to an increase in the reaction rate. In addition, indirect chemiluminescent reactions (Fig. 10.2b) necessarily require the presence of an energy acceptor.

The overall efficiency of a chemiluminescence reaction can be expressed through the luminescence quantum yield (Φ_{CL}) parameter, which is calculated as:

$$\Phi_{CL} = \frac{\text{n}^\circ \text{ emitted photons}}{\text{n}^\circ \text{ consumed reagent molecules}}$$

The quantum efficiency Φ_{CL} for chemiluminescent reactions normally has a value between 0.001 and 0.1. It is closely related to the luminogenic power of the produced molecule and many other factors, such as concentrations of reagents, temperature and pH. The vast majority of the chemiluminescent reactions are characterized by relatively low quantum yield values, amounting to a maximum 0.01–0.02. Only some bioluminescent reactions have significantly higher quantum yield values.

A chemical reaction can directly give rise to an emission of radiation (*direct chemiluminescence*, Fig. 10.2a) only if a number of conditions are fulfilled:

(a) The reaction must be quite exergonic (i.e., it must release enough energy) to obtain an electronically excited singlet state (for an emission in the visible range, 400–750 nm, 40–70 kcal mol^{-1} are required),
(b) There must exist a reaction path leading to the formation of a product in an excited state,
(c) The product in the excited state must be able to emit. This means that the product in the excited state must be fluorescent, or it can populate an excited state by intra- or inter-molecular energy transfer.

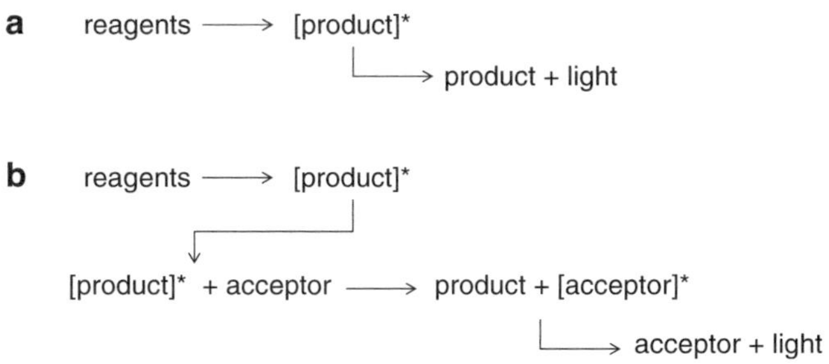

Fig. 10.2 Direct (**a**) and indirect (**b**) chemiluminescent processes

Considering these requirements, the quantum yield of a chemiluminescence reaction can be expressed by the following equation:

$$\Phi_{CL} = \Phi_{CH}\Phi_{CE}\Phi_{FL}$$

In this equation Φ_{CH} is the chemical yield of the reaction (i.e., the fraction of the reactant molecules actually converted into the final product), Φ_{CE} is the chemiexcitation yield (i.e., the fraction of product molecules produced in the excited state) and Φ_{FL} is the fluorescence quantum yield of the reaction product in the excited state. The quantum yield of a chemiluminescent reaction is high only if all of these factors are sufficiently high. In most cases, the low quantum yield of chemiluminescent reactions is mainly due to a low Φ_{CE} value.

In some cases, the reaction product in the excited state is not fluorescent ($\Phi_{FL} \sim 0$), thus the reaction itself is not chemiluminescent. It is however still possible to obtain light emission through the addition of a fluorescent molecule capable of acting as an energy acceptor (*indirect chemiluminescence*, Fig. 10.2b). This molecule is excited through a process of energy transfer from the reaction product, thus becoming the emitting species in the system.

In indirect chemiluminescence the expression of the chemiluminescent quantum yield takes the form:

$$\Phi_{CL} = \Phi_{CH}\Phi_{CE}\Phi_{ET}\Phi_{FL'}$$

where, in addition to the terms previously seen, Φ_{ET} represents the efficiency of the process of energy transfer between the reaction product in the excited state and the fluorescent acceptor, while $\Phi_{FL'}$ represents the fluorescence quantum yield of the acceptor.

Since the quantum yield of each stage of the process is necessarily equal to or less than 1, the quantum yield of the entire chemiluminescence process is also less than 1.

In case when Φ_{CL} is very low because of very low Φ_{CH} or Φ_{CE} values, the yield of the process can be in some cases increased using the appropriate catalyst or by changing experimental conditions (e.g., the concentration of reagents, temperature, pH). If in turn the overall chemiluminescence efficiency is small due to the low Φ_{FL} value, it can be increased by adding to the system an energy acceptor, which can be excited by energy transfer from the excited product of the chemiluminescent reaction and has a high emission quantum yield.

From the analytical chemistry viewpoint, chemiluminescence is a versatile and extremely useful tool for developing sensitive and specific analytical methodologies, with numerous applications in biotechnology, molecular biology, pharmacology, and in the biomedical field.

In the following, the main chemiluminescent systems employed in analytical chemistry and biomedicine are illustrated.

10.3 Chemiluminescent Systems

As previously mentioned, only sufficiently exergonic chemical reactions can give rise to chemiluminescence emission. Many chemiluminescent reactions are oxidation reactions (involving oxygen or hydrogen peroxide as oxidant species) and entail the formation of an organic peroxide, followed by the rupture of the O-O bond. Indeed, O-O bonds are relatively weak and the reorganization of the molecule that follows their rupture releases a considerable amount of energy.

Over the years research activities have been conducted in the attempt to improve the performance of chemiluminescence (as well as other chemical luminescence) detection techniques, in particular through the search for labels with increasingly better photophysical properties.

10.3.1 Luminol

Luminol (5-amino-2,3-dihydro-1,4-phthalazinedione) is capable of emitting light in an alkaline medium and in the presence of an oxidizing agent, such as hydrogen peroxide. After the synthesis of luminol and other derivatives (Fig. 10.3), several research groups studied their use as chemiluminescent markers in bioanalytical methods (e.g., immunoassays).

Luminol is probably the best known chemiluminescent molecule, and it is employed, among other applications, in forensic chemistry for the identification of blood traces (*luminol test*). Luminol chemiluminescent reaction involves its oxidation in alkaline medium, generally carried out by means of hydrogen peroxide, which produces 3-aminophtalate ion in the excited state, with emission at 450 nm wavelength (Fig. 10.4). In normal conditions this reaction is very slow, but many inorganic and organic substances can act as catalysts. For example certain metal ions can have this role, such as iron (indeed, this is the reason why the

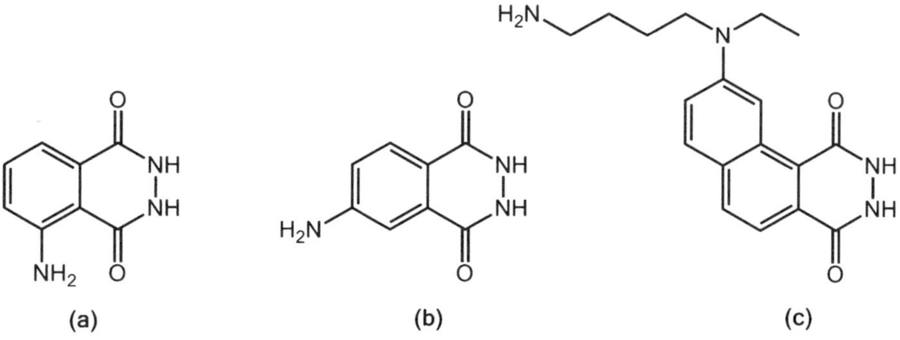

Fig. 10.3 Structure of luminol (**a**) and of some examples of its derivatives: isoluminol (**b**) and 7-[N-(4-aminobutyl)-N-ethyl]naphthalene- 1,2-dicarboxylic acid hydrazide (**c**)

reaction is also catalyzed by hemoglobin), cobalt, copper, manganese. In addition, various enzymes can catalyze the reaction, in particular those belonging to the class of peroxidase, such as horseradish peroxidase (HRP), an enzyme widely used as a biomarker in bioanalytical chemistry.

Recently it has been shown that even noble metal nanoparticles are capable of acting as catalysts in the chemiluminescent reaction of luminol. In particular, gold nanoparticles, with dimensions from 6 to 99 nm, are capable of oxidizing luminol in the presence of a base and of hydrogen peroxide. Gold nanoparticles in the presence of silver nitrate or platinum nanoparticles have also shown the ability to catalyze the reaction of luminol oxidation [7, 8].

The functional group responsible for the chemiluminescent properties of luminol is represented by the amide endocyclic nitrogen atoms, which react with the oxidant to form a very unstable organic peroxide that decomposes immediately with loss of nitrogen, yielding the 5-aminophtalate ion in the excited state (Fig. 10.4). The reaction systems requires energy oxidants when conducted in an aprotic solvent (DMSO or DMF), while in protic solvents (water and alcohols) milder oxidants are sufficient. Luminol quantum yield varies from 5 % (in DMSO) to 1–1.5 % (in water). To obtain an adequate reaction rate, the presence of a catalyst of enzymatic (e.g., peroxidase enzyme), organic (e.g., Heme group) or inorganic (e.g., salts and metal complexes) nature is essential.

To improve the performance of this system, research efforts were directed towards structural modifications of the molecule (e.g., synthesis of derivatives in which the benzene ring is replaced by naphthalene or benzoperilene, as shown in Fig. 10.3c) and the development of "chemical enhancers", i.e., molecules that increase the efficiency of the chemiluminescent reaction.

Indeed, in the 1980s Thorpe et al. discovered that the intensity of light emission obtained by the luminol/peroxide/horseradish peroxidase chemiluminescence system can be increased by the addition of so-called enhancers. The mechanism of enhanced chemiluminescence involves the enhancer oxidation, with generation of enhancer radicals and leads to the increase of horseradish peroxidase turnover.

Several molecules (e.g., phenol, phenylboronic acid, and phenothiazine derivatives) were shown to increase intensity and duration of the chemiluminescence reaction The use of 3-(10-phenothiazinyl)propane-1-sulfonate enhancer combined

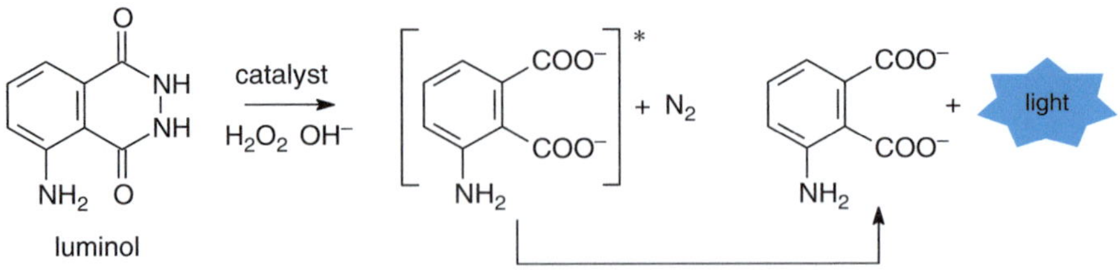

Fig. 10.4 Luminol chemiluminescence reaction in alkaline medium and in the presence of H_2O_2 and a catalyst

with an acylation catalyst (4-dimethylaminopyridine or 4-morpholinopyridine) allowed reaching limits of detection of attomoles of horseradish peroxidase.

The chemiluminescence reaction of luminol has several applications in (bio) analyses. It is used for the quantitative determination of enzymes that produce H_2O_2 (typically oxidases) or their substrates through coupled enzyme reaction systems. It is also used to quantify reactive oxygen species (ROS, e.g., $O_2^{\bullet-}$, $HO^{\bullet}$, $HOO^{\bullet}$) and other reactive species such as ClO^- and $ONOO^-$ produced in biological systems. The luminol system is also used to reveal the reaction catalysts such as Fe^{2+} (reaction at the base of the "luminol test" for the detection of traces of blood) and horseradish peroxidase (used as a marker for chemiluminescence biospecific probes). Luminol derivatives (e.g., isoluminol, shown in Fig. 10.3b, which has the positive feature of being not affected in its chemiluminescence properties upon conjugation with another molecule) are also used for direct marking of biospecific probes, such as antibodies, antigens, or haptens.

10.3.2 Acridinium Esters

Lucigenin (bis-N-methylacridinium nitrate) is the first acridinium salt for which chemiluminescence properties have been demonstrated. This compound is oxidized with formation of N-methylacridone and emission of light at 440 nm in an alkaline medium and in the presence of reactive oxygen species (ROS, especially superoxide anion), as shown in Fig. 10.5.

After the discovery of lucigenin chemiluminescence properties, extensive studies were conducted on classes of structurally related compounds. In the course of these studies, acridinium esters were also identified, as compounds capable of undergoing an oxidative process in the presence of H_2O_2 that produces light (Fig. 10.6).

The reaction mechanism, which leads to chemiluminescence emission by these classes of compounds, is very interesting as it is reminiscent of the firefly luciferase/ luciferin system. The key step is indeed the formation of a very unstable reaction intermediate, consisting of a 1,2-dioxetane derivative (a C_2O_2 four-atom ring) bearing a variety of substituting groups. This decomposes yielding the carbonyl fragments in an electronically excited state, which are (directly or indirectly) responsible for the chemiluminescence emission.

A large number of (bio)analytical applications have been reported for acridinium esters. Their ability to be oxidized by H_2O_2 makes them useful as systems for detecting such molecule. For example, in coupled enzyme reactions, in which a target molecule is detected upon its oxidation by means of a specific oxidase enzyme (e.g., glucose oxidase for glucose detection), which leads to the production of H_2O_2, the latter being detected by means of an acridinium ester.

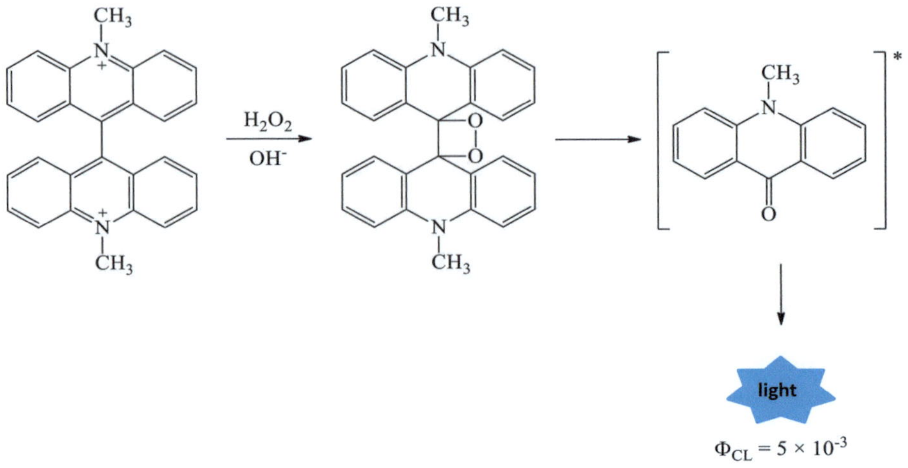

Fig. 10.5 Chemiluminescence reaction of lucigenin oxidation

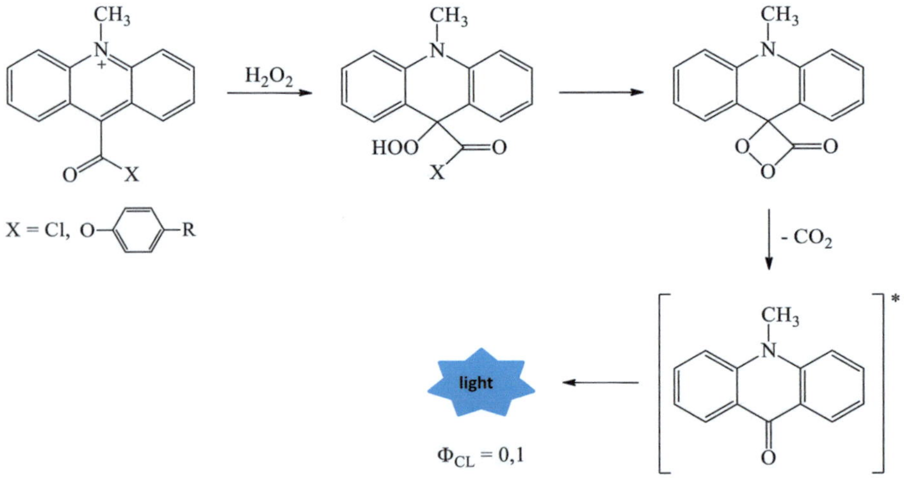

Fig. 10.6 Chemiluminescent oxidation reaction of acridinium esters in the presence of H_2O_2

Acridinium ester derivatives can also be employed as chemiluminescence markers for biospecific molecules (e.g., an antibody or antigen can be chemically labeled with an acridinium ester derivative, to provide a chemiluminescence tracer for immunoassays development). In addition, as they have a high capacity of interaction with single-stranded DNA molecules, acridine derivatives are employed in methods for determining the pairing speed in the processes of hybridization of two DNA strands or the thermodynamics affinity of specific oligonucleotide probes against target molecules.

10.3.3 Acridan Esters

Acridan esters can be employed in chemiluminescence substrates for horseradish peroxidase, as they can be converted into acridinium esters by an enhanced peroxidase-catalyzed oxidation; the formed acridinium ester than reacts with hydrogen peroxide under alkaline conditions to yield the excited state of the corresponding acridanone, as seen above (Fig. 10.7).

With some acridan ester derivatives, the acridinium ester compound that is formed by horseradish peroxidase-catalyzed oxidation at slightly alkaline conditions (pH 8) is resistant to reaction with hydrogen peroxide. This enables accumulation of a large amount of acridinium ester upon reaction with horseradish peroxidase, without observing light emission, even in the presence of hydrogen peroxide; then, by increasing the pH a flash of light is observed. With this systems, subattomole limits of detection (10^{-19} mol) for horseradish peroxidase were achieved, thus providing a major improvement with respect to the enhanced luminol/H_2O_2/horseradish peroxidase system.

Chemiluminescence substrates for horseradish peroxidase based on acridan esters are commercially available for their use in bioanalytical applications involving the use of such enzyme.

10.3.4 1,2-Dioxetanes

1,2-dioxetane and 1,2-dioxetanone derivatives are a class of highly unstable chemiluminescent molecules, able to spontaneously decompose, e.g., as a result of heating. Such decomposition is prevented in commercial chemiluminescent

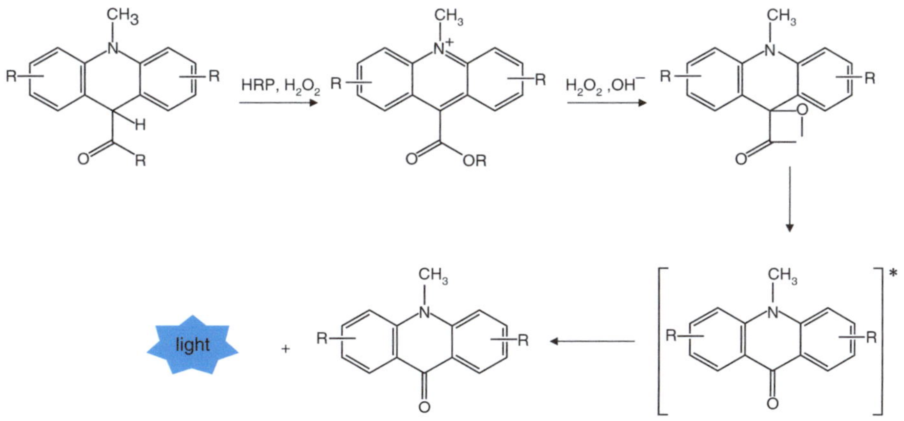

Fig. 10.7 Chemiluminescent reaction of acridan esters in the presence of horseradish peroxidase and hydrogen peroxide

reagents through the introduction of suitable stabilizing groups (e.g., adamantanone that acts as a molecule stabilizing moiety though a steric hindrance effect).

Commercial derivatives also contain suitable protecting groups (e.g., a phosphate or a galactose moiety) that can be removed by chemical or enzymatic hydrolysis when the chemiluminescence reaction needs to be triggered (Fig. 10.8). By varying the nature of these stabilizing groups, specific chemiluminescent reagents for the detection of a given enzyme, such as alkaline phosphatase or β-galactosidase, have been produced. Examples are disodium 3-(4-methoxyspiro {1,2-dioxetane-3,2′-(5′-chloro)tricyclo [3.3.1.13,7]decan}-4-yl)phenyl phosphate (CSPD) and 3-(2′spiroadamantyl)-4-methoxy-4-(3″-phosphoryloxy)-phenyl-1,2-dioxetane (AMPPD) substrates for alkaline phosphatase. Upon enzyme hydrolysis, a negative charge is introduced in the molecule and its delocalization determines destabilization of the molecule, which then splits into two fragments, one of which can be in an electronically excited state. This fragment then decays to the ground state with emission of light or (more frequently) with transfer of its energy to an acceptor, which is then responsible for final emission (this reaction is therefore a classic example of indirect chemiluminescence reaction).

The highest chemiluminescent emission is obtained in alkaline conditions, however a flash-type emission is obtained at high pH values (pH = 11–12). Therefore, mild alkaline conditions (pH = 9–9.5) are preferred, at which the chemiluminescent signal remains nearly constant for minutes or hours in the presence of an excess of substrate.

The sensitivity of the chemiluminescent system alkaline phosphatase-AMPPD is extremely high, enough to detect less than 10–20 mol of enzyme in solution or immobilized on a solid support. Thanks to its exceptional sensitivity, the systems alkaline phosphatase-AMPPD and alkaline phosphatase-CSPD coupled with different fluorescent acceptors, have found applications in clinical diagnostics in the field of immunoassays and molecular biology (in particular nucleic acid hybridization assays) providing among the highest sensitivity performance achievable with these systems.

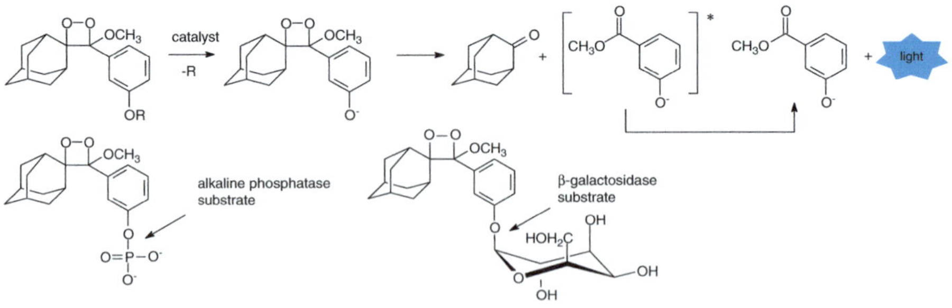

Fig. 10.8 Mechanism of the chemiluminescence reaction of 1,2-dioxetane derivatives employed as enzyme substrates for alkaline phosphatase or β-galactosidase enzymes

10.3.5 Aryloxalates

The oxidation of an aryloxalate is another example of indirect chemiluminescence. Aryloxalates react with oxidizing agents, such as H_2O_2, decomposing and generating a high energy intermediate (a 1,2-dioxetandione derivative) that rapidly decomposes. The product obtained in the excited state is not however fluorescent and therefore radiation emission can only be obtained if the reaction is conducted in the presence of a fluorescent energy acceptor (Fig. 10.9). The emission wavelength depends on the characteristics of the fluorescent acceptor and can therefore be varied by using different acceptors, making these systems very versatile from the point of view of the emission characteristics. However, their applications in bioanalytical chemistry are hampered by the difficulties encountered in the attempts to chemically conjugate aryloxalate derivatives to biological molecules (e.g., antibodies or other proteins) and by their easy hydrolysis in an aqueous environment.

Chemiluminescent aryloxalate derivatives are employed in analytical methods based on the measurement of H_2O_2 produced by chemical reactions (however, these systems are not easily adaptable for use with enzymatic reactions, which require an aqueous reaction medium) and, most frequently, for post-column detection of fluorescent analytes (typically aromatic compounds) in high performance liquid chromatography (HPLC) and capillary electrophoresis (CE).

10.3.6 Other Chemiluminescent Systems

In addition to the chemiluminescent systems described above there are also many others that have found only few application specific fields. Many strong oxidizing agents (e.g., MnO_4^- or Ce^{4+}) can react with organic compounds resulting in a weak chemiluminescence through quite complex mechanisms. These systems are often used to indirectly determine chemical species able to catalyze the chemiluminescent reaction. Examples are detection of drugs in pharmaceutical formulations, as reported in Table 10.2.

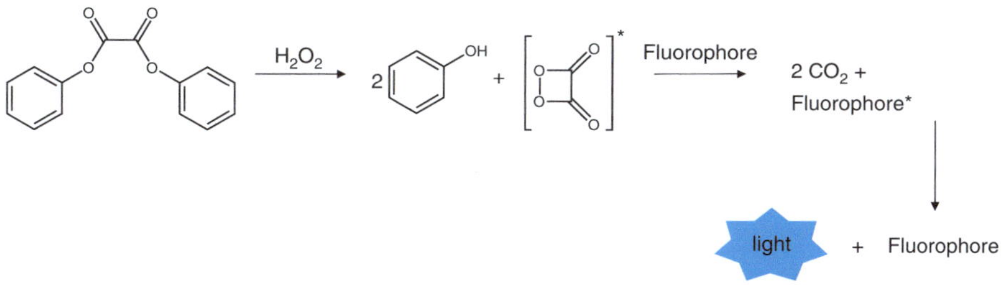

Fig. 10.9 Mechanism of the chemiluminescence reaction of aryloxalate derivatives

10.3.7 Chemiluminescence for the Detection of Enzyme Markers

Many bioanalytical applications of chemiluminescence involve the use of enzyme labels to mark particular biospecific probes. The most commonly used enzymes are horseradish peroxidase and alkaline phosphatase, for both of which chemiluminescence substrates are commercially available, although some applications involving other enzymes such as glucose-6-phosphate-dehydrogenase, β-galactosidase and xanthine-oxidase have also been described.

Since in the presence of an excess substrate the intensity of the chemiluminescent signal is proportional to the concentration of the enzyme, precise and accurate quantitative results can be obtained once the system has been optimized and non-specific background signal has been taken into account. An advantage of the use of enzymes as labels is the possibility to achieve a stationary phase in the chemiluminescence emission kinetics, i.e. a relatively long period of time (generally minutes) in which the signal intensity is fairly constant, which facilitates the signal measurement procedure. Another key feature is the possibility to exploit the enzyme turnover to obtain signal amplification, as many substrate molecules are converted into the emitting reaction intermediate by each enzyme label. This greatly increases signal output, with respect to labeling with a single chemiluminescence molecule (e.g., an antibody labeled with an isoluminol derivative).

Commercial substrates for horseradish peroxidase are generally based on luminol or its derivatives, although commercial products based on acridan esters are also available, while those for alkaline phosphatase are based on 1,2-dioxetane derivatives, particularly CSPD and AMPPD and their derivatives, which offer good performance in terms of signal stability, assay sensitivity, and reproducibility. Generally, alkaline phosphatase-catalyzed chemiluminescence reactions are characterized by a slower kinetics and weaker signals (but higher signal-to-noise ratio) than those of chemiluminescence reactions catalyzed by peroxidase. Peroxidase has the positive feature of a high enzyme turnover number, which allows obtaining relatively intense chemiluminescence signal.

The enzymes horseradish peroxidase and alkaline phosphatase are extensively used as labels in clinical diagnostics (immunoassays) and in molecular biology (DNA hybridization assays) providing excellent analytical performance, particularly a significantly higher detectability as compared with colorimetric and fluorimetric systems (theoretically an amount of enzyme in the order of 10^{-18}–10^{-21} mol can be detected), approaching the detectability obtained with radionuclides.

One of the most important characteristics of these substrates, especially in miniaturized analytical formats such as microarrays or microscopy imaging applications, is the half-life of the reaction product in the electronic excited state, which influences the specificity of the signal. It is indeed important to ensure that this product emits light as quickly as possible so as to identify the precise location in which the reaction occurs. On the contrary, if the half-life of the excited species is long, it can diffuse relatively far from the site of its generation before emitting a photon, thus reducing the ability to precisely localize the target analyte on the microarray or sample surface.

10.4 Instrumentation

Although, at least in principle, a chemiluminescence measurement can be performed with a spectrofluorometer (indeed, in many commercial spectrofluorometers the excitation source can be switched off for this type of measurement), the performance of these instruments does not reach that of luminometers, i.e. instruments specifically designed for chemiluminescence measurements, which main requirement is the ability to collect as much as possible of the emitted radiation.

10.4.1 Luminometers

A luminometer is a much simpler instrument, as compared with a spectrophotometer or with a spectrofluorometer, as no excitation source is required. Nevertheless, due to the low emission intensity of chemiluminescence reactions, highly sensitive light detectors are required. Photodiodes and photomultiplier tubes are the detectors usually found in commercial luminometers. Photomultipliers, thanks to their high sensitivity, are the detectors of choice for measuring extremely low levels of light and they are used in many bench-top luminometers. Photomultipliers are in most cases cooled to further increase the sensitivity by reducing the thermal background noise. On the other hand, the less sensitive photodiodes are generally found in portable instruments, because of their small size and reduced energy consumption.

Some instruments are completely devoid of filters or other wavelength selection systems, which would cause loss of detectability by reducing the fraction of emitted photons that reach the detector. Indeed, as chemical luminescence emission comes from the dark and is highly specific, all emitted photons can be detected and spectral selection of the emitted light is not necessary.

However, some applications may require spectral selection of the light emission, e.g., when two different chemiluminescent reactions with different emission spectra are simultaneously performed in the same sample. In this case, instruments equipped with emission wavelength selection devices are employed. In such case, wide-band interferential optical filters are preferred with respect to grating monochromators, because of their higher light transmission efficiency.

As said above, the main requirement for a luminometer is to achieve maximum light collection efficiency and therefore the instrumentation optics must be designed to reach this goal. As the excitation source is not present in luminometers, the geometrical requirements for the measurement cell are much simpler than those for spectrophotometers and spectrofluorometers. Indeed, chemiluminescence emission can be reliably measured even on samples contained in simple test tubes. Along with luminometers for single tube measurements, instruments for the measurement of microtiter plates (mainly 96- and 384-well microtitre plates) or equipped with flow cells are also available (Fig. 10.10).

Particularly useful is the presence of one or more dispensers for the addition of reagents directly into the measuring cell, which allows the study of chemiluminescent reactions with fast kinetics. In addition, some instruments also offer the possibility to control the temperature of the sample compartment, which increases assay reproducibility, as the speed of chemiluminescent reactions depends on the temperature.

10.4.2 Luminographs

Luminographs are a different type of instruments for chemiluminescence measurements, which have the particular feature of acquiring images of the chemiluminescence emission from the sample (Fig. 10.10). Indeed, they can be used not only to measure the intensity of the signal, but also to evaluate the spatial distribution on the surface of the sample or (by means of an image reconstruction software that combines multiple images taken from different angles) even within the sample itself.

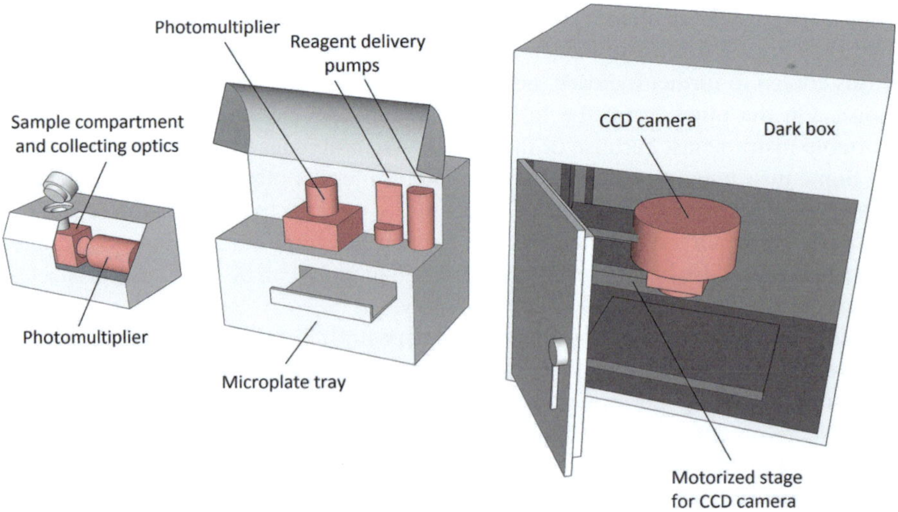

Fig. 10.10 Instrumentation for chemiluminescence measurements: (*left*) luminometer for measurements of test tubes and vials, (*center*) luminometer for measurements of microtiter plates, equipped with reagent dispensers, (*right*) luminograph for general applications equipped with a thermoelectrically-cooled CCD camera

These instruments typically employ digital cameras equipped with highly sensitive CCD (charge-coupled device) sensors, cooled with cryogenic fluids or with a thermoelectric system to reduce thermal background signal and thus improve performance.

Back-illuminated slow-scan CCD cameras are characterized by high detection sensitivity due to their high quantum efficiency and low noise: they allow a quantification of light produced to the level of a single photon. These cameras are suitable for measuring static chemiluminescence signals, as acquisition times are in the order of minutes. Furthermore, slow-scan CCD cameras do not require an image intensification step that could adversely affect the quality of the image and the signal-to-noise ratio.

On the other hand, *intensified CCD cameras* are suitable for real-time capturing dynamic chemiluminescence signals, as acquisition times are very short, in the order of seconds or lower.

For the analysis, the sample is placed in a dark chamber connected to the CCD camera to avoid interference from external light. The instrument is equipped with a suitable support for sample positioning and/or a motorized stage for the CCD camera, which enable regulating the position and the distance of the sample with respect to the CCD camera, and thus the area of the sample to be imaged. The instruments are also accompanied by a specific software for controlling the camera settings and for images elaboration and analysis.

When the analysis is performed on microscopic samples, such as tissue sections or single cells, the objective of the CCD camera is removed and the camera is connected to an optical microscope.

Among commercially available luminographs, some instruments are designed for specific applications, such as gel imaging or in vivo bioluminescence analysis on experimental animals, while others are more versatile as they are designed for general application, thus suitable for the analysis of samples of any nature.

Variables affecting the intensity of the light signal, such as the geometry of the sample, should be considered during the operations of imaging acquisition of the chemiluminescence signal. These effects can be evaluated and corrected using model systems in which, for example, enzymes for chemiluminescence detection are immobilized on a suitable solid support.

10.5 Chemiluminescence Measurements

Chemiluminescence measurements possess significant advantages over other optical detection techniques, such as spectroscopy.

- *Detectability:* the luminescent signal is characterized by a high detectability, due (at least in part) to the absence of the background signals present in other spectroscopic techniques such as fluorescence.

- *Specificity:* a few chemical reactions are luminescent, therefore in many cases the only light emission in the system is due to the presence of the analyte.
- *Quantification:* the chemiluminescence signal can be easily, accurately and reproducibly quantified.
- *Dynamic range:* the dynamic range of chemiluminescence measurements is generally very wide (up to 4–5 orders of magnitude) and therefore the signal is proportional to the concentration of the target analyte in a wide range of values.

In general, in standardized experimental conditions the intensity of the emitted light is proportional to the concentration of analytes, allowing accurate and sensitive quantitative analysis.

However, chemiluminescence signal intensities depend on time, since the intensity of the chemiluminescence signal is proportional to the rate of the reaction. Therefore, the intensity of light emission decreases with time due to the consumption of reagents.

The kinetics of the chemiluminescent reactions are often very fast (*flash-type kinetics*): the emission intensity rapidly increases up to a maximum value, then the signal decays following an exponential function in a time that can vary from fractions of seconds to several seconds, depending on the type of chemiluminescent reaction and experimental conditions (Fig. 10.11a). The management of this type of chemiluminescent signal is quite difficult and a suitable instrumentation is required in order to trigger the chemiluminescent reaction directly in the measurement cell.

Other chemiluminescent reactions are characterized by much slower kinetics (*glow-type kinetics*), in which the emission reaches a constant intensity that is maintained for relatively long time periods, from minutes to hours (Fig. 10.11b). Most of the chemiluminescent reactions catalyzed by enzymes follow this type of kinetics, and the intensity of the light emission is proportional to the activity of the enzyme in the presence of an excess of enzyme substrate.

Chemiluminescent measurements based on glow-type emissions are characterized by high sensitivity and reproducibility; moreover, sample manipulation and luminescence signal management are much easier.

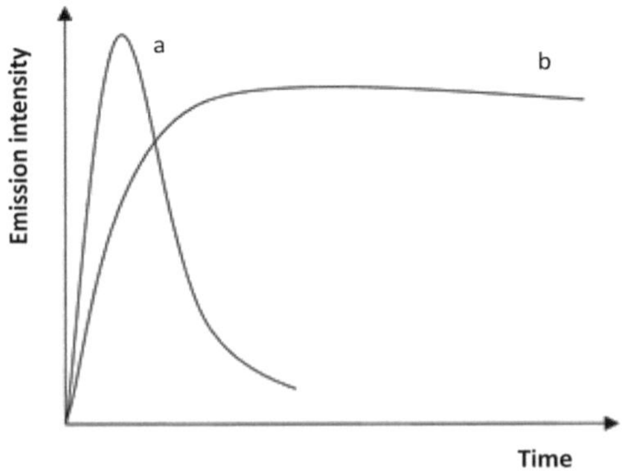

Fig. 10.11 Flash- (**a**) and glow- (**b**) type kinetics for a chemiluminescence reaction

Depending on the type of kinetics, different approaches can be used for signal measurement. For fast kinetics, the maximum intensity or the total signal emission can be measured. The first approach is much faster, but it is often inaccurate because the intensity is affected by the experimental conditions. The second approach guarantees more reproducible results but it could require relatively long measurement times.

For slow kinetics, the measurement of the maximum light intensity in the *plateau* area of the luminescence kinetic profiles usually provides satisfactory results. Nevertheless, to obtain reproducible results all the assay conditions should be standardized; for maximum reproducibility the assay temperature should be also controlled.

10.6 Applications of Chemiluminescence

Some applications of chemiluminescent systems are summarized in Table 10.1.

10.6.1 Flow Injection Analysis

The high specificity of chemiluminescent reactions make them suitable for Flow Injection Analysis (FIA) techniques, in which a precise aliquot of sample is injected in a carrier stream of reagents, and their evolutions, such as Sequential Injection Analysis (SIA) and Multisyringe or Multicommutation Flow Analyses.

Table 10.1 Analytical applications of chemiluminescent systems

System	Application
Luminol	Measurement of enzymes involved in reactions that produce H_2O_2 (e.g., oxidases) and of their substrates by means of coupled enzymatic reactions
	Determination of ROS (reactive oxygen species such as $O_2^{\bullet-}$, $HO^{\bullet}$, $HOO^{\bullet}$) and other reactive species (e.g., ClO^- and $ONOO^-$) in biological systems
	Measurement of catalysts of the chemiluminescent reaction, such as Fe^{2+} (luminol test) and horseradish peroxidase (an enzyme used in bioanalysis as a label for biospecific probes)
	Isoluminol (a luminol derivative) is used in bioanalysis as a label for biospecific probes (e.g., antibodies)
1,2-Dioxetanes	Detection of enzymes, which could be the analytes, labels of biospecific probes, or reporter enzymes in cell biosensors
Aryloxalates	Measurement of H_2O_2 produced by chemical reactions (particularly in organic solvents)
	Post-column detection of fluorescent analytes (e.g., aromatic compounds) in separative techniques such as HPLC and CE

As FIA is not a separative technique, the analyte is detected in the presence of several potential interfering sample matrix components, thus requiring highly specific reactions for its detection. In addition, the simplicity of instrumentation required for chemiluminescence measurements makes them easily integrated in FIA manifolds.

Another key aspect is that automated FIA techniques offer high reproducibility of chemiluminescence measurements, as experimental parameters, such as pH, ionic strength, rapidity and reproducibility of reagents mixing and measurement times, are strictly controlled.

In a typical FIA system (Fig. 10.12), the sample is injected into a continuous flow of reagent mixture and chemiluminescence emission is measured by means of a detector placed immediately downstream of the injection valve and equipped with a flow cell. The result of the analysis consists of a transient signal, the intensity of which depends on the analyte concentration.

In case the reaction of interest is catalyzed by an enzyme, to reduce enzyme consumption a flow reactor containing the immobilized enzyme can be introduced in the system. The signal measurement occurs in a dark chamber containing the flow cell, which is typically a flat spiral of glass or plastic tubing, placed in close proximity to a photomultiplier tube.

Chemiluminescence detection can be employed in FIA exploiting either a direct or indirect approach. In the former, the chemiluminescence signal is generated by the interaction of the analyte with the chemiluminescence reagents; in the latter approach, the analyte acts as a sensitizing or inhibiting agent, which presence influences the intensity of the signal generated by the chemiluminescence reagents. One main limitation is the low selectivity of such assays, due to the fact that usually a class of compounds, rather than a single analyte, are able to elicit the chemiluminescence response.

Chemiluminescence FIA assays have been developed for the quantitative determination of several compounds able to give or catalyze chemiluminescent reactions. Many applications regard the analysis of pharmaceutical formulations [9], for which accurate and reproducible results can be obtained, owing to the relatively simple and standardized sample matrix (Table 10.2). Other applications regard biological, food, and environmental samples.

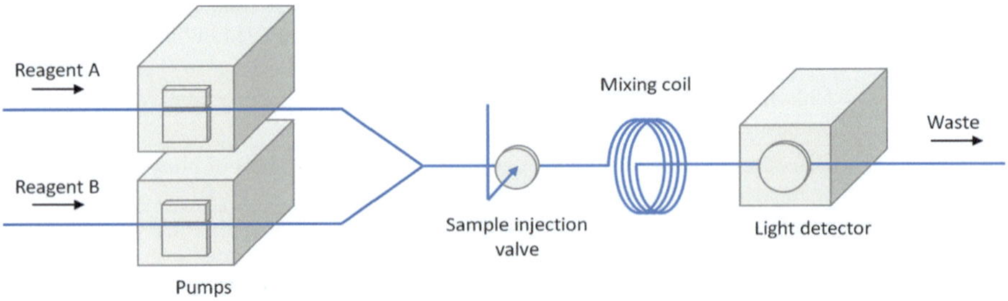

Fig. 10.12 Scheme of the instrumental set-up for a FIA system with chemiluminescence detection

Table 10.2 Examples of FIA assays with chemiluminescence detection in pharmaceutical and food analysis

Analyte	Sample	Chemiluminescence system
Promazine hydrochloride	Serum, pharmaceutical formulations	$Ru(phen)_3^{2+}/Ce^{4+}$ in sulfuric acid medium
Indomethacin	Biological samples, pharmaceutical formulations	$KMnO_4$/formaldehyde in phosphoric acid medium
Retinol, α-tocopherol	Pharmaceutical formulations	$KMnO_4$/formaldehyde in acidic medium
Ciprofloxacin, enrofloxacin	Pharmaceutical formulations	$Ru(bpy)_3^{2+}$
Puerarin	Urine, pharmaceutical formulations	$KMnO_4$/glyoxal in sulfuric acid medium
Clindamycin	Urine, pharmaceutical formulations	Luminol/$K_3Fe(CN)_6$ in alkaline medium
Antioxidants	Pharmaceutical formulations	Luminol/oxidant (O_2^- or NO)
Parathion	Rice	Luminol/H_2O_2
Patulin	Apple juice	Luminol/H_2O_2
Lactic acid	Yoghurt	Lactate oxidase followed by detection of H_2O_2 by luminol/$K_3[Fe(CN)_6]$ system
Ethanol	Wine	Alcohol oxidase followed by detection of H_2O_2 by luminol/$K_3[Fe(CN)_6]$ system

Adapted from Ref. [2]

Chemiluminescence FIA assays offer high sample throughput, possibility of automation, and low consumption of sample and reagents.

Typically limits of detection are in the nM or in the ng/mL range and sample throughput in 100 samples per hour range.

Several miniaturized FIA systems, based on the lab-on-valve or lab-on-chip approaches, with chemiluminescence detection are also increasingly proposed in the scientific literature.

10.6.2 Chromatography and Capillary Electrophoresis

Chemiluminescence reactions are also employed for post-column detection of analytes in chromatography (both gas chromatography and liquid chromatography) and capillary electrophoresis techniques [10, 11]. Analytes can range from molecules that participate in the chemiluminescence process (such as chemiluminescence precursors, reagents, activators, catalysts, inhibitors), that react with other reagents to generate a product that participates in the chemiluminescence reaction, or species that can be derivatized with a CL precursor or a fluorophore.

Several gas-phase chemiluminescence detectors are commercially available for coupling gas chromatography with chemiluminescence detection. This approach, which provides higher selectivity and sensitivity for compounds with particular atoms (e.g., P, S, N) with respect to flame ionization or thermal conductivity detectors, mainly finds applications in environmental analysis (e.g., determination of organophosphorous pesticides, organotin compounds, volatile sulfur compounds, nitrogen-containing organic molecules).

Chemiluminescence detection in liquid chromatography is accomplished employing post-column reaction of the eluting analytes with a reagent that generates the chemiluminescence reaction. In addition, in some cases, analytes derivatization with a fluorophore is required. Therefore, the instrumental assembly is more complex, with the addition of reagents pumping and mixing components.

Chemiluminescence detectors are constituted by a coil of transparent tubing placed in close proximity to a photomultiplier tube, all enclosed in a dark chamber. Instrumental and experimental parameters must be optimized in order to achieve the maximum signal intensity while the solution flows in front of the photomultiplier tube, so that the signal-to-noise ratio is the highest. Reaction temperature should be also controlled for higher reproducibility of the assay.

The solutions composition must be carefully evaluated in order to ensure both an efficient chromatography separation and a sensitive and specific chemiluminescence detection of analytes.

Capillary electrophoresis, which suffers from low detectability due to the very low amounts of sample introduced in the system, can greatly benefit from its coupling with chemiluminescence detection that provides high detectability. Despite this, the technique did not find wide application, yet, mainly because the unavailability of commercial dedicated instrumentation.

Peroxyoxalate and luminol are among those chemiluminescence reagents that are most commonly employed in this field. As concerns peroxyoxalates, analytes derivatization with a fluorophore is required for many classes of compounds; in addition, as the kinetics of the reaction is too slow and causes band broadening, a catalyst is added, such as imidazole. The oxidation reaction of luminol in alkaline medium is also widely employed, with different possible types of oxidants (such as hydrogen peroxide, permanganate, hexacyanoferrate(III), periodate) and of catalysts (such as metal ions, mainly Cu(II) and Co(II), and horseradish peroxidase). In such system, the analyte may be derivatized with a luminol derivative or act as a catalyst, inhibitor, reagent of the chemiluminescence reaction.

Table 10.3 reports some examples of applications of chemiluminescence detection applied to liquid chromatography and capillary electrophoresis.

10.6.3 Measurement of Enzymes and Enzyme Substrates

Chemiluminescence reactions can be exploited to determine enzymes, enzyme cofactors, enzyme substrates and inhibitors of an enzyme reaction. Because of the limited number of chemiluminescent reactions, often this cannot be done directly but through coupled enzyme reactions, in which one of the products of the

Table 10.3 Examples of chromatography and capillary electrophoresis assays with chemiluminescence detection in pharmaceutical and food analysis

Analyte	Sample	Chemiluminescence system
Liquid chromatography		
Catecholamines (nor-epinephrine, epinephrine, dopamine)	Plasma, serum, urine	TDPO[a]/H_2O_2/imidazole, or
		Luminol/$K_3[Fe(CN)_6]$, or
		Luminol/iodine
Phenol, 4-methylphenol	Urine	TCPO[b]/H_2O_2
Amphetamine, methamphetamine, 3,4-methylenedioxy-amphetamine (MDA), 3,4-methylenedioxy-methamphetamine (MDMA)	Hair	CPPO[c]/H_2O_2/imidazole
Estradiol	Plasma	DNPO[d]/H_2O_2/imidazole
Amikacin	Plasma, serum, urine	Luminol/H_2O_2/Cu^{2+}
Oxacillin	Serum	Luminol/H_2O_2/Co^{2+}
Organophosphorous insecticides	Vegetal matrices	Luminol/H_2O_2/CTMAB[e]
N-Methylcarbamate insecticides	Fruit juice, vegetal matrices, environmental water samples	TCPO[b]/H_2O_2, or
		Luminol/$KMnO_4$
Sulfonamides	Milk	TCPO[b]/H_2O_2/imidazole
Aminoglycoside antibiotics	Environmental water samples	Luminol/H_2O_2/Cu(II)
Capillary electrophoresis		
Phenolic compounds	Environmental water samples	TDPO[a]-H_2O_2 derivatization with dansyl chloride
Phenothiazines	Pharmaceutical formulations	Luminol/$K_3[Fe(CN)_6]$ in alkaline medium
Antioxidants	Green tea	Cu(II)/luminol/H_2O_2 in alkaline medium

Adapted from Ref. [2]
[a]TDPO: bis[4-nitro-2-(3,6,9-trioxadecyloxycarbonyl)phenyl]oxalate
[b]TCPO: bis-(2,4,6-trichlorophenyl)oxalate
[c]CPPO: bis-(2,4,5-trichloro-6-carbopentoxyphenyl)oxalate
[d]DNPO: bis-(2,4-dinitrophenyl)oxalate
[e]CTMAB: *N*-cetyl-*N*,*N*,*N*-trimethylammonium bromide

enzymatic reaction is a substrate of a chemiluminescent reaction (Fig. 10.13). In this way, the intensity of the chemiluminescent signal can be correlated to the speed of the enzymatic reaction of interest and thus to the substrate concentration or to the enzyme activity, depending on the assay conditions.

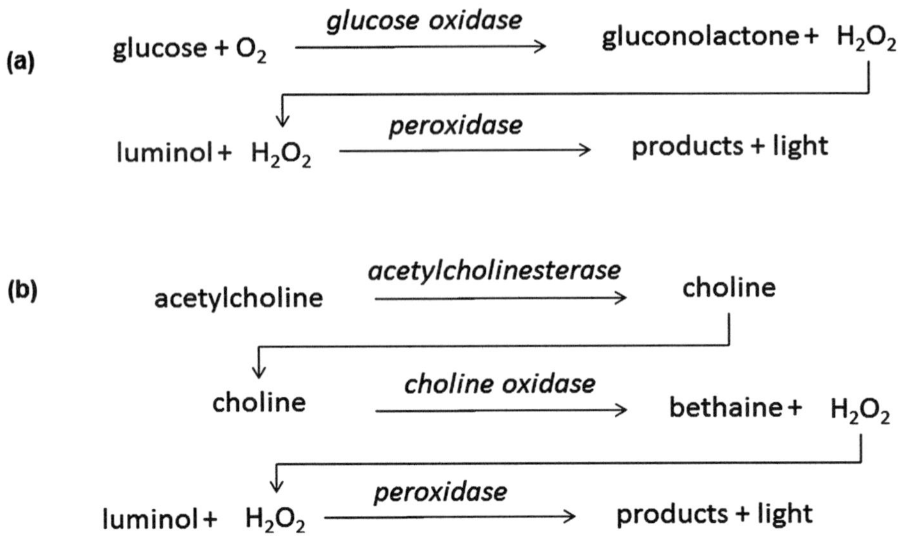

Fig. 10.13 Examples of coupled enzyme reactions with chemiluminescence detection that can be employed for determining enzyme activity or substrate concentration

In flow-based assays, the enzyme(s) necessary for the coupled reactions is(are) most often immobilized in flow reactors.

For example, the luminol/horseradish peroxidase system can be coupled to all oxidase-catalyzed reactions, which produce H_2O_2. With this approach, the analysis of oxidase substrates, such as glucose, cholesterol and phospholipids can be determined. With these measures it is also possible to evaluate the activity of enzyme inhibitors, for example potential drugs that act as inhibitors of an enzyme that is a pharmacological target (e.g., acetylcholinesterase inhibitors).

Similarly, peroxyoxalate compounds can be employed to measure H_2O_2 produced by a target oxidase enzyme reaction.

Any substance that interferes with the chemiluminescent reaction can also be determined. For example, the luminol/horseradish peroxidase system is used to evaluate the antioxidant activity of foods and biological fluids. Indeed, antioxidants react with luminol radicals that are formed during the reaction, inhibiting the emission of light. Under standardized conditions, the time of light emission inhibition is directly related to the antioxidant activity of the sample.

10.6.4 Immunoassays and Gene Probe Hybridization Assays

Analytical methods based on biospecific recognition reactions, such as antigen-antibody (immunoassays) and nucleic acids hybridization reactions (gene probe assays), are particularly useful in clinical applications, as they allow detecting analytes of interest (proteins, peptides, hormones, drugs…) that are present at very low concentrations in complex matrices, such as biological fluids (e.g., plasma, urine…). In addition to sensitivity and selectivity, such analytical methods can offer high analytical throughput, i.e., the analysis of a large number of samples in a short time, and possibility for automation.

Visualization and quantification of the binding of the antibody to the antigen, or of the gene probe to the complementary nucleotide sequence, is usually performed by means of a specific tracer, i.e., a molecule that participates in the reaction and that can be easily quantified using a suitable detection technique. Detection techniques based on radioisotopes were first successfully applied to biospecific recognition assays. Even if these techniques are still used, they are being replaced by non-radioactive techniques, such as those based on spectrophotometry, fluorescence, or luminescence measurements.

Chemiluminescence detection techniques, in particular, have proven to be very sensitive, allowing to obtain analytical performance similar to or better than that achievable using radioisotopes. Chemiluminescent species can be covalently linked to biospecific molecules (antibodies, antigens, nucleic acids) to obtain tracers which participate in the biospecific reaction and are detectable with high sensitivity. In most cases, chemiluminescence-based immune and gene probe assays rely on enzymatic labels (usually horseradish peroxidase or alkaline phosphatase), for which there are various commercially available chemiluminescence substrate formulations, optimized to maximize the signal intensity and duration.

Sensitivity can be further increased using amplified detection systems based on the biotin-streptavidin reaction, which is highly selective and characterized by a very high equilibrium constant (10^{13}–10^{15} L/mol). In such amplified systems, a biotinylated biospecific probe is detected using a reagent containing streptavidin and biotinylated enzyme. Because of the formation of biotin-streptavidin complexes, each biospecific probe binds several enzyme molecules, thus determining an increase in the signal-to-mass ratio of the probe.

Immunoassays are based on the strong and highly specific interaction occurring between antigens (Ag) and antibodies (Ab):

$$\mathrm{Ag} + \mathrm{Ab} \leftrightarrow \mathrm{AgAb}$$

Binding of the antigen with the antibody cannot be easily and conveniently revealed and quantified, therefore a tracer is usually added to the reaction. The quantification of the analyte in the sample is performed by measuring the signal of the tracer bound to the antibody or antigen (depending on the assay format) and by interpolating the result on a dose-response curve obtained by analyzing standard analyte solutions.

Immunoassays can be classified in *homogeneous* or *heterogeneous* assays. In the case of heterogeneous assays, one of the immunoreagents (either the antibody or the antigen) is bound to a solid phase (usually the wells of a microtitre plate or magnetic microspheres). This allows conveniently performing, before the measurement, separation of the tracer free fraction from the bound fraction. In the case of homogeneous assays, the antigen-antibody reaction occurs in solution and there is no separation step of the tracer bound and free fractions before the measurement. As a consequence, these immunoassays can be performed only if the signal of the tracer is changed upon its binding.

Immunoassays can be further classified in *competitive* or *non-competitive* assays. In a typical competitive immunoassay, the antigen Ag (here supposed to be the analyte) and a labeled derivative of the antigen (the tracer, Ag*) compete for a limited number of antibody sites. After the competition has occurred, the fraction of tracer bound to the antibody is inversely proportional to the antigen concentration in the sample. Non-competitive immunoassays for antigens are based on the capture of the antigen Ag by an excess of specific antibodies, then on its quantification by means of a second labeled antibody (the tracer, Ab*) able to bind to a different portion of the antigen. Alternatively, a third antibody, which is labeled with an enzyme and is specific for the species where the detection antibody was produced, can be used to quantify the bound detection antibody. The third antibody represents a "universal" detection reagent, because it can be used to quantify any anti-analyte antibody raised in a specific animal. In non-competitive immunoassays the amount of bound tracer is directly proportional to the antigen in the sample.

Chemiluminescent immunoassays used in clinical chemistry are usually based on the heterogeneous competitive or non-competitive format.

Gene probe assays. The search of specific gene sequences has relevant clinical and diagnostic value in a wide range of diseases. Identification of a gene and its mutation in inherited disorders, e.g., familial hypercholesterolaemia, muscular dystrophy and cystic fibrosis, enables early foetal diagnosis and detection of carriers. Identification of polymorphisms allows the study of their relationship to the susceptibility of individuals to diseases, such as ischaemic heart disease, multiple sclerosis, diabetes and other diseases. Identification of oncogenes and mRNAs for particular polypeptides permits the early detection of cancer cells, while identification of microorganisms (bacteria and viruses) makes possible the early diagnosis of infectious diseases.

Nucleic acid hybridization assays enable the identification of specific sequences of DNA or RNA employing a labeled complementary gene probe. As a hybridization assay is basically a sandwich-type format assay, the sensitivity of which is determined primarily by the detection limit of the label, its analytical performance can be greatly improved by applying chemiluminescence detection. Usually, probes are labeled indirectly with haptens (e.g., digoxigenin, biotin, fluorescein) that are subsequently immunodetected by enzyme-labeled anti-hapten antibodies. Indeed direct labeling of short gene probes (oligonucleotides) with enzymes might affect the kinetics or specificity of the hybridization reaction due to steric hindrance. Amplified labeling systems, such as those based on the biotin-streptavidin reaction, can further improve assay sensitivity. Alternatively, a DNA sequence can be identified using an unlabeled complementary DNA probe and an antibody specific for double-stranded DNA, which binds the hybridization product and is then detected using a labeled anti-species antibody.

Hybrid capture assays, which combine hybridization and immunological reactions, are based on the hybridization of target DNA with a specific RNA probe. Such RNA/DNA hybrids are then captured onto the surface of microtitre plate wells coated with an antibody that selectively recognizes the RNA/DNA hybrids. The bound hybrids are then detected using an enzyme-labeled anti-hybrid antibody and appropriate CL substrate.

10.6.5 Chemiluminescence Imaging

Chemiluminescence low-light imaging allows the analysis of different types of samples, even at the microscopy level (imaging of tissue sections, single cells or devices in microchips), for which the detector is connected to an optical microscope to obtain a spatial resolution of 0.5–1 μM.

The application of such technique enabled the localization and quantification of enzymes, metabolites, antigens and nucleic acids in tissue sections and single cells for a simple and rapid diagnosis of infectious diseases and other diseases, such as genetic disorders and cancer. This technique exceeds the capacity of detection of probes labeled with chromophores or fluorescent probes and it is comparable to detection systems based on radioisotopes.

For ***Southern blot*** analysis, DNA is isolated from cells or tissues, then broken into fragments of different sizes with enzymes that cut DNA at specific sites (i.e., each time a specific nucleotide pattern occurs). These fragments are then separated by size by means of gel electrophoresis. The DNA pieces must be rendered single stranded by using heat or a denaturating agent, in order to react with the probe complementary to them. The single-stranded DNA is transferred to nitrocellulose membrane, the so-called process of Southern blotting, where it sticks firmly. The gene probe is then added, which binds only to DNA complementary to it (i.e., where the adenine-thymine or guanine-cytosine bases match). The unreacted probe is then removed by appropriate washing.

The procedure for Northern blotting is similar to that of Southern blotting, except that it is used to identify RNA instead of DNA.

Western blot enables the identification of a target protein from a complex sample upon a gel electrophoresis separation, transfer on a nitrocellulose membrane and recognition by means of a specific antibody. The antigen-antibody complex is then detected by means of an enzyme labeled probe (e.g., secondary antibody or streptavidin). Chemiluminescence detection can be performed upon addition of the enzyme substrate (e.g., the luminol/H_2O_2/enhancer system for horseradish peroxidase).

The chemiluminescence signal may be acquired employing a very simple detection system based on the use of a photographic film. However, with this this type of instrumentation adequate quantification of signal intensity is not possible. On the contrary, chemiluminescence imaging detection performed with a luminograph equipped with an ultrasensitive CCD camera, immunological or gene hybridization reactions can be directly and quickly detected on the membrane, with results comparable or even better than those obtained with other detection principles (radioisotopes, chromophores, photography).

Immunohistochemistry (IHC) techniques with chemiluminescent detection are important means for localization of specific antigens at both the tissue and cellular levels. Immunohistochemical techniques are widely used in research and in clinical chemistry laboratories in different fields, including cancer detection and infectious disease diagnosis.

Highly specific antibodies, usually monoclonal antibodies, are added on thin tissue sections mounted on glass slides or cell preparations to selectively bind to antigens such as proteins, enzymes and viral or bacterial products. The bound antibody is then determined, usually indirectly, through a secondary (species-specific) antibody conjugated with an enzyme that catalyzes the chemiluminescent reaction (alkaline phosphatase or horseradish peroxidase). This technique allows to determine the spatial distribution of the antigen in the tissue or cell and to evaluate different levels of expression in different samples and/or sample areas, thus providing very informative data on the antigen distribution.

The main advantages of this technique are its specificity and sensitivity, since the interaction between the antigen and primary antibody is completely exclusive. Many chemiluminescence immunohistochemistry methods have been developed, also in multiplex formats, enabling the localization of specific antigens within single cells, primarily for diagnostic and prognostic purposes or for investigating the mechanisms of pathology onset and progression.

In situ hybridization (ISH) is a technique for the detection of genetic material, based on the recognition of a single strand of a target gene sequence (endogenous DNA, mRNA, viral or bacterial nucleic acid present in the sample, such as in a tissue section or in a preparation of chromosomes fixed on a slide) by a complementary specific nucleotide probe, which is labeled, usually with biotin, fluorescein, or digoxigenin. The complementarity between bases allows the formation of covalent bonds with the formation of a double-stranded nucleic acid. The labeled probe, upon hybridizing with the target, enables its detection and localization. Indeed, for the detection, streptavidin, anti-fluorescein or anti-digoxigenin antibodies are added, which are covalently labeled with an enzyme suitable for chemiluminescence detection, such as horseradish peroxidase or alkaline phosphatase.

Chemiluminescence ISH offers high assay sensitivity, thus enabling identification of even very few copies of DNA. In addition, it also allows evaluating whether the genome is integrated into the cell or in episomal form.

Furthermore, the sample is not destroyed or altered in its morphology, thus enabling not only to determine the spatial location of the genome, but also to operate other analyses on the same sample, as the search for target antigens or cellular components (Figs. 10.14 and 10.15).

Isolated perfused organs represent an experimental model widely used in the study of various biological processes, as they represent a relatively simple and reliable method to predict target physiological or pathological events in an intact organism.

Chemiluminescence imaging on organs can be applied in ex vivo studies of all those pathophysiological processes that result in the production of light.

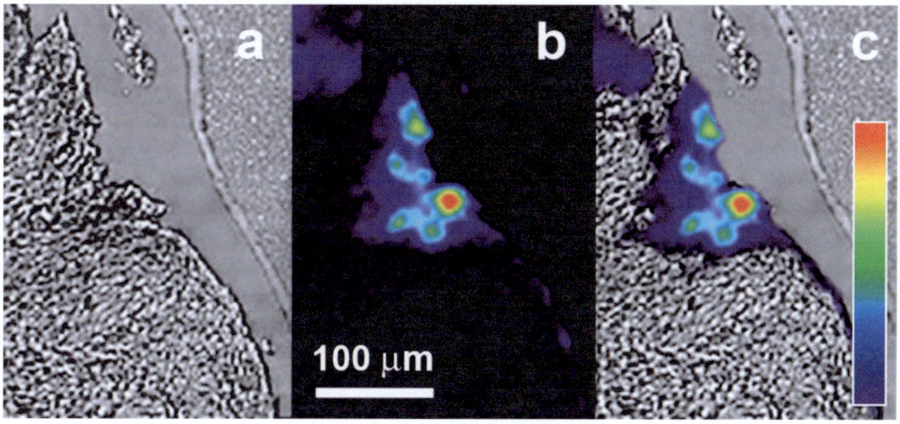

Fig. 10.14 Example of ISH with chemiluminescence detection on a tissue section. Human Papillomavirus 18 DNA was detected by using a digoxigenin-labeled gene probe, followed by a HRP-labeled anti-digoxigenin antibody and a suitable chemiluminescent substrate: (**a**) transmitted light image; (**b**) pseudocolored chemiluminescence image; (**c**) overlay of chemiluminescence image and transmitted light image (Reprinted with permission from [12])

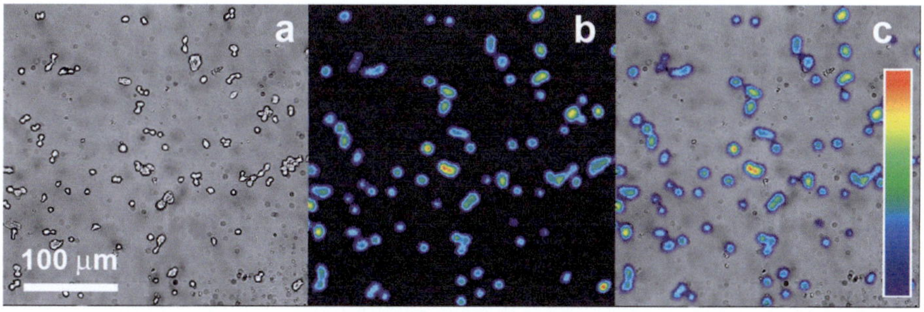

Fig. 10.15 Example of ISH with chemiluminescence detection on fixed cells. Human Papillomavirus 16 DNA was detected in fixed K562 cells by using a digoxigenin-labeled gene probe, followed by a HRP-labeled anti-digoxigenin antibody and a suitable chemiluminescent substrate: (**a**) transmitted light image; (**b**) pseudocolored chemiluminescence image; (**c**) overlay of chemiluminescence image and transmitted light image (Reprinted with permission from [12])

For example, chemiluminescence can be used to study the formation of oxygen free radicals (ROS) in different biological systems, including whole organs, tissue homogenates, isolated cells and intracellular organelles. This is made possible by the weak spontaneous emission of photons associated with oxidative processes related to ROS.

An isolated organ exposed to oxidative damage by ischemia and reperfusion exhibits increased production of ROS, which can be detected by perfusing the organ with lucigenin. This model can be exploited in the study of the pathophysiology of oxidative stress, in the screening of new compounds with antioxidant activity and in assessing the effectiveness of the conditions of preservation of organs before transplantation.

10.6.6 Chemiluminescence-Based Miniaturized Devices

Owing to its high detectability even in low volumes, wide linear range of the response, and simplicity of the instrumental set-up required for its measurement, chemiluminescence is particularly suited as a detection principle in miniaturized analytical devices, such as microfluidic platforms, lab-on-chip, and micro total analysis systems (μTAS) [13, 14]. Ideally, such devices should allow performing all the steps of an analytical procedure employing a portable miniaturized and self-operating system, to enable extra-laboratory analyses. This approach can find applications in different fields, such as clinical chemistry, environmental monitoring, food analysis, and bio-warfare protection.

A variety of detectors have been employed to measure chemiluminescence from microfluidic devices, including photomultiplier tubes, usually placed directly in front of the metering cell or coupled to it by means of optical fibres. This ensures high signal detectability, but at the expense of device portability and system integration. Other approaches rely on imaging, employing CCD or complementary metal-oxide-semiconductor (CMOS) sensors, with the main advantage being the possibility to simultaneously monitor chemiluminescence emission from different spots or channels, thus enabling multiplexing and/or parallel analysis of several samples. These sensors can be employed in a traditional imaging configuration, with a suitable optics for focusing the image of the chip on the sensor surface, or in a "contact" imaging configuration, in which the microfluidic device is placed directly in contact with the sensor surface, thus providing increased light collection efficiency (that leads to improved detectability) and device compactness. Also suitable for multiplexing is the on-chip integration of thin film photosensors, such as amorphous silicon photosensors, organic photodiodes, carbon nanotubes coated with photovoltaic polymers and metal–semiconductor–metal photodetectors. Finally, very recently, the use of a smartphone back side illuminated CMOS sensor for measuring chemiluminescence signals from microfluidic devices has been proposed, as a very convenient solution that eliminates the needs for dedicated

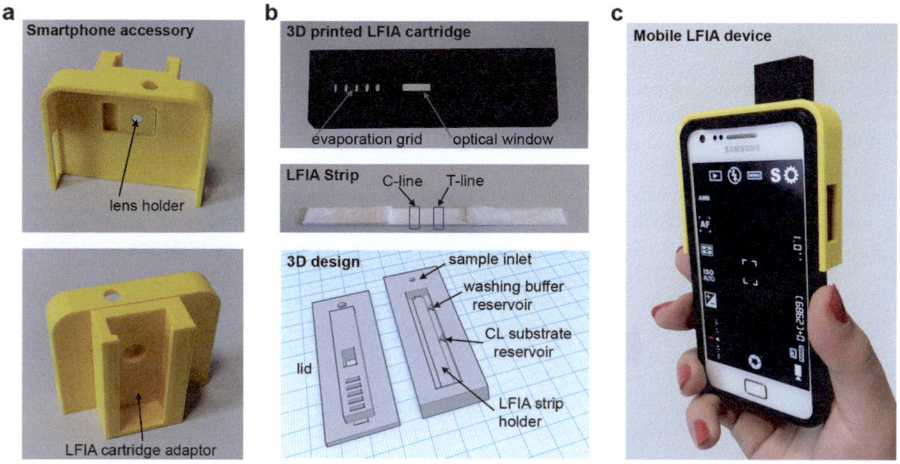

Fig. 10.16 (**a**, **b**) Smartphone accessory cartridge 3D printed in ABS to enable smartphone-based chemiluminescence lateral-flow immunoassay (*LFIA*) for salivary cortisol quantitative detection. (**c**) The integrated cortisol LFIA smartphone-based device with running application for chemiluminescence signal acquisition (Reprinted with permission from [15])

instrumentation for signal detection and enables exploiting smartphone embedded technology and connectivity for a telemedicine approach [15] (Fig. 10.16).

Several microfluidic-based analytical platforms have been described exploiting chemiluminescence detection exploiting different analytical formats, such as microchip flow injection assays, microchannel-based binding assays (immuno- and gene probe assays), microchip capillary electrophoresis, and paper-based assays.

Despite the rapidly expanding number of published miniaturized chemiluminescence-based analytical systems with applications in the clinical diagnostics, food safety control, water process control and forensic analysis fields, most of such systems still need manual operations and external instrumentation to perform the analyses, thus requiring skilled personnel and laboratory environment. Thus, the development of fully automated and integrated platforms, in which all the steps of the analytical procedure are performed, experimental conditions can be carefully controlled and all the device functional modules are miniaturized and integrated is still matter of research [13].

In the view of extremely simple, low-cost, compact and rapid analytical microdevices, paper-based analytical devices (μPAD) are emerging as flexible, lightweight, disposable and self-standing analytical devices on which a variety of bioanalytical assays can be performed in low-resource settings [16].

References

1. Ana M, Garcia-Campana E, Willy RG (2001) Baeyens, chemiluminescence in analytical chemistry. CRC Press, New York
2. Roda A (2010) Chemiluminescence and bioluminescence: past, present and future. RSC Publishing, Cambridge
3. Roda A, Guardigli M (2012) Analytical chemiluminescence and bioluminescence: latest achievements and new horizons. Anal Bioanal Chem 402:69–76
4. Iranifam M (2014) Analytical applications of chemiluminescence methods for cancer detection and therapy. TRAC Trends Anal Chem 59:156–183
5. Rodriguez-Orozco AR, Ruiz-Reyes H, Medina-Serriteno N (2010) Recent applications of chemiluminescence assays in clinical immunology. Mini-Rev Med Chem 10:1393–1400
6. Liu M, Lin Z, Lin J-M (2010) A review on applications of chemiluminescence detection in food analysis. Anal Chim Acta 670:1–10
7. Giokas DL, Vlessidis AG, Tsogas GZ, Evmiridis NP (2010) Nanoparticle-assisted chemiluminescence and its applications in analytical chemistry. TRAC-Trends Anal Chem 29:1113–1126
8. Li N, Liu DQ, Cui H (2014) Metal-nanoparticle-involved chemiluminescence and its applications in bioassays. Anal Bioanal Chem 406:5561–5571
9. Iranifam M (2013) Revisiting flow-chemiluminescence techniques: pharmaceutical analysis. Luminescence 28:798–820
10. Lara FJ, Garcia-Campana AM, Ibanez Velasco A (2010) Advances and analytical applications in chemiluminescence coupled to capillary electrophoresis. Electrophoresis 31:1998–2027
11. Gamiz-Gracia L, Garcia-Campana AM, Huertas-Perez JF, Lara FJ (2009) Chemiluminescence detection in liquid chromatography: applications to clinical, pharmaceutical, environmental and food analysis-A review. Anal Chim Acta 640:7–28
12. Roda A, Guardigli M, Pasini P, Mirasoli M, Michelini E, Musiani M (2005) Bio- and chemiluminescence imaging in analytical chemistry. Anal Chim Acta 541:25–35
13. Mirasoli M, Guardigli M, Michelini E, Roda A (2014) Recent advancements in chemical luminescence-based lab-on-chip and microfluidic platforms for bioanalysis. J Pharm Biomed Anal 87:36–52
14. Seidel M, Niessner R (2014) Chemiluminescence microarrays in analytical chemistry: a critical review. Anal Bioanal Chem 406:5589–5612
15. Zangheri M, Cevenini L, Anfossi L, Baggiani C, Simoni P, Di Nardo F, Roda A (2015) A simple and compact smartphone accessory for quantitative chemiluminescence-based lateral flow immunoassay for salivary cortisol detection. Biosens Bioelectron 64:63–68
16. Ge L, Yu JH, Ge SG, Yan M (2014) Lab-on-paper-based devices using chemiluminescence and electrogenerated chemiluminescence detection. Anal Bioanal Chem 406:5613–5630

Chapter 11
Solar Filters: A Strategy of Photoprotection

Susana Encinas Perea

Abstract Solar filters, as photostable compounds, are used to protect the human skin against harmful UV radiation. Up until recently, sun protection was limited to the use of sunscreens whose active ingredients are UV filters, which reflect or absorb UV rays depending on their chemical nature. Today there is a trend toward higher sun protection factors and the use of additional ingredients in sunscreens aimed at further protecting against the indirect damage primarily caused by the deeply penetrating UVA rays. In fact, these wavelengths have been shown to generate reactive oxygen species, which become mediators of cellular oxidative damage, leading to oxidative stress and immune suppression in skin.

In this chapter a photoprotection strategy has been described based on the development of sunscreens to minimize light absorption by the photosensitizers, and the use of scavengers to eliminate the reactive oxygen species arising therefrom. Advantage has been taken from the photochemical and photophysical techniques, through the combination of steady-state photolysis, photoproducts isolation and characterization with time-resolved spectroscopy, such as laser flash photolysis.

A broadband UV filter, the bis-ethylhexyloxyphenol methoxyphenyl triazine (Tinosorb S®), which provide protection against direct and indirect UV-induced DNA damage by absorbing solar UV and reducing ROS formation in the skin, has been chosen. Based on its characteristics, it has been considered whether it would be feasible and useful to have a compound which would display in the same molecule both UV-absorbing and antioxidant properties. This type-molecules could find potential applications for reducing skin photo-oxidative damage.

S. Encinas Perea
Chemistry Department, Universitat Politècnica de València, Valencia, Spain
e-mail: sencinas@qim.upv.es

G. Bergamini, S. Silvi (eds.), *Applied Photochemistry*,
Lecture Notes in Chemistry 92, DOI 10.1007/978-3-319-31671-0_11

11.1 Ultraviolet Incidence on Earth

The amount of incident ultraviolet (UV) radiation has important implications for human health and terrestrial and aquatic ecosystems. Only a small fraction of the radiation emitted by the Sun that reaches the Earth's surface resides in the UV range. This corresponds to the 6.1 % of the solar electromagnetic spectrum. The UV wavelength range extends from 100 to 400 nm, and it is divided into UVC (100–290 nm), UVB (290–320 nm), and UVA (320–400 nm). UVA is further divided into UVA1 (340–400 nm) and UVA2 (320–340 nm). The shorter the wavelengths, the greater the absorption by the atmosphere, the most energetic UV photons belonging to vacuum UV radiation. Ozone and oxygen completely absorb UVC radiation and absorb the majority (approximately 90 %) of UVB. Therefore, UVA accounts for approximately 95 % of the total UV energy that reaches the Earth's surface, with the remaining 5 % being UVB [1].

11.2 DNA Damage Induced by UVB and UVA

The understanding of UV radiation (UVR) and its effects on the skin is constantly evolving. The UV spectrum has an effect on aging of the skin, sunburn development, the production of precancerous and cancerous lesions, and immunosuppression. The incidence of skin cancer in humans has increased considerably in recent decades. Exposure to solar ultraviolet radiation is certainly involved in numerous skin cancers pathologies. In this context, the role of sunlight UV radiation as an initiator and as a promoter of cancer is no longer in question [2].

The mechanisms involved in the UV damage of biomolecules are largely due to the direct absorption of light to form more reactive, excited molecules. Although UVR represents only a small part of the solar spectrum, it has enormous importance on the structure of the atmosphere, and it has a critical impact on the biosphere. In this sense, the main biological relevance of UV radiation resides in the formation of DNA damage in different aquatic and terrestrial life forms. Because the maximum light absorption by DNA molecule is observed at 260 nm, UVC would be the most effective waveband for the induction of DNA photoproducts; however, these short wavelengths are entirely blocked by the Earth's atmosphere. Additionally, the energy of incident UVB and UVA photons is high enough to generate DNA damage, being UVA capable of penetrating the skin more efficiently than UVB, reaching the basal layers where melanocytes and dividing stem cells are located [3].

The damage mechanisms differ for UVB and UVA radiation. UVB is absorbed directly by the DNA as this sunlight component is considered the most mutagenic and carcinogenic. Although UVA radiation is at least ten times more abundant than UVB in the solar spectrum, it doesn't cause directly damage to DNA but may act through photosensitized reactions due to endogenous or exogenous chromophores contained in several compounds as drugs, cosmetics, etc. This photosensitizing

DNA damage by xenobiotics has attracted considerable attention because it can involve a more extended active fraction of the solar spectrum with carcinogenic potential. In that way, the risk of biomolecules damage is considerably increased [4].

The direct absorption of UVB energy, mainly by the pyrimidine bases, such as cytosine (C) and thymine (T), leads to the formation of pyrimidine dimers. A scheme of the DNA modifications induced by sunlight is shown in Fig. 11.1. The mechanisms underlying the induction of pyrimidine dimers by UV light have been well described previously [5, 6].

Indirect modifications in the DNA molecule can also be derived by UVA radiation. Their mechanisms are mediated by endogenous photosensitizers, such as flavins, nicotinamide adenine dinucleotide phosphate (NADPH) oxidase, heme groups, porphyrins, melanin, and cytochromes, which absorb the UV energy resulting in excited molecules [7–9].

There are two different reactions that can modify biomolecules. The type I reaction involves a complex chain of events starting by charge transfer to a photosensitizer in a triplet excited state, which, in turn, reacts directly with the

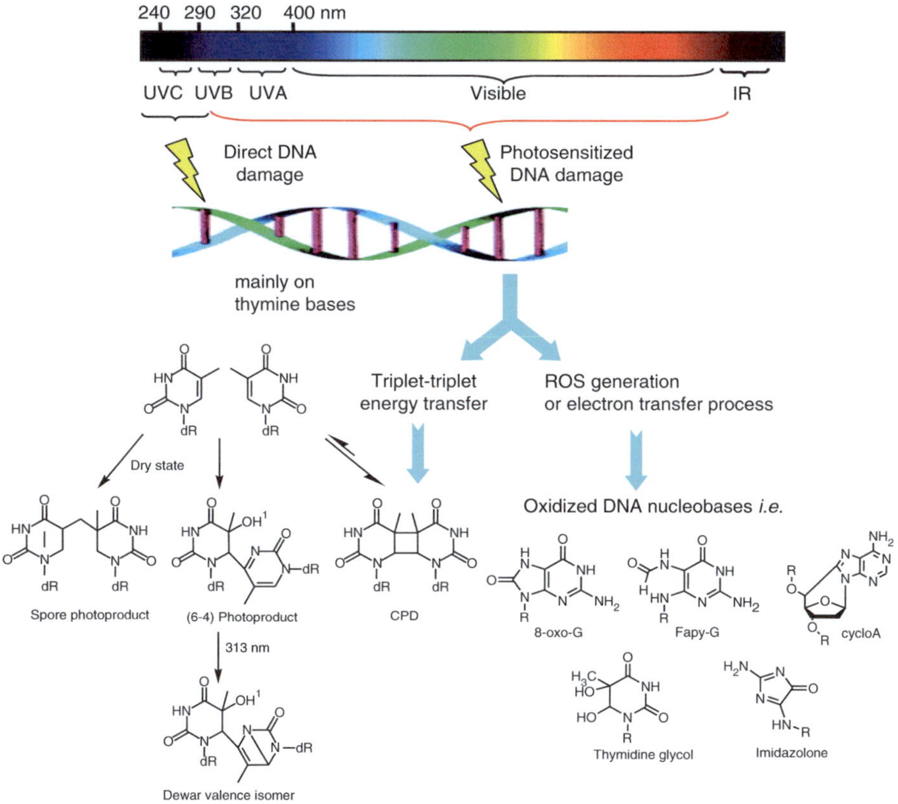

Fig. 11.1 DNA modifications induced by sunlight

substrate in a one-electron transfer reaction. This produces a radical or radical ion in both the sensitizer and the substrate. In the presence of oxygen, these reactions give rise to reactive oxygen species (ROS), such as the superoxide anion radical ($O_2^{\cdot -}$) and the highly reactive hydroxyl radical ($HO^{\cdot}$). Thus, DNA single strand breaks and oxidized purine and pyrimidine bases are the indirect consequences of a type I photosensitization mechanism. On the other hand, the main sensitized oxidation reaction to cellular DNA occurs via the type II reaction, which involves direct energy transfer from the photosensitizer in a triplet excited state to an oxygen molecule, leading to singlet oxygen ($^1O_2$) formation and subsequent DNA oxidation [10].

Consequently, 8-oxo-7,8-dehydroguanine (8-oxoG) is the main oxidized DNA base formed after UVA exposure that can be generated by the attack of several ROS or reactive nitrogen species (RNS), as a result of type I and II photosensitization reactions during UV exposure [11]. In this sense, it is well known that cyclobutane pyrimidine dimers (CPDs) generation and purine bases oxidation are the main reactions responsible for the photoinduced DNA damage due to the fact that the integrity of the genome is provided by the right bases sequence along the double strand (Fig. 11.1).

To sum up, UVB typically induces erythema, acute sunburn and direct DNA damage via pyrimidine dimer formation, whereas UVA mainly associated with generation of ROS contributes to photo-aging and photocarcinogenesis. Therefore, DNA damage is considered to be the main cause for the genetic changes responsible for sunlight-induced skin lesions and carcinogenesis, including malignant melanoma [3].

11.3 Reactive Oxygen Species: Atomic Oxygen

ROS are formed as natural by-products of the oxygen normal metabolism. Environmental stress, as for example, the exposure to UV light, produces large amounts of ROS that can be the origin of the damaging effects to cellular components. In addition, the combination of UVR with photosensitizing compounds has proved to be very effective in the generation of oxidative damage to biomolecules leading, for example, to skin cancer.

The capacity to induce DNA modifications by ROS depends on several factors, such as the production place of reactive species, the relative ability of the biomolecule to be oxidized and the availability of metal ions. In order to combat the attack of these reactive species cells have defenses, small molecules that act as antioxidants or enzymes, which are responsible for limiting the ROS levels. However, even when the cell is operating normally, ROS tend to avoid these defenses and the balance is favorable for oxidative cell damage.

Overall, exposure to solar UVR can produce high amounts of ROS, which become mediators of oxidative DNA damage [12]. In this context, the identity of the reactive intermediates responsible for such damage remains a subject of active

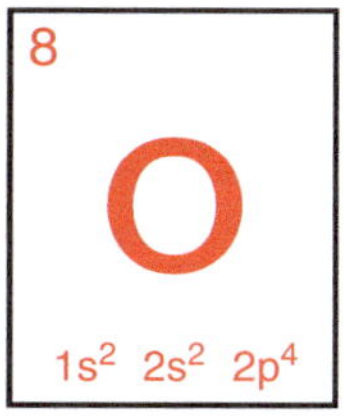

O(^{3}P)

Fig. 11.2 Reactive oxygen species: atomic oxygen

investigation. For instance, among these powerful oxidizing agents, atomic oxygen (Fig. 11.2) in its triplet ground state O(^{3}P) is a species of great relevance in atmospheric and combustion chemistry. Then, its gas-phase chemistry has been the subject of numerous kinetic studies [13]. However, its reactivity in solution and its biological effects remain almost unexplored [14].

11.4 Photoprotection Strategy

As part of a strategy of photoprotection against cell damage is necessary to consider the implementation of several consecutive lines of defense. Among them the development of sunscreens to minimize light absorption by the photosensitizers and the use of scavengers to eliminate the reactive oxygen species arising therefrom.

There are three principal ways to protect the skin against the UV rays: reflect the rays away from the skin; absorb the rays and deactivate them; prevent and repair the damage they cause. The ideal sunscreen should therefore contain chemicals that address all three. Indeed, sunscreens may contain agents which mainly would reflect UV rays, and/or photo-stable UV absorbers which would absorb the rays in different ways according to their chemical structure in order to provide broad spectrum coverage. The third issue may be addressed by using free radical scavengers/antioxidants and agents active in DNA repair to improve photoprotection [15].

11.4.1 Active Sunscreen Ingredients: UV Filters

Sunscreens are used to protect human skin against harmful UV radiation. They are first-line protection from UVR and its negative effects, as it is schematized in Fig. 11.3. Then, the purpose of UV filters used in cosmetic sunscreen formulations is to attenuate the UV radiation of the sun minimizing the extent of UVB and UVA radiation that might reach DNA in cell nuclei [16]. In this sense, sunscreens are widely proved to have positive effects on decreasing the signs of photoaging and the incidence of skin cancers. Thus, because UV radiation is the major cause of the

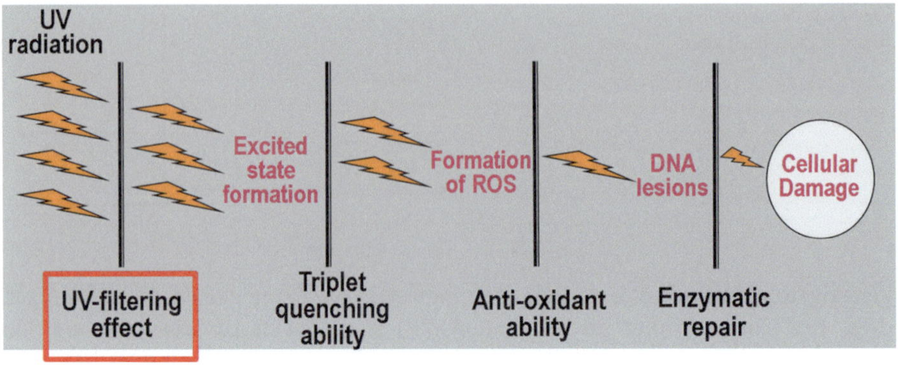

Fig. 11.3 Photoprotection strategy: UV-filtering effect

clinical changes in skin exposed to sunlight, it is extremely important to accurately assess the photoprotection properties of sunscreens in order to prevent harmful consequences caused by prolonged sun exposure, especially for hypersensitive people [2].

In this sense, sunscreens have been available since 1928 and today play a major role in skin cancer prevention and sun protection. Their use as an integral part of the photoprotection strategy has expanded worldwide. Their active ingredients are generally divided into inorganic and organic agents, previously termed physical blockers and chemical absorbers, respectively. On the one hand, inorganic sunscreen ingredients act by reflecting or scattering visible, UV, and infrared radiation over a broad-spectrum. The major inorganic agents used today are zinc oxide and titanium dioxide. On the other hand, organic sunscreen ingredients, many of which specifically filter UVB, act by absorbing UVR and converting it to heat energy [17].

In this context, the fundamental component of a good sunscreen is a broad-spectrum (UVA/UVB) coverage, absorbing over the entire UV spectrum, because of UVA- and UVB-blocking products protect against sunburn as well as more subtle suberythemal skin damage. Only in that way a sunscreen would be able to prevent or significantly reduce the photoinduced biological damage [18].

11.4.1.1 Principles of UV Radiation Absorption by Organic Ultraviolet Filters

The principle of photoprotection in organic sunscreens is the absorption of UVR. In order to absorb the UVR an organic UV light filter must contain a suitable chromophore having conjugated π-electron systems. Increasing the number of conjugated double bonds in the molecule, the absorption maximum would shift to longer wavelengths and also would give rise to a larger absorption cross section and, therefore, stronger absorption. In general, the larger the molecular weight of the chromophore, the more the absorption maximum will be shifted towards longer wavelengths. This is the reason that UVB light filters have smaller molecular weights compared to UVA light or broad-spectrum filters.

Currently, all organic UV absorbers used in sunscreens are aromatic compounds, each containing multiple conjugated π-electron systems (Table 11.1). Furthermore, also the type of substituents and their position at the aromatic ring are important for the UV spectroscopic properties. Especially advantageous are disubstituted systems with an electron-donor and an electron-acceptor group in para position (so-called push-pull systems).

Sunscreens, which mainly absorb UVB, may be less effective in preventing UV radiation-induced immunosuppression than broad-spectrum products [20]. Upon UV radiation, the molecule tends to delocalize the electrons to reach a higher state of energy. Absorption of a UV photon transfers the organic UV absorber into an excited electronic state. The UV filter molecule is switched from a ground state (before absorption) to either a singlet (often short lived) or a triplet excited state (longer lived). The excited molecule may reach a stable equilibrium through reversible isomerization or, under certain conditions, return to its original form (ground state). The energy of the excited electronic state may dissipate after internal conversion into molecular vibrations and further into heat via collisions with surroundings molecules, or have a radiative deactivation (fluorescence from the singlet state or phosphorescence from the triplet state) [21].

On the whole, the faster the rate of internal conversion, the better is the photostability of an absorber molecule. If the excitation energy cannot be disposed off by energy transfer, by emission of light or if the absorbed energy is not sufficiently and speedily dissipated into heat, chemical bonds of the UV absorber molecule may break or rearrange, resulting in degradation of the UV filter. In addition, sunscreens must dissipate the absorbed energy efficiently through photophysical and photochemical pathways that rule out the formation of singlet oxygen, other ROS, and other harmful reactive intermediates [21, 22].

11.4.1.2 Photostability

Nowadays, there is an increasing need for good topical sunscreens to prevent the well-documented damaging effects of UV light on human skin. The ideal sunscreen should be such that no photochemical or photosensitizing transformation of its components occurs within the formulation or on the skin. Photochemical stability is indeed the most important characteristic of an effective UV filter since the light-induced decomposition of the sunscreen agent not only reduces its photoprotective power but can also promote phototoxic or photoallergic contact dermatitis [16].

Nevertheless, several compounds used as UV-filters exhibit some photoreactivity leading to the formation of photoproducts that can absorb in different spectral regions, thus reducing their photoprotective efficacy. Moreover, UV filters can also display some photosensitizing effects: their photochemical intermediates or stable photoproducts can interact with skin components, mediating phototoxic and/or photoallergic processes. Consequently, as new compounds of

Table 11.1 Common UV light filters approved in Australia, Europe, Japan, and the United States [19]

INCI	CE no.[b]	USAN	Trade name[c]	INCI abbreviation
Broad-spectrum and UVA1				
Bis-ethylhexyloxyphenol methoxyphenyl triazine	S 81	Bemotrizinol	Tinosorb S	BEMT
Butyl methoxydibenzoylmethane	S 66	Avobenzone	Parsol 1789	BMDM
Diethylamino hydroxybenzoyl hexyl benzoate	S 83	–	Uvinul A Plus	DHHB
Disodium phenyl dibenzimidazole tetrasulfonate	S 80	Bisdisulizole disodium	Neo Heliopan AP	DPDT
Drometrizole trisiloxane	S 73	–	Mexoryl XL	DTS
Menthyl anthranilate	–	Meradimate	–	MA
Methylene bis-benzotriazolyl tetramethylbutylphenol	S 79	Bisoctrizole	Tinosorb M (active)	MBBT
Terephthalylidene dicamphor sulfonic acid	S 71	Ecamsule	Mexoryl SX	TDSA
Zinc oxide	S 76	Zinc oxide	ZnO (Nanox)	ZnO
UVB and UVA2				
4-Methylbenzylidene camphor	S 60	Enzacamene	Eusolex 6300	MBC
Benzophenone-3	S 38	Oxybenzone	–	BP3
Benzophenone-4	S 40	Sulisobenzone	Uvinul MS40	BP4
Polysilicone-15	S 74	–	Parsol SLX	PS15
Diethylhexyl butamido triazone	S 78	–	Uvasorb HEB	DBT
Ethylhexyltrimethyl PABA[a]	S 08	Padimate O	Eusolex 6007	EHDP
Ethylhexyl methoxycinnamate	S 28	Octinoxate	Uvinul MC 80	EHMC
Ethylhexyl salicylate	S 13	Octisalate	Neo Heliopan OS	EHS
Ethylhexyl triazone	S 69	Octyltriazone	Uvinul T 150	EHT
Homomenthyl salicylate	S 12	Homosalate	Eusolex HMS	HMS
Isoamyl p-methoxycinnamate	S 27	Amiloxate	Neo Heliopan E1000	IMC
Octocrylene	S 32	Octocrylene	Uvinul N 539 T	OCR
Phenylbenzimidazole sulfonic acid	S 45	Ensulizole	Eusolex 232	PBSA
Titanium dioxide	S 75	Titanium dioxide	Eusolex T2000	TiO_2
Tris biphenyl triazine	S 84	–	Tinosorb A2B	TBPT

INCI International Nomenclature for Cosmetic Ingredients
USAN United States Adopted Name
[a]Not being supported in the EU and may be delisted
[b]Cosmetics Europe order number reflects chronology of UV filter development
[c]Trade names are property of their respective manufacturers

unknown toxicology can be formed, evaluation of potential hazards associated with the use of a sunscreen requires a detailed study of its photochemistry [23].

Therefore, the UV absorbers used in sunscreens must be photostable under the conditions of use. Among the most commonly used compounds *para*-Aminobenzoic acid (PABA) was patented in 1943 with the water resistant ability. However, it can elicit photocontact allergies and possibly autoimmune diseases. Moreover, its decomposition is able to produce potentially carcinogenic products. In the 1980s, Benzophenone-3 (BP3) became the most frequently used component of sunscreen formulations and it also reported acute toxic effects [24].

On the other hand, the principle of a fast internal conversion is realized, for instance, in the Menthyl anthranilate (MA) filter because of the orthoamino group, resulting in excellent photostability. In others, this is realized via an orthohydroxy group forming hydrogen bonds (e.g. Bis-ethylhexyloxyphenol methoxyphenyl triazine (BEMT) and Methylene bis-benzotriazolyl tetramethylbutylphenol (MBBT)). In fact, UVA filters that fulfill both efficacy and photostability requirements are rare. For instance, benzophenones provide broad-spectrum UVB and UVA protection; however, they are photolabile and their oxidation can interrupt the antioxidant system. Also Butyl methoxydibenzoylmethane (BMDM), a potent UVA filter, undergoes rapid photodegradation [17, 19].

Furthermore, there are studies demonstrating that some filters react under UV irradiation as it is the case of Phenylbenzimidazole sulfonic acid (PBSA) that generates a variety of free radicals and active oxygen species that may be involved in the photodamage of DNA. It would appear that PBSA has the potential to act as a photosensitizer and interact with DNA, even as it protects skin cells from the effects of direct sunlight [25]. Then, PBSA could behave as a double-edged sword toward DNA questioning its suitability for sunscreen applications [26].

In this context, one approach to enhancing sunscreens is to develop innovative UV filters that are suitable and safe for human use. Most sunscreen agents provide protection in a particular UV range and offer insufficient photoprotection when used alone. In this sense, two factors must be addressed to produce an "ideal" sunscreen. First, it should provide uniform protection across the range of UVB and UVA, which assures that the natural spectrum of sunlight is attenuated in a uniform manner. Then, broad-spectrum (UVB/UVA) products are produced by combining filters with varying UV absorption spectra (see Fig. 11.4). In the second place, the "ideal" sunscreen should also have pleasing sensory and tactile profiles that enhance the user's compliance [19].

Summarizing, an appropriate sunscreen product (with a high sun protection factor, SPF) must fulfil the following critical requirements:

- Provide efficient protection against UVB and UVA radiation
- Be stable to heat and to UV radiation (UVR)
- Be user-friendly (easy and uniform to applicate, nonirritating and non-staining) to encourage frequent application and provide reliable protection
- Be cost-effective [18]

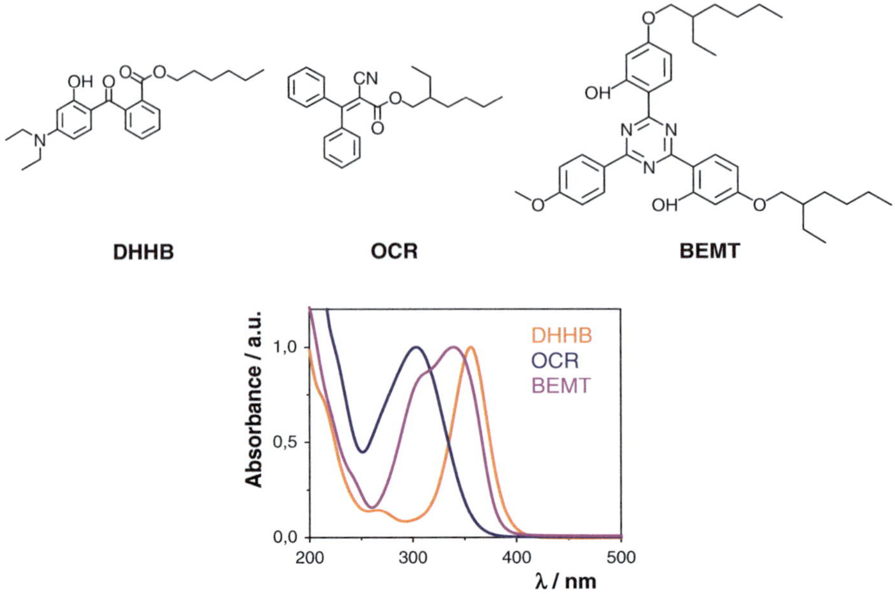

Fig. 11.4 *Top*: Chemical structures of three solar filters. *Bottom*: UV-Vis absorption spectra of the filters with different UV coverage

11.4.1.2.1 Photostabilization by Triplet Quenching: Triazines

Considering the photoprotective strategy described before sunscreen formulations including triplet quenchers (as shown in Fig. 11.3) could provide effective protection from the potential phototoxic and photoallergic effects derived from poor photostability of some filters [23]. Then, photounstable UV absorbers may be additionally stabilized by employing triplet quenchers. In this sense, being aware of these mechanisms and applying them for specific UV filter combinations can help in designing efficient sunscreens.

Among the common UV light filters listed in Table 11.1, BEMT is reported as highly photostable, with a recovery higher than 99 % after irradiation and it is a broad-spectrum (280–380 nm) agent (see Fig. 11.4 bottom). It is oil soluble and has been used successfully to improve the photostability and efficacy of sunscreens containing BMDM and Ethylhexyl methoxycinnamate (EHMC) [27]. Moreover, it has been evidenced that BEMT may act as a triplet quencher and can be used to stabilize photounstable filters. Because of its relatively large size, BEMT filter rarely cause allergic contact dermatitis, systemic absorption, or endocrine-like effects [16].

Given its molecular symmetry (shown in Fig. 11.4 top), the presence of electron-releasing groups such as hydroxyl group substituted on the aromatic rings provides to BEMT an optimal structure for energy dissipative processes allowing electron resonance delocalization upon absorption of a photon, and is most probably able to deactivate sensitizers trough energy transfer (triplet-triplet energy transfer) leading to the isomerization of the acceptor (reversible photoisomerization and deactivating capacity) [21]. To return to its ground state it can efficiently dissipate the accepted energy through intramolecular hydrogen transfer in the excited state followed by internal conversion and thermal deactivation [28].

Very recently, a new related UV filter, known as tris-biphenyl triazine, has just been launched on the market. Its broad spectrum associated with its qualities in terms of efficacy and photostability make it a choice ingredient for the formulation of sun protection products [29].

11.4.2 Antioxidants

The need for photoprotection today is more apparent than ever, with a 20 % of incidence on skin cancer in humans during their lifetime, the majority of these cancers attributable to a lifetime of cumulative UV radiation exposure. The ability for individuals to practice photoprotection is becoming more facile as newer sunscreen agents and technology block UVR more effectively [2].

Until recently, sun protection has been limited to using sunscreens whose active ingredients are UV filters, which reflect or absorb UV rays depending on their chemical composition. Thus, the amount of radiation that penetrates the deeper skin layers is reduced and, consequently the effects of oxidative stress as burns, erythema, aging and cancer decrease. However, today there is a growing interest in the use of additional ingredients in sunscreens for more protection. In addition, the principal damage triggered by the UVA rays in skin is caused by the generation of ROS comprising free radicals.

At present, there is a trend toward higher SPF and extra ingredients are included in the formulation of sunscreens to achieve further protection against the indirect damage primarily caused by the deeply penetrating UVA rays. In that context, vitamins, plant extracts and synthetic antioxidants are being incorporated in sunscreens as an additional measure to delay the aging process and reduce the skin photodamage induced by an excessive exposure to solar radiation [30]. Moreover, the use of antioxidants is reinforced by recent results showing that solar filters can promote DNA damage [23–26].

Actually, topical antioxidants and DNA repair stimulants are being explored as options for expanding the photoprotective abilities of sunscreens, as shown in Fig. 11.5. For the case of antioxidants, they may be a promising option to enhance sunscreen photostability as well as decreasing UV-related skin damage and boosting SPF without introducing additional UV agents [31]. Among these compounds, flavonoids, resveratrol and green tea extracts are reported to probably diminish UV-related skin damage, although they can be unstable and diffuse poorly into the epidermis [17]. Such antioxidants are inefficient UV filters and have low SPF; therefore, they are commonly used in combination with sunscreens to enhance their efficacy [32].

Moreover, these additives have been incorporated into many sunscreen products to neutralize the cytotoxic effects of ROS generated by UV exposure. In this sense,

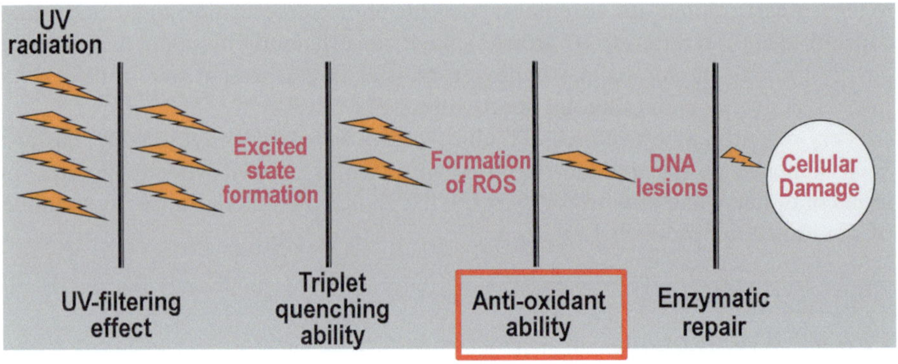

Fig. 11.5 Photoprotection strategy: anti-oxidant ability

they would have the potential of adding protection against the effects of UV radiation. However, a 2011 study concluded that all tested sunscreens had no or minimal antioxidant properties, most likely because of the lack of stability of antioxidants [19]. Thus, a variety of supplements were studied recently in order to improve photoprotection. Nevertheless, many substances added to sunscreens failed to increase the protective effect; as it is the case of dietary β-carotene, known as free radical scavenger, which did not improve the effect of sunscreen use on the development of skin cancer in humans [24].

On the other hand, as a positive example, a synergistic effect of vitamin A palmitate on the phototoxicity of combinations containing BMDM was observed [33]. Additionally, several studies have shown that a combination of UV absorber plus antioxidants is more effective in protecting skin against sun over-exposure than UV absorbers alone [15].

In conclusion, recommended photoprotective measures include sun avoidance during the peak UVR (10 AM–4 PM), the use of photoprotective clothing, wide-brimmed hat, sunglasses, and the use of broad-spectrum sunscreens. There are many other agents with photoprotective properties, which range from antioxidants to plant extracts to DNA repair enzymes. Recently, combinations of UV filters with agents active in DNA repair have been introduced in order to improve photoprotection [24]. Continued investigations in this area should result in the development of even more effective photoprotective agents in the future.

11.4.3 Triazines as UV Filters and Antioxidants

Aromatic compounds with s-triazine (1,3,5-triazine) chromophore have been little studied. Besides being part of numerous pesticides and herbicides commonly used (such as atrazine), reaching to be species with a high concentration in the atmosphere, some of them as BEMT (see in Fig. 11.6) are used as ingredients in sunscreens due to their UVA and UVB absorption. This particular triazine has

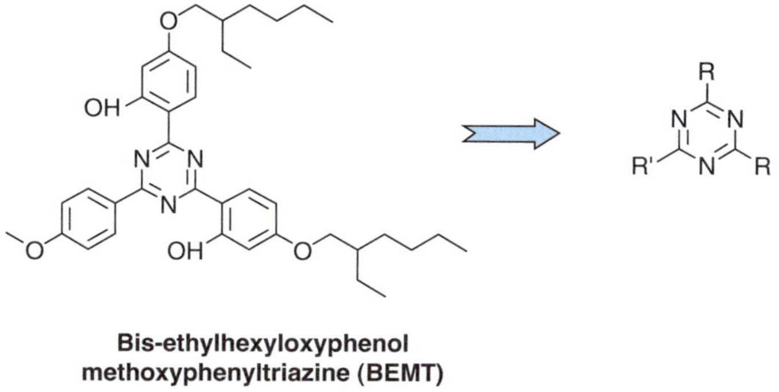

Fig. 11.6 Triazine-containing moiety in the BEMT solar filter

been described as highly photostable; [16] it is capable of dissipating the absorbed energy efficiently through an intramolecular hydrogen transfer in the excited state, followed by a non-radiative internal conversion process and thermal deactivation [28].

It is well known that triazines undergo oxidation reactions. Therefore, they could be compounds potentially capable of reducing the photooxidative cellular damage as an "all in one" acting as UV filters (decreasing the ROS formation) and antioxidants (protecting biomolecules from the generated ROS). Moreover, they are fluorescent compounds that modulate their fluorescent capacity based on their oxidation level and structural rigidity, which would allow its use as fluorescent probes. Thus, it could be opened an improvement on cosmetic formulations leading to significantly reduce the photoinduced carcinogenesis and cellular aging.

With this background, an investigation [34] is presented as an example of the triazine BEMT acting as an all-in-one antioxidant-based UV filter because the molecule possesses both UV-absorbing and antioxidant properties.[1]

11.5 A Filter Moiety Displaying UV-Absorbing and Antioxidant Properties

In order to prove the concept, we have implemented several photophysical and photochemical techniques (taking advantage from the experience in our research group), using heterocyclic *N*-oxides as atomic oxygen generators and the triazine-containing sunscreen BEMT as a potential acceptor.

In this work, as the source for atomic oxygen (one of the ROS that could mediate oxidative DNA damage and also is needing further investigation), we used the

[1] Part of this material has been reproduced from Ref. [34] with permission from the European Society for Photobiology, the European Photochemistry Association, and The Royal Society of Chemistry.

heterocyclic *N*-oxides of pyridazine and 4-nitroquinoline (Fig. 11.7) which display a very broad activity spectrum in biological oxidation and exhibit mutagenic and carcinogenic properties. For these compounds containing the *N*-oxide function, deoxygenation is a very common by process in their photochemistry occurring from the excited triplet state [35], which results in the generation of atomic oxygen $O(^3P)$.

A combination of steady-state photolysis, photoproducts isolation and characterization with time-resolved spectroscopic techniques, such as laser flash photolysis (LFP) has been used in order to selectively generate and sensitively detect the $O(^3P)$ species.

Basically, we have adapted a literature method [36] to reveal $O(^3P)$ generation by means of LFP in acetonitrile. This solvent reacted with the "invisible" atomic oxygen, giving rise to acetonitrile oxide (CH_3CNO) within 500 ns, as a detectable transient according to reaction Eq. 11.1.

$$O(^3P) + CH_3CN \rightarrow CH_3CNO(\lambda_{max} = 340\,nm) \tag{11.1}$$

We have used pyridazine *N*-oxide instead of pyridine *N*-oxide as the source of atomic oxygen, because the former provides a better signal-to-noise ratio at 340 nm. In addition, we have employed 266 nm as the excitation wavelength to maximize light absorption by the *N*-oxide in the presence of BEMT (see UV-spectra in Fig. 11.8).

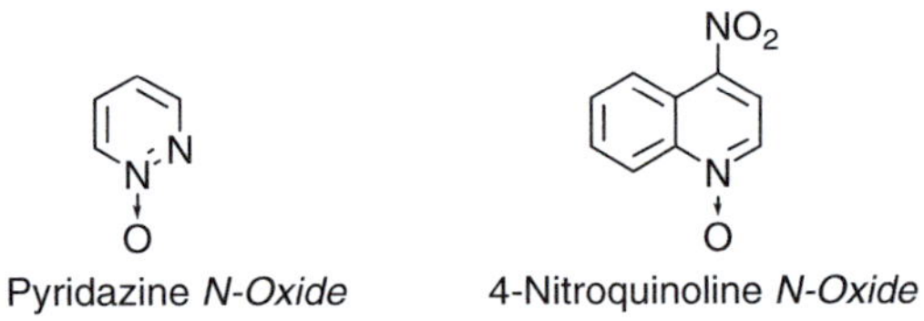

Fig. 11.7 Chemical structures of the *N*-oxides

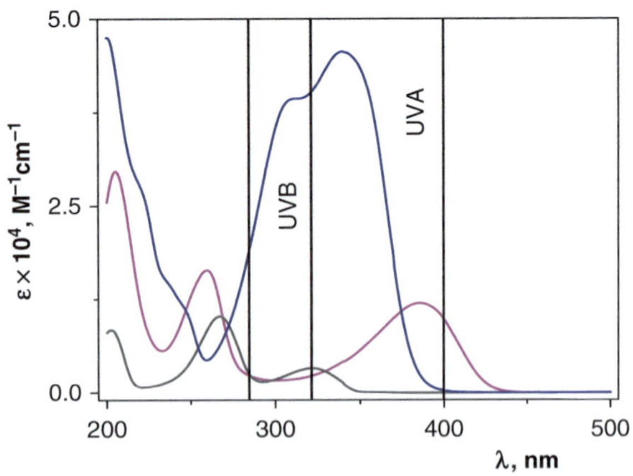

Fig. 11.8 UV-Vis absorption spectra of BEMT (*blue*) and pyridazine (*black*) and 4-nitroquinoline (*magenta*) *N*-oxides. UVB and UVA solar regions are delimited

As a matter of fact, LFP of the pyridazine *N*-oxide under these conditions led to the expected acetonitrile oxide (see the spectrum in Fig. 11.9, bottom). In the presence of increasing amounts of BEMT, a clear decrease in the intensity of this transient at 340 nm was observed (Fig. 11.9, top). From these data, the interaction between O(^{3}P) and the *N*-oxide was confirmed, and its kinetics that follows a Stern-Volmer relationship (Eq. 11.2) pointed out that it is indeed a very effective process (kq *ca.* 10^{11} M^{-1} s^{-1}, Fig. 11.9, bottom).

$$I_0/I = 1 + kq \times \tau_0 \times [BEMT] \tag{11.2}$$

The slight upward curvature of the Stern–Volmer plot suggests that the quenching rate constant could be somewhat overestimated.

Then, to obtain independent information on the order of magnitude of this value, kinetic experiments were performed looking at the formation of phenol from benzene and at the influence of added BEMT on this process. Thus, deaerated solutions of the *N*-oxide in acetonitrile containing benzene were irradiated, in the

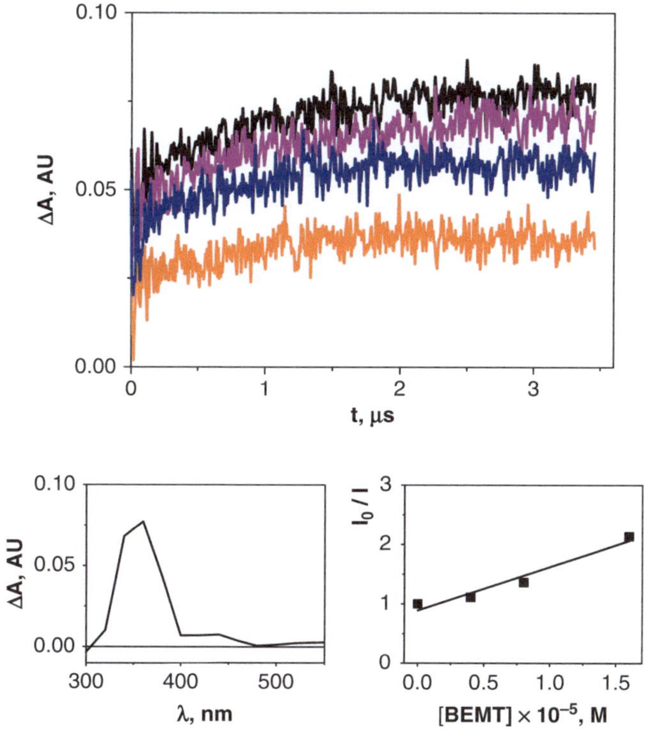

Fig. 11.9 *Top*: Growth traces at 340 nm, recorded after LFP of pyridazine *N*-oxide (4 × 1 0^{-5} M) in CH_3CN under nitrogen, in the presence of several amounts of BEMT (1:0 (*black*); 1:0.1 (*magenta*); 1:0.2 (*blue*); 1:0.4 (*orange*)). *Bottom*: Transient absorption spectrum obtained after LFP of pyridazine *N*-oxide (*left*). Stern–Volmer plot of CH_3CNO absorption intensity versus BEMT concentration (*right*)

absence and in the presence of BEMT. From the relative slopes of the linear plots of phenol formation vs. irradiation time (data not shown) and taking into account the rate constant of atomic oxygen quenching by benzene, a value *ca.* $10^{11}\ M^{-1}s^{-1}$ was again obtained for the solar filter, in agreement with the data obtained by LFP.

At this point, in order to further investigate the atomic oxygen transfer mechanism, the 4-nitroquinoline *N*-oxide was also included in our study. When such *N*-oxide was submitted to LFP, its triplet excited state was observed and characterized with an absorption maximum at 550 nm and a lifetime of 8 μs (as shown in Fig. 11.10, left). This transient decreased concomitantly with the bleaching recovery at ca. 400 nm, the absorption maximum in the ground state.

According to this experiment, in the presence of several amounts of BEMT there was no variation of the triplet excited state formation or decay for the bicyclic *N*-oxide. As the triplet quantum yield of 4-nitroquinoline *N*-oxide remained unchanged (Fig. 11.10, right), it was not affected by a possible screen role by the triazine UV-filter. In addition, there was no direct reaction of excited *N*-oxide with BEMT, as the triplet lifetime exhibited no significant changes (Fig. 11.10, right).

Next, in order to achieve a better understanding of the reaction mechanism, steady-state photolysis studies were performed. Thus, the monocyclic *N*-oxide was irradiated in the presence of the solar filter, under anaerobic atmosphere, leading to two photoproducts that were detected by UPLC-MS/MS. Their exact mass was 644.3691 and 644.3682 corresponding to the molecular formula $C_{38}H_{50}N_3O_6$ (MH^+); both of them matched with the incorporation of one oxygen atom to BEMT. Control experiments confirmed that the presence of both compounds was required for the photochemical formation of these products, which were not detected in the dark.

Finally, to disclose the substructure of the filter responsible for acceptance of $O(^3P)$, several model compounds were used to analyze their behavior under

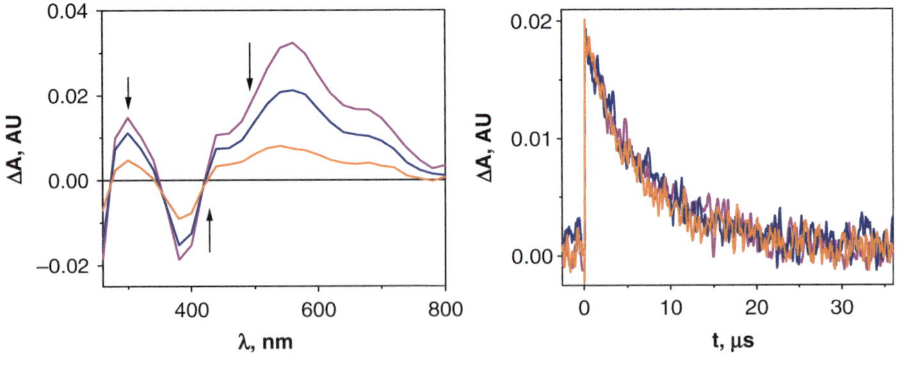

Fig. 11.10 *Left*: Triplet–triplet absorption spectra of 4-nitroquinoline *N*-oxide in CH_3CN under N_2 after LFP. Spectra were taken by increasing the delay time after the laser pulse (0.5, 5 and 10 μs); the time evolution is indicated by *arrows*. *Right*: Decay traces for the *N*-oxide triplet at 550 nm in the presence of several amounts of BEMT (1:0 (*magenta*); 1:0.1 (*blue*); 1:1 (*orange*)) recorded at 0.5 μs after the laser pulse

irradiation in the presence of pyridazine *N*-oxide (as schematized in Fig. 11.11). The selected fragments were related to the substituents of the heterocyclic core (methoxybenzene and m-methoxyphenol) as well as to triazine itself (parent compound and trimethoxy-substituted). The only oxygenated photoproduct was found for methoxybenzene and was identified as o-methoxyphenol.

Overall, these findings support photochemical oxygen transfer from an *N*-oxide to BEMT, confirming the feasibility of this solar filter to act as an oxygen acceptor. Then, two very connected processes are taking place: *N*-oxide deoxygenation and solar filter oxygenation. This occurs by *N*-oxide photocleavage with generation of atomic oxygen, followed by reaction of this species with the solar filter as an acceptor (Fig. 11.12), rather than by direct oxygen transfer in the excited state [37]. Transient absorption spectroscopy, high-resolution mass spectrometry and steady-state photolysis with model compounds strongly support the proposed mechanism.

To sum up, the triazine-containing UV-filter BEMT may act as a scavenger of ROS, specifically atomic oxygen, by means of being a good oxygen acceptor. Combining an antioxidant moiety within a UV-absorber may be an alternative and useful strategy for preventing/reducing photo-oxidative skin damage. This triazine-containing compound would behave as an all-in-one, antioxidant-based UV-filter.

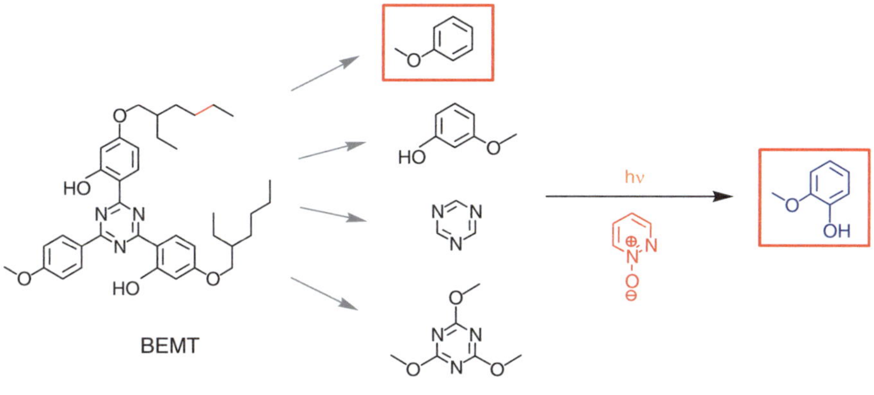

Fig. 11.11 Model fragments study on solar filter oxygenation by steady-state photolysis

Fig. 11.12 Mechanism for the oxygen transfer process between donor and acceptor moieties

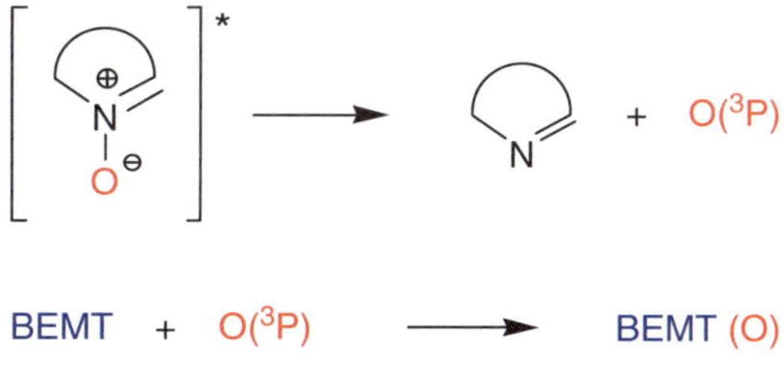

In conclusion, the potential relevance of the results could be associated with the considerable interest of the problem from a fundamental point of view, but also with the practical applicability to minimize the photocarcinogenic and photomutagenic potential of solar radiation. This photobiological risk constitutes currently a public health concern, due to its consequences such as ageing or skin cancer. Thus, the design of a new photoprotection strategy has been envisaged based on the development of sunscreens to minimize light absorption by the photosensitizers and the use of scavengers to eliminate the reactive oxygen species arising therefrom, taking advantage from the use of photochemical and photophysical techniques. As a specific example, the mode of action of a photoactive compound can be tuned by knowledge-based molecular engineering, in order to play a dual role as an antioxidant-based UV-filter and achieve photoprotective ability. In addition to result in an increased photosafety, advances of this type may lead to novel technological applications in pharmaceutical and skin protection industries.

Acknowledgements The author acknowledges M.A. Miranda and N. Duran-Giner for working in this project.

References

1. McKenzie RL, Aucamp PJ, Bais AF, Bjorn LO, Ilyas M, Madronich S (2011) Ozone depletion and climate change: impacts on UV radiation. Photochem Photobiol Sci 10:182–198
2. Palm MD, O'Donoghue MN (2007) Update on photoprotection. Dermatol Ther 20:360–376
3. Passaglia Schuch A, Machado Garcia CC, Makita K, Martins Menck CF (2013) DNA damage as a biological sensor for environmental sunlight. Photochem Photobiol Sci 12:1259–1272
4. Miranda MA (2001) Photosensitization by drugs. Pure Appl Chem 73:481–486
5. Cadet J, Mouret S, Ravanat JL, Douki T (2012) Photoinduced damage to cellular DNA: direct and photosensitized reactions. Photochem Photobiol 88:1048–1065
6. Pfeifer GP, Besaratinia A (2012) UV wavelength-dependent DNA damage and human non-melanoma and melanoma skin cancer. Photochem Photobiol Sci 11:90–97
7. Sage E, Girard PM, Francesconi S (2012) Unravelling UVA-induced mutagenesis. Photochem Photobiol Sci 11:74–80
8. Banyasz A, Vayá I, Changenet-Barret P, Gustavsson T, Douki T, Markovitsi D (2011) Base pairing enhances fluorescence and favors cyclobutane dimer formation induced upon absorption of UVA radiation by DNA. J Am Chem Soc 133:5163–5165
9. Mouret S, Baudouin C, Charveron M, Favier A, Cadet J, Douki T (2006) Cyclobutane pyrimidine dimers are predominant DNA lesions in whole human skin exposed to UVA radiation. Proc Natl Acad Sci U S A 103:13765–13770
10. Foote CS (1991) Definition of type I and type II photosensitized oxidation. Photochem Photobiol 54:659–660
11. Cadet J, Douki T, Ravanat JL, Di Mascio P (2009) Sensitized formation of oxidatively generated damage to cellular DNA by UVA radiation. Photochem Photobiol Sci 8:903–911
12. Wauchope OR, Shakya S, Sawwan N, Liebman JF, Greer A (2007) Photocleavage of plasmid DNA by dibenzothiophene S-oxide under anaerobic conditions. J Sulfur Chem 28:11–16
13. Orrego JF, Truong TN, Mondragon F (2008) A linear energy relationship between activation energy and absolute hardness: a case study with the $O(^3P)$ atom-addition reactions to polyaromatic hydrocarbons. J Phys Chem A 112:8205–8207

14. Zhang M, Ravilious GE, Hicks LM, Jez JM, McCulla RD (2012) Redox switching of adenosine-5′-phospho-sulfate kinase with photoactivatable atomic oxygen precursors. J Am Chem Soc 134:16979–16982
15. Damiani E, Astolfi P, Greci L (2008) Nitroxide-based UV-filters: a new strategy against UV-damage? Househ Pers Care Today 2:20–23
16. Herzog B, Wehrle M, Quass K (2009) Photostability of UV absorber systems in sunscreens. Photochem Photobiol 85:869–878
17. Sambandan DR, Ratner D (2011) Sunscreens: an overview and update. J Am Acad Dermatol 64:748–758
18. Fourtanier A, Moyal D, Seite S (2012) UVA filters in sun-protection products: regulatory and biological aspects. Photochem Photobiol Sci 11:81–89
19. Jansen R, Osterwalder U, Wang SQ, Burnett M, Lim HW (2013) Photoprotection Part II. Sunscreen: development, efficacy and controversies. J Am Acad Dermatol 69:1–14
20. Moyal DD, Fourtanier AM (2008) Broad-spectrum sunscreens provide better protection from solar ultraviolet-simulated radiation and natural sunlight-induced immunosuppression in human beings. J Am Acad Dermatol 58:149–154
21. Chatelain E, Gabard B (2001) Photostabilization of butyl methoxydibenzoylmethane (Avobenzone) and ethylhexyl methoxycinnamate by bis-ethylhexyloxyphenol methoxyphenyl triazine (Tinosorb S), a new UV broadband filter. Photochem Photobiol 74:401–406
22. Serpone N, Dondi D, Albini A (2007) Inorganic and organic UV filters: their role and efficacy in sunscreens and suncare products. Inorg Chim Acta 360:794–802
23. Paris C, Lhiaubet-Vallet V, Jimenez O, Trullas C, Miranda MA (2009) A blocked diketo form of avobenzone: photostability, photosensitizing properties and triplet quenching by a triazine-derived UVB-filter. Photochem Photobiol 85:178–184
24. Maier T, Korting HC (2005) Sunscreens-Which and what for? Skin Pharmacol Physiol 18:253–262
25. Inbaraj JJ, Bilski P, Chignell CF (2002) Photophysical and photochemical studies of 2-phenylbenzimidazole and UVB sunscreen 2-phenylbenzimidazole-5-sulfonic acid. Photochem Photobiol 75:107–116
26. Bastien N, Millau JF, Rouabhia M, Davies RJH, Drouin R (2010) The sunscreen agent 2-phenylbenzimidazole-5-sulfonic acid photosensitizes the formation of oxidized guanines *in cellulo* after UV-A or UV-B exposure. J Invest Dermatol 130:2463–2471
27. Lhiaubet-Vallet V, Marin M, Jimenez O, Gorchs O, Trullas C, Miranda MA (2010) Filter-filter interactions. Photostabilization, triplet quenching and reactivity with singlet oxygen. Photochem Photobiol Sci 9:552–558
28. McGarry PF, Jockusch S, Fujiwara Y, Kaprinidis NA, Turro NJ (1997) DMSO solvent induced photochemistry in highly photostable compounds. The role of intermolecular hydrogen bonding. J Phys Chem A 101:764–767
29. Couteau C, Paparis E, Chauvet C, Coiffard L (2015) Tris-biphenyl triazine, a new ultraviolet filter studied in terms of photoprotective efficacy. Int J Pharm 487:120–123
30. Venditti E, Brugè F, Astolfi P, Kochevar I, Damiani E (2011) Nitroxides and a nitroxide-based UV filter have the potential to photoprotect UVA-irradiated human skin fibroblasts against oxidative damage. J Dermatol Sci 63:55–61
31. Afonso S, Horita K, Sousa e Silva JP, Almeida IF, Amaral MH, Lobao PA, Costa PC, Miranda MS, Esteves da Silva JCG, Sousa Lobo JM (2014) Photodegradation of avobenzone: stabilization effect of antioxidants. J Photochem Photobiol B Biol 140:36–40
32. Kullavanijaya P, Lim HW (2005) Photoprotection. J Am Acad Dermatol 52:937–958
33. Gaspar LR, Tharmann J, Maia Campos PMBG, Liebsch M (2013) Skin phototoxicity of cosmetic formulations containing photounstable and photostable UV-filters and vitamin A palmitate. Toxicol in Vitro 27:418–425
34. Duran-Giner N, Encinas S, Miranda MA (2013) Solar filters as feasible acceptors of atomic oxygen. Photochem Photobiol Sci 12:725–728

35. Albini A, Alpegiani M (1984) The photochemistry of the N-oxide function. Chem Rev 84:43–71
36. Bucher G, Scaiano JC (1994) Laser flash photolysis of pyridine N-oxide: kinetic studies of atomic oxygen [O(^{3}P)] in solution. J Phys Chem 98:12471–12473
37. Ogawa Y, Iwasaki S, Okuda S (1981) A study on the transition state in the photooxygenations by aromatic amine N-oxides. Tetrahedron Lett 22:2277–2280

Chapter 12
Luminescent Chemosensors: From Molecules to Nanostructures

Nelsi Zaccheroni, Francesco Palomba, and Enrico Rampazzo

Abstract Chemical analysis has been improved by the advent of sensory devices based on chemosensors that are able to transform chemical information (composition, concentration, etc.) into an analytically useful signal. These devices, in fact, find applications in many different areas of great social and economic impact. Among all the possible chemosensors, we focus here our attention on luminescent ones as they present many advantages such as sensitiveness, low cost, ease of use and versatility. After a discussion on their pros and cons and of the most important features in the design of new species with customized properties, we present selected examples of different chemosensors that take advantage from various transduction mechanisms always following a supramolecular approach. A step forward in the research and application has been done implementing these structures in nanosized materials to obtain powerful and versatile platforms for addressing crucial issues in sensing, imaging, and molecular testing. In this context we have chosen to restrict the discussion on luminescent multichromophoric silica-based nanoprobes as a virtuous example of how a high design versatility allows the preparation of nanostructures where modulation and multifunctionality offer the possibility to induce collective energy- and electron-transfer processes, that are at the base of signal amplification effects.

12.1 Introduction

The sensing process, in a general acceptation, exploits one or more chemical-physical phenomena to inform about the external environment (sensing domain), and converts the stimulus of the sensed phenomenon/species into a signal or data stream that can be understood and manipulate.

To obtain the detection of the target analyte, therefore, two different processes are needed: molecular recognition and signal transduction. This means that the starting point is always the design of single molecules, or of arrays of molecules,

---

N. Zaccheroni (✉) • F. Palomba • E. Rampazzo
Dipartimento di Chimica "Giacomo Ciamician", Università di Bologna, via Selmi 2, Bologna 40126, Italy
e-mail: nelsi.zaccheroni@unibo.it

G. Bergamini, S. Silvi (eds.), *Applied Photochemistry*,
Lecture Notes in Chemistry 92, DOI 10.1007/978-3-319-31671-0_12

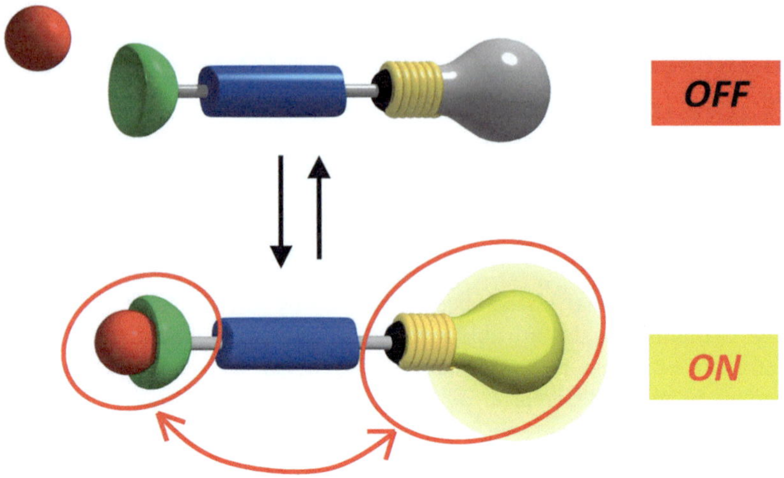

Fig. 12.1 Schematic representation of a luminescent chemosensor

that can specifically recognize an ion or a chemical species in a reversible manner (not always) and in a given concentration range. Molecules of abiotic origin of this kind are called *chemosensors* and, following a supramolecular approach, they can usually be schematized as made of different components (Fig. 12.1): (i) a receptor (responsible for the selective analyte binding), (ii) an active unity (whose properties should change upon recognition) and, in some cases, (iii) a spacer that can change the geometry of the system and tune the electronic interaction between the two former moieties [1].

A valuable feature of chemosensors is that they allow to monitor the analyte presence and concentration fluctuations in real-time and real space. Moreover, in recent times, single analyte sensing has been paralleled by a new kind of systems that are able to detect classes or mixtures of chemicals in a similar manner to which nature has developed human taste or smell.

In this complex framework there are many possibilities to design a chemosensor and the merging of luminescence with the field has brought about so many advantages that the area has boost out and hundreds of species have been proposed and discussed. This interest has been driven by many useful characteristics typical of fluorescence measurements such as their sensitiveness, low cost, ease in be performed, and versatility, with the opportunity to offer sub-micron visualisation and sub-millisecond temporal resolution. Moreover, luminescence can be stimulated or generated, by light energy, e.g. by direct excitation, or via energy or electron transfer processes in donor–acceptor arrays, electrochemically or by chemical inputs, such as ions or neutral molecules. Besides, the photophysical properties of a luminophore can also be modulated by introducing proton-, energy- and electron-transfer processes, or heavy-atoms, by changing the electronic density, or destabilising non emissive $n\pi^*$ excited states. It is not surprising, therefore, how luminescent chemical sensors, presenting such a versatility, play a major role in key fields such as industry, diagnostic and therapeutic medicine, and various kinds of environmental monitoring.

In the case of luminescent chemosensors the need for real-time and in situ measurements of many applications can take advantage of the development of optical fibers. This technology, in fact, has extended the possibility of performing remote measurements by monitoring changes in the absorption or luminescence spectra of chromogenic or fluorescent compounds immobilized on polymer matrix on the tip of a fibre. Microsized optical fibres, therefore, can allow analysis at almost any location in real-time and a tremendous effort has been done for the development of optical sensors (optode) for many different analytes [2, 3].

Another very important feature of chemosensors is typically *reversibility*: this is for sure an essential requirement for continuous or in vivo monitoring, while in most of the on-off cases measurements it's not necessary or at least not a stringent need.

In spite of their applications in the most different fields to detect a variety of analytes, these species share other common critical properties, besides reversibility, to ensure an ideal detection performance.

First of all the receptor must present a suitable selectivity for the target analyte where *selectivity* is defined as the ability of the host to significantly stronger bind a guest (or a class of guests) with respect to all the other species. Selectivity is quantified in terms of ratios among the different values of the equilibrium constants. It can be determined by many factors such as: (i) complementarity of the different host guest sites (receptor and analyte must present mutual spatially and electronically complementary binding sites); (ii) pre-organization in the receptor conformation (when it requires no significant conformational change to bind the analyte of interest); (iii) co-operativity of different ligand functions (when two or more binding sites act in a concerted fashion to produce a combined interaction that is stronger than when the binding sites act independently of each other). It has also to be mentioned that cooperativity is a more general but analogous concept of the chelating effect in coordination chemistry: multi-dentate ligands result in more stable complexes than comparable systems containing multiple uni-dentate ligands, as a result they generally present higher selectivity [4].

The association process must be fast and the chemosensor chemically stable in the work conditions to allow long term monitoring. Moreover, when the analysis is in situ, as already mentioned, reversibility of the association process is necessary to avoid the saturation of the binding sites and the drastic modification of the system in analysis.

The core event is, however, the recognition of the analyte and its signalling via a variation of the photophysical properties of the system, it has to be highlighted that there are many possible mechanisms that enable this as they will be discussed more in detail in the following section. Here, we would only anticipate that not all the variations in luminescence can be considered equivalent from the detection point of view since not all the output signals allow to the same effectiveness, given the same affinity and selectivity of the binding event. For example an "off-on" fluorescent chemosensor can in principle reach lower detection limits than an "on-off" one since the quantification of a light intensity versus the dark is easier and more accurate than the measure of a less intense light versus a more intense one. Pushing further these considerations, an even more valuable structure would be the one in which the analyte recognition would induce the concomitant variation of two (or more) emission signals instead of only one. This is the strategy at the bases of the so called ratiometric luminescent chemosensors, which have the potential to provide more exact and reliable analysis since they are less affected by interferences on the measurement caused by instrumental, environmental and stability factors [5].

As a last consideration, one has always to keep in mind the kind of analysis for which the system has been designed and that many environments introduce additional constrains, as it is the case for bio-medical applications.

To perform intracellular detection, for example, together with all the features typically required for a good performance of a luminescence chemosensor discussed above, it becomes fundamental also the water solubility in physiological conditions, in cell culture media and in body fluids, non-toxicity and suitable cell interaction and internalization. In order to be ideal the probe should also allow precise compartmentalization, without perturbing the equilibrium of the cell itself.

In case of an in vivo analysis other further characteristics become necessary such as: (i) reversibility of the binding event in order to preserve the analyte availability (as already cited above); (ii) convenient time of clearance to avoid toxicity and interference effects; (iii) chemosensors presenting both absorption and emission in the near infrared region (NIR), the so called in vivo transparency window, the only range where light can penetrate deep enough in the tissues.

This last point is of particular importance since the number of known fluorophores is really large, however, very few present a bright emission in the NIR in water. The "window" of increased optical transparency ranges from 650 to 1800 nm but it has to be mentioned that two water absorption peaks (centred around 980 nm and 1450 nm) also fall in this spectral range, therefore it has to be carefully evaluated the possible non-negligible and harmful local heating of biological samples when exciting via laser light in this window. Nevertheless, the use of NIR light absorption and emission ensures a low auto-fluorescence and tissue damage – a possible concern when long-term UV excitation is required – and reduced light scattering.

With the aim of clearness and to help the reader to approach the following sections, in Table 12.1 we have summarized the fundamental features to take into account while designing a luminescent chemosensor that has to be used in a solution, versus systems for in vitro and in vivo applications.

Table 12.1 Summary of fundamental features of a luminescent chemosensor to be used in solution, versus systems for in vitro and in vivo applications

Luminescent chemosensors in water solutions	Luminescent chemosensors for in vitro applications	Luminescent chemosensors for in vivo applications
Water solubility	Water solubility and cell permeability	Water solubility
Target selectivity	Organelle/structure selectivity and target selectivity within the cell	Tissue/organ selectivity and target selectivity
Low detection limit	Low toxicity	Low toxicity
Off-on or ratiometric behaviour (preferable)	Off-on or ratiometric behav-iour (preferable)	Off-on or ratiometric behaviour (preferable)
Photochemical stability	Photochemical stability	Photochemical stability
Good photophysical properties (Φ, ε, Stokes shift)	Good photophysical properties (Φ, ε, Stokes shift) within the cell	Good photophysical properties (Φ, ε, Stokes shift) in physiological environment
	Visible absorption/emission properties	Absorption/emission properties in the red and/or NIR range for penetration

12.2 The Supramolecular Approach in Luminescent Chemosensors

The possible analytes of interest are very different and in great number, among them special attention is often devoted to the development of chemosensors for metal cations, due to their role both beneficial and detrimental in medicine, biochemistry and environmental monitoring. In spite of this, only in the late 1970s, after the development of the supramolecular chemistry and the pioneering work of Pedersen, Cram and Lehn, a systematic research started on the synthesis of abiotic metal ions receptors.

In this contest appeared what we now consider the traditional luminescent chemosensors, that integrate the ion-receptor recognition in systems containing a moiety able to give a binary answer (off-on or on-off) in luminescence, depending on the binding event. This process is usually based on photoinduced electron transfer (PET) occurrence, a mechanism studied since long [6, 7], deeply exploited in this field by professor A. P. De Silva and other groups and then worldwide adopted in sensing processes [8–11]. To this goal, a luminophore and a receptor must be weakly electronically coupled to allow them to maintain their own photophysical and chemical characteristics. When the excited state of the luminescent moiety has a larger energy of the sum of the oxidation and reduction potentials of the receptor and emitter respectively, the electron transfer can occur from the free recognising moiety to the luminescent one. The target binding to the system causes a significant change in the potential of binding moiety preventing the PET with a consequent luminescence recovery [10]. These systems are normally obtained with a fluorophore-linker-receptor architecture, and can be efficiently designed using tools such as electrochemical experiments and orbital energy calculations. The PET mechanism inside a sensor was quite recently demonstrated using a BODIPY linked to the famous Ca^{2+} receptor proposed by Tsien [12].

However, this is not the only mechanism that can be exploited since also fluorescence resonance energy transfer (FRET) [11, 13, 14] or through-bond energy transfer (TBET) [13], internal charge transfer (ICT) [11, 14] and excited state intramolecular proton transfer (ESIPT) can be as much effective as PET [9].

Both energy transfer mechanisms involve two fluorophores with suitable characteristics (significant overlap between the donor emission spectra and the acceptor absorption one) to act as a donor and acceptor couple: the donor emitting at lower wavelengths is able to transfer its energy to the acceptor inducing its emission at longer wavelength. In the case of FRET systems the two units communicate through space while in the TBET the two moieties are connected by electronically conjugate bonds and the overlap constrains are less severe. In the first example the binding event induces a variation in the distance and or the orientation of the couple modulating the efficiency of the energy transfer and therefore signalling the presence of the targeting species. In TBET the presence of the target analyte disrupts or activates the bond conjugation resulting again in a drastic change of the efficiency of the process. ICT systems are instead based on the presence of an electron donating part and an electron accepting one in the luminescent system that has a "push-pull" character of the excited emitting state. In this case when a cation interacts with one of the two parts it modifies their electron-accepting or donating character inducing changes in the absorption spectrum and in the emission quantum yield.

Even if lots of chemosensor have been already developed taking advantage from all these transduction mechanisms, even more valuable systems can be obtained when allowing ratiometric measurements. In this case the analyte recognition induces the concomitant variation of two (or more) emission signals instead of only one and this gives a measurement much less affected by interferences caused by instrumental, environmental and stability factors [5].

The supramolecular approach to sensing applications can be however pushed much further merging the traditional host-guest chemistry with the molecular assembly making a step forward toward the enlarged concept of stimuli induced (cation recognition) formation of supramolecular structures [15]. Since many years it is well known the self-assembling ability of flat polyaromatic fluorophores that give rise to $\pi-\pi$ stacking when in close proximity originating excimeric emission [14]. The excimer complex is, in fact, formed when one excited fluorophore interacts with another one of the same nature but in the ground state, giving rise to a new species that presents different photophysical characteristics from the monomeric one. In particular its emission is generally a broad non-structured band and at lower energy in comparison with the emission of the monomeric, non-complexed structure. When the dimerization or aggregation is induced by the analyte recognition a luminescent ratiometric chemosensor can be obtained. Among the aromatic structures that can be used for excimer-based analyte detection, pyrene is probably the most versatile and simple molecule. Many chemosensors take advantage from the photophysical properties of pyrene to detect ions [14, 16, 17] or molecular species, such as ATP [18], nucleic acids [19, 20] or drugs such as ecstasy (Fig. 12.2) [21].

Also other approaches were investigated and exploited to obtain supramolecular aggregates as, for example, metal-metal interactions (such as aurophilicity) or other non-covalent ones, that can give rise to aggregation-induced emission (AIE). In this systems the aggregation, induced by the analyte (ion-nucleobase, ion- π bond, etc.), prevents interactions such as π–π stacking or internal rotational motions deactivating non-radiative decay pathways.

In the same framework, another signal transduction mechanism has been widely studied: the displacement of a dye from a macrocyclic binding site by an analyte – when this causes a significant change in the dye luminescence (switch on or off) – it is the basic principle of indicator displacement assays (IDA) [22]. This purely supramolecular approach has been extensively used to determine absolute analyte concentrations, primarily of anion targets (Fig. 12.3) [22, 23], but also of cations, small neutral molecules, protein and to carry out chiral sensing [24]. Moreover it can also be exploited to track – in real time – changes in concentrations of species of interest during membrane transport or in enzymatic reactions [25].

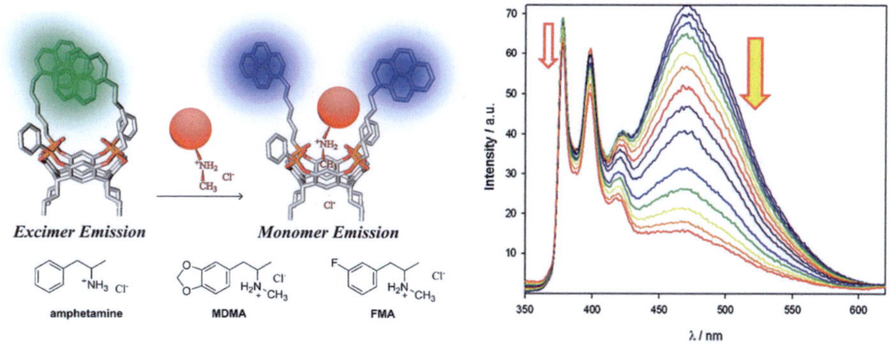

Fig. 12.2 An example of fluorescent chemosensor exploiting pyrene monomer-excimer emission to detect selectively the presence of amphetamine-type-stimulants such as ecstasy (MDMA), developed by L. Prodi et al. Ref. [21]. The figure shows the proposed recognition mode, the chemical structures of the synthesized cavitand receptor and the tested drugs. Fluorescence spectra (λexc = 330 nm) of a dispersion of the chemosensor (1 μM) and of core-shell silica-PEG nanoparticles in water upon addition of increasing amounts of MDMA hydrochlorides (0–2.5 mM) is reported on the *right* (Reproduced from Ref. [21] with permission from The Royal Society of Chemistry)

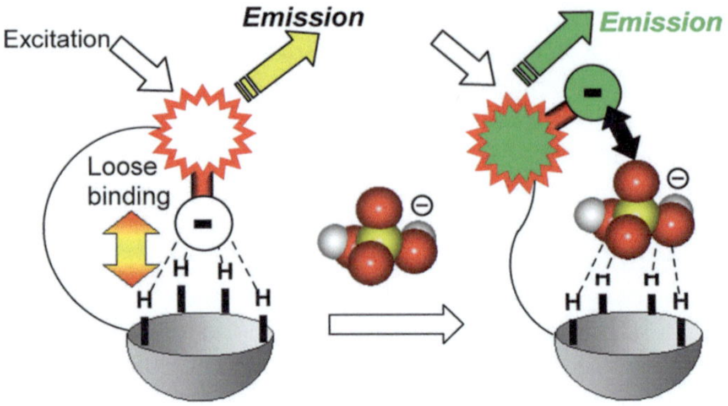

Fig 12.3 Schematization of the indicator displacement assays (IDA) approach for a fluorescent chemosensor (the sensing of an anion, the phosphonate herbicide glyphosate is reported) (Reprinted with permission from Ref. [22]. Copyright 2014 American Chemical Society)

12.3 The Targets

The more interesting examples of fluorescent sensors are developed for biological or environmental applications. They can be designed to monitor several targets, such as protons [26, 27], metal ions [5, 26, 27], anions [5, 28, 29], thiols [30] or to indicate physicochemical properties such as viscosity [31] or membrane potential [32] at organelles or cellular level [33].

From the point of view of the chemosensor architecture, proton is probably the most accessible target, since selectivity can be obtained using a single Brönsted base as a receptor, moreover, protonation can dramatically affect the efficiency of the sensing mechanism (mainly PET) [10]. The importance of proton recognition is well established since proton concentration changes can induce serious health problems (relevant pH deviations have been observed in cancer tissues or in patients with neurodegenerative diseases) and serious environmental concerns (acid rains, as an example) [26, 27].

As already mentioned, among the most studied classes of analytes there are also alkali, alkaline-earth and transition metal ions. Group I and II cations have been chosen during the development of the first example of cation responsive fluorescent sensors, because of the availability of well-established and selective crown ethers or chelating receptors. Systems suitable for practical bio-medical applications were developed to sense K^{+} [34], Ca^{2+} [35], and Mg^{2+} [36] at biological level, since these cations are fundamental in cell metabolism. Many research groups are working in the field and we have contributed to studying a family of fluorescent molecules – diaza-18-crown-6 8-hydroxyquinolines derivatives (DCHQ) – that show a remarkable affinity and specificity for magnesium, higher than all the commercially available probes, thus permitting the detection of the total intracellular magnesium [37]. In particular, the increased conjugation of one component of this family (DCHQ5) took to the most promising characteristics, such as a high fluorescence

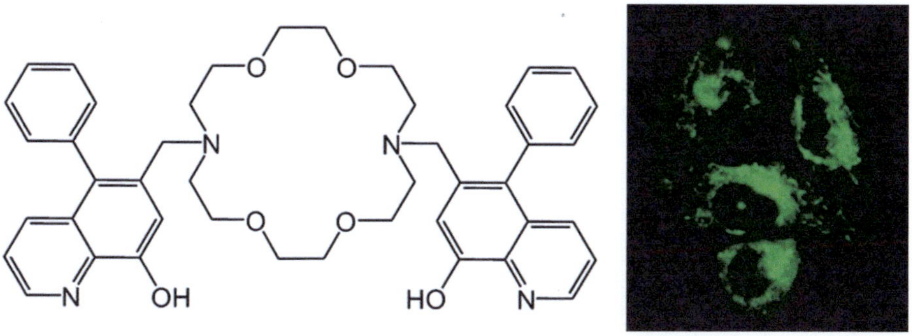

Fig. 12.4 Chemical structure of the diaza-18-crown-6 8-hydroxyquinolines derivative DCHQ5, a highly selective fluorescent chemosensor for Mg^{2+} in cells (*left*). Two-photon fluorescence microscopy images (λexc = 750 nm) of rat osteosarcoma cells stained with [DCHQ5] = 10 mM (*right*) (Reproduced from Ref. [37] with permission from The Royal Society of Chemistry

response location binding, membrane staining and the possibility to be excited both in the UV and visible spectrum (Fig. 12.4).

On the other side, the high toxicity and environmental impact of heavy metals is well known, and this pushed a growing interest in their recognition and quantification, together with the rising awareness of their fundamental role in disease pathologies that are accompanied by metal ion imbalance [38]. Metal homeostasis is apparently affected by diseases resulting in abnormal accumulation of transition metal ions, such as copper [39] and zinc [40], for example in cancer tissues or the brain of subjects presenting neurodegenerative pathologies [38]. It has also to be taken into account that the various oxidation states of the same metal can influence in a different way cell equilibriums. This is, for example, the case of Cu(I) and Cu (II) and a chemical sensor has to be selective over other competitors and specific not only for the metal but also for its oxidation state [39, 41]. Therefore, advances in intracellular quantitative detection, compartimentalization and dynamics of metal ions, that are still mostly unclear, could take to the full understanding of successive illness states. Moreover, some metal ions are present in drugs, for example platinum based ones are central in oncology treatments, and their quantitative revealing would allow pharmacokinetic studies, distribution determination and trafficking [42].

We would like to stress here again that to make intracellular measurements, together with all the features typically required for a good performance of a luminescent chemosensors, other fundamental characteristics are needed (see Table 12.1) such as suitable cell interaction and internalization and, ideally, precise compartmentalization, without perturbing the equilibrium of the cell itself, since the risk of depleting cellular ion pools in the process has been, for example, recently reported [43–45]. Moreover, since the same metal ion in different cells or in different stages of a disease covers a very wide range of concentrations, it is unlikely that one single species presents a suitable fluorescence variation over all the range. For this reason it is often necessary, or preferred, to exploit the synergic action of a group of chemosensors with complementary performances.

It has to be underlined, that the recognition of cations finds its first example in a time that now we can consider almost remote and it is, at least in principle, more accessible than the one of anions, due to their common spherical-like form. The pioneering work on anion chemosensing dates back "only" a couple of decades and it needs different approaches due to the different nature, shape and dimension that anions can have. Traditionally three main approaches have been used to recognize these species: (i) hydrogen bonding; (ii) displacement assays and (iii) chemical reactions generating new species [29]. For their nature anions are hydrated in water (i.e. hindered and already coordinated) and this makes an even more difficult task to detect them in this solvent for medical and environmental applications. On the other side, the bio-imaging of anionic species is gaining more and more interest since many different anions, being involved in most intracellular functions, are of biological significance (F^-, CN^-, I^-; HCO_3^-, ClO^-, HS^- etc.) [28–30].

As a last point, we would like to mention that also the detection of molecules is of great interest such as the colourless, highly toxic hydrogen sulphide [30, 46], synthetic psychoactive compounds that give addictiveness like ecstasy [21], or explosives at trace or ultratrace levels [47], an unfortunately more and more critical challenging issue in these days of political international instability.

12.4 Nanostructured Luminescent Chemosensors

The discussions in the previous sections clearly show that it is not a trivial task to develop efficient and reliable detection systems. For example, the steadiness toward environmental fluctuations, and an high photo-chemical stability are often difficult to achieve in molecular chemosensors. The advent of nanotechnology – for its own nature a multidisciplinary field – has opened up the way to merge different scientific knowledge and new materials, with increasing potentialities to overcome many of the typical limitations of conventional systems. In particular, take advantage of a nanotechnological approach to luminescent signalling can lead to the creation of unique species that could induce great improvements in the technical development of sensing in many areas. Among all new nanomaterials proposed in the last decades, nanoparticles [48] are the most versatile and exploited ones, providing a protected environment for active species from undesirable external triggers and unique features arising from their particular size-dependent opto-electronic properties. There are many classes of nanoparticles [25] such as metallic [49, 50], organic [51, 52], and inorganic ones [53–55], each presenting pros and cons for each specific application. These nanostructures have been implemented via self-organizing synthetic strategies that, even requiring quite moderate synthetic efforts, yield materials able to perform elaborate functions. Nanoparticles have been designed to collect molecular level information about biologically relevant processes through luminescence-related techniques [48], overcoming many of the typical limitations of conventional luminophores (organic dyes) such as poor photostability, low quantum yield, and aggregation in physiological conditions.

In this very complex scenario we will try to give some hints on how the evolution from molecular to nanostructured systems can boost the field of sensing but we will focus our discussion here only on some examples of the development of luminescent (dye-doped) silica nanoparticles (DDSNs) [56]. Two are the reasons of this choice, first of all it is impossible to discuss all the strategies explored in these last years in a single chapter and second, in our opinion, DDSNs are particularly promising and they will play a more and more valuable role in sensing especially in biomedical applications [57]. This can be attributed to many favourable features such as their synthetic versatility, that allows the design of NPs with customized properties for each application, a very important point to limit the degree of nonspecific interactions and environment interference. In particular, crucially positive characteristics for their application are: (i) silica is photophysically inert and for this it is not involved in quenching or photodecomposition processes; (ii) it does not present intrinsic toxicity, even if nanotoxicity is in general difficult to rationalize because it depends by the specific particle and by the biological target; (iii) each silica nanoparticle can contain a large number of photochemically active species allowing bright luminescence and collective processes; [58] (iv) the silica matrix protects the active units segregated inside the nanoparticle, potentially increasing their photostability, fluorescence quantum yield and inferring water solubility [59].

Many scientific publications discuss these advantages of DDSNs that take to their convenient use as fluorescent biosensors [60, 61], and the versatile nature of these systems that allows for different design approaches. Since the photophysical properties of the material are conferred by the doping dyes, one has to take into account also their possible cross-interactions to understand the properties of the final objects, given the tiny dimension that force their proximity. In DDSNs both homo-energy transfer processes and hetero-energy transfer processes, when they are loaded with different dyes, can take place and with a proper structure design they can be intentionally avoided to yield ratiometric systems for quantitative measurements, or optimized to obtain efficient internal FRET yielding high fluorescence intensity and wide absorption with multiple emission colours for applications in multiplex bioassays [58, 62, 63]. Energy transfer processes can therefore induce, in luminescent silica nanoparticles, very valuable collective properties yielding species that present (i) antenna effects – enhanced light-sensitivity obtained by an increase in the overall cross-section for light absorption; [64] (ii) spectral sensitization and (iii) light-energy up-conversion (anti-Stokes luminescence) [65]. These processes can be also designed to obtain remarkable Stokes shifts [63, 66], that typically allow a dramatic increase of the signal-to-noise ratio for more sensitive and precise photoluminescence measurements.

All these classes of systems can be exploited to obtain efficient nanostructured chemosensors, for example the antenna effect gives signal amplification [64, 67] that leads to a large increase of the sensitivity, and, as a consequence, to lower detection limits. To this goal, it is of great importance to know how the different dyes are located and distributed inside the nanoparticle and there are two main possible approaches to the design of these innovative materials: (i) to insert the receptor and luminescent units already covalently linked together to form a chemosensor, or (ii) as separate moieties and use the forced vicinity induced by the nanosize of the particles to obtain their communication [68, 69]. This second approach is probably the most versatile, since it exploits the synthetic advantages offered by self-organization. Several examples have been proposed that follow this strategy and some of the first ones were studied by Tonellato et al., that used the self-organization of small receptors and fluorescent units on or within silica nanoparticles for the sensing of Cu(II) [68–70] or Pb(II) ions (Fig. 12.5) [71]. In addition, the great versatility of this "modular system" opens up also the possibility to adapt these materials for very different applications without requiring each time the design of the synthesis of the chemosensor from scratch.

Another example of nanostructurated off-on Zn(II) selective chemosensor was obtained by Montalti et al. [72] grafting on the suface of silica nanoparticles a derivative (6-methoxy-8-*p*-toluensulfonamide-quinoline) of the well-known Zinquin species. The organization of the chemosensors (in this case the signalling

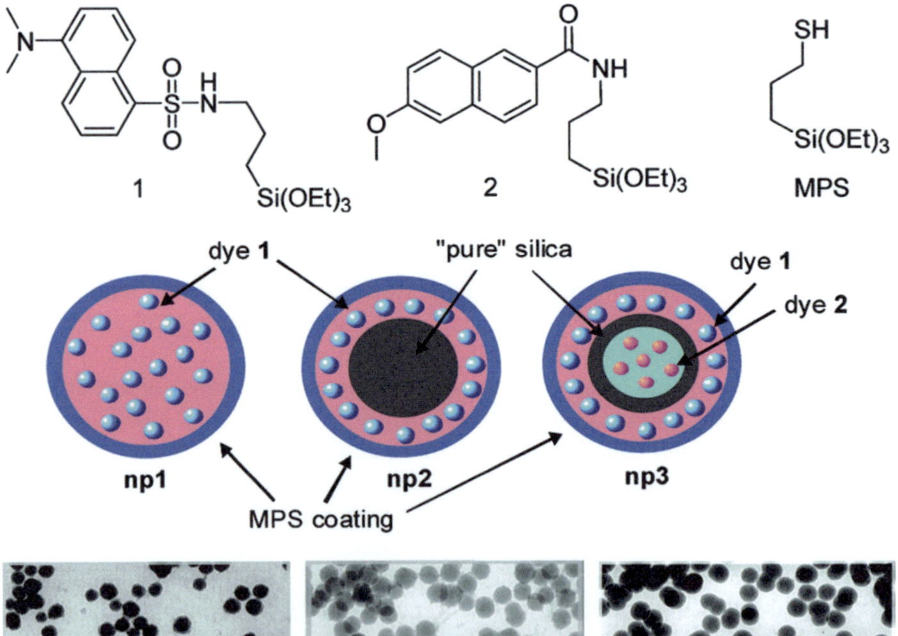

Fig. 12.5 Schematic representation of the main components and of the nanoparticles structures for a ratiometric fluorescent sensor for Pb(II) ions (Reprinted with permission from Ref. [71]. Copyright 2007 American Chemical Society)

and receptor units are are already covalently linked) onto the silica led to a synergic behaviour in the complexation with a large increase in the affinity – sometimes this enhances also selectivity towards the target analyte – thus improving the performance of the system. The receptors proximity and organization concur to confer also other more sophisticated new properties such as signal amplification that is related to the rise of FRET mechanisms (Fig. 12.6) [72].

The overall synthetic versatility of dye doped silica nanoparticles also allows for the possibility to realize onion-like multilayer structures that have been well investigated, and nanoparticles formed by a core, as many layers as desired, and an external modifiable surface (in case, one or more of these parts can be of different materials) have been prepared and characterized. A valuable example showing this versatility was proposed by Prodi et al. [64] (Fig. 12.7). A core-shell silica-PEG nanoparticle covalently doped with 7-(diethylamino)coumarin-3-carboxamide – acting as donor – were able to host a Cu(I) selective fluorescent

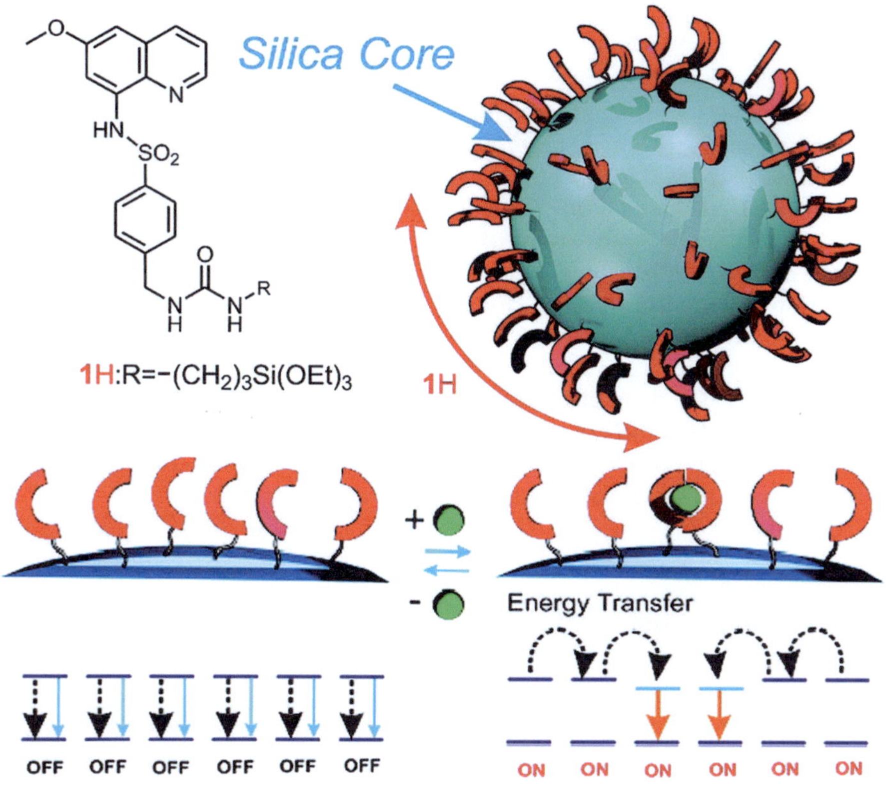

Fig. 12.6 A fluorescent chemosensor obtained with a densely grafted TSQ (6-methoxy-8-*p*-toluensulfonamide-quinoline) derivative onto the surface of preformed silica nanoparticles. The fluorescence transduction occurs in a region of the fluorophore network, that is able to transfer its excitation energy to the complexed units, with amplification of the fluorescence signal (Reprinted with permission from Ref. [72]. Copyright 2008 American Chemical Society)

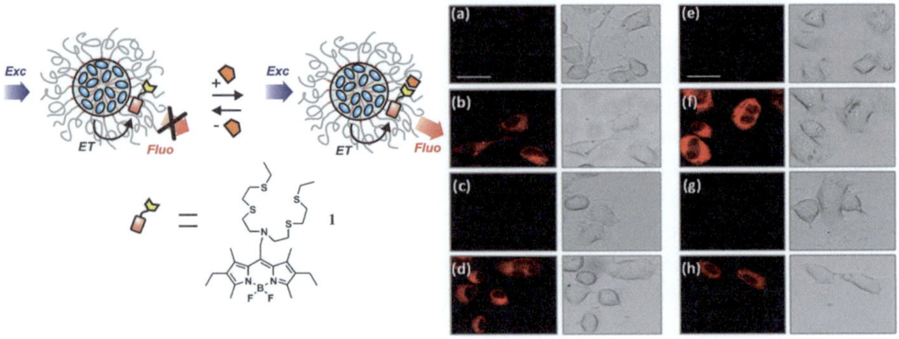

Fig. 12.7 Schematic representation of the proposed transduction mechanism for the Cu(I) off-on chemosensing ensamble. Confocal microscopy emission (λex = 543 nm; λem = 572 nm) and corresponding bright-field images for the SHSY5Y cell line: (**a–d**) cultured in DMEM medium for 24 h, and (**e–h**) supplemented with $CuCl_2$ (100 mm) for 5 h, incubated at 37 °C. (**a**, **e**) control PBS; (**b**, **f**) Cu^+ sensor **1** (5 mM); (**c**, **g**) nanoparticles (5 mM); (**d**, **h**) **1**@NP complex (5 mM; 1:1 molar ratio); scale bar: 30 mm (Adapted from Ref. [64]. Copyright 2011 WILEY-VCH Verlag GmbH & Co. KGaA, Weinheim)

chemosensor [41] – acting as acceptor – in the outer PEG shell. Thanks to a good energy overlap, a very efficient energy transfer process with the molecules hosted in the core was exploited to detect the presence of Cu(I) ions within SHSY5Y cells. The obtained results showed that this approach was straightforward to obtain nanoparticle based chemosensors using an "antenna" behaviour and it took to higher affinity and Stokes shifts and signal amplification.

The same strategy allowed also to obtain a similar system but with another goal, aiming to environmental and industrial applications. A fluorescent species based on the N2S2 pyridinophane macrocycle coupled with the 7-nitrobenzo[1, 2, 5] oxadiazole (NBD) fluorogenic fragment is able to selectively coordinate Pd (II) ions. Loaded on the external PEG shell of dye-doped silica nanoparticles, it can be used as a ratiometric fluorescent chemosensor for Pd(II) in water with high selectivity toward other metal ions including the platinum group ones [73].

12.5 Conclusions

In this chapter we have tried to give an overview of the field of luminescent chemosensors, species that already find applications in many fields and that still display a high potential. We have started from a discussion on their pros and cons and of the most important features to take in mind in designing new species with customerized properties. We have then presented selected examples of different chemosensors that take advantage from various transduction mechanisms always following a supramolecular approach. A very valuable characteristic of these specie, in fact, is the versatility that allows the detection of an astonishing number of possible targets via a proper engineering of the receptor moiety. In recent times, however, the already remarkable performance of molecular luminescent chemosensors has been even improved thanks to their implementation in nanosized materials that have yielded powerful and versatile platforms for addressing crucial issues in sensing, imaging, and molecular testing.

In this context we have again selected some examples trying to evidence the many advantages of luminescent multichromophoric silica-based nanoprobes. These materials are very promising offering remarkable stability and solubility in water even in physiological conditions, combined with low intrinsic toxicity. Their high design versatility also offers the possibility to obtain collective energy- and electron-transfer processes at the base of signal amplification effects. The great added value of silica-based nanosensors can be sketched as the combination of modulation and multifunctionality. This can be obtained following a rational and proper engineering, taking to the optimization of the performance of the whole system with the enhancement of sensitivity, reproducibility and affinity.

In our opinion, however, most of the potentialities of nanostructured luminescent chemosensors have not yet been fully developed and there is still room to further and unforeseen improvements.

References

1. Anthony WC (1993) Fluorescent chemosensors for ion and molecule recognition, vol 538, ACS Symposium Series. American Chemical Society, Washington, DC. doi:10.1021/bk-1993-0538
2. David RW, Christiane M, Ping Y, Shufang L, Steven B (1989) Design, preparation, and applications of fiber-optic chemical sensors for continuous monitoring. In: Chemical sensors and microinstrumentation. ACS Symposium Series, vol 403. American Chemical Society, pp 252–272. doi:10.1021/bk-1989-0403.ch017
3. Wencel D, Abel T, McDonagh C (2014) Optical chemical pH sensors. Anal Chem 86 (1):15–29. doi:10.1021/ac4035168
4. Steed JW, Turner DR, Wallace K (2007) Core concepts in supramolecular chemistry and nanochemistry. Wiley, Chichester
5. Lee MH, Kim JS, Sessler JL (2015) Small molecule-based ratiometric fluorescence probes for cations, anions, and biomolecules. Chem Soc Rev 44(13):4185–4191. doi:10.1039/c4cs00280f
6. Weller A (1968) Electron-transfer and complex formation in the excited state. Pure Appl Chem 16:115–123. doi:10.1351/pac196816010115
7. Marcus RA (1956) On the theory of oxidation-reduction reactions involving electron transfer. I. J Chem Phys 24(5):966–978. doi:http://dx.doi.org/10.1063/1.1742723
8. Richard AB, Silva APd, Gunaratne HQN, Lynch PLM, Colin PM, Glenn EMM, Sandanayake KRAS (1993) Fluorescent photoinduced electron-transfer sensors. In: Fluorescent chemosensors for ion and molecule recognition. ACS Symposium Series, vol 538. American Chemical Society, p 45–58. doi:10.1021/bk-1993-0538.ch004
9. de Silva AP (2011) Luminescent Photoinduced Electron Transfer (PET) molecules for sensing and logic operations. J Phys Chem Lett 2(22):2865–2871. doi:10.1021/jz201311p
10. Daly B, Ling J, de Silva AP (2015) Current developments in fluorescent PET (photoinduced electron transfer) sensors and switches. Chem Soc Rev 44(13):4203–4211. doi:10.1039/c4cs00334a

11. de Silva AP, Gunaratne HQN, Gunnlaugsson T, Huxley AJM, McCoy CP, Rademacher JT, Rice TE (1997) Signaling recognition events with fluorescent sensors and switches. Chem Rev 97(5):1515–1566. doi:10.1021/cr960386p
12. Batat P, Vives G, Bofinger R, Chang R-W, Kauffmann B, Oda R, Jonusauskas G, McClenaghan ND (2012) Dynamics of ion-regulated photoinduced electron transfer in BODIPY-BAPTA conjugates. Photochem Photobiol Sci 11(11):1666–1674. doi:10.1039/c2pp25130b
13. Fan J, Hu M, Zhan P, Peng X (2013) Energy transfer cassettes based on organic fluorophores: construction and applications in ratiometric sensing. Chem Soc Rev 42(1):29–43. doi:10.1039/c2cs35273g
14. Kim JS, Quang DT (2007) Calixarene-derived fluorescent probes. Chem Rev 107 (9):3780–3799. doi:10.1021/cr068046j
15. Yeung MC, Yam VW (2015) Luminescent cation sensors: from host-guest chemistry, supramolecular chemistry to reaction-based mechanisms. Chem Soc Rev 44(13):4192–4202. doi:10.1039/c4cs00391h
16. Yang R-H, Chan W-H, Lee AWM, Xia P-F, Zhang H-K (2003) A ratiometric fluorescent sensor for AgI with high selectivity and sensitivity. J Am Chem Soc 125(10):2884–2885. doi:10.1021/ja029253d
17. Kim SK, Lee SH, Lee JY, Lee JY, Bartsch RA, Kim JS (2004) An excimer-based, binuclear, on–off switchable calix[4]crown chemosensor. J Am Chem Soc 126(50):16499–16506. doi:10.1021/ja045689c
18. Xu Z, Singh NJ, Lim J, Pan J, Kim HN, Park S, Kim KS, Yoon J (2009) Unique sandwich stacking of pyrene-adenine-pyrene for selective and ratiometric fluorescent sensing of ATP at physiological pH. J Am Chem Soc 131(42):15528–15533. doi:10.1021/ja906855a
19. Ostergaard ME, Hrdlicka PJ (2011) Pyrene-functionalized oligonucleotides and locked nucleic acids (LNAs): tools for fundamental research, diagnostics, and nanotechnology. Chem Soc Rev 40(12):5771–5788. doi:10.1039/c1cs15014f
20. Wu J, Zou Y, Li C, Sicking W, Piantanida I, Yi T, Schmuck C (2012) A molecular peptide beacon for the ratiometric sensing of nucleic acids. J Am Chem Soc 134(4):1958–1961. doi:10.1021/ja2103845
21. Masseroni D, Biavardi E, Genovese D, Rampazzo E, Prodi L, Dalcanale E (2015) A fluorescent probe for ecstasy. Chem Commun 51(64):12799–12802. doi:10.1039/c5cc04760a
22. Minami T, Liu Y, Akdeniz A, Koutnik P, Esipenko NA, Nishiyabu R, Kubo Y, Anzenbacher P (2014) Intramolecular indicator displacement assay for anions: supramolecular sensor for glyphosate. J Am Chem Soc 136(32):11396–11401. doi:10.1021/ja504535q
23. Khatua S, Choi SH, Lee J, Kim K, Do Y, Churchill DG (2009) Aqueous fluorometric and colorimetric sensing of phosphate ions by a fluorescent dinuclear zinc complex. Inorg Chem 48 (7):2993–2999. doi:10.1021/ic8022387
24. You L, Zha D, Anslyn EV (2015) Recent advances in supramolecular analytical chemistry using optical sensing. Chem Rev 115(15):7840–7892. doi:10.1021/cr5005524
25. Ghale G, Nau WM (2014) Dynamically analyte-responsive macrocyclic host–fluorophore systems. Acc Chem Res 47(7):2150–2159. doi:10.1021/ar500116d
26. Hamilton GR, Sahoo SK, Kamila S, Singh N, Kaur N, Hyland BW, Callan JF (2015) Optical probes for the detection of protons, and alkali and alkaline earth metal cations. Chem Soc Rev 44(13):4415–4432. doi:10.1039/c4cs00365a
27. Yin J, Hu Y, Yoon J (2015) Fluorescent probes and bioimaging: alkali metals, alkaline earth metals and pH. Chem Soc Rev 44(14):4619–4644. doi:10.1039/c4cs00275j
28. Ashton TD, Jolliffe KA, Pfeffer FM (2015) Luminescent probes for the bioimaging of small anionic species in vitro and in vivo. Chem Soc Rev 44(14):4547–4595. doi:10.1039/c4cs00372a
29. Gale PA, Caltagirone C (2015) Anion sensing by small molecules and molecular ensembles. Chem Soc Rev 44(13):4212–4227. doi:10.1039/c4cs00179f

30. Guo Z, Chen G, Zeng G, Li Z, Chen A, Wang J, Jiang L (2015) Fluorescence chemosensors for hydrogen sulfide detection in biological systems. Analyst 140(6):1772–1786. doi:10.1039/c4an01909a
31. Liu T, Liu X, Spring DR, Qian X, Cui J, Xu Z (2014) Quantitatively mapping cellular viscosity with detailed organelle information via a designed PET fluorescent probe. Sci Rep 4:5418. doi:10.1038/srep05418
32. Miller EW, Lin JY, Frady EP, Steinbach PA, Kristan WB, Tsien RY (2012) Optically monitoring voltage in neurons by photo-induced electron transfer through molecular wires. Proc Natl Acad Sci 109(6):2114–2119. doi:10.1073/pnas.1120694109
33. Roger YT (1993) Fluorescent and photochemical probes of dynamic biochemical signals inside living cells. In: Fluorescent chemosensors for ion and molecule recognition. ACS Symposium Series, vol 538. American Chemical Society, pp 130–146. doi:10.1021/bk-1993-0538.ch009
34. Divakar M, Mariann EL (1993) Fluorescent chemosensors for monitoring potassium in blood and across biological membranes. In: Fluorescent chemosensors for ion and molecule recognition. ACS Symposium Series, vol 538. American Chemical Society, pp 162–182. doi: 10.1021/bk-1993-0538.ch011
35. Michael AK (1993) 1,2-Bis(2-aminophenoxy)ethane- < italic > N,N,N?,N?</italic>,-tetraacetic acid conjugates used to measure intracellular Ca < sup >2+</sup > concentration. In: Fluorescent chemosensors for ion and molecule recognition. ACS Symposium Series, vol 538. American Chemical Society, pp 147–161. doi:10.1021/bk-1993-0538.ch010
36. Farruggia G, Iotti S, Prodi L, Montalti M, Zaccheroni N, Savage PB, Trapani V, Sale P, Wolf FI (2006) 8-hydroxyquinoline derivatives as fluorescent sensors for magnesium in living cells. J Am Chem Soc 128(1):344–350. doi:10.1021/ja056523u
37. Marraccini C, Farruggia G, Lombardo M, Prodi L, Sgarzi M, Trapani V, Trombini C, Wolf FI, Zaccheroni N, Iotti S (2012) Diaza-18-crown-6 hydroxyquinoline derivatives as flexible tools for the assessment and imaging of total intracellular magnesium. Chem Sci 3(3):727–734. doi:10.1039/c1sc00751c
38. Qian X, Xu Z (2015) Fluorescence imaging of metal ions implicated in diseases. Chem Soc Rev 44(14):4487–4493. doi:10.1039/c4cs00292j
39. Cotruvo JA Jr, Aron AT, Ramos-Torres KM, Chang CJ (2015) Synthetic fluorescent probes for studying copper in biological systems. Chem Soc Rev 44(13):4400–4414. doi:10.1039/c4cs00346b
40. Chen Y, Bai Y, Han Z, He W, Guo Z (2015) Photoluminescence imaging of Zn(2+) in living systems. Chem Soc Rev 44(14):4517–4546. doi:10.1039/c5cs00005j
41. Zeng L, Miller EW, Pralle A, Isacoff EY, Chang CJ (2006) A selective turn-on fluorescent sensor for imaging copper in living cells. J Am Chem Soc 128(1):10–11. doi:10.1021/ja055064u
42. Tracey MP, Pham D, Koide K (2015) Fluorometric imaging methods for palladium and platinum and the use of palladium for imaging biomolecules. Chem Soc Rev 44(14):4769–4791. doi:10.1039/c4cs00323c
43. Burns A, Ow H, Wiesner U (2006) Fluorescent core-shell silica nanoparticles: towards "Lab on a Particle" architectures for nanobiotechnology. Chem Soc Rev 35(11):1028–1042. doi:10.1039/b600562b
44. Shi DL (2009) Integrated multifunctional nanosystems for medical diagnosis and treatment. Adv Funct Mater 19(21):3356–3373. doi:10.1002/adfm.200901539
45. Yong KT, Roy I, Swihart MT, Prasad PN (2009) Multifunctional nanoparticles as biocompatible targeted probes for human cancer diagnosis and therapy. J Mater Chem 19(27):4655–4672. doi:10.1039/b817667c
46. Chen Y, Zhu C, Yang Z, Chen J, He Y, Jiao Y, He W, Qiu L, Cen J, Guo Z (2013) A ratiometric fluorescent probe for rapid detection of hydrogen sulfide in mitochondria. Angew Chem Int Ed 52(6):1688–1691. doi:10.1002/anie.201207701

47. Bianchi F, Bedini A, Riboni N, Pinalli R, Gregori A, Sidisky L, Dalcanale E, Careri M (2014) Cavitand-based solid-phase microextraction coating for the selective detection of nitroaromatic explosives in air and soil. Anal Chem 86(21):10646–10652. doi:10.1021/ac5025045
48. Chinen AB, Guan CM, Ferrer JR, Barnaby SN, Merkel TJ, Mirkin CA (2015) Nanoparticle probes for the detection of cancer biomarkers, cells, and tissues by fluorescence. Chem Rev 115(19):10530–10574. doi:10.1021/acs.chemrev.5b00321
49. Zhou W, Gao X, Liu D, Chen X (2015) Gold nanoparticles for in vitro diagnostics. Chem Rev 115(19):10575–10636. doi:10.1021/acs.chemrev.5b00100
50. Yang X, Yang M, Pang B, Vara M, Xia Y (2015) Gold nanomaterials at work in biomedicine. Chem Rev 115(19):10410–10488. doi:10.1021/acs.chemrev.5b00193
51. Renna LA, Boyle CJ, Gehan TS, Venkataraman D (2015) Polymer nanoparticle assemblies: a versatile route to functional mesostructures. Macromolecules 48(18):6353–6368. doi:10.1021/acs.macromol.5b00375
52. Mavila S, Eivgi O, Berkovich I, Lemcoff NG (2015) Intramolecular cross-linking methodologies for the synthesis of polymer nanoparticles. Chem Rev. doi:10.1021/acs.chemrev.5b00290
53. Dong H, Du S-R, Zheng X-Y, Lyu G-M, Sun L-D, Li L-D, Zhang P-Z, Zhang C, Yan C-H (2015) Lanthanide nanoparticles: from design toward bioimaging and therapy. Chem Rev 115 (19):10725–10815. doi:10.1021/acs.chemrev.5b00091
54. Lee N, Yoo D, Ling D, Cho MH, Hyeon T, Cheon J (2015) Iron oxide based nanoparticles for multimodal imaging and magnetoresponsive therapy. Chem Rev 115(19):10637–10689. doi:10.1021/acs.chemrev.5b00112
55. Prodi L, Rampazzo E, Rastrelli F, Speghini A, Zaccheroni N (2015) Imaging agents based on lanthanide doped nanoparticles. Chem Soc Rev 44(14):4922–4952. doi:10.1039/c4cs00394b
56. Montalti M, Prodi L, Rampazzo E, Zaccheroni N (2014) Dye-doped silica nanoparticles as luminescent organized systems for nanomedicine. Chem Soc Rev 43(12):4243–4268. doi:10.1039/c3cs60433k
57. Montalti M, Rampazzo E, Zaccheroni N, Prodi L (2013) Luminescent chemosensors based on silica nanoparticles for the detection of ionic species. New J Chem 37(1):28–34. doi:10.1039/c2nj40673j
58. Genovese D, Rampazzo E, Bonacchi S, Montalti M, Zaccheroni N, Prodi L (2014) Energy transfer processes in dye-doped nanostructures yield cooperative and versatile fluorescent probes. Nanoscale 6(6):3022–3036. doi:10.1039/c3nr05599j
59. Rampazzo E, Voltan R, Petrizza L, Zaccheroni N, Prodi L, Casciano F, Zauli G, Secchiero P (2013) Proper design of silica nanoparticles combines high brightness, lack of cytotoxicity and efficient cell endocytosis. Nanoscale 5(17):7897–7905. doi:10.1039/c3nr02563b
60. Bonacchi S, Genovese D, Juris R, Montalti M, Prodi L, Rampazzo E, Sgarzi M, Zaccheroni N (2011) Luminescent chemosensors based on silica nanoparticles. Top Curr Chem 300:93–138. doi:10.1007/128_2010_104
61. Burns A, Sengupta P, Zedayko T, Baird B, Wiesner U (2006) Core/shell fluorescent silica nanoparticles for chemical sensing: towards single-particle laboratories. Small 2(6):723–726. doi:10.1002/smll.200600017
62. Wang L, O'Donoghue MB, Tan W (2006) Nanoparticles for multiplex diagnostics and imaging. Nanomedicine (London) 1(4):413–426. doi:10.2217/17435889.1.4.413
63. Rampazzo E, Bonacchi S, Genovese D, Juris R, Montalti M, Paterlini V, Zaccheroni N, Dumas-Verdes C, Clavier G, Méallet-Renault R, Prodi L (2014) Pluronic-silica (PluS) nanoparticles doped with multiple dyes featuring complete energy transfer. J Phys Chem C 118(17):9261–9267. doi:10.1021/jp501345f
64. Rampazzo E, Bonacchi S, Genovese D, Juris R, Sgarzi M, Montalti M, Prodi L, Zaccheroni N, Tomaselli G, Gentile S, Satriano C, Rizzarelli E (2011) A versatile strategy for signal amplification based on core/shell silica nanoparticles. Chemistry 17(48):13429–13432. doi:10.1002/chem.201101851

65. Achatz DE, Ali R, Wolfbeis OS (2011) Luminescent chemical sensing, biosensing, and screening using upconverting nanoparticles. Top Curr Chem 300:29–50. doi:10.1007/128_2010_98
66. Genovese D, Bonacchi S, Juris R, Montalti M, Prodi L, Rampazzo E, Zaccheroni N (2013) Prevention of self-quenching in fluorescent silica nanoparticles by efficient energy transfer. Angew Chem Int Ed 52(23):5965–5968. doi:10.1002/anie.201301155
67. Zhu S, Fischer T, Wan W, Descalzo AB, Rurack K (2011) Luminescence amplification strategies integrated with microparticle and nanoparticle platforms. Top Curr Chem 300:51–91. doi:10.1007/128_2010_99
68. Brasola E, Mancin F, Rampazzo E, Tecilla P, Tonellato U (2003) A fluorescence nanosensor for Cu2+ on silica particles. Chem Commun 24:3026–3027. doi:10.1039/b310582b
69. Rampazzo E, Brasola E, Marcuz S, Mancin F, Tecilla P, Tonellato U (2005) Surface modification of silica nanoparticles: a new strategy for the realization of self-organized fluorescence chemosensors. J Mater Chem 15(27–28):2687–2696. doi:10.1039/b502052b
70. Mancin F, Rampazzo E, Tecilla P, Tonellato U (2006) Self-assembled fluorescent chemosensors. Chemistry - A European Journal 12(7):1844–1854. doi:10.1002/chem.200500549
71. Arduini M, Mancin F, Tecilla P, Tonellato U (2007) Self-organized fluorescent nanosensors for ratiometric Pb2+ detection. Langmuir 23(16):8632–8636. doi:10.1021/la700971n
72. Bonacchi S, Rampazzo E, Montalti M, Prodi L, Zaccheroni N, Mancin F, Teolato P (2008) Amplified fluorescence response of chemosensors grafted onto silica nanoparticles. Langmuir 24(15):8387–8392. doi:10.1021/la800753f
73. Arca M, Caltagirone C, De Filippo G, Formica M, Fusi V, Giorgi L, Lippolis V, Prodi L, Rampazzo E, Scorciapino MA, Sgarzi M, Zaccheroni N (2014) A fluorescent ratiometric nanosized system for the determination of PdII in water. Chem Commun 50(96):15259–15262. doi:10.1039/c4cc07969h

Chapter 13
Photochemistry for Cultural Heritage

Maria João Melo, Joana Lia Ferreira, António Jorge Parola, and João Sérgio Seixas de Melo

Abstract Why do certain ancient natural dyes, such as indigo, preserve their colour so well while others, like brazilein, seem to degrade much faster? And how did mauveine change the world of colour? Will modern binding media, as vinyl paints, perform as well as a medieval tempera? Will it be possible to predict their durability? Photochemistry can answer many important questions about materials' stability, providing new tools for the conservation of treasured artworks.

In this chapter, photochemistry emerges as an important contribution to the understanding of those complex processes, providing fascinating insights into a world of colour and light.

13.1 Introduction

> Cultural property constitutes one of the basic elements of civilisation and national culture. (UNESCO 1970)

Conservation science is about art preservation [1]. It is a new science that has developed accurate methodologies for revealing the composition and structure of the materials used by artists to better access the immaterial value of a work of art: to reveal memories, affections and beauty. Conservation science is about a better understanding and access of our culture and our cultural heritage, and for this reason it is also about studying ethnographic and archaeological objects. This chapter will be on the contributions that photochemists have made to this emerging field, in which art walks hand in hand with science.

A main goal of conservation scientists is to stabilize materials (and avoid their degradation whenever possible). For achieving this goal the development of advanced analytical techniques has proven to be essential, according to the editors

M.J. Melo (✉) • J.L. Ferreira • A.J. Parola
LAQV-REQUIMTE, Department of Conservation and Restoration and Department of Chemistry, Faculty of Science and Technology, Universidade NOVA de Lisboa, Lisbon, Portugal
e-mail: mjm@fct.unl.pt

J.S.S. de Melo
CQC, Department of Chemistry, University of Coimbra, 3004-535 Rua Larga, Coimbra, Portugal

G. Bergamini, S. Silvi (eds.), *Applied Photochemistry*,
Lecture Notes in Chemistry 92, DOI 10.1007/978-3-319-31671-0_13

of a special issue of Accounts for Chemical Research [2]: "Such advanced techniques have answered many questions about degradation. As a result, new methods of treatment are being proposed that offer the possibility of attacking the actual cause of the problem, rather than merely the symptoms." The majority of the advanced techniques that Antonio Sgamellotti and Bruno Brunneti refer to are based on light.

As beautifully stated by Vincenzo Balzani and Franco Scandola, light is made of photons, and photons are at the same time energy quanta and information bits [3]. In nature, this double role of light manifests itself in photosynthesis and vision, respectively. This duality in the interaction of light with matter is also present in the conservation of cultural heritage. Light as a bit of information is intrinsically associated with materials characterisation, as case studies will illustrate. As a bit, it may reveal a hidden millenary dye, the degradation mechanisms of a medieval colour, or assess the stability of a twentieth century artistic paint. On the other hand, light as energy may be the cause of degradation of the same beautiful dye and of the paint binder, enhancing the decay of many masterpieces worldwide.

In this chapter we highlight the contributions of photochemistry for a better understanding of cultural heritage and its preservation. We start our journey with the molecules of colour in art, in which we showcase indigo as an ancient textile dye and brazilwood lake pigments as colour paints in fifteenth century manuscript illuminations. Mauveine, the superb purple colour created by Perkin, indirectly responsible for the twilight of natural dye applications, will close this first stage. We will begin a new journey with the invisible in art: the binding medium. The art of polyesters will be examined, with drying linseed oil and vinyl emulsion paints. In the oil case study, we shall focus on how a metal coordinated paint system is made, the development of metal carboxylate aggregates (metal soaps), and how they negatively influence the appearance of a painting surface. The most recent theories predict that self-repair mechanisms are also present in these systems. Poly(vinyl acetate) is at the heart of vinyl paint emulsions, and we shall discuss the importance of measuring quantum yields of photodegradation to accurately determine the stability of a polymer matrix.

13.2 The Molecules of Colour in Art

Natural dyes, discovered through the ingenuity and persistence of our ancestors, can remain bright for centuries or even millennia, and may be found hidden in such diverse places as the roots of a plant, a parasitic insect, and the secretions of a sea snail, Fig. 13.1. In art, natural dyes were used to colour a textile or to paint [4].

Dyeing, in red, blue, purple or yellow, is a complex task that requires skill and knowledge [5]. To be used as a textile dye, the chromophore must also be captured as strongly as possible into the fibres, i.e., it must be resistant to washing. Dyes can bind to the surface of the fibre or be trapped within. They are often bound to the textile with the aid of metallic ions known as **mordants**, which can also play an important role in the final colour obtained. Alum, as a source of aluminium ion,

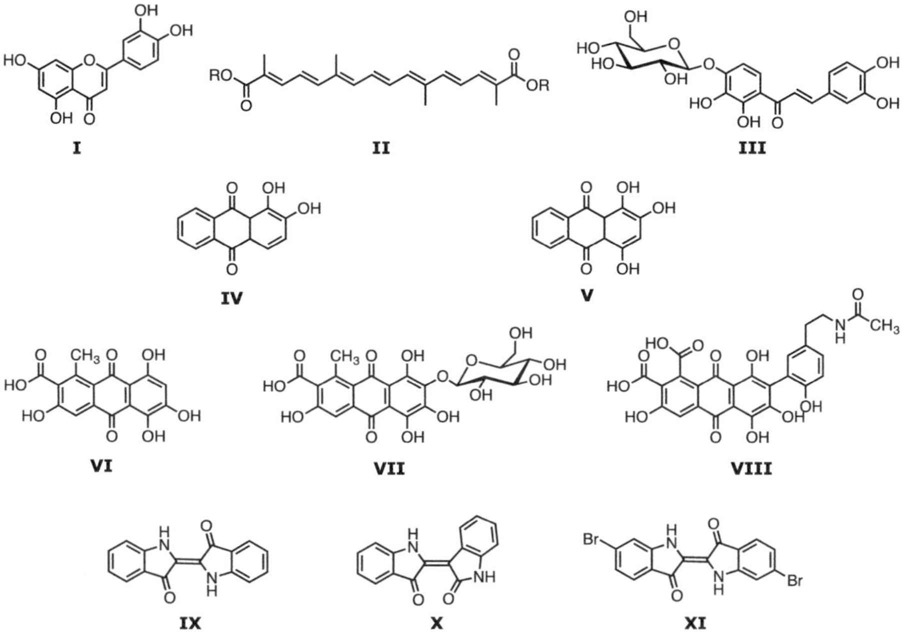

Fig. 13.1 *Yellows*, *reds* and *blues* from natural dyes. The flavone *luteolin* (I), the carotenoid *crocetin* (II) and the chalcone *marein* (III) are examples of yellow dyes. The anthraquinone reds from plants (IV, V) and from insects (VI–VIII) are shown in the 2nd and 3rd rows, respectively: *alizarin* (IV), *purpurin* (V); *kermesic acid* (VI), *carminic acid* (VII) and *laccaic acid A* (VIII). The *blues* and *purple* chromophores are depicted in the fourth row: *indigo* (indigotin, IX), *indirubin* (X) and Tyrian *purple* (XI)

is an important historical mordant and was widely used in the past. Other important mordants were iron, copper, and tin ions [5]. Dyes, like indigo or Tyrian purple, which are trapped in the fibres due to an oxidation-reduction reaction, without the aid of a mordant, are known as **vat dyes**.

It is useful to distinguish between dyes and pigments based on their solubility in the media used to apply the colour; dyes are generally organic compounds that are soluble in a solvent, whereas pigments are usually inorganic compounds or minerals, which are insoluble in the paint medium (oil, water, etc.) and are dispersed in the matrix. A dye may be converted into a **lake pigment** by precipitating the dye extract as a result of complexation with a metal ion such as Al^{3+} [4, 6].

Lake pigments have been widely applied in painting, Fig. 13.2. For example, anthraquinones and their derivatives have been used as red dyes and lake pigments from pre-historic times, and we can find written accounts of their use in ancient Egypt [4, 7]. Pure dyes, such as indigo, were also used as painting materials.

In order to be used as a **colour paint**, for example, to create a medieval illumination on a sheet of parchment, the colours are "tempered" (mixed) with protein binders like glair (egg white), parchment glue (collagen) or even egg yolk [8, 9]. These binding agents allow the colouring to be applied as paint layers, enabling it to adhere to the page. The success of a formulation as well as its

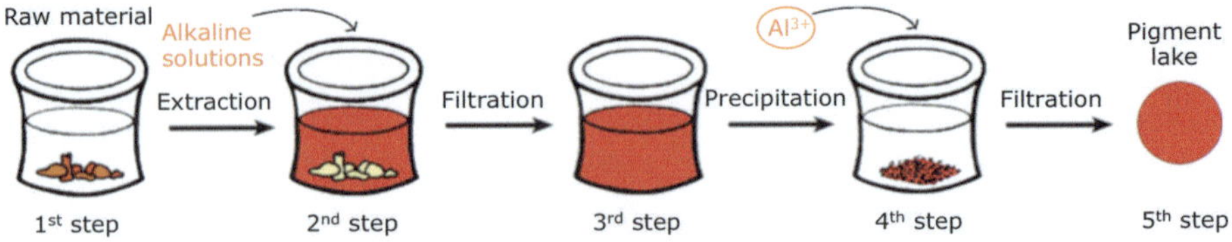

Fig. 13.2 A lake pigment is a metal complex, insoluble in water, prepared by the addition of a metal ion such as Al^{3+} to a an aqueous solution of a dye extracted from a natural source. The complex will precipitate when a neutral molecule is formed, usually between pH 6 and 7. The dye may be extracted in acidic media (in the presence of Al^{3+}), in neutral (in urine) or basic solutions (prepared with the ashes of plants, in which the oxides of alkaline metals are transformed in the corresponding hydroxides by reaction with water)

longevity depends not just on the binder, but also on the additives that (*i*) improve paint application and (*ii*) make it resistant to decay and give it a greater ability to adapt to mechanical stress. In medieval treatises on the art of illumination, ear-wax (human or animal) is an example of the first type, which allows a more fluid application, further minimising the formation of small air bubbles. An example of the second would be binder mixtures (adjusting flexibility), and the use of fillers such as calcium carbonate (increasing mechanical resistance) [9]. The **binding medium** is the "invisible component" of paint, but if the binders fail, then so do the colours.

13.2.1 Andean Textiles: Photophysics and Photochemistry of Indigo Blue

13.2.1.1 Indigo as a Dye

Indigo blue was one of the earliest and most popular dyestuffs known to man, and its longevity as a colouring material has given this molecule an iconic status, Fig. 13.1. Its importance was only surpassed by its brominated derivative Tyrian purple (Box 13.1). The history of indigo as a dye covers over four millennia, and it was ***the*** natural source of blue for textiles. It still is a universally used colour, as the worldwide use of indigo-dyed blue jeans can attest. Indigo sources are found all over the world, and several plants were most likely used in antiquity. Indeed, more than 700 *Indigofera* species can be found in nature [10].

Indigo is one of the most light-stable organic dyes, a characteristic that explains not only its wide use in antiquity and the pre-modern area, but also its longevity as a colourant [10–12]. The stability of indigo is also the reason why it was used in medieval illuminations and by some of the great masters of the seventeenth and eighteenth centuries, such as Rubens [4]. The explanation of indigo's blue colour and its derivatives has also been an intriguing and fascinating subject, and was explored in great detail and ingenuity in the 1970s and 1980s, as described in Sect. 13.2.1.2.

Box 13.1: Ancient Purple (Tyrian Purple)

Purple was the most important colour in antiquity [4] and, possibly, was produced since the mid-second millennium BC. Purple was a status-symbol for the Roman emperors as well as, in later years, for the powerful representatives of the Catholic Church. It was the Persian monarchs who, as early as the ninth century BC, made purple the official mark of royalty. The kings of Israel also wore purple. Imperial or divine purple was used for powerful emotional moments, where authority was conveyed, in rituals of high visual impact. Purple was also the colour of the mantle Christ used in its sacrifice and as so it is an important liturgical colour: a sacred colour (Cavallo 1998). The original colour was the most expensive dye in antiquity, but even its imitations (a combination of indigo with madder or other similar red) could express its status-symbol [5].

The true purple dye, Tyrian purple, was obtained from the Mediterranean shellfish of the genera *Purpura*[1] (McGovern 1990). As one of the luxury goods traded by the Phoenicians, it was named "Tyrian purple" after the Phoenician city, Tyre. The precursors of Tyrian purple are stored in the hypobranchial glands of live muricids and can be released after death or when the mollusc is pressed and rubbed. Piles of purple-yielding mollusc shells on Crete and along the Levantine coast, and elsewhere in the Mediterranean, have provided the primary archaeological evidence for the extent of this industry (McGovern 1990).

It is worth pointing out that indigo is blue, whereas its brominated derivatives are violet and indirubin is reddish. The shade of purple obtained depends on the species of mollusc and on the dyeing process used. For more on the production of purple please see [4]. The story of purple was to be continued in the nineteenth century with the discovery of mauve by Perkin, and the rebirth of the ancient *purple mania* as *mauve mania*.

For more details see Guglielmo Cavallo, La porpora tra scienze e culture. Una introduzione, *in* La porpora, realtà e immaginario di un colore simbolico, Oddone Longo ed., Istituto Veneto di Scienze, Lettere ed Arti, Venezia 1998 and P. E. McGovern, R. H. Michel, Royal purple dye: the chemical reconstruction of the ancient Mediterranean industry, *Acc. Chem. Res.*, 23, 152–158 (1990)

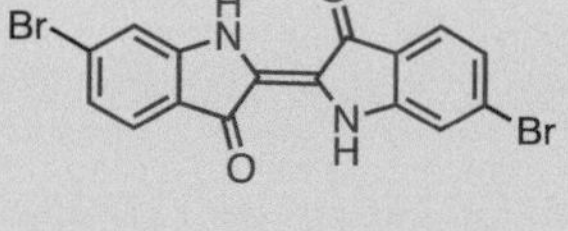

Figure Box 13.1 6,6′-Dibromoindigotin (Tyrian purple)

[1] The Mediterranean purple molluscs, *Murex brandaris*, *Murex trunculus* and *Purpura haemastoma*, have recently been renamed *Bolinus brandaris*, *Hexaplex trunculus* and *Stramonita haemastoma*, respectively, with the result that *Murex* and *Purpura* had disappeared from the name, and with it the memory of their importance as historic dyes was cancelled

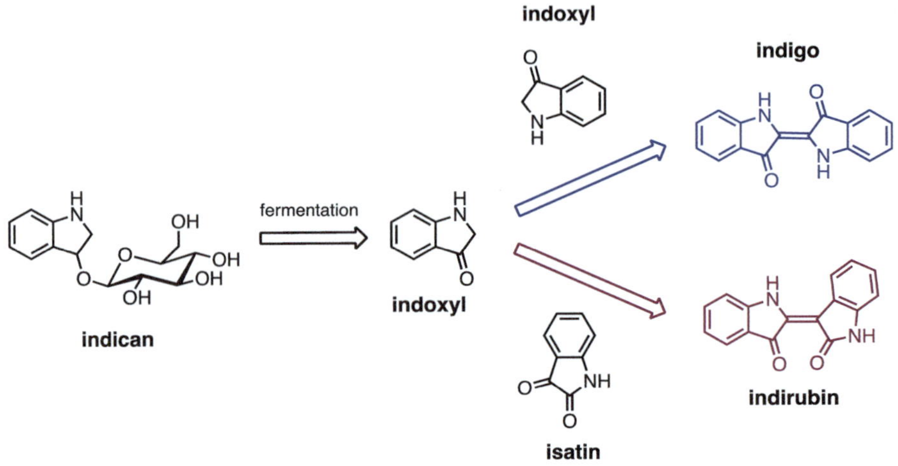

Fig. 13.3 Indoxyl, obtained from fermentation of the water-soluble indican, is the common precursor for the blue indigo and the reddish indirubin

The precursors of indigo (=indigotin) can be extracted entirely from plants. The extract, a solution of the water-soluble indoxyl glucoside, indican, is hydrolysed via fermentation to give indoxyl, which will further react with another indoxyl molecule producing indigo; or, in a side reaction, with isatin, resulting in indirubin, a reddish dye [4], Fig. 13.3. Indirubin, also known as the red shade of indigo, is formed in the interface where more oxygen is available [13]. Indigo derivatives, in their oxidized forms, are insoluble in water [4, 5]. Extraction from plants was this dye's unique source until Bayer opened the way for an industrial process that became economically feasible. The structural determination and synthesis of indigo [14] were triumphs of the growing German chemical industry of the second half of the nineteenth century, and Adolf von Baeyer was awarded the 1905 Nobel Prize in Chemistry in recognition of his contributions to chemical synthesis, where his synthesis of indigo is specifically cited [15]. Bayer considered the *cis*- configuration to be the most stable, and it was only in 1928 that Reis and Schneider [16], using X-ray crystal, established the correct structure to be the *trans*- isomer, Fig. 13.1.

13.2.1.2 Indigo: Colour and Photophysics

13.2.1.2.1 Colour

Some relevant characteristics of indigo are [17–25]: (*i*) its blue colour (with an unusual long wavelength of absorption), (*ii*) the effect of substitution on its colour, Box 13.1, (*iii*) its redox chemistry that interconverts indigo *keto* form into *leuco* (reduced) and *dehydroindigo* (oxidized) species, and (*iv*) its high photochemical stability resulting from a fast excited state proton transfer from the N-H to the C= O group.

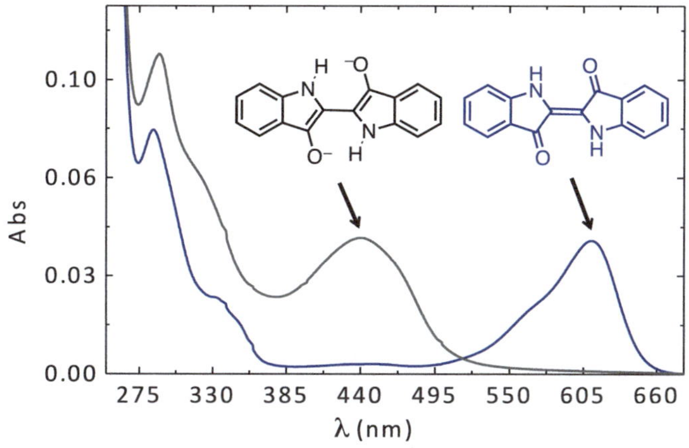

Fig. 13.4 Absorption spectra of the leuco (λ_{max} ~ 440 nm, gray trace) and keto (λ_{max} ~ 610 nm, blue trace) forms of indigo

Indigo's intense blue colour is due to its absorption at long wavelengths, Fig. 13.4, an intriguing[2] feature for a relatively small molecule [17–19]. The causes of colour in indigo were investigated in the 1970s with the synthesis of several derivatives – Fig. 13.5 [20] – proving that the fundamental chromophore in indigo consists of a central C=C bond (connecting the two rings) together with the nitrogen and carbonyl groups; neither the benzene rings nor the double bonds in the five-member rings are responsible for the characteristic indigo spectrum and blue colour. In summary, ressonance is not the cause of indigo's blue colour [19–23]. Indigo is best described as a donor-acceptor dye,[3] where the amino groups act as the electron donors and the carbonyl groups as the electron acceptors, Fig. 13.5. More recently, TD-DFT calculations supported this description, showing that the HOMO is essentially located in the central C=C bond and the nitrogen atoms and the LUMO in the central C=C bonds and oxygen atoms [24]. The donor groups raise the π orbital of the C=C double bond (HOMO), while the acceptor groups lower the π * orbital (LUMO). In a simplified approximation, the absorption of visible light may be explained by a narrow HOMO-LUMO separation [19].

[2] Merocyanines structures, even with 6 conjugated double bonds, display an absorption wavelength maximum of only 510 nm (with a shift of 90–100 nm compared to indigo); see Fig. 13.5.

[3] It will be useful to describe a dye as a two constituent system: the chromophore and the auxochrome. The chromophore group is responsible for basic colour, whereas the auxochromes allow the "enrichment" of the colour. Dilthey and Wizinger extended this concept defining the chromophore as an electron accepting group and the auxochromes as electron donating, linked to each other by conjugated bonds. This gave rise to the concept of the donor/acceptor dye type.

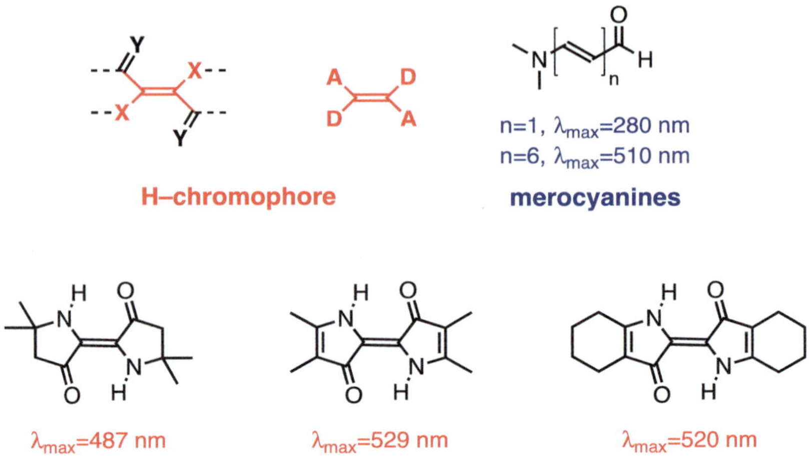

Fig. 13.5 H-chromophore structure and absorption maxima wavelengths of merocyanines and of indigo pseudo-derivatives (lacking the aromatic unit)

13.2.1.3 Indigo: Photoprotection by Excited Proton Transfer

George Wyman, in a seminal work published in 1994, brought to the photophysical scene indigo and the amazing fact that, in contrast with the strongly fluorescent thioindigos, it showed no detectable emission [18]. It was later determined that indigo had a fluorescence quantum yield of $\phi_F = 0.0023$, a fluorescence lifetime $\tau_F = 0.14$ ns, and an energy for the first lying singlet excited state of 15728 cm^{-1} [11, 25]. With such a weak emitter, it was a challenge to obtain the yields (and rate constants) of the radiationless deactivation channels (internal conversion, ϕ_{IC} and singlet to triplet intersystem crossing, ϕ_{ISC}). This was done by obtaining the triplet energy with pulse radiolysis energy transfer experiments ($E_T = 1.0 \pm 0.1$ eV), followed by a determination of the ϕ_{ISC} (=0.0066) from photoacoustic calorimetry [26]. These numbers show that over 99.99 % of the quanta are lost through non-radiative channels (internal conversion).

In theory, the efficiency of the internal conversion deactivation channels may result from a proton transfer in the excited state (ESPT) or be the consequence of the golden rule for radiationless transitions: when the energy difference between the ground and first singlet ($\Delta E = S_1 – S_0$) is very small, which is the case of indigo, it favours an efficient coupling between the S_0 and S_1 vibronic modes [11, 25].

It was demonstrated in both experiments [26–28] and theoretical calculations [24] that the high stability of indigo has its mechanism associated with a fast proton transfer in the excited state (ESPT), although there is still a dispute about whether the excited proton transfer involves a concerted proton transfer between the two N-H and two C=O groups or from a single group, and if it is intra or inter molecular. This ESPT is schematically depicted in Fig. 13.6: the *keto* excited species is converted into its *enol* isomer with a rate constant of ~8×10^{10} s^{-1}, in

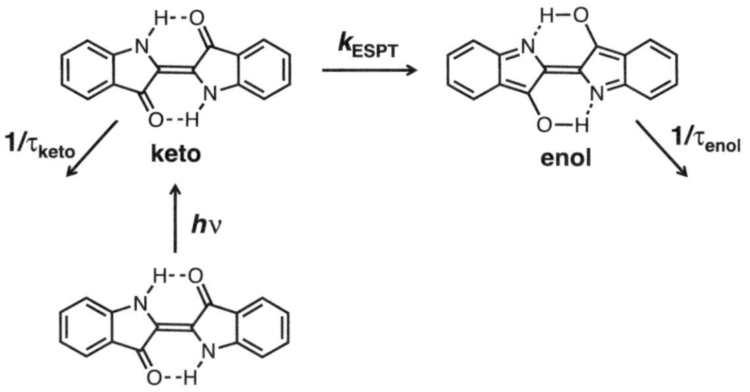

Fig. 13.6 Kinetic scheme with the keto (ground and excited state) and enol (excited state) forms of indigo

which the existence of these two excited species is in agreement with the biexponential decay observed in indigo (and indigocarmine[4]) [27].

It was fascinating to discover that excited state proton transfer plays such a crucial role in the photochemical stability of this extraordinary dye. This fast and highly efficient internal conversion provides a 'self-protective channel', minimizing access to photodegradation. This photoprotective process is shared with many other historic dyes and with DNA bases (purines and pyrimidines) [29]. For the latter, De Vries proposed that this mechanism has been preserved as a molecular fossil: "Evolution requires the existence of self-replicating molecules; the selection of the nucleobases as building blocks of those macromolecules would thus logically have taken place prior to any biological processes. Therefore, if the excited state properties have played a role in a chemical selection of today's nucleobases, they may be a relic of prebiotic chemistry on an early earth. **These photochemical properties then would be molecular fossils, so to speak, of four-billion-year-old chemistry**" [29]. *Molecular fossil* is a powerful expression that enriches our chemical vocabulary, and disclose indigo both a millenary dye and a molecular fossil!

13.2.1.4 Andean Textiles and Indigo Photochemistry

Following the detailed characterization of indigo photophysics we started the study of its degradation mechanisms, having selected as case study the fabulous Andean textiles in the collection of the Museum of Fine Arts-Boston [12, 30].

Pre-Columbian textiles represent the longest continuous textile record in world history and are precious and unique works of art, Fig. 13.7 [31]. Fortunately, in

[4] For the water soluble 5,5′-disulfonate derivative of indigo, indigocarmine, in DMF, the keto form is instantaneously formed with an intermolecular rate constant of ~1.4×10^{11} s^{-1} (by proton transfer to the solvent), whereas in methanol it is ~1.2×10^{11} s^{-1}, making it more likely to be intramolecular.

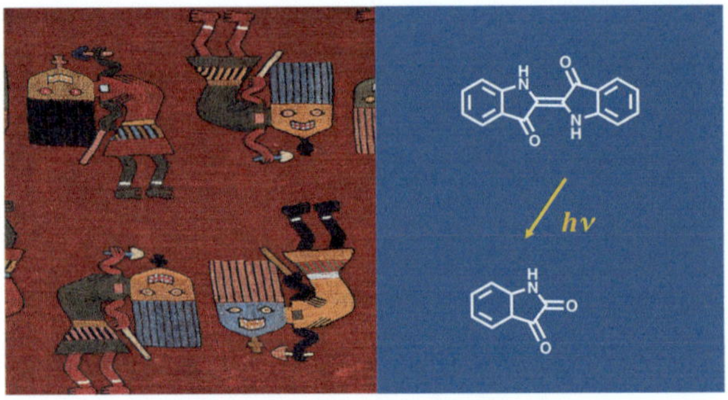

Fig. 13.7 In the extraordinary Andean textiles, in the collection of the Museum of Fine Arts Boston, colours are beautifully preserved. Nevertheless, the degradation product of indigo was detected in all the samples studied. For more details please see [12]

extremely arid archeological sites, the cultural heritage of different Andean cultures such as Paracas, Nasca, Chancay and Lambayeque has been preserved.

The red[5] and blue colours were fully characterized in microsamples from different Andean textiles, dated from 200 B.C. to A.D. 1476 [12, 30]. In this section we will center our attention in the blues, particularly in the degradation products found in these Andean textiles naturally aged over centuries. These will be compared with what observed in our simulation studies, for which we selected transparent gels (collagen and CMC) to reproduce the dye environment in textiles, wool (protein) and cotton (cellulose). As these are aqueous gels, and indigo is insoluble in water, in these media it was replaced by its analogous the 5,5′-indigodisulfonic acid sodium salt (indigo carmine) [12].

Quantum yields of reaction were obtained with excitation at 335 nm and 610 nm in organic (THF, DMF) and aqueous solutions as well as in the transparent gels, Box 13.2. In solution, the photodegradation quantum yields (Φ_R) were in the order of 10^{-4} and dependent on the irradiation wavelength; with the exception of indigo carmine in water where a $\Phi_R = 9 \times 10^{-6}$ was obtained; in the case of indigo carmine the Φ_R values were found to suffer a twofold increase upon going from water to gels [12].

Box 13.2: Measuring a Reaction Quantum Yield in a Transparent Gel
Fernando Pina
In a photochemical reaction in solution as represented in Eq. (13.1) the quantum yield is defined by means of Eq. (13.2)

$$R \xrightarrow{h\nu} P \tag{13.1}$$

(continued)

[5] Microspectrofluorimetry was used to analyze the red chromophores and SEM-EDX screening enabled to confirm the use of aluminium ion, Al^{3+}, as a mordant and also to conclude that all the red samples studied were made of camelid fibers. Emission and excitation spectra were obtained in a 30 μm spot. The fluorophores were identified by comparing their spectra with those from historical reconstructions assembled in a database. In the Paracas and Nasca textiles, dated from 200 B.C. to A.D.1476, purpurin and pseudopurpurin were the red dyes used. Carminic acid was detected in textiles dated close to the Inca Empire, A.D. 1000–1476 [31].

Box 13.2 (continued)

$$\Phi = \frac{number\ of\ P\ molecules\ formed}{number\ of\ absorbed\ photons}$$

$$= \frac{\frac{\Delta A_{\lambda m}}{\Delta \varepsilon_{\lambda m}} v_{irr}}{I_0\left(1 - 10^{-A_{\lambda_{irr}}}\right) \frac{A_{\lambda_{irr}(R)}}{A_{\lambda_{irr}(R)} + A_{\lambda_{irr}(P)}} \Delta t} \tag{13.2}$$

where ΔA_{λ_m} is the mean absorbance variation at the monitoring wavelength λ_m, $\Delta \varepsilon_{\lambda_m}$ is the difference between the molar absorption coefficients of reagent and product at the same wavelength, I_0 is the intensity of the light, and $A_{\lambda_{irr}} = A_{\lambda_{irr}(R)} + A_{\lambda_{irr}(P)}$ is the total absorbance of the solution at the irradiation wavelength, where $A_{\lambda_{irr}(R)} = \varepsilon_{\lambda_{irr}(R)} b[R]$ and $A_{\lambda_{irr}(P)} = \varepsilon_{\lambda_{irr}(P)} b[P]$ are the contributions from reagent and product, respectively. The irradiation path length is b, while v_{irr}, the irradiated volume, is the product of the illuminated surface of the cell and the optical path. The time interval over which the absorption changes are measured is Δt. The term $I_0\left(1 - 10^{-A_{\lambda_{irr}}}\right)$ accounts for the total light absorbed by the system (both reagent and product), while the fraction $\frac{A_{\lambda_{irr}(R)}}{A_{\lambda_{irr}(R)} + A_{\lambda_{irr}(P)}}$ gives the fraction that is absorbed by the reagent responsible for the photochemical events.

One interesting aspect of photochromic gels is the formation of a gradient of product concentration that can be observed at naked eye in concentrated solutions absorbing in the visible. This effect is due to the penetration of light through the optical path. The gradient formation does not affect the measurement of the quantum yields, if the irradiated volume is defined by the area of the cell under light and the optical path. For more details see F. Pina, A. T. Hatton, "Photochromic Soft Materials: Flavylium compounds Incorporated in Pluronic F-127 hydrogel Matrixes", *Langmuir*, 2008, *24*, *2356–2364*.

In conclusion, Eq. (13.2) can be used indistinctly in a transparent gel or in a stirred solution. However in the case of a gel the irradiated volume should be defined by the irradiated face and the optical pathway and the incident light should be homogeneously distributed through the surface.

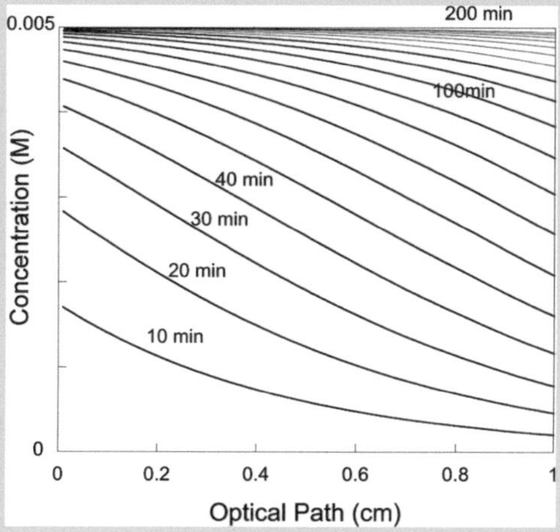

Figure Box 13.2 Simulation of the product distribution (P) resulting from a photochemical reaction $R \xrightarrow{h\nu} P$ in a gel. Calculation carried out with the parameters: $I_0 = 8 \times 10^{-7}$ Einstein min^{-1}; $\Phi = 0.3$; $v_{irr} = 3$ ml: $[C_0] = 0.005$ M; $\varepsilon_R = 180\ M^{-1}cm^{-1}$; $\varepsilon_P = 10\ M^{-1}cm^{-1}$

The reaction kinetics, in the absence and presence of oxygen, was followed by UV-visible absorption and the resulting photoproducts were characterized by HPLC-DAD-MS. Isatin was the major photodegradation product observed in all the media studied [12].

The data obtained was in agreement with the analysis of blues from millenary Andean textiles, dated from BC 200 to 200 AD. These textiles were kept for millennia protected from light in a very dry environment [31], in a region named Paracas (South Peru). In these Paracas blues, isatin was also found as the major photodegradation product, in low to moderate relative amount ($\leq$10 % and $\leq$20 %), indicating that indigo is a very stable molecule in certain environments.

This was the part of the story that was straightforward. During our experiments it was also observed that the quantum yield for the photodegradation of indigo is different for DMF freshly distilled and DMF (from open bottles with one or more days), which in conjunction with the data obtained in water enables to consider that $OH^{\cdot}$ like radicals or other oxygen based reactive species are involved in this degradation mechanism, as others suggested [32]; the hydroxyl radical attacks the central double bond and the result is the formation of isatin. Others have also found that superoxide dismutase converts indigo carmine into isatin sulfonic acid prompting once more for the involvement of $O_2^{-\cdot}$ and $OH^{\cdot}$ radicals [33]. It turned out that the unexpected behaviour of indigo in DMF solutions provide us crucial evidence into the fundamental photodegradation step. In both cases isatin is produced from a reaction of indigo with a radical, resulting from an attack to the central double bond.

This type of mechanism is very plausible when considering the electrochemical studies that lead to the conversion of the keto into the leuco form, Scheme 13.1, namely the mechanism involved in the disruption of the central double bond connecting the two indole moieties [34]. A stopped flow study on indigo carmine reduction by dithionite ($S_2O_4^{2-}$) concluded that the mechanism involved a two-step reduction, and that the reduction potential, $E^{1/2}_{red}$, was pH dependent, Scheme 13.1; therein it was suggested the possibility of a protonation step following charge transfer to be rate-limiting [34]. Further studies on the electrochemical reduction of indigo dissolved in organic solvents (DMSO, DMF) [35] confirmed the general reduction mechanism. This reduction mechanism shows the formation of a radical in one of the carbon atoms that are part of the central double bond; this radical will react very promptly with other radicals present in solution, in our case, with radicals present in DMF solutions not freshly distilled.

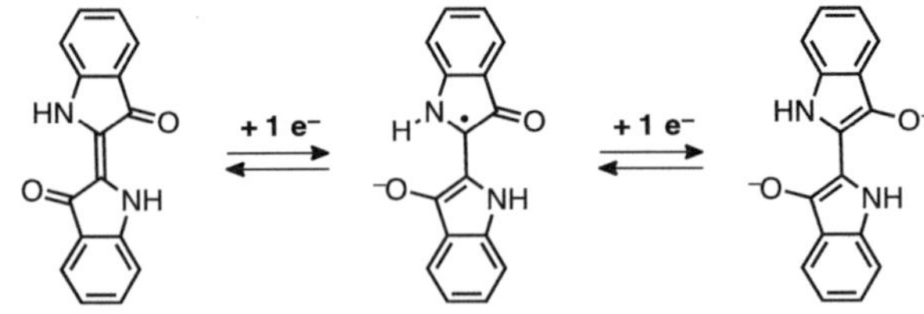

Scheme 13.1 Mechanistic scheme for the reduction of indigo in non acidic media (Adapted from [34])

Based on all the experimental evidence collected, we propose that the central double bond is at the same time a photoprotector mechanism and a weak point. The central double bond is the backbone of the photoprotection mechanism as it enables a very fast non-radiative deactivation process, from the excited to the fundamental state, "locking" the molecule. But, the central double bond is also the most reactive bond of indigo; in our photodegradation experiments, we have proved that its disruption could occur via a mechanism similar to the one proposed for indigo reduction, Scheme 13.1. Finally, on the basis of the obtained Φ_R, we propose that photodegradation quantum yields as found in water, below 10^{-6}, could be representative for indigo in many works of art, and that the values ranging from 10^{-4} to 10^{-3} could be representative of environments where the presence of oxygen based radicals or reducing species could be present.

13.2.2 A World of Light: Brazilwood in Medieval Manuscripts and in Medieval Culture

Brazilwood was an important source for pink and red colours in the Middle Ages [4, 5]. By the nineteenth century, despite being considered less-stable to light than the anthraquinone reds (carminic acid, laccaic acid, alizarin, purpurin), it was still used in oil painting [4, 36].

In the early Middle Ages, it began to be imported into Europe on a large scale from Sri Lanka, India and Southeast Asia [5, 6]. The coloured heartwood of the trunk and larger branches of the trees were used, cut up into chips, and ground into powder [6] as source of dye for inks and to colour textiles, as well as to prepare lake pigments for manuscript illumination [36].

By the fourteenth and fifteenth centuries brazilwood pinks had become a fashionable colour among the aristocracy, as one can appreciate in the lavish illuminations from *Ajuda songbook* (*Cancioneiro da Ajuda*) [37]. In this songbook, lyrics and music from the medieval troubadours were collected and an illumination introduced each poet/author. The book remained incomplete, as the music is missing, but it is still possible to see how much the noblemen were fond of this pink colour [37]. Brazilwood pinks were present in their dressing, and brazilwood lakes were used to colour them in the songbook. Together with the precious lapis lazuli blues, brazilwood lakes were also the distinctive colours of the luxurious Books of Hours, a mandatory item among the aristocracy.

13.2.2.1 The Chromophore: Brazilein

Brazilwood lakes may be easily obtained by extraction and precipitation from aqueous solutions, with the colour being modulated by the metal ion (Al^{3+}, Ca^{2+}, Pb^{2+}) used to complex with the dye [36]. It has been confirmed by HPLC analysis that, in the pigment, the main red chromophore is brazilein [36]. Brazilein is an organic acid, and the first deprotonation occurs with pK_{a1} of 7, in a water-methanol solution (70:30; v:v) and 6.5 in water, Fig. 13.8.

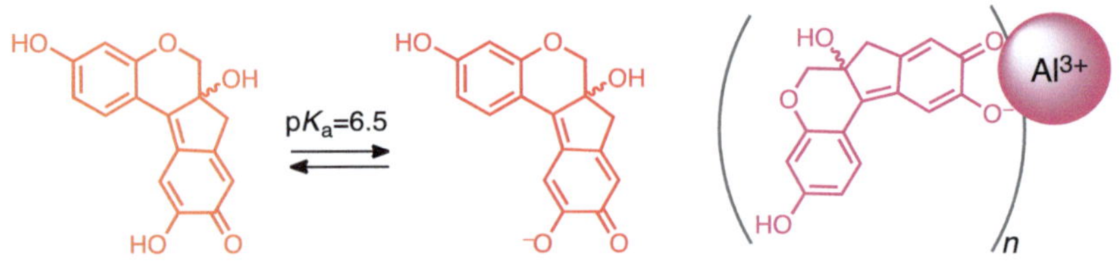

Fig. 13.8 Brazilein has a pK_a of 6.5 in water. Addition of an aluminium salt precipitates the corresponding lake pigment whose structure is still to be fully disclosed

13.2.2.2 Microspectrofluorimetry and Fibre Optic Reflectance spectroscopy for the Characterisation of Brazilwood and Its Lake Pigments

Brazilwood lakes are often considered in literature to be rather unstable to light. For this reason, in a systematic study of the materials and methods used to produce the illuminations of French books of hours in Portuguese collections, special attention was paid to brazilwood lakes.[6] This research is being carried out by an interdisciplinary team, integrating contributions from art history, conservation and chemistry.

Box 13.3: Works of Art Through the Eye of Science
A multi-analytical approach provides a full characterization of the color paints, including, pigments, additives and binders. Micro-spectroscopic techniques, based on infrared, UV-VIS and X-ray radiation, enable to disclose the palette of these medieval manuscripts. Inorganic pigments are fully characterized by combining microXRF (Energy-Dispersive X-Ray Fluorescence Spectroscopy), microFTIR (Fourier Transform Infrared spectroscopy) and microRaman. Infrared analysis proved invaluable for a full characterization of the additives that were applied as fillers or whites (chalk, gypsum and white lead) as well as the proteinaceous and polysaccharide binders that were found pure or in mixture. It is also the only technique that provides a semiquantitave analysis of the overall paint formulation.

Generally, dyes are a major challenge in the study of works of art, and particularly in illuminations. Contrary to textiles, rare are the case studies in which there is a sufficient amount of sample available for analysis by HPLC-MS-DAD. To overcome this obstacle new advanced techniques have been

(continued)

[6] This research was launched within the master thesis of Rita Araújo and Tatiana Vitorino
- Rita Araújo, *Os Livros de Horas (séc. XV) na colecção do Palácio Nacional de Mafra: estudo e conservação*, 5 December 2012, [http://hdl.handle.net/10362/9329].
- Tatiana Vitorino, A Closer Look at Brazilwood and its Lake Pigments, 5 December 2012, [http://run.unl.pt/bitstream/10362/10179/1/Vitorino_2012.pdf].

Box 13.3 (continued)
recently developed and tested [9, 30, 36]: SERS (surface-enhanced Raman spectroscopy), microspectrofluorimetry and FORS (UV-Visible-NIR Fibre Optic Reflectance Spectroscopy). SERS is the only of the above techniques that allows acquiring a "molecular fingerprint" of the main dye present.

On the other hand, combined, microspectrofluorimetry and FORS proved to be the only techniques that could identify, in situ or in micro-samples, the chromophores responsible for the organic pinks, purples and reds of medieval illuminations [36]. This characterization was possible by comparison with a database of references for colorants, binders and colour paints. For more details on historically accurate reconstructions see Box 13.4.

The possibility of in situ analysis of colorants, offered by these two techniques, is a considerable advantage when considering the minute details of the paintings in books of hours. Moreover, fluorimetry offers high sensitivity and selectivity, and the importance of sensitivity is clear when considering that the organic chromophores are usually present in very low concentrations, because they may have faded or may have been applied as very thin coats over, or mixed with an inorganic pigment or extender.

Microfluorescence and UV-Visible-NIR Fibre Optic Reflectance spectroscopy were used to identify the chromophore(s) and Fourier Transform Infrared spectroscopy to characterize the binding media and additives, such as chalk and gypsum, present in the paint colours, Box 13.3. The results obtained for reference pigments and paints were compared with data obtained for brazilein lakes and for pink and red colours in manuscript illuminations from fifteenth century French books of hours (Palácio Nacional de Mafra, Portugal) [35], Fig. 13.9. For all medieval and reference pigments, FORS and microspectrofluorimetry consistently identified the same brazilwood pigment. Nevertheless, when comparing the absorption and excitation spectra with those from the main chromophore, the brazilein-Al^{3+} complex,[7] on filter paper, a shift of *ca.* 30 nm is observed [35]. This may indicate that FORS and microspectrofluorimetry are sensing a different chromophore environment than that present in the brazilein-Al^{3+} complex, Fig. 13.10, consequence of ageing or of a different production method. Combined these two techniques will enable us to collect evidence on the technology used to produce these medieval paints, advancing our knowledge on these ancient dyes and paving the way for a sustainable conservation of our cultural heritage.

[7] Colour coordinates for the brazilein-Al^{3+} complex, on filter paper: $L^* = 89$, $a^* = 15$, $b^* = 4$.

Fig. 13.9 Books of Hours were books of prayer for the laymen, usually of small dimensions. Lavishly illuminated and treasured by their owners, they were very expensive objects in the Middle Ages. Ms n°24 (197 × 135 × 50 mm), fifteenth century; Portugal, Palácio Nacional de Mafra

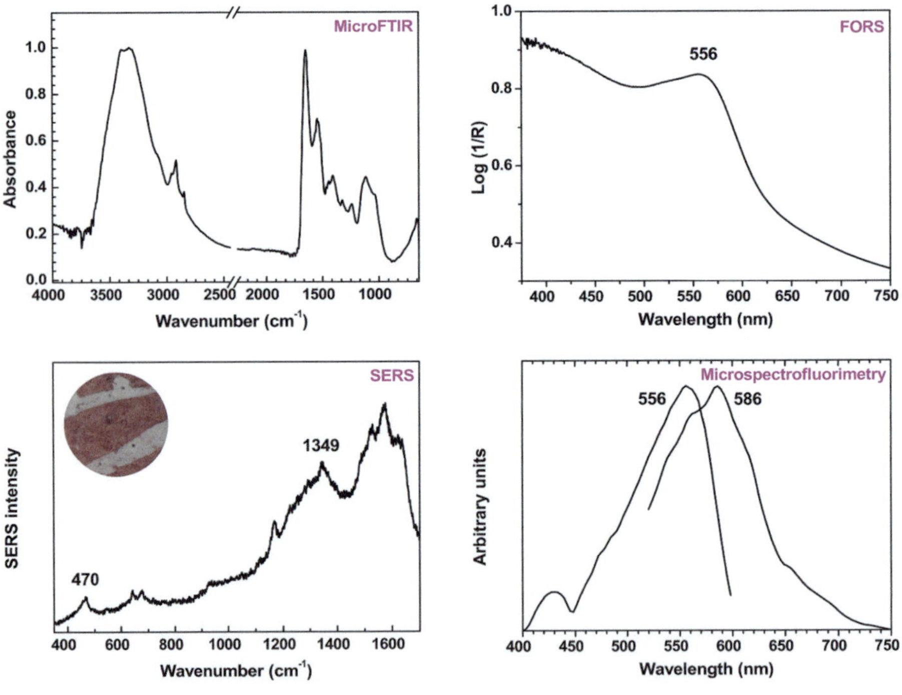

Fig. 13.10 Characterizing brazilwood lake pigments with light. FORS and microspectrofluorimetry enable an identification of the chromophore (by comparison with a reference database of historical reconstructions). SERS provides a molecular fingerprint for the chromophore, but it is not always possible to record good spectra. Infrared spectroscopy gives invaluable information on the paint formulation, binder and additives. The spectra displayed were obtained for the pink colour found in a Book of Hours in Palácio Nacional de Mafra collection (ms22, f76v&r). In the microspectrofluorimetry data, the emission spectra was obtained using $\lambda_{exc} = 490$ nm and the excitation spectrum was collected at $\lambda_{em} = 610$ nm

13.2.2.3 Photostability in Brazilwood Lake Pigments

The mechanisms associated with the exceptional photostability of ancient dyes have been linked to an ultrafast excited state proton transfer (ESPT). This has been already described for indigo (Sect. 13.2.1.2) and also observed for brazilwood [38] and dragon's blood chromophores (*Dracoflavylium*) [39, 40]. However, detailed mechanisms governing photostability, thermal stability and stability towards acid/alkaline attack, in these chromophores, are aspects that still lack a complete explanation. To fully understand the degradation mechanisms involved in such a complex material as a work of art, we started a detailed research on the production of historically accurate reference materials, Box 13.4.

Box 13.4: Historically Accurate Reconstruction Techniques
Vanessa Otero
The HART (Historically Accurate Reconstruction Techniques) methodology enables the production of reference materials with as much historical accuracy as possible. These references are validated by its "closeness to the true value", the historic material or artwork under study.

When studying historic materials it is essential to have access to their methods of production since today's materials do not represent the formulations used in the past. For this reason, the HART approach is a three-part model which combines documentary research of sources of technical information contemporary to the historic material, preparation of historically accurate reconstructions according to the latter and a multi-analytical characterization of both the historic materials and the reconstructions, which certifies the HART references. These can be used to study the history of use, the chemistry of the materials as well as their chemical and physical evolution over time. Ultimately, these studies will allow assessing the original appearance and condition of the artworks, advancing knowledge towards their conservation.

For brazilwood lake pigments reconstructions were prepared from two different sources: medieval recipes from the fifteenth century *Livro de como se fazem as cores* (*The Book on How to Make Colours*), and recipes from the nineteenth century archive of the artists' colourman, Winsor & Newton (W&N) [41]. By reconstructing the recipes using ingredients and methods appropriate to their period, it was possible to study the effect of the different procedures described, and of the different ingredients called for, such as precipitating agents, chelating ions, extenders or buffers, and the role of pH.

These reconstructions enabled us to reach an important conclusion: the colours of the reconstructed pigments were consistent with the pink and red brazilwood colours found in French books of hours [35]. This comparison showed that, despite their reputation for being relatively unstable and subject to fading, the colours derived from brazilwood in the books of hours appear to be very well preserved (possibly, because they were kept protected from light).

13.2.3 *A Colour That Changed the World: Mauveine*

The synthesis of mauveine by W. H. Perkin in 1856 constitutes a landmark in the history of modern chemistry. Variously known as aniline purple (1857), Tyrian purple (1858), mauve dye (from 1859), phenamine or indisine [42], Perkin's dye achieved great success as a fashionable colour between 1858 and 1863. With his serendipitous discovery, a new era of synthetic organic dyes was initiated with natural dyes being almost completely replaced by synthetic dyes [43–45]. The original mauve is believed to have been synthesized by Perkin on 23 March 1856 and patented in 1856 (granted on 26 August 1856 and sealed on 20 February 1857) [45]. Mauve was manufactured on a small industrial scale at Greenford, Middlesex, by William Perkin, his father George Fowler Perkin and his brother Thomas Dix Perkin, between late 1857 and the end of 1873 [45].

In 1856 the 18 years-old young William Perkin was asked by Hofmann, director of the "Royal College", to synthesise quinine which he could not succeed [42–46]. Indeed, what Perkin synthesized by reacting aniline, 2-methylaniline and 4-methylaniline together was a complex mixture of products [42, 46], as shown in Fig. 13.11.

Recently, we had the opportunity to study several historical samples that included both a set of powders and textiles from museum collections in the quest for the original Mauve [42]. We were able to show that the two major chromophores present are mauveines A (C_{26}) and B (C_{27}), Fig. 13.11. It was also possible to disclose, by the chromophores distribution, the two processes used by Perkin to produce the dye. In the first commercial product, mauveine A is the major chromophore (50 %) followed by mauveines C_{25} (15–20 %), Fig. 13.11. Later, Perkin changed the process to obtain a more soluble mauve with acetate as counter-ion. In this second process, a higher concentration of toluidine as starting material was used, leading to higher percentages of mauveine B, Fig. 13.11. We could also confirm that mauve dyed silk very easily and no mordant was necessary to bind the colouring matter to the textile.

Again we observed for mauve a photoprotective mechanism based on a very fast internal conversion channel. This was proved with the photophysical studies on two relevant chromophores (mauveines B2 and C), which showed a predominance of the $S_1\sim\sim\text{->}S_0$ radiationless deactivation over all other excited state deactivation pathways [47].

13.3 The Invisible in Paint: The Binding Medium

Oil painting and tempera painting are commonly employed for describing a painting technique in museums and exhibitions. These descriptions hide more than they tell, because an oil paint formulation in the fourteenth or fifteenth century is only remotely similar to what we may find nowadays in commercial artist's materials. Nevertheless, we need to acknowledge the tremendous impact binding media have had on the history of art; proteinaceous temperas were used in the twelfth century

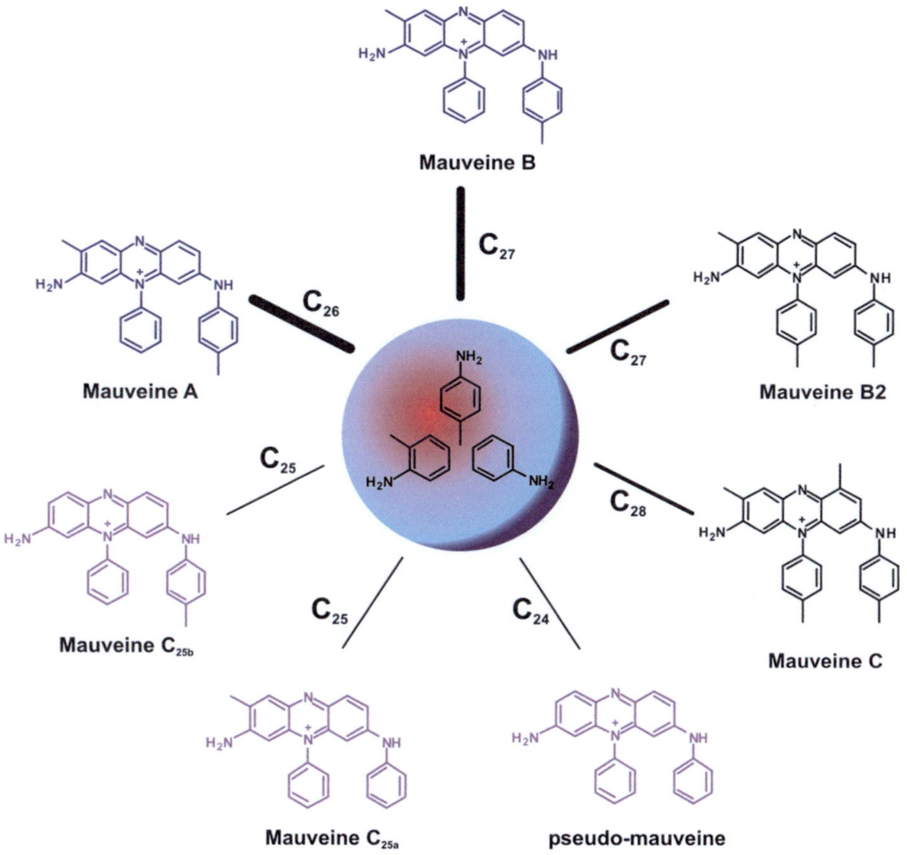

Fig. 13.11 Mauveine structures isolated from historic samples (From Ref. [42] with permission from Wiley)

when the great art of colour was in the codex, illuminating God's word; the Flemish school masterfully used linseed oil to promote a revolution in the way nature and humans were portrayed; the twentieth and twenty-first century have inherited the new direction that Impressionism motivated, and welcomed experimentation with new materials never seen before. Oil painting is still important, but in the twenty-first century water emulsions are ubiquitous. So in a certain way, we may say that with vinyl or acrylic emulsion pints we are closer to the twelfth century medium.

The two binding media selected as case studies share a common functional group: the ester. However, the mechanisms of formation of the two polyester matrixes are as different as are their photostability and durability. When we say an oil paint dries, we are conveying a cascade of chemical reactions that will result in an oil paint: a slow process that requires patience. On the other hand, when a vinyl emulsion dries, the artists only need to wait for water to evaporate to find a pristine paint film (as the polymer is already formed in the formulation): it is rapid and it is modern.

13.3.1 Oil Paint Systems: Self-Repair or Self-Destruction?

A drying oil is a natural product mainly composed of triglycerides, i.e., esters of glycerol linked to three carbon chains (fatty acids). The fatty acid portion of the drying oils generally used in oil paints is mostly composed of C_{18} polyunsaturated acids: oleic acid (C18:1), linoleic (C18:2) and linolenic (C18:3); and a small percentage of saturated fatty acids with 12–18 carbon atoms, e.g., palmitic (C16:0) and stearic (C18:0) [48], Box 13.5.

Box 13.5: Drying of a Sective Oil (e.g., Linseed and Poppy Oil)

A drying oil is a natural product mainly composed of triglycerides, i.e., esters of glycerol linked to three long carbon chains (Figure Box 13.3). The fatty acid portion of the drying oils generally used in oil paints, is mostly composed of C_{18} polyunsaturated acids: oleic acid (C18:1), linoleic (C18:2) and linolenic (C18:3); and a small percentage of saturated fatty acids with 12–18 carbon atoms, e.g., palmitic (C16:0) and stearic (C18:0).

Drying of the oil results from chemical reactions. The double bonds of the unsaturated fatty acids promote the initiation of these reactions, which can be light catalyzed, and will led to cross-linking of the fatty acid chains. This process may also be catalysed by the presence of metal ions from pigments. In the mature stage, a **three-dimensional polyanionic network** is formed by carboxylate anions and the pigment metal cations.

In the curing and ageing of an oil paint, at the same time that cross-linking of the unsaturated fatty acids result in the three-dimensional polyester network, hydrolysis of the ester functions to carboxylic acids occurs. Auto and photo oxidation leads to the formation of hydroperoxides, which will evolve to radicals such as alkoxy ($RO^{\bullet}$) and carbonyl moieties that, in turn, will initiate a cascade of Norrish reactions, leading both to chain scission and cross-linking.

The diacids that may be formed as end products, such as azelaic acid (C9) may also contribute to the stabilisation of the network due to their chain building ability in three dimensions, contributing to the paint system *self-repair*.

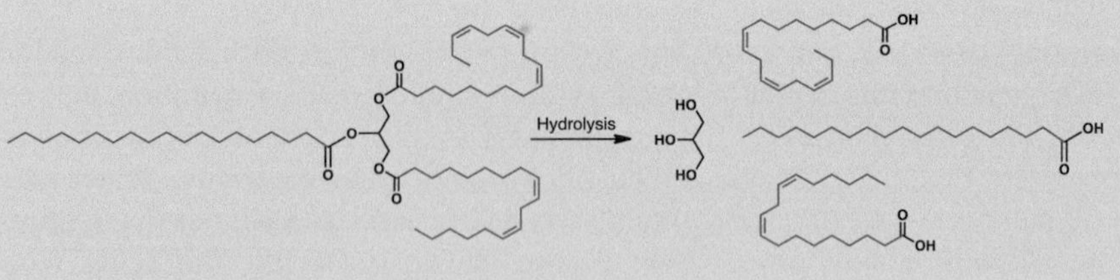

Figure Box 13.3 Hydrolysis of a triglyceride

A paint oil dries primarily as a result of photo-oxidation and oxidation reactions. The reactivity of the double bonds of the unsaturated fatty acids promotes the initiation of these reactions. This process may be catalysed by the presence of metal ions from pigments and driers (e.g. lead oxide or acetate) [48]. Jaap Boon and co-workers described the end result as a complex three-dimensional polyanionic network based on the glycerol ester with a cross-linked fraction of carboxylate anions stabilized by metal cations [49]. These same researchers have recently described the paint system as being capable of *self-repair*, preventing paint failure [49]; the loose ends would self-organize in a way similar to crystal liquids, thus preventing migration of these components within the paint and to the surface. Diacids, such as azelaic acid (C9), would thus contribute to the stabilisation of the network due to their chain-building ability in three dimensions. Mono and dicarboxylic acids that are not incorporated in the polyanionic network may migrate to react with metal cations leading to the formation of an increased proportion of metal carboxylates (metal soaps). In some cases, metal soaps form aggregates within the paint film.

The presence of metal soap aggregates in high proportions can negatively influence the appearance of oil paintings and have a dramatic aesthetic impact, as in the case when an opaque white lead colour is transformed into a translucent whitish soap [49]. Dissolution of opaque lead white, a basic lead carbonate, leads to a dramatic decrease in reflectance, thus inducing a darkening. As lead white is a dominant colour in old master paintings, pure or admixed, the colour balance of the painted surface may be severely lost [49].

The relevant contribution of the research led by Jaap Boon, which we briefly highlighted, proved invaluable in our understanding of the complexity of an oil paint matrix. We are now able to diagnose numerous degradation situations and recognize their impact in the perception of precious works of art. This was a scientific achievement as it has enabled us to "see" phenomena not known before nor considered in a conservation intervention. How the formation of paint soaps correlates with paint composition, environmental conditions, and conservation history of a painting is a hot topic in conservation and will lay the groundwork for the development of innovative sustainable conservation treatments.

13.3.2 The Preservation of Contemporary Art: The Case of Vinyl Paints

13.3.2.1 The 'Plastic Age'

After the Stone Age, Bronze Age and Iron Age, the development of synthetic polymers in the early 1900s revolutionized the life of modern society. In 1979 the volume of plastics produced worldwide surpassed that of steel. This event was the landmark for what we now consider the 'Plastic Age' [50]. The novelty of synthetic polymers fascinated artists and they have been experimenting with these new

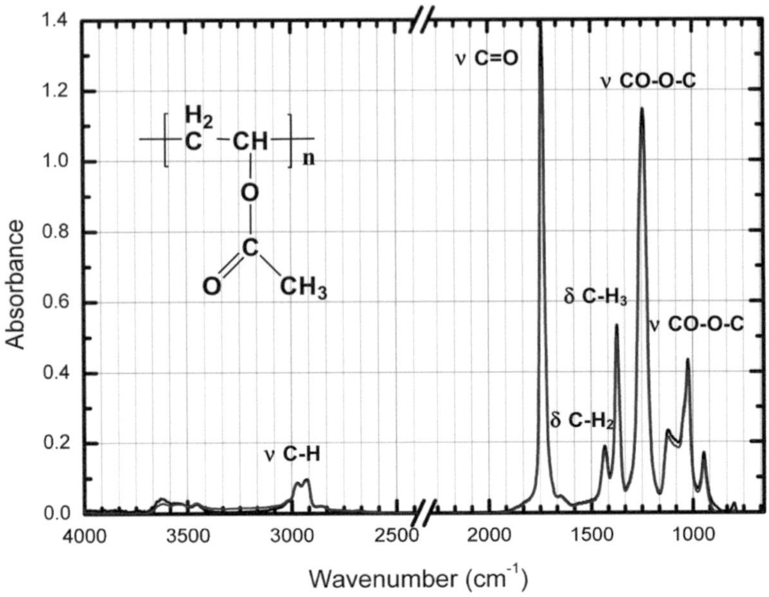

Fig. 13.12 Infrared spectra of PVAc homopolymer: *black line* – t = 0 h; *blue line* – 5000 h of irradiation with a Xenon-arc lamp ($\lambda \geq 300$ nm; 800 W/m^2); the *black line* is the upper line at ca. 1100 cm^{-1}

materials ever since [51–54]. During the 1940s and 1950s, the first artist's formulations based on vinyl[8] and acrylic emulsions were commercialized [51] and replaced the primacy of oil paints.

13.3.2.2 Vinyl Paints in Art

Vinyl paints have been used by important artists mainly as emulsions, e.g. Bridget Riley (1931-) and the Portuguese artists Joaquim Rodrigo (1912–1997) and Ângelo de Sousa (1938–2011) [51, 54], and in fewer cases in the form of a solvent-based solution, e.g. Yves Klein (1928–1962) [55].

Artists' quality paints are similar to household ones in their general formulation, but they are likely to have higher quality pigments and higher application viscosity. Also, these paints have been formulated with less filler, more pigment and thickener [56]. According to Crook and Learner, the first Polymer Tempera was an artists' PVAc emulsion paint developed in 1946 by the paint manufacturer Borden Co. with the collaboration of the American artist Alfred Duca (1920–1997). PVAc resin[9] was used by Yves Klein in his famous paint formulation, patented as IKB – International

[8] Poly(vinyl acetate), PVAc (Fig. 13.12), was introduced in the market in 1928, and it has been widely used for household paints and adhesives in the form of aqueous emulsions.

[9] Only soluble in organic media.

Fig. 13.13 *Left*: *Os quintais* (The backyards), Joaquim Rodrigo (1989); *right: Plantas* (Plants), Ângelo de Sousa (1962). National Museum of Modern Art- Museu do Chiado (Lisbon) and artist collection (Porto), respectively

Klein Blue, with which he produced the Blue Monochrome Series between 1955 and 1962. IKB consists of artificial ultramarine blue and poly(vinyl acetate) binder [55].

The preservation condition of selected vinyl paintings by Portuguese artists Joaquim Rodrigo and Ângelo de Sousa, Fig. 13.13, has been assessed [54] in comparison with photodegradation studies in model samples, which will be described in Sect. 13.3.2.3. The models studied are based on dated paint samples: paints prepared by Joaquim Rodrigo and the vinyl aqueous emulsion paints, Sabu, produced by A Favrel Lisbonense. These models were prepared with titanium white, iron oxide, and ultramarine blue pigments, along with the filler calcium carbonate, widely used in paint formulations. Rodrigo prepared his own paints by mixing the vinyl emulsion with a restricted palette based on red and yellow iron oxides, titanium dioxide, and a 'suitable' black [54]. This approach allows a correlation to be made between paint samples aged naturally and those aged under artificial conditions. The conclusion is that the selected works of art are in a good state of preservation, with no signs of physical changes or molecular degradation.

13.3.2.3 PVAc in Art: Reference Samples, Photochemistry and Degradation Mechanisms

Photo-oxidation is considered the main factor affecting polymer weathering [57–61]. Works of art made with vinyl paints are mainly exposed to radiation filtered by glass windows ($\lambda \geq 300$ nm), and consequently no direct absorption of light by the polymer carbonyl group is expected. When studying the stability of polymers, an important factor is the structure of a commercial polymer, which is likely to be more complex than indicated by its molecular formula. Commercial polymers typically have structural irregularities introduced during polymerisation and storage. Radiation at $\lambda \geq 300$ nm will most likely be absorbed by chromophores in the PVAc matrix, such as hydroperoxide groups formed during synthesis or processing [60]. Even present in very low amounts, and therefore absorbing a very small fraction of light, they may act as initiation sites for polymer degradation.

For a better understanding of the photo-oxidative mechanisms of poly(vinyl acetate) **homopolymer**, the main polymer identified in selected paintings works by Joaquim Rodrigo and Ângelo de Sousa, studies under both polychromatic (Xenon-arc lamp at $\lambda \geq 300$ nm) and monochromatic (at 313 nm) radiation have been carried out. As the first indications for the molecular deterioration of the polymer are given by changes in the molecular weight distribution (e.g., scission or crosslinking), the molecular evolution is followed by infrared spectroscopy (FTIR) and size exclusion chromatography (SEC). Reaction quantum yields (Φ_R) constitute a universal and straightforward parameter to assess and compare polymer photostability and as such they were also calculated [57, 62, 63], Box 13.6. In polymer systems, Φ_R is defined as the ratio between the number of molecules undergoing chain scission, crosslinking, or any other relevant photodegradation aspect per photon absorbed [61, 62], Box 13.6.

Box 13.6: Measuring a Reaction Quantum Yield (Φ_R) in a Solid Polymer Matrix

The Φ_R for PVAc photodegradation is considered as the rate of scissions per chain, S′, per mole of photons absorbed, I_A Eq. (13.4). To minimize experimental error in using all the absorption spectra for $\lambda \geq 300$ nm, Φ_R is calculated for 313 nm, a wavelength in which it is still possible to measure the polymer matrix absorption with acceptable accuracy [57]. As Φ_R is dimensionless, the units for both S′ and I_A should be the same (as detailed).

I_0* and *I_A

I_0 at 313 nm is obtained using potassium hexacyanocobaltate(III) as actinometer and Eq. (13.1); I_0 is expressed as mol.cm^{-2}. min^{-1}, i.e., the mole of photons per unit area per unit time.

$$I_0 = \frac{V_{sol} \times \left(\frac{\Delta A}{\Delta \varepsilon}\right)}{1000 \times \Phi_R \times \Delta t} = 1.16 \times 10^{-7} \text{mol.L}^{-1}\text{.cm.min}^{-1}$$
$$= 1.16 \times 10^{-7} \times 10^{3} \text{mol.cm}^{-2}\text{.min}^{-1} \qquad (13.1)$$

Where, V_{sol} is the volume of irradiated solution in ml (3 ml); ΔA is the change in absorbance at the monitoring wavelength (380 nm) over the irradiation time period, corrected by the light absorption of the reagent at 313 nm; $\Delta\varepsilon$ is the difference between the molar absorption coefficients of reagent (ε(R) = 10 M^{-1}cm^{-1}) and product (ε(P) = 280 M^{-1}cm^{-1}) at the monitoring

(continued)

Box 13.6 (continued)
wavelength, in L.mol^{-1}.cm^{-1}; 1,000 corresponds to ml; Δt to irradiation period in min; and Φ_R is the quantum yield of the reaction ($\Phi_R = 0.31$) [57].

The number of photons absorbed is:

$$I_A(PVAc) = I_0 x\left(1 - 10^{-A}\right) = 1.16 \times 10^{-4} \times \left(1 - 10^{-0,0155}\right)$$
$$= 4.1 \times 10^{-6}\text{mol.cm}^{-2}\text{.min}^{-1} \qquad (13.2)$$

Rate of scissions per chain

The number of scissions per chain per unit time (S') was determined as 2.0×10^{-6}. To be used to calculate Φ_R this number must refer to the moles irradiated per area, i.e., it must be multiplied by the weight of polymer irradiated (w, g) and divided by M_{n0} and irradiated area (a, cm^2), as indicated below:

N° of moles undergoing chain scission per unit time per area (PVAc) =

$$\frac{S' \times w}{Mn_0 \times a} = \frac{2.0 \times 10^{-6} \times 0.01981}{39982 \times 3.3} = 3.0 \times 10^{-13}\text{mol.cm}^{-2}\text{.min}^{-1} \qquad (13.3)$$

Quantum yield

The Φ_R may now be obtained by dividing the moles of polymer undergoing chain scission Eq. (13.3) by the moles of photons absorbed, both by unit area and unit time:

$$\Phi_R(PVAc) = \frac{3.0 \times 10^{-13}}{4.1 \times 10^{-6}} = 7.40 \times 10^{-8} \qquad (13.4)$$

13.3.2.3.1 Molecular Evolution

The photo-oxidative mechanisms of poly(vinyl acetate) homopolymer are discussed in Box 13.7. Infrared monitoring has shown that molecular changes in the polymer, are small, almost negligible, Fig. 13.12. After prolonged irradiation, the intensity of the PVAc spectrum is lowered by a maximum of 10 %, and its shape is maintained when compared to non-irradiated samples.[10] Furthermore no PVAL formation could be detected [57]. Changes in molecular weight distributions are present after 500 h under irradiation. With no indication for crosslinking, the serial decrease in weight average molecular weight (Mw) and the slight increase in its

[10] For all samples, Gaussian decomposition of the spectra in the carbonyl region (c.a. 1840–1660 cm^{-1}) as well as the $\delta CH_3/C{=}O$ and $\nu C\text{-}O/C{=}O$ ratios (absorption and peak area) show less than 5.5 % variations upon irradiation, indicating that the photochemical reactions involving the carbonyl group, if present, are still not relevant as detected by IR.

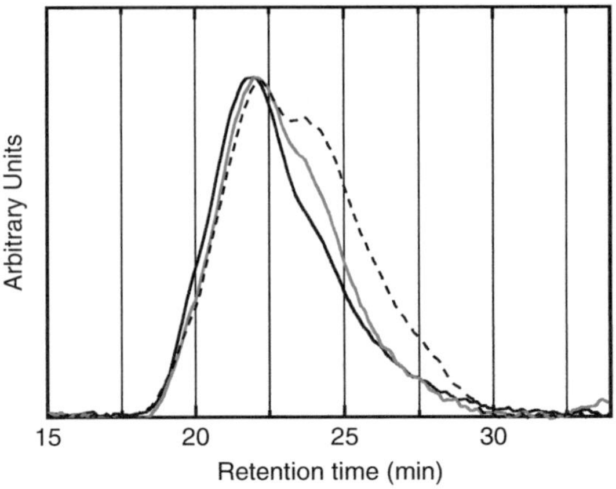

Fig. 13.14 SEC chromatograms of PVAc before and after irradiation with a Xenon-arc lamp (λ ≥300 nm; 800 W/m^2): *black line* – 0 h; *grey line* – 1500 h; dashed line – 3500 h

dispersity indicate that PVAc undergoes chain scission from the beginning of irradiation, Fig. 13.14.

Box 13.7: Norrish Reaction Mechanisms in PVAc Photodegradation

The formation of acetic acid, carbon monoxide and other volatile molecules observed upon irradiation at 254 nm [61] may be explained based on an heterolytic process or/and via an homolytic mechanism based on Norrish type I reactions, Figure Box 13.4. Heterolytic bond scission may occur as a Norrish type II mechanism or an ester hydrolysis that will lead to the formation of acetic acid and poly(vinyl alcohol).

It has been shown that upon PVAc irradiation at 254 nm acetic acid is formed as the main product with a quantum yield of reaction of 10^{-2} [61]. This would be extremely dangerous in a museum environment, because acetic acid reacts with innumerous materials in works of art, such as pigments and metals where lead is present. As photochemists we are aware that a reaction activated by light at 254 nm may or may not occur when the same sample is irradiated at 313 nm. A work of art may be found indoor or outdoor and in neither case it will be possible for a material to absorb at 254 nm, as this part of the solar spectrum does not reach the Earth's surface [54]. In our set-up we mimic the photodegradation that may occur upon natural ageing by irradiating PVAc paints with light filtered at 300 nm or monochromatic 313 nm radiation. We now will take a closer look on the processes leading to the formation of acetic acid to demonstrate how they may be monitored by infrared spectroscopy.

Acetic acid may be formed by heterolytic (non-photochemical) acidic hydrolysis or homolytic Norrish type I reactions, resulting in the formation of poly(vinyl alcohol), which, according to our previous studies, is detected in the infrared spectrum for a concentration of PVAL greater than 10 %

(continued)

Box 13.7 (continued)

[57]. The appearance of an O-H stretching from the alcohol group present in PVAL together with the ratios $\nu CH_3/C = O$, $\nu CH_2/C = O$, $\delta CH_3/C = O$ and $\nu C\text{-}O/C = O$ provide relevant information about the extent and type of the ester fragmentation.

In a Norrish Type II mechanism, Figure Box 13.4, acetic acid release is accompanied by the formation of a double bond in the main chain. Due to its low molar absorptivity the formation of these double bonds might not be detected in the infrared spectrum, but the extensive production of acetic acid would be reflected in a decrease of polymer weight; it would also be detectable in the infrared spectra by a loss of intensity as well as a broadening of the ester absorption band.

In a Norrish type I, besides the changes in the carbonyl absorption just described, the side chain reactions are readily observed in the infrared spectrum; the loss of a CH_3 group would be monitored both in the $C = O$/ C-H ratios and in alteration of the CH_3 stretching and CH_3 bending, at approximately 2970 cm^{-1} and 1373 cm^{-1}, respectively.

Figure Box 13.4 Photodegradation pathways for PVAc and proposed intermediates leading to the formation of volatile fragments, such as CO, CH_4 and acetic acid (Adapted from [57])

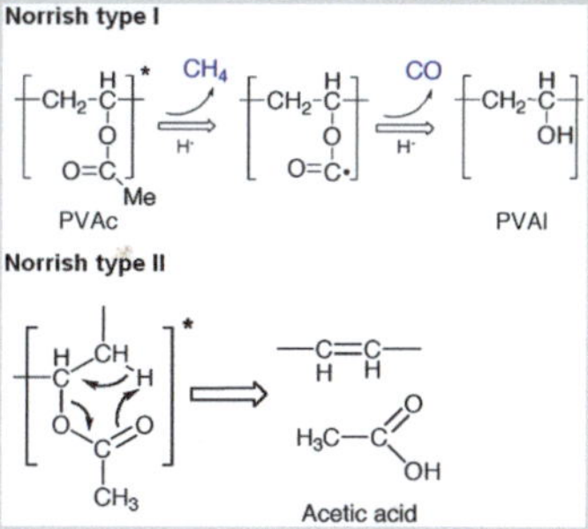

An important aspect to consider when analysing the degradation of paints is the effect of pigments and fillers on the polymer stability [57]. Looking at the scission process in more detail, the number of scissions per chain, S, may be calculated by $S = Mn_0/Mn_t - 1$[11] and it is possible to separate the pigments into two groups: stabilising and sensitising, Fig. 13.15. Titanium dioxide and iron oxide display a protective effect on paint films shown by a decrease in the scission rate when compared to PVAc alone. On the other hand, ultramarine blue and calcium carbonate in particular (frequently present in paints as filler) promote an increase in the polymer scission rate.

[11] As no relevant weight losses relating to the release of volatiles are found, the number of scissions per chain, defined as $S = Mn_0(1\text{-}x)/Mn_t - 1$, where Mn is the number average molecular weight before (Mn_0) and after irradiation (Mn_t), and x is the fraction of volatilized polymer, may be calculated through the simplified expression.

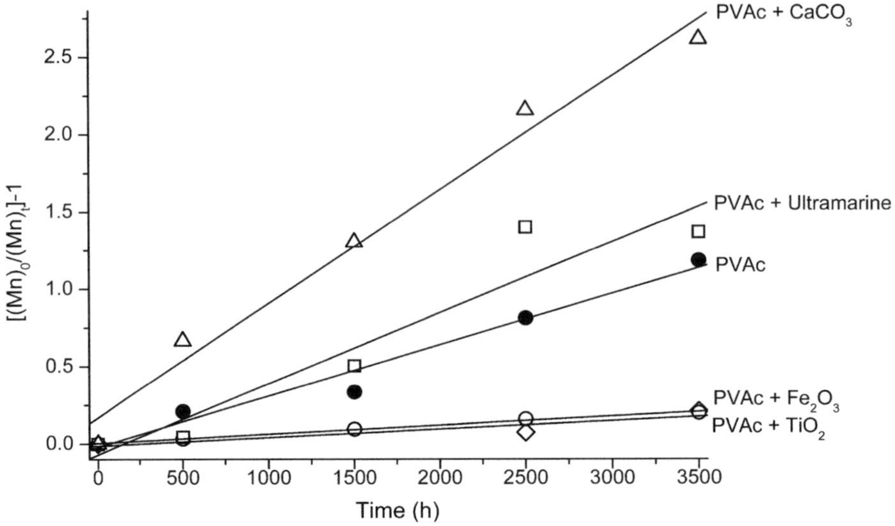

Fig. 13.15 Scission per chain as a function of irradiation time for PVAc and its mixtures with pigments. Where Mn_0 = original number average molecular weight; Mn_t = number average molecular weight after irradiation (From [57] with permission from Elsevier)

As results cannot be explained based on the photocatalytic properties of the pure pigments, where both Fe^{3+} and Ti^{4+} are well-known photosensitizers, it is possible that TiO_2 and Fe_2O_3 are sold with a protective coating, i.e. encapsulated with inert materials. For a discussion of the results, it is necessary to take into account the partition of light absorption by the composite system, i.e. between pigment and polymer. If the pigment is able to compete for light absorption, without promoting any secondary photochemical reactions, it will display a protective effect; the larger the fraction of light absorbed, especially for lower wavelengths, the more efficient this effect will be. Both iron oxide and titanium dioxide absorb completely (cut off) radiation below 400 nm [57] and are possibly encapsulated. Therefore, a protective effect is observed. In turn, the results obtained for calcium carbonate do not present an immediate explanation, but some hypothesis may be formulated. Calcium carbonate does not display a sharp cut-off, and between 300 and 350 nm both reflects and scatters light [57]. In this region, both the scattered and reflected light could be re-absorbed by the polymer (internal scattering), intensifying the light absorption [65]. A similar reasoning might explain the observation for ultramarine [57], but in this case due to the presence of kaolin as an impurity. Finally, the ultramarine pigment is an aluminosilicate matrix where colour is provided by S_2^- and S_3^- ions entrapped in the lattice. Therefore, it is a pigment where no catalytic activity would be predicted, but recent studies proved that in paintings this is not always the case.

The Φ_R for PVAc homopolymer at 313 nm is 7.40×10^{-8}. When compared with the values published in literature for irradiation involving 254 nm [64], the Φ_R for

PVAc is five orders of magnitude lower, which indicates that a different photochemical mechanism is involved [57, 64, 65].

Given that no changes have been detected in the infrared spectra, the discussion that follows only entails hypotheses. Main chain scission may be the result of excitation of hydroperoxides, present in very low amounts in the polymer matrix and not detected in the infrared spectrum ($\leq 3\,\%$). The absorption of light by these groups results in main chain scission and possibly the formation of functional groups, like double bonds and other carbonyls, and the release of volatile moieties (also $\leq 3\,\%$). Or, the first cleavage may be accompanied by some depolymerisation, and this could explain both the unchanged IR spectra after irradiation and the slight decrease in intensity by polymer loss. Quantitative measurements of volatiles will be necessary to gain further insight into this aspect.

13.4 Overview

With a "little help" from photochemistry much may be learned from medieval paintings and textiles; possibly, beyond what could have been told by the makers themselves. Our wish is that the reader has appreciated learning about the photochemical aspects underlined by historic dyes, and may agree with us that it is fascinating "seeing" indigo as a "molecular fossil" [29], sharing with DNA bases an ancestral photoprotective mechanism; in indigo, the excited state properties that provide an highly efficient internal conversion are based on a very fast ESPT.

Conservation of contemporary art is a challenge our field is facing in this new century, and binders are at the very centre of paint stability. Taking into account that Φ_R is a fundamental parameter for quantifying photostability and understanding the reaction mechanisms in modern vinyl paints, we have demonstrated the essential contributions of photochemistry for the lifetime prediction and material stabilization of valuable artworks [54, 57].

In the very near future, the bright light made available by synchrotron facilities will provide the molecular detail that is still lacking for an accurate description of the complexity of oil paint systems as found in works of art. This is an emerging field where important breakthroughs will enable a better preservation of masterpieces [66].

We hope you have enjoyed this tour into a world of light, molecules and art. Now, it is up to you to engage with and contribute to the safeguard of our cultural heritage!

References

1. Wouters J (2008) Protecting cultural heritage: reflections on the position of science in multidisciplinary approaches. Chem Int 30:4–7
2. Brunetti BG, Sgamellotti A, Clark AJ (2010) Advanced techniques in art conservation (Editorial). Acc Chem Res 43:693–694
3. Balzani V, Scandola F (1991) Supramolecular photochemistry. Ellis Horwood, Chichester
4. Melo MJ (2009) History of natural dyes in the ancient mediterranean world. In: Bechtold T, Mussak R (eds) Handbook of natural colorants. Wiley, Chichester, pp 3–18
5. Cardon D (2007) Natural dyes. Sources, tradition, technology and science. Archetype Publications, London
6. Vitorino T, Melo MJ, Carlyle L, Otero V (2015) New insights into brazilwood manufacture through the use of historically accurate reconstructions. Stud Conserv. doi:10.1179/2047058415Y.0000000006
7. Halleux R (ed) (2002) Les alchimistes grecs: papyrus de Leyde, papyrus de Stockholm, recettes. Les Belles Lettres, Paris
8. Kroustallis S (2011) Binding media in medieval manuscript illumination: a source of research. Rev Hist Arte FCSH-UNL Sér W 1:113–125
9. Melo MJ, Castro R, Miranda A (2014) Colour in medieval portuguese manuscripts: between beauty and meaning. In: Sgamellotti A, Brunetti BG, Miliani C (eds) Science and art: the painted surface. The Royal Society of Chemistry, London
10. Balfour-Paul J (2000) Indigo. British Museum Press, London
11. Seixas de Melo JS, Moura AP, Melo MJ (2004) Photophysical and spectroscopic studies of indigo derivatives in their keto and leuco forms. J Phys Chem A 108:6975–6981
12. Sousa MM, Miguel C, Rodrigues I, Parola AJ, Pina F, Seixas de Melo JS, Melo MJ (2008) A photochemical study on the blue dye indigo: from solution to ancient Andean textiles. Photochem Photobiol Sci 7:1353–1359
13. Meijer L, Guyard N, Skaltsounis LA, Eisenbrand G (eds) (2006) Indirubin, the red shade of indigo. Life in Progress, Roscoff
14. Baeyer A, Drewson V (1882) Darstellung von Indigblau aus Orthonitrobenzaldehyd. Ber Dtsch Chem Ges 15(2):2856–2864
15. de Meijere A (2005) Adolf von Baeyer: winner of the nobel prize for chemistry 1905. Angew Chem Int Ed 44(48):7836–7840
16. Reis A, Schneider W (1928) On the crystal structure of indigo and fumaric acid. Z Kristall 68 (6):543–566
17. Cooksey CJ (2001) Tyrian purple: 6,6′-dibromoindigo and related compounds. Molecules 6:736–769
18. Wyman GM (1994) Reminescences of an accidental photochemist. EPA News Lett 50:9–13
19. Bauer H, Kowski K, Kuhn H, Luttke W, Rademacher P (1998) Photoelectron spectra and electronic structures of some indigo dyes. J Mol Struct 445(1–3):277–286
20. Wille E, Luttke W (1971) Theoretical and spectroscopic studies on Indigo Dyes. 9. 4,4,4′,4-′-Tetramethyl-Delta2,2′-Bipyrrolidine-3,3′-Dione, a compound having basic chromophore system of Indigo. Angew Chem Int Ed 10(11):803–804
21. Elsaesser T, Kaiser W, Luttke W (1986) Picosecond spectroscopy of intramolecular hydrogen bonds in 4,4′,7,7′-tetramethyllndigo. J Phys Chem 90:2901–2905
22. Klessinger M (1982) The origin of the color of indigo dyes. Dyes Pigm 3(2–3):235–241
23. Klessinger M (1980) Captodative substituent effects and the chromophoric system of indigo. Angew Chem Int Ed Engl 19(11):908–909
24. Jacquemin D, Preat J, Wathelet V, Perpete EA (2006) Substitution and chemical environment effects on the absorption spectrum of indigo. J Chem Phys 124(7):074104
25. Miliani C, Romani A, Favaro G (1998) A spectrophotometric and fluorimetric study of some anthraquinoid and indigoid colorants used in artistic paintings. Spectecrochim Acta A Mol Biomol Spectr 54:581–588

26. Seixas de Melo JS, Serpa C, Burrows HD, Arnaut LG (2007) The triplet state of indigo. Angew Chem Int Ed Engl 46:2094–2096
27. Seixas de Melo JS, Rondão R, Burrows HD, Melo MJ, Navaratnam S, Edge R, Voss G (2006) Spectral and photophysical studies of substituted Indigo derivatives in their Keto forms. Chem Phys Chem 7:2303–2311
28. a) Nagasawa Y, Taguri R, Matsuda H, Murakami M, Ohama M, Okada T, Miyasaka H (2004) The effect of hydrogen-bonding on the ultrafast electronic deactivation dynamics of indigo carmine. Phys Chem Chem Phys 6(23):5370–5378; b) Iwakura I, Yabushita A, Kobayashi T (2011) Transition state in a prevented proton transfer observed in real time. Bull Chem Soc Jap 84(2):164–171
29. Kleinermanns K, Nachtigallová D, de Vries MS (2013) Excited state dynamics of DNA bases. Int Rev Phys Chem 32(2):308–342
30. a) Claro A, Melo MJ, Seixas de Melo JS, van den Berg KJ, Burnstock A, Montague M, Newman R (2010) Identification of red colorants in cultural heritage by microspectrofluorimetry. J Cult Herit 11:27–34; b) Melo MJ, Claro A (2010) Bright light: microspectrofluorimetry for the characterization of lake pigments and dyes in works of art Acc Chem Res 43: 857–866
31. Paul A (ed) (1991) Paracas art & architecture, object and context in South Coastal. University of Iowa Press, Peru
32. Frank AT, Adenike A, Aebisher D, Greer A, Gao R, Liebman JF (2007) Paradigms and paradoxes: energetics of the oxidative cleavage of indigo and of other olefins. Struct Chem 18(1):71–74
33. Kettle AJ, Clark BM, Winterbourn CC (2004) Superoxide converts indigo carmine to isatin sulfonic acid. Implications for the hypothesis that neutrophils produce ozone. J Biol Chem 279:18521–18525
34. Srividya N, Paramavisan G, Seetharaman K, Ramamurthy P (1994) Two-step reduction of indigo carmine by dithionite: a stopped-flow study. J Chem Soc Faraday Trans 90:2525–2530
35. Bond AM, Marken F, Hill E, Compton RG, Hügel H (1997) The electrochemical reduction of indigo dissolved in organic solvents and as a solid mechanically attached to a basal plane pyrolytic graphite electrode immersed in aqueous electrolyte solution. J Chem Soc Perkin Trans 2:1735–1742
36. Melo MJ, Otero V, Vitorino T, Araújo R, Muralha VSF, Lemos A, Picollo M (2014) Three books of hours from the 15th century: a multi-analytical and interdisciplinary approach. Appl Spec 68:434–444
37. http://cantigas.fcsh.unl.pt/manuscritos.asp
38. Rondão R, Seixas de Melo JS, Melo MJ, Vitorino T, Parola AJ (2013) Brazilwood Reds: the (Photo)chemistry of Brazilin and Brazilein. J Phys Chem A 117:10650–10660
39. Melo MJ, Sousa MM, Parola AJ, Seixas de Melo JS, Catarino F, Marçalo J, Pina F (2007) Identification of 7,4′-dihydroxy-5-methoxyflavylium in "Dragon's blood". To be or not to be an anthocyanin. Chem Eur J 13:1417–1422
40. Pina F, Melo MJ, Laia CAT, Parola AJ, Lima JC (2012) Chemistry and applications of Flavylium compounds: a handful of colours. Chem Soc Rev 41:869–908
41. Otero V, Carlyle L, Vilarigues M, Melo MJ (2012) Chrome yellow in nineteenth century art: historic reconstructions of an artists' pigment. RSC Adv 2:1798–1805
42. Sousa MM, Melo MJ, Parola AJ, Morris PJT, Rzepa HS, Seixas de Melo JS (2008) A study in Mauve: unveiling Perkin's Dye in historic samples. Chem Eur J 14:8507–8513
43. Perkin WH (1906) Address of Sir Silliam Henry Perkin. Science 24:488–493
44. Garfield S (2002) MAUVE. How one man invented a color that changed the world. W. W. Norton & Company, New York
45. Travis AS (2007) Mauve, its impact, and its anniversaries. Bull Hist Chem 32:35–44
46. Meth-Cohn O, Smith M (1994) What did Perkin, W.H. Actually make when he oxidized aniline to obtain mauveine. J Chem Soc Perkin Trans 1:5–7
47. Seixas de Melo J, Takato S, Sousa MM, Melo MJ, Parola AJ (2007) Revisiting Perkin's dyes (s): the spectroscopy and photophysics of two new mauveine compounds (B2 and C). Chem Commun 2624–2626

48. Otero V, Sanches D, Montagner C, Lopes JA, Vilarigues M, Carlyle L, Melo MJ (2014) In situ characterisation of metal carboxylates by Raman and infrared spectroscopy in works of art. J Raman Spectrosc 45:1197–1206
49. Boon J, Hoogland FG, Keune K (2007) Chemical processes in aged oil paints affecting metal soap migration and aggregation. In: Mar Parkin, H (ed) 34th annual meeting of the American Institute for Conservation of Historic & Artistic Works, Providence June 2006. AIC Paintings Specialty Group Postprints, vol 19. American Institute for Conservation, Washington p 16
50. Morgan J (1993) A joint project on the conservation of plastics by the conservation unit and the Plastics Historical Society. In: Grattan DW (ed) Saving the twentieth-century: the conservation of modern materials: Proceedings of Symposium 9, Ottawa, September 1991. Canadian Conservation Institute, Otawa, p 43
51. Crook J, Learner T (2000) The impact of modern paints. Tate Gallery Publishing Ltd, London
52. Sonoda N, Rioux JP (1990) Identification des matériaux synthétiques dans les peintures modernes. I. Vernis et liants polyméres. Stud Conserv 35:189–204
53. Chiantore O, Rava A (2012) Conserving contemporary art – issues, methods, material and research. The Getty Conservation Institute, Los Angeles
54. Ferreira JL, Ávila MJ, Melo MJ, Ramos AM (2013) Early aqueous dispersions paints: Portuguese artist's use of poly(vinyl acetate) 1960s–1990s. Stud Conserv 58:211–225
55. Mancusi-Ungaro C (1982) A technical note on IKB. In: Yves Klein, 1928–1962: a retrospective, exhibition catalogue. Institute for the Arts, Rice University, Houston p 258
56. Croll S (2007) Overview of developments in the paint industry since 1930. In: Learner TJS, Smithen P, Krueger JW, Schilling MR (eds) Proceedings from the Symposium Modern Paints Uncovered, London May 2006. Getty Publications, Los Angeles, p 17
57. Ferreira JL, Ramos AM, Melo MJ (2010) PVAc paints in works of art: a photochemical approach – part 1. Polym Degrad Stabil 95:453–461
58. Lemaire J, Gardette J, Lacoste J, Delprat P, Vaillant D (1996) Mechanisms of photo-oxidation of polyolefins: prediction of lifetime in weathering conditions. In: Clough RL, Billingham NC, Gillen KT (eds) Polymer durability: degradation, stabilization and lifetime predictions. American Chemical Society, Boston, pp 577–598
59. Pospíšil J, Pilař J, Billingham NC, Marek A, Horák Z, Nešpůrek S (2006) Factors affecting accelerated testing of polymer stability. Polym Degrad Stabil 91:417–422
60. Allen NS, Edge M (1992) Fundamentals of polymer degradation and stabilization. Elsevier, London
61. Rabek J (1995) Polymer photodegradation: mechanisms and experimental methods. Chapman & Hall, London
62. Fox RF (1997) Photodegradation of high polymers. In: Jenkins AD (ed) Progress in polymer science, vol 1. Pergamon, London, p 45
63. Montalti M, Credi A, Prodi L, Gandolfi MT (2006) Handbook of photochemistry, 3rd edn. Taylor & Francis, Boca Raton
64. David C, Borsu M, Geuskens G (1970) Photolysis and radiolysis of polyvinyl acetate. Eur Polym J 6:959–963
65. Buchanan KJ, McGill WJ (1980) Photodegradation of poly(vinyl esters) – II: volatile product formation and changes in the absorption spectra and molecular mass distributions. Eur Polym J 16:313–318
66. Cotte M, Susini J, Dik J, Janssens K (2010) Synchrotron-based X-ray absorption spectroscopy for art conservation: looking back and looking forward. Acc Chem Res 43:705–714

Index

G. Bergamini, S. Silvi (eds.), *Applied Photochemistry*, Lecture Notes in Chemistry 92, DOI 10.1007/978-3-319-31671-0

MIX
Papier aus verantwortungsvollen Quellen
Paper from responsible sources
FSC® C105338

If you have any concerns about our products,
you can contact us on
ProductSafety@springernature.com

In case Publisher is established outside the EU,
the EU authorized representative is:
Springer Nature Customer Service Center GmbH
Europaplatz 3, 69115 Heidelberg, Germany

Printed by Libri Plureos GmbH
in Hamburg, Germany